Crystal Plasticity

Crystal Plasticity

Editor

Wojciech Polkowski

MDPI • Basel • Beijing • Wuhan • Barcelona • Belgrade • Manchester • Tokyo • Cluj • Tianjin

Editor
Wojciech Polkowski
Łukasiewicz Research Network—Krakow Institute of Technology
Poland

Editorial Office
MDPI
St. Alban-Anlage 66
4052 Basel, Switzerland

This is a reprint of articles from the Special Issue published online in the open access journal *Crystals* (ISSN 2073-4352) (available at: https://www.mdpi.com/journal/crystals/special_issues/Crystal_Plasticity).

For citation purposes, cite each article independently as indicated on the article page online and as indicated below:

LastName, A.A.; LastName, B.B.; LastName, C.C. Article Title. *Journal Name* **Year**, *Volume Number*, Page Range.

ISBN 978-3-0365-0838-2 (Hbk)
ISBN 978-3-0365-0839-9 (PDF)

© 2021 by the authors. Articles in this book are Open Access and distributed under the Creative Commons Attribution (CC BY) license, which allows users to download, copy and build upon published articles, as long as the author and publisher are properly credited, which ensures maximum dissemination and a wider impact of our publications.

The book as a whole is distributed by MDPI under the terms and conditions of the Creative Commons license CC BY-NC-ND.

Contents

About the Editor ix

Preface to "Crystal Plasticity" xi

Wojciech Polkowski
Crystal Plasticity
Reprinted from: *Crystals* 2021, *11*, 44, doi:10.3390/cryst11010044 1

Aleksander Zubelewicz
Mechanisms-Based Transitional Viscoplasticity
Reprinted from: *Crystals* 2020, *10*, 212, doi:10.3390/cryst10030212 5

Umer Masood Chaudry, Kotiba Hamad and Jung-Gu Kim
Ca-induced Plasticity in Magnesium Alloy: EBSD Measurements and VPSC Calculations
Reprinted from: *Crystals* 2020, *10*, 67, doi:10.3390/cryst10020067 25

David Bürger, Antonin Dlouhý, Kyosuke Yoshimi and Gunther Eggeler
How Nanoscale Dislocation Reactions Govern Low- Temperature and High-Stress Creep of
Ni-Base Single Crystal Superalloys
Reprinted from: *Crystals* 2020, *10*, 134, doi:10.3390/cryst10020134 33

Faisal Qayyum, Sergey Guk, Matthias Schmidtchen, Rudolf Kawalla and Ulrich Prahl
Modeling the Local Deformation and Transformation Behavior of Cast X8CrMnNi16-6-6
TRIP Steel and 10% Mg-PSZ Composite Using a Continuum Mechanics-Based Crystal
Plasticity Model
Reprinted from: *Crystals* 2020, *10*, 221, doi:10.3390/cryst10030221 49

**Mohamed H. Mussa, Ahmed M. Abdulhadi, Imad Shakir Abbood, Azrul A. Mutalib and
Zaher Mundher Yaseen**
Late Age Dynamic Strength of High-Volume Fly Ash Concrete with Nano-Silica and
Polypropylene Fibres
Reprinted from: *Crystals* 2020, *10*, 243, doi:10.3390/cryst10040243 75

Uttam Bhandari, Congyan Zhang and Shizhong Yang
Mechanical and Thermal Properties of Low-Density $Al_{20+x}Cr_{20-x}Mo_{20-y}Ti_{20}V_{20+y}$ Alloys
Reprinted from: *Crystals* 2020, *10*, 278, doi:10.3390/cryst10040278 99

**Rostislav Kawulok, Ivo Schindler, Jaroslav Sojka, Petr Kawulok, Petr Opěla, Lukáš Pindor,
Eduard Grycz, Stanislav Rusz and Vojtěch Ševčák**
Effect of Strain on Transformation Diagrams of 100Cr6 Steel
Reprinted from: *Crystals* 2020, *10*, 326, doi:10.3390/cryst10040326 109

**Werner Skrotzki, Aurimas Pukenas, Eva Odor, Bertalan Joni, Tamas Ungar, Bernhard Völker,
Anton Hohenwarter, Reinhard Pippan and Easo P. George**
Microstructure, Texture, and Strength Development during High-Pressure Torsion of
CrMnFeCoNi High-Entropy Alloy
Reprinted from: *Crystals* 2020, *10*, 336, doi:10.3390/cryst10040336 125

Dmitriy Panov, Olga Dedyulina, Dmitriy Shaysultanov, Nikita Stepanov, Sergey Zherebtsov and Gennady Salishchev
Mechanisms of Grain Structure Evolution in a Quenched Medium Carbon Steel during Warm Deformation
Reprinted from: *Crystals* **2020**, *10*, 554, doi:10.3390/cryst10070554 139

Ning Zhang, Li Meng, Wenkang Zhang and Weimin Mao
Study on Texture and Grain Orientation Evolution in Cold-Rolled BCC Steel by Reaction Stress Model
Reprinted from: *Crystals* **2020**, *10*, 680, doi:10.3390/cryst10080680 155

Myoungjae Lee, In-Su Kim, Young Hoon Moon, Hyun Sik Yoon, Chan Hee Park and Taekyung Lee
Kinetics of Capability Aging in Ti-13Nb-13Zr Alloy
Reprinted from: *Crystals* **2020**, *10*, 693, doi:10.3390/cryst10080693 169

Stanislav Minárik and Maroš Martinkovič
On the Applicability of Stereological Methods for the Modelling of a Local Plastic Deformation in Grained Structure: Mathematical Principles
Reprinted from: *Crystals* **2020**, *10*, 697, doi:10.3390/cryst10080697 177

Jesús Galán-López and Javier Hidalgo
Use of the Correlation between Grain Size and Crystallographic Orientation in Crystal Plasticity Simulations: Application to AISI 420 Stainless Steel
Reprinted from: *Crystals* **2020**, *10*, 819, doi:10.3390/cryst10090819 205

Alexey Shveykin, Peter Trusov and Elvira Sharifullina
Statistical Crystal Plasticity Model Advanced for Grain Boundary Sliding Description
Reprinted from: *Crystals* **2020**, *10*, 822, doi:10.3390/cryst10090822 225

Hiroaki Kosuge, Tomoya Kawabata, Taira Okita and Hidenori Nako
Accurate Estimation of Brittle Fracture Toughness Deterioration in Steel Structures Subjected to Large Complicated Prestrains
Reprinted from: *Crystals* **2020**, *10*, 867, doi:10.3390/cryst10100867 243

Chun-Yu Ou, Rohit Voothaluru and C. Richard Liu
Fatigue Crack Initiation of Metals Fabricated by Additive Manufacturing—A Crystal Plasticity Energy-Based Approach to IN718 Life Prediction
Reprinted from: *Crystals* **2020**, *10*, 905, doi:10.3390/cryst10100905 261

Boris Straumal, Natalia Martynenko, Diana Temralieva, Vladimir Serebryany, Natalia Tabachkova, Igor Shchetinin, Natalia Anisimova, Mikhail Kiselevskiy, Alexandra Kolyanova, Georgy Raab, Regine Willumeit-Römer, Sergey Dobatkin and Yuri Estrin
The Effect of Equal-Channel Angular Pressing on Microstructure, Mechanical Properties, and Biodegradation Behavior of Magnesium Alloyed with Silver and Gadolinium
Reprinted from: *Crystals* **2020**, *10*, 918, doi:10.3390/cryst10100918 271

Xufei Fang, Lukas Porz, Kuan Ding and Atsutomo Nakamura
Bridging the Gap between Bulk Compression and Indentation Test on Room-Temperature Plasticity in Oxides: Case Study on $SrTiO_3$
Reprinted from: *Crystals* **2020**, *10*, 933, doi:10.3390/cryst10100933 287

Faisal Qayyum, Aqeel Afzal Chaudhry, Sergey Guk, Matthias Schmidtchen, Rudolf Kawalla and Ulrich Prahl
Effectof 3D Representative Volume Element (RVE) Thickness on Stress and Strain Partitioning in Crystal Plasticity Simulations of Multi-Phase Materials
Reprinted from: *Crystals* **2020**, *10*, 944, doi:10.3390/cryst10100944 303

Jae Hyung Kim, Taekyung Lee and Chong Soo Lee
Microstructural Influence on Stretch Flangeability of Ferrite–Martensite Dual-Phase Steels
Reprinted from: *Crystals* **2020**, *10*, 1022, doi:10.3390/cryst10111022 321

Bjørn Holmedal
Regularized Yield Surfaces for Crystal Plasticity of Metals
Reprinted from: *Crystals* **2020**, *10*, 1076, doi:10.3390/cryst10121076 331

Oleg Matvienko, Olga Daneyko and Tatiana Kovalevskaya
Mathematical Modeling of Plastic Deformation of a Tube from Dispersion-Hardened Aluminum Alloy in an Inhomogeneous Temperature Field
Reprinted from: *Crystals* **2020**, *10*, 1103, doi:10.3390/cryst10121103 349

Mahesh R.G. Prasad, Siwen Gao, Napat Vajragupta, and Alexander Hartmaier
Influence of Trapped Gas on Pore Healing under Hot Isostatic Pressing in Nickel-Base Superalloys
Reprinted from: *Crystals* **2020**, *10*, 1147, doi:10.3390/cryst10121147 367

Sari Yanagida, Takashi Nagoshi, Akiyoshi Araki, Tso-Fu Mark Chang, Chun-Yi Chen, Equo Kobayashi, Akira Umise, Hideki Hosoda, Tatsuo Sato and Masato Sone
Heterogeneous Deformation Behavior of Cu-Ni-Si Alloy by Micro-Size Compression Testing
Reprinted from: *Crystals* **2020**, *10*, 1162, doi:10.3390/cryst10121162 383

Khanh Nguyen, Meijuan Zhang, Víctor Jesús. Amores[1], Miguel A. Sanz and Francisco J. Montáns
Computational Modeling of Dislocation Slip Mechanisms in Crystal Plasticity: A Short Review
Reprinted from: *Crystals* **2021**, *11*, 42, doi:10.3390/cryst11010042 391

About the Editor

Wojciech Polkowski (born 1985) graduated from the Faculty of Advanced Technologies and Chemistry of the Military University of Technology in Warsaw. In the years 2013 to 2016 he worked at the Faculty as an academic teacher (Assistant Professor), where he was involved in organizing and conducting classes, as well as in planning and (co)supervising of students' diploma works. After receiving PhD diploma he has left the Military University of Technology in Warsaw, and since October 2016 he is employed on a permanent position as a Postdoctoral Researcher in the Centre for High Temperature Studies of the Foundry Research Institute in Krakow. After a restructuration into Łukasiewicz Research Network (April 2019) he is playing the role of the Head of Centre for Materials Research and the Leader of High Temperature Research Area at Łukasiewicz-Krakow Institute of Technology. He is an author or co-author of 108 scientific disseminated works including: 45 publications in journals from Journal Citation Report list; 44 conference communicates and publications in proceedings; and 3 authored chapter in a book. For his scientific work and achievements, he has received three awards from the Rector of Military University of Technology (including the Award for the Best Thesis of the Year, 2009). In the years 2013 to 2014 years he was the laureate of the Scholarship for the Best PhD students founded by the Marshal of Mazovia. In 2017, he was awarded by the three year Scholarship of Ministry of Science and Higher Education for Young Outstanding Scientists.

Preface to "Crystal Plasticity"

The term "Crystal Plasticity" builds a bridge between pure crystallography, material science, and industrial processing of commonly applied material pieces (sheets, plates, wires, etc.).

As material scientists and technologists working in the field of (poly)crystals plasticity, we all tend to provide valuable quantitative and qualitative indicators that describe the process→(crystalline) structure→properties relationship. Generally, our efforts are focused on recognizing possible ways to improve materials' behavior under predicted operational conditions and applied mechanical or thermal external loadings. However, we all know that this goal can be achieved only by having well-established knowledge on crystal structure evolution regarding mechanical and plastic deformation processing.

Nowadays, the research on crystal plasticity-related phenomena is of high practical importance in the view of the following:

- The on-going progress in conventional fabrication techniques (as a forging or a cold rolling processes);

- The design of new processing methods (e.g., various complex severe plastic deformation techniques);

- The development of novel materials (e.g., high-entropy alloys, inter-metallics, bulk metallic glasses, ultra-fine-grained alloys, nano-steel, etc.).

In this regard, this book is especially dedicated to present theoretical and experimental research works providing new insights and practical findings in the field of crystal plasticity-related topics.

Wojciech Polkowski
Editor

Editorial

Crystal Plasticity

Wojciech Polkowski

Łukasiewicz Research Network–Krakow Institute of Technology, Zakopiańska 73 Str., 30-418 Krakow, Poland; wojciech.polkowski@kit.lukasiewicz.gov.pl; Tel.: +48-12-26-18-115

Received: 5 January 2021; Accepted: 5 January 2021; Published: 6 January 2021

The Special Issue on "Crystal Plasticity" is a collection of 25 original articles (including one review paper) dedicated to theoretical and experimental research works providing new insights and practical findings in the field of crystal plasticity-related topics.

Crystal plasticity is an inherently multi-scale process starting at the atomic scale (dislocation cores) towards substructural dislocation arrangements in a single grain and up to the macroscopic mechanical response of the material. Its multi-dimensional nature and a high practical importance build a space for scientists and engineers working within various methodological domains.

The main intention of this Special Issue was to present a wide spectrum of the Crystals Plasticity area, i.e., to combine a mathematical modeling with experimental investigations on the processing/structure/property relationship, to show its practical importance in examining both "traditional" and novel materials (e.g., steels and high-entropy alloys) and processing techniques (e.g., hot-rolling and additive manufacturing). After collecting all the papers, I am extremely happy that a great contribution of researchers all over the world (from 18 different countries!) allowed the attainment of this goal. All the papers can be virtually divided into three groups, namely (i) "modelling and simulation"; (ii) "methodological aspects" and (iii) "experiments on process/structure/properties relationship".

In terms of more theoretical works in which crystal plasticity model attempts were evaluated to reproduce the complex deformation processes of polycrystalline metals, Zubelewicz [1] has provided an extensive review on mechanisms-based transitional viscoplasticity and Nguyen et al. [2] have made a short review of computational modeling of dislocation slip mechanisms. Masood Chaudry et al. [3] have compared the results of Electron Backscatter Diffraction measurements and viscoplastic self-consistent calculations upon analysis of Ca-induced plasticity in Mg alloys, while a similar approach was also applied by Galán-López and Hidalgo [4] to stainless steel. Other examples of such works are modeling the local deformation and transformation behavior of cast metal matrix composite by using a continuum mechanics-based crystal plasticity model (Qayyum et al. [5]) or mathematical modeling of plastic deformation of a tube from dispersion-hardened aluminum alloy in an inhomogeneous temperature field (Matvienko et al [6]). Ou et al. [7] have incorporated a crystal plasticity finite element model for predicting a crack initiation in additively manufactured IN718 alloy, while Kosuge et al. [8] have utilized strain gradient plasticity theory based on the finite element method to estimate a behavior of steel structures subjected to large complicated pre-strains. Further examples on crystal plasticity modeling involve a reaction stress model (Zhang et al. [9]), statistical crystal plasticity constitutive models of polycrystalline metals and alloys (Shveykin et al. [10]) or 3D representative volume element approach (Qayyum et al. [11]). Furthermore, a theoretical work related to yield surfaces and slip systems has been provided by Holmedal [12].

The second group of papers is dedicated towards the development of new metodological aspects related to examining crystal plasticity issues (also at the microscale). In this field, Yanagida et al. [13] have investigated the characteristic deformation behavior of a precipitation strengthening-type Cu-Ni-Si alloy by microcompression of Focused Ion Beam produced specimens, while Fang et al. [14] have made an attempt to bridge the gap between bulk compression and indentation tests on room-temperature plasticity in oxides. Minárik and Martinkovič [15] have designed a computationally low consuming procedure for quantification of local deformation in a grained structure based on the distortion of the image of this structure in a cross-sectional view.

Finally, the last group of papers deals with an experimental assessment of structural and mechanical properties evolution upon plastic deformation of various metallic materials. The following subjects are described:

- an influence of trapped gas on pore healing under hot isostatic pressing in nickel-base superalloys (Prasad et al. [16]);
- a microstructural influence on stretch flangeability of ferrite–martensite dual-phase steels (Kim et al. [17]);
- an effect of equal-channel angular pressing on microstructure, mechanical properties, and biodegradation behavior of Mg alloyed with Ag and Gd (Straumal et al. [18]);
- kinetics of capability aging in Ti-13Nb-13Zr alloy (Lee et al. [19]);
- mechanisms of grain structure evolution in a quenched medium carbon steel during warm deformation (Panov et al. [20]);
- a microstructure, texture, and strength development during high-pressure torsion of CrMnFeCoNi high-entropy alloy (Skrotzki et al. [21]);
- an effect of strain on transformation diagrams of 100Cr6 steel (Kawulok et al. [22]);
- mechanical and thermal properties of low-density Al20+xCr20-xMo20-yTi20V20+y alloys (Bhandari et al. [23]);
- the late age dynamic strength of high-volume fly ash concrete with nano-silica and polypropylene fibres (Mussa et al. [24]);
- dislocation reactions governing low-temperature and high-stress creep of ni-base single crystal superalloys (Burger et al [25]).

I hope that this collection of papers will meet expectations of readers looking for new advances in the Crystal Plasticity field, as well as it bringing inspirations for further research work.

Funding: The author is thankful to the Polish National Science Center for the financial support of his past and present research projects related to crystal plasticity topics (e.g., under Grants no. UMO-2013/09/N/ST8/04366 and UMO-2016/23/D/ST8/01269), that to some extent have brought him to being a Guest Editor of this Special Issue.

Acknowledgments: A contribution of all authors is gratefully acknowledged. The author would like to express his thanks to the *Crystals* Editorial Office, and on top of that to Oscar Guo (a Technical Coordinator of the Issue) for the excellent communication, support, friendly and fully professional attitude.

Conflicts of Interest: The author declares no conflict of interest.

References

1. Zubelewicz, A. Mechanisms-Based Transitional Viscoplasticity. *Crystals* **2020**, *10*, 212. [CrossRef]
2. Nguyen, K.; Zhang, M.; Amores, V.J.; Sanz, M.A.; Montáns, F.J. Computational Modeling of Dislocation Slip Mechanisms in Crystal Plasticity: A Short Review. *Crystals* **2021**, *11*, 42. [CrossRef]
3. Masood Chaudry, U.; Hamad, K.; Kim, J.-G. Ca-induced Plasticity in Magnesium Alloy: EBSD Measurements and VPSC Calculations. *Crystals* **2020**, *10*, 67. [CrossRef]
4. Galán-López, J.; Hidalgo, J. Use of the Correlation between Grain Size and Crystallographic Orientation in Crystal Plasticity Simulations: Application to AISI 420 Stainless Steel. *Crystals* **2020**, *10*, 819.
5. Qayyum, F.; Guk, S.; Schmidtchen, M.; Kawalla, R.; Prahl, U. Modeling the Local Deformation and Transformation Behavior of Cast X8CrMnNi16-6-6 TRIP Steel and 10% Mg-PSZ Composite Using a Continuum Mechanics-Based Crystal Plasticity Model. *Crystals* **2020**, *10*, 221.

6. Matvienko, O.; Daneyko, O.; Kovalevskaya, T. Mathematical Modeling of Plastic Deformation of a Tube from Dispersion-Hardened Aluminum Alloy in an Inhomogeneous Temperature Field. *Crystals* **2020**, *10*, 1103.
7. Ou, C.-Y.; Voothaluru, R.; Liu, C.R. Fatigue Crack Initiation of Metals Fabricated by Additive Manufacturing—A Crystal Plasticity Energy-Based Approach to IN718 Life Prediction. *Crystals* **2020**, *10*, 905. [CrossRef]
8. Kosuge, H.; Kawabata, T.; Okita, T.; Nako, H. Accurate Estimation of Brittle Fracture Toughness Deterioration in Steel Structures Subjected to Large Complicated Prestrains. *Crystals* **2020**, *10*, 867. [CrossRef]
9. Zhang, N.; Meng, L.; Zhang, W.; Mao, W. Study on Texture and Grain Orientation Evolution in Cold-Rolled BCC Steel by Reaction Stress Model. *Crystals* **2020**, *10*, 680.
10. Shveykin, A.; Trusov, P.; Sharifullina, E. Statistical Crystal Plasticity Model Advanced for Grain Boundary Sliding Description. *Crystals* **2020**, *10*, 822.
11. Qayyum, F.; Chaudhry, A.A.; Guk, S.; Schmidtchen, M.; Kawalla, R.; Prahl, U. Effect of 3D Representative Volume Element (RVE) Thickness on Stress and Strain Partitioning in Crystal Plasticity Simulations of Multi-Phase Materials. *Crystals* **2020**, *10*, 944. [CrossRef]
12. Holmedal, B. Regularized Yield Surfaces for Crystal Plasticity of Metals. *Crystals* **2020**, *10*, 1076. [CrossRef]
13. Yanagida, S.; Nagoshi, T.; Araki, A.; Chang, T.-F.M.; Chen, C.-Y.; Kobayashi, E.; Umise, A.; Hosoda, H.; Sato, T.; Sone, M. Heterogeneous Deformation Behavior of Cu-Ni-Si Alloy by Micro-Size Compression Testing. *Crystals* **2020**, *10*, 1162. [CrossRef]
14. Fang, X.; Porz, L.; Ding, K.; Nakamura, A. Bridging the Gap between Bulk Compression and Indentation Test on Room-Temperature Plasticity in Oxides: Case Study on $SrTiO_3$. *Crystals* **2020**, *10*, 933. [CrossRef]
15. Minárik, S.; Martinkovič, M. On the Applicability of Stereological Methods for the Modelling of a Local Plastic Deformation in Grained Structure: Mathematical Principles. *Crystals* **2020**, *10*, 697.
16. Prasad, M.R.G.; Gao, S.; Vajragupta, N.; Hartmaier, A. Influence of Trapped Gas on Pore Healing under Hot Isostatic Pressing in Nickel-Base Superalloys. *Crystals* **2020**, *10*, 1147. [CrossRef]
17. Kim, J.H.; Lee, T.; Lee, C.S. Microstructural Influence on Stretch Flangeability of Ferrite–Martensite Dual-Phase Steels. *Crystals* **2020**, *10*, 1022. [CrossRef]
18. Straumal, B.; Martynenko, N.; Temralieva, D.; Serebryany, V.; Tabachkova, N.; Shchetinin, I.; Anisimova, N.; Kiselevskiy, M.; Kolyanova, A.; Raab, G.; et al. The Effect of Equal-Channel Angular Pressing on Microstructure, Mechanical Properties, and Biodegradation Behavior of Magnesium Alloyed with Silver and Gadolinium. *Crystals* **2020**, *10*, 918. [CrossRef]
19. Lee, M.; Kim, I.-S.; Moon, Y.H.; Yoon, H.S.; Park, C.H.; Lee, T. Kinetics of Capability Aging in Ti-13Nb-13Zr Alloy. *Crystals* **2020**, *10*, 693. [CrossRef]
20. Panov, D.; Dedyulina, O.; Shaysultanov, D.; Stepanov, N.; Zherebtsov, S.; Salishchev, G. Mechanisms of Grain Structure Evolution in a Quenched Medium Carbon Steel during Warm Deformation. *Crystals* **2020**, *10*, 554. [CrossRef]
21. Skrotzki, W.; Pukenas, A.; Odor, E.; Joni, B.; Ungar, T.; Völker, B.; Hohenwarter, A.; Pippan, R.; George, E.P. Microstructure, Texture, and Strength Development during High-Pressure Torsion of CrMnFeCoNi High-Entropy Alloy. *Crystals* **2020**, *10*, 336. [CrossRef]
22. Kawulok, R.; Schindler, I.; Sojka, J.; Kawulok, P.; Opěla, P.; Pindor, L.; Grycz, E.; Rusz, S.; Ševčák, V. Effect of Strain on Transformation Diagrams of 100Cr6 Steel. *Crystals* **2020**, *10*, 326. [CrossRef]
23. Bhandari, U.; Zhang, C.; Yang, S. Mechanical and Thermal Properties of Low-Density Al20+xCr20-xMo20-yTi20V20+y Alloys. *Crystals* **2020**, *10*, 278. [CrossRef]
24. Mussa, M.H.; Abdulhadi, A.M.; Abbood, I.S.; Mutalib, A.A.; Yaseen, Z.M. Late Age Dynamic Strength of High-Volume Fly Ash Concrete with Nano-Silica and Polypropylene Fibres. *Crystals* **2020**, *10*, 243. [CrossRef]
25. Bürger, D.; Dlouhý, A.; Yoshimi, K.; Eggeler, G. How Nanoscale Dislocation Reactions Govern Low-Temperature and High-Stress Creep of Ni-Base Single Crystal Superalloys. *Crystals* **2020**, *10*, 134.

© 2021 by the author. Licensee MDPI, Basel, Switzerland. This article is an open access article distributed under the terms and conditions of the Creative Commons Attribution (CC BY) license (http://creativecommons.org/licenses/by/4.0/).

Review

Mechanisms-Based Transitional Viscoplasticity

Aleksander Zubelewicz

Civil Engineering Department, University of New Mexico, Albuquerque, New Mexico, NM 87131, USA; alek@unm.edu

Received: 10 February 2020; Accepted: 16 March 2020; Published: 18 March 2020

Abstract: When metal is subjected to extreme strain rates, the conversation of energy to plastic power, the subsequent heat production and the growth of damages may lag behind the rate of loading. The imbalance alters deformation pathways and activates micro-dynamic excitations. The excitations immobilize dislocation, are responsible for the stress upturn and magnify plasticity-induced heating. The main conclusion of this study is that dynamic strengthening, plasticity-induced heating, grain size strengthening and the processes of microstructural relaxation are inseparable phenomena. Here, the phenomena are discussed in semi-independent sections, and then, are assembled into a unified constitutive model. The model is first tested under simple loading conditions and, later, is validated in a numerical analysis of the plate impact problem, where a copper flyer strikes a copper target with a velocity of 308 m/s. It should be stated that the simulations are performed with the use of the deformable discrete element method, which is designed for monitoring translations and rotations of deformable particles.

Keywords: high strain rate; Hall–Petch relation; Taylor–Quinney coefficient; transitional viscoplasticity; ductile damage; OFHC copper; deformable discrete element method

1. Introduction

Metals subjected to extreme dynamics experience rapid microstructural evolution. Dislocations are generated, become entangled and form fine structures [1]. As a result, microscopic plastic flow is frequently interrupted and rerouted. The complex motion of dislocations at quasi-static conditions becomes increasingly sophisticated at high strain rates. When energy is delivered to metals with rates faster than the rate at which the energy is converted to plastic work and damages, then there is uncompensated energy, which is partly stored in the newly created dislocation structures and the rest of it activates micro-dynamic excitations. Among direct consequences of the excitations are dynamic strengthening [2–4] and the magnification of plasticity-induced heating [5]. In the Taylor–Quinney interpretation, a significant portion of plastic work is converted to heat [6–9], while the remaining work is stored in dislocation structures. Dislocations nucleate, move through the material, are piling up and become entangled. Thus, the majority of plastic work aids configurational entropy of the material, while plasticity-induced heating quantifies the efficiency of the reconfigurations. The grain size dependence is an integral part of the picture. Hall and Petch along with Armstrong [10–13] explored the grain size sensitivities and proposed the generally accepted Hall–Petch relation. The relation successfully survived for nearly sixty years. We are learning that this relation does not work in nanocrystalline metals [14,15]. In fact, when grains are smaller than 15–20 nanometers, the Hall–Petch effect is reversed [16,17]. The relation is not quite applicable to metals composed of very large grains. Thus, there is still much to learn about the phenomena.

The proposed concept is quite different from the celebrated strength models introduced by Zerilli and Armstrong [18], Johnson and Cook [19], Follansbee and Kocks [20], Preston, Tonks and Wallace [21], among others [22]. The models are based on a salient assumption that the yield stress carries all relevant material information, such as sensitivities to strain rate, temperature, size effect, etc. In a typical scenario, the models adapt the strain rate-insensitive solving scheme, where radial return seems to be the favorite procedure. In consequence, plastic strain is merely a byproduct of the requirement that stress must adhere to the yield surface. An interesting technique suitable for generating random dislocation arrangements is proposed by Gurrutxanga-Lerma [23]. The main concern of the study is focused on how a given population of dislocation structures influences the dominant features of the stress field. The stochastic analysis provides justification for Taylor's equation and the Hall–Petch relation. The study, however, is not concerned with how the dislocation structures are formed in the first place. Another approach to crystal plasticity is proposed by Langer [24]. His thermodynamics dislocation theory (TDT) is based on the observation that dislocation experience complex chaotic motions and the motion is responsible for the field fluctuations. One should expect the frequency of the fluctuations to be many orders of magnitude lower in comparison with the frequency of thermal excitations. According to Langer, the non-deterministic nature of dislocation dynamics makes the behaviors suitable for statistical and thermodynamics treatment. Brown [25] offered an interesting idea that plastic flow is responsible for the formation of dislocation structures in the state of self-organized criticality. In this interpretation, the dislocation structures evolve until reaching the state of maximum structural adaptiveness such that every site of the structure has a statistically equal chance to act as a nucleation site of further deformation. At the initial stages of plastic deformation, flow is initiated at points of weakness. However, as the process proceeds, the initiation sites become inactive and are replaced by newly created ones in the evolving system. Consequently, there are no fixed sites in time and space. The state of self-organized criticality is dynamic, the fluctuations are inherently erratic and carry a measurable amount of energy, where the energy is fed back into the transient structures. The fluctuations are micro-mechanical and, while allowing to explore the neighboring states, drive the system back into the state of self-organized criticality. The scale-free patterning of self-organized dislocations is further tested in ice [26]. Again, the results confirm the existence of scale-free dislocation assemblies in single crystals. The power-law distribution of dislocation avalanches is insensitive to temperature. Unlike single crystals, polycrystals display a strong deviation from the power-law distribution. Zaiser and Nikitas [27] offered yet another clarification. In their view, the dislocation avalanches exhibit lamellar geometry and, therefore, a very small volume of the material is actively involved in the deformation process. Avalanches produce small macroscopic plastic strain, thus, suggesting an extrinsic limit to the avalanche size. The explanation supports the idea of self-organized criticality.

Here, the proposed concept is aimed to address several questions raised by Brown, Langer, Zaiser and Nikitas. In this study, the mechanism of plastic flow is introduced first, followed by a discussion on the thermal activation of plastic processes. The following section is focused on flow complexities. Slip bursts are shown to activate microscale dynamic excitations. At high strain rates, the excitations are responsible for the dynamic upturn of stress and magnify plasticity-induced heating. Lastly, I construct an energy-based Hall–Petch relation. In contrast to the accepted stress-based interpretation, the proposed Hall–Petch relation is directly linked to the flow constraints. I argue that the stress-based and kinematics-based interpretations represent the upper and lower bound estimates of the grain size effect. For clarity, I made an effort to write each section in a semi-autonomous manner.

All the mechanisms are incorporated into a unified constitutive model. The model reproduces metal responses from quasi-static loading to extreme dynamics and at temperatures from cryogenic environments to melting points. The average size of crystals is shown to affect the responses in a manner that is consistent with experimental observations. The model is calibrated for OFHC copper and has been tested against available experimental data. Lastly, the model is implemented to a code of deformable discrete element method and is validated in a plate impact analysis.

2. Mechanism of Plastic Flow

In metals, the macroscopic plastic flow $\dot{H}^p$ results from slip events along many θ planes defined by two orthogonal unit vectors, n^θ and s^θ. The planes are somewhat misoriented with respect to the plane of the dominant slip (Figure 1). Crystal-to-crystal misorientations, grain boundaries and other obstacles broaden the distribution of active slip systems. When slip is arrested along one plane, other less favorable pathways are created. At advanced deformation, the rerouting is linked to dislocation interactions, includes grain rotations, causes lattice distortions, and the processes are generally dynamic. A material point in a continuum description contains a large number of grains and, given grain-to-grain misorientations, there is a much larger number of slip planes. While the distribution remains discrete, the continuum approximation seems to be a justifiable assumption. For this reason, one can assume that the θ-planes experience continuous reorientations, such that

$$\left\{\begin{array}{c}\dot{n}^\theta\\ \dot{s}^\theta\end{array}\right\}=\left[\begin{array}{cc}0 & \dot{\theta}\\ -\dot{\theta} & 0\end{array}\right]\left\{\begin{array}{c}n^\theta\\ s^\theta\end{array}\right\}. \tag{1}$$

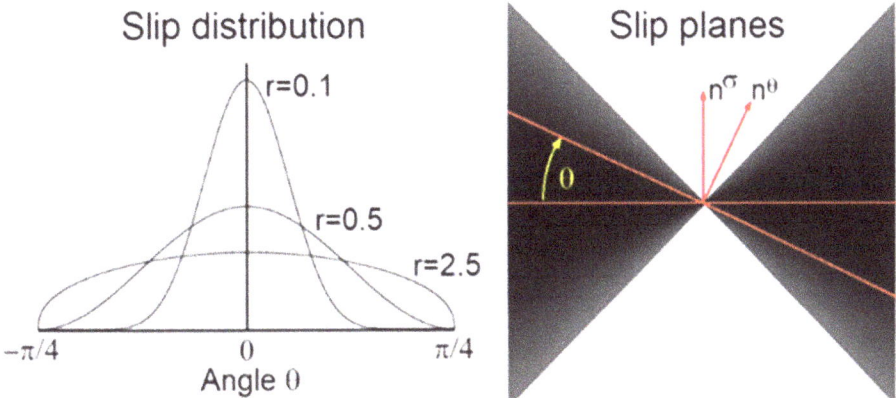

Figure 1. Slip-events distribution around the dominant plane $\theta = 0$.

The dominant slip is defined at $\theta = 0$. At each angle, slip events occur with frequency $f_\theta(\theta)$, where the angle is taken between $\pm\pi/4$. The distribution $f_\theta(\theta) = \frac{2\,\Gamma(1+1/2r)}{\sqrt{\pi}\,\Gamma(1/2+1/2r)}\cos^{1/r} 2\theta$ is constructed in such a manner that $\int_{-\pi/4}^{\pi/4} f_\theta(\theta)d\theta = 1$ and Γ is a gamma function. As shown in Figure 1, the exponent r controls the shape of the distribution. When r is approaching zero, slip occurs along the dominant direction. On the other hand, when r is a large number, slip is distributed indiscriminately in all orientations between $\pm\pi/4$.

The rate of macroscopic plastic strain $\dot{H}^p$ results not only from local slip events, but also reflects the process of slip reorganizations (Equation (1)), hence

$$\dot{H}^p = \frac{1}{2}\int_{-\pi/4}^{\pi/4}\frac{\partial\left[\left(n^\theta s^\theta + s^\theta n^\theta\right)e^\theta\right]}{\partial t}d\theta + \dot{H}^d. \tag{2}$$

The volumetric strain rate $\dot{H}^d$ quantifies slip incompatibilities in tension. Plastic strain e^θ is assigned to each θ-slip plane. The θ-strains e^θ are weighted $\dot{e}^\theta = f_\theta(\theta)\,\dot{e}^p$, where $\dot{e}^p$ is the rate of macroscopic strain. Since the reorientations are local events, therefore $\dot{\theta} = A_T(T)\left(\partial\theta/\partial e^\theta\right)\dot{e}^\theta$. The coefficient A_T makes the reorganization a thermally activated process. The concept of thermal activation is introduced in the next section. Slip systems aligned closely with the dominant direction ($\theta = 0$) do not rotate [28]. The slip reorganizations are maximized at $\theta = \pm\pi/4$. The relation $e^\theta\left(\partial\theta/\partial e^\theta\right) = 2\sin 2\theta$ properly reflects the trend. The dominant slip plane is defined in terms of two orthogonal unit vectors n^σ and s^σ and the flow tensor is a dyadic product of the two $N = (n^\sigma s^\sigma + s^\sigma n^\sigma)$.

Since plastic flow is stimulated by stress σ, therefore, the tensor must be expressed in terms of the current stress $N = N^\sigma(\sigma)$. The representation $N^\sigma = \alpha_0\, \mathbf{1} + \alpha_1\, S + \alpha_2\, S^2$ of the dyadic product is constructed on the basis of stress deviator $S = \sigma - \mathbf{1}\, tr\sigma/3$ and the three variables α_0, α_1 and α_2 are functions of stress invariants. The representation must reproduce invariants of the original flow tensor $(n^\sigma s^\sigma + s^\sigma n^\sigma)$, thus $trN^\sigma = 0$, $tr(N^\sigma)^2 = 2$ and $N^\sigma = (N^\sigma)^3$. There are three solutions, where the representations are constructed on the planes of principal stresses $\{\sigma_1, \sigma_2\}$, $\{\sigma_2, \sigma_3\}$ and $\{\sigma_1, \sigma_3\}$ and where $\sigma_1 > \sigma_2 > \sigma_3$. I select the solution which reproduces the maximum shear mechanism [29], hence

$$N^\sigma = \frac{2\cos\frac{\varphi}{3}}{\sqrt{3}\cos\varphi}\left(1 - \frac{3}{2 J_2} S^2\right) + \frac{\cos\frac{2\varphi}{3}}{\sqrt{J_2}\cos\varphi} S. \tag{3}$$

The angle $\varphi = \sin^{-1}\left(J_3\sqrt{27/4}/J_2^3\right)$ and two invariants J_2 and J_3 of stress deviator S complete the relation. It turns out that $N^\sigma : \sigma/2$ is the maximum shear stress. It is worth mentioning that several tensor representations have been developed and some of them are reported in [30]. In the current (Tresca) description, the θ-vectors n^θ and s^θ are co-rotational with the dominant slip plane $n^\theta = n^\sigma\cos\theta + s^\sigma\sin\theta$ and $s^\theta = s^\sigma\cos\theta - n^\sigma\sin\theta$. As a result, the plastic flow (Equation (2)) becomes

$$\dot{H}^p = \frac{1}{2} N^\sigma M \dot{e}^p + \dot{H}^d, \tag{4}$$

where the average Schmid factor $M = M_s - M_r$ specifies the current slip arrangement M_s and captures the contribution of the flow reorganizations M_r. The term $M_s = \frac{\Gamma^2\left(1+\frac{1}{2r}\right)}{\Gamma\left(\frac{1}{2}+\frac{1}{2r}\right)\Gamma\left[\frac{1}{2}\left(3+\frac{1}{r}\right)\right]}$ is closely approximated by $M_s \cong \frac{1+r}{1+\pi\, r/2}$, where the error of approximation does not exceed one percent. The second term $M_r = \frac{4\, A_T(T)}{\pi(1+2r)}$ is takes as derived. Note that the Schmid factor is a function of the exponent r. Advances of plastic deformation broaden the spectrum of the active planes and, therefore, $r = r(e^{pe})$. The exponent $r = r_0\left(1 + e^{pe}/e_r^p\right)$ reflects the slip organization, where r_0 and e_r^p are constants. Here, e^{pe} is the magnitude of effective plastic strain.

As stated earlier, ductile damage results from slip incompatibilities [31,32]. In tension, the compatibility is restored by the stress-favorable nucleation and growth of voids. The damage predominantly occurs on the plane of maximum shear, as shown in [24], and is $\dot{H}^d = \frac{q}{2}(N^\sigma)^2 M \dot{e}^{pe}$. The rate of damage is controlled by the parameter q.

3. Thermal Activation

Thermal activation (TA) in fcc metals results from the dislocation interactions, where the intersections of dislocations are the controlling mechanism [33]. In bcc metals, TA is determined by Peierls' mechanism, where much stronger temperature dependence of the yield stress is observed at low temperatures. In a continuum-level description, the activation processes can be described in terms of free activation energy $F_a = E_a - S_a T$, where E_a and S_a represent activation energy and activation entropy [34]. Note that the entropic contribution $S_a T$ weakens the energy barriers F_a, where entropy S_a is nearly independent of temperature. I introduce a non-dimensional function $\xi_a = F_a/k_b T$, where k_b is Boltzmann constant and T is temperature. Thermal activation is stochastic with respect to ξ_a. When dislocation glide is the dominant mechanism, the temperature has a minor influence on the formation of dislocation structures [25]. The distribution of dislocation avalanches is shown to be independent of temperature [26]. For these reasons, temperature in ξ_a is just a scaling factor. One can visualize a local landscape of activation energy that contains energetically weakest sites. In the state of self-organized criticality, some of the sites expire and new ones are created, thus it is an evolving landscape. A macroscopic material point contains a very large number of the sources and, therefore, a continuum-type distribution would be an acceptable assumption. The Weibull distribution $f_T = k_a\, \xi_a^{k_a-1}\, \exp\left[-\xi_a^{k_a}\right]$ seems to fit the description. The material is expected to increase the number of activation sites with frequency $f_T(T)$ and the exponent k_a determines the rate of the process.

The number of the sites at temperature T includes all the sites which are theoretically available up to this temperature. Therefore, the actual thermal resistance to plastic flow A_T is integrated over all frequencies taken in the thermal domain that has not been explored, that is between T_c/T_m and T_c/T, hence $A_T = \int_{T_c/T_m}^{T_c/T} f_T(s) ds$. I introduce the transition temperature T_c. At temperatures above T_c, the dislocation mobility becomes more probable. Here, activation energy E_a is expressed in terms of $k_b T_c$ such that $E_a = g_a^{1/k_a} k_b T_c$. Activation entropy is estimated at melting point $F_a(T_m) = 0$ and we obtain $S_a = E_a/T_m$. The activation factor becomes

$$A_T(T) = 1 - \exp\left[-g_a\left(\frac{T_c}{T} - \frac{T_c}{T_m}\right)^{k_a}\right], \tag{5}$$

where g_a determines the magnitude of activation energy. At melting temperature, when $A_T(T_m) \to 0$, the dislocation activities are replaced by diffusional flow. At cryogenic temperatures, the factor A_T is slightly smaller than one and the material exhibits the strongest resistance to plastic deformation.

4. Rerouting of Plastic Flow and Consequences

At small scales, plastic flow is an erratic process made of irreversible bursts of slip and includes other forms of local instabilities. The sources are activated, deactivated, are spatially fluctuating and undergo continuous reorganizations [35]. As Brown stated [25], the dynamic reconfigurations drive the system toward the state of self-organized criticality. One can visualize the process by monitoring the trajectory of a selected material particle (Figure 2).

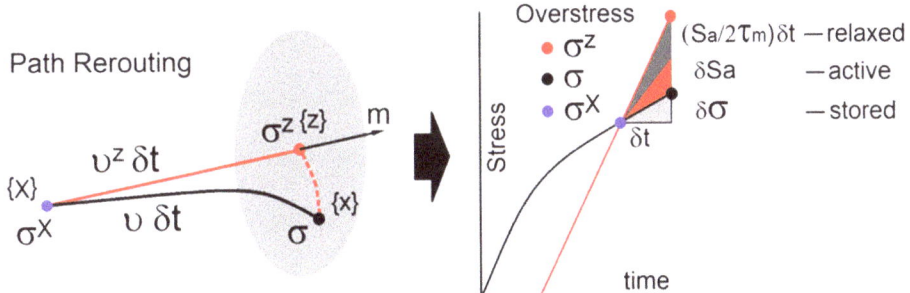

Figure 2. Path rerouting activates dynamic excitations. Note that overstress is only partly stored in newly created dislocation structures.

Initially, the particle resides in position $\{X\}$ and stress is σ^X. Shortly later (δt), the particle is forced to change its trajectory and arrives at position $\{x\}$, where stress is σ. Should the original path be available, the particle would move to position $\{z\}$ with velocity v^z and the stress would be σ^z. In all the scenarios, equations of motion $\nabla \cdot \sigma = \rho \dot{v}$ must be preserved. The particle acceleration and mass density are denoted by $\dot{v}$ and ρ. The rerouting alters tractions projected onto the surface ∂V_0 that is normal to the direction of the particle velocity m. The difference in tractions ($\sigma^z \cdot m \neq \sigma \cdot m$) between positions $\{z\}$ and $\{x\}$ triggers stress perturbations $\delta \sigma = \frac{l_r}{V_0} \int_{\partial V_0} (\sigma - \sigma^z) \cdot (m\,m) dS$. The perturbations are stretched over distance l_r. Next, the volume V_0 is reduced to a material point and stress perturbations become $\delta \sigma = l_r \psi$, where $\psi = \rho(\delta v\, m + m\, \delta v)/2$ is the momentum tensor. The path rearrangements affect particle acceleration $\delta \dot{v} = (\dot{v} - \dot{v}^z)$, while the overstress ($\sigma^z - \sigma$) is partly stored in dislocation structures (Figure 1). The uncompensated stress δS^a in Figure 1 activates dynamic excitations. Given sufficient time, the overstress is fully relaxed.

5. Dynamic Overstress

The stress perturbations $\delta\sigma$ slow down plastic flow $\dot{H}^t = \dot{H}^e + (\dot{H}^p - \delta\sigma/\kappa)$, where $\dot{H}^t, \dot{H}^e$ and $\dot{H}^p$ are the rates of total, elastic and drag-free plastic strains. The term $\delta\sigma/\kappa$ represents the drag on dislocations and the drag coefficient is κ. Stress is partly stored in the lattice $\dot{\sigma} = C \cdot \dot{H}^e$, where C is the elastic matrix. During an active plastic process, stress $C \cdot \dot{H}^t$ exceeds the elastic limit and its uncompensated part triggers dynamic excitations $\dot{\sigma} + \kappa \, \dot{\psi}/\rho \, l_r = C \cdot \dot{H}^t$. The excitations give rise to the viscous overstress $\kappa \dot{\psi}/\rho \, l_r$. The overstress can be expressed in terms of stress perturbations $\kappa \dot{\psi}/\rho \, l_r = R_k^{-2}\delta\sigma/\tau$, where the resistance to plastic flow $R_k = l_r/\tau v_s$ carries information on the size of the perturbed domain l_r and specifies the relaxation time $\tau = \kappa/\mu$ required for the equilibration of the dislocation structures. Here, $v_s = \sqrt{\mu/\rho}$ is shear velocity and μ is shear modulus. As a result, we have $\dot{\sigma} + R_k^{-2}\delta\sigma/\tau = C \cdot \dot{H}^t$. Once again, the resistance to plastic flow $R_k = v_r/v_s$ arises in the actively perturbed domain l_r, such that $v_r = l_r/\tau$. The excitations are redistributed into the surroundings with velocity slower than shear velocity v_s, thus the resistance R_k cannot be greater than one. In summary, the material description is given in terms of three relations

$$\dot{H}^t = \dot{H}^e + \left(\dot{H}^p - \delta\sigma/\kappa\right)$$
$$\dot{\sigma} + R_k^{-2}\delta\sigma/\tau = C \cdot \dot{H}^t \qquad (6)$$
$$\dot{\sigma} = C \cdot \dot{H}^e$$

Stress perturbations slow down the slip process, while the part of stress which exceeds the elastic limit is converted to viscous overstress. The relations are further rearranged

$$\dot{H}^t = \dot{H}^e + \dot{H}^{pe}$$
$$\dot{\sigma} + C \cdot \dot{H}^{pe} = C \cdot \dot{H}^t \qquad (7)$$
$$\delta\sigma = \tau \, R_k^2 \, C \cdot \dot{H}^{ep}$$

The rate of effective (true) plastic strain $\dot{H}^{pe} = P^{-1} \cdot \dot{H}^p$ is scaled by the fourth-order drag tensor $P = I + R_k^2 \, C/\mu$. In textured metals, the drag tensor P channels plastic deformation and damages along crystallographically favorable pathways. In an isotropic metal, the drag tensor $P^{-1} = I/(1 + 2 R_k^2)$ and the rate of plastic strain $\dot{H}^{pe} = \dot{H}^p/(1 + 2R_k^2)$ take much simpler forms. Comparison of Equations (6_2) and (7_2) $\left(C \cdot \dot{H}^{pe} = R_k^{-2}\delta\sigma/\tau\right)$ confirms that the viscous overstress in Equation (6) is acting on the plastic strain. Relaxation of the viscous overstress is best described by Maxwell's process $\dot{S}^a + S^a/2\tau = C \cdot \dot{H}^{pe}$, where S^a is the active overstress (Figure 2). Rerouting of plastic flow is enabled by the excess of energy $\dot{W}_l = S^a : \dot{S}^a/2\mu$, where $trS^a = 0$. During slow processes, a large portion of W_l is dissipated. At high strain rates, W_l explicitly affects the plastic flow, increases storage of energy, intensifies plasticity-induced heating and influences the damage mechanism.

The initial resistance to plastic flow R_k^0 is a consequence of pre-existing defects. At dynamic conditions, the excess of energy W_l further amplifies the resistance. Consistently with the description of thermal resistance (Section 3), the relation $R_k = R_k^0 + (1 - R_k^0)(1 - e^{-(W_l/G_l)^{n_r}})$ is constructed on the basis of the weakest link hypothesis. The constants G_l and n_r specify dynamic constraints. The process occurs along the lowest energy pathways, while the frequency of the rerouting events is controlled by the exponent n_r. The dynamic resistance to plastic flow R_k is responsible for the stress upturn (Figure 3). It is shown that the upturn becomes much less pronounced in metals composed of fine grains, where the initial value of R_k^0 is closer to one. A full justification of the result is provided in Section 7 (Equation (13)). In Figure 3, the data points marked in red are collected from multiple sources [5,36–44]. Experiments performed on fine-grained tantalum [14] and copper [15] confirm the predictions constructed for microcrystalline copper.

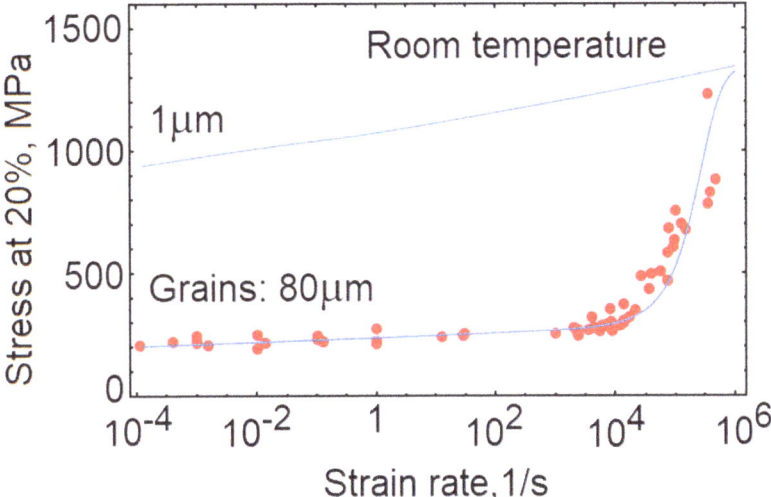

Figure 3. Stress at strain 20% is plotted as a function of strain rate. The data points are collected from [3,36–44].

Ductile damage results from slip incompatibilities. In tension, as stated earlier, kinematical compatibility is restored by the stress-favorable nucleation and growth of voids. In the Tresca description, the growth of damage is defined on the plane of dominant shear $(N^\sigma)^2 = (n^\sigma n^\sigma + s^\sigma s^\sigma)$ and is $\dot{H}^d = \frac{q}{2}(N^\sigma)^2 M \dot{e}^{pe}$. Subsequently, the volumetric change is $tr\dot{H}^d = qM\dot{e}^{pe}$. The relation was derived in [24]. The internal friction parameter q scales the damage process. When $q = 1$, voids nucleate and grow on the plane of maximum tensile stress. In compression, q is equal to zero. Here, the flow incompatibilities in tension $q = 1 - e^{-e_d^{pe}/e_d}$ are controlled by ductility e_d, where plastic strain e_d^{pe} is accumulated during tensile loading only. The damage is responsible for the degradation of the apparent bulk and shear moduli. The scaling parameter is $\left(1 - e^{-H_{kk}^d/H_{kk}^0}\right)$. Here, H_{kk}^0 specifies the strain at which the damage process is macroscopically evident. Consequently, in a damage-free metal, the scaling factor $\left(1 - e^{-H_{kk}^d/H_{kk}^0}\right)$ is equal to zero. The factor is equal to one in a completely fractured material.

6. Plasticity-Induced Heating

In 1934, Taylor and Quinney calculated plasticity-induced heating and arrived at the conclusion that about 90% of plastic work is turned into heat. Recent estimates contradict the assessment [45]. Moreover, plasticity-induced heat is found to be a function of plastic strain, strain rate and type of loading. The strongest generation of heat occurs at the initial stage of plastic deformation. Further increase of plastic strain slows down the process, and then, the rate increases again at an even more advanced stage of deformation. In the Taylor and Quinney interpretation, a part of plastic power is converted to hear $\dot{Q} = (\beta \sigma_{eq}) \dot{e}^{pe}$, where the Taylor–Quinney coefficient β quantifies the conversion process. I challenge the Taylor–Quinney interpretation and postulate that plastic work contributes to the increase of configurational entropy of the material, while heat is a measure of the process inefficiencies caused by the unavoidable rerouting of the deformation pathways.

The argument is constructed as follows. First, we note that stress perturbations $\delta\sigma = l_r \dot{\psi}$ are expressed in terms of the plastic strain rate $\delta\sigma = \tau R_k^2 (C \cdot \dot{H}^{pe})$. Thus, the change in momentum is $\dot{\psi} = R_k\, C\cdot\dot{H}^{pe}/v_s$. In Maxwell's process ($\dot{S}^a + S^a/2\tau = C\cdot\dot{H}^{pe}$), the part ($S^a/2\tau$) in $C\cdot\dot{H}^{pe}$ quantifies the already dissipated excitations

$$\dot{\psi}_d = R_k\, S^a/2v_s\tau. \tag{8}$$

The energy of the excitations is absorbed in dislocation structures. Still, there a fraction ξ_{ef} of $\dot{\psi}_d$ which gives rise to phonon vibrations $\dot{\psi}_Q = \xi_{ef}\dot{\psi}_d$. The term $S^a/2\tau$ in Equation (8) is replaced by $(C\cdot\dot{H}^{pe} - \dot{S}^a)$, where the phonon-scale overstress $\dot{S}^a$ is a vanishing quantity. Therefore, the heat generating excitations $\dot{\psi}_Q \cong \xi_{ef}\, R_k(C\cdot\dot{H}^{pe})/v_s$ are activated by the overstress $C\cdot\dot{H}^{pe}$ and are affected by the flow constraints R_k. As plastic deformation advances, heat sources continue to evolve $\psi_Q = \psi_p + \xi_{ef}\, R_k C\cdot H^{pe}/v_s$. In this expression, momentum ψ_p is linked to shear stress capable of triggering path rearrangements.

Heat is produced when excitations $\dot{\psi}$ act on the existing sources of heat ψ_Q, such that $\dot{Q} = \psi_Q : \dot{\psi}/\rho$. Consequently, plasticity-induced heat is generated with rate

$$\dot{Q} = 2R_k\big[R_k\,\xi_{ef}\, C\cdot H^{pe} + v_s\,\psi_p\big] : \dot{H}^{pe}. \tag{9}$$

The relation is further simplified $\dot{Q} = 2R_k\big[R_k\,\xi_{ef}\mu\, e^{ep} + \sigma_y\big]\dot{e}^{pe}$, where equivalent plastic strain e^{pe} has already been defined and σ_y is yield stress. The Taylor–Quinney coefficient $\beta = \dot{Q}/\dot{W}^p$ becomes $\beta = 2R_k\big[R_k\xi_{ef}\mu\, e^{pe} + \sigma_y\big]/\sigma_{eq}$. The heat sources are formed discretely both in time and space, where spatial resolution of the sources is estimated to be in a sub-micrometer range. Experimental techniques are yet to be developed for the measurement of such small-scale temperature fluctuations. Large-scale atomistic simulations might provide invaluable insight into the phenomena. It should be stated that the contours in Figure 4a,b are constructed under the assumption that the copper specimen is subjected to uniaxial compression and the temperature rise is depicted at strain 20%. The coefficient ξ_{ef} is calibrated using experimental data presented in [46]. In Figure 4a, the temperature rise is calculated as a function of grain size d and initial temperature T_0/T_m. The samples are deformed at a strain rate of $10^4/s$. The contours indicate that plasticity-induced heating is significantly stronger in microcrystalline copper. The environment T_0 affects the rate of heating. In large-grain copper, high strain rates noticeably magnify plasticity-induced heating, while small grains make the strain rate sensitivity an irrelevant factor (see Equation (13) and Figure 4b). For example, at a strain rate of $2.5\,10^5/s$ the same temperature rise is predicted in copper with average grains 2.5 μm and 77 μm. It would be very interesting and, perhaps, technologically important to experimentally validate the theoretical predictions. The grain size dependence is discussed in the next section.

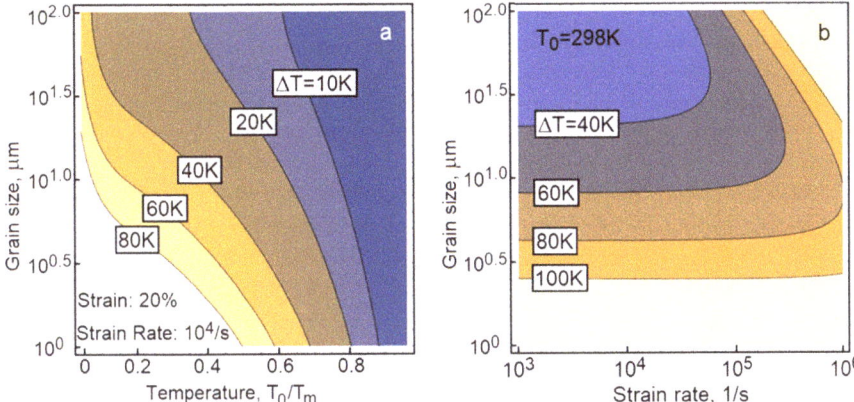

Figure 4. Contours of temperature rise are plotted as a function of grain size. (**a**) Temperature rise is calculated at strain rate 10^4/s and in a broad range of initial temperatures T_0. (**b**) Temperature rise is plotted at room temperature and for strain rates between 10^3/s and 10^6/s. The coefficient ξ_{ef} is estimated based on experimental measurements in [46].

7. Hall–Petch Relation

In early 1950s, Hall [10] and Petch [11] introduced a relation, which correlates yield stress σ_y with an average size of grains d

$$\sigma_y = \sigma_y^p + k_y\, d^{-1/2}. \tag{10}$$

In this construction, σ_y^p is the size-independent strength, k_y is a constant and d is the average size of grains. In this notation, yield stress corresponds to equivalent stress $\sigma_y = \sigma_{eq}$. The relation is generally applicable to metals and other polycrystalline materials. A few years later, Armstrong [12] proposed a dislocation pileup interpretation and further explained the mechanism in numerous writing [47]. In this construction, the parameter k_y is considered a Griffith type stress intensity factor, where the grain size strengthening results from the dislocation pileup along slip bands blocked at a grain boundary, thus behaving similarly to a shear crack. Among other descriptions, it is worth mentioning the concept by Ashby [48]. In Ashby's formula, the constant k_y includes the additional term $\left(\mu b^{1/2}\right)$. It will be shown that the term is reproduced in the proposed energy-based formula. Strong experimental evidence exists that dislocation pileups are responsible for the observed heterogeneous distribution of stresses. The supporting evidence was obtained in micro-diffraction (DAXM) experiments further supported by high-resolution electron backscatter diffraction (HR-EBSD) measurements [49]. It is worth noting that nanoindentation tests [50] have shown that the material hardness increases in the proximity of grain boundaries. Thus, the experimental results convey a clear message that dislocation pileups near grain boundaries are indeed responsible for the stress and energy heterogeneities.

The Hall–Petch interpretation is based on the postulate that, at a given rate of plastic strain, the upsurge of plastic power is controlled by the increases of yield stress $\dot{W}_{\dot{\varepsilon}^p}^p = \left(1 + 2R_k^2\right) \sigma_y^p M \dot{\varepsilon}^p$. Here, the constraints R_k magnify the effective yield stress $\sigma_y = \sigma_y^p \left(1 + 2R_k^2\right)$. When assuming that the Hall–Petch relation prevails, one can obtain $R_k^2 = k_y\, d^{-1/2}/2\sigma_y^p$. Next, I construct a kinematics-based interpretation. At the prescribed stress σ_y, the resistance R_k is responsible for slowing down the plastic flow. For this reason, we write $\dot{W}_{\sigma_y}^p = \sigma_y M \dot{\varepsilon}^p / \left(1 + 2R_k^2\right)$. The two interpretations represent the upper and lower bound estimates of the size effect.

7.1. Energy-Based Hall–Petch Relation

I propose an energy-based interpretation of the Hall–Petch strengthening. As stated above, the Hall–Petch size effect is mostly caused by grain boundaries, which slow down the plastic flow.

Suppose that polycrystalline metal is subjected to stress σ_y^p. This stress activates slip inside grains, but the deformation is blocked at grain boundaries. Grain boundaries are energy barriers defined in terms of surface energy g_{gb}. At stress σ_y^p, the barriers are partly climbed g_{gb}^y. As loading advances, blocked dislocations magnify the pileup stress. It should be stated that dislocations are emitted in other areas, such as grain boundary corners. After some time (the delay time), the pileups accumulate sufficient energy for a dislocation to cross the barriers g_{gb}. In essence, it is a stochastic process. Stress concentrations appear in different locations and grow at different rates. Large grains contain a large reservoir of energy which, in turn, feeds the energy concentrations more efficiently. The pileup-stimulated increase of internal energy is taken per unit grain boundary area S_{gb} and becomes $E_{gr} = (V_s/S_{gb}) \int_{\sigma_y^p}^{\sigma_y} w(\sigma_s) \sigma_s d\sigma_s / 2\mu$. The average shear stress σ_s continues to rise and becomes the macroscopic yield stress σ_y. At this point, dislocations cross the boundaries in large numbers. The average rate of energy accumulation slows down. The correction function $w = (\sigma_y - \sigma_y^p)/2\sigma_s$ captures the slowing down effect. When energy E_{gr} overcomes the energy barrier $(g_{gb} - g_{gb}^a)(S_{gb}/V_s)$, the macroscopic plastic flow is initiated

$$(g_{gb} - g_{gb}^a)(S_{gb}/V_s) = \frac{(\sigma_y - \sigma_y^p)^2}{4\mu}. \tag{11}$$

Grain boundary area per unit volume S_{gb}/V_s characterizes granular structure of the polycrystal. This ratio is sensitive to the grain size, shape and number of grains in a sample. The Hall–Petch relation becomes

$$\sigma_y = \sigma_y^p + k_{yS} \left(\frac{S_{gb}}{V_s}\right)^{1/2}. \tag{12}$$

In this expression, the Hall–Petch mechanism is controlled by grain boundary area per unit volume. The parameter $k_{yS} = 2\sqrt{\mu(g_{gb} - g_{gb}^a)}$ is closely related to constant k_y in Equation (10), such that $k_{yS} \cong k_y/2$. It has been shown [51] that the grain boundary energy is linearly dependent on shear modulus multiplied by a length parameter, where the parameter specifies the grain boundary type. Thus, the relation $(g_{gb} - g_{gb}^a) = \mu(\alpha_{yS}^2 b)$ suggests that $k_{yS} = 2\mu \alpha_{yS} b^{1/2}$ and α_{yS} carries information on the metal crystallographic structure. It turns out that the parameter α_{yS} is a material constant in fcc and bcc metals. Based on the data collected in [52], the mean value of α_{yS} for fcc metals is $\alpha_{yS} = 0.04604 \pm 0.00616$. In bcc metals, the values are found to be $\alpha_{yS} = 0.1089 \pm 0.0353$. Estimates are also made for hcp metals, but the data scatter is much larger. The hcp deformation twins might be responsible for the scatter. As usual, the magnitude of the Burgers vector is denoted by b.

In Equation (12), the grain boundary area per unit volume is the size variable. The variable can be estimated with the use of stereology methods [53]. In bulk samples composed of nearly equiaxed grains, the average size of grains is about $d \sim 4V_s/S_{gb}$. However, the validity of the approximation can be questioned in samples with a small number of grains. The grain shape matters too. In one case, yield stress is significantly reduced in thin films of nickel, where the film's thickness is comparable with the size of grains [54]. In the case of mild steel, Armstrong [12] made similar observations.

7.2. Kinematics-Based Construction of Hall–Petch Relation

At constant stress, energy barriers slow down plastic flow and, in this manner, affect plastic power $\dot{W}_{\sigma_y}^p = \sigma_y M \dot{e}^p / (1 + 2R_k^2)$. In the energy-based construction, the term $(1 + 2R_k^2)$ in $\dot{H}^{pe} = \dot{H}^p / (1 + 2R_k^2)$ becomes $\left[1 + \frac{2\mu \alpha_{yS} b^{1/2}}{\sigma_y^p} (S_{gb}/V_s)^{1/2}\right]$. The pre-existing resistance to plastic flow $R_k = R_k^0$ becomes a quantifiable material constant

$$R_k = \sqrt{\frac{k_{yS}}{2\sigma_y^p}} (S_{gb}/V_s)^{1/4}. \tag{13}$$

Now, it becomes clear that the resistance to plastic flow is strongly dependent on the size of grains. In microcrystalline copper, the value of R_k^0 is approaching one, thus the size effect becomes less pronounced. As shown in Section 5, the resistance $R_k = l_r/(v_s \tau)$ is a function of relaxation time $\tau = l_r/(R_k v_s)$. The length l_r is directly related to the grain boundary area per unit volume and is $l_r = (S_{gb}/V_s)^{-1}$. Based on the above, the relaxation time becomes the size-dependent material property

$$\tau = \frac{(S_{gb}/V_s)^{-5/4}}{v_s}\sqrt{\frac{2\,\sigma_y^p}{k_y s}}. \tag{14}$$

It is worth stating that relaxation time τ (Equation (14)) matches earlier estimates made for OFHC copper [55,56]. In nanocrystalline copper, relaxation time is in the range of hundred picoseconds. The kinematically-based Hall–Petch relation fits well the existing experimental data in [52] (Figure 5). Similar trends are reported for other metals [52,57] as well.

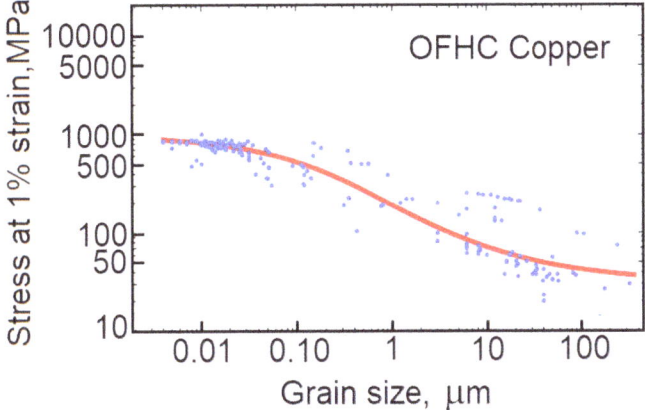

Figure 5. Grain size strengthening depicted at strains approximately one percent. Experimental data (blue points) gathered from [52]. The red line represents the model predictions. The middle part of the log–log plot complies with the Hall–Petch relation.

8. Transitional Viscoplasticity

Plastic power $\sigma : \dot{H}^{pe} = \sigma_{eq}\dot{e}^{pe}$ is expressed in terms of equivalent stress $\sigma_{eq} = \frac{1}{2}N^\sigma : \sigma$ and strain rate $\dot{e}^{pe}$. During active deformation processes, plastic power $\dot{W}^p$ aids to configurational entropy (suppleness [25]) of the material. The rate of true plastic strain $\dot{H}^{pe} = \frac{1}{2}[N^\sigma + q(N^\sigma)^2]\dot{e}^{pe}$ is defined in terms of the effective rate $\dot{e}^{pe} = \frac{M\dot{e}^p}{1+2R_k^2}$, where the resistance-free strain rate $\dot{e}^p$ is a function of equivalent stress and the relation is

$$\dot{e}^p = \dot{\Lambda}_p \cdot (\sigma_{eq}/\sigma_y^p)^{n_p}. \tag{15}$$

The power-law viscoplasticity was first introduced in [58,59]. Strength $\sigma_y^p = \sigma_0\,A_T(T)$ is determined at 0.2% strain, where σ_0 is athermal strength. Exponent n_p determines the elastic-plastic transition. What makes the relation uncommon is that the strain rate sensitivity is further tuned by the rate factor $\dot{\Lambda}_p = \Lambda_0\,\dot{e}_N^0[\dot{e}_N^t/\dot{e}_N^0]^{\omega_p}$, where Λ_0 and ω_p are constants and $\dot{e}_N^0 = 1/s$. The normalized rate of total strain $\dot{e}_N^t = \sqrt{\dot{H}^t : \dot{H}^t}/2$ completes the expression. As shown earlier, the average Schmid factor $M = \frac{1+r}{1+\pi r/2} - \frac{4\,A_T}{\pi\,(1+2\,r)}$ reflects slip plane misorientations, where the exponent r controls the evolution of the process (Figure 1). Advances of plastic deformation broaden the spectrum of active planes and, therefore, $r = r(e^{pe})$.

At the initial stage of deformation, the rate of elastic strain dominates ($\dot{H}^e \to \dot{H}^t$) and, therefore, we have $\dot{\varepsilon}^p \propto \left(\sigma_{eq}/\sigma_y^p\right)^{n_p}$. At advanced deformation, plastic flow takes over ($\dot{H}^p \to \dot{H}^t$) and the mechanism evolves $\dot{\varepsilon}^p \propto \left(\sigma_{eq}/\sigma_y^p\right)^{n_p/(1-\omega_p)}$, accordingly. Thus, there is a smooth transition from diffusional flow $\dot{\varepsilon}^p \propto \left(\sigma_{eq}/\sigma_y^p\right)^{n_p}$ to dislocation glide $\dot{\varepsilon}^p \propto \left(\sigma_{eq}/\sigma_y^p\right)^{n_p/(1-\omega_p)}$, where we should note that $n_p/(1-\omega_p) \gg n_p$.

9. OFHC Copper

The constitutive model captures copper responses in a broad range of temperature and strain rates, predicts ductile damage, accounts for the size of grains, and includes the description of plasticity-induced heating. There are three groups of parameters. Parameters readily available in the literature are listed in Table 1. Among the constants are elastic properties, mass density, yield stress at 0.2% strain, Burgers vector, melting point and specific heat.

Table 1. Text-book material constants.

Bulk Modulus	Shear Modulus	Mass Density	Yield Stress, 298 K	Burgers Vector	Melting Point	Specific Heat, 298 K
B GPa	μ GPa	ρ kg/m^3	$\sigma_{y0.2\%}$ MPa	b nm	T_m K	C_p J/(kg K)
138	56	8930	84	0.2555	1356	385

Several parameters display a marginal variability. Among the parameters are the quasi-static strain-rate sensitivity ω_p, stress exponent n_p, thermal efficiency of plastic flow ξ_{ef}, crystallographic constant α_{yS}, the initial distribution of slip planes r_0, the transition temperature T_c and the rate of defect formation n_r. The parameters are shown in Table 2. In the absence of experimental data, the constants should properly approximate responses of fcc metals.

Table 2. Material constants with marginal variability.

Strain Rate Exponent	Stress Exponent	Heat Coefficient	Crystallographic Constant	Schmid Factor	Transition Temperature	Overstress Exponent
ω_p	n_p	ξ_{ef}	α_{yS}	r_0	T_c	n_r
–	–	–	–	–	K	–
0.99	0.56	0.031	0.044	0.3	596.6	0.5

The material-specific parameters are listed in Table 3. Among them are the activation energy constants (g_a, k_a), the strength pre-factor Λ_0, critical energy G_I, hardening strain e_r^p and damage parameters (e_d, H_{kk}^0). In this group, the activation energy constants show low variability and, therefore, there are only five fully tunable constants. One may argue that the total number of parameters is large. It is worth noting that the parameters have well-defined interpretations and most of them are material constants. I emphasize that the constitutive model captures material responses in a very broad range of temperature and strain rates, predicts ductile damage, accounts for the size of grains, and includes plasticity-induced heating.

Table 3. Material-specific parameters.

Activation Energy Factor	Thermal Activation	Stress Pre-Factor	Critical Energy	Schmid Factor	Ductility	Damage Strain
g_a	k_a	Λ_0	G_I	e_r^p	e_d	H_{kk}^0
–	–	–	MJ/m^3	–	–	–
1.13	0.087	6.8	5.0	0.016	0.148	0.0296

The model capabilities are summarized in the deformation map (Figure 6), where stress at 20% strain is plotted as a function of strain rate and temperature. Two maps are constructed. In Figure 6a, the average grain size is 80 μm.

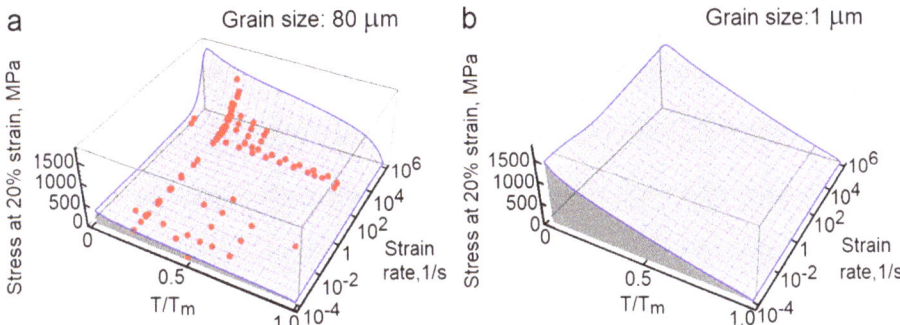

Figure 6. Deformation maps for copper presented in terms of stress at 20% strain. The blue mesh represents the model predictions taken in a broad range of temperatures and strain rates. (**a**) Experimental data collected from References [5,35–43,60–62] and the points are marked in red. (**b**) Microcrystalline copper (grains size 1 μm) does not exhibit the stress upturn.

The red points depict experimental data collected from several sources [5,35–43,60–62] and the blue mesh represents the model predictions. The model correctly predicts copper responses (grains 80 μm) at temperatures from cryogenic environments up to melting points and strain rates from quasi-static loading to extreme strain rates. The second plot in Figure 6b is constructed for microcrystalline copper, where the grain size is 1 μm. Note that in the second case the stress upturn is nearly gone. This result matches experimental measurements performed on small-grained tantalum [14] and copper [15]. The strain rate sensitivity was also studied in nickel [63], where ultrafine-grained nickel exhibits a diminishing rate-sensitivity.

For completeness, I included selected stress–strain responses (Figure 7). The predictions (solid lines) are calculated at three temperatures and three strain rates. The experimental data points (red, blue and black dots) are collected from [37,61]. In all cases, I included the contribution of plasticity-induced heating, where specific heat varies with temperature.

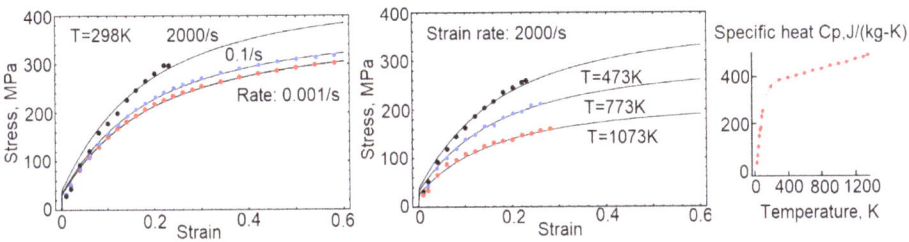

Figure 7. Selected stress–strain responses are constructed for three strain rates and three temperatures. The average size of grains is assumed to be 80 μm. The data is collected from [37,61].

10. Plate Impact Problem

The constitutive model is implemented to a code of deformable discrete element method (DDEM). In DDEM, each discrete element represents a crystallographic grain. The grains experience large deformation and are subjected to translation and rotation, where the rigid motions are directly measurable quantities. A more detailed description of the method is presented in Appendix A. In the DDEM construction shown in Figure 8, the Cu flyer is discretized with the use of 7200 elements, while the Cu target is composed of 14,400 elements. The thicknesses of the flyer and target are 1.9 mm and 3.9 mm, respectively. The flyer travels with velocity 308 m/s and, after striking the target, shock waves are sent into both of the samples.

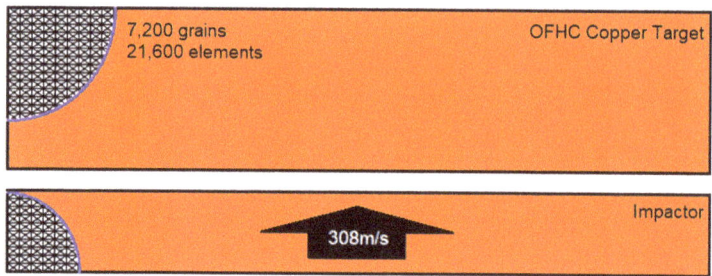

Figure 8. Copper target is struck by a copper flyer with velocity 308 m/s. The entire system is constructed with the use of 7200 triangle particles, where each particle is composed of three elements. Artificial viscosity is not used in this simulation.

A simple Mie Gruneisen equation of state for copper is used. The equation is further modified to account for the contribution of dynamic excitations. It is worth stating that the calculations are made without the use of artificial viscosity. In Figure 9, VISAR measurements [64] are compared with the results of this simulation. Note that the constitutive model correctly predicts the entire impact responses, and this includes the pull-back signals. Four points, A, B, C and D, are selected for plotting contours of damage. The damage is scaled between zero and one, as described in Section 5. The scaling factor refers to the degradation of elastic properties from zero (no damage) to one (open crack). The damage factor $\left(1 - e^{-H_{kk}^d / H_{kk}^0}\right)$ is determined in terms of plasticity-induced dilatation H_{kk}^d. As discussed earlier, the volume change is triggered by slip incompatibilities in tension. The contact between flyer and target experiences periodic separations. The spallation damage is clearly seen in the middle section of the target. Less severe damage is noticed in other areas of the samples. The damages might be responsible for the observed ringing in the VISAR plot.

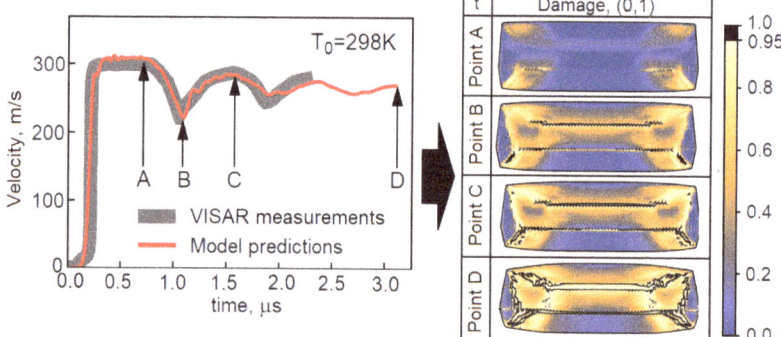

Figure 9. VISAR measurement (thick gray line) reported in [64] is compared with results of the numerical simulation (red line). Damage levels from zero (no damage) to one (fully developed crack) are shown at VISAR points A, B, C and D. The contours are rescaled to fit the window.

11. Conclusions

The paper summarized a five-year effort aimed for the development of a mechanisms-based viscoplasticity concept for metals. The constitutive description comprises of several novel ideas.

1. The macroscopic plastic flow results from plastic slippages and slip reorganizations. The description is constructed on the basis of the tensor representation concept. It is my conviction that tensor representations derived for generic dyads represent useful tools in the hands of a modeler.
2. In the proposed model, thermally activated processes are considered stochastic. The concept explains the transition of flow mechanisms from power-law creep to high strain rate dislocation glide.
3. The proposed description of plasticity-induced heating is based on the hypothesis that plasticity-induced heating quantifies the efficiency of the plastic flow process, while plastic work aids in configurational entropy (suppleness) of the material.
4. Drag on dislocations is activated by dynamic excitations. As shown, the excitations result from the kinematically-necessary readjustments of flow pathways.
5. The stress–strain relations are constructed in the framework of transitional viscoplasticity. The power-law relations enable a smooth elastic-plastic transition during loading and unloading processes.
6. I developed an energy-based Hall–Petch relation, where the commonly known stress-based relation is replaced by its kinematics-based counterpart. The proposed Hall–Petch concept was born out of extensive discussions with Ron Armstrong, who walked me through the sixty years of Hall–Petch interpretations, for which I am grateful.
7. The model is calibrated for OFHC copper, implemented to a deformable discrete element code (my Ph.D. thesis) and validated in simulations of a plate impact problem. The method itself describes a semi-Cosserat medium, where grain translations and rotations are accounted for.

The most important conclusion of the study is that the flow mechanisms, transitional viscoplasticity, thermal activation, drag on dislocations and size effect are inseparable parts of the constitutive description.

Funding: This research received no external funding.

Acknowledgments: The work was supported by Alek and Research Associates, LLC. I wish to acknowledge Prof. Ron Armstrong for sharing his inside into the six decades of the Hall–Petch relation. It was a unique learning experience, for which I am grateful. I am thankful to Prof. Mick Brown for explaining his prospective on the role of self-organized criticality in plastic flow. I would like to thank Prof. Dany Rittel for offering constructive comments and suggestions.

Conflicts of Interest: The author declare no conflict of interest.

Appendix A. A Short Note on Deformable Discrete Element Method

The deformable discrete element method (DDEM) was first reported in [65]. I believe that this might be one of the first if not the first DEM approach used for the prediction of fracture processes in otherwise continuum media. A complete description of the method can be found in [66]. Later, the method was reduced to a system of rigid blocks and springs, while the description of deformable particles was forgotten. Here, I describe the method as originally developed. Each particle is subjected to translations and rotations (Figure A1). In addition, each particle experiences deformation as defined in the classic finite element method. In a general case, particle rotations act on moments and, in this way, the system represents a pseudo-Cosserat medium. In this approach, mass and moments of inertia are placed in the particle mass centers. Some portion of mass is also lumped into external nodes of each particle. As a result, equations of motion are constructed in internal and external nodes.

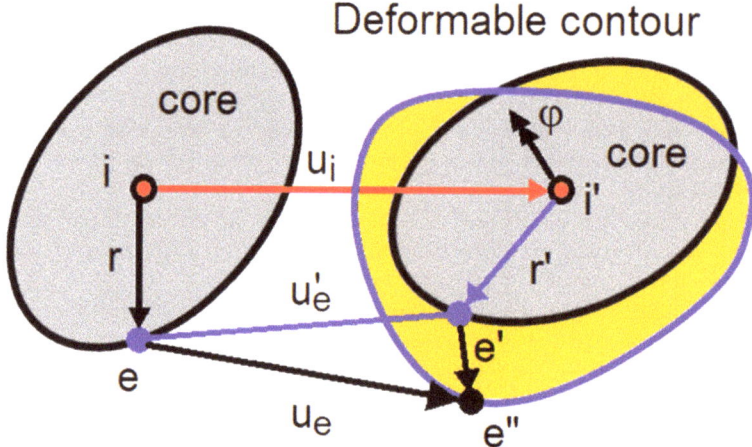

Figure A1. Motion and deformation of discrete element.

The motion of deformable particles includes translations u_i and rotations φ of mass centers (i-nodes), while particle-to-particle interactions are defined in e-external nodes. The particle deformation is measured in terms of relative displacements $\Delta u_{ei} = u_e - u'_e$ between points $e\prime$ and $e\prime\prime$. The relations become $\Delta u_{ei} = u_e - (u_i + r \times \varphi)$. In this expression, vectors r connect selected external node with the mass center. The deformation matrix B links strains H_{ie} and relative displacements

$$H_{ie} = B \cdot \Delta u_{ei}. \tag{A1}$$

The rotational deformation is defined with respect to the mass center and is

$$H^i_\varphi = B_\varphi : \Delta u_{ei}, \tag{A2}$$

where $H^i_\varphi = \left(\sum_{N_i} \Delta u^i_\varphi / r_i\right)/N_i$. The relative displacements Δu^i_φ represent rotational deformation subtracted from the rigid core rotations. Each internal node is coupled with N_i- external nodes. The rotation-induced strain is responsible for additional storage of elastic energy $U^\varphi = \mu\, \beta_\varphi \left(H^i_\varphi\right)^2 /2$, where the parameter β_φ scales the rotational constraints. Nodal forces in each element are collected from all sub-elements and are

$$f_{ie} = \int \left(B^T \cdot \sigma + B^T_\varphi\, \mu\, \beta_\varphi H^i_\varphi\right) dV^i_s. \tag{A3}$$

Stress σ is calculated in deformable particles and is $\sigma = C \cdot H^e_{ie}$. Lastly, equations of motion are constructed in external nodes and mass centers (internal nodes)

$$\begin{aligned} F_e &= \sum_{N_i} f_{ei} + M_e\, \ddot{u}_e \\ F_i &= \sum_{N_{ie}} f_{ei} + M_i\, \ddot{u}_i \\ M_\varphi &= \sum_{N_e} r_i \times f_{ie} + J_M\, \ddot{\varphi}_i \end{aligned} \tag{A4}$$

During an active deformation process, particles may become separated, and then, the number of interconnected nodes may vary. Forces in mass centers are collected from N_{ie} external nodes. Moments are calculated as described in Equation (A4). In this notation, mass in external nodes is denoted M_e, mass and moment of inertia in internal nodes are M_i and J_M. In DDEM, mass is redistributed between external and internal nodes in such a manner that inertia forces due to rotations and translations are decoupled. Finite elements (particles) can be subjected to large reshaping and repositioning. Thus, the

DDEM method is not just a numerical solver, but it carries information on the size and potential shape of grains.

References

1. Meyers, M.S.; Jarmakani, H.; Bringa, E.M.; Remington, B.A. Dislocations in shock compression and release. In *Dislocations in Solids*; Hirth, J.P., Kubin, L., Eds.; North-Holland: Amsterdam, The Netherland, 2009; Volume 15, pp. 91–197.
2. Follansbee, P.S. High-strain-rate deformation of FCC metals and alloys. In Proceedings of the EXPLOMET '85—International Conference on Metallurgical Applications of Shock Wave and High Strain-Rate Phenomena, Portland, OR, USA, 28 July 1985.
3. Kumar, A.; Kumble, R.G. Viscous drag on dislocations at high strain rates in copper. *J. Appl. Phys.* **1969**, *40*, 3475–3480. [CrossRef]
4. Regazzoni, G.; Kocks, U.F.; Follansbee, P.S. Dislocation kinetics at high strain rates. *Acta Metall.* **1987**, *35*, 2865–2875. [CrossRef]
5. Bragov, A.; Igumnov, L.; Konstantinov, A.; Lomunov, A.; Rusin, E. Effects of high strain rate on plastic deformation of metal materials under fast compression loading. *J. Dyn. Behav. Mater.* **2019**, *5*, 309–319. [CrossRef]
6. Dodd, B.; Bai, Y. *Adiabatic Shear Localization: Frontiers and Advances*, 2nd ed.; Elsevier: Amsterdam, The Netherlands, 2012; pp. 2–18.
7. Bever, M.B.; Holt, D.L.; Titchener, A.L. The stored energy of cold work. *Prog. Mater Sci.* **1973**, *17*, 5–177. [CrossRef]
8. Farren, W.S.; Taylor, G.I. The heat developed during plastic extension of metals. *Proc. R. Soc. A* **1952**, *5*, 398–451.
9. Taylor, G.I.; Quinney, H. The latent energy remaining in a metal after cold working. *Proc. R. Soc. A* **1934**, *143*, 307–326.
10. Hall, E.O. The deformation and aging of mild steel. *Proc. Phys. Soc. Lond.* **1951**, *64*, 747–753. [CrossRef]
11. Petch, N.J. The cleavage strength of polycrystals. *J. Iron Steel Inst.* **1953**, *174*, 25–28.
12. Armstrong, R.W. On size effect in polycrystal plasticity. *J. Mech. Phys. Solids* **1961**, *9*, 196–199. [CrossRef]
13. Armstrong, R.W.; Codd, L.; Douthwaite, R.M.; Petch, N.J. The plastic deformation of polycrystalline aggregates. *Philos. Mag.* **1962**, *7*, 45–58. [CrossRef]
14. Park, H.-S.; Rudd, R.E.; Cavallo, R.M.; Barton, N.R.; Arsenlis, A.; Belof, J.L.; Blobaum, K.J.M.; El-Dasher, B.S.; Florando, J.N.; Huntington, C.M.; et al. Grain-size-independent plastic flow at ultrahigh pressures and strain rates. *Phys. Rev. Lett.* **2015**, *114*, 065502. [CrossRef]
15. Mao, Z.N.; An, X.H.; Liao, X.Z.; Wang, J.T. Opposite grain size dependence of strain rate sensitivity of copper at low vs high strain rates. *Mater. Sci. Eng. A* **2018**, *738*, 430–438. [CrossRef]
16. Arzt, E. Size effects in materials due to microstructural and dimensional constraints: A comparative review. *Acta Mater.* **1998**, *46*, 5611–5626. [CrossRef]
17. Pande, C.S.; Cooper, K.P. Nanomechanics of Hall-Petch relationship in nanocrystalline materials. *Prog. Mater. Sci.* **2009**, *54*, 689–706. [CrossRef]
18. Zerilli, F.J.; Armstrong, R.W. Dislocation-mechanics-based constitutive relations for material dynamics calculations. *J. Appl. Phys.* **1987**, *61*, 1816–1825. [CrossRef]
19. Johnson, G.R.; Cook, W.H. A constitutive model and data for metals subjected to large strains, high strain rates and high. In Proceedings of the 7th International Symposium, Ballistics, The Hague, 19–21 April 1983; pp. 541–547.
20. Follansbee, P.S.; Kocks, U.F. A constitutive description of the deformation of copper based on the use of the mechanical threshold. *Acta Metall.* **1988**, *36*, 81–93. [CrossRef]
21. Preston, D.L.; Tonks, D.L.; Wallace, D.C. Model of plastic deformation for extreme loading conditions. *J. Appl. Phys.* **2003**, *93*, 211–220. [CrossRef]
22. Salvado, F.C.; Teixeira-Dias, F.; Walley, S.M.; Lea, L.J.; Cardoso, J.B. A review on the strain rate dependency of the dynamic viscoplastic response of fcc metals. *Prog. Mater. Sci.* **2017**, *88*, 186–231. [CrossRef]
23. Gurrutxanga-Lerma, B. A stochastic study of the collective effect of random distribution of dislocations. *J. Mech. Phys. Solids* **2019**, *124*, 10–34. [CrossRef]

24. Langer, J.S. Statistical thermodynamics of crystal plasticity. *J. Stat. Phys.* **2019**, *175*, 531–541. [CrossRef]
25. Brown, L.M. Power laws in dislocation plasticity. *Philos. Mag.* **2016**, *96*, 2696–2713. [CrossRef]
26. Richeton, T.; Weiss, J.; Louchet, F. Dislocation avalanches: Role of temperature, grain size and strain hardening. *Acta Mater.* **2005**, *53*, 4463–4471. [CrossRef]
27. Zaiser, M.; Nikitas, N. Slip avalanches in crystal plasticity: Scaling of avalanche cutoff. *J. Stat. Mech. Theory Exp.* **2007**, *2007*, P04013. [CrossRef]
28. Pantheon, W. Distribution in dislocation structures: Formation and spatial correlation. *J. Mater. Res.* **2002**, *17*, 2433–2441. [CrossRef]
29. Zubelewicz, A. Micromechanical study of ductile polycrystalline materials. *J. Mech. Phys. Solids* **1993**, *41*, 1711–1722. [CrossRef]
30. Zubelewicz, A. *Tensor Representations in Application to Mechanisms-Based Constitutive Modeling*; Technical Report ARA-2; Alek & Research Associates, LLC: Los Alamos, NM, USA, 2015. [CrossRef]
31. Zubelewicz, A. Overall stress and strain rates for crystalline and frictional materials. *Int. J. Non Linear Mech.* **1991**, *25*, 389–393. [CrossRef]
32. Zubelewicz, A. Metal behavior at extreme loading rates. *Mech. Mat.* **2009**, *41*, 969–974. [CrossRef]
33. Armstrong, R.W.; Walley, S.M. High strain rate properties of metals and alloys. *Int. Mater. Rev.* **2008**, *53*, 105–128. [CrossRef]
34. Ryu, S.; Kang, K.; Cai, W. Entropic effect on the rate of dislocation nucleation. *Proc. Natl. Acad. Sci. USA* **2011**, *108*, 5174–5178. [CrossRef]
35. Kubin, L.P. Dislocation patterns: Experiment, theory and simulations. In *Stability of Materials*; Plenum Press: New York, NY, USA, 1996; pp. 99–135.
36. Gray, G.T.; Chen, S.R.; Wright, W.; Lopez, M.F. *Constitutive Equations for Annealed Metals under Compression at High Strain Rates and High Temperature*; IS-4 Report, LA-12669-MS; Los Alamos National Laboratory: Los Alamos, NM, USA, 1994.
37. Gray, G.T., III. High strain rate deformation: Mechanical behavior and deformation substructures induced. *Annu. Rev. Mater. Res.* **2012**, *42*, 285–303. [CrossRef]
38. Gao, C.Y.; Zhang, L.C. Constitutive modelling of plasticity of fcc metals under extremely high strain rates. *Int. J. Plast.* **2012**, *32–33*, 121–133. [CrossRef]
39. Huang, S.H.; Clifton, R.J. *Macro and Micro-Mechanics of High Velocity Deformation and Fracture*; Kawata, K., Shioiki, J., Eds.; IUTAM: Tokyo, Japan, 1985; pp. 63–74.
40. Tong, W.; Clifton, R.J.; Huang, S.H. Pressure-shear impact investigation of strain rate history effects in oxygen-free high-conductivity copper. *J. Mech. Phys. Solids* **1992**, *40*, 1251–1294. [CrossRef]
41. Nemat-Nasser, S.; Li, Y. Flow stress of fcc polycrystals with application to OFHC Cu. *Acta Mater.* **1988**, *46*, 565–577. [CrossRef]
42. Tanner, A.B.; McGinty, R.D.; McDowell, D.L. Modeling temperature and strain rate history effects in OFHC Cu. *Int. J. Plast.* **1999**, *15*, 575–603. [CrossRef]
43. Baig, M.; Khan, A.S.; Choi, S.-H.; Jeong, A. Shear and multiaxial responses of oxygen free high conductivity (OFHC) copper over wide range of strain-rates and temperatures and constitutive modeling. *Int. J. Plast.* **2013**, *40*, 65–80. [CrossRef]
44. Jordan, L.J.; Siviour, C.R.; Sunny, G.; Bramlette, C.; Spowart, J.E. Strain rate-dependent mechanical properties of OFHC copper. *J. Mater. Sci.* **2013**, *48*, 7134–7141. [CrossRef]
45. Rittel, D.; Zhang, L.H.; Osovski, S. The dependence of the Taylor-Quinney coefficient on the dynamic loading mode. *J. Mech. Phys. Solids* **2017**, *107*, 96–114. [CrossRef]
46. Nieto-Fuentes, J.C.; Rittel, D.; Osovski, S. On a dislocation-based constitutive model and dynamic thermomechanical considerations. *Int. J. Plast.* **2018**, *108*, 55–69. [CrossRef]
47. Armstrong, R.W. Size effect on material yield strength/deformation/fracturing properties. *J. Mater. Res.* **2019**, *34*. [CrossRef]
48. Ashby, M.F. The deformation of plasticity non-homogeneous materials. *Philos. Mag.* **1970**, *21*, 399–424. [CrossRef]
49. Guo, Y.; Collins, D.M.; Tarleton, E.; Hofmann, F.; Tischler, J.; Liu, W.; Xu, R.; Wilkinson, A.J.; Britton, T.B. Measurements of stress fields near a grain boundary: Exploring blocked arrays of dislocations in 3D. *Acta Mater.* **2015**, *96*, 229–236. [CrossRef]

50. Voyiadjis, G.Z.; Zhang, C. The mechanical behavior during nanoindentation near the grain boundary in a bicrystal fcc metal. *Mater. Sci. Eng. A* **2015**, *621*, 218–228. [CrossRef]
51. Udler, D.; Seidman, D.N. Grain boundary and surface energies of fcc metals. *Phys. Rev. B* **1996**, *54*, 133–136. [CrossRef]
52. Cordero, Z.C.; Knight, B.E.; Schuh, C.A. Six decades of the Hall-Petch effect—A survey of grain-size strengthening studies on pure metals. *Int. Mater. Rev.* **2016**, *61*, 495–512. [CrossRef]
53. Van Vlack, L.H. *Elements of Materials Science and Engineering*, 6th ed.; Addison Wesley: Boston, MA, USA, 1989.
54. Keller, C.; Hug, E.; Retoux, R.; Feaugas, X. TEM study of dislocation patterns in near-surface and core regions of deformed nickel polycrystals with few grains across the cross section. *Mech. Mater.* **2010**, *42*, 44–54. [CrossRef]
55. Zubelewicz, A. Metal behavior in the extremes of dynamics. *Sci. Rep.* **2018**, *8*, 5162. [CrossRef]
56. Zubelewicz, A. Century-long Taylor-Quinney interpretation of plasticity-induced heating reexamined. *Sci. Rep.* **2019**, *9*, 9088. [CrossRef]
57. Ravi Chandran, K.S. A new exponential function to represent the effect of grain size on the strength of pure iron over multiple length scales. *J. Mater. Res.* **2019**, *34*, 2315–2324. [CrossRef]
58. Zubelewicz, A.; Addessio, F.L.; Cady, C. Constitutive model for uranium-niobium alloy. *J. Appl. Phys.* **2006**, *100*, 013523. [CrossRef]
59. Zubelewicz, A.; Zurek, A.K.; Potocki, M.L. Dynamic behavior of copper under extreme loading rates. *J. Phys. IV* **2006**, *134*, 23–27. [CrossRef]
60. Chen, S.R.; Kocks, U.F. On the strain rate dependence of dynamic recrystallization in copper polycrystals. In Proceedings of the International Conference Recrystallization 92, San Sebastian, Spain, 31 August–4 September 1992.
61. Freed, A.D.; Walker, K.P. High temperature constitutive modeling: Theory and applications. In Proceedings of the Winter Annual Meeting of the American Society of Mechanical Engineers, Atlanta, GA, USA, 1–6 December 1991.
62. Banerjee, B. An evaluation of plastic flow stress models for the simulation of high-temperature and high-strain-rate deformation of metals. *arXiv* **2005**, arXiv:cond-mat/0512466. [CrossRef]
63. Selyutina, N.S.; Borodin, E.N.; Petrov, Y.V. Structural-temporal peculiarities of dynamic deformation of nanostructured and nanoscaled metals. *Phys. Solid State* **2018**, *60*, 1813–1820. [CrossRef]
64. Thomas, S.A.; Veeser, L.R.; Turley, W.D.; Hixson, R.S. Comparison of CTH simulations with measured wave profiles for simple flyer plate experiments. *J. Dyn. Behav. Mater.* **2016**, *2*, 365–371. [CrossRef]
65. Zubelewicz, A. Iterative method of finite elements. In Proceedings of the 2nd Conference on Computational Methods in Mechanics of Structures, Gdansk, Poland, 24–26 November 1975.
66. Zubelewicz, A. A Certain Version of Finite Element Method. Ph.D. Thesis, Warsaw University of Technology, Warsaw, Poland, 1979.

© 2020 by the author. Licensee MDPI, Basel, Switzerland. This article is an open access article distributed under the terms and conditions of the Creative Commons Attribution (CC BY) license (http://creativecommons.org/licenses/by/4.0/).

Article

Ca-induced Plasticity in Magnesium Alloy: EBSD Measurements and VPSC Calculations

Umer Masood Chaudry, Kotiba Hamad * and Jung-Gu Kim *

School of Advanced Materials Science & Engineering, Sungkyunkwan University, Suwon 16419, Korea; umer@skku.edu
* Correspondence: hamad82@skku.edu (K.H.); kimjg@skku.ac.kr (J.-G.K.)

Received: 3 January 2020; Accepted: 23 January 2020; Published: 24 January 2020

Abstract: In the present work, Ca-induced plasticity of AZ31 magnesium alloy was studied using electron backscattered diffraction (EBSD) measurements supported by viscoplastic self-consistent (VPSC) calculations. For this purpose, alloy samples were stretched to various strains (5%, 10%, and 15%) at room temperature and a strain rate of 10^{-3} s^{-1}. The EBSD measurements showed a higher activity of non-basal slip system (prismatic slip) as compared to that of tension twins. The VPSC confirmed the EBSD results, where it was found that the critical resolved shear stress of the various slip systems and their corresponding activities changed during the stretching of the alloy samples.

Keywords: AZ31-Ca magnesium alloy; plasticity; slip systems; EBSD; VPSC; microstructure; texture

1. Introduction

Due to its low density and high natural abundance, magnesium (Mg) is considered as one of the most promising metals for structural applications. On the other hand, the naturally intrinsic brittleness of Mg, which is related to the hexagonal close packed (HCP) structure, limits its applicability as a structural material. In such structures, various slip families with several slip systems are usually available; those families are basal, prismatic, and pyramidal [1,2]. In spite of the presence of various slip families, the number of independent slip systems that can operate at room temperature are limited to those only belonging to the basal family, and this is due to the high critical resolved shear stress (CRSS) of prismatic and pyramidal slip systems (~40 and ~100 times higher than that of basal systems) [3]. One way to enhance the poor ductility of Mg-based materials is by weakening the basal texture observed after a primary processing (casting and rolling, for example). The evolution of weak texture leads to a higher activation of the non-basal systems due to the higher shear stress that can be resolved in the slip plane. Texture weakening in Mg alloys can be achieved through a thermomechanical processing of these alloys. For example, a weak basal-textured AZ31 alloy has been fabricated using a warm severe plastic deformation through differential speed rolling (DSR) process [4,5]. The results showed that the processing by DSR could optimize both strength and ductility in AZ31 Mg alloy, and this was related to the role of DSR in grain refinement and texture weakening of the AZ31 alloy. In addition, intermetallic compounds (IMC) formed due to the alloying of Mg with other elements can improve the ductility of Mg-based materials, where these compounds have a dual effect on the ductility. One is that these IMCs can contribute to the texture weakening during the primary processing, and hence, improve the ductility [6,7]. Another effect comes from the role of IMCs in changing the activity of the various slip systems, and this is related to the strengthening effect of the IMC in the basal plane [2,8]. Very recently, it has been reported that addition of 0.5wt. % calcium (Ca) can efficiently enhance the ductility and stretch formability of AZ31 Mg alloy [8], and this was attributed to the texture weakening induced by the (Mg,Al)$_2$Ca particles. Viscoplastic self-consistent (VPSC) calculations conducted on sample strained to 10% showed that the Ca addition can change the activity of the various slip systems. To gain further understanding of the Ca-induced plasticity in the AZ31 Mg alloy, more experiments and investigations are needed. Hence, the present study investigates the microstructural, textural evolution and deformation mechanisms of AZ31-0.5Ca as a result of various stretching conditions. In addition, particular focus was given to the slip behaviors and their related activities at various loading conditions. Accordingly, in the present work, VPSC calculations will be conducted on samples strained at various conditions, and these calculations will be supported by electron-back scattered diffraction (EBSD) measurements.

2. Experiments and VPSC Procedure

The composition of AZ31-0.5Ca Mg alloy investigated in the present work was 3.12 wt. % Al, 0.76 wt. % Zn, 0.5 wt. % Ca, 0.3 wt. % Mn and the balance is Mg. Samples (120 mm × 40 mm) cut from the alloy were stretched at room temperature and a strain rate of 10^{-3} s^{-1}. For EBSD measurements, specimens cut from the TD-RD (direction-rolling direction) plane of the stretched alloy samples were mechanically ground and polished using a cross-sectional polisher. The EBSD data were analyzed by using TSL OIM 6.1.3 software (EDAX Corporate, New Jersey, NY, USA). Electron probe micro-analyzing (EPMA) was also carried out on the non-stretched alloy.

To identify the deformation modes, which might be activated during the pre-stretching experiment, plasticity simulations based on a viscoplastic self-consistent (VPSC) model with full constraints (FC) approach were performed. The FC approach was used to account for the average overall response of the grains as a result of deformation. It is already established that a VPSC model can successfully predict the relative activities of various slip systems by comparing the simulated texture to the experimentally received texture based on the Voce hardening equation [9,10].

$$\tau^s = \tau_0^s + (\tau_1^s + \theta_1^s \Gamma)\left(1 - \exp\left(-\Gamma \left|\frac{\theta_0^s}{\tau_1^s}\right|\right)\right) \tag{1}$$

where Γ is accumulated shear strain and s, θ_0^s, θ_1^s, τ_0^s and $(\tau_0^s + \tau_1^s)$ are the slip system, initial and final slopes of the hardening curve, initial critical resolved shear stress, and the back-extrapolated CRSS, respectively. VPSC simulations begin with initial texture (~10,000 orientations) of AZ31-0.5Ca alloy that was used to predict the relative activities of slip systems during various stretching conditions. In this regard, CRSS and hardening parameters of each deformation mode were adjusted until the simulated textures matched with the experimentally received textures for all stretching conditions.

3. Results and Discussion

Figure 1 shows the microstructural features, texture and room temperature tensile properties of the AZ31-0.5Ca Mg alloy, investigated in the present work. It can be seen that this alloy has a fine-grained structure with weak basal texture, and interestingly, some twin boundaries (~4%) were observed in this alloy (tension twins at 86° as shown by Figure 1c). In addition, Figure 1d shows that this alloy exhibited an improved tensile ductility, which was higher than that of conventional AZ31 alloy. The enhanced ductility of the alloy is mainly related to the Ca addition and the intermetallic compounds (IMCs) that formed as a result of this addition. The EPMA maps of the alloy presented in Figure 1e confirmed the formation of the IMCs in the form of particles (~1 μm) and those particles are composed of Mg, Al and Ca, as shown by the EPMA maps associated with these elements.

For a further understanding on how such addition can contribute to the plasticity of the Mg alloy, EBSD observations were carried out on alloy samples stretched at various strains (5%, 10%, and 15%) and the counterpart VPSC calculations were then conducted. Figure 2 shows the inverse pole figure (IPF) maps of strained samples and the related microstructural and textural features. It can be noted that the straining of the alloy samples mainly lead to a further weakening in the basal texture, where most of orientations were tilted away from the normal direction pole and distributed along the rolling direction and this effect became more significant by increasing the amount of strain from

5% to 15% [11]. The grain size was also influenced by the magnitude of stretching where, after 15%, the structure was slightly coarser as compared to those after 5% and 10%; this is shown by arrows in Figure 2b. A more important trend was noticed from the misorientation angle distribution of the strained alloy samples, as shown in Figure 3c. In this trend, it can be noted that the fraction of the tension twin boundaries (at 86°) was higher as compared to that of the non-strained sample (Figure 1c) and this fraction decreased by increasing the magnitude of stretching, where fractions of 10%, 8% and 6% were observed in the samples strained at 5%, 10% and 15%, respectively. Even in such behavior, one can suggest the contribution of twins as a support mechanism to enhance the plasticity of this alloy; however, the fraction of the twins evolved in this alloy upon the stretching is still much smaller than that observed in conventional Mg alloy (~40% after 10% deformation) [12,13]. The lower activity of twin mode in this alloy can be attributed to the contribution of the more non-basal slip systems in the plasticity process. This can be roughly explored through the evolution of a larger fraction of low-angle grain boundaries (LAGBs) (<15%), where fractions of LAGBs of 30%, 40% and 55% were evolved in the samples strained at 5%, 10% and 15%, respectively. It is well known that boundaries with a low angle of misorientation—less than 15°—are considered as arrays of parallel dislocation lines. The higher activity of non-basal slip systems expected in this alloy based on the EBSD measurements is mainly attributed to the effect of Ca addition, which can be clarified through the VPSC calculations under the same stretching conditions. As aforementioned, initial orientations (~10,000 orientations) from the AZ31-0.5Ca alloy were introduced into the VPSC model, which were used to simulate the final textures received after the various pre-stretching conditions (5%, 10%, 15%). For each pre-stretching condition, the latent and voce hardening parameters were adjusted until the simulated pole figures (PFs) were almost identical to the experimentally obtained PFs. Figure 3 shows the simulated and received (0001) PFs for 5%, 10% and 15% stretched samples. It can be seen that VPSC modeling reproduced the PFs for all the various conditions successfully, indicating that the parameters used for conducting simulations are reliable. The CRSS values for various deformation modes for all the stretched conditions (5%, 10%, 15%) are listed in Table 1. The received values were found to be higher as compared to a Mg single crystal. It can be attributed to the formation of $(Mg,Al)_2Ca$ intermetallic particles as a result of Ca addition to AZ31 which, in turn, can act as an obstacle to plastic deformation. In addition, higher CRSS values of pyramidal slip were recorded for all the stretched conditions. However, some interesting differences in the CRSS of other deformation modes can be noticed. For instance, the CRSS of basal slip was received to be almost double to prismatic slip (50 vs 95) for the 5% stretched sample, while on the other hand, a smaller difference in CRSS of the basal slip and prismatic slip (57 vs 91) was observed for the 10% stretched sample. This difference was received to be further reduced as a result of 15% stretching (Figure 3b). It is already established that as the ratio ($CRSS_{Prismatic}/CRSS_{Basal}$) is reduced, it can lead to the higher activity of prismatic slip system which, in turn, can enhance the performance of the Mg alloy [14–17]. Figure 3c demonstrates the activities of prismatic slip, basal slip and tension twins in AZ31-0.5Ca, which was stretched by 15%, along the rolling direction. As is evident from Figure 3c, at the beginning of the deformation process, the basal slip acted as the dominating slip system. As the deformation process continued, higher activity of the prismatic slip was noticed, which eventually surpassed the basal slip for higher stretching conditions. In addition, a very low activity of tension twins was recorded during the deformation process.

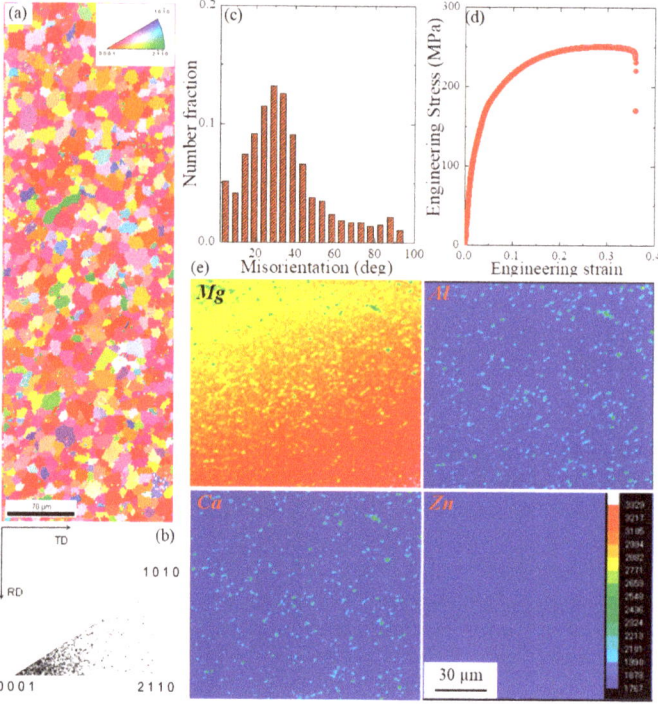

Figure 1. (**a**), (**b**) and (**c**) Inverse pole figure map, texture inverse pole figure, and misorientation angle distribution of the AZ31-Ca Mg alloy, respectively. (**d**) Room temperature tensile curve of the alloy. (**e**) Electron probe microanalysis maps of the alloy.

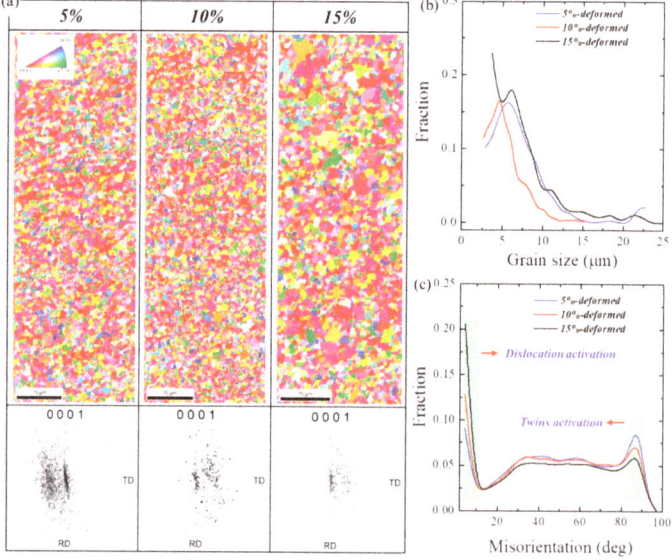

Figure 2. (**a**) Inverse pole figure maps and related 0001 pole figures of the AZ31-Ca Mg alloy samples stretched to 5%, 10% and 15%. (**b**) and (**c**) grain size and misorientation profile of the stretched alloy samples, respectively.

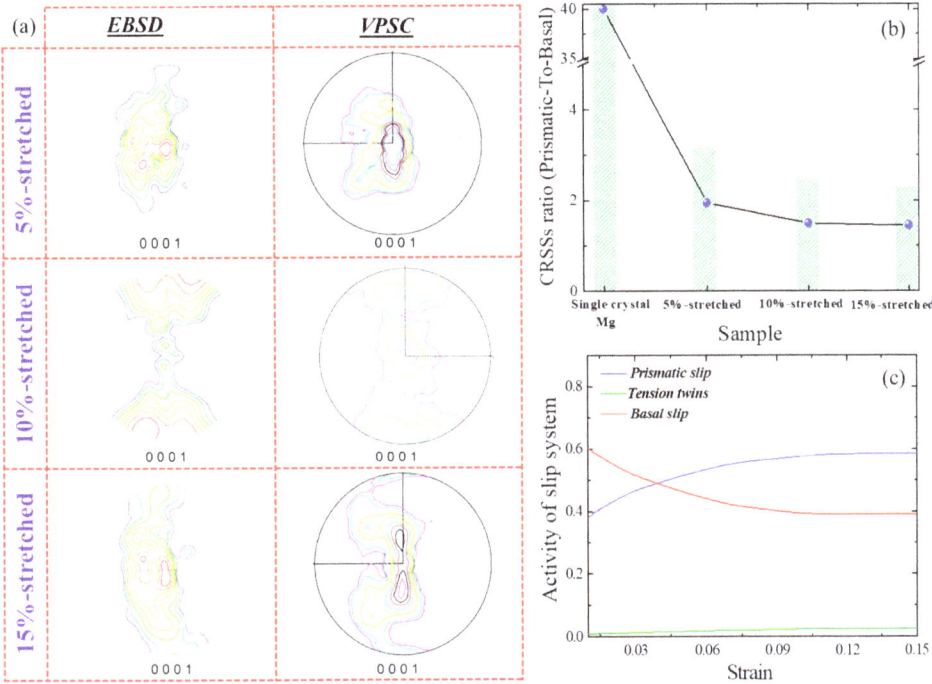

Figure 3. (a) Pole figures of the stretched samples obtained by the EBSD measurements and by VPSC modeling. (b) CRSSs ratio (prismatic/basal) of single crystal Mg and the stretched samples. (c) The activity of basal slip, prismatic slip and tension twins of the AZ31-Ca alloy investigated in the present work.

Table 1. CRSS values and hardening parameters for VPSC modeling of the samples (5%, 10%, 15% stretched).

Alloy Sample	Deformation Mode	τ_0	τ_1	θ_0	θ_1	$CRSS_P/CRSS_B$
5%	<a> basal slip	50	12	99	55	1.90
	<a> prismatic slip	95	29	135	10	
	<a+c> pyramidal slip	315	38	57	19	
	Tension twin	14	170	105	50	
10%	<a> basal slip	57	9	115	62	1.50
	<a> prismatic slip	91	37	148	8	
	<a+c> pyramidal slip	305	44	64	22	
	Tension twin	15	185	110	55	
15%	<a> basal slip	65	15	125	45	1.47
	<a> prismatic slip	96	35	137	9	
	<a+c> pyramidal slip	302	195	58	33	
	Tension twin	20	178	112	58	

In-crystal misorientation axis (ICMA) measurement carried out by EBSD is a useful tool to determine the type of slip system which might operate during deformation of metallic materials [18,19].

Generally, dislocations generated during the early stage of plastic deformation bend the crystal, leading to the so-called in-crystal misorientation, and through the characteristics of this misorientation axis, the type of dislocation can be determined. For example, basal dislocations induce the crystal bending around $\langle 10\bar{1}0 \rangle$, whereas an axis parallel to $\langle 0001 \rangle$ is obtained in the case of prismatic dislocation-plasticized crystals [20,21]. The ICMA measurements carried out on the stretched alloys samples are shown in Figure 4. The ICMA was conducted on three randomly selected grains from each sample. Figure 4 shows that the number of misorientation axes that come to be parallel with $\langle 0001 \rangle$ increased upon increasing the magnitude of stretching, suggesting that the crystal plasticity of this alloy is mainly aided by the prismatic dislocations rather than the activation of tension twins. This is consistent with the misorientation angle distributions and VPSC calculations presented in Figures 2c and 3, respectively. The higher activity of prismatic slip systems is mainly related to the Ca addition, which led to the formation of $(Mg,Al)_2Ca$ particles, as shown by the EPMA maps in Figure 1e. During the primary processing of the Ca-added Mg alloy, these particles can weaken the basal texture through inducing the recrystallization [22,23]. Accordingly, one way to explain the high activity of the prismatic slip is the particle-induced basal texture weakening, where in such a texture, a higher amount of stress can be resolved in the prismatic plane when comparing with strong basal-textured Mg, leading to the high activity of the prismatic slip. Another approach to figure out such high activity of prismatic slip is related to the strengthening effect of these particles in the basal plane which, in turn, lead to reducing the ratio between the CRSS of prismatic and CRSS of basal (Table 1). The intrinsically hard-to-operate prismatic slip becomes preferred relative to the extrinsically hardened basal slip, leading to the enhanced plasticity of the alloy investigated in the present work. Further transmission electron microscopy-based observations are required to investigate the nature and characteristics of the interaction between basal dislocations and particles in the AZ31- 0.5Ca Mg alloy.

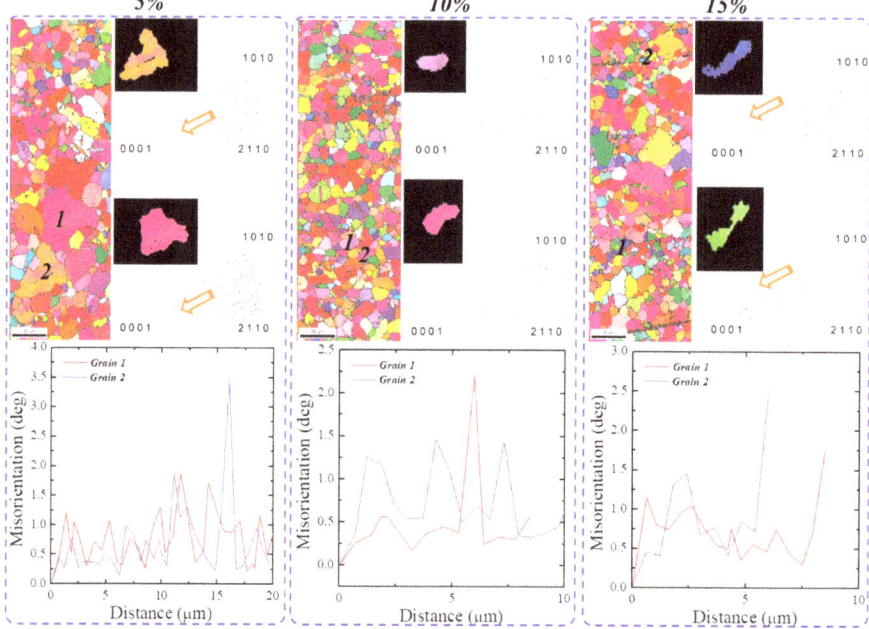

Figure 4. In-crystal misorientation axis (ICMA) measurements of the stretched samples. The misorientation profiles presented in this figure were taken from the grains used for the ICMA measurements.

4. Conclusions

In the present work, EBSD measurements supported by VPSC calculations were used to explain the high plasticity recorded in AZ31-0.5Ca alloys. The results showed that the plastic deformation in the alloy is mainly aided by the operation of dislocations rather than the activation of twinning. This is related to the texture weakening induced by the intermetallic particles formed in the AZ31 alloy due to the Ca addition. The strengthening effect of these particles on the basal dislocations, additionally, reduced the ratio $CRSS_P/CRSS_B$, leading to the higher contribution of prismatic slip in the plasticity of this alloy.

Author Contributions: K.H. and J.-G.K designed the experiments. U.M.C. carried out the experiments and VPSC simulations. K.H. and U.M.C. analyzed the data and wrote the manuscript. K.H., J.-G.K. and U.M.C. revised the manuscript. All authors have read and agreed to the published version of the manuscript.

Funding: This research was funded by National Research Foundation (NRF) of South Korea (2017R1C1B5017204).

Acknowledgments: The authors express their gratitude to *C. N. Tomé* (Material Science and Technology Division, Los Alamos National Laboratory) for providing the VPSC software to conduct the simulations. This research was supported by National Research Foundation (NRF) of Korea (2017R1C1B5017204).

Conflicts of Interest: The authors declare no conflict of interest.

References

1. Miller, V.M.; Berman, T.D.; Beyerlein, I.; Jones, J.W.; Pollock, T.M. Prediction of the plastic anisotropy of magnesium alloys with synthetic textures and implications for the effect of texture on formability. *Mater. Sci. Eng. A* **2016**, *675*, 345–360. [CrossRef]
2. Chaudry, U.M.; Hamad, K.; Kim, J.G. On the ductility of magnesium based materials: A mini review. *J. Alloys Compd.* **2019**, *792*, 652–664. [CrossRef]
3. Agnew, S.R.; Duygulu, O. Plastic anisotropy and the role of non-basal slip in magnesium alloy AZ31B. *Int. J. Plast.* **2005**, *21*, 1161–1193. [CrossRef]
4. Luo, D.; Luo, D.; Wang, H.; Zhao, L.; Wang, C.; Liu, G.; Liu, Y.; Jiang, Q. Effect of differential speed rolling on the room and elevated temperature tensile properties of rolled AZ31 Mg alloy sheets. *Mater. Charac.* **2017**, *124*, 223–228. [CrossRef]
5. Hamad, K.; Ko, Y.G. A cross-shear deformation for optimizing the strength and ductility of AZ31 magnesium alloys. *Sci. Rep.* **2016**, *6*, 1–8. [CrossRef] [PubMed]
6. Samman, T.Al.; Li, X. Sheet texture modification in magnesium-based alloys by selective rare earth alloying. *Mater. Sci. Eng. A* **2011**, *528*, 3809–3822. [CrossRef]
7. Sandlobes, S.; Pie, Z.; Friak, M.; Zhu, L.F.; Wang, F.; Zaefferer, S.; Raabe, D.; Neugebauer, J. Ductility improvement of Mg alloys by solid solution: Ab initio modeling, synthesis and mechanical properties. *Acta. Mater.* **2014**, *70*, 92–104. [CrossRef]
8. Chaudry, U.M.; Kim, T.H.; Park, S.D.; Kim, Y.S.; Hamad, K.; Kim, J.G. On the high formability of AZ31-0.5Ca magnesium alloy. *Materials.* **2018**, *11*, 1–15.
9. Agnew, S.R.; Yoo, M.H.; Tomé, C.N. Application of texture simulation to understanding mechanical behavior of Mg and solid solution alloys containing Li or Y. *Acta. Mater.* **2001**, *49*, 4277–4289. [CrossRef]
10. Lebensohn, R.A.; Tomé, C.N. A self-consistent anisotropic approach for the simulation of plastic deformation and texture development of polycrystals: Application to zirconium alloys. *Acta. Metal et Mater.* **1993**, *41*, 2611–2624. [CrossRef]
11. Khan, A.S.; Pandey, A.; Herold, T.G.; Mishra, R.K. Mechanical response and texture evolution of AZ31 alloy at large strains for different strain rates and temperatures. *Int. J. Plast.* **2011**, *27*, 688–706. [CrossRef]
12. Sandlobes, S.; Friak, M.; Kerzel, S.K.; Pie, Z.; Neugebauer, J.; Raabe, D. A rare-earth free magnesium alloy with improved intrinsic ductility. *Sci. Rep.* **2017**, *7*, 1–8. [CrossRef] [PubMed]
13. He, J.; Jiang, B.; Yang, Q.; Li, X.; Xia, X.; Pan, F. Influence of pre-hardening on microstructure evolution and mechanical behavior of AZ31 magnesium alloy sheet. *J. Alloys Compd.* **2015**, *621*, 301–306. [CrossRef]

14. Martin, R.S.; Prado, M.T.P.; Segurado, J.; Bohlen, J.; Urrutia, I.G.; Llorca, J.; Aldareguia, J.M.M. Measuring the critical resolved shear stress in Mg alloys by instrumented nanoindentation. *Acta. Mater.* **2014**, *71*, 283–292. [CrossRef]
15. Manrique, H.P.; Yi, S.B.; Bohlen, J.; Letzig, D.; Perez, P.M.T. Effect of Nd additions on extrusion texture development and slip activity in a Mg-Mn alloy. *Metall. Mater. Trans. A* **2013**, *44*, 4819–4829. [CrossRef]
16. Chaudry, U.M.; Kim, T.H.; Park, S.D.; Kim, Y.S.; Hamad, K.; Kim, J.G. Effect of calcium on the activity of slip systems in AZ31 magnesium alloy. *Mater. Sci. Eng. A* **2019**, *739*, 289–294. [CrossRef]
17. Angew, S.R.; Tome, C.N.; Brown, D.W.; Holden, T.M.; Vogel, S.C. Study of slip mechanisms in a magnesium alloy by neutron diffraction and modeling. *Scripta. Mater.* **2003**, *48*, 1003–1008.
18. Suh, B.C.; Kim, J.H.; Bae, J.H.; Hwang, J.H.; Shim, M.S.; Kim, N.J. Effect of Sn addition on the microstructure and deformation behavior of Mg-3Al alloy. *Acta. Mater.* **2017**, *124*, 268–279. [CrossRef]
19. Chun, Y.B.; Battaini, M.; Davies, C.H.J.; Hwang, S.K. Distribution characteristics of in-grain misorientation axes in cold-rolled commercially pure titanium and their correlation with active slip modes. *Metall. Mater. Trans. A* **2010**, *41*, 3473–3487. [CrossRef]
20. Chaudry, U.M.; Kim, Y.S.; Hamad, K. Effect of Ca addition on the room-temperature formability of AZ31 magnesium alloy. *Mater. Lett.* **2019**, *238*, 305–308. [CrossRef]
21. Yuasa, M.; Miyazawa, N.; Hayashi, M.; Mabuchi, M.; Chino, Y. Effects of group II elements on the cold stretch formability of Mg-Zn alloys. *Acta. Mater.* **2015**, *83*, 294–303. [CrossRef]
22. Trang, T.T.T.; Zhang, J.H.; Kim, J.H.; Zargaran, A.; Hwang, J.H.; Suh, B.C.; Kim, N.J. Designing a magnesium alloy with high strength and high formability. *Nat. Comm.* **2018**, *9*, 1–6. [CrossRef] [PubMed]
23. Kim, J.H.; Kang, N.E.; Yim, C.D.; Kim, B.K. Effect of calcium content on the microstructural evolution and mechanical properties of wrought Mg-3Al-1Zn alloy. *Mater. Sci. Eng. A* **2009**, *525*, 18–29. [CrossRef]

© 2020 by the authors. Licensee MDPI, Basel, Switzerland. This article is an open access article distributed under the terms and conditions of the Creative Commons Attribution (CC BY) license (http://creativecommons.org/licenses/by/4.0/).

Article

How Nanoscale Dislocation Reactions Govern Low-Temperature and High-Stress Creep of Ni-Base Single Crystal Superalloys

David Bürger [1,*], Antonin Dlouhý [2], Kyosuke Yoshimi [3] and Gunther Eggeler [1]

1 Institute for Materials, Ruhr-Universität Bochum, Universitätsstr. 150, 44801 Bochum, Germany; gunther.eggeler@rub.de
2 Institute of Physics of Materials, Zizkova 22, 616 62 Brno, Czech Republic; dlouhy@ipm.cz
3 Department of Materials Science, Tohoku University, 6-6-02 Aramaki Aza Aoba, Aoba-ku, Sendai 980-8579, Japan; yoshimi@material.tohoku.ac.jp
* Correspondence: david.buerger@rub.de

Received: 27 January 2020; Accepted: 20 February 2020; Published: 22 February 2020

Abstract: The present work investigates γ-channel dislocation reactions, which govern low-temperature (T = 750 °C) and high-stress (resolved shear stress: 300 MPa) creep of Ni-base single crystal superalloys (SX). It is well known that two dislocation families with different b-vectors are required to form planar faults, which can shear the ordered γ'-phase. However, so far, no direct mechanical and microstructural evidence has been presented which clearly proves the importance of these reactions. In the mechanical part of the present work, we perform shear creep tests and we compare the deformation behavior of two macroscopic crystallographic shear systems $[01\bar{1}](111)$ and $[11\bar{2}](111)$. These two shear systems share the same glide plane but differ in loading direction. The $[11\bar{2}](111)$ shear system, where the two dislocation families required to form a planar fault ribbon experience the same resolved shear stresses, deforms significantly faster than the $[01\bar{1}](111)$ shear system, where only one of the two required dislocation families is strongly promoted. Diffraction contrast transmission electron microscopy (TEM) analysis identifies the dislocation reactions, which rationalize this macroscopic behavior.

Keywords: dislocation reactions; single crystal Ni-base superalloys; shear creep testing; transmission electron microscopy; nucleation of planar fault ribbons

1. Introduction

Creep, the time-dependent plastic deformation of materials, limits the service life of single crystal Ni-base superalloys (SX) [1,2]. During low-temperature (T < 800 °C) and high-stress (σ > 600 MPa) ⟨100⟩ tensile creep of SX, dislocation ribbons with a ⟨112⟩ displacement vectors move through the γ/γ'-microstructure [3–9]. The nucleation of such ribbons requires the reaction between ordinary a/2⟨110⟩γ-channel dislocations. In the early stages of low-temperature and high-stress primary tensile creep in the ⟨100⟩ direction, two creep rate minima are observed [10,11]. Dislocation processes, which rationalize this phenomenon, were studied. The increase of creep rate after the first minimum was interpreted as a result of an increasing intensity of reactions between two a/2⟨110⟩ γ-channel dislocations with different Burgers vectors. While this reaction could be modelled using a simplified two-dimensional discrete dislocation dynamics (2D DDD) model [11], no direct experimental evidence for this reaction has been provided. In the present work, we combine high precision shear creep testing [12] with diffraction contrast scanning transmission electron microscopy (STEM) to show that these dislocation reactions govern low-temperature and high-stress SX creep.

2. Materials and Methods

2.1. Material and Shear Creep Testing

The material investigated in the present work is the single crystal superalloy ERBO-1C (CMSX-4 type), which was available as a 140 mm × 100 mm × 20 mm slab. The chemical composition is given in Table 1 and all other details concerning its heat treatment and its microstructure have been published elsewhere [10,11,13,14].

Table 1. Chemical composition of ERBO/1 in weight-%.

Element	Al	Co	Cr	Hf	Mo	Re	Ta	Ti	W	Ni
weight-%	5.5	9.7	6.4	0.1	0.6	2.9	6.5	1.0	6.4	Bal.

The average spacing between primary dendrites was close to 500 μm [10,13]. After heat treatment and prior to creep, the average γ'-volume fractions were 72% and 77% in the dendritic and interdendritic regions, respectively [13,14]. In the present work, we used a shear creep test technique, which was recently documented [12]. The SX material was first mounted onto a goniometer and oriented into a specific crystallographic direction using a Laue system from Multiwire Laboratories (Lansing, NY, USA), equipped with an electronic position control unit and a sensitive real-time back reflection camera of type MWL120. The goniometer with the oriented specimen was subsequently fitted into an electro discharge machine (EDM) of type Charmilles Robofil 240 (Schorndorf, Germany), where the specimen was spark erosion machined as described in [15]. Figure 1 shows a photograph of the specimen together with a drawing which illustrates dimensions (shear width L = 2 mm, height H = 8 mm, and depth D = 3 mm) and crystallographic directions. Note that two macroscopic crystallographic shear systems (MCSS) are considered: $[01\bar{1}](111)$ and $[11\bar{2}](111)$. During shear creep testing, the outer two loading wings were pulled upwards while the central loading section was pushed down. During creep, the shear displacement Δ was measured and the engineering shear γ was obtained by

$$\gamma = \frac{\Delta}{L} \quad (1)$$

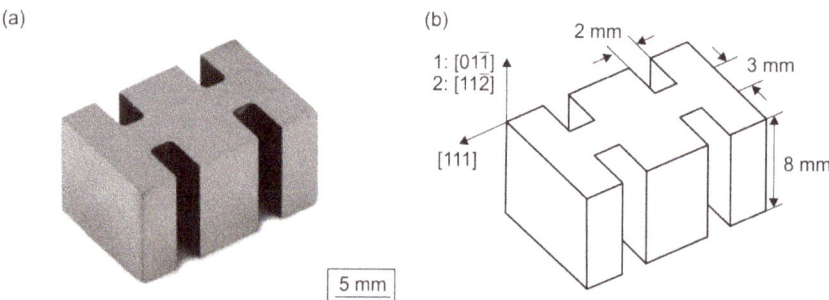

Figure 1. Double shear creep specimen. (a) Specimen after manufacturing. (b) Schematic drawing showing specimen dimensions and the shear crystallography.

Creep testing was performed using a creep machine of type Denison Mayes (Hunslet, Leeds, United Kingdom), equipped with a vertically movable three-zone resistance furnace. The individual zones were controlled by Eurotherm controllers (Neuss, Germany). The double shear creep specimen was positioned in the temperature constant zone of the furnace (at 750 °C: 100 mm). In addition to three control thermocouples, two measurement thermocouples were fixed to the upper and lower part of the shear specimens. The precision of the temperature measurement at 750 °C was

±2 °C. Displacements were measured using a ceramic rod in tube extensometry, which transferred displacements to electromechanical sensors (linear variable differential transformers, LVDTs) outside the furnace.

After mounting the specimens into the creep test rig, they were heated up to temperature within 2 to 3 hours under a small pre-load, which corresponds to 5% of the end load. Loads were applied over a 15:1 lever arm within 1 minute. As the creep specimens accumulated strain, the lever arm was kept in a horizontal position by an electromechanical control system. In the tensile studies which motivated the present work [10,11,15], the applied tensile stress in the ⟨100⟩ direction was 800 MPa. Comparable shear experiments on a macroscopic {111} shear system must therefore be performed at a directly applied shear stress of 328 MPa (800 MPa × 0.41). This shear stress yielded too high creep rates and therefore a stress of 300 MPa was chosen for the present work. Figure 2 shows a [01$\bar{1}$](111) shear creep experiment, which was performed at a test temperature of 750 °C and 300 MPa. The experiment was interrupted after 5% shear strain, which corresponds to a shear displacement of 100 µm. This extensometer displacement result was verified in the SEM, as shown in Figure 2b,c. Creep tests were performed at a shear stress similar to the resolved shear stress which governed [001] tensile creep testing at 800 MPa (creep condition considered in [10,11]).

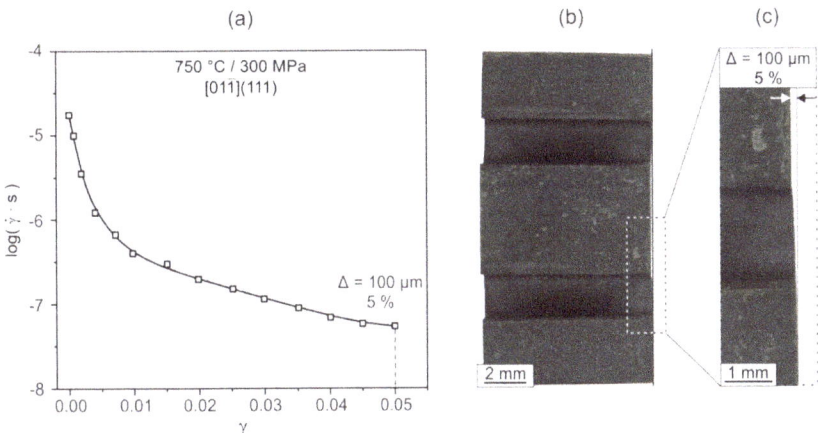

Figure 2. Shear creep displacements in the macroscopic crystallographic shear system [01$\bar{1}$](111). (**a**) Creep curve presented as the logarithmic shear rate log($\dot{\gamma}$) as a function of engineering shear γ. The experiment was interrupted at γ = 5% corresponding to a displacement of 100 µm. (**b**) and (**c**): Corresponding specimen after 5% shear creep deformation, and 100 µm shear displacement detected in the SEM.

2.2. Scanning Transmission Electron Microscopy (STEM)

STEM was performed on crept specimens using a Jeol JEM-2100 F (Tokyo, Japan) operating at 200 kV. To prepare TEM specimens, 300 µm slices were cut out parallel to the shear planes of the double shear creep specimen using an Accutom 5 cutting disk from Struers. These were polished down to a thickness of 100 µm using emery paper of mesh size 4000. Electron transparent thin foils were obtained by electrochemical thinning using a TenuPol-5. An electrolyte consisting of 70 vol.-% methanol, 20 vol.-% glycerin, and 10 vol.-% perchloric acid yielded good thinning results at 2 °C, a voltage close to 14 V, and a flow rate of 16. All details describing TEM specimen preparation and the standard TEM procedures used in the present work have been described elsewhere [10,11,16–19]. The microstructure of the material prior to creep is shown in Figures 2 (SEM) and 3 (TEM) of [10]. No rafting is observed after 750 °C creep exposure [10]. It was shown how γ-channel widths (15–125 nm, average: 65 nm) and γ'-cube sizes (50–750 nm, average: 442 nm) scatter [10]. In the present work, we evaluated

STEM micrographs, which were taken using two beam contrast conditions. Table 2 summarizes corresponding diffraction vectors together with the figure number of the STEM micrograph presented in this work.

Table 2. The operating diffraction vectors used in this work with the appropriate figure numbers for conditions shown in the last row. Note that in Figure 5c the condition minus g9 is used.

g-vector	g1	g2	g3	g4	g5	g6	g7	g8	g9	g10	g11
Type	$(1\bar{1}\bar{1})$	$(1\bar{1}\bar{1})$	$(0\bar{2}2)$	$(\bar{2}20)$	$(\bar{2}02)$	$(\bar{1}1\bar{1})$	$(\bar{1}1\bar{1})$	$(\bar{1}\bar{1}1)$	$(3\bar{1}\bar{1})$	$(1\bar{3}1)$	$(0\bar{2}0)$
Fig. no.	4b	5b,6a	6d	6e	6f	6b	4a, 5a	6c, 7a	5c, 6g	6h	6i

In the present work, dislocation densities were determined using Ham's intercept method [20], following the procedure described by Wu et al. [10]. The thicknesses of TEM foils were measured using a stereo technique [19,21]. Dislocation densities were evaluated from 4 to 5 TEM micrographs per material state (g: $(1\bar{1}\bar{1})$). The micrographs were taken from TEM foils parallel to (111). Foil regions of about 10×10 μm² were considered. Sixteen reference lines (8 lines parallel and 8 horizontal to shear direction) were evaluated for one dislocation density value. The measured dislocation densities represent mean values obtained from all micrographs. Error bars show the highest and lowest dislocation density measured on the micrographs for one material state.

2.3. Calculation of Peach Köhler Glide and Climb Forces (PKFg and PKFc)

In order to rationalize dislocation reactions, Peach Köhler glide and climb forces were calculated, using the 2D DDD methodology described previously [11,18,22,23]. We considered an actual dislocation substructure, which was characterized using STEM and used the input parameters listed in [11] (see Table 3 in [11]).

3. Results

3.1. Creep in the Two MCSSs

In Figure 3, we compare the creep behavior of the macroscopic shear systems $[01\bar{1}](111)$ and $[11\bar{2}](111)$. Two creep curves are shown for the two systems, which both show reproducible creep behavior. In Figure 3a, shear strains are plotted as a function of time t. Figure 3b shows the logarithmic creep rates as a function of strain. Early on, the two shear systems deform at a similar rate. However, starting at shear strains of the order of 0.5%, the creep rates of the two shear systems start to deviate from each other. At shear strains close to 1%, the $[11\bar{2}](111)$ shear system deforms four times faster than the $[01\bar{1}](111)$ shear system. This ratio increases to 25 at a strain of 4.25%, as shown in Figure 3b.

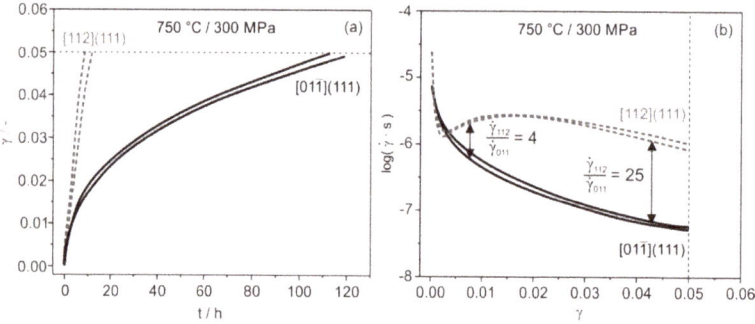

Figure 3. Direct comparison of the creep behavior of the two macroscopic shear systems $[01\bar{1}](111)$ and $[11\bar{2}](111)$ at 750 °C and 300 MPa. (a) Creep curves presented as shear strain γ vs. time t. (b) Creep data presented as log($\dot{γ}$) vs. γ.

3.2. Dislocations and Planar Faults

Overview micrographs of dislocations and planar faults that formed during shear creep deformation at 750 °C and 300 MPa are shown in Figure 4a,b. They illustrate the state of the microstructure after 1% shear deformation in the macroscopic $[01\bar{1}](111)$ (Figure 4a, $g7=(\bar{1}11)$) and $[11\bar{2}](111)$ (Figure 4b, $g1=(1\bar{1}1)$) shear systems. Cube projections in the upper left corners of both STEM micrographs indicate the foil orientations. At the equivalent contrast condition of type {111}, the two microstructures show clear differences, as shown in Figure 4a,b. A high density of gamma channel dislocations is observed in Figure 4a. These dislocations enter the γ-channels in the early stages of creep and, depending on the channel type, deposit 60° and screw dislocation segments at the γ/γ'-interfaces. In Figure 4a, dislocations do not cut into the γ'-particles and no planar faults in the γ'-phase are observed. In contrast, numerous γ'-stacking faults (either superlattice intrinsic (SISF) or extrinsic (SESF)) have formed after 1% shear in the $[11\bar{2}](111)$ shear system, as shown in Figure 4b. One of the early γ'-cutting events is marked by a white arrow in Figure 4b.

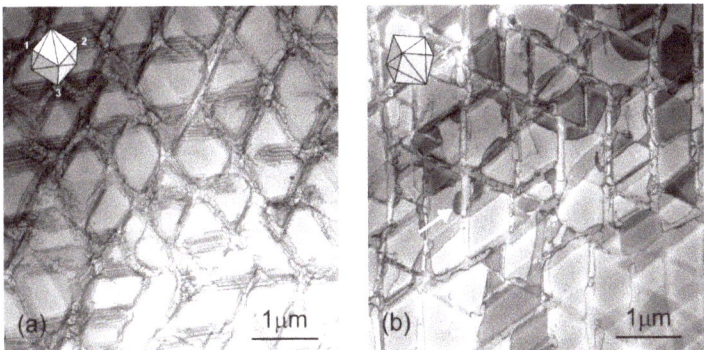

Figure 4. Scanning transmission electron microscopy (STEM) micrographs showing dislocation microstructures, which formed during shear creep deformation at 750 °C and 300 MPa to 1% strain. (a) Macroscopic shear system $[01\bar{1}](111)$. (b) Macroscopic shear system $[11\bar{2}](111)$.

Figure 5 shows three STEM micrographs taken after 1% $[01\bar{1}](111)$ shear creep deformation. These micrographs document full visibility for $g7=(\bar{1}11)$, as shown in Figure 5a, and two effective invisibilities for $g2=(1\bar{1}1)$, as shown in Figure 5b, and $-g9 = (\bar{3}11)$, as shown in Figure 5c. The effective $g·b = 0$ invisibility criterion suggests that the Burgers vectors of these dislocations are of type $\pm 1/2[01\bar{1}]$. Several of the leading $\pm 1/2[01\bar{1}]$ segments shown in Figure 5a (three of which are highlighted by white arrows pointing up) decompose into classical stacking fault coupled Shockley partials $\pm 1/6[\bar{1}2\bar{1}]$ and $\pm 1/6[11\bar{2}]$ on the (111) plane. The white arrow pointing down marks a contamination, which serves as a reference location in all three micrographs of Figure 5. The locations of the three Shockley partials are also highlighted by white arrows in Figure 5b.

Nine STEM images of the cutting event highlighted with the white arrow in Figure 4b are shown in Figure 6a–i. They were taken using different contrast conditions (g-vectors) after macroscopic $[11\bar{2}](111)$ shear creep deformation to 1% (750 °C, 300 MPa). Stereo microscopy was used (images not shown) in order to clarify the spatial arrangement of individual dislocation segments [19,21].

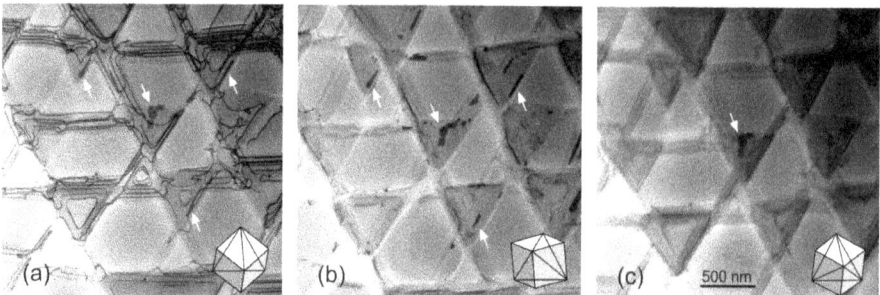

Figure 5. STEM micrographs taken after 1% strain accumulated in the $[01\bar{1}](111)$ shear creep system at 750 °C and 300 MPa. (a) **g7** $=(\bar{1}11)$: $1/2[01\bar{1}]$-type dislocations and stacking faults (SFs) in full contrast. (b) **g2** $= (1\bar{1}\bar{1})$: $1/2[01\bar{1}]$-type dislocations out of contrast, SFs in contrast. (c) **−g9** $=(\bar{3}11)$: $1/2[01\bar{1}]$-type dislocations and SFs out of contrast.

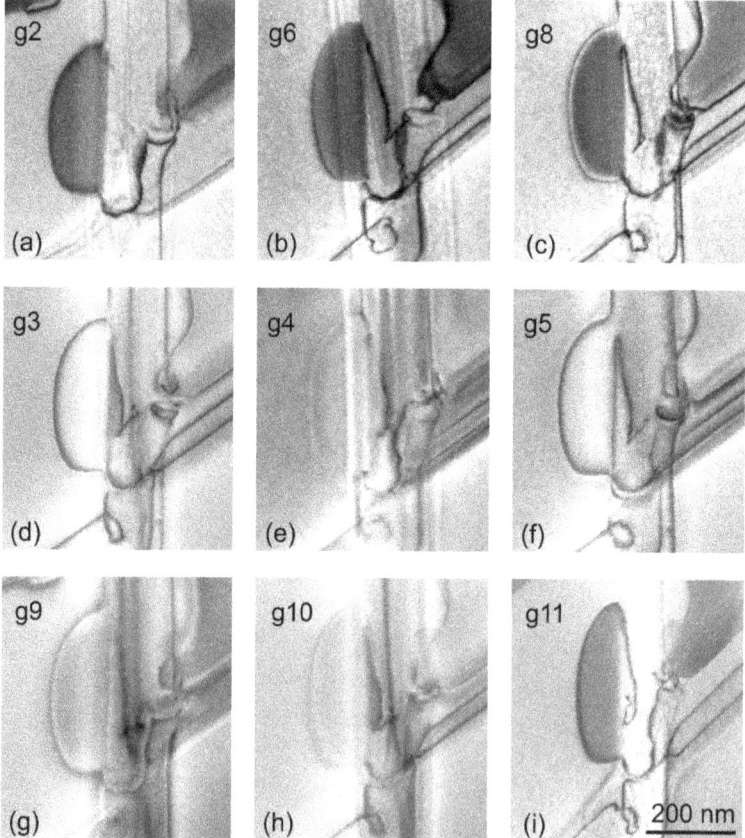

Figure 6. Nine STEM micrographs taken at different g-vectors of the cutting event marked with a white arrow in Figure 7b. (a) **g2** $=(1\bar{1}\bar{1})$. (b) **g6** $=(\bar{1}1\bar{1})$. (c) **g8** $=(\bar{1}\bar{1}1)$. (d) **g3** $=(0\bar{2}2)$. (e) **g4** $=(\bar{2}20)$. (f) **g5** $=(\bar{2}02)$. (g) **g9** $=(\bar{3}1\bar{1})$. (h) **g10** $=(1\bar{3}1)$. (i) **g11** $=(0\bar{2}0)$.

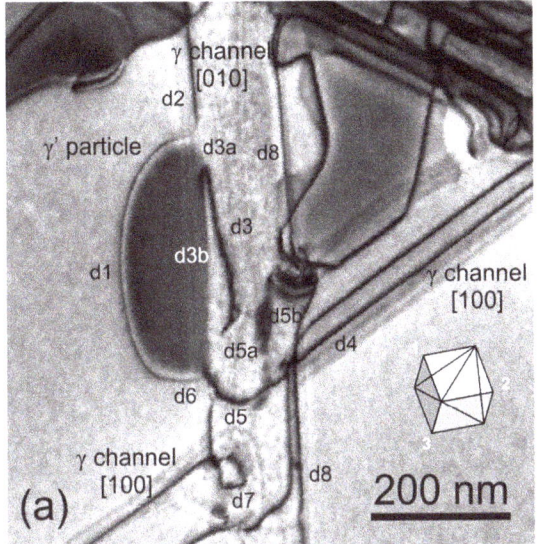
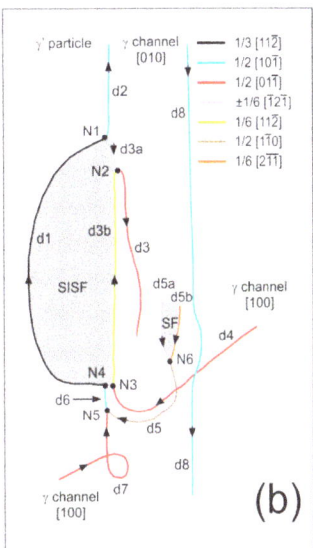

Figure 7. STEM results and schematic illustrations of dislocation reactions, which are involved in forming planar fault. (**a**) Experimental STEM image. (**b**) Identification of Burgers vectors and orientations of dislocation line segments (d1 to d8). Nodes N1 to N6 indicate locations where dislocation reactions have taken place.

Figure 7a shows the microstructure presented in Figure 6c at a higher magnification. Figure 7a and the associated schematic drawing in Figure 7b identify individual dislocation segments. The results of the effective visibility/invisibility Burgers vector analysis are compiled in Table 3 (effective visibilities/invisibilities for **g2** to **g6**) and Table 4 (effective visibilities/invisibilities for **g8** to **g11**). The last column of Table 4 shows the resulting b-vectors.

Table 3. Effective visibilities (+) and invisibilities (-) for g-vectors 2 to 6 from Table 2.

Dislocation	g2: $(\bar{1}\bar{1}1)$		g3:$(0\bar{2}2)$		g4: $(\bar{2}20)$		g5: $(\bar{2}02)$		g6: $(\bar{1}11)$	
d1	+	2/3	+	−2	res	0	+	−2	+	2/3
d2	+	1	+	−1	+	−1	+	−2	−	0
d3	−	0	+	−2	+	1	+	−1	+	1
d3a	−	−1/3	+	−1	+	1	−	0	+	2/3
d3b	−	−1/3	w	−1	−	0	+	−1	−	1/3
d4	−	0	+	−2	+	1	+	−1	+	1
d5	+	1	+	1	+	−2	+	−1	+	−1
d5a	−	1/3	+	1	+	−1	−	0	+	−2/3
d5b	+	2/3	−	0	w	−1	+	−1	−	−1/3
d6	+	1	+	−1	?	−1	+	−2	−	0
d7	−	0	+	−2	+	1	+	−1	+	1
d8	+	1	+	−1	w	−1	+	−2	−	0

Table 4. Effective visibilities (+) and invisibilities (−) for g-vectors 8 to 11 from Table 2.

Dislocation	g8: (111)		g9: (311)		g10: (131)		g11: (020)		b/a
d1	+	−4/3	+	4/3	+	−4/3	+	−2/3	1/3[11$\bar{2}$]
d2	+	−1	+	2	−	0	−	0	1/2[10$\bar{1}$]
d3	+	−1	res	0	+	−2	+	−1	1/2[01$\bar{1}$]
d3a	−	−1/3	−	−2/3	+	−4/3	+	−2/3	1/6[$\bar{1}$2$\bar{1}$]
d3b	w	−2/3	+	2/3	−	−2/3	−	−1/3	1/6[11$\bar{2}$]
d4	+	−1	res	0	+	−2	+	−1	1/2[01$\bar{1}$]
d5	−	0	+	2	?	2	+	1	1/2[1$\bar{1}$0]
d5a	−	1/3	?	2/3	?	4/3	+	2/3	1/6[1$\bar{2}$1]
d5b	−	−1/3	+	4/3	?	2/3	−	−1/3	1/6[2$\bar{1}\bar{1}$]
d6	+	−1	+	2	−	0	−	0	1/2[10$\bar{1}$]
d7	+	−1	res	0	+	−2	+	−1	1/2[01$\bar{1}$]
d8	+	−1	+	2	−	0	−	0	1/2[10$\bar{1}$]

Figure 8 shows how dislocation densities evolve as a function of shear strain during macroscopic [01$\bar{1}$](111) and [11$\bar{2}$](111) shear loading. The dislocation density data obtained after shear creep testing in the present work are plotted as a function of shear strain γ. For comparison, the [001] tensile creep results from Wu et al. [10] are plotted as a function of strain ε into the same diagram (same x-axis). Figure 8a shows the increase of the overall dislocation density. As can be seen in Figure 8b (channel dislocation densities), there is a higher increase of dislocation density after 1% in the macroscopic [01$\bar{1}$](111) shear system than in [11$\bar{2}$](111), while the opposite holds for 5%. Figure 8c shows the dislocation density evolution in the γ'-phase. Here, it is important to highlight that after 1% strain, more dislocations are detected in the γ'-phase after [11$\bar{2}$](111) than after [01$\bar{1}$](111) shear loading.

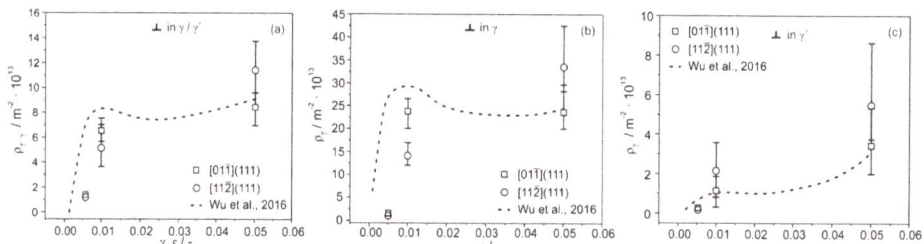

Figure 8. Evolution of dislocation densities during [01$\bar{1}$](111) and [11$\bar{2}$](111) shear creep testing at 750°C and 300 MPa (dislocation densities vs. shear strain γ). Dislocation densities from [001] tensile creep tests reported by Wu et al. [10] (dislocation densities vs. tensile strain ε). (a) Overall dislocation density. (b) γ-channel dislocation density. (c) Dislocation density in the γ'-phase.

4. Discussion

4.1. Mechanical and Micro Mechanical Analysis

As a striking new result, it was found that from 0.5% onwards, the [11$\bar{2}$](111) shear system deforms significantly faster than the shear system [01$\bar{1}$](111), as shown in Figure 3 (see also strain vs. time data in [12]). The corresponding logarithmic shear rate plotted as a function of shear strain is presented in Figure 3b. As a first step in our analysis, we took a look at the twelve basic microscopic crystallographic glides systems of type $\langle 110 \rangle \{111\}$, which govern plasticity in cubic face-centered and related lattices. Table 5 compiles loading modes and Schmid factors [24], which scale as the resolved shear stresses, which, in turn, drive the dislocation activities in the associated microscopic crystallographic slip systems. We keep in mind that the present work was motivated by mechanical and microstructural findings related to ⟨001⟩ tensile loading (T). As can be seen in the fourth column of Table 5, precisely oriented ⟨001⟩ tensile loading [10,11,15] results in activation of eight out of twelve slip systems. This makes it difficult to single out the role of individual dislocation families and to analyze their reactions. In contrast, one microscopic crystallographic slip system promoted during ⟨110⟩{111} shear creep (Schmid factor 1 is highlighted in bold in Table 5) but two during ⟨112⟩{111} loading (Schmid factors 0.87 are highlighted in bold in Table 5). The corresponding Schmid factors for these two shear loading modes are listed in columns 5 and 6 of Table 5. Our mechanical results, presented in Figure 3, show that the macroscopic crystallographic shear system $[11\bar{2}](111)$, where two slip systems have a Schmid factor of 0.87, deforms significantly faster than the $[01\bar{1}](111)$ shear system, where one dislocation family is promoted by the strongest resolved shear stress (Schmid factor: 1).

Figure 3 shows that the creep rates in both shear systems decrease during early primary creep, which is due to the filling of γ-channels by dislocations. In the $[01\bar{1}](111)$ shear system this decrease of creep rate continues throughout, up to 5% shear, where the test was terminated. In contrast, the $[11\bar{2}](111)$ shear system shows an increase of creep rate towards an intermediate maximum, very similar to what was reported by Wu et al. [10] for uniaxial [001] testing in the same temperature range. In this study, we show that this increase of creep rate is associated with the start of γ'-cutting. The fact that creep rates during [001] tensile testing decrease much more rapidly in the early stages of creep [10] is due to the interaction of multiple slip systems. This also explains why γ-channel dislocation densities increase more strongly during [001] creep (see data from [10] shown as dashed line in Figure 8). Figure 8 also shows that at shear strains of the order of 1%, where γ'-cutting starts, γ-channel dislocation densities in the $[11\bar{2}](111)$ shear systems are lower than in the $[01\bar{1}](111)$ shear system. The fact that γ'-cutting can result in a decrease of dislocation density has been explained in [10]. The continuous decrease of creep rate for shear strains > 2% has been attributed to the increase of γ-channel dislocation density (growing population of 1/2⟨110⟩ dislocations [6]).

Table 5. Activation of twelve microscopic $a/2\langle110\rangle\{\bar{1}11\}$ crystallographic slip systems during [001] tensile testing (T) and shear creep testing (S) in the two macroscopic crystallographic shear systems $[01\bar{1}](111)$ and $[11\bar{2}](111)$.

System	Plane	Direction	T: [001]	S: [011̄](111)	S: [112̄](111)
1	(111)	[1̄10]	0	0.5	0
2	(111)	[101̄]	−0.41	0.5	0.87
3	(111)	[011̄]	−0.41	1	0.87
4	(1̄11)	[110]	0	0.17	−0.19
5	(1̄11)	[1̄01̄]	−0.41	0.17	0.48
6	(1̄11)	[011̄]	−0.41	0.33	0.29
7	(11̄1)	[1̄10]	0	0.50	0.19
8	(11̄1)	[101̄]	−0.41	0.17	0.29
9	(11̄1)	[011̄]	−0.41	0.67	0.48
10	(111̄)	[1̄10]	0	0.17	0
11	(111̄)	[101]	−0.41	0.50	0.67
12	(111̄)	[011]	−0.41	0.67	0.67

4.2. Analysis of STEM Overview Micrographs

The shear test technique used in the present work not only facilities micro mechanical analysis. As compared to ⟨001⟩ tensile experiments, which are commonly used to study these phenomena, it also has the advantage that it is easy to obtain TEM foils, which contain the glide planes of the activated slip systems. Therefore, all STEM micrographs presented in the present work contain long dislocation segments, which facilitate the analysis of dislocation reactions. The STEM image presented in Figure 4a shows that dislocations do not enter γ'-particles after 1% shear strain in the macroscopic $[01\bar{1}](111)$ shear system. It is generally assumed that dislocation plasticity is confined to the γ-channels and that dislocations do not easily enter the γ′-phase because of the high antiphase boundary (APB) energy associated with the classical pairwise cutting of the ordered L1$_2$ phase [2,25–30]. The STEM micrographs shown in Figure 5a–c prove, that only the 1/2[01$\bar{1}$] slip system (SF=1, Table 5) carries high temperature plasticity in the early stages of [01$\bar{1}$](111) shear creep deformation. In contrast, in the macroscopic [11$\bar{2}$](111) shear system, which after 1% shear deforms significantly faster than the [01$\bar{1}$](111) shear system, frequent cutting events are observed, which manifest themselves by the presence of planar faults in the γ′-phase, as shown in Figure 4b. This represents direct experimental evidence for a rate controlling dislocation reaction of type

$$a/2 \cdot [10\bar{1}]_\gamma + a/2 \cdot [01\bar{1}]_\gamma \rightarrow a/3 \cdot [11\bar{2}]_{\gamma'} + a/6 \cdot [11\bar{2}]_{\gamma'/\gamma} \qquad (2)$$

which accounts for the formation of the leading half of a dislocation ribbon (e.g., [3–11,31–34]). In combination with our mechanical results, this suggests that this reaction is promoted (and γ′-cutting more frequent) when the two dislocation families which participate in the reaction are equally strongly activated. While this result supports views suggested in the literature [3–11,31–34], the specific experimental evidence, proving the character of the reacting dislocation families and characterizing the reaction event itself, has previously not been given.

4.3. Analysis of a Local Dislocation Reaction

The results shown in Figures 6 and 7 and Tables 3 and 4 can be interpreted in terms of the mechanism, which was proposed by Wu et al. [11]. The simplified scenario in Figure 9 helps to understand how the cutting configuration of Figure 7 evolved during low-temperature and high-stress creep, as the creep exposure time increases from t1 to t4. We keep in mind that Figures 7b and 9 show [111] projections of the γ/γ′-microstructure ([111] pointing into the image planes). The microstructure shown in Figure 9 is subjected to an external shear stress acting in the (111) plane in the [11$\bar{2}$] direction.

At t1, in the upper left scenario of Figure 9, the blue dislocation loop with a Burgers vector of [10$\bar{1}$] (dislocation family 1, glide plane: (111)) has expanded from the top to the bottom of the vertical grey γ-channel (channel directions given in Figure 9). From the right side, the red loop (Burgers vector: a/2[01$\bar{1}$], dislocation family 2, glide plane (111)) has arrived emerging from the green [100] channel. Our stereo TEM method [19] revealed that at the channel crossing the blue and red dislocation lines are not in the same (111) plane, the red dislocation being located slightly above the blue (not shown here).

At t2, the red dislocation segment d3 has moved by glide and climb. It has reached the γ′/γ-interface at the left side of the [010] γ-channel and has reached the same glide plane as the blue dislocation segment d2, which has locally entered the γ′-phase and created a thin APB ribbon. As suggested in the 2D DDD model of Wu et al. [11], the compact d3 segment dissociates into two Shockley partials. The leading Shockley partial is located at the γ/γ′-interface, while the trailing partial remains a short distance from the interface. The two Shockley partials are coupled by a stacking fault.

At t3, the leading Shockley partial has joined the blue dislocation segment resulting in the formation of the black dislocation segment d1. This fast reaction converts the high energy APB fault into a superlattice intrinsic stacking fault (SISF), while the trailing partial closes up to the γ/γ′-interface. Dislocation segment d1 bows out in between the two black nodes N1 and N4 at the γ/γ′-interface. This represents the nucleation event of the first half of the cutting ribbon. Finally, at t4, the experimentally observed configuration has been established. Segment d1 has paved its way into the γ′-phase, while the upper node N1 is dragged upwards along the interface.

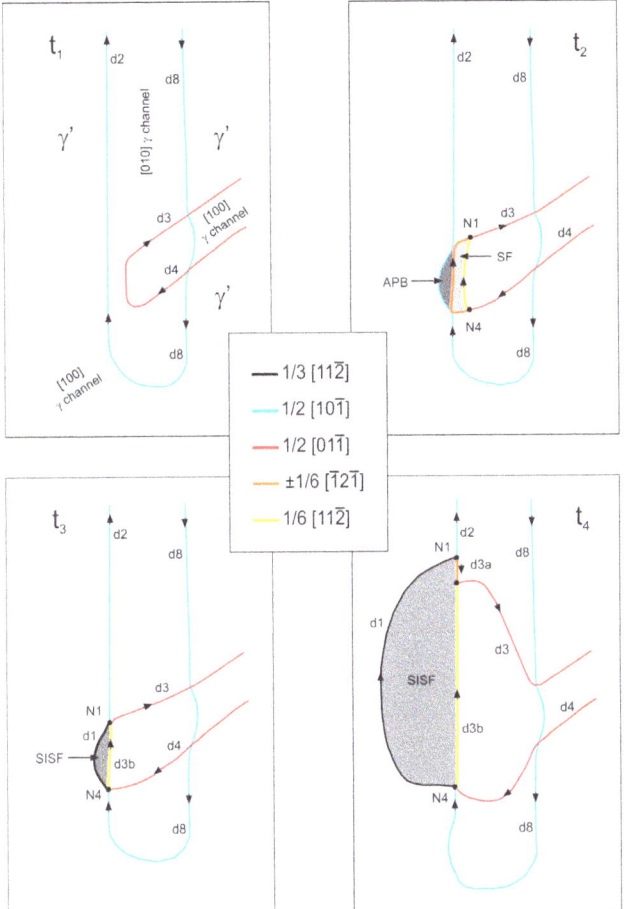

Figure 9. Simplified schematic drawings illustrating how the dislocation substructure shown in the STEM micrograph of Figure 7 has evolved. Creep exposure time increases from t1 (starting configuration) to t4 (final configuration). For details see text.

An effort was made to support this scenario by calculating the forces, which act on the dislocation segments. Using the methodology and input parameters outlined in [11], one can calculate Peach Köhler glide and climb forces (PKFg and PKFc), which act on the blue and red dislocations highlighted in Figure 9. These PKF were calculated for a local stress state, which accounts for the applied shear stress and the superimposed coherency stress fields [11]; the results are presented in Tables 6 and 7. The calculations were performed for three types of γ-channels [100], [010], and [001] (present in the upper right corner of Figure 7a but not shown in Figures 7b and 9). As can be seen in Tables 6 and 7, the character of the interface dislocation segments and the Peach Köhler glide and climb forces depend on the type of channel. From the results presented in Tables 6 and 7, one can conclude that the dislocations with Burgers vectors 1/2[10$\bar{1}$] (like segments d2 and d8 in Figure 7) prefer [010] channels. Dislocations with Burgers vectors 1/2[01$\bar{1}$] (like dislocation segment d4 in Figure 7), on the other hand, experience higher Peach Köhler glide forces in [100] channels. The results for the channel in which our reaction takes places is highlighted in grey in Tables 6 and 7. They show that both dislocation families experience positive glide forces, which pushes them towards the γ/γ'-interfaces. Moreover, the red

60° dislocation also experiences a positive climb force, which enables climb down onto the common (111) glide plane. Thus, the forces acting on the dislocations highlighted in Figure 9 are in line with the suggested scenario.

Table 6. Normalized Peach Köhler glide (PKFg) and climb forces (PKFc) for dislocation family 1 (blue dislocation in Figure 9, Burgers vector: 1/2[10$\bar{1}$]) in the three types of γ-channels.

Channel	Line Direction	Dislocation Type	PKFg	PKFc
[100]	[0$\bar{1}$1]	60°	0.190	0.217
[010]	[10$\bar{1}$]	screw	0.612	0.0
[001]	[$\bar{1}$10]	60°	1.035	−0.217

Table 7. Normalized Peach Köhler glide (PKFg) and climb forces (PKFc) for dislocation family 2 (red dislocation in Figure 9, Burgers vector: 1/2[01$\bar{1}$]) in the three types of γ-channels.

Channel	Line Direction	Dislocation Type	PKFg	PKFc
[100]	[0$\bar{1}$1]	screw	0.612	0
[010]	[10$\bar{1}$]	60°	0.190	0.217
[001]	[$\bar{1}$10]	60°	1.035	−0.217

4.4. Relation to Previous Work

The dislocation reaction (Equation (2)) has been considered by several authors throughout the last five decades, and the fact that it is crucial to form the first half of a dislocation ribbon, which can propagate to the γ/γ'-microstructure, is well appreciated (e.g., [4–6,8,9]). Early primary creep phenomena were recently related to the opening of a window of opportunity [6,8,10,11,31] (the Rae-window [10]), where, after an incubation period, good conditions for the onset of γ'-cutting are established. However, only recently the dislocation reactions, which lead to the nucleation of the first half of the dislocation ribbon which shears the γ/γ'-microstructure, have received full attention [11].

A 2D DDD model helped to rationalize the underlying processes. While isolated elements of the overall scenario were presented in the literature (e.g., ribbon formation and peculiar primary creep behavior [5], observation of two γ-channel dislocation families reacting [32–34]), no clear link between the macroscopic mechanical data from the low-temperature high-stress creep regime and the underlying elementary dislocation processes was established. While we see the same dislocation pair as Sass et al. [34] (one screw and one 60° dislocation reacting at a γ/γ'-interface), our calculation suggests that the 60° dislocation climbs while the screw dislocation is sessile. Our screw dislocation does not cross slip, as suggested by Sass et al. [34]. Moreover, Sass et al. [34] did not rationalize their scenario by providing the Peach Köhler forces, which drive their reaction. During [001] and [011] tensile testing, eight and four microscopic crystallographic slip systems are activated, respectively, which makes it difficult to clarify the impact of individual dislocation processes on macroscopic creep rates.

There are other complicating factors associated with standard ⟨001⟩ tensile creep testing, such as (i) tensile specimens not always being precisely oriented, (ii) other mechanisms of microstructural evolution like rafting interfering [35], and (iii) modelling efforts not being guided by experimentally observed dislocation scenarios in realistic microstructures (with γ/γ'-misfit) [35,36]. The direct shear loading performed in the present work helps to overcome most of these issues. In the present study, it could be clearly shown that when only one microscopic crystallographic slip system is activated, ribbon formation is hampered, and strain rates keep decreasing. In contrast, when just the two dislocation families required for the formation of the first half of a shearing dislocation ribbon (Equation (2)) are activated, cutting starts and has the accelerating effect on creep rates described in this work. Evidence for these two different types of microstructural evolution are presented in the overview STEM micrographs shown in Figures 4 and 5. Most importantly, a local dislocation scenario was analyzed after 1% shear creep deformation of the macroscopic crystallographic shear system [11$\bar{2}$](111), which rationalizes the macroscopic behavior and provides experimental proof for the 2D DDD model presented by Wu et al. [11]. In agreement with what was recently reported, we detected creep anisotropy [37]. However, it is important to keep in mind that microscopic ⟨112⟩{111} slip systems depend on the continuous operation of ⟨110⟩{111} slip systems in the γ-channels.

5. Conclusions

The present work provides direct experimental evidence for the importance of γ-channel dislocation reactions which govern low-temperature (T = 750 °C) and high-stress (resolved shear stress: 300 MPa) creep of Ni-base single crystal superalloys (SX). Two dislocation families with different Burgers vectors must react to form the first half of a dislocation ribbon, which can shear the γ'-phase. For the first time, the present work provides direct mechanical and microstructural proof, which documents the importance of this reaction. From the results obtained in the present work, the following conclusions can be drawn:

(1) Shear creep experiments can help to identify specific dislocation reactions, which are more difficult to study using standard ⟨001⟩-tensile creep testing. They allow for the direct activation of specific microscopic crystallographic slip systems, which facilitates micromechanical analysis. Shear experiments also provide easier access to TEM foils, which are parallel to the activated glide planes and contain long dislocation segments. This facilitates defect analysis in the TEM.

(2) The macroscopic crystallographic shear system $[11\bar{2}](111)$ creeps significantly faster than the macroscopic crystallographic shear system $[01\bar{1}](111)$. This represents direct mechanical proof for the fact that the formation of the first half of a dislocation ribbon which cuts the γ'-phase is rate controlling. In contrast to the multiple slip conditions associated with ⟨001⟩ or ⟨011⟩ tensile loading, shear loading of the macroscopic crystallographic shear system $[11\bar{2}](111)$ allows for the specific activation of the two dislocation families, which are required for this reaction, with equally high Schmid factors of 0.81. The higher activation of only one of the two partners during macroscopic $[01\bar{1}](111)$ shear creep deformation cannot produce equally high creep rates.

(3) Overview TEM micrographs of the material states, which were interrupted after 1% strain, show that dislocation activity is confined to the γ-channels during $[01\bar{1}](111)$ shear creep deformation, where no planar faults are observed. In contrast, planar faults are frequently observed after 1% shear strain in specimens subjected to $[11\bar{2}](111)$ shear creep deformation. This is one element of microstructural proof required to explain the difference in creep rates in the two shear system systems.

(4) Another element of microstructural proof is provided using higher magnification STEM diffraction contrast analysis in combination with micromechanical DDD modelling. This allows for the detection of the two dislocation families, which are required to form the first half of a dislocation ribbon.

(5) The present work provides experimental proof for the scenario suggested in the 2D DDD model by Wu et al. [11] and thus fully rationalizes the peculiar creep behavior (incubation times, intermediate creep rate maxima) which have been reported for low-temperature and high-stress creep of single crystal superalloys.

Author Contributions: D.B. performed the experimental work as part of his Dr.-Ing. Thesis, A.D. brought in his expertise in TEM defect analysis and 2D micromechanical modelling, G.E. wrote the paper during his time at Tohoku University (host: K.Y.), and all authors contributed to the discussion. All authors have read and agreed to the published version of the manuscript.

Funding: This research was funded by the Deutsche Forschungsgemeinschaft (DFG) through the collaborative research center SFB/TR 103.

Acknowledgments: The authors acknowledge funding through projects A1 and A2 of the collaborative research center SFB/TR 103 on super alloy single crystals funded by the Deutsche Forschungsgemeinschaft (DFG). G.E. acknowledges funding by the Japanese Society for the Promotion of Science (JSPS). A.D. acknowledges financial support from MSMT through a project CEITEC 2020 no. LQ1601.

Conflicts of Interest: The authors declare no conflict of interest. The funders had no role in the design of the study; in the collection, analyses, or interpretation of data; in the writing of the manuscript, or in the decision to publish the results.

References

1. Durand-Charre, M. *The Microstructure of Superalloys*; CRC Press: Boca Raton, FL, USA, 1997.
2. Reed, R.C. *The Superalloys: Fundamentals and Applications*; Cambridge University Press: Cambridge, UK, 2006.
3. Kear, B.H.; Leverant, G.R.; Oblak, J.M. An analysis of creep-induced intrinsic/extrinsic fault pairs in a precipitation hardened Nickel-base alloy. *Trans. ASM* **1969**, *62*, 639–650.
4. Kear, B.H.; Oblak, J.M.; Giamei, A.F. Stacking faults in gamma prime (Ni3(Al,Ti) precipitation hardened Nickel-base alloys. *Metall. Trans.* **1970**, *1*, 2477–2486.
5. Matan, N.; Cox, D.C.; Rae, C.M.F.; Reed, R.C. Creep of CMSX-4 single crystals: Effect of misorientation and temperature. *Acta Mater.* **1999**, *47*, 1549–1563. [CrossRef]
6. Rae, C.M.F.; Matan, N.; Reed, R.C. The role of stacking fault shear in the primary creep of [001]-oriented single crystal superalloys at 750 °C and 750 MPa. *Mater. Sci. Eng.* **2001**, *300*, 125–134. [CrossRef]
7. Drew, G.L.; Reed, R.C.; Kakehi, K.; Rae, C.M.F. Single crystal superalloys: The transition from primary to secondary creep. In *Superalloys 2004*; Green, K.A., Ed.; TMS: Warrendale, PA, USA, 2004; pp. 127–136.
8. Rae, C.M.F.; Reed, R.C. Primary creep in single crystal superalloys: Origins, mechanisms and effects. *Acta Mater.* **2007**, *55*, 1067–1081. [CrossRef]
9. Rae, C.M.F.; Zhang, L. Primary creep in single crystal superalloys: Some comments on effects of composition and microstructure. *Mater. Sci. Technol.* **2009**, *25*, 228–235. [CrossRef]
10. Wu, X.; Wollgramm, P.; Somsen, C.; Dlouhy, A.; Kostka, A.; Eggeler, G. Double minimum creep of single crystal Ni-base superalloys. *Acta Mater.* **2016**, *112*, 242–260. [CrossRef]
11. Wu, X.; Dlouhy, A.; Eggeler, Y.M.; Spiecker, E.; Kostka, A.; Somsen, C.; Eggeler, G. On the nucleation of planar faults during low temperature and high stress creep of single crystal Ni-base superalloys. *Acta Mater.* **2018**, *144*, 624–655. [CrossRef]
12. Eggeler, G.; Wieczorek, N.; Fox, F.; Berglund, S.; Bürger, D.; Dlouhy, A.; Wollgramm, P.; Neuking, K.; Schreuer, J.; Agudo Jácome, L.; et al. On shear testing of single crystal Ni-base superalloys. *Metall. Mater. Trans.* **2018**, *49*, 3951–3962. [CrossRef]
13. Parsa, A.P.; Wollgramm, P.; Buck, H.; Somsen, C.; Kostka, A.; Povstugar, I.; Choi, P.P.; Raabe, D.; Dlouhy, A.; Müller, J.; et al. Advanced scale bridging microstructure analysis of single crystal Ni-base superalloys. *Adv. Eng. Mater.* **2015**, *17*, 216–230. [CrossRef]
14. Yardley, V.; Povstugar, I.; Choi, P.P.; Raabe, D.; Parsa, A.B.; Kostka, A.; Somsen, C.; Dlouhy, A.; Neuking, K.; George, E.; et al. On phase equilibria and the appearance of nanoparticles in the microstructure of single crystal Ni-base superalloys. *Adv. Eng. Mater.* **2016**, *18*, 1556–1567. [CrossRef]
15. Wollgramm, P.; Bürger, D.; Parsa, A.B.; Neuking, K.; Eggeler, G. The effect of stress, temperature and loading direction on creep behavior of Ni-base single crystal superalloy miniature tensile creep specimens. *Mater. High Temp.* **2016**, *33*, 346–360. [CrossRef]
16. Agudo Jácome, L.; Göbenli, G.; Eggeler, G. Transmission electron microscopy study of the microstructural evolution during high-temperature and low stress shear creep deformation of the superalloy single crystal CMSX 4. *J. Mater. Res.* **2017**, *32*, 4492–4502. [CrossRef]
17. Jácome, L.A.; Nörtershäuser, P.; Heyer, J.K.; Lahni, A.; Frenzel, J.; Dlouhy, A.; Somsen, C.; Eggeler, G. High-temperature and low-stress creep anisotropy of single-crystal superalloys. *Acta Mater.* **2013**, *61*, 2926–2943. [CrossRef]
18. Jácome, L.A.; Nörtershäuser, P.; Somsen, C.; Dlouhy, A.; Eggeler, G. On the nature of γ' phase cutting and its effect on high temperature and low stress creep anisotropy of Ni-base single crystal superalloys. *Acta Mater.* **2014**, *69*, 246–264. [CrossRef]
19. Jácome, L.A.; Eggeler, G.; Dlouhy, A. Advanced scanning transmission stereo electron microscopy of structural and functional engineering materials. *Ultramicroscopy* **2012**, *122*, 48–59. [CrossRef]
20. Ham, R.K. The determination of dislocation densities in thin films. *Philos. Mag.* **1961**, *6*, 1183–1184. [CrossRef]
21. Dlouhy, A.; Pesicka, J. Estimate of foil thickness by stereo TEM microscopy. *Chech. J. Phys.* **1990**, *40*, 539–584.
22. Eggeler, G.; Dlouhy, A. On the formation of <010> dislocations in the gamma prime phase of superalloy single crystals during high temperature and low stress creep. *Acta Mater.* **1997**, *45*, 4251–4262. [CrossRef]
23. Probst Hein, M.; Dlouhy, A.; Eggeler, G. Interface dislocation in superalloy single crystals. *Acta Mater.* **1999**, *47*, 2497–2510. [CrossRef]
24. Schmid, E.; Boas, W. *Kristallplastizität*; Springer Verlag: Berlin, Germany, 1935.

25. Williams, R.O. Origin of strengthening of precipitation: Ordered particles. *Acta Metall.* **1957**, *5*, 241–244. [CrossRef]
26. Gleiter, H.; Hornbogen, E. Aushärtung durch kohärente geordnete Teilchen. *Acta Metall.* **1965**, *13*, 576–578. [CrossRef]
27. Gleiter, H.; Hornbogen, E. Theorie der Wechselwirkung von Versetzungen mit kohärenten geordneten Zonen I. *Phys. Stat. Sol.* **1965**, *12*, 235–250. [CrossRef]
28. Gleiter, H.; Hornbogen, E. Theorie der Wechselwirkung von Versetzungen mit kohärenten geordneten Zonen II. *Phys. Stat. Sol.* **1965**, *12*, 251–264. [CrossRef]
29. Tituts, M.; Moturra, A.; Viswanathan, G.B.; Suzuki, A.; Mills, M.J.; Pollock, T.M. High resolution energy dispersive mapping of planar defects in L12-containing Co-based superalloy. *Acta Mater.* **2015**, *82*, 423–437. [CrossRef]
30. Titus, M.; Eggeler, Y.M.; Suzuki, M.; Pollock, T.M. Creep-induced planar defects in L12-containing Co- and CoNi-based superalloys. *Acta Mater.* **2015**, *82*, 530–539. [CrossRef]
31. Pollock, T.M.; Field, R.D. Dislocations and high-temperature plastic deformation of superalloy single crystals. In *Dislocations in Solids*; Nabarro, F.R.N., Duesbery, M.S., Eds.; Elsevier: Amsterdam, The Netherlands, 2002; pp. 547–618.
32. Feller-Kniepmeier, M.; Kuttner, T. [011] creep in a single crystal Ni-base superalloy at 1037 K. *Acta Metall. Mater.* **1994**, *42*, 3167–3174. [CrossRef]
33. Sass, V.; Glatzel, U.; Feller-Kniepmeier, M. Anisotropic creep properties of the Ni-base superalloy CMSX-4. *Acta Mater.* **1996**, *44*, 1967–1977. [CrossRef]
34. Sass, V.; Feller-Kniepmeier, M. Orientation dependence of dislocation structures and deformation mechanisms in creep deformed CMSX-4 single crystals. *Mater. Sci. Eng.* **1998**, *245*, 19–28. [CrossRef]
35. Svoboda, J.; Lukas, P. Model of creep in <001>-oriented superalloy single crystals. *Acta Mater.* **1998**, *46*, 3421–3431. [CrossRef]
36. Haghighat, S.M.H.; Eggeler, G.; Raabe, D. Effect of climb on dislocation mechanisms and creep rates in gamma prime strengthened superalloy single crystals: A discrete dislocation dynamics study. *Acta Mater.* **2013**, *61*, 3709–3723. [CrossRef]
37. Li, Y.; Wang, L.; Zhang, G.; Zhang, J.; Lou, L. Creep anisotropy of a 3rd generation Ni-base single crystal super alloy at 850 °C. *Mater. Sci. Eng. A* **2019**, *760*, 26–36. [CrossRef]

© 2020 by the authors. Licensee MDPI, Basel, Switzerland. This article is an open access article distributed under the terms and conditions of the Creative Commons Attribution (CC BY) license (http://creativecommons.org/licenses/by/4.0/).

Article

Modeling the Local Deformation and Transformation Behavior of Cast X8CrMnNi16-6-6 TRIP Steel and 10% Mg-PSZ Composite Using a Continuum Mechanics-Based Crystal Plasticity Model

Faisal Qayyum *, Sergey Guk, Matthias Schmidtchen, Rudolf Kawalla and Ulrich Prahl

Institute of Metal Forming, Technische Universität Bergakademie Freiberg 09599, Germany; sergey.guk@imf.tu-freiberg.de (S.G.); Matthias.Schmidtchen@imf.tu-freiberg.de (M.S.); Rudolf.Kawalla@imf.tu-freiberg.de (R.K.); ulrich.prahl@imf.tu-freiberg.de (U.P.)
* Correspondence: Faisal.qayyum@student.tu-freiberg.de; Tel.: +49-3731-39-4073

Received: 28 January 2020; Accepted: 18 March 2020; Published: 20 March 2020

Abstract: A Transformation-Induced Plasticity (TRIP) steel matrix reinforced with magnesium-partially stabilized zirconia (Mg-PSZ) particles depicts a superior energy absorbing capacity during deformation. In this research, the TRIP/TWIP material model already developed in the framework of the Düsseldorf Advanced Material Simulation Kit (DAMASK) is tuned for X8CrMnNi16-6-6 TRIP steel and 10% Mg-PSZ composite. A new method is explained to more accurately tune this material model by comparing the stress/strain, transformation, twinning, and dislocation glide obtained from simulations with respective experimental acoustic emission measurements. The optimized model with slight modification is assigned to the steel matrix in 10% Mg-PSZ composite material. In the simulation model, zirconia particles are assigned elastic properties with a perfect ceramic/matrix interface. Local deformation, transformation, and the twinning behavior of the steel matrix due to quasi-static tensile load were analyzed. The comparison of the simulation results with acoustic emission data shows good correlation and helps correlate acoustic events with physical attributes. The tuned material models are used to run full phase simulations using 2D Electron Backscatter Diffraction (EBSD) data from steel and 10% Mg-PSZ zirconia composites. Form these simulations, dislocation glide, martensitic transformation, stress evolution, and dislocation pinning in different stages of deformation are qualitatively discussed for the steel matrix and ceramic inclusions.

Keywords: TRIP Steel; zirconia composite; numerical simulation; crystal plasticity; local deformation behavior

1. Introduction

Transformation-Induced Plasticity (TRIP) steel magnesium-partially stabilized zirconia (Mg-PSZ) composites are of great interest for various applications due to their high energy absorbing capacity, which they owe to transformation in present phases during deformation. The metastable austenitic phase in these materials transforms into martensite under applied strain, and the Mg-PSZ particles transform into a monoclinic phase from the tetragonal phase under applied stress [1–4]. This transformation in the material into more compact and harder phases during deformation strengthens the material by increased hardening while allowing the material to deform under applied external load [5,6].

Researchers in the past have worked extensively on the experimental investigation and understanding of the transformation phenomena in low and medium alloy steels. Due to low alloying, the Stacking Fault Energy (SFE) of the stabilized austenitic phase in such steels is low at room temperature, i.e., in the range of 10–12 mJ/m^2 [7]. When these materials with low SFEs are deformed at room temperature, the stabilized Face Centered Cubic (FCC) austenitic phase transforms into the Body-Centered Cubic (BCC) ά-martensitic phase through the highly deformed Hexagonal Closest Packed (HCP) ε-martensitic phase [8,9]. Based on the experimental investigations, researchers have reported properties of the cast X8CrMnNi16-6-6 steel stacking faults [10], the martensitic lath thickness [11], the transformation behavior of the material [12], the nano hardness measurement of different phases [13], the evolution of the hardening coefficient of the material, and critical strains for transformation [9,14,15].

On the one hand, this transformation of the soft phase into harder phases during deformation is of interest for high-end applications. On the other hand, its dependence on many chemical and physical material parameters adds complexity to developing a complete understanding of the phenomena. It hence restricts the material scientists from obtaining full control over material engineering during manufacturing [16–24]. Many factors affect the formability of these materials; some of the major ones are the composition of the composite, grain size, orientation distribution, the employed manufacturing technique, working temperatures, the rate of deformation degree, and the strain rate adopted during deformation [25–30].

Researchers [31] in an earlier in-situ tensile deformation analysis of X5CrNi18.10 and 5% Mg-PSZ composites showed that the zirconia particles are loosely attached to the matrix. During deformation, the interface decohesion occurs at as low as 1.6% of the total strain with eventual cracking of some particles at around 6% of the total stain. Other researchers [9,32] analyzed the behavior of X5CrMnNi15-6-6 and 5 vol. % Mg-PSZ composites under compressive load. The EBSD and X-Ray Diffraction (XRD) analysis of the tested samples at different compressive strains showed that the martensitic transformation within the TRIP steel is mainly responsible for the shape of the flow curve and higher plasticity, and they explained that martensite formation in the TRIP steel majorly occurs near the poles of embedded zirconia particles. The interdependence and complete understanding of these parameters are still not entirely understood, therefore, until now, the only reliable way to investigate the mechanical and structural properties of these components is through extensive experimentation, which is an expensive and slow path.

2. Micro-Mechanical Modeling

The development of an accurate numerical model incorporated with physical-based microstructural attributes can be beneficial [27–30]. Such a model can be used to understand the structure-property relationship of the material better. An open-source, crystal plasticity-based simulation framework, Düsseldorf Advanced Material Simulation Kit (DAMASK), was developed by the Max Planck Institute for Iron Research (MPIE) [33]. The usage of the Fast Fourier Transform (FFT) approach for modeling and the prediction of micromechanical fields in polycrystalline materials was proposed [34], and the DAMASK framework was extended to solve elastic–viscoplastic dislocation-based boundary value problems using the spectral solver in combination with various constitutive models [35]. The spectral methods were developed further to predict the micro-mechanical behavior of plastically deforming heterogeneous materials [36], i.e., multi-phase steels and composites.

A constitutive, physically based material model to capture the Transformation Induced Plasticity and Twinning Induced Plasticity (TRIP/TWIP) effect in high manganese steels at different temperatures was developed by Wong et al. [37]. The model was incorporated in already developed crystal plasticity-based numerical simulation tool DAMASK. The SFE of high Manganese steels at room temperature is higher than 20 mJ/m^2 at room temperature. Therefore, they exhibit TRIP/TWIP phenomena, and with increasing temperature, the SFE increases more than 45 mJ/m^2, which results in dislocation glide only. The model was further developed by Madivala, and he showed the model to be working in the range of SFE 13–118 mJ/m^2 [38]. The challenge is that the proposed model comprises of ~50 physical-based and fitting parameters. Although the coefficient values for the high manganese TRIP/TWIP steel model have been proposed and validated with experimental results, the identification and calibration of the fitting parameters for other materials of the same class is a challenge.

It is essential to identify the occurrence of each deformation and transformation phenomena independently for better understand and tuning of the material model. Recently, researchers showed the use of different characterization techniques for collecting and validating such data for X8CrMnNi16-6-6 cast TRIP steel [12]. In the list of proposed methods, Acoustic Emission analysis (AE-analysis) of such materials is a viable technique for the in-situ measurement of twining, transformation, and dislocation glide data. These different microstructural deformation mechanisms were independently analyzed and compared with experimental observations [39].

In this research, a small size benchmark Representative Volume Element (RVE) of $10 \times 10 \times 10$ units containing 1000 elements having randomly assigned orientations were used for the parametric calibration of steel and composite. This technique was recently established [40]. The identification of the TRIP/TWIP material model was carried out by matching simulation results with experimental observations. The flow curves, hardening coefficient, and martensitic transformation in the material observed from DAMASK numerical simulations were compared. Strain evolution from different phenomena is compared with AE cluster data previously published [39].

The difference in the mechanical properties of martensite and austenite is substantial [13]; therefore, in current research, martensite is considered as an elastic phase without differentiating á-martensite and ε-martensite. The evolution of both phases is studied without differentiation, which makes it impossible to analyze the volume change of the individual phase. This assumption makes it easier to use the already developed TRIP/TWIP material model and tune the model parameters for low SFE and ΔG to match the numerical simulation results with experimental observations. The tuned material model is further used to run simulations for actual EBSD data of X8CrMnNi16-6-6 TRIP steel to analyze the evolution of local stresses, strains, and transformation in the microstructure qualitatively.

It has been pointed out by previous researchers that the mechanical properties of the Metal Matrix Composite (MMC) reinforced with Mg-PSZ are mainly controlled by the deformation of the matrix material, and zirconia particles undergo significantly less strain [9]. Therefore, the zirconia particles in the current simulation models were assumed to be elastic and non-transformative to reduce the complexity of the model and bring down the computation time. The tuned material model for steel with slight modification was adopted for the simulation modeling of 10% composite. Full phase simulations with 2D EBSD data of MMC composite are run for applied tensile load in the horizontal direction to analyze the evolution of stress/strain partitioning, martensitic transformation, and dislocation density to better understand the interplay of these attributes at the local scale.

Details of material data, sample preparation, and testing to obtain experimental stress-strain (σ–ε) and transformation data are provided in Sections 3 and 4. Section 5 deals with the methodology, details of the numerical simulation model, and parametric identification method. Section 6 presents the results of the study in comparison with experimental observations. It also contains full phase simulation results from 2D EBSD data. In Section 7, results are discussed in comparison with state of the art, and insight into the outlook is provided. The study is concluded in Section 8.

3. Material Data

High-alloyed Cr–Mn–Ni cast steel was used to manufacture steel samples, the chemical composition and SFE are shown in Table 1. On the other hand, 10% powder of MgO-stabilized ZrO_2 ceramic (d10 = 25 µm, d50 = 35 µm and d90 = 48 µm) with the chemical composition presented in Table 2 was mixed with the steel powder before sintering to manufacture composite samples.

The manufactured material was 99% dense with a homogeneous distribution of monoclinic zirconia particles in the matrix. The sintered plates were heat-treated at 1330K for 30 min, followed by cooling to room temperature. The plates were cut using a water jet cutting machine into corresponding specimen shapes for metallographic and tensile testing.

Table 1. Chemical composition and Stacking Fault Energy (SFE) of tested cast X8CrMnNi16-6-6 Steel.

Chemical Composition (wt.%)							Characteristics
C	Cr	Mn	Ni	Si	Al	N	SFE (mJ/m^2) [41]
0.08	16.0	6.0	6.1	1.0	0.05	0.05	9.6–15.6

Table 2. Chemical composition of Mg–PS–ZrO$_2$.

Chemical Composition (wt.%)						
ZrO$_2$	MgO	Na$_2$O	CaO	TiO$_2$	Fe$_2$O$_3$	SiO$_2$
0.08	16.0	6.0	6.1	1.0	0.05	0.05

4. Experimentation

For tensile testing, flat samples with a total length of 50mm, a gauge length of 12 mm, and a cross-section of 5 × 2.5 mm^2 were manufactured. The samples were taken from the center area of the cast plate with the orientation perpendicular to the top surface. The samples were ground from both sides up to 250µm and polished to a final grade of 1µm before tensile testing. The macro tensile test was carried out on a screw-driven loading device at room temperature and a strain rate of 1×10^{-3} s^{-1}.

To determine the volume fraction of deformation-induced martensite, the samples were analyzed using the technique of measuring the ferromagnetic phase fraction with a Ferritscope. Then, the volume fraction of martensite was calculated by multiplying the Fe% by the factor of 1.7 and by using the calibrated initial value. Details of the sample mentioned above preparation and characterization techniques are published in detail elsewhere [12].

Acoustic emission data for the same material was measured and published earlier [39]. Therefore, the data is being adopted to calibrate the numerical model and for validating the trends observed in the numerical simulation results. A differential wideband AE sensor 1045D by Fuji Ceramics (Japan) was attached to the shoulder part of the specimen using a rubber band. The signals from the sensor were amplified by 60 dB. A two-channel AE setup based on the 18-bit PCI-2 data acquisition board (Physical Acoustic Corporation, 195 Clarksville Rd, Princeton Jct, NJ 08550, USA) was used in a continuously streaming mode operating at 2 MHz. The data were further processed and clustered based on the frequency and the amplitude of the signal. The details of the method are published elsewhere [39].

For EBSD analysis, the steel and composite samples were prepared. The sample surfaces for EBSD measurements were carefully ground and polished up to 1 µm grade. The final preparation step was vibration polishing (Buehler) for four hours using a colloidal SiO$_2$ suspension with 0.02 µm particle size. EBSD scanning was done using a high-magnification field-emission SEM (MIRA 3 XMU, Tescan, Czech Republic), which is equipped with a retractable four-quadrant BSE detector and an EBSD/EDX system of EDAX/TSL (Ametek).

5. Numerical Simulation

The simulations were performed using the spectral solver implemented in DAMASK, which is an open-source crystal plasticity simulation framework [33]. In the current work, the constitutive model developed using DAMASK, incorporating the dislocation glide, twinning, and martensite phase transformation was used. A description of the constitutive law is outlined with detailed theoretical discussion elsewhere [37,38].

A periodic representative volume element (RVE) with 10 × 10 × 10 voxels with each voxel assigned a random orientation distribution was constructed from a Voronoi tessellation on a grid. Such an RVE for single or multiphase materials has been shown to behave isotopically when loaded in any direction [40]. The boundary conditions were chosen to represent uniaxial tension along the horizontal direction, which was prescribed through mixed boundary conditions shown in Equations (1) and (2).

$$\dot{\bar{F}}_{ij} = \begin{bmatrix} 1 & 0 & 0 \\ 0 & * & 0 \\ 0 & 0 & * \end{bmatrix} \times 10^{-3}.s^{-1} \quad (1)$$

Similarly, the first Piola-Kirchhoff stress corresponding to uniaxial tension in x-direction:

$$\bar{P}_{ij} = \begin{bmatrix} * & * & * \\ * & 0 & * \\ * & * & 0 \end{bmatrix} Pa \quad (2)$$

The boundary conditions imposed on the RVE were similar to the mechanical loading in the uniaxial tension experiments with the strain rate of 1×10^{-3} s^{-1}, which is in the quasi-static strain range. The averaged stress and strain values at each increment obtained from the RVE simulations were compared with the experimental stress–strain (σ–ε) and the strain-hardening (dσ/dε–ε) behavior. The procedure for obtaining optimized constitutive model parameters for dislocation glide, twinning, and martensite transformation is explained elsewhere [38]. The tuned material model is used to run numerical simulations on actual EBSD data of the material. In full phase simulations, the RVEs are simulated in-plane stress conditions. Under this assumption, the three-dimensional crystal model is compelled to undergo two-dimensional deformation. Due to this assumption, quantitative discussions of the results are not possible.

For a steel sample, the EBSD data were obtained in the middle of the prepared and polished sample. The data were later cleaned using HKL channel five software to remove any unidentified points considering five neighboring points and is shown in Figure 1. The data were converted from tabular format into a geometry file using an already available subroutine.

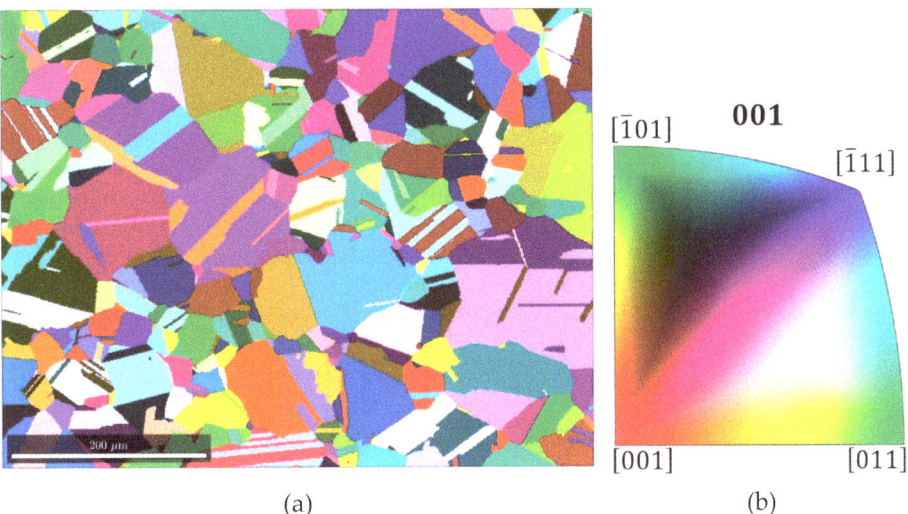

Figure 1. Two-dimensional (**a**) Electron backscatter diffraction (EBSD) maps of X8CrMnNi16-6-6 Transformation-Induced Plasticity (TRIP) steel, which are used for simulations colored with an IPF scheme and an (**b**) IPF orientation scale.

The elasticity parameters used for austenite and martensite phases are shown in Table 3. Due to the low Stacking Fault Energy of the current alloy, the austenitic iron phase (γ) passes through the dislocation-density-rich phase generally regarded as ϵ-martensite [8] and transforms into ά-martensite. However, considering the problem complexity and model limitations, for simplicity, both ϵ-martensite and ά-martensite are treated as a single phase. As compared to γ-austenite, both these phases have high stiffness (at least 24% higher) [13]. Therefore, the already developed model is adopted, which considers the HCP nature of martensitic transformation, and five elastic constants are considered to define HCP ϵ-martensite.

Table 3. Single-crystal elastic constants for austenite, martensite, and zirconia incorporated in the model during simulations.

Austenite [37]	Martensite [42]	Zirconia [43]	Units
$C_{11} = 175.0$	$C_{11} = 242.3$	$C_{11} = 191.0$	GPa
$C_{12} = 115.0$	$C_{12} = 117.7$	$C_{12} = 80.0$	GPa
$C_{44} = 135.0$	$C_{13} = 45.0$	$C_{44} = 40.0$	GPa
	$C_{33} = 315.0$		GPa
	$C_{44} = 40.5$		GPa

The physical-based parameters for the steel matrix were adopted from literature and proposed by Wong et al. [37]. The solid solution strength was tuned to match the experimentally observed yielding. The calibration of fitting parameters is a challenging task and was carried out manually by adjustment of the flow curve and transformation volume evolution by iteratively comparing simulation results with experimental observations. The calibrated parameters for the steel matrix are presented in Table 4. It is essential to mention that without physical sense, there might be limitless possible combinations of presented fitting parameters, which might yield a good correlation factor.

For the composite sample, the EBSD data were obtained in the middle of the prepared and polished sample. The data were later cleaned using HKL channel five software to remove any unidentified points considering five neighboring points and is shown in Figure 2. High-magnification EBSD was recorded and processed from the same sample at a different location to run simulations; it is shown in Figure 2b. A medium magnification simulation is better for getting a general overview of local material behavior. In contrast, high-magnification simulations are helpful in closeup analysis while keeping the computation time to optimal.

For modeling the deformation behavior of 10% Mg-PSZ composite, the zirconia particles were assigned elastic properties using three cubic elastic contacts, as shown in Table 3. The fitting parameters already tuned for the austenite matrix were used, except a few dislocation glide and transformation parameters, which are presented in Table 5. These few parameters need tuning when incorporating the second phase due to the assumption of perfect ceramic/matrix interface and incapability of the model to calculate the solid solution strengthening of the composite material internally.

The Orowan equation [44] provides the shear rate on the slip system α in a dislocation-based system as:

$$\dot{\gamma}^\alpha = \rho_e b_s v_0 \exp\left[-\frac{Q_s}{k_b T}\left\{1 - \left(\frac{|\tau^\alpha_{eff}|}{\tau_{sol}}\right)^p\right\}^q\right] sign(\tau^\alpha) \quad (3)$$

The mean free path reduces due to the pileup of dislocation, twins, and martensitic transformation as follows:

$$\frac{1}{\Lambda^\alpha_s} = \frac{1}{d} + \frac{1}{\lambda^\alpha_{slip}} + \frac{1}{\lambda^\alpha_{sliptwin}} + \frac{1}{\lambda^\alpha_{sliptrans}} \quad (4)$$

where

$$\frac{1}{\lambda^\alpha_{slip}} = \frac{1}{i_{slip}}\left(\sum_{\acute{\alpha}=1}^{N_s} \xi_{\alpha\acute{\alpha}}\left(p^{\acute{\alpha}}_{edge} + p^{\acute{\alpha}}_{dipole}\right)\right)^{\frac{1}{2}} \quad (5)$$

The mean free bath between two obstacles seen by growing twin is computed as:

$$\frac{1}{\Lambda_{tw}^{\beta}} = \frac{1}{i_{tw}}\left(\frac{1}{d} + \sum_{\beta=1}^{N_{tw}} \varepsilon_{\beta\dot{\beta}} f^{\beta} \frac{1}{t_{tw}(1-f_{tw})}\right) \quad (6)$$

The mean free bath for martensite is computed as:

$$\frac{1}{\Lambda_{tr}^{\chi}} = \frac{1}{i_{tr}}\left(\frac{1}{d} + \sum_{\dot{\chi}=1}^{N_{tr}} \varepsilon_{\chi\dot{\chi}} f^{\dot{\chi}} \frac{1}{t_{tr}(1-f_{tr})}\right) \quad (7)$$

The probability of twin nuclei to bow out and form a twin is:

$$p_{tw} = \exp\left[-\left(\frac{\hat{\tau}_{tw}}{\tau^{\beta}}\right)^{A}\right] \quad (8)$$

The probability of martensite lath to evolve is computed as:

$$p_{tr} = \exp\left[-\left(\frac{\hat{\tau}_{tr}}{\tau^{\chi}}\right)^{B}\right] \quad (9)$$

For a description of the model in detail and a better understanding of the assumptions and limitations, readers are encouraged to read the constitutive law outlined with detailed theoretical discussion elsewhere [37,38].

Table 4. Optimized constitutive model parameters for X8CrMnNi16-6-6 TRIP steel.

	Symbol	Description	Value	Unit	
martensite transformation parameters	$C_{threshhold-trans.}$	Adj. parameter controlling trans threshold stress	0.5	–	
	L_{tr}	Width of martensite lath during nucleation	5.0×10^{-8}	m	
	t_{tr}	Average martensite thickness	5.0×10^{-6}	m	[45]
	B	ß-exponent in transformation formation probability	3.0	–	
	i_{tr}	Adj. parameter controlling trans mean free path	10	–	
Twinning formation parameters	A	r-exponent in twin formation probability	1.0	–	
	i_{tw}	Adj. parameter controlling twin mean free path	5.0	–	
	$C_{threshhold-TWIN.}$	Adj. parameter controlling twin threshold stress	1.3	–	
	Γ_{sf}	Stacking fault energy	10	mJ/m²	[7]
	ΔG	Change in Gibbs free energy	-2.54×10^7	J/m³	[46]
Dislocation glide parameters	τ_{sol}	Solid solution strength	5×10^7	Pa	
	ρ_{ini}	Initial dislocation density	1.0×10^{12}	m^{-2}	[37]
	i_{slip}	Adj. parameter controlling dislocation mean free path	55	–	
	d	Avg. grain size	50×10^{-6}	m	[3]

It is essential to mention here that the EBSD data presented in Figures 1 and 2 is of sintered X8CrMnNi16-6-6 TRIP steel and 10% Mg-PSZ composite, but is adopted to simulate and show how these materials behave locally under applied deformation. The EBSD data were cleaned considering the phase and orientation of five neighboring pixels. After cleaning the data and allocating unidentified points to a specific phase, the porosity of the samples is eliminated, which yields a cast-like material microstructure. A substantial number of annealing twins are observed in the initial microstructure, i.e., in Figures 1 and 2. For the simplicity of the simulation model, they were assumed to be part of the austenitic matrix with default orientation.

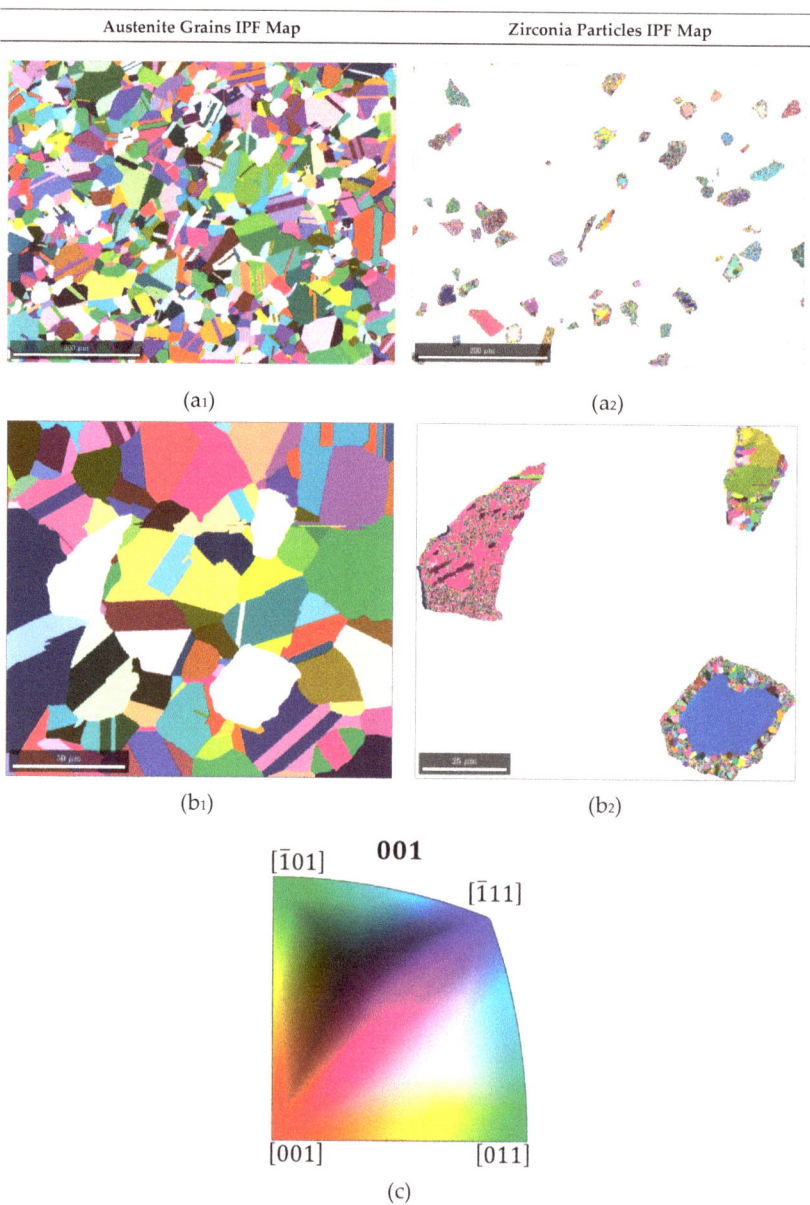

Figure 2. Two-dimensional EBSD maps of X8CrMnNi16-6-6 TRIP steel 10% Mg-PSZ composite, which are used for simulations, (**a**) medium magnification, and (**b**) high magnification. The white space in a_1 and b_1 corresponds to the embedded Zirconia particles shown in a_2 and b_2, (**c**) IPF orientation scale.

Table 5. Optimized constitutive model parameters for the X8CrMnNi16-6-6 TRIP steel matrix in 10% Mg-PSZ MMC.

	Symbol	Description	Value	Unit
Dislocation glide parameters	τ_{sol}	Solid solution strength	8.5×10^7	Pa
	i_{slip}	Adj. parameter controlling dislocation mean free path	350	–
Martensite transformation parameters	i_{tr}	Adj. parameter controlling trans mean free path	5.0	–
	B	ß-exponent in transformation formation probability	3.0	–

It is also observed that the grain size in the cast samples is relatively larger when compared with sintered materials. The model is based on length scale-dependent physical parameters, i.e., average grain size, the width of martensitic lath, which are used in the model for calculating the volume percent evolution of martensite and twins and the eventual drop in the mean free path. Such calculations are carried out locally for each gaussian point, but the model remains unaware of the global dimensional scale of the input geometry. Due to the absence of dimensional scale in the simulation model and consideration of very small areas under analysis, this assumption does not play a very profound role. It will not significantly affect the output of the study

All the simulations were run by applying a horizontal tensile load using a small increment step size of 0.01 deformation/inc. The global simulations for steel and composite were run for 42% and 21% of true stain, whereas the full phase simulations for steel and composite were run for 33% and 17% of true strain, respectively. The simulation data were post-processed to analyze the stain due to dislocation glide, transformation, and twinning. The evolution of the stress, transformation, and dislocation density of the material was investigated. During post-processing, logarithmic strain tensor is calculated from the total deformation gradient by using already available subroutines in the DAMASK package. The strain is described as deformation in terms of the relative displacement of particles in the RVE. The strain tensors for each Lagrangian point are averaged using von Mises' criteria. These averaged values of strain are plotted or presented graphically in the figures.

6. Results

In the first part of this research, the fitting parameters of the TRIP/TWIP material model were tuned for the X8CrMnNi16-6-6 cast TRIP steel. The simulation results match with the global stress/strain and transformation behavior of the material. The results of strain due to dislocation glide and transformation were also compared with already published acoustic emission data of the same material. This model was used to simulate the 2D EBSD data of X8CrMnNi16-6-6 cast TRIP steel to get a better understanding of the local stress, strain, dislocation density, and transformation behavior of the material. These obtained results are presented and discussed in the first part.

In the second part of this research, the tuned model for the matrix combined with elastic zirconia particles was used to model the 10% MMC behavior. The simulation's result nicely matches the experimental observations. This tuned model was used to run full phase simulations on a 2D EBSD map of MMC at different magnifications, and the local evolution of stress and strain in both phases is presented in the second part. The local dislocation glide and transformation behavior of the matrix, especially in the close vicinity of the ceramic particles, is critically analyzed.

6.1. Deformation Behavior of Steel Samples

In this section, the results from global and local deformation behavior of steel samples are analyzed by comparing with experimentally observed trends from tensile and AE tests.

6.1.1. Global Behavior

The experimental flow curve and measured transformation data [39] in comparison with the averaged stress/strain and transformation data from the 1000-grain benchmark simulations are presented in Figure 3a. It is observed that the tuned simulation model accurately captures the stress/strain behavior of the material with less than 1% error. The onset of martensitic evolution, its increase in volume, and the overall trend of evolution also correlate well with experimental observations. It is important to note here that the method of measuring martensite volume with a ferritoscope is not the most accurate. Hence, some errors might occur due to miscalculations during experimental measurements. Figure 3b, the strain hardening rate is plotted against true strain in the material, and it is observed that the material hardens due to martensitic evolution after 20% of the true strain. This increase of hardening is associated with local dislocation pinning and martensitic growth (conversion of austenite to a harder phase) in the material, as can be observed in Figure 3c. The dislocation density is observed to increase quadratically with increasing strain.

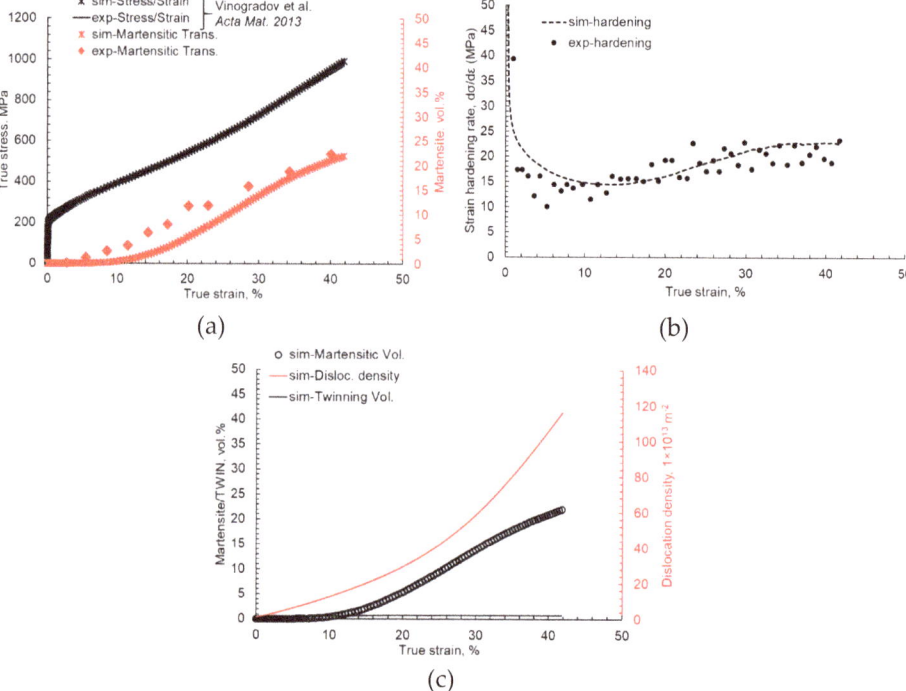

Figure 3. Numerical simulation results of the tuned model compared with experimental results of cast X8CrMnNi16-6-6 steel, (**a**) Comparison of experimental and numerical simulation stress – strain and martensite volume evolution, (**b**) Comparison of experimental and numerical simulation strain hardening curve, (**c**) Evolution of martensite volume percent, TWIN volume fraction and dislocation density with increasing true strain.

It is easy to understand that the dislocation density linearly increases up to 20% strain, after which there is an exponential increase in the dislocation density due to the high transformation which reduces mean free path for the dislocations to glide. This interplay of different phenomena in the material is better understood by full phase simulation results, which are discussed in Section 6.1.2.

In the DAMASK TRIP/TWIP material model, total deformation is composed of an elastic part and a plastic part. The total plastic deformation is calculated by adding strain due to dislocation glide, twinning, and transformation. In the model, each deformation phenomenon is governed by its own set of deformation equations comprising of physical and fitting parameters. Therefore, the results of the simulation can be post-processed to decompose the total deformation into its components and can be plotted to see the evolution of different phenomena in the material. Such a plot for an accurately tuned model is presented in Figure 4. As there is no twinning in the material, accumulated strain due to partial dislocations is almost zero. Strain accumulation due to transformation starts at 15% true strain and tends to saturate with increasing deformation. As observed, the strain due to dislocation glide starts with plastic deformation and increases almost linearly with total strain in the material to a maximum of 1.2.

Recently, researchers [12,39] developed and tuned the technique of using acoustic emission data to simultaneously record and analyze the deformation in TRIP steels in situ. This technique was shown to successfully yield useful information about transformation and dislocation dynamics in the form of increasing AE events. The limitation of this technique is that the output is in the form of AE events of different amplitudes and frequencies, which are amplified, filtered, clustered and added for every strain increment, it is the statistical data that do not correspond to the physical material deformation parameter. In Figure 4, the numerical simulation results of the strain occurrence due to dislocation glide, transformation, and twinning is compared with the acoustic emission data of the same material. The onset of transformation at 15% of strain is accurately captured by the simulation model. The correlation between two measurements is based on the singular starting point and scaling of axis such that there is maximum correlation factor in strain due to dislocation motion.

The value of strain due to transformation with the same trend is under-predicted by the simulation model. This reduced strain due to transformation is compensated by the linear trend of strain (by dislocation motion). The slope of dislocation slip, AE elements, and strain due to dislocation glide in simulation is the same up to 23% of the true strain, after which the simulation model slightly over predicts this value.

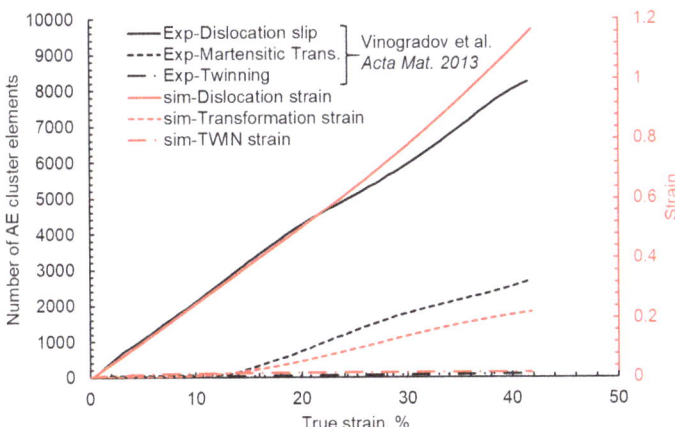

Figure 4. Numerical simulation results of the tuned model compared with experimental results of cast X8CrMnNi16-6-6 steel.

6.1.2. Local Behavior

Based on the presented global results, it is evident that the numerical simulation with a tuned material model can be used to model material behavior in a broad strain range. Using this material model, full phase numerical simulations were run on EBSD data containing 310 grains. The orientation and position of information from 2D EBSD data presented in Figure 1 were adopted for these simulations. The local strain evolution and its components at 5%, 15%, and 30% of true strain are shown in Figure 5.

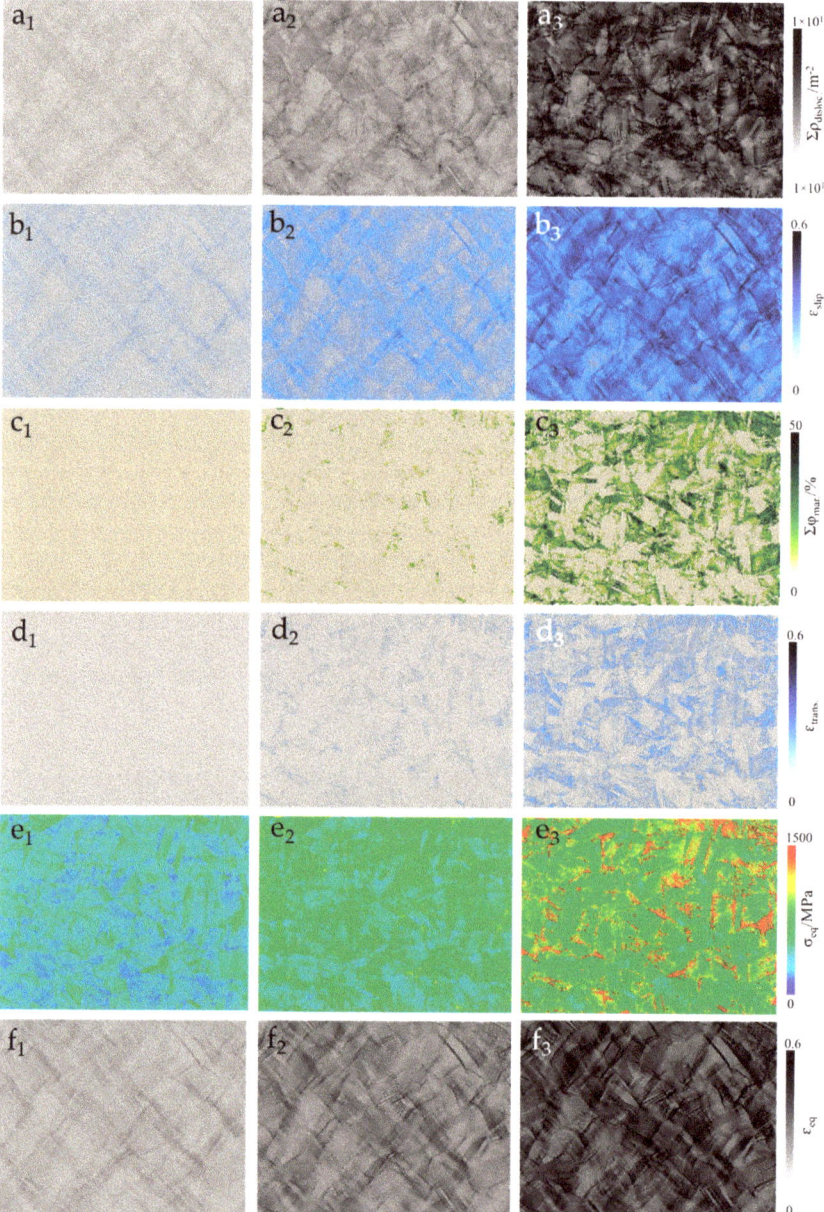

Figure 5. Simulation results at different levels of applied tensile deformation in principal x-direction (left to right: 5%, 15%, and 30%) for X8CrMnNi16-6-6 TRIP steel. The evolution of the (**a**) total dislocation density of the microstructure ($\Sigma\rho_{disloc.}$) calculated by adding edge and dipole dislocation densities, the (**b**) accumulated strain due to slipping in all planes (ϵ_{slip}), the (**c**) total martensitic volume percent ($\Sigma\phi_{mar.}$), the (**d**) accumulated strain due to martensitic transformation ($\epsilon_{trans.}$), (**e**) VON MISES' stress (σ_{eq}), and (**f**) VON MISES' strain (ϵ_{eq}).

The analysis provided in Figure 5 reveals that a substantial heterogeneity develops in the strain distribution (ranging from local strain values of 0.0–0.6) depending upon the grain orientation and position according to the neighboring grains, which is per earlier experimentally observed trends [47]. It is observed in Figure 5f that intensely strained zones in the microstructure enter into the plastic deformation regime first, followed by the formation of high strain channels between these zones.

For better understanding, the total strain component has been divided into its components. The strain due to twinning is neglected due to its small contribution. Total strain due to dislocation glide and strain due to martensitic transformation is plotted in Figures 5b and 5d, respectively. It is observed that the primary contributing mechanism for the strain is dislocation glide, i.e., accumulated shear due to slip in Figure 5b corresponds well with the true strain map in Figure 5f.

Although martensitic transformation also contributes to the total strain in the material, the value is small (Figure 5d). Moreover, those positions after transformation restrict the dislocation motion and hence are observed in areas with low total strain. In the material under consideration, at room temperature, the austenitic matrix starts transforming into martensite around 15% of the true strain (Figure 5(c_2)). This martensitic evolution reduces the available mean free path for the dislocation glide, and hence the dislocation pinning occurs near these martensitic islands. This increasing dislocation pinning results in higher dislocation densities, which further reduces the plastic flow of the material in these zones and results in the global hardening of the material. These zones with a high transformed martensitic volume and a high dislocation density also bear the highest amount of stress during deformation. They are the most susceptible zones for void formation [17–19,48].

The evolution of dislocation density in the full phase model is shown in Figure 5a. When compared with martensitic volume in Figure 5c, it is observed that the high dislocation density zones and high martensite evolution zones are mostly the same. The dislocation density reached up to 1×10^{15} m^{-2} in areas of high martensite volume. Von Mises' true stress evolution during deformation is shown in Figure 5e. The stress is higher in areas of high martensite volume. The study of stress evolution reveals that, as expected, the martensite phase is carrying most of the applied stress when compared with austenitic zones.

6.2. Deformation Behavior of Composite Samples

In this section, the global and local deformation behavior in both phases of analyzed composite material is presented in comparison with experimental observations.

6.2.1. Global Behavior

The true stress/strain and transformation results obtained from the RVE simulations of the composite are compared with experimental observations in Figure 6a. Due to accurate identification of the material parameters, the onset of yielding and plastic flow of the material is accurately captured with less than 2% error. The developed model does not have inclusive failure criteria; therefore, it is not able to capture the final necking and failure of the sample, as is observed in experimental data. The hardening of the material is found to drop exponentially by up to 7% percent of the strain and stays constant, as shown in Figure 6b.

To understand this constant hardening in the material, TWIN/TRIP vol.% and the dislocation density evolution of the material are plotted with true strain in Figure 6c. It is observed that the dislocation density increases from 1.2×10^{12} m^{-2} (assigned initial dislocation density) to 17×10^{13} m^{-2} up to 20% of the strain, after which the material fails.

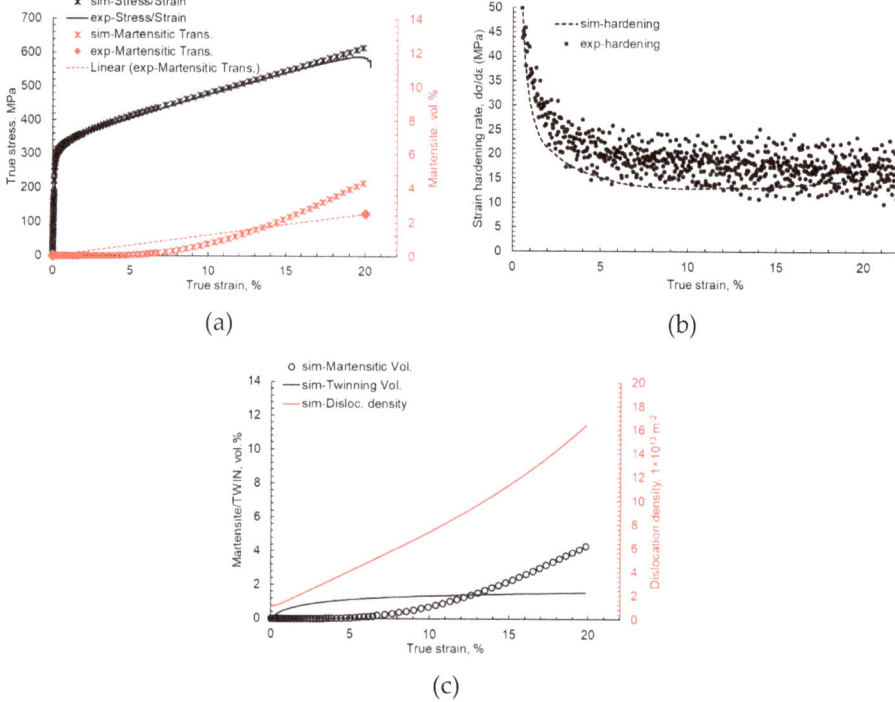

Figure 6. Numerical simulation results of the tuned model compared with experimental results of cast 10% Mg-PSZ composite, (**a**) Comparison of experimental and numerical simulation stress—strain and martensite volume evolution, (**b**) Comparison of experimental and numerical simulation strain hardening curve, (**c**) Evolution of martensite volume percent, TWIN volume fraction and dislocation density with increasing true strain.

6.2.2. Local Behavior

To better understand this phenomenon, the full phase simulation results of the 2D EBSD data of the composite with moderate and high magnifications were carried out. The results for the steel matrix are separated from the composite particles for better visualization of stress and strain partitioning in each phase. The results showing local evolution of the total dislocation density accumulated strain due to slipping in all planes, total martensitic volume percent, accumulated strain due to martensitic transformation, equivalent stress, and equivalent strain at 4%, 10%, and 16% of true strain are presented in Figures 7 and 8a–f. It is observed that substantial heterogeneity in the local distribution of these parameters is present at the microstructural scale. The results of the steel matrix and zirconia particles are separately analyzed for better scaling and visualization.

Zirconia particles in the composite, being highly stiff, are observed to carry most of the stress and undergo significantly less strain during the overall deformation of the composite, as shown in Figure 7. Although the zirconia particles are assumed to be perfectly elastic in these simulations, it is observed in Figure 7(a$_3$) at 16% of the global strain that the stress distribution inside the particles is heterogeneous. It is difficult to analyze this heterogeneity at low-magnification simulations. Therefore, a high-magnification simulation model was run to get a closer look at the overall parametric distribution in the microstructure at elevated strains.

Steel matrix in the composite, at room temperature, starts transforming into martensite around 7% of the true strain. This martensitic evolution reduces the available mean free path for the dislocation glide, and hence the dislocation pinning occurs near these martensitic islands. This increasing dislocation pinning results in higher dislocation densities, which further reduces the plastic flow of the material in these zones and results in the global hardening of the material. These zones with high transformed martensitic volume and high dislocation density also bear the highest amount of stress during deformation [17–19,48]. The evolution of dislocation density in the full phase model is shown in Figure 8a. When compared with martensitic volume in Figure 8c, it is observed that the high dislocation density zones and high martensite evolution zones are the same. It is seen that the dislocation density reached up to 5×10^{14} m^{-2} in areas of high martensite volume.

For a better understanding, the total strain component has been divided into its components. The strain due to twinning is neglected due to its small contribution. Total strain due to dislocation glide and strain due to martensitic transformation is plotted in Figures 8b and 8d, respectively. It is observed that the primary contributing mechanism for the strain is dislocation glide the local map of strain due to accumulated shear due to slip in Figure 8b corresponds well with the true strain map in Figure 8f. In the 10% composite simulation, martensite starts evolving at around 7% of the true strain the material. It is observed that although martensitic transformation also contributes to the total stain in the material, the value is small. Moreover, those positions after transformation restrict the dislocation motion and hence are observed in areas with low total strain and high total stress.

In the high-magnification simulation model, it is observed that martensite formation in the TRIP steel occurs close to the ceramic particles and at the ceramic/matrix interface, as shown in Figure 9c. The degree of plasticity in the austenite grains above and below a ceramic particle depends on the crystal orientation, dislocation motion, and stacking fault generation. Consequently, more dislocations are generated, or glide systems of higher-order are activated in the favorably oriented grains near the TRIP steel/ceramic interfaces. Hence, an enhanced martensite formation is observed in such regions with a concentrated strain. A different situation is seen at the flanks of the Mg-PSZ particles (Figure 5), where the deformation in the TRIP steel is limited by the strain in the ceramic particle. In these regions, the martensite formation is retarded. The martensitic evolution leads to dislocation pinning and increased dislocation density in these regions, as shown in Figure 9c. In the steel, matrix martensite bears the most stress.

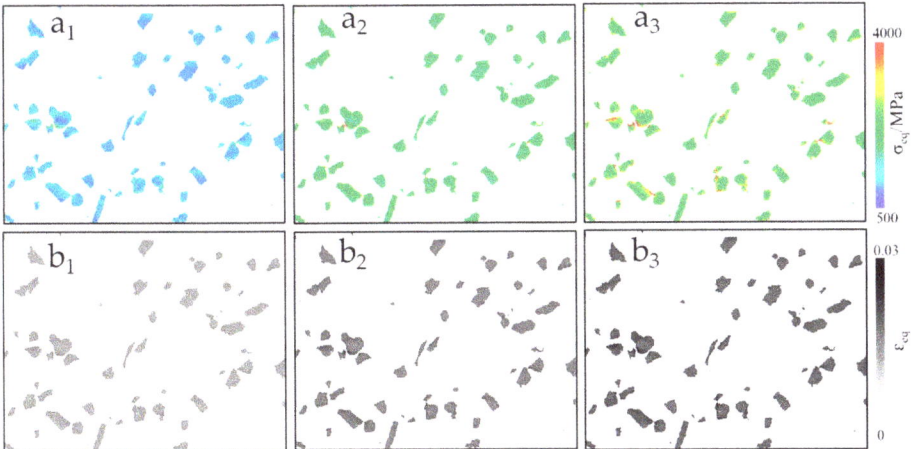

Figure 7. Simulation results of ceramic particles at different levels of applied tensile deformation in principal x-direction (left to right: 4%, 10%, and 16%) in zirconia particles of X8CrMnNi16-6-6 TRIP steel and 10% Mg-PSZ composite. The evolution of the (**a**) VON MISES' stress (σ_{eq}) and the (**b**) VON MISES' strain (ϵ_{eq}).

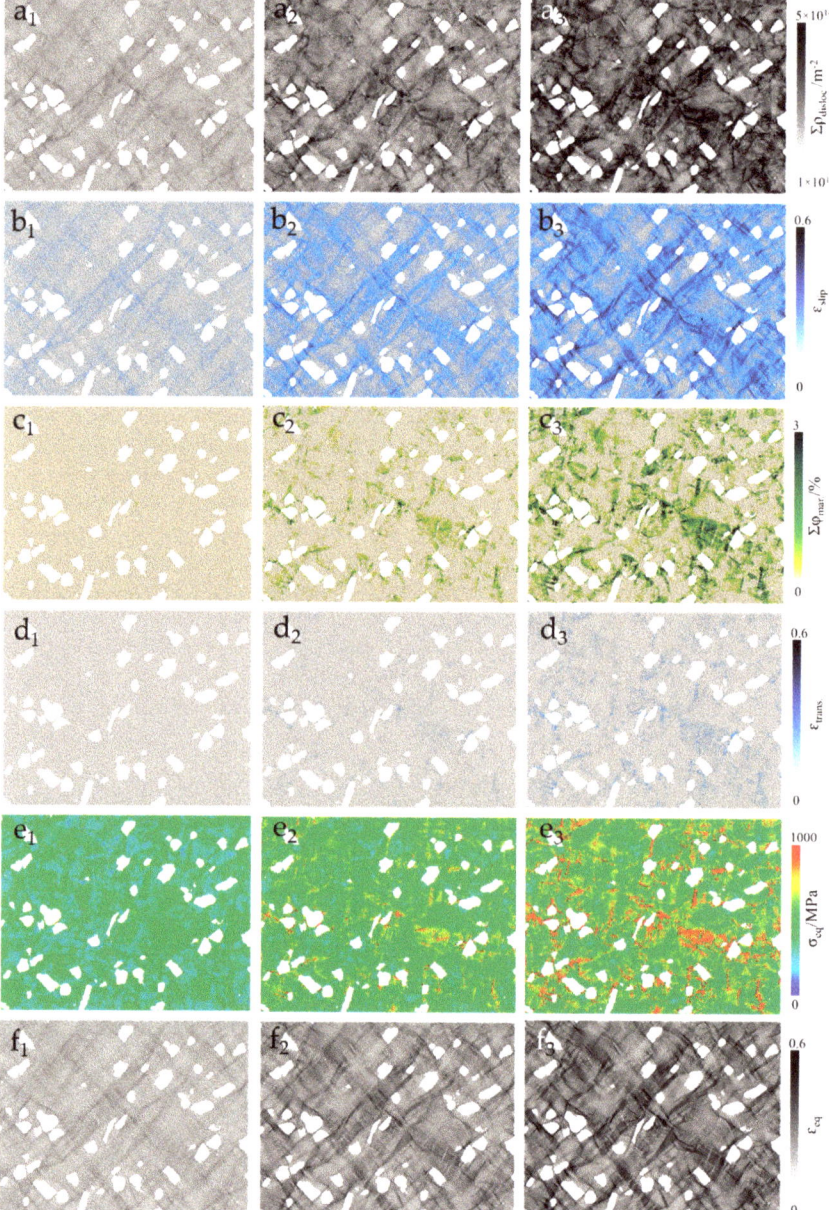

Figure 8. Simulation results at different levels of applied tensile deformation in principal x-direction (left to right: 4%, 10%, and 16%) in the steel matrix of X8CrMnNi16-6-6 TRIP steel and 10% Mg-PSZ composite. The evolution of the (**a**) total dislocation density in the microstructure ($\Sigma\rho_{disloc.}$) calculated by adding edge and dipole dislocation densities, the (**b**) accumulated strain due to slipping in all planes (ϵ_{slip}), the (**c**) total martensitic volume percent ($\Sigma\phi_{mar.}$), the (**d**) accumulated strain due to martensitic transformation (ϵtrans.), (**e**) VON MISES' stress (σ_{eq}), and (**f**) VON MISES' strain (ϵ_{eq}).

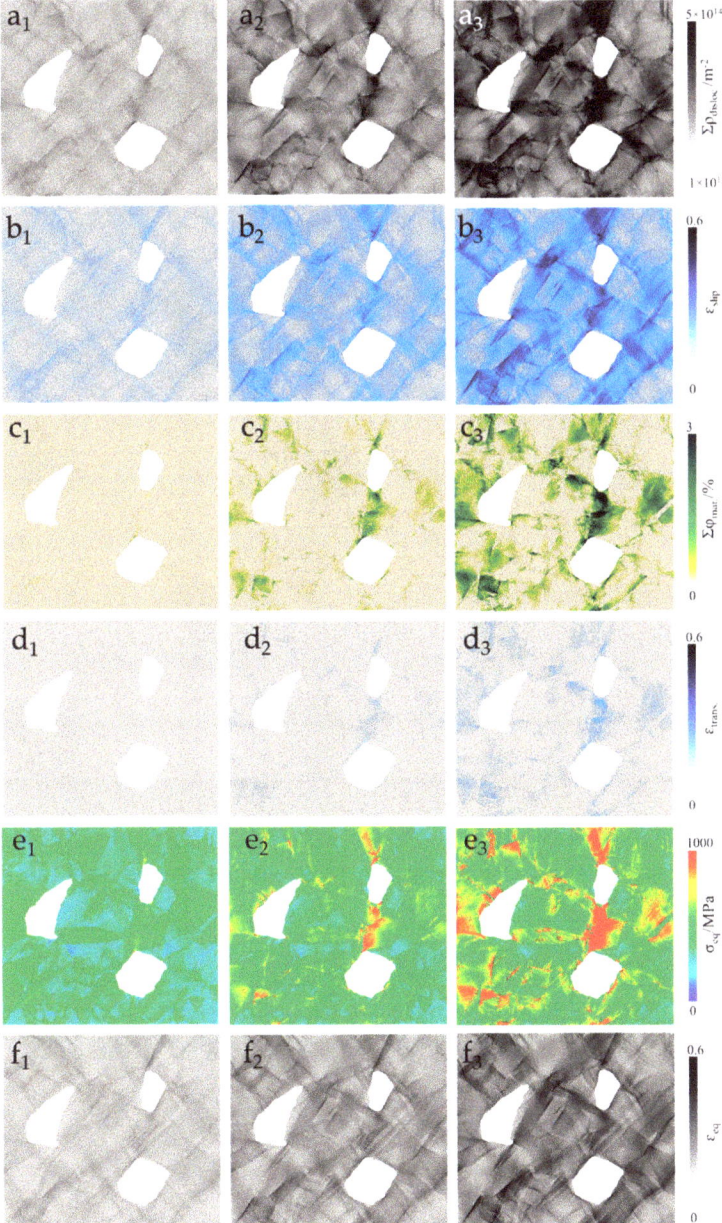

Figure 9. Simulation results of the steel matrix at different levels of applied tensile deformation in principal x-direction (left to right: 4%, 10%, and 16%) in the steel matrix of X8CrMnNi16-6-6 TRIP steel and 10% Mg-PSZ composite at high magnification. The evolution of the (**a**) total dislocation density in the microstructure ($\Sigma\rho_{disloc.}$) calculated by adding edge and dipole dislocation densities, the (**b**) accumulated strain due to slipping in all planes (ϵ_{slip}), the (**c**) total martensitic volume percent ($\Sigma\phi_{mar.}$), (**d**) the accumulated strain due to martensitic transformation (ϵtrans.), (**e**) VON MISES' stress (σ_{eq}), and (**f**) VON MISES' strain (ϵ_{eq}).

Due to the higher fracture strength of the ceramic particles compared to the low yield strength of the austenitic steel, the applied strain is absorbed mainly via the plastic deformation of the TRIP steel, and minor strain occurs in the zirconia particles, as shown in Figure 10b. The local stress distribution in the zirconia particles is very high compared to the steel matrix, yet the distribution is not homogeneous. In Figure 10(a_3), it is observed that the stress evolution in the zirconia particle is higher near the interface. This heterogeneous distribution of stress in zirconia particles is due to three reasons. Firstly, it is due to a variation in the orientation distribution assignment (see Figure 2(a_1,b_1)). Secondly, it is due to dependence on the neighbor grain orientations. Thirdly, it is due to dependence on particle size.

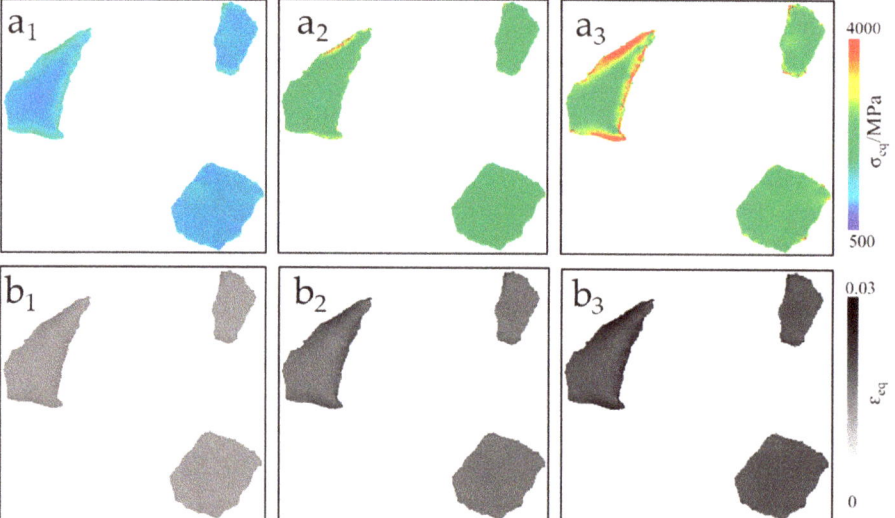

Figure 10. Simulation results of ceramic particles at different levels of applied tensile deformation in principal x-direction (left to right: 4%, 10%, and 16%) in zirconia particles of X8CrMnNi16-6-6 TRIP steel and 10% Mg-PSZ composite at high magnification. The evolution of (**a**) VON MISES' stress (σ_{eq}) and (**b**) VON MISES' strain (ϵ_{eq}).

The simulation results reveal how the orientation, size, and neighborhood of a grain largely influence the martensitic evolution in TRIP steels and composites. This evolution is further facilitated by the presence of ceramic particles inside the matrix, which act as barriers for free dislocation motion. The dislocations start accumulating on the edges of zirconia particles, preferably perpendicular to the applied load, resulting in high local strain accumulation.

To put deformation into perspective, Figure 11 represents the deformed stress and strain geometries of all simulated RVEs in comparison with initial geometry (grey outline). The presented deformed states are scaled to 100% deformation.

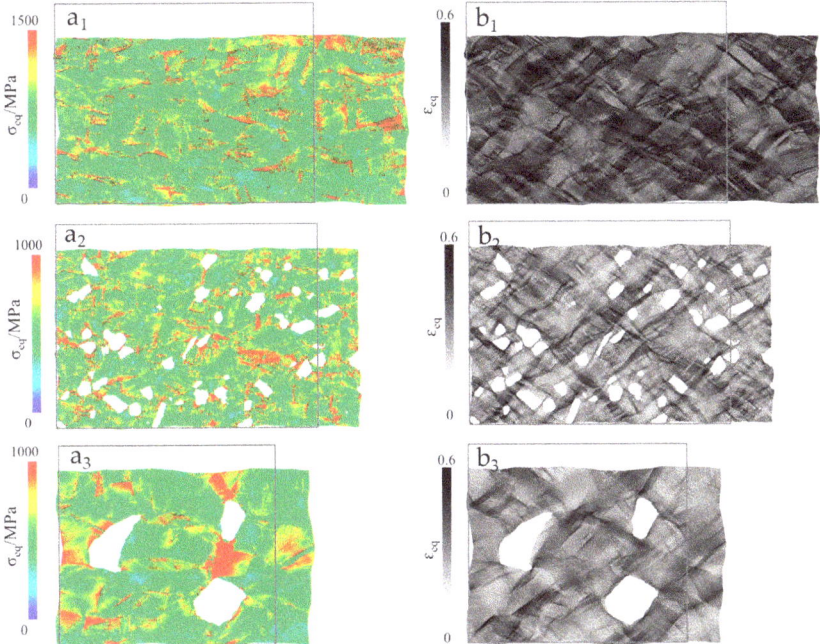

Figure 11. The deformed states of representative volume elements (RVEs) compared with undeformed shapes (grey outline). (**a**) VON MISES' stress (σ_{eq}) states at maximum deformation and (**b**) VON MISES' strain (ε_{eq}) states as maximum deformation. (1) is for X8CrMnNi16-6-6 TRIP steel at 30% true strain, (2) is for steel matrix in 10% Mg-PSZ composite at high magnification at 16% true strain, and (3) is for steel matrix in 10% Mg-PSZ composite at high magnification at 16% true strain.

7. Discussion

In this research, the already developed TRIP/TWIP material model for high manganese steels was adopted and calibrated using a small size benchmark RVE of 10 × 10 × 10 units containing 1000 elements having randomly assigned orientations for the medium alloyed X8CrMnNi16-6-6 TRIP steel. The physical parameters were adopted from literature, and fitting parameters were calibrated by comparing the stress/strain behavior of the tuned material model with experimental observations. The comparison reveals a good fit of stress/strain and twinning volume percent data with <1% error. But regarding the martensitic volume percent, the simulation model slightly under predicts the evolution for low strains but nicely captures the trend for higher strain values. This mismatch is possibly due to a slight error in the ferritoscope measurements.

The strain due to individual phenomena, when compared with acoustic emission data, shows that the trends from the experimental and numerical techniques are comparable. However, quantitatively, there is a slight mismatch in the values. It is still unknown if the simulation trend is better or the AE data as the signal measurement and clustering is an approximate technique and can yield varying results depending on subjective choices. Given the accurate output of the simulation model when comparing the stress/strain and transformation behavior in Figure 3a, a deviation in the amount of strain due to individual phenomena when compared with AE data is unlikely. The analysis of this deviation needs further detailed study with critical AE data measurement, clustering, and analysis, which is a task for future research.

The calibration and validation of fitting parameters for such models is a big challenge these days as a model can yield the same results by using an infinite number of fitting parameter combinations where each combination will yield effective minima of the correlation function. It is evident that the demonstrated simulation model and the in-situ AE analysis complement each other and should be combined in future studies to get a better insight into the local material behavior. Combining AE with such detailed crystal plasticity models can be very helpful in the accurate calibration or validation of fitting parameters for many different deformation phenomena occurring at the same time. The method can also help obtain a unique solution regarding fitting parameter identification. This has been conclusively demonstrated in this research.

The tuned material model for steel with minor modification was adopted for the simulation modeling of 10% composite. Due to the consideration of plain stress condition and non-periodic EBSD data used for simulations, the results of the study cannot be quantitatively accurate and need careful attention for a comparative qualitative discussion. Due to non-periodicity, the stress/strain evolution on the RVE edges is ignored. It is observed in the case of steel and composite that localized plasticity initiates in austenitic grains more favorably aligned in the deformation direction to load. The deformation propagates in the form of narrow strain bands throughout the microstructure. More interestingly, nearly undeformed regions are observed in between these highly strained zones and narrow strain bands, even at high average strain levels (Figure 5(f_3)). Thus, it appears that the full strain hardening capacity of the microstructure is not exploited during the deformation. In the case of the composite depending upon the ceramic particle distribution, the heterogeneity of plastic strain accommodation differs.

It is observed that the results of stress/strain and transformation are accurate for the steel matrix. Still, the simulation model over-predicts the transformation in the matrix by 51% in the case of the composite. This can be due to the assumption of an ideal steel/matrix interface taken for simulations, which acts as a source of high dislocation density and transformation initiation site. Whereas, in reality, this interface is fragile, and the interface decohesion of some particles has been reported to start at as low as 1.6% of the global strain [47,49–53]. This needs further investigation in future research by modeling the ceramic/matrix interface in the simulation model and comparing the results.

The reinforcing mechanism of Mg-PSZ particles in highly-alloyed TRIP steel was studied experimentally by Martin et al. in detail [32]. The results obtained in the current analysis for high-magnification simulation correspond well with his experimental observations. It is observed that the transformation occurs mostly at the poles of the zirconia particles if the austenitic grain orientation is more favorable. In the steel matrix, the deformation due to dislocation motion starts with the applied strain and generally increases with increasing strain. Martensitic evolution in the matrix starts around 8% of global stain and increases linearly.

The simulations were run for medium magnification geometries in case of steel and composite for the better overall analysis, and a high-magnification simulation was run for the composite for detailed local analysis. Such a methodology of decoupled low- and high-magnification analysis helps reduce the computation time significantly. The correct identification of the parameters and the calibration of the model for this specific material class can help researchers in better understanding and prediction of the material behavior; hence, it will improve the chances of cost-effectively tuning the material properties for particular applications.

8. Conclusions

In this research, the TRIP/TWIP material model already developed in the framework of DAMASK is tuned for X8CrMnNi16-6-6 TRIP steel and 10% Mg-PSZ composite. A new method to more accurately tune this material model is explained by comparing stress/strain, transformation, twinning, and dislocation glide obtained from simulations with experimental acoustic emission measurements. The calibrated model for a steel matrix with slight modification is used to model a 10% Mg-PSZ composite matrix. In the simulation model, zirconia particles are assigned elastic properties with a perfect ceramic/matrix interface. The local deformation, transformation, and twinning behavior of the materials due to quasi-static tensile load were analyzed. The global simulation results were compared with experimental observations, and the full phase simulations were run to on 2D EBSD maps to see the evolution of local stresses, strains, and transformation in the microstructure. The research can be concluded as follows:

- The developed material model with identified fitting parameters presented in this research and for the X8CrMnNi16-6-6 TRIP steel matrix and 10% Mg-PSZ composite can accurately predict the deformation behavior of steel matrix and ceramic particles with less than 2% error.
- The individual deformation phenomena of dislocation glide and transformation in the steel matrix for the current model correspond well with in-situ AE analysis data and help in associating the number of AE events with physical attributes.
- There is considerable heterogeneity in dislocation motion and strain evolution in the steel matrix with or without ceramic particles, which results in intense stress partitioning.
- Most of the local strain evolution in the steel matrix is due to dislocation motion, and the strain due to partial dislocation motion or transformation is significantly less, contributing very little to the overall strain in the material.
- It is observed that the zirconia particles take the most stress and undergo significantly less strain during deformation, which leads to high stress/strain partitioning at the ceramic/matrix interface, which is the possible reason for interface decohesion in this material.

The model and the methodology presented here for full phase simulation modeling is a useful tool for assisting development and microstructural optimization of cast X8CrMnNi16-6-6 TRIP steel 10% Mg-PSZ composites due to an improved understanding of the phenomena at the microstructural level accurately. Further research is needed to establish a relationship between AE data and numerical simulation results. Similar to global deformation behavior, validation of the local deformation behavior of steel and composite will be carried out in the future by comparing the results of in-situ deformation tests.

Author Contributions: F.Q. is the principal author of this article who carried out most of this study for his doctoral studies. S.G. assisted in acquiring material, preparing samples, executing tests, and analyzing experimental data. M.S. assisted in analyzing the simulation data and organizing the results. R.K. and U.P. are supervising this research project. They helped in acquiring the right computation resources and assisted in finalizing this research. All co-authors helped in scripting the article and finalizing it. All authors have read and agreed to the published version of the manuscript.

Funding: This research received no external funding.

Acknowledgments: The authors acknowledge the DAAD Faculty Development for Balochistan, 2016 (57245990) - HRDI-UESTP's/UET's funding scheme in cooperation with the Higher Education Commission of Pakistan (HEC) for sponsoring the stay of Faisal Qayyum at IMF TU Freiberg. The authors acknowledge the support of DFG funded collaborative research group SFB799 (TRIP matrix composites) for sharing knowledge and helpful discussions. The authors gratefully acknowledge the support of Martin Diehl & Franz Roters (MPIE, Düsseldorf) for help regarding the functionality of DAMASK, Anja Weidner (IWT, TU Freiberg) & Stephan Martin (IWW, TU Freiberg) for valuable feedback regarding current work.

Conflicts of Interest: The authors declare no conflict of interest.

Abbreviations

Acronyms:

SSS	Solid Solution Strength
SFE	Stacking Fault Energy
TRIP	Transformation Induced Plasticity
MMC	Metal Matrix Composite
Mg-PSZ	Magnesium partially stabilized Zirconia
EBSD	Electron Back Scatter Diffraction
XRD	X-ray Diffraction
AE	Acoustic Emission
IPF	Inverse Pole Figure

Symbols:

ϵ	True strain
σ	True stress
γ	Austenite phase (FCC)
ρ	Dislocation density
ΔG	Change in Gibbs free energy
τ_{eff}^{α}	effective resolved shear stress o slip system α
τ_{sol}	Solid Solution Strength
b_s	Length of Burgers vector for slip
v_0	Dislocation glide velocity
Q_s	The activation energy for dislocation slip
k_b	Boltzmann constant
T	Temperature, in the current case, is 300 K
p	Fitting parameter for strain hardening
q	Fitting parameter for strain hardening
d	Average grain size
λ_{slip}^{α}	dislocation pile-up
$\lambda_{sliptwin}^{\alpha}$	twin pile up
$\lambda_{sliptrans}^{\alpha}$	martensite pile up
f_{tw}	the total twin volume fraction
f_{tr}	the total transformation volume fraction
t_{tw}	average twin thickness
t_{tr}	average martensite thickness
i_{slip}	fitting parameter for slip mean free path
i_{tw}	fitting parameter for twin mean free path
i_{tr}	fitting parameter for transformation mean free path
$\xi_{\alpha\acute{\alpha}}$	Interaction matrix between different slip systems α and $\acute{\alpha}$
$\xi_{\beta\acute{\beta}}$	Interaction matrix between different slip systems β and $\acute{\beta}$
$\xi_{\chi\acute{\chi}}$	Interaction matrix between different slip systems χ and $\acute{\chi}$
$f^{\acute{\beta}}$	the volume fraction of twins for the twin system $\acute{\beta}$
$f^{\acute{\chi}}$	the volume fraction of twins for the twin system $\acute{\chi}$
$\hat{\tau}_{tw}$	the critical stress for twining
$\hat{\tau}_{tr}$	the critical stress for transformation
τ^{β}	resolved shear stress on twin system β
τ^{χ}	resolved shear stress on transformation system χ
A	fitting parameter for twin probability
B	fitting parameter for transformation probability

References

1. Herrera, C.; Ponge, D.; Raabe, D. Design of a novel Mn-based 1 GPa duplex stainless TRIP steel with 60% ductility by a reduction of austenite stability. *Acta Materialia* **2011**, *59*, 4653–4664. [CrossRef]
2. Grässel, O.; Krüger, L.; Frommeyer, G.; Meyer, L. High strength Fe–Mn–(Al, Si) TRIP/TWIP steels development—properties—application. *Int. J. Plast.* **2000**, *16*, 1391–1409. [CrossRef]
3. Guk, S.; Milisova, D.; Pranke, K. Influence of deformation conditions on the microstructure and formability of sintered Mg-PSZ reinforced TRIP-matrix-composites. *Key Eng. Mater.* **2016**, *684*. [CrossRef]
4. Guk, S.; Pranke, K.; Müller, W. Flow Curve Modelling of an Mg-PSZ Reinforced TRIP-matrix-composite. *ISIJ Int.* **2014**, *54*, 2416–2420. [CrossRef]
5. Olson, G. *Transformation Plasticity and the Stability of Plastic Flow*; ASM: Washington, DC, USA, 1984.
6. Olson, G.; Cohen, M. Kinetics of strain-induced martensitic nucleation. *Metall. Trans. A* **1975**, *6*, 791. [CrossRef]
7. Rafaja, D.; Krbetschek, C.; Ullrich, C.; Martin, S. Stacking fault energy in austenitic steels determined by using in situ X-ray diffraction during bending. *J. Appl. Crystallogr.* **2014**, *47*, 936–947. [CrossRef]

8. Martin, S.; Ullrich, C.; Rafaja, D. Deformation of Austenitic CrMnNi TRIP/TWIP Steels: Nature and Role of the ε− martensite. *Mater. Today Proc.* **2015**, *2*, S643–S646. [CrossRef]
9. Martin, S.; Decker, S.; Krüger, L.; Martin, U.; Rafaja, D. Microstructure Changes in TRIP Steel/Mg-PSZ Composites Induced by Low Compressive Deformation. *Adv. Eng. Mater.* **2013**, *15*, 600–608. [CrossRef]
10. Weidner, A.; Martin, S.; Klemm, V.; Martin, U.; Biermann, H. Stacking faults in high-alloyed metastable austenitic cast steel observed by electron channelling contrast imaging. *Scripta Materialia* **2011**, *64*, 513–516. [CrossRef]
11. Weidner, A.; Segel, C.; Biermann, H. Magnitude of shear of deformation-induced α'-martensite in high-alloy metastable steel. *Mater. Lett.* **2015**, *143*, 155–158. [CrossRef]
12. Weidner, A.; Biermann, H. Combination of different in situ characterization techniques and scanning electron microscopy investigations for a comprehensive description of the tensile deformation behavior of a CrMnNi TRIP/TWIP steel. *JOM* **2015**, *67*, 1729–1747. [CrossRef]
13. Weidner, A.; Hangen, U.D.; Biermann, H. Nanoindentation measurements on deformation-induced α'-martensite in a metastable austenitic high-alloy CrMnNi steel. *Philos. Mag. Lett.* **2014**, *94*, 522–530. [CrossRef]
14. Decker, S.; Krüger, L.; Richter, S.; Martin, S.; Martin, U. Strain-Rate-Dependent Flow Stress and Failure of an Mg-PSZ Reinforced TRIP Matrix Composite Produced by Spark Plasma Sintering. *Steel Res. Int.* **2012**, *83*, 521–528. [CrossRef]
15. Martin, S.; Wolf, S.; Martin, U.; Krüger, L. Influence of temperature on phase transformation and deformation mechanisms of cast CrMnNi-TRIP/TWIP steel. *Solid State Phenomena* **2011**, *172–174*, 172–177. [CrossRef]
16. Ullah, M.; Pasha, R.A.; Chohan, G.Y.; Qayyum, F. Numerical Simulation and Experimental Verification of CMOD in CT Specimens of TIG Welded AA2219-T87. *Arab. J. Sci. Eng.* **2015**, *40*, 935–944. [CrossRef]
17. Qayyum, F.; Kamran, A.; Ali, A.; Shah, M. 3D numerical simulation of thermal fatigue damage in wedge specimen of AISI H13 tool steel. *Eng. Fract. Mech.* **2017**, *180*, 240–253. [CrossRef]
18. Ullah, M.; Wu, C.S.; Qayyum, F. Prediction of crack tip plasticity induced due to variation in solidification rate of weld pool and its effect on fatigue crack propagation rate (FCPR). *J. Mech. Sci. Technol.* **2018**, *32*, 3625–3635. [CrossRef]
19. Mukhtar, F.; Qayyum, F.; Anjum, Z.; Shah, M. Effect of chrome plating and varying hardness on the fretting fatigue life of AISI D2 components. *Wear* **2019**, *418*, 215–225. [CrossRef]
20. Guk, S.; Yanina, A.; Müller, W.; Kawalla, R. Mathematical-physical model of resistance sintering with current conducting electrode punches of steel ceramic composites. *Materialwissenschaft und Werkstofftechnik* **2010**, *41*, 33–38. [CrossRef]
21. Weigelt, C.; Aneziris, C.; Yanina, A.; Guk, S. Ceramic Processing for TRIP-Steel/Mg-PSZ Composite Materials for Mechanical Applications. *Steel Res. Int.* **2011**, *82*, 1080–1086. [CrossRef]
22. Yanina, A.; Guk, S.; Müller, W.; Kawalla, R.; Weigelt, C. Dynamic and Static Softening of Sintered MgO-PSZ/TRIP-matrix Composites with up to 10 vol.-% ZrO_2. *Steel Res. Int.* **2011**, *82*, 1158–1165. [CrossRef]
23. Bokuchava, G.; Gorshkova, Y.E.; Papushkin, I.; Guk, S.; Kawalla, R. Investigation of Plastically Deformed TRIP-Composites by Neutron Diffraction and Small-Angle Neutron Scattering Methods. *J. Surf. Investig. X-Ray Synchrotron Neutron Tech.* **2018**, *12*, 227–232. [CrossRef]
24. Kirschner, M.; Guk, S.; Kawalla, R.; Prahl, U. Further Development of Process Maps for TRIP Matrix Composites during Powder Forging. *Mater. Sci. Forum* **2019**, *949*, 15–23. [CrossRef]
25. Khan, U.; Hussain, A.; Shah, M.; Shuaib, M.; Qayyum, F. Investigation of mechanical properties based on grain growth and microstructure evolution of alumina ceramics during two step sintering process. *IOP Conf. Ser. Mater. Sci. Eng.* **2016**, *146*, 012046.
26. Qayyum, F.; Shah, M.; Muqeet, A.; Afzal, J. The effect of anisotropy on the intermediate and final form in deep drawing of SS304L, with high draw ratios: Experimentation and numerical simulation. *IOP Conf. Ser. Mater. Sci. Eng.* **2016**. [CrossRef]
27. Khan, F.; Qayyum, F.; Asghar, W.; Azeem, M.; Anjum, Z.; Nasir, A.; Shah, M. Effect of various surface preparation techniques on the delamination properties of vacuum infused Carbon fiber reinforced aluminum laminates (CARALL): Experimentation and numerical simulation. *J. Mech. Sci. Technol.* **2017**, *31*, 5265–5272. [CrossRef]

28. Asghar, W.; Nasir, M.A.; Qayyum, F.; Shah, M.; Azeem, M.; Nauman, S.; Khushnood, S. Investigation of fatigue crack growth rate in CARALL, ARALL and GLARE. *Fatigue Fract. Eng. Mater. Struct.* **2017**, *40*, 1086–1100. [CrossRef]
29. Butt, Z.; Anjum, Z.; Sultan, A.; Qayyum, F.; Ali, H.M.K.; Mehmood, S. Investigation of electrical properties & mechanical quality factor of piezoelectric material (PZT-4A). *J. Electr. Eng. Technol.* **2017**, *12*, 846–851.
30. Elahi, H.; Eugeni, M.; Gaudenzi, P.; Qayyum, F.; Swati, R.F.; Khan, H.M. Response of piezoelectric materials on thermomechanical shocking and electrical shocking for aerospace applications. *Microsyst. Technol.* **2018**, *24*, 3791–3798. [CrossRef]
31. Weidner, A.; Berek, H.; Segel, C.; Aneziris, C.G.; Biermann, H. In situ tensile deformation of TRIP steel/Mg-PSZ composites. *Mater. Sci. Forum* **2013**. [CrossRef]
32. Martin, S.; Richter, S.; Decker, S.; Martin, U.; Krüger, L.; Rafaja, D. Reinforcing Mechanism of Mg-PSZ Particles in Highly-Alloyed TRIP Steel. *Steel Res. Int.* **2011**, *82*, 1133–1140. [CrossRef]
33. Roters, F.; Eisenlohr, P.; Kords, C.; Tjahjanto, D.; Diehl, M.; Raabe, D. DAMASK: the Düsseldorf Advanced MAterial Simulation Kit for studying crystal plasticity using an FE based or a spectral numerical solver. *Procedia IUTAM* **2012**, *3*, 3–10. [CrossRef]
34. Lebensohn, R.A.; Kanjarla, A.K.; Eisenlohr, P. An elasto-viscoplastic formulation based on fast Fourier transforms for the prediction of micromechanical fields in polycrystalline materials. *Int. J. Plast.* **2012**, *32*, 59–69. [CrossRef]
35. Eisenlohr, P.; Diehl, M.; Lebensohn, R.A.; Roters, F. A spectral method solution to crystal elasto-viscoplasticity at finite strains. *Int. J. Plast.* **2013**, *46*, 37–53. [CrossRef]
36. Shanthraj, P.; Eisenlohr, P.; Diehl, M.; Roters, F. Numerically robust spectral methods for crystal plasticity simulations of heterogeneous materials. *Int. J. Plast.* **2015**, *66*, 31–45. [CrossRef]
37. Wong, S.L.; Madivala, M.; Prahl, U.; Roters, F.; Raabe, D. A crystal plasticity model for twinning-and transformation-induced plasticity. *Acta Materialia* **2016**, *118*, 140–151. [CrossRef]
38. Madivala, M.; Schwedt, A.; Wong, S.L.; Roters, F.; Prahl, U.; Bleck, W. Temperature dependent strain hardening and fracture behavior of TWIP steel. *Int. J. Plast.* **2018**, *104*, 80–103. [CrossRef]
39. Vinogradov, A.; Lazarev, A.; Linderov, M.; Weidner, A.; Biermann, H. Kinetics of deformation processes in high-alloyed cast transformation-induced plasticity/twinning-induced plasticity steels determined by acoustic emission and scanning electron microscopy: Influence of austenite stability on deformation mechanisms. *Acta Materialia* **2013**, *61*, 2434–2449. [CrossRef]
40. Qayyum, F.; Guk, S.; Prüger, S.; Schmidtchen, M.; Saenko, I.; Kiefer, B.; Kawalla, R.; Prahl, U. Investigating the local deformation and transformation behavior of sintered X3CrMnNi16-7-6 TRIP steel using a calibrated crystal plasticity-based numerical simulation model. *Int. J. Mater. Res. (formerly: Zeitschrift für Metallkunde)* **2020**. [CrossRef]
41. Dai, Q.; Wang, A.; Cheng, X. Stacking fault energy of cryogenic austenitic steel. *Gangtie Yanjiu Xuebao (J. Iron Steel Res.)* **2002**, *14*, 34–37.
42. Music, D.; Schneider, J.M. The correlation between the electronic structure and elastic properties of nanolaminates. *JOM* **2007**, *59*, 60–64. [CrossRef]
43. Pabst, W.; Ticha, G.; Gregorová, E. Effective elastic properties of alumina-zirconia composite ceramics-Part 3. Calculation of elastic moduli of polycrystalline alumina and zirconia from monocrystal data. *Ceram. Silikaty* **2004**, *48*, 41–48.
44. Orowan, E. Zur kristallplastizität. i. *Zeitschrift für Physik* **1934**, *89*, 605–613. [CrossRef]
45. Ackermann, S.; Martin, S.; Schwarz, M.R.; Schimpf, C.; Kulawinski, D.; Lathe, C.; Henkel, S.; Rafaja, D.; Biermann, H.; Weidner, A. Investigation of phase transformations in high-alloy austenitic TRIP steel under high pressure (up to 18 GPa) by in situ synchrotron X-ray diffraction and scanning electron microscopy. *Mater. Trans. A* **2016**, *47*, 95–111. [CrossRef]
46. Andersson, J.-O.; Helander, T.; Höglund, L.; Shi, P.; Sundman, B. Thermo-Calc & DICTRA, computational tools for materials science. *Calphad* **2002**, *26*, 273–312. [CrossRef]
47. Weidner, A.; Yanina, A.; Guk, S.; Kawalla, R.; Biermann, H. Microstructure and Local Strain Fields in a High-Alloyed Austenitic Cast Steel and a Steel-Matrix Composite Material after in situ Tensile and Cyclic Deformation. *Steel Res. Int.* **2011**, *82*, 990–997. [CrossRef]
48. Qayyum, F.; Guk, S.; Kawalla, R.; Prahl, U. Experimental Investigations and Multiscale Modeling to Study the Effect of Sulfur Content on Formability of 16MnCr5 Alloy Steel. *Steel Res. Int.* **2018**, *90*. [CrossRef]

49. Krüger, L.; Decker, S.; Ohser-Wiedemann, R.; Ehinger, D.; Martin, S.; Martin, U.; Seifert, H. Strength and Failure Behaviour of Spark Plasma Sintered Steel-Zirconia Composites Under Compressive Loading. *Steel Res. Int.* **2011**, *82*, 1017–1021. [CrossRef]
50. Poklad, A.; Motylenko, M.; Klemm, V.; Schreiber, G.; Martin, S.; Decker, S.; Abendroth, B.; Haverkamp, M.; Rafaja, D. Interface Phenomena Responsible for Bonding between TRIP Steel and Partially Stabilised Zirconia as Revealed by TEM. *Adv. Eng. Mater.* **2013**, *15*, 627–637. [CrossRef]
51. Guk, S.; Müller, W.; Pranke, K.; Kawalla, R. Mechanical Behaviour Modelling of an Mg-Stabilized Zirconia Reinforced TRIP-Matrix-Composite under Cold Working Conditions. *Mater. Sci. Appl.* **2014**, *5*, 812–822. [CrossRef]
52. Hensl, T.; Mühlich, U.; Budnitzki, M.; Kuna, M. An eigenstrain approach to predict phase transformation and self-accommodation in partially stabilized zirconia. *Acta Materialia* **2015**, *86*, 361–373. [CrossRef]
53. Weigelt, C.; Berek, H.; Aneziris, C.G.; Wolf, S.; Eckner, R.; Krüger, L. Effect of minor titanium additions on the phase composition of TRIP steel/magnesia partially stabilised zirconia composite materials. *Ceram. Int.* **2015**, *41*, 2328–2335. [CrossRef]

© 2020 by the authors. Licensee MDPI, Basel, Switzerland. This article is an open access article distributed under the terms and conditions of the Creative Commons Attribution (CC BY) license (http://creativecommons.org/licenses/by/4.0/).

Article

Late Age Dynamic Strength of High-Volume Fly Ash Concrete with Nano-Silica and Polypropylene Fibres

Mohamed H. Mussa [1,2,*], Ahmed M. Abdulhadi [2], Imad Shakir Abbood [3], Azrul A. Mutalib [1] and Zaher Mundher Yaseen [4]

[1] Department of Civil Engineering, Faculty of Engineering & Built Environment, Universiti Kebangsaan Malaysia, Bangi 43600 UKM, Selangor, Malaysia; azrulaam@ukm.edu.my
[2] Department of Civil Engineering, University of Warith Al-Anbiyaa, Karbala 56001, Iraq; ahmedmouse@uowa.edu.iq
[3] Engineering Affairs Department, Sunni Endowment Diwan, Baghdad, Iraq; imadshakirabbood@gmail.com
[4] Sustainable Developments in Civil Engineering Research Group, Faculty of Civil Engineering, Ton Duc Thang University, Ho Chi Minh City, Vietnam; yaseen@tdtu.edu.vn
* Correspondence: eng.mhmussa@siswa.ukm.edu.my or dr.mhmussa@uowa.edu.iq; Tel.: +964-7735047594

Received: 1 March 2020; Accepted: 23 March 2020; Published: 26 March 2020

Abstract: The dynamic behaviour of high-volume fly ash concrete with nano-silica (HVFANS) and polypropylene fibres at curing ages of 7 to 90 days was determined by using a split Hopkinson pressure bar (SHPB) machine. At each curing age, the concrete samples were laboratory tested at different temperatures conditions under strain rates reached up to 101.42 s^{-1}. At room temperature, the results indicated that the dynamic compressive strength of plain concrete (PC) was slightly higher than HVFANS concrete at early curing ages of 7 and 28 days, however, a considerable improvement in the strength of HVFANS concrete was noted at a curing age of 90 days and recorded greater values than PC owing to the increase of fly ash reactivity. At elevated temperatures, the HVFANS concrete revealed a superior behaviour than PC even at early ages in terms of dynamic compressive strength, critical strain, damage and toughness due to increase of nano-silica (NS) activity during the heating process. Furthermore, equations were suggested to estimate the dynamic increase factor (DIF) of both concretes under the investigated factors.

Keywords: curing age; HVFANS concrete; SHPB test; DIF; toughness; critical damage

1. Introduction

Nowadays, the production process of one ton Portland cement could release about one ton of greenhouses gases to the air which mainly consisted of carbon dioxide CO_2 [1]. The scientific community reported that CO_2 contributes about 65% of global warming which means that the temperature of the planet will be raised between 1.4 and 5.8 °C during the next century so that several natural disasters might occur [2]. Therefore, scholars suggested several alternative binders as instant (fly ash (FA), silica fume, nano-silica (NS), slag, metakaolin, etc.) to reduce the reliance on Portland cement in construction.

Huge amounts of FA are produced annually worldwide about one billion tonnes as reported by Bakharev [3]. However, only a few amounts are used approximately 20% to 30% and the rest is landfilled with potential risks of environmental pollution [4]. Fly ash is gathered from thermal power plants during the pulverised coal combustion. Most of the FA particles are solid spheres with sizes between 1 to 100 µm, and a specific gravity of 1.9 to 2.8 with a surface area of 300 to 500 m^2/kg. The classification of FA depends on its chemical composition and typically classified as Class F or Class C based on the percentages of iron, silica and aluminium oxide, which have to be a minimum of 70% or 50%, respectively. Recently, numerous studies have been conducted to investigate the ability to use a high quantity of fly ash in concrete as a cement replacement because of its beneficial effects on concrete behaviour and low cost [5].

The main disadvantage of using high-volume fly ash (HVFA) was the sharp reduction in the strength of concrete particularly at primary ages owing to the slow reactivity of fly ash which contained a high amount of oxide calcium (CaO) [6]. Montgomery et al. [7] revealed that CaO apparently had not reacted at early ages during the normal curing circumstances. The differences in strength between the plain concrete (PC) and HVFA concrete are diminished with the increase of curing age. At late ages, the strength of HVFA concrete might equalise or exceed the plain concrete (PC) strength according to the quantity, fineness and reactivity of FA, water to cementations material ratio (w/(c + p)) and curing conditions such as humidity and temperature.

Several scholars aimed to develop the static strength of HVFA concrete and mortar at early curing ages by adding nano-silica (NS) material due to its high pozzolanic reactivity and pore-filling effect. Li [8] revealed that using 4% of NS in concrete consisted of 50% of FA could increase the static compressive strength by 68.57%, 39.34%, 18.72% and 7.58% at curing ages of 7, 28, 56, and 112 days, respectively. In the same context, Zhang and Islam [9] stated that the strength of mortar samples consisted of 50% FA and 1% NS improved by 62.16%, 24.31%, 17.1%, 6.86% and 4.35% during curing ages of 1, 3, 7, 28 and 91 days, respectively. Furthermore, the fire resistance of HVFA concrete inclusion with NS was studied. Ibrahim [10,11] reported that high-volume fly ash nano-silica (HVFANS) concrete could maintain about 94.54% of its strength under a temperature of 700 °C at a curing age of 28 days via substituting the cement with 2.5% and 52.5% of nano-silica and fly ash, respectively.

Several structures as instant tunnels may be subject to accidental or man-made explosions during the service time [12–19]. It leads to the fact that the concrete material is in danger of fire exposure and dynamic loads simultaneously. The previous studies mainly focused on determining the response of normal strength concrete (NSC) under the combined effect of temperature and strain rate at a curing age of 28 days using the SHPB machine [20–25]. These studies concluded that the concrete strength reduced with temperature particularly above 400 °C and the dynamic increase factor (DIF) showed a linear relationship with strain rate. Chen et al. [26] investigated the behaviour of NSC containing 15% to 20% of FA under high temperatures reaching up to 950 °C and strain rate range between 10 and 205 s^{-1}. The results showed a remarkable relation between the dynamic properties of concrete and strain rates at elevated temperature. Whereas, the failure of concrete samples under high strain rates were considerably different than those at room temperature. On the other hand, limited research was performed to investigate the dynamic response of high strength concrete (HSC) contained a high-volume of FA and nano-silica materials under the combined effect of temperatures and strain rate. Most of these studies were conducted at a curing age of 28 days. Li et al. [27] tested the HSC concrete consisting from 4.5% to 23% of silica fume and fly ash at a strain rate range between 33.70 and 194.12 s^{-1} and elevated temperatures reached up to 800 °C. The results revealed a linear increase of dynamic strength with the strain rate which obviously reduced after 400 °C. The tested concrete was able to maintain 51.85% and 24.3% of its original strength at 600 and 800 °C, respectively.

In the same context, Su et al. [28] studied the behaviour of HSC concrete consisting of 20% and 5% of FA and silica fume, respectively, under temperatures reached up to 800 °C within strain rate range between 27.5 to 121.6 s^{-1}. The results indicated that the concrete strength was increased by 14% at 400 °C as compared to the ambient temperature. However, its strength was significantly decreased by 16% and 48% at temperatures of 600 and 800 °C, respectively. Furthermore, the concrete toughness showed a notable increase at temperatures of 200 and 400 °C by 31 and 95, respectively, as compared to ambient temperature but its toughness considerably declined at 800 °C. According to the previous studies, the curing age had a clear effect on the static properties of fly ash concrete with nano-silica materials. Nevertheless, the dynamic behaviour of this concrete type is still not further investigated at early or late curing age while most of the prior researchers determined its properties at 28 days only. Therefore, understanding the behaviour of HVFANS concrete under various curing ages will increase the dependability on this concrete in construction. The proportions of HVFANS concrete proposed by Ibrahim [10,11] were adopted in this study due to its high-fire resistance.

2. Materials and Equipment

2.1. Components of Concrete Mixture

The proportions of concrete samples with a target strength of 60MPa are selected based on (ACI 211.4R-93) with water to binder materials ratio of 0.29 as described in Table 1 [29]. The percentages of FA and NS as cement replacement were 52.5% and 2.5%, respectively. These percentages were chosen owing to the fire properties of its produced concrete at temperatures reached up to 700 °C [10,11].

Table 1. Proportions of concrete samples.

Samples	Cement	Sand	Gravel	Water	Fly Ash	Nano-Silica	PPF %
PC	531.30	780.06	942.64	170.16	-	-	1
HVFANS	225.80	682.01	942.64	154.85	278.93	26.56	1

The Portland cement (Type I) manufactured by Tasek Cement Company according to Malaysian standard (MS 522) was used [30]. The Fly ash class F was collected from a local power plant with 44.16% of silicon dioxide (SiO_2), 24.6% of aluminium oxide (Al_2O_3) and 12.5% of ferric oxide (Fe_2O_3) was adopted as a cement replacement [31]. Colloidal nano-silica (Cembinder W8) was manufactured via AkzoNobel Company was used to improve the early compressive strength of concrete, thereby shortening the setting time of the cement slurry [10,11]. The surface area of NS particles was equal to 80 m^2/g with an average size of 35 nm and a silica concentration of 50% with a bulk density of 1050 kg/m^3 and pH of 10. Natural river sand was used as a fine aggregate passed via a sieve size of 4.75 mm with fineness modulus of 2.98 [32,33]. The bulk specific gravity of the sand was 2.53 with a compacted and loose bulk density of 1721.83 and 1510.18 kg/m^3 [34]. Coarse aggregate with a maximum size of 10 mm with a specific gravity of 2.07 was used in the mix design.

Fibrillated Polypropylene fibres (PPF) with a 1% dosage were added to the concrete samples exposed to the high temperature to increase the ductility of concrete and eliminate the surface spalling by providing new pores after melting which work as channels to relieve the internal energy of water vapour pressure [35–40]. These fibres manufactured by Timuran Company with the length of 12 mm and a specific gravity of 0.9 have a white colour with a melting point ranging between 160 and 170 °C and tensile strength of 0.36 kN/mm^2. Furthermore, superplasticiser with a dosage of 1% was added to provide the required workability and improve the compressive strength of concrete via decreasing the amount of water in the mixture.

2.2. Samples Preparation and Curing Conditions

The feeding sequence of concrete materials into a horizontal pan mixer was carried out according to American Society for Testing and Materials (ASTM) C192 [41]. However, cement balls were formed mainly in the case of HVFANS concrete samples as shown in Figure 1. Therefore, a part of coarse aggregate with fine aggregate was firstly mixed with fly ash and cement to avoid the above problem. Afterwards, the water and colloidal NS were added to the mixture. Lastly, the other parts of the aggregate and fibres were added to destroy any noodles of a mixture [42]. The concrete mixture was cast into three layers inside cubic and cylindrical steel moulds with dimensions of 150 × 150 × 150 mm and 50 × 50∅ mm, respectively, and then each layer was compacted by a vibration table for ten seconds. Before the curing process, the casted samples were covered by a plastic membrane to reduce the moisture evaporation. The PC samples remained in moulds for 24 h, while the HVFANS concrete samples were de-moulded after 48 h. Afterwards, all the casted cubes and cylinders are cured into a water tank for 7, 28 and 90 days within room temperature. Water with a pH of 8 was used for making the mixes and curing the specimens. The water was free of chloride and deleterious salts according to the American Society for Testing and Materials (ASTM) specifications for mixing water used in the production of hydraulic cement concrete [43].

Figure 1. Formation of cement balls during the mixing process.

2.3. Laboratory Tests

2.3.1. Equipment

An electric kiln with a maximum temperature of 1000 °C and capacity of three concrete cubes was used to carry out the heating process according to ISO 834 fire curve [44]. After two hours, the heating process was stopped and the specimens were left to gradually cool down within the room temperature. The average static compressive strength of three cubic samples for both concrete was recorded at curing ages of 7, 28 and 90 days within room temperatures of 400 and 700 °C according to British standards by using a compression machine with a maximum load capacity of 5000 kN [45]. The SHPB machine was used to perform the dynamic test at strain rates between 30.09 and 102.48 s^{-1} as shown in Figure 2 [46,47]. The machine bars were manufactured from a stainless steel material with a diameter of 50 mm which was selected to reduce the wave dispersion and increase the accuracy of the recorded signals [46]. A hand grinder was used to decrease and achieve good contact between the bars and concrete samples. Moreover, an alignment by a laser with a precision reached up to 0.01 mm was used to accomplish a good coaxial and fulfil the force equilibrium condition [46]. The average dynamic strength of five cylindrical samples (350 samples) for both concrete was recorded at each exposing temperature and curing age.

Figure 2. Split Hopkinson pressure bar (SHPB) machine components.

2.3.2. Recorded Signals of Dynamic Test

In the current study, cylinder specimens with a ratio of length to the diameter of 1 were used during the dynamic test. Davies and Hunter [48] indicated that there is no clear influence for the size of a concrete cylinder sample if its length to diameter ratio kept within 0.31 to 1.55 [46,47]. Besides, Lindholm [49] reported that the length of concrete specimens should be short enough to achieve a uniform state of stress throughout the test. During the test, the incident bar was a strike and a uniaxial incident pulse within amplitude primarily depends on the velocity of the striker bar that is calculated by Newton's law [50,51]. A part of the incident pulse was reflected as a reflected pulse along the incident bar because of the impedance mismatch between the concrete sample and incident bar while the rest of the pulse was moved to the transmitter bar and recorded as a transmitted pulse as shown in Figure 3.

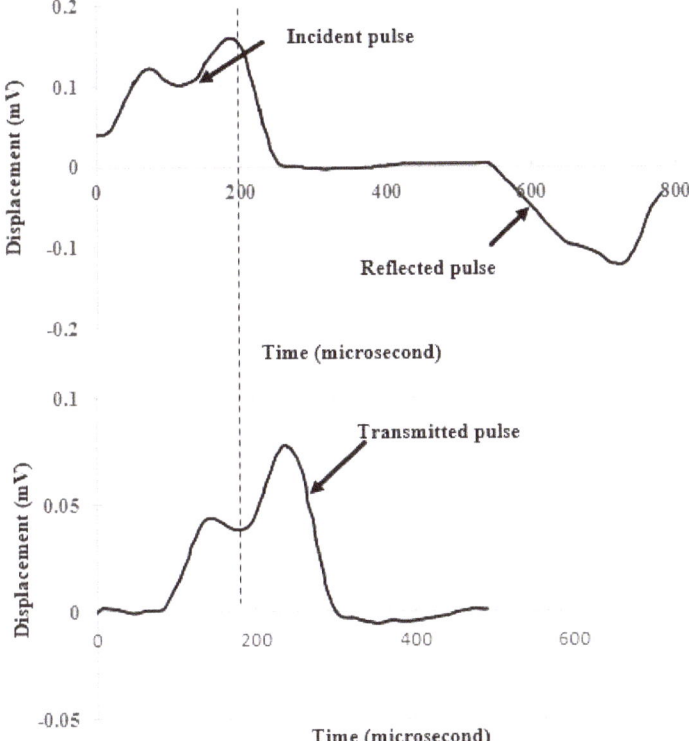

Figure 3. Displacement time-history records of the incident, reflected and transmitted pulses during the test.

Quarterly connected strain gauges (SGD-3/350-LY11) were utilised to record the elongation of incident and transmitter bars during the test. These gauges were connected to bridge sensors model (OM2-163) to amplify the voltage change. The voltage signals were converted into digital data that can be saved into a computer and analysed via MATLAB software by using a data logger (OMB-DAQ-3000). All the recorded pulses were used to calculate the stress $\sigma(t)$, strain rate $\dot{\varepsilon}(t)$, and strain $\varepsilon(t)$, respectively, according to the following equations [47,52–54]:

$$\sigma(t) = \frac{AE}{2A_s}[\varepsilon_i(t) + \varepsilon_r(t) + \varepsilon_t(t)] \tag{1}$$

$$\dot{\varepsilon}(t) = \frac{C_0}{L_s}[\varepsilon_i(t) - \varepsilon_r(t) - \varepsilon_t(t)] \qquad (2)$$

$$\varepsilon(t) = \frac{C_0}{L_s}\int_0^t [\varepsilon_i(t) - \varepsilon_r(t) - \varepsilon_t(t)]\,dt \qquad (3)$$

where $\varepsilon_i(t)$ is the incident pulse, $\varepsilon_r(t)$ reflected pulse and $\varepsilon_t(t)$ is the transmitted pulse, A is the area of the bar cross-section, E is the bar modulus of elasticity equals to 210 GPa, C_0 is the bar wave velocity equalling 5190 m/s and l_s, A_s are the sample length and area.

3. Results of Laboratory Tests

3.1. Static Test

Table 2 revealed that the static compressive strength of HVFANS concrete was slightly lesser than PC at early curing ages of 7 and 28 days within room temperature by 6.87% and 4.21%, respectively. This behaviour attributed to the presence of fly ash (class F) within high quantities in the mixture which has a low amount of cement-like properties (CaO), besides, the slow pozzolanic reactivity of FA at an early age [55]. Montgomery et al. [7] observed that CaO was not significantly able to react at an early age under normal curing conditions. However, the differences between both concrete at room temperature were diminished at late curing age of 90 days and the strength of HVFANS concrete exceeded the PC by 2.24% due to the increase of FA reactivity that expressively enhanced the calcium silicate hydrate (C–S–H) structure [56].

Table 2. Curing age effect on the response of plain concrete (PC) and high-volume fly ash concrete with nano-silica (HVFANS) concrete at 25 °C.

Sample No.	Time (days)					
	7		28		90	
	PC	HVFANS	PC	HVFANS	PC	HVFANS
1	50.81	49.12	61.82	57.95	61.48	67.82
2	49.62	45.34	62.66	61.55	66.86	66.54
3	51.74	47.91	59.08	56.68	65.72	64.06
Average (MPa)	50.72	47.46	61.19	58.72	64.69	66.14
Maximum relative error (%)	2.17	4.47	3.45	4.82	4.96	3.14

The HVFANS concrete showed an excellent performance than PC at all the investigated curing ages within elevated temperatures of 400 and 700 °C as shown in Tables 3 and 4. This performance, owing to the filler effect of nano-silica on the concrete mixture, caused a great increase in the C–S–H content [55]. Whereas the maximum differences between HVFANS concrete and PC were observed at 70 °C. by 34.07%, 26.01% and 39.37% at curing ages of 7, 28 and 90 days, respectively. As can be noted that the compressive strength of both concrete was considerably reduced in most curing age cases under elevated temperatures particularly at 700 °C. due to the massive growth of vapour pressure into the samples which resulted in numerous cracks beside that the cement paste binder products could significantly dehydrate at this temperature [10,11]. The relative error was less than 7.78% and fairly reflected the reliability of the laboratory records.

Table 3. Curing age effect on the response of PC and HVFANS concrete at 400 °C.

Sample No.	Time (days)					
	7		28		90	
	PC	HVFANS	PC	HVFANS	PC	HVFANS
1	42.31	44.37	58.61	67.79	59.67	64.81
2	40.48	41.81	56.48	61.01	62.31	67.12
3	43.13	43.50	59.48	64.86	61.15	65.60
Average (MPa)	41.97	43.23	58.19	64.55	61.04	65.84
Maximum relative error (%)	3.55	3.28	2.94	5.48	2.24	1.94

Table 4. Curing age effect on the response of PC and HVFANS concrete at 700 °C.

Sample No.	Time (days)					
	7		28		90	
	PC	HVFANS	PC	HVFANS	PC	HVFANS
1	28.91	40.14	38.87	51.6	39.81	53.12
2	30.68	38.50	33.54	52.13	37.11	50.44
3	27.22	37.77	36.71	46.26	35.81	49.91
Average (MPa)	28.94	38.80	36.37	49.99	37.58	51.16
Maximum relative error (%)	5.94	3.45	7.78	7.46	5.93	3.83

British standards (BS 1881-101) [57] defined the satisfactory failure modes of cubic concrete samples under compression load as shown in Figure 4a. The results appeared a good agreement with the British standard for both concrete samples that mainly failed by cracking the four faces of concrete cube equally within slight damage in its connected face with plates as shown in Figure 4b.

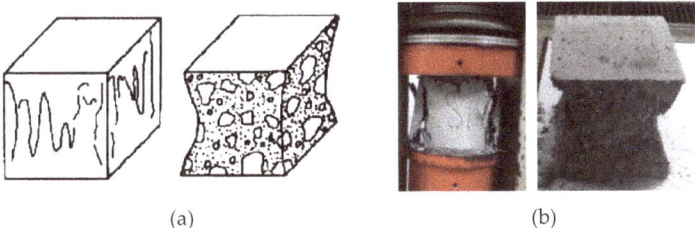

(a)   (b)

Figure 4. Comparison of failure modes between (a) British standard and (b) static test results.

3.2. Dynamic Test

3.2.1. Dynamic Compressive Strength (f_{cd})

The peak value of the stress–strain diagrams is considered as the ultimate dynamic strength (f_{cd}) of both concrete samples. The downward part of the diagrams signified the strain relaxing and fragmentation of concrete specimens after the ultimate stress [58]. The average of five strain rates is recorded at each curing age tested within different temperatures and strain rates as shown in Figures 5–7. The diagrams indicated that the dynamic strength of both concrete linearly increased with the strain rate at all the investigated curing age as shown in Table 5. The dependency of dynamic strength with strain rate was noted by several scholars on concrete and other materials such as steel [59], ceramics [60], composites [61] and rocks [62,63]. Grady [64] attributed this behaviour to the growing of tensile micro-cracks on the rocks. While Ross et al. [65] revealed that the moisture of concrete could add inertia forces and increase the fracture strength of wet concrete compared with dry.

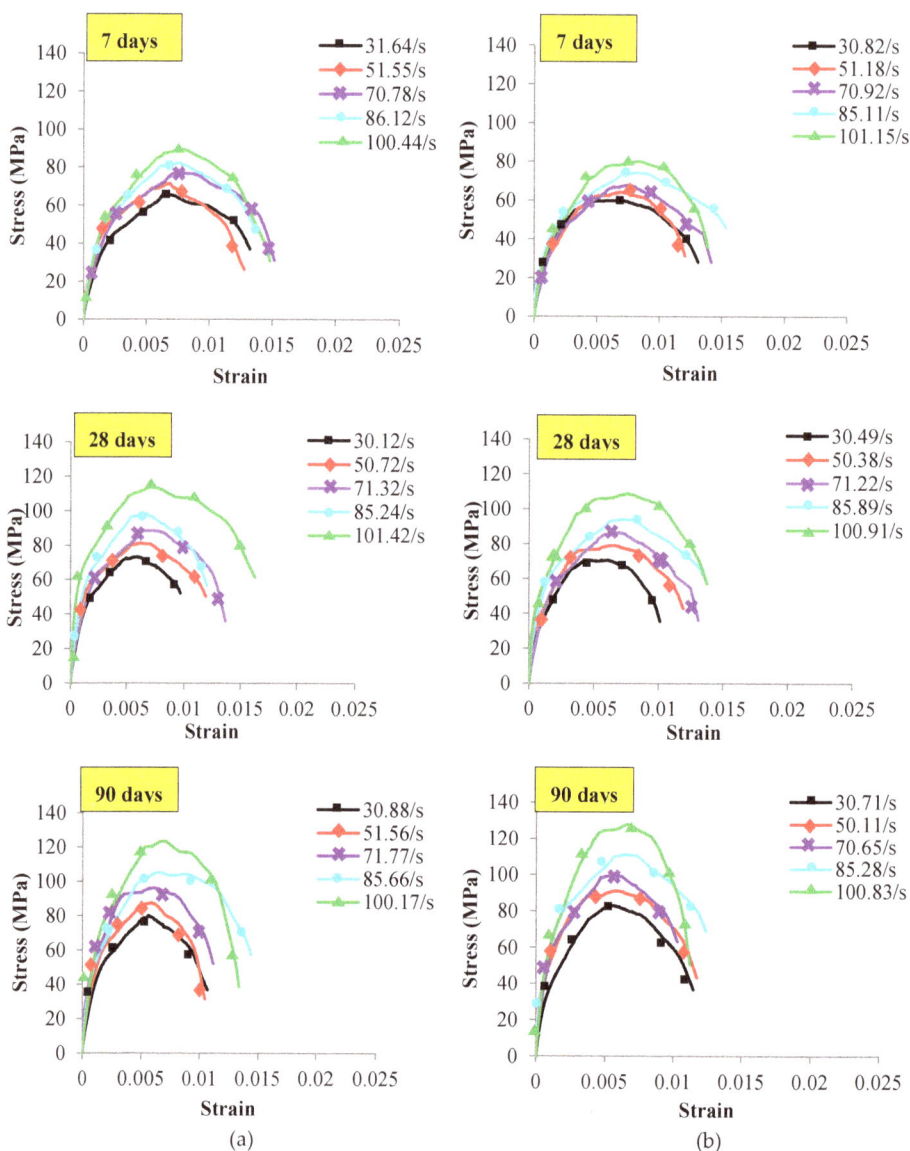

Figure 5. Stress–strain diagrams at 25 °C: (**a**) PC and (**b**) HVFANS concrete.

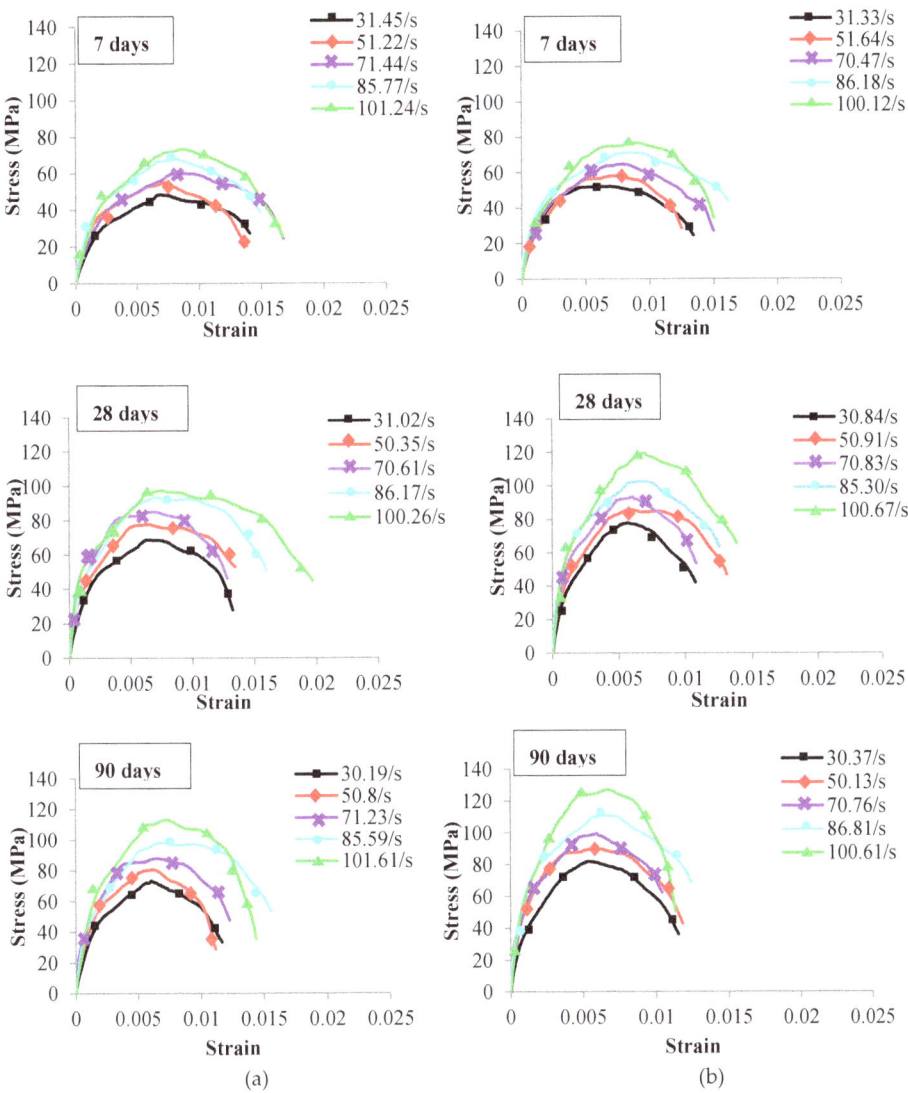

Figure 6. Stress–strain diagrams at 400 °C: (a) PC and (b) HVFANS concrete.

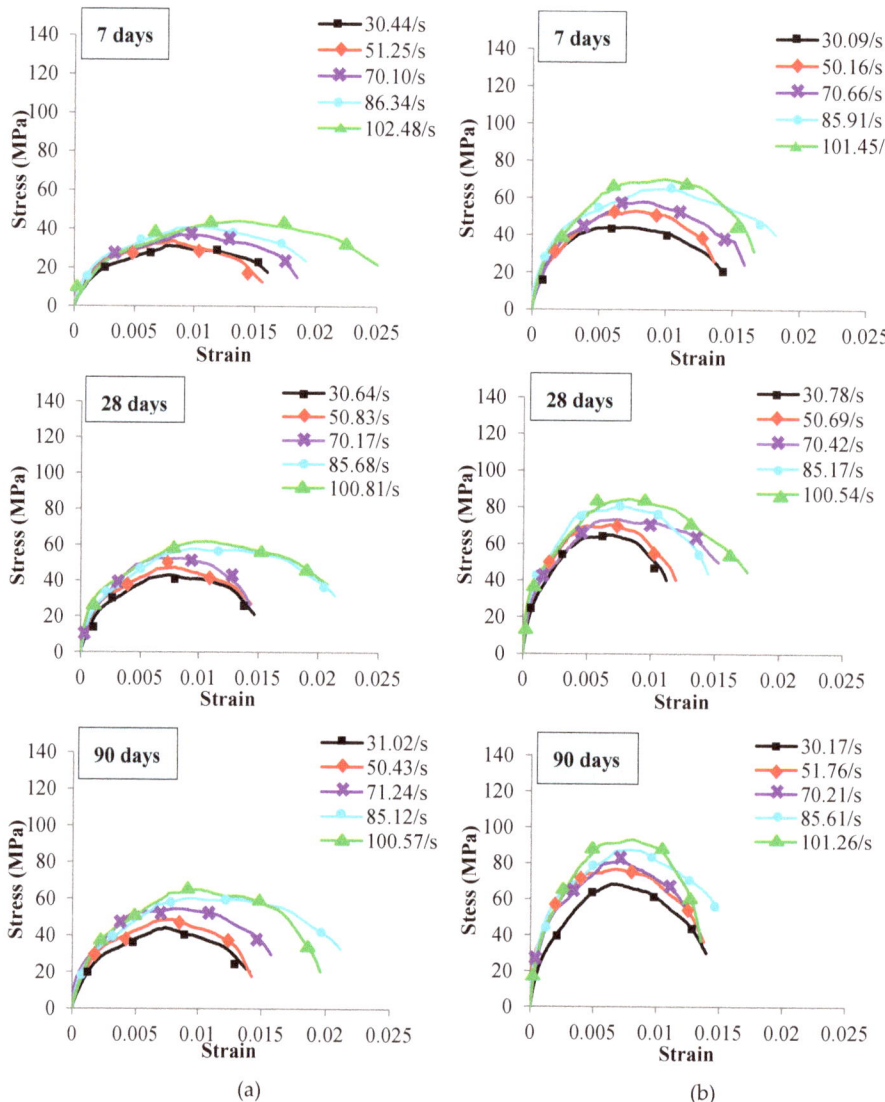

Figure 7. Stress–strain diagrams at 700 °C: (**a**) PC and (**b**) HVFANS concrete.

The results indicated that the curing age has a significant effect on the dynamic behaviour of both concrete, particularly on HVFANS concrete samples. At room temperature, the dynamic strength of HVFANS concrete was smaller than PC at curing ages of 7 and 28 days under all the investigated strain rates and the maximum difference is observed at strain rates above 100 s^{-1} by 11.78% and 4.55%, respectively, due to the slow reactivity of FA at these ages. However, the strength of HVFANS concrete is significantly increased at a curing age of 90 days under room temperature and recorded higher values than PC at all the investigated strain rates owing to the improvement of FA reactivity at the late ages. At elevated temperatures, the dynamic compressive strength of both concrete was declined at all the investigated curing ages as compared to room temperature due to the gradual dehydration of the binder products and also massive induced vapour pressure into the concrete samples [10,11].

Nevertheless, the HVFANS concrete showed a superior behaviour than PC at all the studied curing ages and strain rates owing to the filler effect of nano-silica material [55]. The highest differences between both concretes were noted at exposing temperature of 700 °C under strain rates above 100 s^{-1}. Whereas the strength of HVFANS concrete was higher than PC by 59.59%, 36.68% and 43.59% at curing ages of 7, 28 and 90, respectively. These results proved that HVFANS concrete has superior fire and explosion resistance than PC even at early curing ages and reflected the possibility of using this concrete in the structure that might be exposed to these kinds of threats during service life.

Table 5. Effect of curing age on dynamic compressive strength of PC and HVFANS concrete at different temperatures and strain rates.

Concrete	Temperature (°C)	Curing Age (days)	Average Strain Rate (s^{-1})					Dynamic Compressive Strength (MPa)				
			C1	C2	C3	C4	C5	C1	C2	C3	C4	C5
PC	25	7	31.64	51.55	70.78	86.12	100.44	65.77	71.32	76.9	82.42	89.56
		28	30.12	50.72	71.32	85.24	101.42	73.39	81.25	88.66	98.61	113.77
		90	30.88	51.56	71.77	85.66	100.17	79.88	87.61	96.23	105.33	123.6
HVFANS	25	7	30.82	51.18	70.92	85.11	101.15	59.89	64.11	67.77	74.16	80.12
		28	30.49	50.38	71.22	85.89	100.91	70.68	79.2	86.63	94.18	108.82
		90	30.71	50.11	70.65	85.28	100.83	83.11	91.23	100.65	111.24	127.87
PC	400	7	31.45	51.22	71.44	85.77	101.24	48.43	55.12	60.12	67.88	73.15
		28	31.02	50.35	70.61	86.17	100.26	68.69	77.86	85.14	93.39	97.46
		90	30.19	50.8	71.23	85.59	101.61	73.11	80.88	87.95	98.11	113.2
HVFANS	400	7	31.33	51.64	70.47	86.18	100.12	52.13	58.2	64.87	71.23	76.79
		28	30.84	50.91	70.83	85.3	100.67	77.73	85.8	93.38	102.82	119.43
		90	30.37	50.13	70.76	86.81	100.61	81.76	90.16	99.31	110.98	127.11
PC	700	7	30.44	51.25	70.1	86.34	102.48	31.2	33.52	36.66	41.12	43.78
		28	30.64	50.83	70.17	85.68	100.81	42.86	47.18	52.39	57.72	61.77
		90	31.02	50.43	71.24	85.12	100.57	44.13	48.58	54.43	60.15	64.86
HVFANS	700	7	30.09	50.16	70.66	85.91	101.45	44.12	52.78	57.9	64.76	69.83
		28	30.78	50.69	70.42	85.17	100.54	64.88	70.3	73.28	80.31	84.43
		90	30.17	51.76	70.21	85.61	101.26	68.31	76.45	81.22	87.4	93.13

3.2.2. Critical Strain (ε_{cr})

The values of critical strain (ε_{cr}) described in Figures 5–7 clarify the ductility of both concrete at curing ages of 7, 28 and 90 days within exposing temperatures of 25, 400 and 700 °C under different ranges of strain rate. The results showed that the strain increased linearly with the strain rate at all the investigated cases owing to the lateral confinement that could form high quantities of micro-cracks and prevent the formation of macro cracks [66,67]. Moreover, the outcomes revealed the important role of PP fibres under dynamic loads due to its ability to improve the concrete ductility as well as produce channels to release vapour pressure at high temperatures which clearly reduced the critical strain values particularly at temperatures of 25 and 400 °C [68–70]. The highest values of critical strain for both concretes are recorded at a curing age of seven days and exposing temperature 700 °C, however, HVFANS concrete showed smaller values as compared to PC within a maximum difference of 43.56% at a strain rate above 100 s^{-1} owing to improve nano-silica at elevated temperatures [68,69]. On the other hand, the critical strain showed a positive relationship with the increase of curing age and recorded the smallest values at 90 days at the same exposing temperature due to the increase of fly ash reactivity and hydration process which enhance the concrete structure.

3.2.3. Toughness

The toughness of both concretes was represented in term of energy absorption (SEA) calculated as the following [27,71,72]:

$$\text{SEA} = \frac{AEc}{A_s l_s} \int_0^T [\varepsilon_i^2(t) - \varepsilon_r^2(t) - \varepsilon_t^2(t)] dt \qquad (4)$$

where c is the bar's wave speed and T is the failure moment of the concrete sample. The results proved that the toughness of both concretes improved linearly with strain rate as shown in Figure 8. This behaviour was noted by other scholars who tested the concrete via SHPB machine under room or high temperatures [72,73].

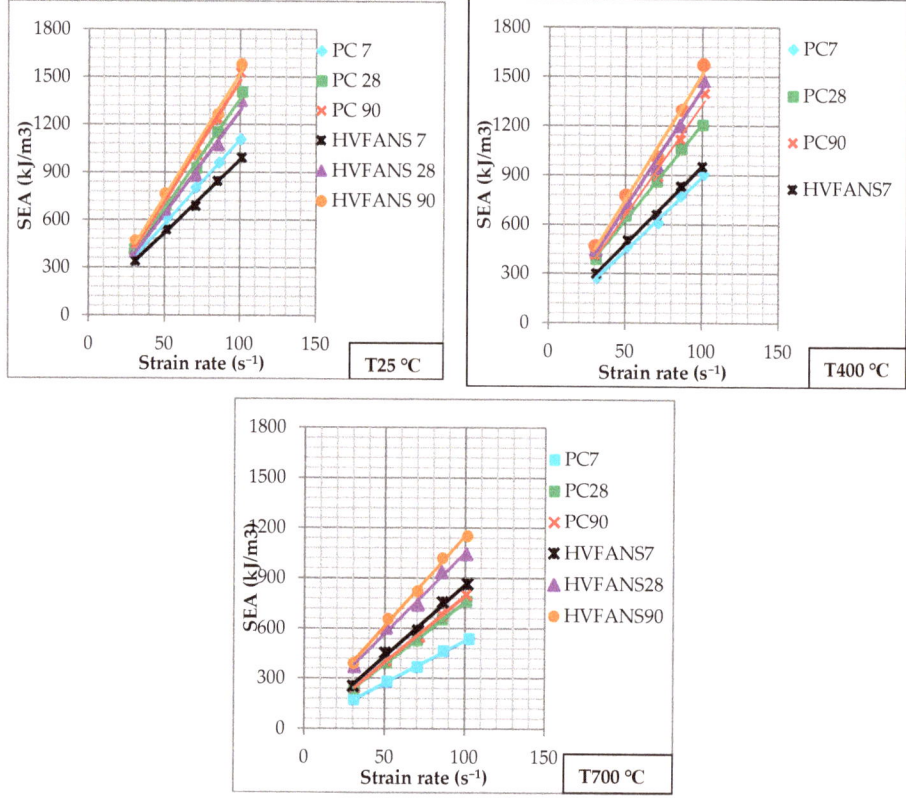

Figure 8. Effect of curing age on the toughness of concrete specimens at different temperatures and strain rates.

The results indicated that HVFANS concrete has lower toughness than PC at early ages of 7 and 28 days within room temperature condition. However, the toughness of HVFANS concrete significantly increased than PC at the late age of 90 days and recorded a difference of 11.68% at strain rates above 100 s^{-1}. At elevated temperatures, the toughness of HVFANS concrete considerably increased even at an early age and recorded values higher than PC that attribute to the high pozzolanic reactivity of NS as well as its filler ability. The lowest toughness values of both concrete are recorded at exposing temperature of 700 °C, nevertheless, the HVFANS concrete showed strong resistance to dynamic loads at all the investigated curing ages and strain rates as compared to PC as can be clearly noted at average strain rate above 100 s^{-1} at curing ages of 7, 28 and 90 by differences of 59.57%, 36.74% and 43.66%, respectively.

3.2.4. Critical Damage

The critical damage of both concretes is determined by using mechanics theory of continuous damage based on the variances of stress–strain diagrams with strain rate. A variable (D_{cr}) is used to define the critical damage as following [74–76]:

$$D_{cr} = \begin{cases} 0 & \text{For } \varepsilon = 0 \\ 1 - \dfrac{\sigma_p}{E_o \varepsilon_{cr}} & \text{For } \varepsilon > 0 \end{cases} \quad (5)$$

where σ_p is the peak stress, and E_o is the initial Young's modulus. The above formula represented the damage of concrete samples in terms of a kinematic variable according to the voids density as well as the cracks in a certain area that could cause a gradual demolition in the concrete sample [77,78]. Moreover, the state of $D_{cr} = 1$ signifies that the concrete sample is not capable of sustaining the applied dynamic loads. The experimental results indicated that the damage of both concretes improved linearly with strain rate as shown in Figure 9. At room temperature, the damage of HVFANS concrete at early ages of 7 and 28 days was higher than PC; however, the increase of NS reactivity at a late age of 90 days reduced the damage of HVFANS concrete and recorded lower values as compared to PC at all investigated strain rates. At elevated temperatures, the failure of HVFANS concrete was lesser as compared to PC even at early ages owing to the increase of NS reactivity and hydration process [55].

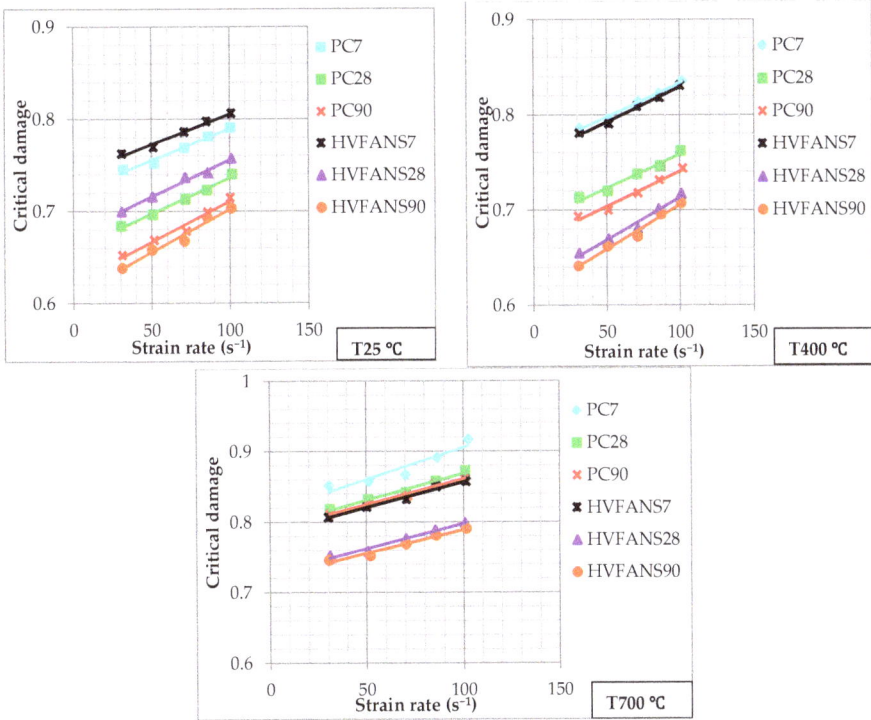

Figure 9. Effect of curing age on the critical damage of concrete specimens at different temperatures and strain rates.

Figure 10 showed that the failure of both concretes was violent whereas the cylindrical specimens were destroyed into pieces at high strain rates. Three crack patterns were observed for both concrete with the increase of curing age. The cracks were clearly straight and passed through a large area that has more broken aggregate that occurred at a curing age of seven days with a strain rate of 30 s^{-1} as well as at 28 and 90 days within strain rates above 70 s^{-1}. The number of cracks in concrete samples was considerably increased and resulted in serious destruction into small pieces to release the internal energy that occurred at a curing age of seven days with strain rate above 50 s^{-1} as well as at curing ages of 28 and 90 days within strain rates above 85 s^{-1}. While the crack path was passed through the concrete mortar only at curing ages of 28 and 90 days within strain rates between 30 and 70 s^{-1}.

Curing age seven days at $\dot{\varepsilon}_r \geq 30$ s^{-1}
Curing age 28 and 90 days at $\dot{\varepsilon}_r \geq 70$ s^{-1}

Curing age seven at $\dot{\varepsilon}_r \geq 50$ s^{-1}
Curing age 28 and 90 at $\dot{\varepsilon}_r \geq 85$ s^{-1}

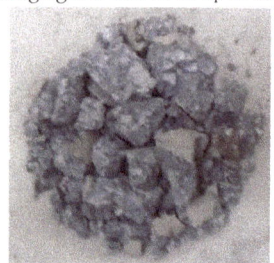

Curing age 28 and 90 at $\dot{\varepsilon}_r \geq 30$ s^{-1}

Figure 10. Effect of curing age on the damage mode of PC and HVFANS concrete specimens at different curing ages.

3.2.5. Dynamic Increase Factor (DIF)

The effect of strain rate on the dynamic behaviour of both concrete is described in terms of dynamic increase factor (DIF) which represents the ratio of the dynamic to static strength. The DIF curves showed linear increases with strain rate values at all the investigated curing ages and temperatures as shown in Figure 11. Design-Expert Software was used to statistically analyse the experimental data and derive a formula gather between curing age (CA), temperature (T) and strain rate ($\dot{\varepsilon}_d$) by using response surface methodology (RSM) [79]. The 2FI model with inverse transformation was found to be the most appropriate to fit the DIF data of both concrete as follows:

$$\frac{1}{DIF} = 0.905844 + 0.000385 CA + 0.000156T - 0.003549 \dot{\varepsilon}_d$$
$$-1.19369 * 10^{-6} CA * T - 3.94641 * 10^{-6} CA * \dot{\varepsilon}_d \quad \text{For PC} \quad (6)$$
$$-3.1153 * 10^{-7} T * \dot{\varepsilon}_d$$

$$\frac{1}{DIF} = 0.952904 - 0.000460 CA - 0.000033T - 0.003811 \dot{\varepsilon}_d$$
$$-2.40444 * 10^{-7} CA * T + 1.56509 * 10^{-6} CA * \dot{\varepsilon}_d \quad \text{For HVFANS concrete} \quad (7)$$
$$+3.94112 * 10^{-7} T * \dot{\varepsilon}_d$$

where CA is the curing age between 7 and 90 days, T is the exposing temperature range between 25 and 700 °C, and $\dot{\varepsilon}_d$ is the strain rate between 30.09 to 102.48 s^{-1}.

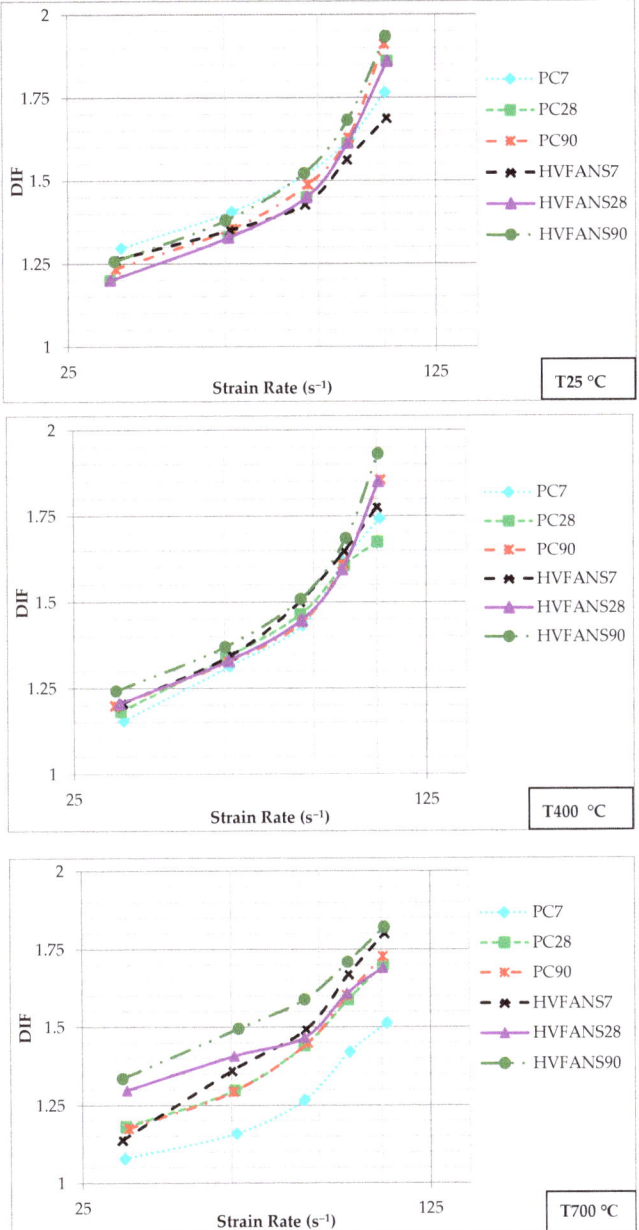

Figure 11. Effect of curing age on the dynamic increase factor (DIF) of concrete specimens at different temperatures and strain rates.

The outcomes of ANOVA variance analysis indicated an excellent accuracy for the adopted model to estimate the DIF of both concretes within correlation coefficient (R^2) values close to one as indicated in Table 6 [80,81]. Furthermore, the differences between Predicted and Adjusted R^2" was less than 0.2 within Adequate Precision (AP) of more than four that measured the noise-to-signal ratio [82].

Table 6. ANOVA analysis results.

Concrete Type	Transformation	Model	R^2	Adjusted R^2	Predicted R^2	AP
PC	inverse	2FI	0.9534	0.9460	0.9341	36.9194
HVFANS	Inverse	2FI	0.9557	0.9487	0.9283	35.9574

The normal probability and predicted vs. actual values plots of both concrete types were determined to confirm the accuracy of the selected model to provide an adequate approximation of the real system. The results clearly indicated that the data were normally distributed since almost all the points followed the straight line as described in Figure 12. Therefore, the predicted values can be considered in good agreement with actual values.

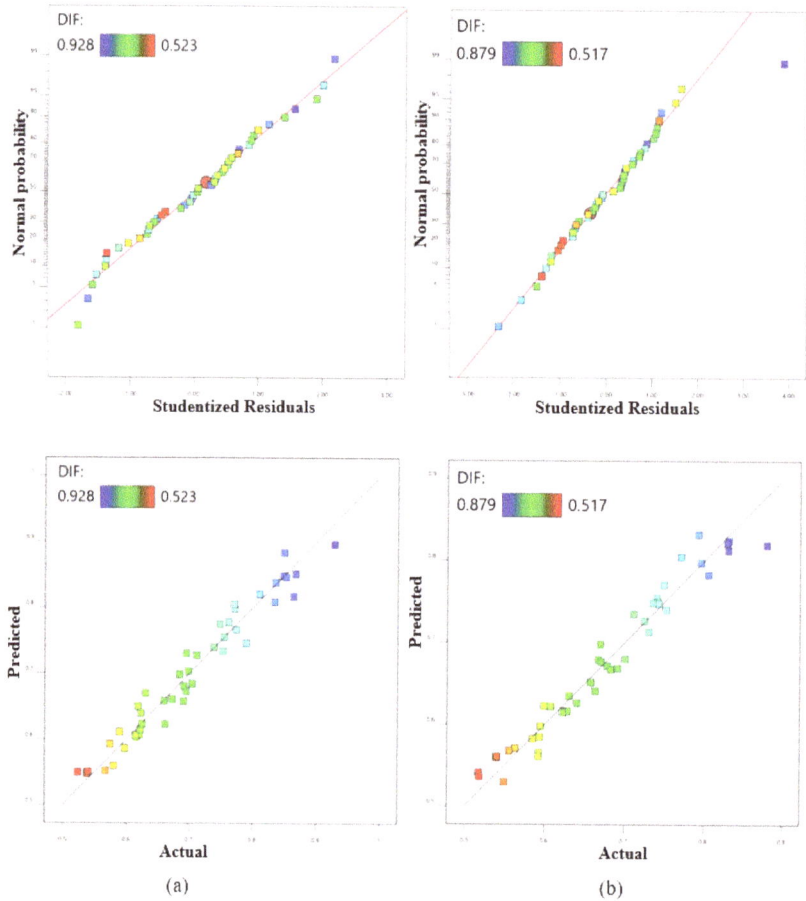

Figure 12. Plots of normal probability and predicted vs. actual values: (a) PC and (b) HVFANS concrete.

3.2.6. Comparison of DIF with Previous Recommended Expressions

Prior scholars widely utilised the DIF curves to calculate the effects of strain rate on the compression strength of concrete [66,83]. Therefore, several equations were recommended to estimate the DIF diagrams:

(a) In 1993, CEB [84] proposed a formula to estimate the DIF of concrete in compression within a strain rate range reached up to 300 s^{-1}:

$$DIF_{CEB} = \begin{cases} (\dot{\varepsilon}_d/\dot{\varepsilon}_s)^{1.026\alpha_s} & \text{For } \dot{\varepsilon}_d \leq 30 \text{ s}^{-1} \\ \gamma_s(\dot{\varepsilon}_d/\dot{\varepsilon}_s)^{1/3} & \text{For } \dot{\varepsilon}_d > 30 \text{ s}^{-1} \end{cases} \qquad (8)$$

where $\dot{\varepsilon}_s$ is static strain rate that equals to 30×10^{-6}s^{-1}, $\alpha_s = 1/(5+9f_{cs}/10)$, $\log \gamma_s = 6.156\alpha - 2$, and f_{cs} is static compressive strength in MPa.

(b) In 1997, Tedesco et al. [85] reported that the transition from low strain rate to higher values occurred at 63.1 s^{-1} as follows:

$$DIF_{Tedesco} = \begin{cases} 0.000965 \log \dot{\varepsilon}_d + 1.058 & \text{For } \dot{\varepsilon}_d \leq 63.1 \text{ s}^{-1} \\ 0.758 \log \dot{\varepsilon}_d - 0.289 & \text{For } \dot{\varepsilon}_d > 63.1 \text{ s}^{-1} \end{cases} \qquad (9)$$

(c) In 2008, Zhou and Hao [86] determined an empirical formula to predict the DIF diagrams for the concrete-like materials as follows:

$$DIF_{ZhouandHao} = \begin{cases} 0.0225 \log \dot{\varepsilon}_d + 1.12 & \text{For } \dot{\varepsilon}_d \leq 10 \text{ s}^{-1} \\ 0.2713(\log \dot{\varepsilon}_d)^2 - 0.3563 \log \dot{\varepsilon}_d + 1.2275 & \text{For } \dot{\varepsilon}_d > 10 \text{ s}^{-1} \end{cases} \qquad (10)$$

The DIF diagrams of both concrete compared with the outcomes of the above equations at different curing ages and exposing temperatures within strain rate range reached up to 102.48 s^{-1} as shown in Figure 13. At room temperature, the CEB expression was more accurate to estimate the DIF of both concrete within differences range between 0.092% and 25.30% particularly observed at higher strain rate above 30 s^{-1} within curing ages of 7 and 28 days. At 400 °C, the results of the proposed expression by Zhou and Hao [86] were more close to the determined DIF especially at early ages of 7 and 28 days. However, some variances were noted at an age of 90 days in the case of HVFANS concrete and the comparison proved that the CEB model is more appropriate to predict the performance of this concrete type at 400 °C that may attribute to the increase of the FA and NS reactivity at late age during the heating process. At 700 °C, Tedesco's expression was more appropriate to estimate the behaviour of PC at the early age of seven days. Zhou and Hao's expression revealed good ability to predict the performance of PC with the increase of curing age to 28 and 90 days as well as in the case of HVFANS concrete at seven days. On the other hand, the CEB model showed the good capability to estimate the DIF of HVFANS concrete at curing ages of 28 and 90 days as compared to other expressions due to the low degradation in strength at 700 °C.

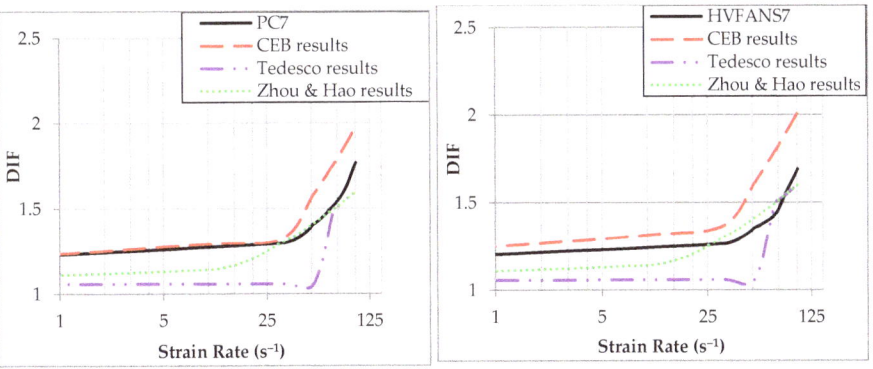

Figure 13. Cont.

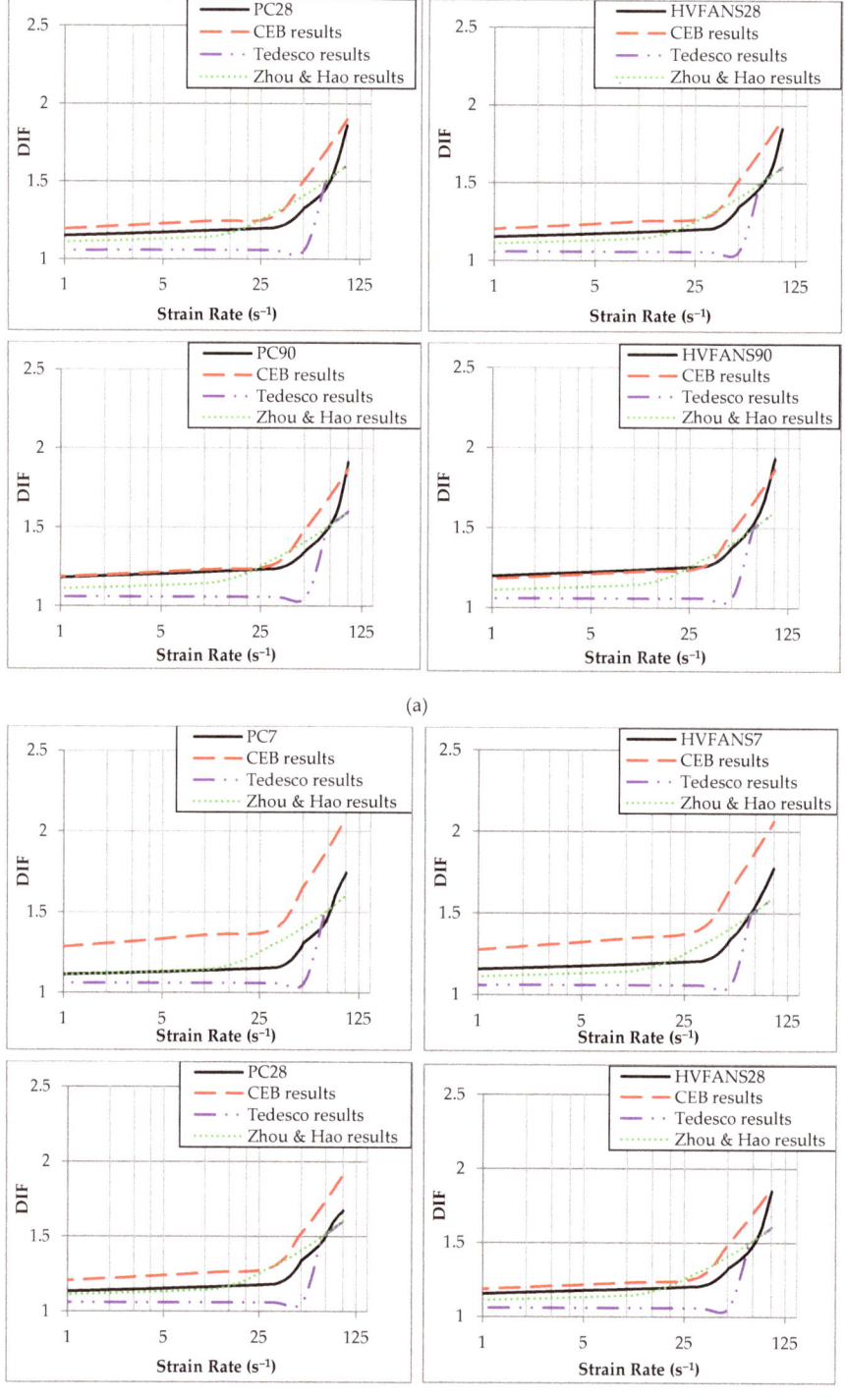

(a)

Figure 13. Cont.

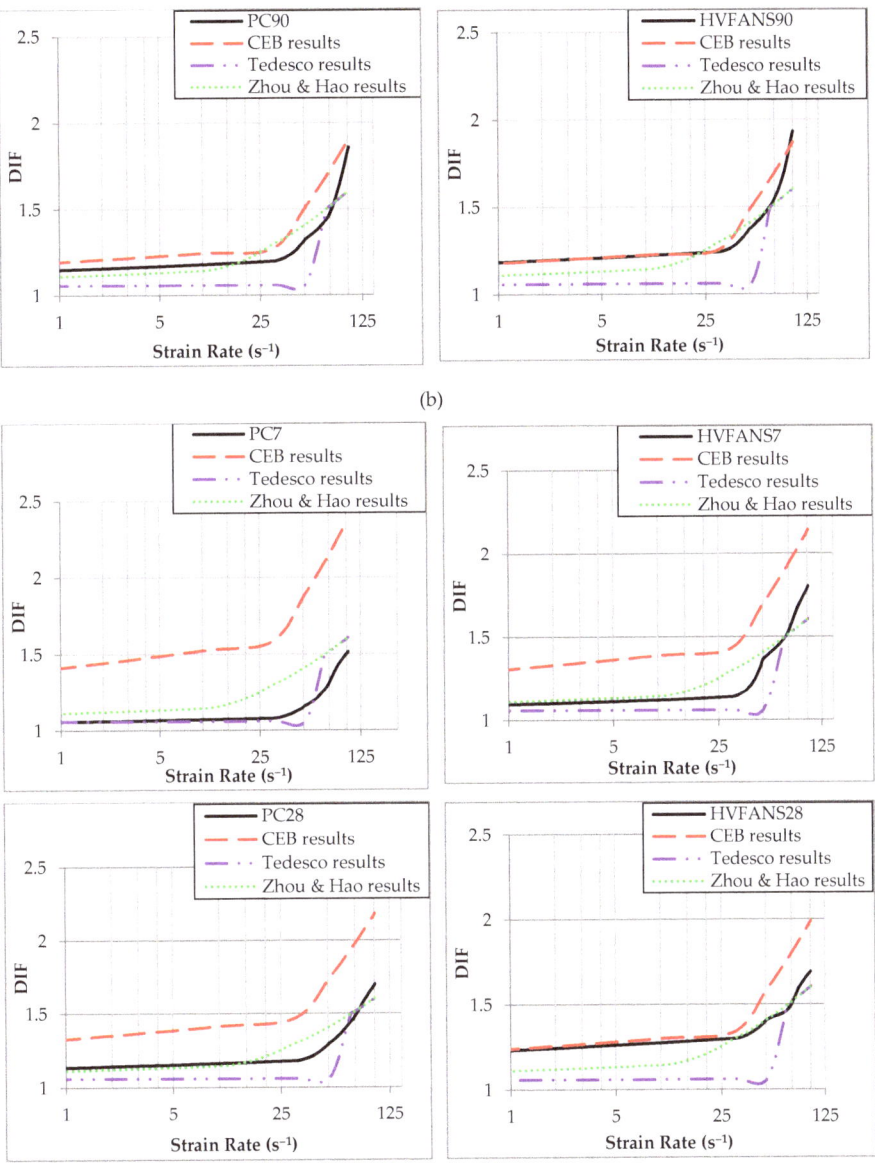

(b)

Figure 13. Cont.

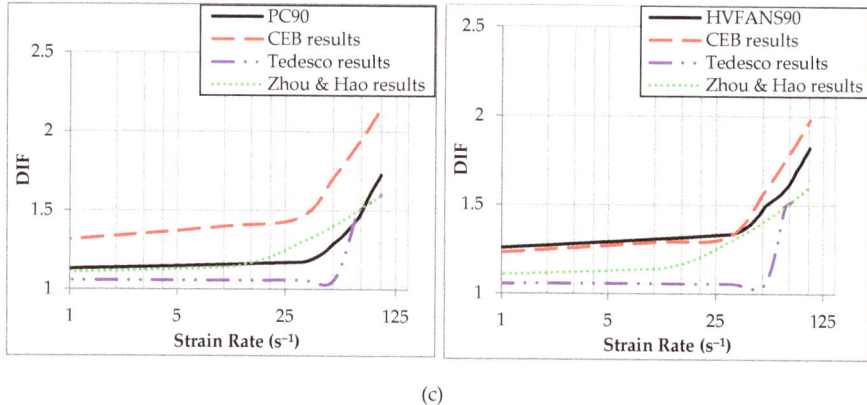

(c)

Figure 13. Comparison of DIF with the previously recommended expressions at different curing ages and temperatures of (**a**) 25 °C, (**b**) 400 °C and (**c**) 700 °C.

4. Conclusions

The results of the current study can be concluded as follows:

1. At room temperature, the PC concrete showed better static and dynamic behaviour than HVFANS concrete at curing ages of 7 and 28 days. However, the behaviour of HVFANS concrete was significantly improved at a curing age of 90 days and recorded superior performance than PC during the static and dynamic tests.
2. At elevated temperatures, the dynamic strength of HVFANS concrete exceeded the PC at all the studied curing ages. The maximum differences between both concretes were noted at 700 °C by 59.59%, 36.68% and 43.59% at curing ages of 7, 28 and 90, respectively. Accordingly, the HVFANS concrete recorded lower strain values than PC within a maximum difference of 43.56% at a strain rate above 100 s^{-1}.
3. The formulas were suggested to estimate the dynamic increase factors of PC and HVFANS concrete at a curing age between 7 and 90 days under temperatures range of 25 to 700 °C subjected to strain rates reached up to 102.48 s^{-1}.
4. The CEB expression was more accurate to estimate the DIF of HVFANS concrete at different curing age within room temperature as well as at elevated temperatures of 400 and 700 °C at curing ages of 28 and 90 days, respectively. While the proposed expression by Zhou and Hao was more appropriate to determine the DIF of HVFANS concrete at a curing age of seven under 400 and 700 °C.

The above results indicated that the dynamic resistance of the HVFANS concrete significantly improved with the increase of curing age under room and elevated temperatures as compared with PC. This behaviour revealed a good possibility to utilize this concrete to protect the engineering structures exposed to a massive fire and dynamic loads.

Author Contributions: Conceptualization, M.H.M.; methodology, M.H.M.; investigation, M.H.M.; formal analysis, M.H.M.; supervision, A.A.M.; project administration, A.A.M. and M.H.M.; Funding acquisition, I.S.A. and A.A.M.; resources, A.M.A. and Z.M.Y.; data curation, M.H.M.; visualization, M.H.M.; writing—original draft preparation, M.H.M.; writing—review and editing, M.H.M. All authors have read and agreed to the published version of the manuscript.

Funding: This research was funded by the Research University Grant, GUP-2018-029.

Conflicts of Interest: The authors declare that they have no conflict of interest.

References

1. Guo, X.; Shi, H.; Dick, W.A. Compressive strength and microstructural characteristics of class C fly ash geopolymer. *Cem. Concr. Compos.* **2010**, *32*, 142–147. [CrossRef]
2. Rehan, R.; Nehdi, M. Carbon dioxide emissions and climate change: Policy implications for the cement industry. *Environ. Sci. Policy.* **2005**, *8*, 105–114. [CrossRef]
3. Bakharev, T. Geopolymeric materials prepared using class F fly ash and elevated temperature curing. *Cem. Concr. Res.* **2005**, *35*, 1224–1232. [CrossRef]
4. Fernández-Jiménez, A.; Palomo, A.; Criado, M. Microstructure development of alkali-activated fly ash cement: A descriptive model. *Cem. Concr. Res.* **2005**, *35*, 1204–1209. [CrossRef]
5. Jaturapitakkul, C.; Kiattikomol, K.; Sata, V.; Leekeeratikul, T. Use of ground coarse fly ash as a replacement of condensed silica fume in producing high-strength concrete. *Cem. Concr. Res.* **2004**, *34*, 549–555. [CrossRef]
6. Jiang, L.; Guan, Y. Pore structure and its effect on strength of high-volume fly ash paste. *Cem. Concr. Res.* **1999**, *29*, 631–633. [CrossRef]
7. Montgomery, D.; Hughes, D.; Williams, R. Fly ash in concrete—A microstructure study. *Cem. Concr. Res.* **1981**, *11*, 591–603. [CrossRef]
8. Li, G. Properties of high-volume fly ash concrete incorporating nano-SiO_2. *Cem. Concr. Res.* **2004**, *34*, 1043–1049. [CrossRef]
9. Zhang, M.H.; Islam, J. Use of nano-silica to reduce setting time and increase early strength of concretes with high volumes of fly ash or slag. *Constr. Build. Mater.* **2012**, *29*, 573–580. [CrossRef]
10. Ibrahim, R.K. The Strength and Micro Structures of Sustainable High Strength High-Volume Fly Ash Concrete with Nano Materials Exposed to High Temperature. Ph.D. Thesis, Universiti Kebangsaan Malaysia (UKM), Bangi, Malaysia, 2013.
11. Ibrahim, R.K.; Hamid, R.; Taha, M.R. Fire resistance of high-volume fly ash mortars with nanosilica addition. *Constr. Build. Mater.* **2012**, *36*, 779–786. [CrossRef]
12. Mussa, M.H.; Mutalib, A.A.; Hamid, R.; Naidu, S.R.; Radzi, N.A.M.; Abedini, M. Assessment of damage to an underground box tunnel by a surface explosion. *Tunn. Undergr. Space Technol.* **2017**, *66*, 64–76. [CrossRef]
13. Mussa, M.H.; Mutalib, A.A.; Hamid, R.; Raman, S.N. Blast damage assessment of symmetrical box-shaped underground tunnel according to peak particle velocity (PPV) and single degree of freedom (SDOF) criteria. *Symmetry* **2018**, *10*, 158. [CrossRef]
14. Abedini, M.; Mutalib, A.A.; Raman, S.N.; Akhlaghi, E.; Mussa, M.H.; Ansari, M. Numerical investigation on the non-linear response of reinforced concrete (RC) columns subjected to extreme dynamic loads. *J. Asian Sci. Res.* **2017**, *7*, 86. [CrossRef]
15. Saadun, A.; Mutalib, A.A.; Hamid, R.; Mussa, M.H. Behaviour of polypropylene fiber reinforced concrete under dynamic impact load. *J. Eng. Sci. Technol.* **2016**, *11*, 684–693.
16. Abedini, M.; Mutalib, A.A.; Mehrmashhadi, J.; Raman, S.N.; Alipour, R.; Momeni, T.; Mussa, M.H. Large Deflection Behavior Effect in Reinforced Concrete Columns Exposed to Extreme Dynamic Loads. 2019. Available online: https://engrxiv.org/6n5fs/ (accessed on 25 July 2019).
17. Abedini, M.; Khlaghi, E.A.; Mehrmashhadi, J.; Mussa, M.H.; Ansari, M.; Momeni, T. Evaluation of concrete structures reinforced with fiber reinforced polymers bars: A review. *J. Asian Sci. Res.* **2017**, *7*, 165. [CrossRef]
18. Mutalib, A.A.; Mussa, M.H.; Hao, H. Effect of CFRP strengthening properties with anchoring systems on PI diagrams of RC panels under blast loads. *Constr. Build. Mater.* **2019**, *200*, 648–663. [CrossRef]
19. Abbood, I.S.; Mahmod, M.; Hanoon, A.N.; Jaafar, M.S.; Mussa, M.H. Seismic response analysis of linked twin tall buildings with structural coupling. *Int. J. Civ. Eng. Technol.* **2018**, *9*, 208–219.
20. Ziyan, L.; Yuzhuo, G.W.G. Dynamic compression behavior of heated concrete. *China Civ. Eng. J.* **2011**, *4*, 013.
21. Huo, J.S.; He, Y.M.; Xiao, L.P.; Chen, B.S. Experimental study on dynamic behaviours of concrete after exposure to high temperatures up to 700 °C. *Mater. Struct.* **2013**, *46*, 255–265. [CrossRef]
22. Huo, J.; Wang, P.; Yu, Q.; He, Y. Dynamic behaviour of normal-strength carbonate aggregate concrete at temperatures up to 800 °C. *Mag. Concr. Res.* **2014**, *66*, 975–990. [CrossRef]
23. He, Y.; Huo, J.; Xiao, Y. Experimental study on dynamic behavior of concrete at elevated temperatures. *Adv. Sci. Lett.* **2011**, *4*, 1128–1131. [CrossRef]
24. Zhai, C.; Chen, L.; Fang, Q.; Chen, W.; Jiang, X. Experimental study of strain rate effects on normal weight concrete after exposure to elevated temperature. *Mater. Struct.* **2017**, *50*, 40. [CrossRef]

25. Mussa, M.H.; Mutalib, A.A. Effect of geometric parameters (β and τ) on behaviour of cold formed stainless steel tubular X-joints. *Int. J. Steel Struct.* **2018**, *18*, 821–830. [CrossRef]
26. Chen, L.; Fang, Q.; Jiang, X.; Ruan, Z.; Hong, J. Combined effects of high temperature and high strain rate on normal weight concrete. *Int. J. Impact Eng.* **2015**, *86*, 40–56. [CrossRef]
27. Li, Z.; Xu, J.; Bai, E. Static and dynamic mechanical properties of concrete after high temperature exposure. *Mater. Sci. Eng. A* **2012**, *544*, 27–32. [CrossRef]
28. Su, H.; Xu, J.; Ren, W. Experimental study on the dynamic compressive mechanical properties of concrete at elevated temperature. *Mater. Des.* **2014**, *56*, 579–588. [CrossRef]
29. Committee, A. *Guide for Selecting Proportions for High-Strength Concrete with Portland Cement & Fly ASH-ACI 211.4 R-93*; American Concrete Institute: Farming-ton Hills, MI, USA, 1998.
30. Standard, M. *Portland Cement (Ordinary and Rapid-Hardening): Part 1*, 2nd ed.; MS: Bangi, Malaysia, 2003; p. 522.
31. Standard, B. *Fly ASH for Concrete—Part 1: Definition, Specifications and Conformity Criteria*; European Committee for Standardization: Brussels, Belgium, 2005.
32. Standard, A. *C136 Standard Test Method for Sieve Analysis of Fine and Coarse Aggregates*; ASTM International: West Conshohocken, PA, USA, 2006.
33. ASTM, A. *C128-07a Standard Test Method for Density, Relative Density (Specific Gravity), and Absorption of Fine Aggregate*; ASTM International: West Conshohocken, PA, USA, 2007.
34. ASTM, C. *Standard Test Method for Bulk Density ("Unit Weight") and Voids in Aggregate*; American Society for Testing and Materials, Annual Book: West Conshohocken, PA, USA, 2009.
35. Nishida, A. Study in the Properties of High Strength Concrete with Short Polypropylene Fiber for Spalling Reasistance, Concrete under Severe Condetions. In Proceedings of the International Conference on Concrete under Severe Conditions, Sapporo, Japan, 2–4 August 1995; Volume 2, pp. 1141–1150.
36. Kalifa, P.; Chene, G.; Galle, C. High-temperature behaviour of HPC with polypropylene fibres: From spalling to microstructure. *Cem. Concr. Res.* **2001**, *31*, 1487–1499. [CrossRef]
37. Noumowe, A. Mechanical properties and microstructure of high strength concrete containing polypropylene fibres exposed to temperatures up to 200 °C. *Cem. Concr. Res.* **2005**, *35*, 2192–2198. [CrossRef]
38. Zeiml, M.; Leithner, D.; Lackner, R.; Mang, H.A. How do polypropylene fibers improve the spalling behavior of in-situ concrete? *Cem. Concr. Res.* **2006**, *36*, 929–942. [CrossRef]
39. Bilodeau, A.; Kodur, V.; Hoff, G. Optimization of the type and amount of polypropylene fibres for preventing the spalling of lightweight concrete subjected to hydrocarbon fire. *Cem. Concr. Compos.* **2004**, *26*, 163–174. [CrossRef]
40. Mutalib, A.A.; Mussa, M.; Mohd Taib, A. Behaviour of prestressed box beam strengthened with CFRP under effect of strand snapping. *Gradevinar* **2020**, *72*, 103–113.
41. Standard, A. *C192 Standard Practice for Making and Curing Concrete Test Specimens in the Laboratory (ASTM C192-07)*; ASTM International: West Conshohocken, PA, USA, 2007.
42. Neville, A.M. *Properties of Concrete*; Longman Publisher: London, UK, 1995.
43. ASTM, C. *1602/C 1602M-12 Standard Specification for Mixing Water Used in the Hydraulic Cement Concrete*; ASTM International: West Conshohocken, PA, USA, 2012.
44. Standard, M.J. *Thermal Insulation—Determination of Steady-State Thermal Resistance and Related Properties—Guarded Hot Plate Apparatus (ISO 8302: 1991, IDT)*; Department of Standards Malaysia: Cyberjaya, Malaysia, 2003.
45. EN, B. Testing Hardened Concrete—Part 3: Compressive Strength of Test Specimens. Available online: http://home.aktor.qa/External%20Documents/Intenational%20Specifications/British%20Standards/BS%20EN/BS%20EN%2012390-3-2009.pdf (accessed on 31 May 2009).
46. Al-Masoodi, A.H.H.; Kawan, A.; Kasmuri, M.; Hamid, R.; Khan, M.N.N. Static and dynamic properties of concrete with different types and shapes of fibrous reinforcement. *Constr. Build. Mater.* **2016**, *104*, 247–262. [CrossRef]
47. Chien Yet, T.; Hamid, R.; Kasmuri, M. Dynamic stress-strain behaviour of steel fiber reinforced high-performance concrete with fly ash. *Adv. Civ. Eng.* **2012**, *2012*, 907431. [CrossRef]
48. Davies, E.; Hunter, S. The dynamic compression testing of solids by the method of the split Hopkinson pressure bar. *J. Mech. Phys. Solids.* **1963**, *11*, 155–179. [CrossRef]
49. Lindholm, U.S. High strain rate tests. *Meas. Mech. Prop.* **1971**, *5*, 199–271.

50. Kirby, M.S.P. Designing a Data Acquisition System for a Split Hopkinson Pressure Bar. Available online: https://pdfs.semanticscholar.org/36d6/e475d65a484381a88dfb064e8d99b8d389b5.pdf (accessed on 14 June 2015).
51. Dyab, M.M. Providing Learning Opportunities by Designing a Split Hopkinson Pressure Bar. *Age* **2013**, *23*, 1.
52. Lok, T.; Zhao, P.; Lu, G. Using the split Hopkinson pressure bar to investigate the dynamic behaviour of SFRC. *Mag. Concr. Res.* **2003**, *55*, 183–191. [CrossRef]
53. Lu, Y.; Li, Q. Appraisal of pulse-shaping technique in split Hopkinson pressure bar tests for brittle materials. *Int. J. Prot. Struct.* **2010**, *1*, 363–390. [CrossRef]
54. Mussa, M.H.; Mutalib, A.A.; Hamid, R.; Raman, S.N. Dynamic properties of high volume fly ash nanosilica (HVFANS) concrete subjected to combined effect of high strain rate and temperature. *Lat. Am. J. Solids Struct.* **2018**, *15*. [CrossRef]
55. Berndt, M. Properties of sustainable concrete containing fly ash, slag and recycled concrete aggregate. *Constr. Build. Mater.* **2009**, *23*, 2606–2613. [CrossRef]
56. Ji, T. Preliminary study on the water permeability and microstructure of concrete incorporating nano-SiO_2. *Cem. Concr. Res.* **2005**, *35*, 1943–1947. [CrossRef]
57. Institutions, B.S. *Method for Determination of Compressive Strength of Concrete Cubes*; BS: London, UK, 1881.
58. Miller, O.; Freund, L.; Needleman, A. Modeling and simulation of dynamic fragmentation in brittle materials. *Int. J. Fract.* **1999**, *96*, 101–125. [CrossRef]
59. Cadoni, E.; Fenu, L.; Forni, D. Strain rate behaviour in tension of austenitic stainless steel used for reinforcing bars. *Constr. Build. Mater.* **2012**, *35*, 399–407. [CrossRef]
60. Johnstone, C.; Ruiz, C. Dynamic testing of ceramics under tensile stress. *Int. J. Solids Struct.* **1995**, *32*, 2647–2656. [CrossRef]
61. Lee, O.; Kim, G.H. Thickness effects on mechanical behavior of a composite material (1001P) and polycarbonate in split Hopkinson pressure bar technique. *J. Mater. Sci. Lett.* **2000**, *19*, 1805–1808. [CrossRef]
62. Zhao, J.; Li, H. Experimental determination of dynamic tensile properties of a granite. *Int. J. Rock Mech. Min. Sci.* **2000**, *37*, 861–866. [CrossRef]
63. Cadoni, E. Dynamic characterization of orthogneiss rock subjected to intermediate and high strain rates in tension. *Rock Mech. Rock Eng.* **2010**, *43*, 667–676. [CrossRef]
64. Grady, D.E.; Kipp, M.E. The micromechanics of impact fracture of rock. In *International Journal of Rock Mechanics and Mining Sciences & Geomechanics Abstracts*; Elsevier: Amsterdam, The Netherlands, 1979.
65. Ross, C.A.; Jerome, D.M.; Tedesco, J.W.; Hughes, M.L. Moisture and strain rate effects on concrete strength. *Mater. J.* **1996**, *93*, 293–300.
66. Bischoff, P.; Perry, S. Compressive behaviour of concrete at high strain rates. *Mater. Struct.* **1991**, *24*, 425–450. [CrossRef]
67. Lai, J.; Sun, W. Dynamic behaviour and visco-elastic damage model of ultra-high performance cementitious composite. *Cem. Concr. Res.* **2009**, *39*, 1044–1051. [CrossRef]
68. Köksal, F.; Altun, F.; Yiğit, İ.; Şahin, Y. Combined effect of silica fume and steel fiber on the mechanical properties of high strength concretes. *Constr. Build. Mater.* **2008**, *22*, 1874–1880. [CrossRef]
69. Shannag, M. High strength concrete containing natural pozzolan and silica fume. *Cem. Concr. Compos.* **2000**, *22*, 399–406. [CrossRef]
70. Nili, M.; Afroughsabet, V. The effects of silica fume and polypropylene fibers on the impact resistance and mechanical properties of concrete. *Constr. Build. Mater.* **2010**, *24*, 927–933. [CrossRef]
71. Li, W.; Xu, J. Mechanical properties of basalt fiber reinforced geopolymeric concrete under impact loading. *Mater. Sci. Eng. A* **2009**, *505*, 178–186. [CrossRef]
72. Li, W.; Xu, J. Impact characterization of basalt fiber reinforced geopolymeric concrete using a 100-mm-diameter split Hopkinson pressure bar. *Mater. Sci. Eng. A* **2009**, *513*, 145–153. [CrossRef]
73. Wang, S.; Zhang, M.-H.; Quek, S.T. Effect of specimen size on static strength and dynamic increase factor of high-strength concrete from SHPB test. *J. Test. Eval.* **2011**, *39*, 1–10.
74. Clayton, J.D. A model for deformation and fragmentation in crushable brittle solids. *Int. J. Impact Eng.* **2008**, *35*, 269–289. [CrossRef]
75. Wang, Z.L.; Liu, Y.; Shen, R. Stress—Strain relationship of steel fiber-reinforced concrete under dynamic compression. *Constr. Build. Mater.* **2008**, *22*, 811–819. [CrossRef]
76. Hamdi, E.; Romdhane, N.B.; le Cléac'h, J.M. A tensile damage model for rocks: Application to blast induced damage assessment. *Comput. Geotech.* **2011**, *38*, 133–141. [CrossRef]

77. Addessio, F.L.; Johnson, J.N. A constitutive model for the dynamic response of brittle materials. *J. Appl. Phys.* **1990**, *67*, 3275–3286. [CrossRef]
78. Wang, Z.L.; Li, Y.C.; Wang, J. A damage-softening statistical constitutive model considering rock residual strength. *Comput. Geosci.* **2007**, *33*, 1–9. [CrossRef]
79. Vaughn, N.; Polnaszek, C. *Design-Expert® Software*; Stat-Ease, Inc.: Minneapolis, MN, USA, 2007.
80. Joglekar, A.; May, A. Product excellence through design of experiments. *Cereal Foods World* **1987**, *32*, 857.
81. Noordin, M.Y.; Venkatesh, V.C.; Sharif, S.; Elting, S.; Abdullah, A. Application of response surface methodology in describing the performance of coated carbide tools when turning AISI 1045 steel. *J. Mater. Process. Technol.* **2004**, *145*, 46–58. [CrossRef]
82. Montgomery, D.C. *Design and Analysis of Experiments 6th Edition with Design Expert Software*; John Wiley & Sons: Hoboken, NJ, USA, 2004.
83. Malvar, L.J.; Ross, C.A. Review of strain rate effects for concrete in tension. *Mater. J.* **1998**, *95*, 735–739.
84. MC90, C. *Design of Concrete Structures. CEB-FIP Model Code 1990*; Thomas, T., Ed.; British Standard Institution: London, UK, 1993.
85. Tedesco, J.W.; Powell, J.C.; Ross, C.A.; Hughes, M.L. A strain-rate-dependent concrete material model for ADINA. *Comput. Struct.* **1997**, *64*, 1053–1067. [CrossRef]
86. Zhou, X.; Hao, H. Modelling of compressive behaviour of concrete-like materials at high strain rate. *Int. J. Solids Struct.* **2008**, *45*, 4648–4661. [CrossRef]

 © 2020 by the authors. Licensee MDPI, Basel, Switzerland. This article is an open access article distributed under the terms and conditions of the Creative Commons Attribution (CC BY) license (http://creativecommons.org/licenses/by/4.0/).

Article

Mechanical and Thermal Properties of Low-Density $Al_{20+x}Cr_{20-x}Mo_{20-y}Ti_{20}V_{20+y}$ Alloys

Uttam Bhandari, Congyan Zhang and Shizhong Yang *

Department of Computer Science, Southern University and A&M College, Baton Rouge, LA 70813, USA; uttam_bhandari_00@subr.edu (U.B.); congyan_zhang@subr.edu (C.Z.)
* Correspondence: shizhong_yang@subr.edu

Received: 18 March 2020; Accepted: 2 April 2020; Published: 7 April 2020

Abstract: Refractory high-entropy alloys (RHEAs) $Al_{20+x}Cr_{20-x}Mo_{20-y}Ti_{20}V_{20+y}$ ((x, y) = (0, 0), (0, 10), and (10, 15)) were computationally studied to obtain a low density and a better mechanical property. The density functional theory (DFT) method was employed to compute the structural and mechanical properties of the alloys, based on a large unit cell model of randomly distributed elements. Debye–Grüneisen theory was used to study the thermal properties of $Al_{20+x}Cr_{20-x}Mo_{20-y}Ti_{20}V_{20+y}$. The phase diagram calculation shows that all three RHEAs have a single body-centered cubic (BCC) structure at high temperatures ranging from 1000 K to 2000 K. The RHEA $Al_{30}Cr_{10}Mo_5Ti_{20}V_{35}$ has shown a low density of 5.16 g/cm^3 and a hardness of 5.56 GPa. The studied RHEAs could be potential candidates for high-temperature application materials where high hardness, ductility, and low density are required.

Keywords: density functional theory; phase diagram; high-entropy alloy; mechanical and thermal properties

1. Introduction

The operating temperatures of aircraft engines and nuclear reactors are very high. Currently, most widely used alloys are nickel-based with working temperatures of 1200 K–1600 K. The performance of traditional nickel-based superalloys is thus limited to high-temperature applications (>1900 K), due to their low melting temperature and high density [1,2]. In past years, several high-temperature nickel-based superalloys such as Inconel 625 [3], SSR 99 [4], and Co–Ti–Cr [5] have been widely investigated. These reports show that while they have excellent mechanical properties, they also have high density. Ideally, the structural components of aircraft engines need low-density materials with high toughness and a high melting temperature, which motivates researchers in trying titanium-mixed refractory high-entropy alloys (RHEAs) [6–8]. RHEAs have attracted the interest of researchers due to their excellent mechanical property, high melting point, hardness, ductility, wear resistance, and high-temperature strength [9–12]. Studies on CrNbTiVZr [13], MoNbTaVW [14,15], and MoNbTaW [16] have shown that these RHEAs could be potential future candidate materials for high-temperature applications in aircraft engines and nuclear reactors. However, these RHEAs either have a relatively high density (>10 g/cm^3) or are brittle at room temperature, which limits their applications. Recently, Xu et al. [6] designed three RHEAs $Ti_{50-x}Al_xV_{20}Nb_{20}Mo_{10}$ (x = 10, 15, 20) with low densities ranging from 6.01 g/cm^3 to 5.876 g/cm^3. These alloys have high strength and good plasticity at high temperatures. So far, Vickers hardness of these alloys has not been reported. Khaled group [7] also designed a low-density nanocrystalline high-entropy alloy $Al_{20}Li_{20}Mg_{10}Sc_{20}Ti_{30}$ with a hardness of 5.9 GPa. However, this alloy has a low melting point which is not appropriate for high-temperature applications. Kang et al. [8] prepared an equimolar low-density RHEA AlCrMoTiV. They found a single body-centered cubic (BCC) phase in AlCrMoTiV using the X-ray diffraction (XRD) method. The measured Vickers hardness of AlCrMoTiV is 5.54 GPa. No computer simulation is found on this RHEA AlCrMoTiV. The density and other physical properties of this alloy can be further tuned by optimizing the component concentration. It is highly desirable to develop a better RHEA based on this AlCrMoTiV, with a lower density, a higher melting point, and enhanced hardness and ductility.

The first-principles density functional theory (DFT) method was used to optimize the HEA AlCrMoTiV. We investigated a novel RHEA $Al_{30}Cr_{10}Mo_5Ti_{20}V_{35}$ with the aim of decreasing the density for a more balanced mechanical property when compared to the previously discussed RHEAs. The molar percentages of Mo (10.28 g/cm^3) and Cr (7.14 g/cm^3) were reduced, as they both have high density. The molar percentages of the low-density elements V (6.11 g/cm^3) and Al (2.7 g/cm^3) were increased to ensure the low density of the alloy. As the melting point of Ti is lower than that of V, it was kept constant to keep the alloy highly solidus at high temperatures.

2. Computational Methods

Thermo-Calc 2019 software [17] was used to predict the phase equilibria of the RHEAs. The database used in this software was Thermo-Calc's high-entropy alloy (TCHEA1) database [18]. The phase prediction based on the TCHEA1 database was in good agreement with many experimental reports [19–22]. The first-principles method used in this study was based on DFT [23,24]. The computations were performed with the Vienna Ab initio Simulation Package (VASP 5.4) [25] as implemented in the MedeA software environment [26]. In this study, the electron-ion interactions were described by the projector augmented wave (PAW) [27]. The generalized gradient approximation (GGA) [28] of Perdew–Burke–Ernzerhof (PBE) [29] was used as an exchange-correlation function to optimize the structure. The BCC structure of the alloys was constructed with the help of the Knuth shuffle model [30] using Python code. The distribution of the component atoms was randomized in a 100-atom supercell model, which is shown in Figure 1a. To confirm the random distribution of the elements inside the supercell, we plotted the total and partial pair distribution functions (PDF) of the supercell, as shown in Figure 1b. In the BCC structure, each element had eight nearest adjacent elements, and six second nearest adjacent elements. In this case, there were a total of 400 pairs combined by the first adjacent elements, and 700 pairs by the second adjacent elements in the 100-atom supercell. In this supercell, there were 12 elements out of 400 pairs which were combined by the first nearest neighboring elements, and 22 elements out of 700 pairs were connected by the second nearest adjacent element. This demonstrates that the elements inside the supercell were uniformly distributed, providing an accurate high-entropy model.

The elastic constants were calculated using Le Page and Saxe's stress-strain method [31,32] on MedeA software. The stress tensor was calculated using analytic expressions. To ensure that the equilibrium structure corresponded to zero strain, the initial optimization of the theoretical cell parameters was made with high accuracy. The k-spacing was set as 0.2 per angstrom. Plane-wave cut-off energy of 500 eV was used. The geometry convergence criterion was set to 0.02 eV/Å, and the SCF convergence criterion to 10×10^{-5} eV. The k-mesh was forced to be centered on the gamma point. After calculating the elastic constant, the Voight–Reuss–Hill approximation [33,34] was used to calculate the mechanical properties of the RHEAs. The mathematical relations for calculating bulk modulus (B), shear modulus (G), Young's modulus (E), and Poisson ratio (ν) are listed in the following equations:

$$B = \frac{1}{3}(C_{11} + 2C_{12}), \quad G = \frac{1}{2}\left(G_{Voight} + G_{Reuss}\right) \quad (1)$$

where $G_{Voight} = \frac{1}{5}(C_{11} - C_{12} + 3C_{44})$ and $G_{Reuss} = \frac{5}{4(S_{11}-S_{12})+3S_{44}}$, S is compliance matrix,

$$E = \frac{9BG}{3B+G}, \quad \text{and } \nu = \frac{3B-2G}{2(3B+G)}. \quad (2)$$

The thermodynamics calculations used in this study were based on the Debye theory [35]. In this theory, only the elastic constant and averaged Grüneisen constant values are needed to compute the thermal expansion coefficient for a cubic material. The Grüneisen constant (γ_G) was derived from the pressure–volume equation by Mayer et al. [36]. The Debye temperature was computed by using the formula of the mean sound velocity (V_m) as follows:

$$\theta_D = \frac{\hbar}{K}\left(\frac{6\pi^2 q}{V_0}\right)^{\frac{1}{3}} V_m \qquad (3)$$

where q is the number of atoms in the unit cell, V_0 its volume, and $\hbar$ and K_B are Planck and Boltzmann constants, respectively. The lattice contribution to the specific heat capacity, C_V, as a function of temperature, T, was evaluated. The linear thermal expansion coefficient (α) was calculated using the relation in Reference 37 [36,37].

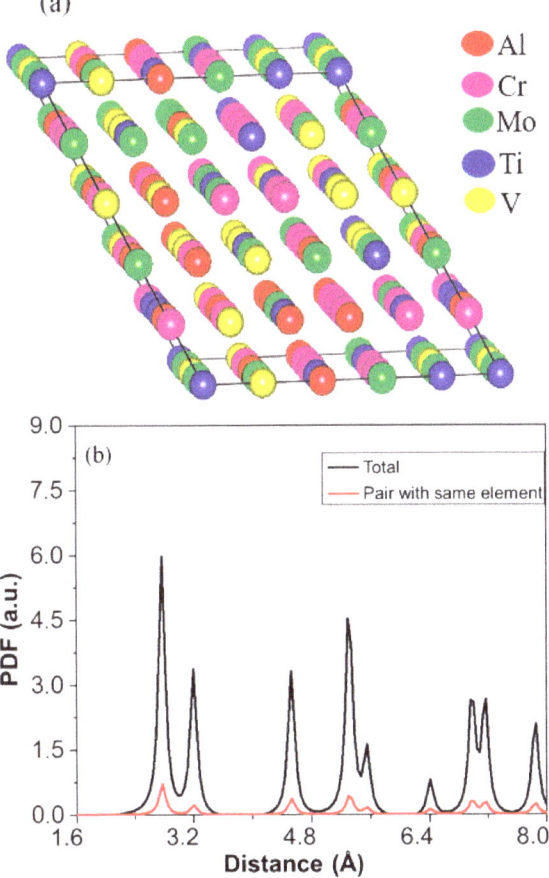

Figure 1. (a) Schematic of a random 100-atom unit cell model used for densitiy functional theory (DFT) calculations. (b) Plot between total pair distribution function (PDF) of all pairs in the supercell and total PDF of pairs as a function of distance.

3. Results and Discussion

3.1. Structural Properties

The phase formation and crystal structures of the design alloys were predicted by calculating the entropy of mixing (ΔS_{mix}) [38], mixing enthalpy (ΔH_{mix}) [39], atomic size differences (δ) [40], and valence electron concentration (VEC) [41,42]. For the current RHEAs, corresponding values of ΔS_{mix}, ΔH_{mix}, VEC, and δ were calculated using the following formula, and the results are listed in Table 1:

$$\Delta S_{mix} = -R \sum_{i=1}^{n} x_i \ln x_i \quad (4)$$

$$\Delta H_{mix} = 4 \times \sum_{i=1, i \neq j}^{n} \Delta H_{ij}^{mix} x_i x_j \quad (5)$$

$$\delta = 100 \times \sqrt{\sum_{i=1}^{n} c_i \left(1 - \frac{r_i}{\bar{r}}\right)^2}, \quad \bar{r} = \sum_{i=1}^{n} x_i r_i \quad (6)$$

$$VEC = \sum_{i=1}^{n} C_i (VEC)_i \quad (7)$$

where R is ideal gas constant, x_i and x_j are the atomic percentages of the ith and j[th] elements, respectively, r_i is the radius of the ith element, $\bar{r}$ is averaged atomic radius, and $(VEC)_i$ is the valence electron concentration of the ith element.

Table 1. Calculated values of entropy of mixing (ΔS_{mix}), mixing enthalpy (ΔH_{mix}), valence electron concentration (VEC), and atomic size differences (δ) of the refractory high-entropy alloys (RHEAs).

Name of Alloys	ΔS_{mix} (J/K.mol)	ΔH_{mix} (kJ/mol)	δ (%)	VEC
$Al_{20}Cr_{20}Mo_{20}Ti_{20}V_{20}$	13.38	−12.16	4.83	4.8
$Al_{20}Cr_{20}Mo_{10}Ti_{20}V_{30}$	12.94	−13.04	4.92	4.7
$Al_{30}Cr_{10}Mo_{5}Ti_{20}V_{35}$	11.89	−16.98	4.46	4.3

To form a sold solution, Zhang et al. [38] proposed that ΔH_{mix} should be $-15 \leq \Delta H_{mix} \leq 5$ kJ/mol, while Guo et al. [41] reported that ΔH_{mix} should be $-22 \leq \Delta H_{mix} \leq 7$ kJ/mol. These conditions were statistically calculated, and the data had some deviations in the previous findings [38,42–44]. Other criteria to form a stable solid solution are $12 \leq \Delta S_{mix} \leq 17.5$ and $\delta \leq 6.6\%$. Another parameter to predict the crystal structure of HEAs is VEC. If VEC is ≥ 8, the alloy will form a face-centered cubic (FCC) crystal, whereas if VEC is < 6.87, it will form a BCC crystal. Our calculated data in Table 1 show that the present RHEAs could form solid solutions with stable BCC crystal structures.

To provide more information on our initial structural prediction of the candidate alloys, the CALPHAD based Thermo-Calc software with the TCHEA1 database was employed to compute the phase diagram. The calculated equilibrium phase diagrams of the designed RHEAs are shown in Figure 2. The range of calculated temperature in the graph was from 900 K to 3000 K. As can be seen from the diagram, all the designed alloys keep the stable BCC phase at a high temperature of more than 2000 K, with no sign of phase transformation. According to the Thermo-Calc software, the melting point of $Al_{20}Cr_{20}Mo_{20}Ti_{20}V_{20}$, $Al_{20}Cr_{20}Mo_{10}Ti_{20}V_{30}$, and $Al_{30}Cr_{10}Mo_5Ti_{20}V_{35}$ were 2190 K, 2137 K, and 2025 K, respectively. The XRD patterns reported by Kang et al. [8] show the existence of a single BCC crystal structure in $Al_{20}Cr_{20}Mo_{20}Ti_{20}V_{20}$. This experimental finding agrees with our prediction for $Al_{20}Cr_{20}Mo_{20}Ti_{20}V_{20}$. Considering the experimental agreement and validation of the TCHEA1 database prediction according to a previous publication [19–22], we expect that future experimental results may confirm our phase prediction findings for $Al_{20}Cr_{20}Mo_{10}Ti_{20}V_{30}$ and $Al_{30}Cr_{10}Mo_5Ti_{20}V_{35}$.

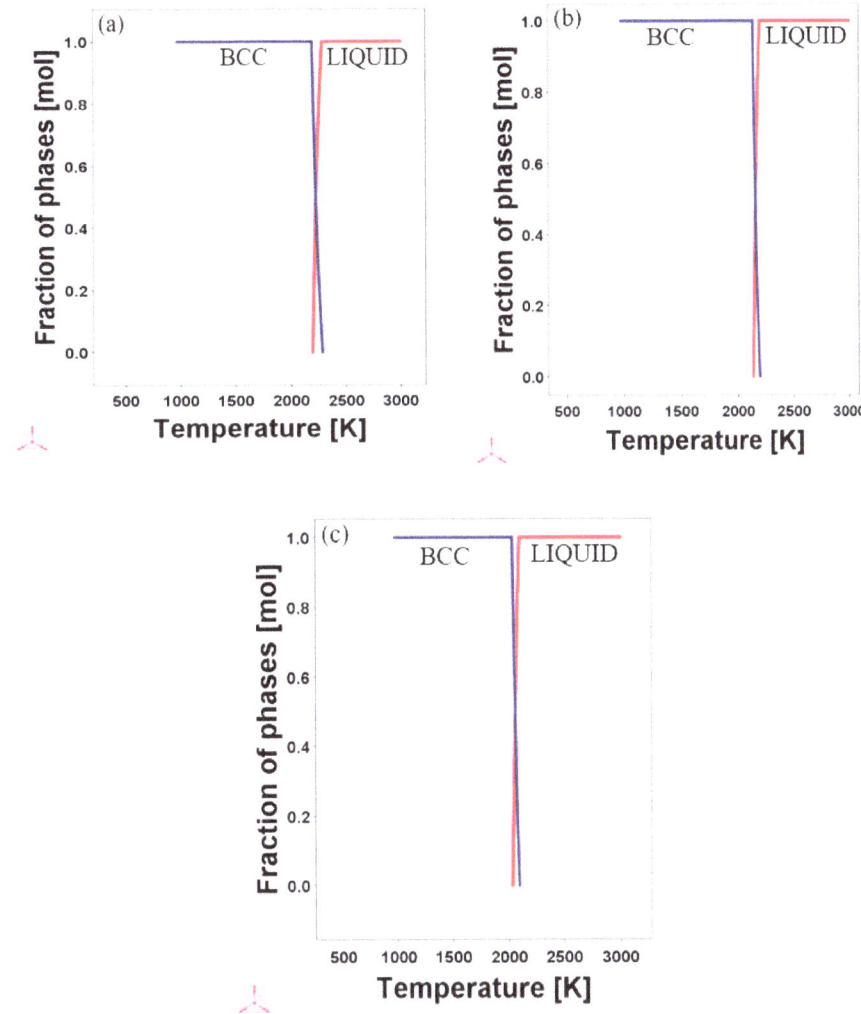

Figure 2. Calculated equilibrium phase diagrams of the RHEAs at high temperature: (a) $Al_{20}Cr_{20}Mo_{20}Ti_{20}V_{20}$; (b) $Al_{20}Cr_{20}Mo_{10}Ti_{20}V_{30}$; (c) $Al_{30}Cr_{10}Mo_5Ti_{20}V_{35}$.

3.2. Mechanical Properties

The structural optimization was performed with a supercell of 100 atoms to avoid the chance of lattice distortion among its constituent elements. The calculated elastic constants and other mechanical properties of the alloys are listed in Table 2. The optimized volumes of the 100-atoms models of $Al_{20}Cr_{20}Mo_{20}Ti_{20}V_{20}$, $Al_{20}Cr_{20}Mo_{10}Ti_{20}V_{30}$, and $Al_{30}Cr_{10}Mo_5Ti_{20}V_{35}$ were 1447.42 Å^3, 1425.36 Å^3, and 1461.51 Å^3, respectively. From these optimized volumes, lattice constants and densities were calculated. From the calculation, $Al_{30}Cr_{10}Mo_5Ti_{20}V_{35}$ was found to have a low density of 5.16 g/cm^3.

Table 2. Calculated three independent elastic constants (GPa); C_{11} ; C_{12} ; C_{44}, Cauchy pressure C_{11}–C_{44} (GPa), bulk modulus B (GPa), shear modulus G (GPa), Young's modulus E (GPa), Poisson's ratio (ν), Pugh's ratio (B/G), lattice constant a (Å), density ρ (gm/cm^3) and hardness H_v (GPa) at zero pressure and Kelvin, respectively.

Alloys	C_{11}	C_{12}	C_{44}	C_{11}–C_{44}	B	G	E	ν	B/G	ρ	a	H_v
$Al_{20}Cr_{20}Mo_{20}Ti_{20}V_{20}$	255	119	69	186	165	61	163	0.33	2.72	6.28	3.07	5.48
$Al_{20}Cr_{20}Mo_{10}Ti_{20}V_{30}$	247	111	64	183	157	60	161	0.32	2.62	5.85	3.05	5.66
$Al_{30}Cr_{10}Mo_{5}Ti_{20}V_{35}$	246	90	63	183	143	56	149	0.32	2.71	5.16	3.08	5.56

The elastic constant of materials can yield information about their bonding and mechanical characteristics. All the given alloys satisfy the mechanically stable conditions as (C_{44} >0, C_{11} > |C_{12}|, and $C_{11} + 2C_{11} > 0$) [45]. The ratio of bulk modulus (B) to shear modulus (G) can be used as an indicator to find the ductility in materials using the Pugh criterion [46]. The Pugh criterion states that a material with a B/G ratio greater than 1.75 behaves as ductile. The Poisson's ratio (ν) can be used as a descriptor for ductility. Gu et al. [47] reported that bulk metallic glasses with $\nu > 31$ are very ductile. Moreover, the positive sign value of Cauchy pressure ($C_{11} - C_{44}$) is associated with the metallic nature of polycrystalline material [48]. All three designed RHEAs show metallic nature, as all of them have a positive value of Cauchy pressure. Based on the computed outcomes of the Cauchy pressure, the Poisson ratio, and the Pugh criterion, we conclude that the designed RHEAs are ductile. The Vickers hardness (H_v) of the designed alloy was calculated using Tian's model [49] as $H_v = 0.92 \times K^{1.137} \times G^{0.708}$ where $K = G/B$. The calculated H_v of $Al_{20}Cr_{20}Mo_{20}Ti_{20}V_{20}$ is 5.48 GPa while the experimentally measured H_v value of $Al_{20}Cr_{20}Mo_{20}Ti_{20}V_{20}$ is 5.55 GPa [8]. The excellent agreement of calculated hardness with the experimental result confirms the reliability of our predicted value of bulk modulus and shear modulus. The excellent agreement of calculated hardness with the experimental result also confirms the reliability of our predicted value of elastic properties. The randomized model of the supercell used in our calculation may be used in predicting the mechanical and thermal properties of other future RHEAs. Our prediction may stimulate further experimental investigations of $Al_{30}Cr_{10}Mo_5V_{35}Ti_{20}$, as it has a low density, is very ductile, and has a high melting point.

3.3. Thermal Properties

The thermal coefficient of linear expansion, alpha (α), as a function of temperature for $Al_{20}Cr_{20}Mo_{20}Ti_{20}V_{20}$, $Al_{20}Cr_{20}Mo_{10}Ti_{20}V_{30}$ and $Al_{30}Cr_{10}Mo_5Ti_{20}V_{35}$ RHEAs is shown in Figure 3a. The calculated Debye temperature (θ_D) of $Al_{20}Cr_{20}Mo_{20}Ti_{20}V_{20}$, $Al_{20}Cr_{20}Mo_{10}Ti_{20}V_{30}$, and $Al_{30}Cr_{10}Mo_5Ti_{20}V_{35}$ were 428 K, 443 K, and 451 K, respectively. The higher value of θ_D in materials reflects the strength of the covalent bond component in materials [50]. This reveals that the covalent bond component inside $Al_{30}Cr_{10}Mo_5Ti_{20}V_{35}$ could be stronger than the other two RHEAs, as it has a higher value of θ_D. We found that the α increases rapidly as temperature increases from 0 K to 200 K, and slowly becomes linear at temperatures above 600 K for the given RHEAs. The value of α for $Al_{30}Cr_{10}Mo_5Ti_{20}V_{35}$ is similar to that of $Al_{20}Cr_{20}Mo_{20}Ti_{20}V_{20}$ and $Al_{20}Cr_{20}Mo_{10}Ti_{20}V_{30}$ at temperatures below 200 K, but its α increases rapidly at temperatures above 298 K. This variation of the thermal coefficient of linear expansion indicates that the α of $Al_{30}Cr_{10}Mo_5Ti_{20}V_{35}$ is very sensitive at temperatures above 298 K. It may be due to the addition of low-density elements such as Al and V in the RHEAs. Figure 3b shows the specific heat capacity at constant volume, C_v, as a function of temperature between $Al_{20}Cr_{20}Mo_{20}Ti_{20}V_{20}$, $Al_{20}Cr_{20}Mo_{10}Ti_{20}V_{30}$, and $Al_{30}Cr_{10}Mo_5Ti_{20}V_{35}$. At low temperatures below 50 K, the C_v of all three RHEAs has identical behavior. At higher temperatures above 50 K, the C_v of $Al_{30}Cr_{10}Mo_5Ti_{20}V_{35}$ > C_v of $Al_{20}Cr_{20}Mo_{10}Ti_{20}V_{30}$ > C_v of $Al_{20}Cr_{20}Mo_{20}Ti_{20}V_{20}$. This trend implies that the lowering of the high-density elements Mo and Cr increases the C_v of $Al_{30}Cr_{10}Mo_5Ti_{20}V_{35}$. Since the specific heat capacity of $Al_{30}Cr_{10}Mo_5Ti_{20}V_{35}$ is higher, it can absorb more heat than the other two RHEAs. We are not aware of the experimental report for the thermal properties of the present RHEAs, and future experiments will likely confirm our thermal properties results.

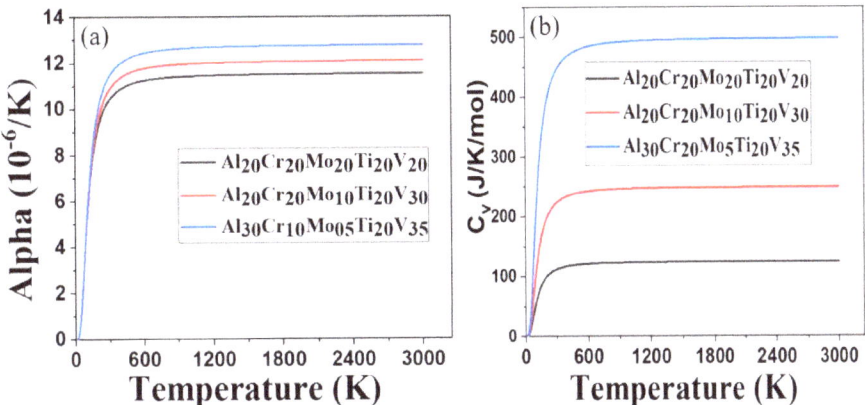

Figure 3. (a) Thermal coefficient of linear expansion as a function of temperature; (b) Specific heat capacity at constant volume as a function of temperature.

4. Conclusions

Based on the first principles DFT calculation, CALPHAD method, and Debye theory method, structural, mechanical, and thermal properties of three RHEAs were studied. Novel $Al_{30}Cr_{10}Mo_5Ti_{20}V_{35}$ RHEA has a low density of 5.16 g/cm^3, a hardness of 5.56 GPa, and maintains a stable BCC structure at high temperatures of 1000–2000 K. It also has a B/G ratio of 2.7, which indicates its ductile nature. Calculations of thermal properties show that the addition of elements Al and V increases the covalent bond component, thermal expansion coefficient, and the specific heat capacity of $Al_{30}Cr_{10}Mo_5Ti_{20}V_{35}$ when compared with the previous RHEA $Al_{20}Cr_{20}Mo_{20}Ti_{20}V_{20}$. The present study reveals that based on structural and physical property calculations, $Al_{30}Cr_{10}Mo_5Ti_{20}V_{35}$ could be a potential candidate for high-temperature applications. It is expected that further experimental exploration may validate our results on the structural, mechanical and thermal properties of these RHEAs.

Author Contributions: U.B. performed the computer simulation and wrote the original manuscript. C.Z. tested and verified 100 atom random unit cell model. U.B., C.Z. and S.Y. contributed to the discussion and writing. All authors have read and agreed to the published version of the manuscript.

Funding: This research is partially supported by the National Science Foundation (NSF) through the collaborative consortium CIMM and Department of Defense (DoD) through AFOSR.

Acknowledgments: This research is partially supported by NSF EPSCoR CIMM project under Award # OIA-1541079 and DoD support under contract W911NF1910005. The computational simulations were supported by the Louisiana Optical Network Infrastructure (LONI) with the supercomputer allocation loni_mat_bio12.

Conflicts of Interest: The authors declare no conflict of interest. The funders had no role in the design of the study; in the collection, analyses, or interpretation of data; in the writing of the manuscript, or in the decision to publish the results.

References

1. Perepezko, J.H. The Hotter the Engine, the Better. *Science* **2009**, *326*, 1068–1069. [CrossRef] [PubMed]
2. Pollock, T.M. Alloy design for aircraft engines. *Nat. Mater.* **2016**, *15*, 809–815. [CrossRef] [PubMed]
3. Albert, B.; Völkl, R.; Glatzel, U. High-temperature oxidation behavior of two nickel-based superalloys produced by metal injection molding for aero engine applications. *Metall. Mater. Trans. A* **2014**, *45*, 4561–4571. [CrossRef]

4. Kuhn, H.A.; Biermann, H.; Ungár, T.; Mughrabi, H. An X-ray study of creep-deformation induced changes of the lattice mismatch in the γ′-hardened monocrystalline nickel-base superalloy SRR 99. *Acta Metallurgica et Materialia* **1991**, *39*, 2783–2794. [CrossRef]
5. Zenk, C.H.; Povstugar, I.; Li, R.; Rinaldi, F.; Neumeier, S.; Raabe, D.; Göken, M. A novel type of Co–Ti–Cr-base γ/γ′ superalloys with low mass density. *Acta Mater.* **2017**, *135*, 244–251. [CrossRef]
6. Xu, Z.Q.; Ma, Z.L.; Wang, M.; Chen, Y.W.; Tan, Y.D.; Cheng, X.W. Design of novel low-density refractory high entropy alloys for high-temperature applications. *Mater. Sci. Eng. A* **2019**, *755*, 318–322. [CrossRef]
7. Youssef, K.M.; Zaddach, A.J.; Niu, C.; Irving, D.L.; Koch, C.C. A Novel Low-Density, High-Hardness, High-entropy Alloy with Close-packed Single-phase Nanocrystalline Structures. *Mater. Res. Lett.* **2015**, *3*, 95–99. [CrossRef]
8. Kang, M.; Lim, K.R.; Won, J.W.; Lee, K.S.; Na, Y.S. Al-ti-containing lightweight high-entropy alloys for intermediate temperature applications. *Entropy* **2018**, *20*, 355. [CrossRef]
9. Li, C.; Li, J.C.; Zhao, M.; Jiang, Q. Effect of alloying elements on microstructure and properties of multiprincipal elements high-entropy alloys. *J. Alloy. Compd.* **2009**, *475*, 752–757. [CrossRef]
10. Zhang, Y. Mechanical properties and structures of high entropy alloys and bulk metallic glasses composites. *Mater. Sci. Forum* **2010**, *654*, 1058–1061. [CrossRef]
11. Zhou, Y.J.; Zhang, Y.; Wang, Y.L.; Chen, G.L. Solid solution alloys of AlCoCrFeNiTix with excellent room-temperature mechanical properties. *Appl. Phys. Lett.* **2007**, *90*, 181904. [CrossRef]
12. Jung, C.; Kang, K.; Marshal, A.; Pradeep, K.G.; Seol, J.B.; Lee, H.M.; Choi, P.P. Effects of phase composition and elemental partitioning on soft magnetic properties of AlFeCoCrMn high entropy alloys. *Acta Mater.* **2009**, *171*, 31–39. [CrossRef]
13. Senkov, O.N.; Senkova, S.; Woodward, C.; Miracle, D.B. Low-density, refractory multi-principal element alloys of the Cr-Nb-Ti-V-Zr system: Microstructure and phase analysis. *Acta Mater.* **2013**, *61*, 1545–1557. [CrossRef]
14. Senkov, O.N.; Wilks, G.; Scott, J.; Miracle, D.B. Mechanical properties of $Nb_{25}Mo_{25}Ta_{25}W_{25}$ and $V_{20}Nb_{20}Mo_{20}Ta_{20}W_{20}$ refractory high entropy alloys. *Intermetallics* **2011**, *19*, 698–706. [CrossRef]
15. Senkov, O.; Wilks, G.; Miracle, D.; Chuang, C.; Liaw, P. Refractory high-entropy alloys. *Intermetallics* **2010**, *18*, 1758–1765. [CrossRef]
16. Zou, Y.; Maiti, S.; Steurer, W.; Spolenak, R. Size-dependent plasticity in an $Nb_{25}Mo_{25}Ta_{25}W_{25}$ refractory high-entropy alloy. *Acta Mater.* **2014**, *65*, 85–97. [CrossRef]
17. Thermo-Calc Software. Available online: https://www.thermocalc.com/ (accessed on 7 April 2020).
18. Andersson, J.-O.; Helander, T.; Höglund, L.; Shi, P.; Sundman, B. Thermo-Calc & DICTRA, computational tools for materials science. *Calphad* **2002**, *26*, 273–312.
19. Lederer, Y.; Toher, C.; Vecchio, K.S.; Curtarolo, S. The search for high entropy alloys: A high-throughput ab-initio approach. *Acta Mater.* **2018**, *159*, 364–383. [CrossRef]
20. Chen, H.L.; Mao, H.; Chen, Q. Database development and Calphad calculations for high entropy alloys: Challenges, strategies, and tips. *Mater. Chem. Phys.* **2018**, *210*, 279–290. [CrossRef]
21. Mao, H.; Chen, H.L.; Chen, Q. TCHEA1: A thermodynamic database not limited for "high entropy" alloys. *J. Phase Equilibria Diffus.* **2017**, *38*, 353–368. [CrossRef]
22. Gao, M.C.; Zhang, B.; Yang, S.; Guo, S. Senary refractory high-entropy alloy HfNbTaTiVZr. *Metall. Mater. Trans. A* **2016**, *47*, 3333–3345. [CrossRef]
23. Hohenberg, P.; Kohn, W. Inhomogeneous electron gas. *Phys. Rev.* **1964**, *136*, B864. [CrossRef]
24. Kohn, W.; Sham, L.J. Self-Consistent Equations Including Exchange and Correlation Effects. *Phys. Rev.* **1965**, *140*, A1133–A1138. [CrossRef]
25. Kresse, G.; Furthmüller, J. Efficient iterative schemes for ab initio total-energy calculations using a plane-wave basis set. *Phys. Rev. B* **1996**, *54*, 11169–11186. [CrossRef] [PubMed]
26. MedeA® Software. Available online: https://www.materialsdesign.com/medea-software (accessed on 7 April 2020).
27. Blöchl, P.E. Projector augmented-wave method. *Phys. Rev. B* **1994**, *50*, 17953. [CrossRef] [PubMed]
28. Perdew, J.P.; Burke, K.; Ernzerhof, M. Generalized Gradient Approximation Made Simple. *Phys. Rev. Lett.* **1996**, *77*, 3865–3868. [CrossRef] [PubMed]
29. Monkhorst, H.J.; Pack, J.D. Special points for Brillouin-zone integrations. *Phys. Rev. B* **1976**, *13*, 5188. [CrossRef]
30. Knuth, E.D. *The Art of Computer Programming*, 3rd ed.; Addison-Wesley: Massachusetts, MA, USA, 1965.

31. Le Page, Y.; Saxe, P. Symmetry-general least-squares extraction of elastic coefficients from ab initio total energy calculations. *Phys. Rev. B* **2001**, *63*, 174103. [CrossRef]
32. Le Page, Y.; Saxe, P. Symmetry-general least-squares extraction of elastic data for strained materials from ab initio calculations of stress. *Phys. Rev. B* **2002**, *65*, 104104. [CrossRef]
33. Anderson, O.L. A simplified method for calculating the Debye temperature from elastic constants. *J. Phys. Chem. Solids* **1963**, *24*, 909–917. [CrossRef]
34. Andrews, K. Elastic moduli of polycrystalline cubic metals. *J. Phys. D: Appl. Phys.* **1978**, *11*, 2527. [CrossRef]
35. Grüneisen, E. *Zustand Thermische Eigenschaften der Stoffe*; Springer: Berlin, Germany, 1926.
36. Mayer, B.; Anton, H.; Bott, E.; Methfessel, M.; Sticht, J.; Harris, J.; Schmidt, P. Ab-initio calculation of the elastic constants and thermal expansion coefficients of Laves phases. *Intermetallics* **2003**, *11*, 23–32. [CrossRef]
37. Ashcroft, N.W.; Mermin, N.D. *Solid State Physics*; Holt-Saunders Int.: Philadelphia, PA, USA, 1976.
38. Zhang, Y.; Zhou, Y.J.; Lin, J.P.; Chen, G.L.; Liaw, P.K. Solid-solution phase formation rules for multi-component alloys. *Adv. Eng. Mater.* **2008**, *10*, 534–538. [CrossRef]
39. Takeuchi, A.; Inoue, A. Classification of bulk metallic glasses by atomic size difference, heat of mixing and period of constituent elements and its application to characterization of the main alloying element. *Mater. Trans.* **2005**, *46*, 2817–2829. [CrossRef]
40. Zhang, Y.; Zuo, T.T.; Tang, Z.; Gao, M.C.; Dahmen, K.A.; Liaw, P.K.; Lu, Z.P. Microstructures and properties of high-entropy alloys. *Prog. Mater. Sci.* **2014**, *61*, 1–93. [CrossRef]
41. Guo, S.; Ng, C.; Lu, J.; Liu, C. Effect of valence electron concentration on stability of fcc or bcc phase in high entropy alloys. *J. Appl. Phys.* **2011**, *109*, 103505. [CrossRef]
42. Wang, Z.; Huang, Y.; Yang, Y.; Wang, J.; Liu, C. Atomic-size effect and solid solubility of multicomponent alloys. *Scripta Materialia* **2015**, *94*, 28–31. [CrossRef]
43. Sheng, G.; Liu, C.T. Phase stability in high entropy alloys: Formation of solid-solution phase or amorphous phase. *Prog. Nat. Sci. Mater. Int.* **2011**, *21*, 433–446.
44. Yeh, J.W.; Chang, S.Y.; Hong, Y.D.; Chen, S.K.; Lin, S.J. Anomalous decrease in X-ray diffraction intensities of Cu-Ni-Al-Co-Cr-Fe-Si alloy systems with multi-principal elements. *Mater. Chem. Phys.* **2007**, *103*, 41–46. [CrossRef]
45. Wallace, D.C. *Thermodynamics of Crystals*; Wiley: Hoboken, NJ, USA, 1972.
46. Pugh, S. XCII. Relations between the elastic moduli and the plastic properties of polycrystalline pure metals. *Lond. Edinb. Dublin Philos. Mag. J. Sci.* **1954**, *45*, 823–843. [CrossRef]
47. Gu, X.; McDermott, A.; Poon, S.J.; Shiflet, G.J. Critical Poisson's ratio for plasticity in Fe-Mo-C-B-Ln bulk amorphous steel. *Appl. Phys. Lett.* **2006**, *88*, 211905. [CrossRef]
48. Nguyen-Manh, D.; Mrovec, M.; Fitzgerald, S.P. Dislocation driven problems in atomistic modelling of materials. *Mater. Trans.* **2008**, *49*, 2497–2506. [CrossRef]
49. Tian, Y.; Xu, B.; Zhao, Z. Microscopic theory of hardness and design of novel superhard crystals. *Int. J. Refract. Met. Hard Mater.* **2012**, *33*, 93–106. [CrossRef]
50. Wang, S.; Zhao, Y.; Hou, H.; Wen, Z.; Zhang, P.; Liang, J. Effect of anti-site point defects on the mechanical and thermodynamic properties of MgZn$_2$, MgCu$_2$ Laves phases: A first-principle study. *J. Solid State Chem.* **2018**, *263*, 18–23. [CrossRef]

© 2020 by the authors. Licensee MDPI, Basel, Switzerland. This article is an open access article distributed under the terms and conditions of the Creative Commons Attribution (CC BY) license (http://creativecommons.org/licenses/by/4.0/).

Article

Effect of Strain on Transformation Diagrams of 100Cr6 Steel

Rostislav Kawulok [1,*], Ivo Schindler [1], Jaroslav Sojka [1], Petr Kawulok [1], Petr Opěla [1], Lukáš Pindor [2], Eduard Grycz [2], Stanislav Rusz [1] and Vojtěch Ševčák [1]

[1] VŠB-TU Ostrava, Faculty of Materials Science and Technology, VSB-Technical University of Ostrava, 70800 Ostrava, Czech Republic; ivo.schindler@vsb.cz (I.S.); jaroslav.sojka@vsb.cz (J.S.); petr.kawulok@vsb.cz (P.K.); petr.opela@vsb.cz (P.O.); stanislav.rusz.fmmi@vsb.cz (S.R.); vojtech.sevcak@vsb.cz (V.Š.)
[2] Třinecké Železárny, a.s., 73961 Třinec, Czech Republic; lukas.pindor@trz.cz (L.P.); eduard.grycz@trz.cz (E.G.)
* Correspondence: rostislav.kawulok@vsb.cz

Received: 5 March 2020; Accepted: 18 April 2020; Published: 21 April 2020

Abstract: Based on dilatometric tests, the effect of various values of previous deformation on the kinetics of austenite transformations during the cooling of 100Cr6 steel has been studied. Dilatometric tests have been performed with the use of the optical dilatometric module of the plastometer Gleeble 3800. The obtained results were compared to metallographic analyses and hardness measurements HV30. Uniaxial compression deformations were chosen as follows: 0, 0.35, and 1; note that these are true (logarithmic) deformations. The highly important finding was the absence of bainite. In addition, it has been verified that with the increasing amount of deformation, there is a further shift in the pearlitic region to higher cooling rates. The previous deformation also affected the temperature martensite start, which decreased due to deformation. The deformation value of 1 also shifted the critical cooling rate required for martensite formation from the 12 °C/s to 25 °C/s.

Keywords: transformation diagrams; 100Cr6 steel; dilatometric test

1. Introduction

Transformation diagrams can be divided into two types. The first type is the time-temperature-transformation (TTT) diagram, which describes the austenite transformations during an isothermal dwell. The second type is the continuously cooling transformation (CCT) diagram that describes the effect of cooling rate on austenite transformation [1,2]. In addition, this type of diagram can be modified to a deformation continuously cooling transformation (DCCT) type, which also includes the effect of previous deformation. Such diagrams then find their practical application in the case of the controlling of steel forming processes (rolling, forging), and their findings often lead to an improvement in the economics of the production process [3–7].

Austenite transformation kinetics during cooling is influenced by many factors. The most important factors are the chemical composition and the rate of cooling of the steels. Further, the transformation of supercooled austenite is influenced by austenitization temperature, initial structure, austenitic grain size, and previous deformation [8–17]. The effects of all these factors are summarized in the schematic CCT diagram in Figure 1 [8].

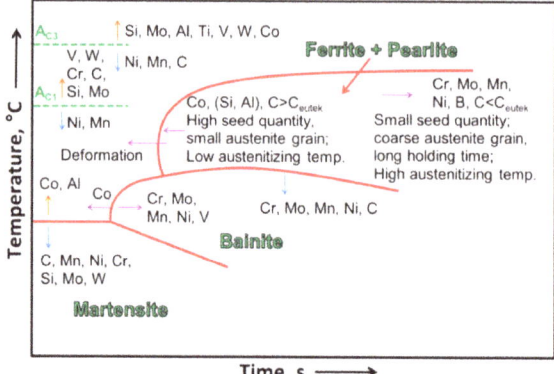

Figure 1. Influence of alloying, thermomechanical factors, and structure state on transformation kinetic [8].

The effect of previous deformation itself on different types of steel is evaluated in a number of research studies [18–28]. From the obtained findings, it can be claimed that the previous deformation, in combination with the chemical composition and other factors, has different effects on individual transformations or transformation products. The authors of these works assume that the deformation accelerates diffusion-controlled transformations, specifically the transformation of austenite to ferrite and pearlite [18–26]. Due to the deformation, the number of lattice defects increases, which promotes the diffusion of all atoms in the solid solution and leads to a faster nucleation and growth of the new phase nuclei—this leads to the acceleration of both transformations [23–26]. The accelerating effects of increasing deformation and strain rate on ferritic transformation are shown in Figures 2 and 3, respectively (where temperature A_{r3} is the transformation temperature of austenite to ferrite and A_{r1} is the transformation temperature of austenite to pearlite by the cooling of hypoeutectoid steels). In addition, from Figure 2, it is evident that with increasing deformation (up to about 0.4), the area of ferritic transformation is narrowed—thus, the pearlitic transformation is accelerated [26]. This thesis confirms that even our previous research detected that the effect of previous deformation leads to an increased fraction of pearlite and ferrite [11,20].

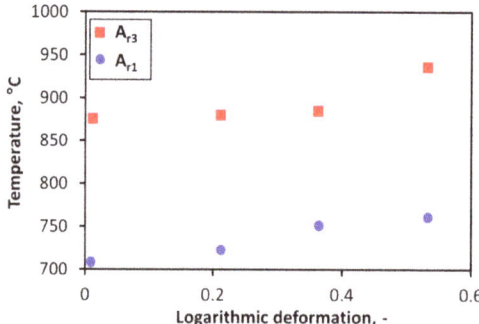

Figure 2. Effect of deformation on the ferritic transformation of HSLA (High-Strength Low-Alloy) steel [26]. The temperatures of Ar1 and Ar3 represent the temperatures of the transition during cooling.

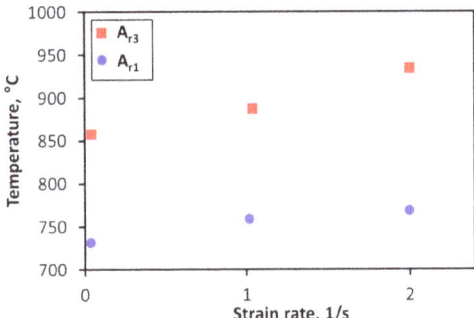

Figure 3. Effect of strain rate on the ferritic transformation of HSLA steel [26]. The temperatures of Ar1 and Ar3 represent the temperatures of the transition during cooling.

In the case of assessing the effect of plastic deformation on the transformation of austenite to bainite and martensite, this effect is ambiguous and, therefore, cannot be generalized as in the previous case. When austenite is deformed, a dense dislocation network is formed and inhibits the progress of the phase boundaries, and despite a large number of nuclei, the share of the new phase is lower than that of non-deformed austenite, especially at higher cooling rates. The role of dislocation on the martensite start temperature is more controversial. The large amount of dislocations generated by a large plastic deformation of austenite prior to the martensite may stabilize the glissile embryo–austenite interface, leading to a decrease in martensite (Ms)-temperature (temperature of starting martensite transformation). This is known as the dislocation stabilization mechanism. However, there is sometimes an opposite phenomenon; accumulated lattice defects initiate martensite formation and enable its formation at higher temperatures than in the case of non-deformed austenite transformation. An example of this is the graph in Figure 4, which shows that the Ms temperature rises after a height deformation of 30% and 60% compared to the Ms temperature of the undeformed specimens of selected steels [18,23,24,29–36].

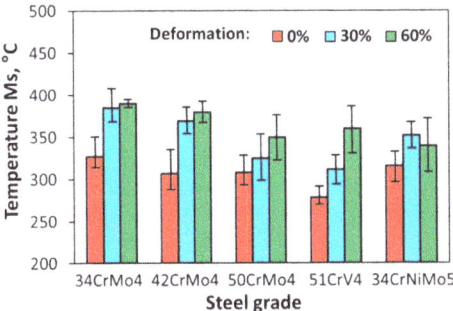

Figure 4. Influence of the previous deformation on martensite (Ms) temperature in selected low-alloyed steels [23].

All diagram types can be constructed on the basis of mathematical simulations using specialized programs (JMatPRO, QTSteel, etc.) or by physical tests on dilatometers or dilatometric modules of universal plastometers. In the case of the mathematical calculation of the diagrams, the diagrams are calculated on the basis of the equations compiled for the selected steel chemical composition range, but their accuracy is not always optimal. For this reason, it is certainly more appropriate to design diagrams on the basis of dilatometric tests that are performed on specimens of specific qualities [37–41].

The subject of this article was to evaluate the influence of the previous deformation of two different true strains, logarithmic deformation (e = 0.35 and 1), on the construction of (D)CCT diagrams of 100Cr6 steel, thus contributing to the extension of knowledge of the austenite transformation kinetics during cooling of the bearing steels. The 100Cr6 steel has good hot formability, is suitable for direct quenching, and, in a soft annealed condition, is reasonably machinable and suitable for components with a very hard and wear-resistant surface. Primarily, this steel is intended for the production of bearing balls up to 25 mm in diameter, and rollers and taper roller bearings up to a 16 mm wall thickness [42–44]. This experiment was realized in order to verify the effect of the previous deformation on the final structure of the hypereutectoid steel.

2. Materials and Methods

As can be seen from the chemical composition of the 100Cr6 steel, which is presented in Table 1, it is a hyper-eutectoid high-carbon and low-alloyed steel.

Table 1. Chemical composition of investigated 100Cr6 steel in wt.%.

C	Mn	Si	P	S	Cr
0.994	0.38	0.324	0.011	0.001	1.45

For the purpose of the experiment, simple cylindrical specimens with a diameter of 6 mm and length of 86 mm were prepared from the 100Cr6 steel. The initial state of the investigated steel was deformed as it was prepared from the rolled rods 12 mm in diameter.

The dilatometric experiments were performed with the use of the optical dilatometric module (The Model 39112 Scanning Non-Contact Optical Dilatometer and Extensometer System with Green LED Technology (Dynamic Systems Inc., Poestenkill, NY, USA) of the Gleeble 3800 plastometer. This optical dilatometric module is based on the measurement cross-section (radial components of strain) of samples with a repeatability of ±0.3 µm by a frequency of 2400 Hz with a maximum temperature of 1200 °C [18–20].

The first step was to determine the temperatures of A_{c1} and A_{cm}, which represent the temperatures of the transition during heating. In this case, the specimen was heated at 5 °C/s to 500 °C, and the heating rate was then slowed down to 10 °C/min (0.167 °C/s) to locate the transformation area. Evaluation of the measured data was carried out using the semi-automatic CCT software (Dynamic Systems Inc., Poestenkill, NY, USA), which uses the tangential method in combination with the derivation of the dilatation curve to determine the transformation temperatures. The result of this test is shown in Figure 5.

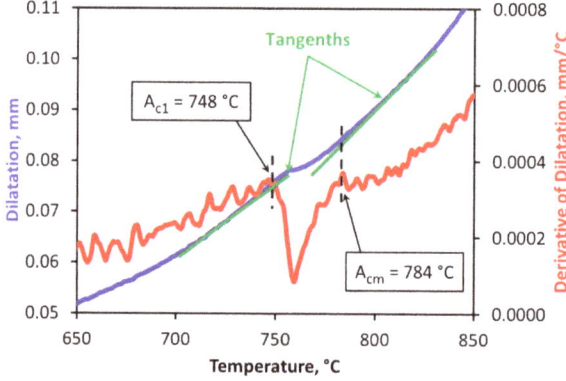

Figure 5. Determination of transformation temperatures A_{c1} and A_{cm} during heating.

Based on the previous experiment (i.e., determination of A_{c1} and A_{cm} temperatures), other specimens were uniformly preheated at 850 °C with a subsequent 10 min dwell at this temperature in order to construct (D)CCT diagrams. After that (in the case of the CCT diagram), the specimens were continuously cooled to room temperature by the constant cooling rates in the range of 0.2–150 °C/s. To achieve cooling rates above 25 °C/s, the specimens (with cooling rates of 60 and 150 °C/s = only for construction of CCT diagram) had to have a special hollow-head structure for high-speed cooling by air nozzles. Unfortunately, the disadvantage of these specially designed specimens is the inability to perform a deformation [45,46]. The temperature of heating at 850 °C is normal for the construction of CCT diagrams of 100Cr6 steel [42–44,47,48].

In the case of the construction of both variants of DCCT diagrams, the continuous cooling was preceded by the uniaxial compression deformation, which was executed directly after the dwell at the austenitization temperature. The magnitude of the true (logarithmic) deformation in the first and second case was equal to 0.35 and 1, respectively. In both cases, the deformation was performed at the strain rate of 1/s. The limiting element in the case of the construction of the DCCT diagrams was the creation of specimens that could allow a maximum cooling rate of 35 °C/s to be achieved [45] after the previous deformation; thus, in both cases of DCCT diagrams, the cooling rates were selected in the range of 0.2–35 °C/s.

Examples of dilatometrically tested specimens without deformation and with the deformation of 0.35 and 1 are given in Figure 6.

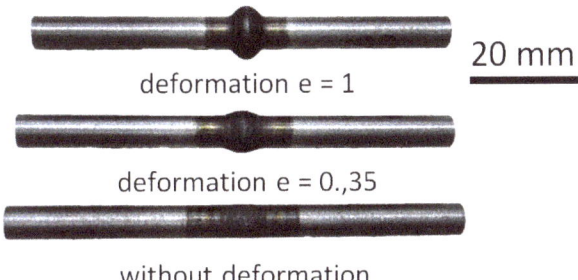

Figure 6. Examples of tested specimens without deformation and after the deformation of 0.35 and 1.

The obtained results were compared to metallographic analyses—scanning electron microscopy (SEM), light-microscopy, and HV30 hardness measurement. Samples intended for metallographic analysis have been prepared by means of mechanical grinding and polishing. The microstructure was revealed via etching in the 2% picric acid solution [20].

As mentioned in the introduction, transformation of austenite depends on the size of austenitic grains as well; therefore, for every (D)CCT diagram besides chemical composition and other thermomechanical parameters, it should be mentioned for what size of austenitic grains this transformation diagram is valid. Therefore, the investigated steel was tested for determination of the average size of austenitic grains (AGS). Other heating parameters, like the heating rate and temperature dwell time, were the same as in the case of the dilatometric experiment, except the fact that instead of deformation after the dwell time, the samples were directly water-quenched to fix the origin austenitic structure. The scheme of this heat-treatment experiment is evident from Figure 7a. In order to evaluate the austenitic grain size, classical optical metallography has been applied when using the method of origin-austenitic-grain revealing. The resulting microstructure of this additional experiment is included in Figure 7b. It was found that all 3 transformation diagrams, which are presented in this article, were constructed for AGS = 9.8 μm. AGS was evaluated by the means of the specialized software, Quick PHOTO INDUSTRIAL 3.2 (PROMICRA s.r.o., Prague, Czech Republic).

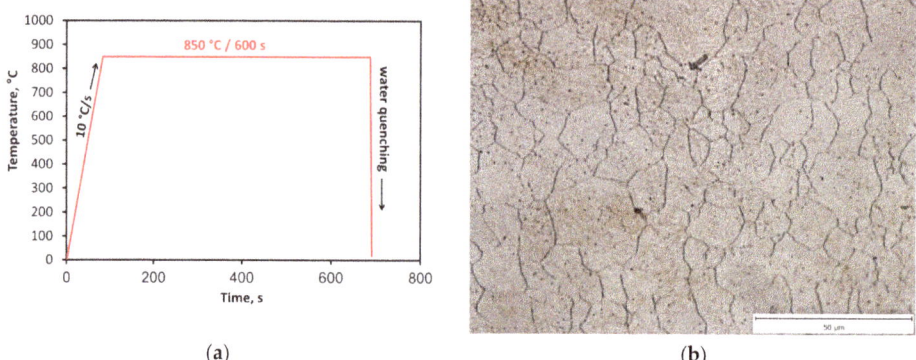

Figure 7. The analysis of the influence of temperature on the austenitic grain size of the 100Cr6 steel. (a) scheme of austenitic grain size (AGS) experiment; (b) Microstructure of austenite, AGS = 9.8 μm.

3. Results and Discussion

3.1. CCT Diagram

As the investigated steel is of hyper-eutectoid type, only austenite–pearlite and austenite–martensite transformations were dilatometrically detected and localized. The example of the dilatation curves with the localized transformation temperatures is given in Figure 8.

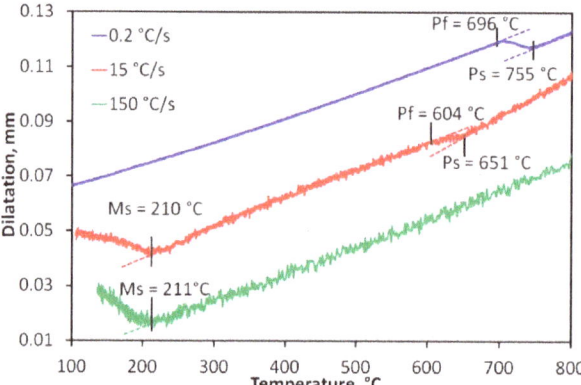

Figure 8. Selected examples of dilatation curves, including determination of transformation points by use of the tangential method.

Based on the analysis of the dilatation curves, the CCT diagram has been compiled (see Figure 9). The constructed CCT diagram consists of only the pearlitic and martensitic regions. Other structural components detected in the samples of the investigated steel were carbides, which do not transform either during heating or during cooling. It is clear that the austenite is transformed to pearlite at cooling rates up to 18 °C/s. The transformation of austenite to martensite proceeds above the cooling rate of 12 °C/s.

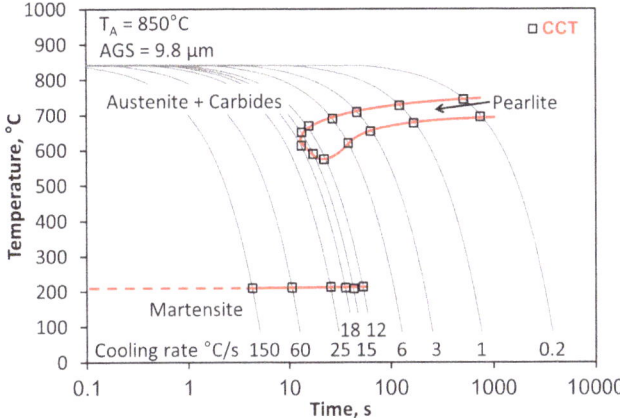

Figure 9. Continuously cooling transformation (CCT) diagram of the 100Cr6 steel.

The constructed CCT diagram was compared to a scanning electron microscopy (SEM) and HV30 hardness measurement. The microstructure of the non-deformed state has been documented at four selected cooling rates, namely 1, 12, 25, and 150 °C/s (Figure 10a–d).

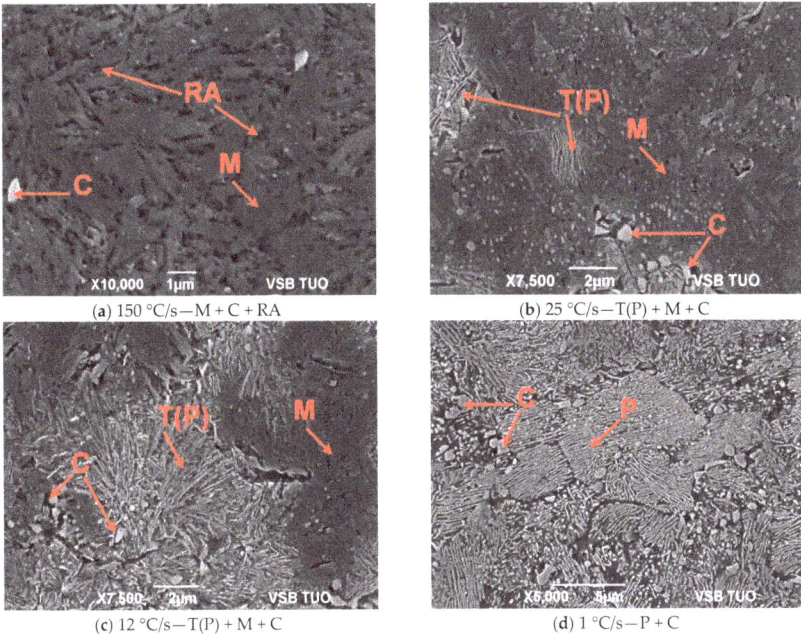

Figure 10. SEM microstructures of selected samples, without deformation. M, martensite; C, carbides; P, pearlite; T(P), troostite; RA, retained austenite.

The presence of retained austenite was confirmed by X-ray diffraction analysis that was performed using a Co Kα source ($\lambda = 0.17902$ nm) by means of the Bruker-AXS D8 (Bruker GmbH, Karlsruhe, Germany) Advance apparatus. The specimen with a cooling rate of 150 °C/s (Figure 10a) contained approximately 6.3% of retained austenite (RA).

The microstructure of pearlitic blocks can be considered fine-grained with the average size of pearlitic blocks of about 6 μm. For measurement of the size of pearlitic blocks (grains), specialized software Quick PHOTO INDUSTRIAL (PROMICRA s.r.o., Prague, Czech Republic) was used, as well as in the determination of AGS.

In all cases, carbides of different sizes from particles smaller than 0.1 μm to particles reaching up to around 1 μm were observed in the microstructure. This confirms that the structure of the steel at the end of the heating dwell consisted of the austenite and carbides. The microstructure of the sample, which was cooled by the rate of 1 °C/s, consists of lamellar pearlite P and carbides (Figure 10d). Precipitation of fine carbides C along the boundaries of former austenitic grains was observed only to a very small extent. The microstructure of the samples, which were cooled by the intermediate rates, 12 and 25 °C/s, consists of pearlite. With respect to an inter-lamellar distance and orientation of pearlitic colonies, the morphology of this pearlite can be identified as troostite T(P) [49] (Figure 10b,c). The structure of these samples also contains martensite M, whose amount rises with the cooling rate. Carbides were observed in these structures as well. The sample cooled by the high rate of 150 °C/s (Figure 10a) consists of the microstructure that is given only by martensite with a certain share of retained austenite RA (see small darker areas in Figure 10a) and carbides.

By comparing the experimentally obtained CCT diagram (Figure 9) to the similar diagrams obtained from available literature (Figures 11 and 12) [41,49–51], it is quite clear that the basic difference between the compared diagrams is the absence of bainite in the experimentally obtained diagram (Figure 9), which was confirmed by the metallographic analyses—see Figure 10. This finding could be caused by the increased content of Si in the investigated steel, which is, in this case, near maximum for this grade of bearing steel. It is known that in the case of high-carbon steels with increasing content of Si, the kinetics of bainitic transformation slows down [51–54].

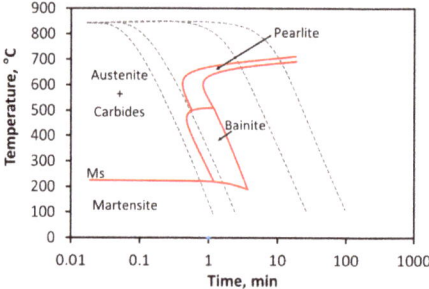

Figure 11. CCT diagram of 100Cr6 steel—according to [51].

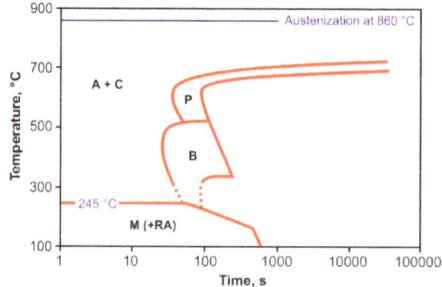

Figure 12. CCT diagram of 100Cr6 steel—according to [42].

From the point of view of the localization of the pearlitic transformation and the temperature of the beginning of martensite formation on the temperature axis, there was a good match. A good match was also obtained when comparing all three CCT diagrams, namely the time domain of pearlite formation, also taking into account slightly different temperatures of austenitization (850 °C vs. 860 °C).

However, martensite appeared in the structure at a cooling rate of 12 °C/s, which is a higher rate than in the case of the compared CCT diagrams from the literature.

3.2. DCCT Diagram—Deformation e = 0.35

While it is possible to encounter a large number of CCT diagrams of the investigated 100Cr6 steel in the literature [42–44,51–54], there is a minimum of information about the effect of prior deformation on this steel. In the case of the presented research, the investigated steel was dilatometrically tested after two levels of the previous deformation. The DCCT diagram in Figure 13 documents the influence of the previous deformation of e = 0.35.

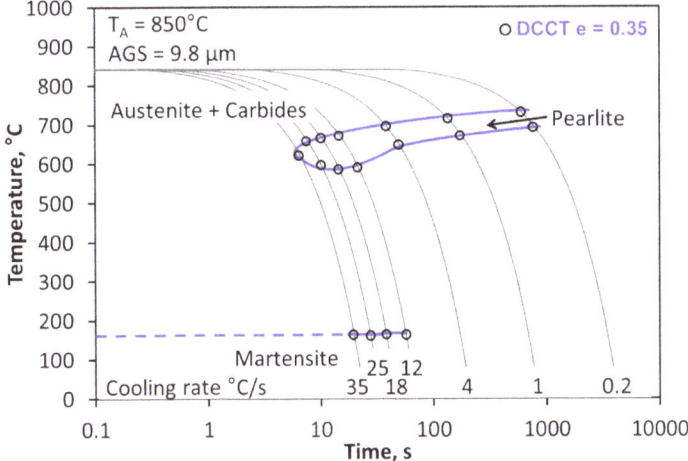

Figure 13. Deformation continuously cooling transformation (DCCT) diagram of the 100Cr6 steel—after the deformation of e = 0.35.

Only the austenite–pearlite and austenite–martensite transformations were found in this DCCT diagram (Figure 13). However, it is evident that the previous deformation accelerated the pearlitic transformation; in other words, due to the previous deformation, the pearlitic nose has been shifted to the left, i.e., to shorter times and thus to higher cooling rates—from 18 °C/s (maximum for CCT diagram) to 35 °C/s. The martensitic transformation was influenced by the previous deformation in the sense of lowering the martensite-start temperature by approximately 50 °C (from 220 °C for CCT to 170 °C for DCCT 0.35).

In order to obtain a completely martensitic-carbidic structure, it was necessary to quench the specimens after the deformation by the use of water jets. This, however, makes it impossible to install the dilatometric module. The course of quenching after the 0.35 deformation is shown in Figure 14. The cooling rate between 810 °C and 100 °C was close to 3500 °C/s.

The microstructure of all samples that were deformed by the strain of 0.35 was observed by light-microscopy and SEM methods (Figure 15). Observed microstructures coincided with those described above for the material state without deformation. An increased troostite share at the expense of martensite was observed in the case of deformed samples, which were cooled by intermediate rates. Documentation via the SEM method has been performed only in the case of the sample, which was cooled at the rate of 25 °C/s (see Figure 15a,b), because light-microscopy would not clearly distinguish the individual structural components, especially troostite.

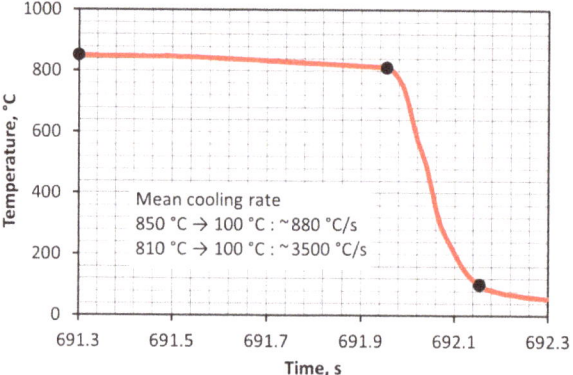

Figure 14. Recording of cooling rate during quenching with the use of water jets.

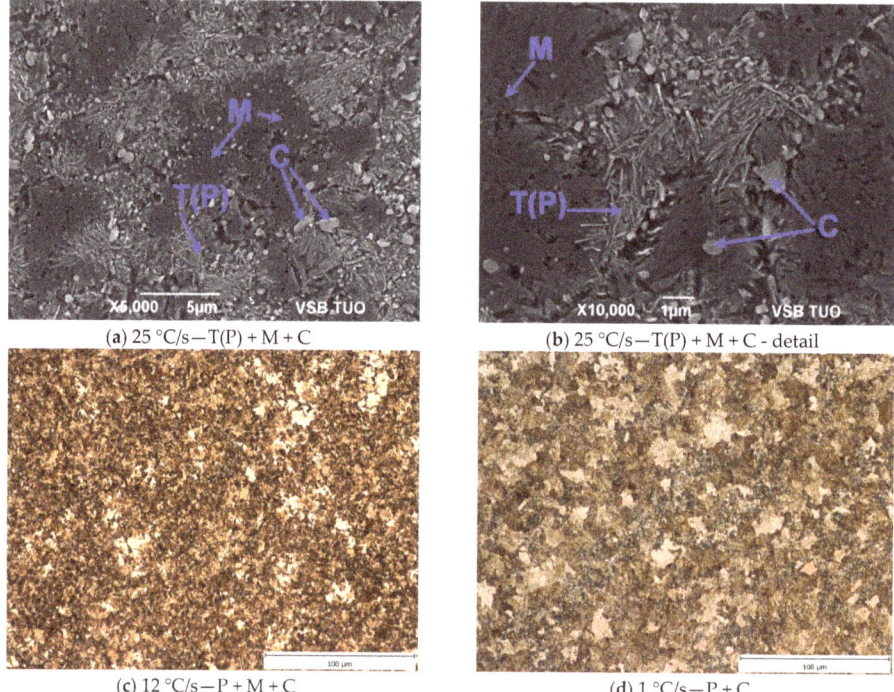

Figure 15. The microstructures (SEM and light-microscopy) of selected samples and after the deformation of e = 0.35. M, martensite; C, carbides; T(P), troostite.

From the point of view of the content of individual structural components, there was a consensus in the case of the representatives of the structures after the deformation of 0.35. The structure of specimens after the quenching consisted only of martensite, carbides, and the occasional occurrence of retained austenite (RA) (see in Figure 15).

3.3. DCCT Diagram—Deformation e = 1

The previous deformation (e = 1) meant a further acceleration of the pearlitic transformation, which unfortunately cannot be precisely quantified, due to the limiting cooling rate (35 °C/s) for the specimen type used for dilatometric deformation tests. The higher deformation did not significantly influence the martensite-start temperature (average martensite-start temperature is about 160 °C). However, the martensite region was shifted to higher cooling rates (25 °C/s). This is shown in Figure 16.

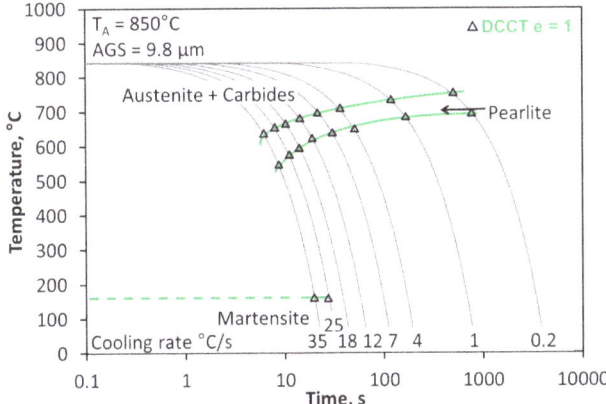

Figure 16. DCCT diagram of the 100Cr6 steel, after the deformation of e = 1.

Metallographic analyses of selected samples after the deformation of 1 revealed that the structure after the quenching contained only martensite, carbides, and retained austenite. In addition, pearlite was detected in the structure of the sample cooled at 25 °C/s, while the fraction of martensite was minor. Unlike the undeformed or deformed specimen (e = 0.35), the structure was solely formed of pearlite and carbides after the cooling rate of 12 °C/s. The mixture of pearlite and carbides was also evident in the sample structure after the cooling rate of 1 °C/s (Figure 17a). The microstructure obtained by SEM is shown in Figure 17b for the specimen cooled at 25 °C/s. It was a mixture of troostite, martensite, and carbides.

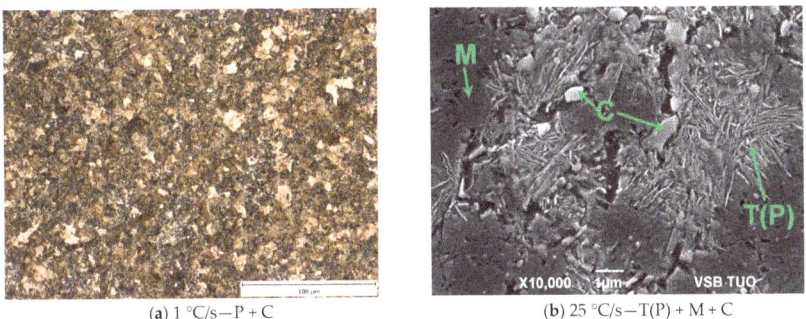

(a) 1 °C/s—P + C   (b) 25 °C/s—T(P) + M + C

Figure 17. The microstructures (light-microscopy and SEM) of selected samples and after the deformation of e = 1. M, martensite; C, carbides; T(P), troostite.

3.4. Comparison of the (D)CCT Diagrams

Comparisons of all three (D)CCT diagrams after different experimental-deformation modes are shown in Figure 18. The shift in the pearlitic region to the left due to the deformation is relatively clear and noticeable, supporting the notion that deformation increases the number of lattice defects, which promotes the diffusion of all atoms in solid solution and leads to faster nucleation and growth of new phase nuclei. Unfortunately, a significant acceleration of the pearlitic transformation due to the pre-deformation of 1 prevented the localization of the pearlitic nose on the selected experimental device. The deformation lowered the martensite-start temperature and, in the case of the deformation of 1, shifted the region to higher cooling rates. A similar conclusion was reported by Nikravesh et al. [55] and Maalekian et al. [56] They reported that the starting temperature of martensite (Ms) slightly decreased with hot deformation. Although the hot deformation of austenite promotes the formation of ferrite and pearlite, this process enhances the stability of austenite against bainitic and martensitic transformation. This phenomenon is known as 'mechanical stabilization' [55–57].

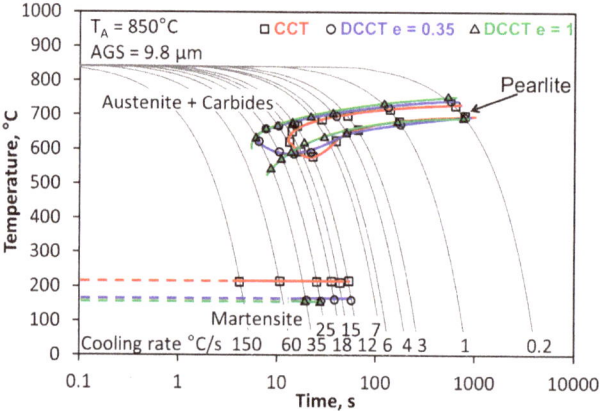

Figure 18. Comparison of all three types of (D)CCT diagrams of the 100Cr6 steel.

3.5. Comparison of the Hardness HV30

Comparison of hardness HV30 of the selected samples after all dilatometric deformation regimes can be seen in Figure 19. As expected, the hardness increased with the cooling rate when martensite in the structure had a significant effect in the sense of hardness increase. The comparison with respect to the effect of the previous deformation on the measured hardness clearly supported the correctness of the construction of the transformation diagrams, as the apparent acceleration of the pearlitic transformation due to the increasing deformation can also be observed. This phenomenon is documented in Figure 19 by the lower hardness compared to undeformed samples, indicating a higher amount of pearlite and carbides in the steel.

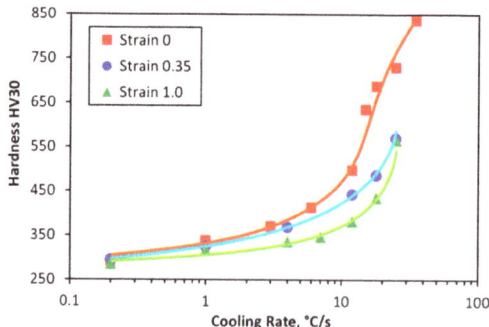

Figure 19. Comparison of measured HV30 hardness in dependence on cooling rate and deformation value.

The influence of the martensite fraction on the measured hardness HV30 is essential and can be documented by the graph in Figure 20. Moreover, the obtained results make it possible to predict the hardness of HV30 of the investigated steel depending on the martensite content (this was determined using specialized software Quick PHOTO INDUSTRIAL (this software works on the basis of structure differentiation by means of its own database, where the shares of individual components are evaluated on the basis of their color spectrum)), which is also presented in Figure 20. It is evident from Figure 20 that hardness HV30 depends linearly on martensite fraction, which can be caused by the relatively narrow range of martensite fraction (10–45%). However, from the literature, we can find that in the wider range of martensite fraction, it has an exponential dependence, where steepness grows with higher carbon content [58,59].

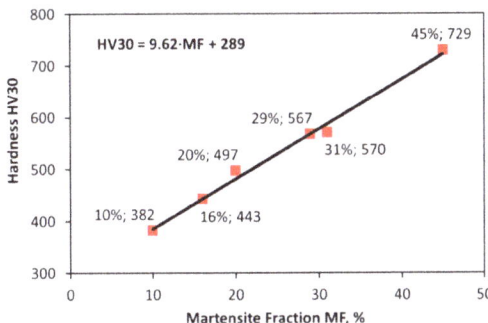

Figure 20. Influence of martensite content on hardness value HV30.

4. Conclusions

Based on the experiments, three variants of (D)CCT diagrams were constructed for the 100Cr6 steel, which can contribute to effective production, in terms of controlled forming and heat treatment.

The study of the influence of the previous deformation on the transformation diagrams of the investigated bearing steel has resulted in several fundamental findings.

The most important finding was the absence of bainite, which is very difficult to detect in such a fine-grained structure, and dilatation curves did not confirm its presence. In this case, the absence of bainite is caused by a higher content of Si, which generally (in the case of high-carbon steels) slows down the kinetics of bainite transformation. The absence of bainite in this steel could be disadvantageous by means of its hardness decreasing, which is one of the most important mechanical properties in the case of bearing steels.

Furthermore, the thesis on the acceleration of diffusion-controlled transformations, including the transformation of austenite to pearlite, was confirmed. In addition, it has been verified that with the increasing amount of deformation, there is a further shift in the pearlitic region to higher cooling rates. Another gained knowledge in the case of the deformation effect on pearlitic transformation is the enlargement of the temperature interval of this transformation due to the increasing deformation, especially in the area of slow deformation rates.

The previous deformation also affected the martensitic transformation that caused its decrease. The deformation of 1 also shifted the critical cooling rate required for martensite formation from 12 °C/s to 25 °C/s.

All these findings were subsequently reflected in hardness, which decreased for specific cooling rates due to increasing deformation.

All three transformation diagrams were compiled for the same heating conditions, i.e., for the same average sizes of origin austenitic grain—9.8 µm, specifically.

Author Contributions: All authors have read and agree to the published version of the manuscript. Conceptualization, R.K. and I.S.; methodology, R.K.; software, R.K.; validation, R.K., P.K., P.O. and J.S.; formal analysis, R.K. and V.Š.; investigation, R.K., P.K., P.O., S.R., L.P., E.G. and J.S.; resources, R.K. and I.S.; data curation, R.K.; writing—original draft preparation, R.K. and V.Š.; writing—review and editing, R.K., P.K., I.S., P.O. and J.S.; visualization, R.K.; supervision, I.S., L.P. and E.G.; project administration, I.S.; funding acquisition, I.S.

Funding: This paper was created at the Faculty of Materials Science and Technology within Project No. CZ.02.1.01/0.0/0.0/17_049/0008399 funded by the Ministry of Education, Youth and Sports of the Czech Republic; and within the students' grant project SP2020/88 supported at the VŠB—TU Ostrava by the Ministry of Education of the Czech Republic.

Conflicts of Interest: The authors declare no conflict of interest.

References

1. Verlinden, B.; Driver, J.; Samajdar, I.; Doherty, R.D. *Thermo-Mechanical Processing of Metallic Materials*, 1st ed.; Elsevier: Oxford, UK, 2007; 528p.
2. Kawulok, R.; Schindler, I.; Kawulok, P.; Rusz, S.; Opela, P.; Kliber, J.; Solowski, Z.; Čmiel, K.M.; Podolinsky, P.; Mališ, M.; et al. Transformation kinetics of selected steel grades after plastic deformation. *Metalurgija* **2016**, *55*, 357–360.
3. Timoshenkov, A. Influence of Deformation on Phase Transformation and Precipitation of Steels for Oil Country Tubular Goods. *Steel Res. Int.* **2014**, *6*, 954–967. [CrossRef]
4. Totten, G.; Xie, L.; Funatani, K. *Modeling and Simulation for Material Selection and Mechanical Design*; Marcel Dekker: Basel, Switzerland, 2004; 166p.
5. Kruglova, A.A.; Orlov, V.V. Effect of hot plastic deformation in the austenite interval on structure formation in low-alloyed—Carbon steels. *Met. Sci. Heat Treat.* **2007**, *12*, 556–560. [CrossRef]
6. Kawulok, R.; Schindler, I.; Kawulok, P.; Rusz, S.; Opěla, P.; Solowski, Z.; Čmiel, K.M. Effect of deformation on the CCT diagram of steel 32CrB4. *Metalurgija* **2015**, *54*, 473–476.
7. Grajcar, A.; Kuziak, R.; Zalecki, W. Designing of cooling conditions for Si-Al microalloyed TRIP steel on the basis of DCCT diagrams. *J. Achiev. Mater. Manuf. Eng.* **2011**, *45*, 115–124.
8. QForm Heat Treatment Process Analysis Software, for Forging. Available online: https://www.indiamart.com/proddetail/heat-treatment-6322730912.html (accessed on 25 February 2019).
9. Jagiełło, A.; Trzaska, J.; Dobrzański, L.A. Computer Software for modelling CCT Diagrams. *Czas. Technol. Mech.* **2008**, *105*, 87–94.
10. Domański, T.; Piekarska, W.; Kubiak, M.; Saternus, Z. Determination of the final microstructure during processingcarbon steel hardening. *Procedia Eng.* **2016**, *136*, 77–81. [CrossRef]
11. Kawulok, R.; Schindler, I.; Mizera, J.; Kawulok, P.; Rusz, S.; Opěla, P.; Podolinsky, P.; Čmiel, K.M. Transformation diagrams of selected steel grades with consideration of deformation effect. *Arch. Metall. Mater.* **2018**, *63*, 55–60.
12. Xie, H.J.; Wu, X.C.; Min, Y.A. Influence of Chemical Composition on Phase Transformation Temperature and Thermal Expansion Coefficient of Hot Work Die Steel. *J. Iron Steel Res. Int.* **2008**, *15*, 56–61. [CrossRef]
13. Calvo, J.; Jung, I.H.; Elwazri, A.M.; Bai, D.; Yue, S. Influence of the chemical composition on transformation behaviour of low carbon microalloyed steels. *J. Mater. Sci. Eng. A* **2009**, *520*, 90–96. [CrossRef]
14. Liu, S.K.; Yang, L.; Zhu, D.G.; Zhang, J. The Influence of the Alloying Elements upon the Transformation Kinetics and Morphologies of Ferrite Plates in Alloy Steels. *Metall. Mater. Trans. A* **1994**, *25*, 1991–2000. [CrossRef]
15. Mun, D.J.; Shin, E.J.; Choi, Y.W.; Lee, S.J.; Koo, Y.M. Effects of cooling rate, austenitizing temperature and austenite deformation on the transformation behavior of high-strength boron steel. *Mater. Sci. Eng. A* **2012**, *545*, 214–224. [CrossRef]
16. Zhizhong, H. *The Handbook of Steel and Its Heat Treatment Curve*; Defence Industry Press: Beijing, China, 1987.
17. Aranda, M.; Kim, B.; Rementeria, R.; Capdevila, C.; García de Andrés, C. Influence of prior austenit grain size to pearlite transformation in a hypoeuctectoid Fe-C-Mn steel. *Metall. Mater. Trans. A* **2014**, *4*, 1778–1786. [CrossRef]
18. Kawulok, P.; Podolinsky, P.; Kajzar, P.; Schindler, I.; Kawulok, R.; Ševčák, V.; Opěla, P. The influence of deformation and austenitization temperature on the kinetics of phase transformations during cooling of high-carbon steel. *Arch. Metall. Mater.* **2018**, *63*, 1743–1748.

19. Kawulok, P.; Schindler, I.; Mizera, J.; Kawulok, R.; Rusz, S.; Opěla, P.; Olszar, M.; Čmiel, K.M. The influence of a cooling rate on the evolution of microstructure and hardness of the steel 27MnCrB5. *Arch. Metall. Mater.* **2018**, *63*, 907–914.
20. Kawulok, R.; Kawulok, P.; Schindler, I.; Opěla, P.; Rusz, S.; Ševčák, V.; Solowski, Z. Study of the effect of deformation on transformation diagrams of two low-alloy manganese-chromium steels. *Arch. Metall. Mater.* **2018**, *63*, 1735–1741.
21. Grajcar, A.; Zalecki, W.; Skrzypczyk, P.; Kilarski, A.; Kowalski, A.; Kołodziej, S. Dilatometric study of phase transformations in advanced high-strength bainitic steel. *J. Therm. Anal. Calorim.* **2014**, *118*, 739–748. [CrossRef]
22. Grajcar, A.; Morawiec, M.; Zalecki, W. Austenite decomposition and precipitation behavior of plastically deformed low-Si microalloyed steel. *Metals* **2018**, *8*, 1028. [CrossRef]
23. Nürnberger, F.; Grydin, O.; Schaper, M.; Bach, F.W.; Koczurkiewicz, B.; Milenin, A. Microstructure transformations in tempering steels during continuous cooling from hot forging temperatures. *Steel Res. Int.* **2010**, *81*, 224–233. [CrossRef]
24. Jandová, D.; Vadovicová, L. Influence of deformation on austenite decomposition of steel 0.5C-1Cr-0.8Mn-0.3Si. In *Metal*; TANGER: Ostrava, Czech Republic, 2004.
25. Khlestov, V.M.; Konopleva, E.V.; McQueen, H.J. Effects of deformation and heating temperature on the austenite transformation to pearlite in high alloy tool steels. *Mater. Sci. Technol.* **2002**, *18*, 54–60. [CrossRef]
26. Yin, S.B.; Sun, X.J.; Liu, Q.Y.; Zhang, Z.B. Influence of Deformation on Transformation of Low-Carbon and High Nb-Containing Steel during Continuous Cooling. *J. Iron Steel Res. Int.* **2010**, *17*, 43–47. [CrossRef]
27. Opiela, M.; Zalecki, W.; Grajcar, A. Influence of plastic deformation on CCT-diagrams of new-developed microalloyed steel. *J. Achiev. Mater. Manuf. Eng.* **2012**, *51*, 78–89.
28. Mohamadizadeh, A.; Zarei-Hanzaki, A.; Heshmati-Manesh, S.; Imandoust, A. The effect of strain induced ferrite transformation on the microstructural evolutions and mechanical properties of aTRIP-assisted steel. *Mater. Sci. Eng. A* **2014**, *607*, 621–629. [CrossRef]
29. Du, L.X.; Yi, H.L.; Ding, H.; Liu, X.H.; Wang, G.D. Effects of Deformation on Bainite Transformation During Continuous Cooling of Low Carbon Steels. *J. Iron Steel Res.* **2006**, *13*, 37–39. [CrossRef]
30. Nadeiri, M. Influence of Hot Plastic Deformation and Cooling Rate on Martenzite and Bainite Start Temperatures in 22MnB5 steel. *Mater. Sci. Eng.* **2012**, *30*, 24–29.
31. Rusz, S.; Schindler, I.; Kawulok, P.; Kawulok, R.; Opěla, P.; Kliber, J.; Solowski, Z. Phase transformation and cooling curves of the mild steel influenced by previous hot rolling. *Metalurgija* **2016**, *55*, 655–658.
32. Yamamoto, S. Effects of austenite Grain Size and Deformation in the Unrecrystallized Austenite Region on Bainite Transformation Behavior and Microstructure. *ISIJ Int.* **1995**, *15*, 1020–1026. [CrossRef]
33. Kawata, H.; Fujiwara, K.; Takahashi, M. Effect of Carbon Content on Bainite Transformation Start Temperature in Low Carbon Fe–9Ni–C Alloys. *ISIJ Int.* **2017**, *57*, 1866–1873. [CrossRef]
34. Liu, Z.; Yao, K.F.; Liu, Z. Quantitative research on effects of stresses and strains on bainitic transformation kinetics and transformation plasticity. *Mater. Sci. Technol.* **2000**, *16*, 643–647. [CrossRef]
35. Xu, Y.; Xu, G.; Mao, X.; Zhao, G.; Bao, S. Method to Evaluate the Kinetics of Bainite Transformation in Low-Temperature Nanobainitic Steel Using Thermal Dilatation Curve Analysis. *Metals* **2017**, *7*, 330. [CrossRef]
36. He, B.B.; Xu, W.; Huang, M.X. Increase of martensite start temperature after small deformation of austenite. *Mater. Sci. Eng. A* **2014**, *609*, 141–146. [CrossRef]
37. Trzaska, J.; Dobrzañski, L.A. Modelling of CCT diagrams for engineering and constructional steels. *J. Mater. Process. Technol.* **2007**, *192–193*, 504–510. [CrossRef]
38. Vermeulen, W.; Zwaag, S.; Morris, P.; Weijer, T. Prediction of the continuous cooling transformation diagram of some selected steels using artificial neural networks. *Mater. Technol.* **1997**, *68*, 72–79. [CrossRef]
39. Motyčka, P.; Kövér, M. Evaluation methods of dilatometer curves of phase transformations. In *Comat*; Tanger Ltd.: Pilsen, Czech Republic, 2012; p. 1237.
40. Qiu, C.; Zwaag, S. Dilatation measurements of plain carbon steels and their thermodynamic analysis. *Steel Res. Int.* **1997**, *68*, 32–38. [CrossRef]
41. Rożniata, E.; Dziurka, R. The Phase transformations in hypoeutectoid steels Mn-Cr-Ni. *Arch. Metall. Mater.* **2015**, *60*, 497–502. [CrossRef]

42. OVAKO—Material Data Sheet of Steel Grade 100Cr6. Available online: https://steelnavigator.ovako.com/steel-grades/100cr6/ (accessed on 25 February 2019).
43. IMS—Data Sheet of 100Cr6 Steel—Normativa di Riferimento UNI 3097. Available online: https://www.ims.it/files/100Cr6.pdf (accessed on 20 February 2019).
44. Durand-Chare, M. *Microstructure of Steels and Cast Irons*, 1st ed.; Springer: Berlin, Germany, 2004; pp. 297–304.
45. Schindler, I.; Kawulok, P. Application possibilities of the plastometer Gleeble 3800 with simulation model Hydrawedge II at the VŠB-TU Ostrava. *Hut. Listy* **2013**, *66*, 85–90.
46. Mandziej, S.T. Physical simulation of metallurgical processes. *Mater. Technol.* **2010**, *44*, 105–119.
47. Alvarez, W.S. Microstructural Degradation of Bearing Steels. Ph.D. Thesis, Department of Materials Science and Metallurgy, University of Cambridge, Cambridge, UK, December 2014.
48. AUSA—SPECIAL STEELS—Material data sheet of Steel grade 100Cr6. Available online: https://www.ausasteel.com/fichas/Bearing-Steel-100Cr6-AUSA.pdf (accessed on 14 October 2019).
49. Glazunov, A. On the structure of troostite. *Collect. Czech. Chem. Commun.* **1933**, *5*, 76–83. [CrossRef]
50. Müştak, O. Characterization of SAE 52100 Bearing Steel for Finite Element Simulation of through-Hardening Proces. Master's Thesis, Master of Science in Metallurgical and Materials Engineering Department, Middle East Technical University, Ankara, Turkey, September 2014.
51. Bhadeshia, H.K.D.H. Steels for bearings. *Prog. Mater. Sci.* **2012**, *57*, 268–435. [CrossRef]
52. Ellerman, A.; Scholtes, B. The strength differential effect in different heat treatment conditions of the steels 42CrMoS4 and 100Cr6. *Mater. Sci. Eng. A* **2015**, *620*, 262–272. [CrossRef]
53. Perez, M.; Sidoroff, C.; Vincent, A.; Esnouf, C. Microstructural evolution of martensitic 100Cr6 bearing steel during tempering: From thermoelectric power measurements to the prediction of dimensional changes. *Acta Mater.* **2009**, *57*, 3170–3181. [CrossRef]
54. Ryttberg, K.; Wedel, M.K.; Recina, V.; Dahlman, P.; Nyborg, L. The effect of cold ring rolling on the evolution of microstructure and texture in 100Cr6 steel. *Mater. Sci. Eng. A* **2010**, *527*, 2431–2436. [CrossRef]
55. Nikravesh, M.; Naderi, M.; Akbari, G.H. Influence of hot plastic deformation and cooling rate on martensite and bainite start temperatures in 22MnB5 steel. *Mater. Sci. Eng. A* **2012**, *540*, 24–29. [CrossRef]
56. Maalekian, M.; Lendinez, M.L.; Kozeschnik, E.; Brantner, H.P.; Cerjak, H. Effect of hot plastic deformation of austenite on the transformation characteristics of eutectoid carbon steel under fast heating and cooling conditions. *Mater. Sci. Eng. A* **2007**, *454–455*, 446–452. [CrossRef]
57. Wang, H.-Z.; Yang, P.; Mao, W.-M.; Lu, F.-Y. Effect of hot deformation of austenite on martensitic transformation in high manganese steel. *J. Alloys Compd.* **2013**, *558*, 26–33. [CrossRef]
58. Mola, J.; Ren, M. On the hardness of high carbon ferrous martensite. *IOP Conf. Ser. Mater. Sci. Eng.* **2018**, *373*, 012004. [CrossRef]
59. Xu, X.; Zwaag, S.; Xu, W. The effect of martensite volume fraction on the scratch and abrasion resistance of a ferrite-martensite dual phase steel. *Wear* **2016**, *348–349*, 80–88.

© 2020 by the authors. Licensee MDPI, Basel, Switzerland. This article is an open access article distributed under the terms and conditions of the Creative Commons Attribution (CC BY) license (http://creativecommons.org/licenses/by/4.0/).

Article

Microstructure, Texture, and Strength Development during High-Pressure Torsion of CrMnFeCoNi High-Entropy Alloy

Werner Skrotzki [1,*], Aurimas Pukenas [1], Eva Odor [2], Bertalan Joni [2], Tamas Ungar [2,3], Bernhard Völker [4], Anton Hohenwarter [4], Reinhard Pippan [5] and Easo P. George [6,7]

1. Institute of Solid State and Materials Physics, Technische Universität Dresden, D-01062 Dresden, Germany; aurimas.pukenas@tu-dresden.de
2. Department of Materials Physics, Eötvös University, H-1117 Budapest, Hungary; odoreva94@gmail.com (E.O.); jonibertalan@gmail.com (B.J.); ungar@ludens.elte.hu (T.U.)
3. Materials Performance Centre, School of Materials, The University of Manchester, Manchester M13 9PL, UK
4. Department of Materials Science, Chair of Materials Physics, Montanuniversität Leoben, A-8700 Leoben, Austria; bernhard.voelker@mcl.at (B.V.); anton.hohenwarter@unileoben.ac.at (A.H.)
5. Erich Schmid Institute of Materials Science, Austrian Academy of Sciences, 8700 Leoben, Austria; reinhard.pippan@oeaw.ac.at
6. Materials Science and Technology Division, Oak Ridge National Laboratory, Oak Ridge, TN 37831, USA; georgeep@ornl.gov
7. Department of Materials Science and Engineering, University of Tennessee, Knoxville, TN 37996, USA
* Correspondence: werner.skrotzki@tu-dresden.de; Tel.: +49-351-463-35144; Fax: +49-351-463-37048

Received: 27 March 2020; Accepted: 20 April 2020; Published: 24 April 2020

Abstract: The equiatomic face-centered cubic high-entropy alloy CrMnFeCoNi was severely deformed at room and liquid nitrogen temperature by high-pressure torsion up to shear strains of about 170. Its microstructure was analyzed by X-ray line profile analysis and transmission electron microscopy and its texture by X-ray microdiffraction. Microhardness measurements, after severe plastic deformation, were done at room temperature. It is shown that at a shear strain of about 20, a steady state grain size of 24 nm, and a dislocation density of the order of 10^{16} m^{-2} is reached. The dislocations are mainly screw-type with low dipole character. Mechanical twinning at room temperature is replaced by a martensitic phase transformation at 77 K. The texture developed at room temperature is typical for sheared face-centered cubic nanocrystalline metals, but it is extremely weak and becomes almost random after high-pressure torsion at 77 K. The strength of the nanocrystalline material produced by high-pressure torsion at 77 K is lower than that produced at room temperature. The results are discussed in terms of different mechanisms of deformation, including dislocation generation and propagation, twinning, grain boundary sliding, and phase transformation.

Keywords: high-entropy alloy; high-pressure torsion; microstructure; texture; phase transformation; strength

1. Introduction

High-entropy alloys (HEAs) represent a new class of single-phase multi-element (≥5) solid solution alloys with near-equiatomic concentrations of the individual elements [1]. In some cases, due to the large number of constituent elements the contribution of configurational entropy to the Gibbs free energy is high enough to suppress compound formation and phase separation. Among the wide variety of reported HEAs with simple crystal structures, such as face-centered cubic (FCC), body-centered cubic (BCC), and hexagonal close-packed (HCP), the most thoroughly investigated alloy is the quinary equiatomic FCC HEA CrMnFeCoNi [2] often referred to as Cantor alloy. This alloy is stable as an FCC single-phase solid solution at high temperatures above about 1073 K [3,4], but decomposes into several different metallic and intermetallic phases during annealing at intermediate temperatures [4–6]. Application of hydrostatic pressure at room temperature (RT) transforms it to the HCP structure, see recent review [7]. The transformation is sluggish and occurs over a large pressure range. During depressurization part of the HCP phase transforms back to FCC. The back-transformation is suppressed up to a high annealing temperature. Although the onset pressure of the HCC phase transformation reported in the literature varies much, some trends are evident and may account for the large scatter. With decreasing hydrostaticity, i.e., increasing deviatoric stress, the onset pressure decreases, while for decreasing grain size it increases. These results are in agreement with finite-temperature ab initio calculations showing that the HCP structure is energetically favored at low temperatures [8–10].

In the FCC solid-solution state the Cantor alloy exhibits certain noteworthy mechanical properties, including simultaneous strength and ductility increase with decreasing temperature leading to outstanding fracture toughness at cryogenic temperatures, see recent review [11]. To unravel the deformation mechanisms of these advanced alloys numerous investigations of microstructure and texture have been carried out on polycrystalline samples deformed by tension, wire drawing, swaging, compression, rolling, and high-pressure torsion (HPT), for a recent review see [12]. It is found that at RT and below, above a certain stress (equivalently strain) in addition to dislocation slip mechanical twinning contributes to deformation. Twinning is more pronounced at cryogenic temperatures. There is texture formation during deformation, but its intensity is quite low. The observed dislocation dissociation and texture are typical of medium/low stacking fault energy (SFE) metals and alloys.

The present paper extends recent work of the authors on microstructure and texture development of CrMnFeCoNi HEA processed by HPT at RT [13]. Here a similar detailed study was conducted at liquid nitrogen temperature (LNT, 77 K). Moreover, the microhardness of the HPT deformed materials was measured at RT. Emphasis is put on the differences in microstructure, texture and strength observed for the two HPT deformation temperatures.

2. Experimental

The CrMnFeCoNi HEA was synthesized from high-purity elements (>99.9 wt.%) by arc melting and drop casting under pure argon atmosphere into cylindrical molds (diameter: 25.4 mm, length: 127 mm). The drop-cast ingots were encapsulated in evacuated quartz ampules and homogenized for 48 h at 1200 °C. Discs with a radius $r = 4$ mm and an initial thickness $t_i \approx 0.8$ mm were cut from the cast and homogenized ingots and deformed by HPT [14]. During HPT the shear strain along the radius is approximately given by $\gamma = 2\pi rn/<t>$, where n is the number of rotations (Rot n) and $<t> = (t_i + t_f)/2$ with t_f = final thickness. With this approximation the maximum error in shear strain is less than 15%. HPT under a quasi-hydrostatic pressure of 7.8 GPa was conducted at RT and LNT at a nominal speed of 0.2 rotations/min yielding a maximum initial shear strain rate of 10^{-1} s^{-1} at the outer radius. The initial grain size was several hundred micrometers, while the saturation microstructure after HPT at RT was estimated by transmission electron microscopy to consist of grains with a size of about 50 nm. During HPT at RT the alloy does not decompose as confirmed by 3D atom probe tomography [4].

To investigate the microstructure, X-ray diffraction (XRD) measurements were carried out in a special high-resolution diffractometer dedicated to line-profile-analysis using Cu $K\alpha_1$ radiation, for details see [13,15]. The measurements are done along the radius of the HPT discs, i.e., at positions where the shear strains are different. The line profiles are evaluated by using the convolutional multiple whole profile (CMWP) procedure [16]. The measured diffraction pattern is matched by the theoretically calculated and convoluted profile functions accounting for the effects of size, distortion, planar defects and instrumental effects, while the background is determined separately. Because of the double-crystal high resolution diffractometer used, the instrumental effect is neglected [15]. The parameters obtained by using the CMWP method characterizing the substructure are the area average crystallite (subgrain) size $<x>_{area}$, dislocation density ρ, dislocation character q (edge versus screw), dislocation arrangement parameter $M = R_e \sqrt{\rho}$ (R_e = effective outer cut-off radius of dislocations), twin density β (number of twin boundary planes within hundred {111} lattice planes parallel to the twin boundary) and average distance between adjacent twin boundaries $d_{Tw} = 100\, d_{\{111\}}/\beta$. The M parameter is closely related to the dipole character of the dislocation arrangements [16]. In plastic deformation there are usually equal numbers of plus and minus dislocations, where the plus-minus pairs form dislocation dipoles. When the two dislocations in the pairs are close or far from each other the dipole character is strong or weak and the corresponding M parameter close to unity or considerably larger than one, respectively. The CMWP procedure can handle more than one phase [16]. This option is used to evaluate the dislocation, size and planar defect parameters in the FCC and HCP phases at the same time. The two separate phases are identified unequivocally by indexing and using the non-overlapping peaks in the evaluation process. The overlapping peaks are also evaluated as two separate peaks, where the intensities are given according to the volume fractions of the two phases deduced from the non-overlapping peaks.

In addition, standard bright-field images and diffraction patterns were taken with a JEOL 2100F image-side C_S-corrected transmission electron microscope (JEOL Ltd., Tokyo, Japan) at an acceleration voltage of 200 kV. Specimen preparation for transmission electron microscopy (TEM) was done by conventional electropolishing.

In order to resolve the local texture of the deformed discs along the radial direction, i.e., as a function of increasing shear strain from the middle to the edge of the disc, two-dimensional X-ray micro-diffraction was performed using the system D8 Discover (BRUKER AXS GmbH, Karlsruhe, Germany). The system is equipped with an Euler cradle including x-y-z-stage, a laser-video adjustment system, a low-power micro-focus X-ray tube IµS (Cu $K\alpha$ radiation, spot size about 100 µm) and a two-dimensional detector VÅNTEC 2000. The intensities of the measured Debye-Scherrer rings were integrated along their curvature to calculate the pole figures (200, 220, and 111) on a 5° × 5° grid. For calculation of the orientation distribution function (ODF) with these pole figures Multex 3 [17] and LaboTex software [18] were used. The Euler angles given are in the Bunge notation [19] with crystal and sample reference systems defined as $x\|$ shear direction, $y\|$ shear plane normal and $z\|$ transverse direction yielding an ODF representation appropriate for simple shear [20]. The textures are represented by $\varphi_2 = 45°$ ODF-sections, which for FCC metals contain all major shear components.

Vickers hardness (HV0.5) values were measured at RT using a Zwick/Roell-ZHµ-Indentec microhardness tester (Zwick/Roell GmbH & Co. KG, Ulm, Germany). The measurements were done along the radius of the disc-shaped samples from the center to the edge, i.e., as a function of shear strain. The hardness values presented here correspond to those radial positions where the XRD measurements were made.

3. Results

The most striking feature observed by XRD is the phase transformation from FCC to HCP at LNT (Figure 1). Unfortunately, the XRD pattern at LNT is not suitable to quantitatively estimate the volume fraction of the HCP phase. A very rough estimate made by comparing the intensities of non-overlapping HCP and FCC peaks yields a volume fraction of about 30%. To overcome this problem, diffraction of high-energy synchrotron radiation and Rietveld analysis is under way.

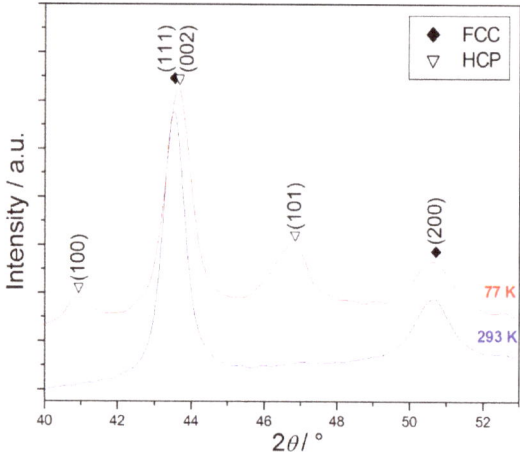

Figure 1. Typical X-ray diffractograms after high-pressure torsion (HPT) at room temperature (293 K) and 77 K (shear strain $\gamma \approx 130$) clearly showing the phase transformation at 77 K.

The results of the CMWP analysis show that during HPT of CrMnFeCoNi HEA a very fast refinement of the FCC microstructure takes place. The crystallite size ($<x>_{area}$) after HPT at RT and 77 K reaches a very low steady-state value of 24 nm after a shear strain of about 20 (Figure 2). Simultaneously, the dislocation density saturates at 3×10^{16} m^{-2} (Figure 3a). Surprisingly, it is lower for HPT at 77 K (10^{16} m^{-2}). The dislocation character q, after HPT at RT, changes from near edge-type ($q_{edge} = 1.4$) to near screw-type ($q_{screw} = 2.4$) and then slightly decreases to 2, which is the saturation value after HPT at 77 K (Figure 3b). The dipole character M is quite weak ($M > 1$) and after HPT at RT saturates at about 6 (Figure 3c). It is even weaker after HPT at 77 K, about twice as large. The twin density after HPT at RT reaches a maximum value of 2% at $\gamma \approx 20$ (Figure 4a) leading to a mean twin separation distance d_{Tw} smaller than $<x>_{area}$ for $10 < \gamma < 60$ (Figure 4b). After HPT at 77 K instead of twinning a phase transformation from FCC to HCP is observed (Figures 1 and 5a). There are clear diffraction peaks of the HCP phase ((100) and (101)) found in the X-ray diffractogram (Figure 1) and the TEM diffraction pattern (Figure 5b). Unfortunately, the HCP diffraction spots cannot be selected separately to locally identify the HCP phase by dark-field imaging. However, in comparison with high resolution TEM images of CrFeCoNi and CrCoNi medium-entropy alloys (MEAs) [21,22] tensile tested at cryogenic temperatures the lamellar features observed may be reasonably attributed to the HCP phase. The spotty TEM diffraction pattern for HPT at LNT in comparison to that at RT [4] indicates that with decreasing temperature deformation becomes more heterogeneous. It should be mentioned that despite of differences in the production of the starting HEA material resulting in different grain sizes and probably slightly different textures, and differences in HPT parameters (sample size, hydrostatic pressure, rotation speed), the steady-state crystallite sizes and dislocation and twin densities as well as microhardness values reported agree quite well with those of the present study, both for HPT at RT [23,24] and LNT [23]. An exception are the results given in [25], where evidently the steady-state has not been reached.

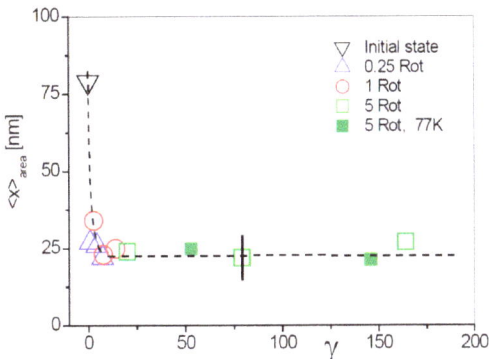

Figure 2. Grain size <x_{area}> after HPT at RT (open symbols) and at 77 K (filled symbols) versus shear strain γ.

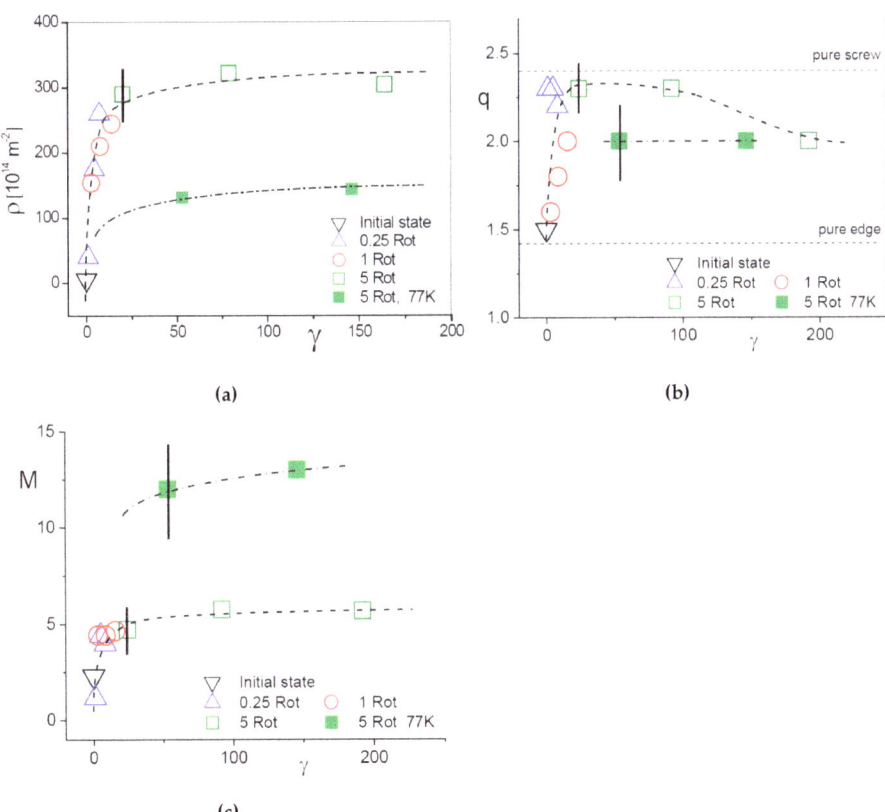

Figure 3. (a) Dislocation density ρ, (b) dislocation character q, and (c) dislocation arrangement parameter M versus shear strain γ. Open and filled symbols refer to HPT at RT and 77 K, respectively.

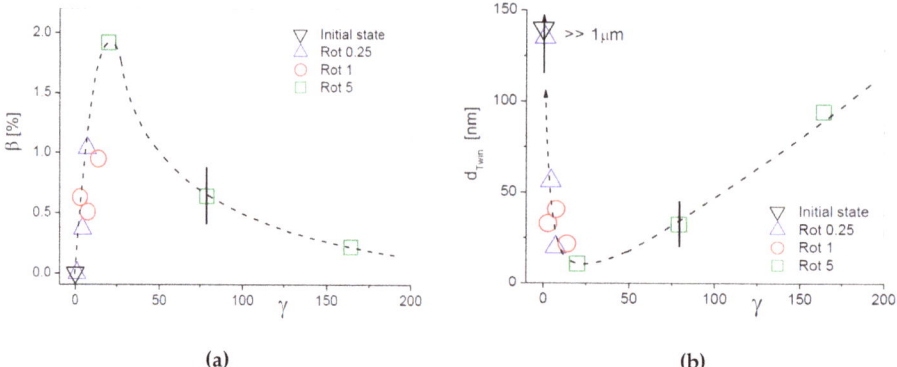

Figure 4. (a) Twin density β and (b) average distance between adjacent twin boundaries d_{Tw} versus shear strain γ after HPT at RT.

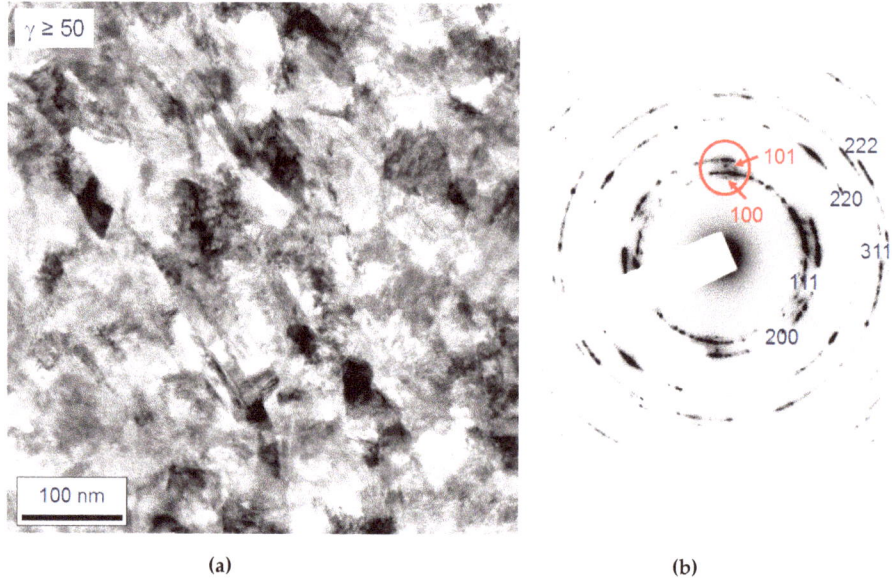

Figure 5. (a) Bright-field transmission electron microscopy (TEM) image representing the microstructure in the steadystate regime after HPT at liquid nitrogen temperature (LNT, 77 K), (b) corresponding diffraction pattern clearly showing 100 and 101 diffraction spots belonging to the hexagonal close-packed (HCP) phase (indicated by circle and arrows in red).

The texture observed after HPT is the typical shear texture observed in FCC metals and alloys. Because of the initial coarse-grain structure the texture measured by micro-diffraction is statistically reliable only after a shear strain of about 5. As shown in Figure 6a, the texture of CrMnFeCoNi HEA after HPT at RT is quite weak (below 2.5 mrd). The main texture components developed by dislocation slip and mechanical twinning are {111}<112> (A_1^*/A_2^*) and {112}<110> (B/$\bar{B}$), with (B/$\bar{B}$) dominating (Figure 6b). With increasing twinning activity the volume fraction of these components decreases. When twinning ceases texture intensity initially increases but then decreases again at high shear strains. The texture is less developed after HPT at 77 K.

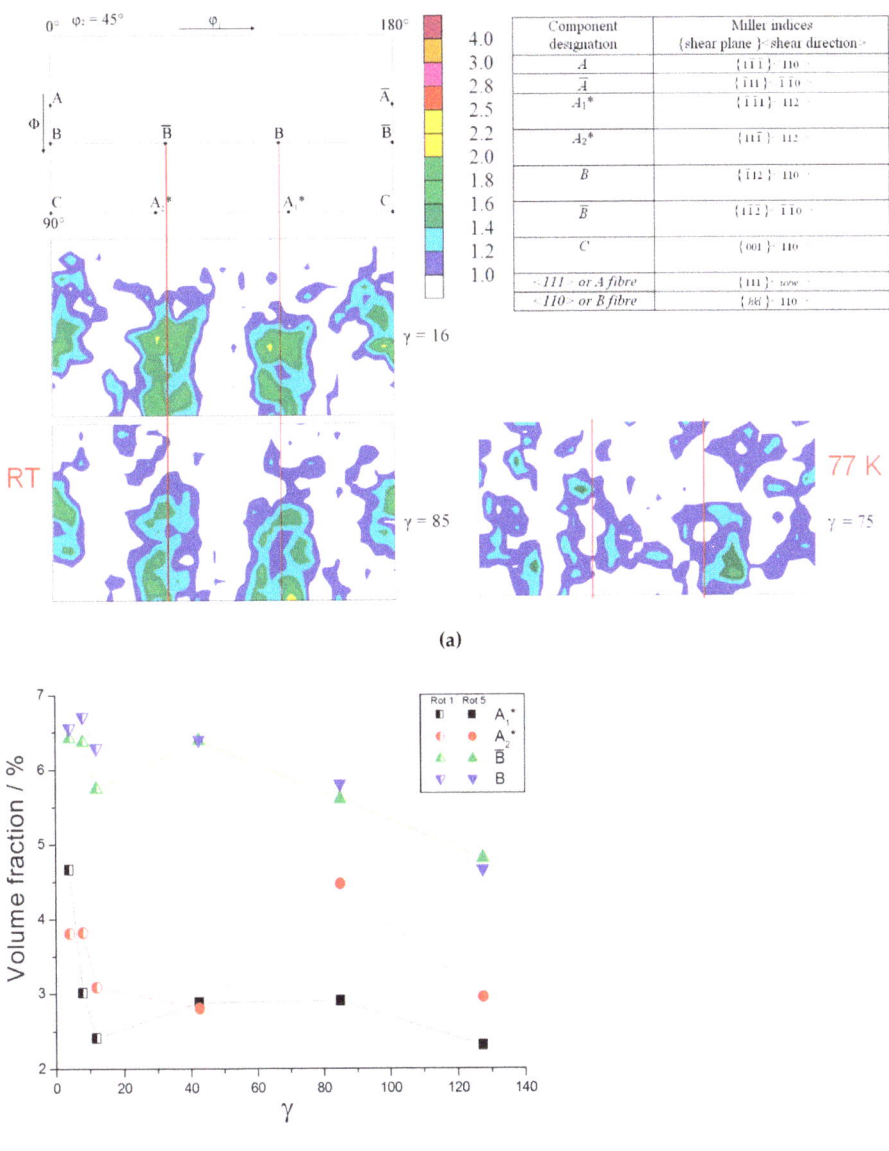

Figure 6. (a) Texture after HPT at RT and 77 K represented as $\varphi_2 = 45°$ orientation distribution function (ODF)-sections at different shear strains γ. Intensities are given in multiples of a random orientation distribution. The key figure shows the main texture components (for component designation see table) that develop during simple shear of face-centered cubic (FCC) metals. (b) Volume fraction of texture components (15° spread from ideal positions) after HPT at RT versus shear strain γ.

The Vickers hardness measured at RT increases with shear strain and saturates at about the same strain level as the crystallite size and dislocation density (Figure 7). The saturation hardness is about 500 HV and 450 HV for samples HPT-deformed at RT and 77 K, respectively. Reasons for this anomalous strength behaviour are discussed below.

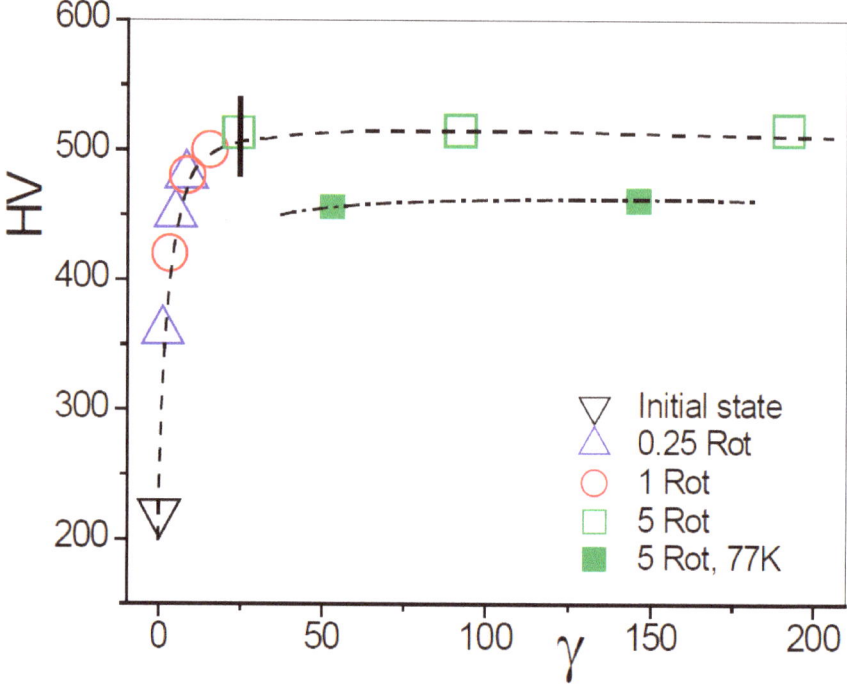

Figure 7. Microhardness measured at RT of samples HPT-deformed at RT (open symbols) and 77 K (filled symbols) versus shear strain γ.

4. Discussion

4.1. Microstructure Development

The strong grain refinement of the FCC HEA during HPT at RT is correlated to the high mechanical twinning activity in the medium SFE ($\leq$30 mJm^{-2} [8,26,27]) alloy [11]. Based on TEM investigations a mechanism of fast grain refinement via primary and secondary twinning has been proposed by Wang et al. [28] for brass. Primary twins by accumulation of a high density of dislocations evolve into curved high angle grain boundaries from which secondary twins are emitted. The emission of secondary twins further refines the grains and transforms the elongated grains into equiaxed nanograins. At a certain shear strain twinning ceases, may be because grain boundary sliding takes over as the deformation mechanism. A supporting indication of this is the further randomization of texture [29,30]. The generally low texture strength may be also caused by twinning. In addition to twinning, in CrFeCoNi MEA Wu et al. [31] observed nanobands and attributed the significant grain refinement to concurrent nanoband subdivision and deformation twinning. Instead of mechanical twinning, during HPT at LNT a strain-induced phase transformation takes place. Nano HCP lamellae seem to have a similar effect on grain refinement as twin lamellae. Surprisingly, despite of phase transformation and in particular the 220 °C lower HPT temperature, the steady state crystallite size in the FCC phase is not smaller than that after HPT at RT. Reasons for this finding may be: (i) dynamic recrystallization, and/or (ii) static recrystallization/grain growth, and/or (iii) back-transformation of part of the HCP phase to FCC during pressure release, heating up to RT, and storage at this temperature. All these processes depend on purity of the material, SFE, temperature and degree of deformation, and annealing temperature. A clear distinction between dynamic and static behaviour can be only made by in-situ studies. Evidence for process (ii) has been provided for pure metals HPT-deformed at cryogenic temperature (100 K) [32], but also for $Cr_{26}Mn_{20}Fe_{20}Co_{20}Ni_{14}$ HPT deformed at LNT [33]. This HEA has a low SFE (3.5 mJm^{-2}) and a moderate melting temperature T_m = 1557 K. Thus, HPT at LNT (homologous temperature T/T_m = 0.05) leads to an increased stored energy of deformation promoting "self-annealing" at RT (T/T_m = 0.19). In contrast, the SFE of the Cantor alloy is much higher (see above) and therefore, "self-annealing" is less likely and so far to the best of our knowledge has not been reported for this alloy. Moreover, neither in the Cantor alloy nor in other HEAs nothing is known yet about process (iii).

It is also surprising that the dislocation density after HPT at LNT is lower than that after HPT at RT. A plausible reason would be recovery during "self-annealing", as was found for $Cr_{26}Mn_{20}Fe_{20}Co_{20}Ni_{14}$ [33] in combination with a lowering of the microhardness. However, the same effect has not been observed for the Cantor alloy [23]. Therefore, it has been suggested, that the shear produced by the deformation-induced HCP phase formation leads to a reduction in strain and, thus, to a reduction in dislocation density in the FCC phase.

The medium SFE of the Cantor alloy leads to widely dissociated dislocations (screws: ≅4 nm, edges: ≅6 nm) [26]. After relatively low compressive strains at 77 K, dislocations with long screw segments and large kinks having mixed character are seen on the {111} planes suggesting that the mobilities of edge and screw dislocations are not significantly different [26]. Here, we also find that during HPT at RT and 77 K the dislocation character tends to become more screw-like with increasing shear strain taking a value of $q \approx 2$ at large shear strain. It is somewhat surprising that the steady-state dipole character of dislocations observed is low and becomes even lower at LNT (RT: $M \approx 5$; 77 K: $M \approx 12$). This may be caused by the wide dislocation dissociation suppressing thermally activated edge dislocation climb and screw dislocation cross slip.

Hydrostatic pressure applied to the Cantor alloy stabilizes the HCP structure [7]. To transform the FCC phase to HCP a certain onset pressure is necessary depending on hydrostaticity and grain size. On the one hand, by using different media to apply pressure, the onset pressure at RT decreases from 22.1 GPa (helium) to 6.9 (silicone oil) and 2.2–6.6 GPa (amorphous boron), i.e., with increasing non-hydrostaticity [34,35] (see also effect on iron [36]). On the other hand, for a given medium, silicone oil, decreasing the grain size from 5 µm to about 0.01 µm increases the onset pressure from 6.9 to 12.3 GPa [35]. As the FCC to HCP martensitic transformation takes place by slip of 1/6 <112> partial dislocations on every second {111} plane [37], it is surprising, that in the present case under high shear stress, HPT at RT under a pressure of 7.8 GPa did not lead to the transformation. The reason may be the fast grain refinement into the nano range suppressing the transformation similar to deformation twinning [38]. However, lowering the HPT temperature to 77 K, i.e., lowering the SFE [8,10], favours the transformation at the pressure applied.

4.2. Texture Formation

The brass-type texture in shear at RT with strong B/$\overline{B}$ components agrees with results of texture simulations on HPT samples of nanocrystalline Pd-10at.%Au alloy [39] and rolling of nanocrystalline Ni-18.7at.%Fe alloy [40] despite the much lower SFE of the Cantor alloy. These simulations strongly suggest that nanoplasticity is determined by slip of partial dislocations emitted from the grain boundaries also leading to twinning. Twinning may have led to the almost equal intensities of the twinning related A_1* and A_2* shear components. Recent simulations also show that grain boundary sliding leads to randomization of the texture while maintaining the texture signature [29,30]. Thus, grain boundary sliding in the nano-grained HEA during RT-HPT may have led to the weak texture observed.

The texture after HPT at LNT is characterized by many low intensity maxima spread almost randomly in the ODF (Figure 6a). This may be caused by the heterogeneous deformation at LNT indicated by the spotty TEM diffraction pattern of Figure 5b. According to the discussion above, recrystallization by "self-annealing" may be excluded.

4.3. Strength Development

The enormous grain refinement during HPT of the Cantor alloy leads to a nanocrystalline material with a crystallite size of about 24 nm and a dislocation density of the order of 10^{16} m^{-2}. These values are comparable to those of nanocrystalline Pd-10at.%Au produced by a bottom–up technique (consolidation of inert gas condensated powder) and subjected to HPT [39]. In that case grain coarsening from an initial crystallite size of 16 nm led to a higher steady-state value, while the high initial dislocation density decreased to a lower steady-state value. Thus, HPT leads to a steady-state microstructure by either grain refinement or grain coarsening. In the following, we will pursue the issue of which mechanisms determine the strength of the nanocrystalline Cantor alloy measured by microhardness at RT.

The transition from grain size softening to grain size hardening takes place at a grain size of about 20 nm [39]. From the microhardness of 510 HV a maximum stress $\sigma_{max} \approx 10$ HV/3 = 1.7 GPa for the RT HPT-deformed material (Figure 7) can be estimated, which represents about the maximum strength achievable at RT in the polycrystalline Cantor alloy. This value is about 1/20 of the theoretical strength $\sigma_{th} \approx M_T G/2\pi$, with the shear modulus G = 79 GPa [26] and the Taylor factor M_T = 3.07 for a random orientation distribution. In the case of nanocrystalline Pd-10at.%Au a slightly higher value 1/15 was measured [39]. The strength determined here from microhardness is about 15% lower than that measured in tension (1.95 GPa [4]) and compression (2.0 GPa [41]). This discrepancy may be due to the generally used formula for conversion of microhardness to strength, but does not change the conclusions drawn below.

For the material deformed by HPT at LNT the strength at RT (1.5 GPa) is about 10% lower than for that deformed by HPT at RT. Scheriau et al. [42] made a similar observation for a modified 316L austenitic steel. To search for the reason, it is necessary to look for the differences in texture and microstructure after HPT at the two temperatures. The texture is quite weak, almost random, after large-strain HPT at RT and LNT. Therefore, the small texture difference observed may not account for the strength anomaly. With regard to microstructure the estimate of the crystallite size is about the same, while the dislocation density is lower and the dipole character is weaker after HPT at LNT compared to HPT at RT. In addition, based on TEM and texture investigations the microstructure is quite inhomogeneous. The large scatter in local nanohardness measured by Podolskiy et al. [23] supports this statement. Moreover, the material consists of a certain volume fraction of HCP phase after HPT at LNT. In an HEA of different composition ($Cr_{20}Mn_6Fe_{34}Co_{34}Ni_6$), which is prone to the same phase transformation, compression of micro pillars shows that the HCP phase is harder than the FCC one [43]. This also explains the higher strength measured for the dual phase alloy $Cr_{10}Mn_{30}Fe_{50}Co_{10}$ [44]. Here, this harder phase cannot be responsible for the lower strength measured, rather it may have indirectly led to the inhomogeneous microstructure with lower average dislocation density in the FCC phase evidently determining the strength anomaly. However, the inhomogeneity of the microstructure may be just a low temperature effect. For more speculations about the strength anomaly and special experiments to prove its existence under in-situ HPT conditions, the reader is referred to the recent work of Podolskiy et al. [23].

For the nanocrystalline Cantor alloy, as dislocation sources in the nanograins are unlikely, the strength may be determined either by nucleation of full dislocations and/or partial dislocations at grain boundaries (i), or by propagation of the dislocations through the nano grains and getting absorbed by the opposing grain boundary. In the latter case, the dislocations experience lattice friction (ii), and interac with solutes (iii), remnant dislocations (iv), SFs, twin boundaries, and HCP lamellae (v). The resulting yield stresses for mechanisms (i) and (iv) are approximately given by the following equations:

Full dislocation emission [45]:

$$\sigma_L = M_T \frac{Gb}{d} \qquad (1)$$

Partial dislocation emission [45]:

$$\sigma_P = M_T \left(\frac{\alpha - 1}{\alpha} \frac{\gamma_{SF}}{b} + \frac{1}{3} \frac{Gb}{d} \right) \qquad (2)$$

Dislocation passing [46]:

$$\sigma_\rho = M_T \frac{Gb}{2\pi} \sqrt{\rho} \qquad (3)$$

With $b = b (1/2 <110>) = 0.255$ nm [26], $\alpha = d/\delta$ ($\delta \approx \frac{b^2}{12\pi G \gamma_{SF}} = 18\,b$ equilibrium dislocation splitting [45]), $\gamma_{SF} = 30$ mJ/m² SFE [26], for the grain size d = 24 nm and $\rho = 3 \times 10^{16}$/m² at RT the equations above yield $\sigma_L = 2.6$ GPa, $\sigma_P = 1.1$ GPa, and $\sigma_\rho = 1.7$ GPa. The friction stress (Peierls stress at 0 K) has been estimated by density functional theory calculations to be $\sigma_F = M_F\,0.178$ GPa $= 0.546$ GPa [47]. Due to thermally activated kink formation it is negligible at RT [41]. The solid solution strengthening at 0 K is $\sigma_{SS} \approx 0.4$ GPa [41]. Due to thermal activation the stress contribution at RT is negligible, too [41]. The density of SFs and twin lamellae is low at RT due to nano grain size (Figure 4). Consequently, their strengthening effect can be neglected. Hence, the stress for propagation of dislocations at RT is $\sigma_\rho \approx 1.7$ GPa. For the material that is HPT-deformed at LNT, because of a lower dislocation density, only the interaction stress of dislocations σ_ρ changes to 1.2 GPa. Moreover, there is an additional contribution from the interaction of dislocations with HCP lamellae. Comparing these values with σ_{max} it is concluded that the yield stress of the nanocrystalline Cantor alloy at RT is predominantly determined by dislocation–dislocation interaction. A similar conclusion has been drawn by Podolskiy et al. [23] and Heczel et al. [24]. Moreover, Podolskiy et al. [23] argued that the smaller strength at cryogenic HPT processing can be understood as an indirect consequence of the shear-induced FCC to HCP martensitic phase transformation. This transformation provides a significant part of the shear strain, which must not be supplied by dislocation activity in the FCC phase and, thus, may allow for a reduction in dislocation density. The corresponding reduction of Taylor hardening is almost equal to the reduction in macroscopic strength so that other contributions to it appear to be negligible. To study the effect of the FCC to HCP martensitic phase transformation on the strength of the Cantor alloy in more detail, HPT at different pressures at LNT producing different volume fractions of HCP phase is under way.

The strengthening mechanism of the nanocrystalline Cantor alloy differs from that of nanocrystalline Pd-10at.%Au alloy, where the strength is mainly determined by the emission of full and partial dislocations from grain boundaries [39]. However, the effect of partial dislocation slip on texture development is the same. This also holds for grain boundary sliding as a strain contributing mechanism for grain sizes at the transition to grain boundary softening.

Apparently, the classical strength approach describes quite well the results on nanoplasticity presented here. However, considering all structure and strength phenomena taking place in HEAs and austenitic steels during severe plastic deformation at different temperatures and thermal treatments, there is no conclusive picture yet [4,23,33,42,48]. This requires in-depth studies.

5. Conclusions

Based on investigations of microstructure, texture and strength on the HPT-deformed prototype CrMnFeCoNi HEA the following conclusions are drawn:

(1) HPT at RT and LNT leads to strong grain refinement of this HEA. Severe mechanical twinning at RT and martensitic phase transformation from FCC to HCP structure at LNT in addition to dislocation slip may cause this.
(2) Correlated with a steady-state grain size in the nano range is a high remnant dislocation density. These dislocations are predominantly screw-type and show a weak dipole character due to a wide dissociation into Shockley partials. HPT at LNT leads to an inhomogeneous microstructure with a lower dislocation density compared to HPT at RT.

(3) The weak texture developed at RT is a dominant brass-type shear texture indicating that the deformation during SPD mainly occurs by partial dislocation slip accompanied by twinning and grain boundary sliding. Due to an inhomogeneous deformation at LNT coupled with phase transformation the texture becomes almost random.

(4) The strength of the nanocrystalline HEA is mainly controlled by dislocation–dislocation interaction. For the material HPT-deformed at LNT with a lower dislocation density there is an additional contribution from HCP lamellae which act as further dislocation barriers.

Author Contributions: Conceptualization, W.S., T.U., A.H., R.P. and E.P.G.; methodology, W.S., T.U. and A.H.; software, T.U.; validation, formal analysis, A.P., E.O., B.J., T.U. and A.H.; investigation, A.P., E.O., B.J., B.V. and A.H.; writing—review and editing, W.S., T.U., A.H., R.P. and E.P.G.; supervision, W.S., T.U., A.H. and R.P.; funding acquisition, A.H. and R.P. All authors have read and agreed to the published version of the manuscript.

Funding: This research was funded by European Research Council (ERC), grant number 340185 USMS, Austrian Science Fund (FWF), project number P26729-N19 and Open Access Funding by the Publication Fund of the TU Dresden.

Acknowledgments: Funding for this work (R.P.) has been provided by the European Research Council under ERC Grant Agreement No. 340185 USMS and by the Austrian Science Fund (FWF) in the framework of research project P26729-N19 (A.H. and B.V.). E.P.G. is supported by the U.S. Department of Energy, Office of Science, Basic Energy Sciences, Materials Sciences and Engineering Division.

Conflicts of Interest: The authors declare no conflict of interest.

Notice of Copyright: This manuscript has been co-authored by UT-Battelle, LLC under Contract No. DE-AC05-00OR22725 with the U.S. Department of Energy. The United States Government retains and the publisher, by accepting the article for publication, acknowledges that the United States Government retains a non-exclusive, paid-up, irrevocable, worldwide license to publish or reproduce the published form of this manuscript, or allow others to do so, for United States Government purposes. The Department of Energy will provide public access to these results of federally sponsored research in accordance with the DOE Public Access Plan (http://energy.gov/downloads/doe-public-access-plan).

References

1. Yeh, J.W.; Chen, S.K.; Lin, S.J.; Gan, J.Y.; Chin, T.S.; Shun, T.T.; Tsau, C.H.; Chang, S.Y. Nanostructured high-entropy alloys with multiple principal elements: Novel alloy design concepts and outcomes. *Adv. Eng. Mater.* **2004**, *6*, 299–303. [CrossRef]
2. Cantor, B.; Chang, I.T.H.; Knight, P.; Vincent, A.J.B. Microstructural development in equi atomic multicomponent alloys. *Mater. Sci. Eng. A* **2004**, *375–377*, 213–218. [CrossRef]
3. Otto, F.; Yang, Y.; Bei, H.; George, E.P. Relative effects of enthalpy and entropy on the phase stability of equiatomic high-entropy alloys. *Acta Mater.* **2013**, *61*, 2628–2638. [CrossRef]
4. Schuh, B.; Mendez-Martin, F.; Völker, B.; George, E.P.; Clemens, H.; Pippan, R.; Hohenwarter, A. Mechanical properties, microstructure and thermal stability of a nanocrystalline CoCrFeMnNi high-entropy alloy after severe plastic deformation. *Acta Mater.* **2015**, *96*, 258–268. [CrossRef]
5. Otto, F.; Dlouhy, A.; Pradeep, K.G.; Kubenova, M.; Raabe, D.; Eggeler, G.; George, E.P. Decomposition of the single-phase high-entropy alloy CrMnFeCoNi after prolonged anneals at intermediate temperatures. *Acta Mater.* **2016**, *112*, 40–52. [CrossRef]
6. Pickering, E.J.; Munoz-Moreno, R.; Stone, H.J.; Jones, N.G. Precipitation in the equiatomic high-entropy alloy CrMnFeCoNi. *Scr. Mater.* **2016**, *113*, 106–109. [CrossRef]
7. Zhang, F.; Lou, H.; Cheng, B.; Zeng, Z.; Zeng, Q. High-pressure induced phase transitions in high-entropy alloys: A review. *Entropy* **2019**, *21*, 239. [CrossRef]
8. Huang, S.; Li, W.; Lu, S.; Tian, F.; Shen, J.; Holmström, E.; Vitos, L. Temperature dependent stacking fault energy of FeCrCoNiMn high entropy alloy. *Scr. Mater.* **2015**, *108*, 44–47. [CrossRef]
9. Ma, D.; Grabowski, B.; Körmann, F.; Neugebauer, J.; Raabe, D. *Ab initio* thermodynamics of the CoCrFeMnNi high entropy alloy: Importance of entropy contributions beyond the configurational one. *Acta Mater.* **2015**, *100*, 90–97. [CrossRef]
10. Zhao, S.; Stocks, G.M.; Zhang, Y. Stacking fault energies of face-centered cubic concentrated solid solution alloys. *Acta Mater.* **2017**, *134*, 334–345. [CrossRef]

11. George, E.P.; Curtin, W.A.; Tasan, C.C. High entropy alloys: A focused review of mechanical properties and deformation mechanisms. *Acta Mater.* **2020**, *188*, 435–474. [CrossRef]
12. Sathiaraj, G.D.; Pukenas, A.; Skrotzki, W. Texture formation in face-centered cubic high-entropy alloys. *J. Alloy. Compd.* **2020**, *826*, 15183. [CrossRef]
13. Skrotzki, W.; Pukenas, A.; Joni, B.; Odor, E.; Ungar, T.; Hohenwarter, A.; Pippan, R.; George, E.P. Microstructure and texture evolution during severe plastic deformation of CrMnFeCoNi high-entropy alloy. *IOP Conf. Ser. Mater. Sci. Eng.* **2017**, *194*, 012028. [CrossRef]
14. Pippan, R.; Scheriau, S.; Hohenwarter, A.; Hafok, M. Advantages and limitations of HPT: A review. *Mater. Sci. Forum* **2008**, *584–586*, 16–21. [CrossRef]
15. Ungar, T.; Ott, S.; Sanders, P.G.; Borbely, A.; Weertman, J.R. Dislocations, grain size and planar faults in nanostructured copper determined by high resolution X-ray diffraction and a new procedure of peak profile analysis. *Acta Mater.* **1998**, *46*, 3693–3699. [CrossRef]
16. Ribarik, G.; Joni, B.; Ungar, T. Global optimum of microstructure parameters in the CMWP line-profile-analysis method by combining Marquardt-Levenberg and Monte-Carlo procedures. *J. Mater. Sci. Technol.* **2019**, *35*, 1508–1514. [CrossRef]
17. *Multex Manual*; Multex 3; Bruker AXS GmbH: Karlsruhe, Germany, 2008.
18. Pawlik, K. Determination of the orientation distribution function from pole figures in arbitrarily defined cells. *Phys. Stat. Sol. B* **1986**, *134*, 477–483. [CrossRef]
19. Bunge, H.-J. Zur Darstellung allgemeiner Texturen. *Z. Met.* **1965**, *56*, 872–874.
20. Toth, L.S.; Molinari, A. Tuning a self consistent viscoplastic model by finite element results—I. Modeling. *Acta Metall. Mater.* **1994**, *42*, 2453–2458. [CrossRef]
21. Miao, J.; Slone, C.E.; Smith, T.M.; Niu, C.; Bei, H.; Ghazisaeidi, M.; Pharr, G.M.; Mills, M.J. The evolution of the deformation substructure in a Ni-Co-Cr equiatomic solid solution alloy. *Acta Mater.* **2017**, *132*, 35–48. [CrossRef]
22. Lin, Q.; Liu, J.; An, X.; Wang, H.; Zhang, Y.; Liao, X. Cryogenic-deformation-induced phase transformation in an FeCoCrNi high-entropy alloy. *Mater. Res. Lett.* **2018**, *6*, 236–243. [CrossRef]
23. Podolskiy, A.V.; Shapovalov, Y.O.; Tabachnikova, E.D.; Tortika, A.S.; Tikhonovsky, M.A.; Joni, B.; Odor, E.; Ungar, T.; Maier, S.; Rentenberger, C.; et al. Anomalous evolution of strength and microstructure of high-entropy alloy CoCrFeNiMn after high-pressure torsion at 300 and 77 K. *Adv. Eng. Mater.* **2020**, *22*, 1900752. [CrossRef]
24. Heczel, A.; Kawasaki, M.; Labar, J.L.; Jang, J.; Langdon, T.G.; Gubicza, J. Defect structure and hardness in nanocrystalline CoCrFeMnNi high-entropy alloy processed by high-pressure torsion. *J. Alloy. Compd.* **2017**, *711*, 143–154. [CrossRef]
25. Zherebtsov, S.; Stepanov, N.; Ivanisenko, Y.; Shaysultanov, D.; Yurchenko, N.; Klimova, M.; Salishchev, G. Evolution of microstructure and mechanical properties of a CoCrFeMnNi high-entropy alloy during high-pressure torsion at room and cryogenic temperatures. *Metals* **2018**, *8*, 123. [CrossRef]
26. Okamoto, N.L.; Fujimoto, S.; Kambara, Y.; Kawamura, M.; Chen, Z.M.; Matsunshita, H.; Tanaka, K.; Inui, H.; George, E.P. Size effect, critical resolved shear stress, stacking fault energy, and solid solution strengthening in the CrMnFeCoNi high-entropy alloy. *Sci. Rep.* **2016**, *6*, 35836. [CrossRef] [PubMed]
27. Yeh, J.W. Physical metallurgy of high-entropy alloys. *JOM* **2015**, *67*, 2254–2261. [CrossRef]
28. Wang, Y.B.; Liao, X.Z.; Zhao, Y.H.; Lavernia, E.J.; Ringer, S.P.; Horita, Z.; Langdon, T.G.; Zhu, Y.T. The role of stacking faults and twin boundaries in grain refinement of a Cu-Zn alloy processed by high-pressure torsion. *Mater. Sci. Eng. A* **2010**, *527*, 4959–4966. [CrossRef]
29. Zhao, Y.; Toth, L.S.; Massion, R.; Skrotzki, W. Role of grain boundary sliding in texture evolution for nanoplasticity. *Adv. Eng. Mater.* **2017**, *20*, 1–9. [CrossRef]
30. Toth, L.S.; Skrotzki, W.; Zhao, Y.; Pukenas, A.; Braun, C.; Birringer, R. Revealing grain boundary sliding from textures of a deformed nanocrystalline Pd-Au alloy. *Materials* **2018**, *11*, 190. [CrossRef]
31. Wu, W.; Song, M.; Ni, S.; Wang, J.; Liu, Y.; Liu, B.; Liao, X. Dual mechanisms of grain refinement in a FeCoCrNi high-entropy alloy processed by high-pressure torsion. *Sci. Rep.* **2017**, *7*, 46720. [CrossRef]
32. Edalati, K.; Cubero-Sesin, J.M.; Alhamidi, A.; Mohamed, I.F. Influence of severe plastic deformation at cryogenic temperature on grain refinement and softening of pure metals: Investigation using high-pressure torsion. *Mater. Sci. Eng. A* **2014**, *613*, 103–110. [CrossRef]

33. Moon, J.; Qi, Y.; Tabachnikova, E.; Estrin, Y.; Choi, W.-M.; Joo, S.-H.; Lee, B.-J.; Podolskiy, A.; Tikhonovsky, M.; Kim, H.S. Microstructure and mechanical properties of high-entropy alloy $Co_{20}Cr_{26}Fe_{20}Mn_{20}Ni_{14}$ processed by high-pressure torsion at 77 K and 300 K. *Sci. Rep.* **2018**, *8*, 11074. [CrossRef] [PubMed]
34. Zhang, F.; Wu, Y.; Lou, H.B.; Zeng, Z.D.; Prakapenka, V.B.; Greenberg, E.; Yan, J.Y.; Okasinski, J.S.; Liu, X.J.; Liu, Y.; et al. Polymorphism in a high-entropy alloy. *Nat. Commun.* **2017**, *8*, 15687. [CrossRef] [PubMed]
35. Zhang, F.; Lou, H.; Chen, S.; Chen, X.; Zeng, Z.; Yan, J.; Zhao, W.; Wu, Y.; Lu, Z.; Zeng, Q. Effects of non-hydrostaticity and grain size on pressure-induced phase transformation of the CoCrFeMnNi high-entropy alloy. *J. Appl. Phys.* **2018**, *124*, 115901. [CrossRef]
36. Von Bergen, N.; Boehler, R. Effect of non-hydrostaticity on the—Transition in iron. *High Press. Res.* **1990**, *6*, 133–140.
37. Yang, X.-S.; Sun, S.; Ruan, H.-H.; Shi, S.-Q.; Zhang, T.Y. Shear and shuffling accomplishing polymorphic fcc γ→hcp ε→bct α martensitic phase transformation. *Acta Mater.* **2017**, *136*, 347–354. [CrossRef]
38. Wu, S.W.; Wang, G.; Yi, J.; Jia, Y.D.; Hussain, I.; Zhai, Q.J.; Liaw, P.K. Strong grain-size effect on deformation twinning of an $Al_{0.1}$CoCrFeNi high-entropy alloy. *Mater. Res. Lett.* **2017**, *5*, 276–283. [CrossRef]
39. Skrotzki, W.; Eschke, A.; Joni, B.; Ungar, T.; Toth, L.S.; Ivanisenko, Y.; Kurmanaeva, L. New experimental insight into the mechanisms of nanoplasticity. *Acta Mater.* **2013**, *61*, 7271–7284. [CrossRef]
40. Li, L.; Ungar, T.; Toth, L.S.; Skrotzki, W.; Wang, J.D.; Ren, J.; Choo, H.; Fogarassy, Z.; Zhou, X.Z.; Liaw, P.K. Shear-coupled grain growth and texture development in a nanocrystalline Ni-Fe alloy during cold rolling. *Metall. Mat. Trans. A* **2016**, *47*, 6632–6644. [CrossRef]
41. Podolskiy, A.V.; Schafler, E.; Tabachnikova, E.D.; Tikhonovsky, M.A.; Zehetbauer, M.J. Thermally activated deformation of nanocrystalline and coarse grained CoCrFeNiMn high entropy alloy in the temperature range 4.2–350 K. *Low Temp. Phys.* **2018**, *44*, 1245–1253. [CrossRef]
42. Scheriau, S.; Zhang, Z.; Kleber, S.; Pippan, R. Deformation mechanisms of modified 316L austenitic steel subjected to high pressure torsion. *Mater. Sci. Eng. A* **2011**, *528*, 2776–2786. [CrossRef]
43. Chen, S.; Oh, H.S.; Gludovatz, B.; Kim, S.J.; Park, E.S.; Zhang, Z.; Ritchie, R.O.; Yu, Q. Real-time observations of TRIP-induced ultrahigh strain hardening in a dual-phase CrMnFeCoNi high-entropy alloy. *Nat. Commun.* **2020**, *11*, 826. [CrossRef] [PubMed]
44. Li, Z.; Pradeep, K.G.; Deng, Y.; Raabe, D.; Tasan, C.C. Metastable high-entropy dual-phase alloys overcome the strength-ductility trade-off. *Nature* **2016**, *534*, 227–231. [CrossRef] [PubMed]
45. Asaro, R.S.; Suresh, S. Mechanistic models for the activation volume and rate sensitivity in metals with nanocrystalline grains and nano-scale twins. *Acta Mater.* **2005**, *53*, 3369–3382. [CrossRef]
46. Seeger, A. *Handbuch Der Physik*; Springer: Berlin/Heidelberg, Germany, 1958; Volume VII/2.
47. Patriarca, L.; Ojha, A.; Sehitoglu, H.; Chumlyakov, Y.I. Slip nucleation in single crystal FeNiCoCrMn high entropy alloy. *Scr. Mater.* **2016**, *112*, 54–57. [CrossRef]
48. Renk, O.; Hohenwarter, A.; Eder, K.; Kormout, K.S.; Cernay, J.M.; Pippan, R. Increasing the strength of nanocrystalline steels by annealing: Is segregation necessary? *Scr. Mater.* **2015**, *95*, 27–30. [CrossRef]

© 2020 by the authors. Licensee MDPI, Basel, Switzerland. This article is an open access article distributed under the terms and conditions of the Creative Commons Attribution (CC BY) license (http://creativecommons.org/licenses/by/4.0/).

Article

Mechanisms of Grain Structure Evolution in a Quenched Medium Carbon Steel during Warm Deformation

Dmitriy Panov, Olga Dedyulina, Dmitriy Shaysultanov, Nikita Stepanov, Sergey Zherebtsov * and Gennady Salishchev

Laboratory of Bulk Nanostructured Materials, Belgorod State University, 85 Pobeda Str., 803015 Belgorod, Russia; Panov_D@bsu.edu.ru (D.P.); dedyulina@bsu.edu.ru (O.D.); shaysultanov@bsu.edu.ru (D.S.); stepanov@bsu.edu.ru (N.S.); salishchev@bsu.edu.ru (G.S.)
* Correspondence: zherebtsov@bsu.edu.ru; Tel.: +74-7225-854-16

Received: 13 June 2020; Accepted: 26 June 2020; Published: 29 June 2020

Abstract: The as-quenched medium-carbon low-alloy Fe-0.36wt.%C-1wt.%Cr steel was subjected to warm deformation via uniaxial compression at temperatures of 400–700 °C and strain rates of 10^{-4}–10^{-2} s^{-1}. At low temperatures (400–550 °C), the microstructure evolution was mainly associated with dynamic recovery with the value of activation energy of 140 ± 35 kJ/mol. At higher temperatures (600–700 °C), dynamic recrystallization was developed, and activation energy in this case was 243 ± 15 kJ/mol. The presence of nanoscale carbide particles in the structure at temperatures of 400–600 °C resulted in the appearance of threshold stresses. A two-component <001>//compression direction (CD) and <111>//CD deformation texture was formed during deformation. Deformation at the low temperatures resulted in the formation of elongated ferritic grains separated mainly by high-angle boundaries (HAB) with a strong <001>//CD texture. The grains with the <111>//CD orientation were wider in comparison with those with the <001>//CD orientation. The development of substructure in the form of low-angle boundaries (LAB) networks was also observed in the <111>//CD grains. The development of dynamic recrystallization restricted the texture formation. The processing map for warm deformation of the 0.36C-1Cr steel was constructed.

Keywords: deformation behavior; warm deformation; texture; microstructure; dynamic recrystallization; processing map

---

1. Introduction

Medium-carbon low-alloy (MCLA) steels are a workhorse material of modern industry due to their low cost and high performance. Generally, MCLA steels are used as structural materials for the machine components' production. The main problem of these steels is the lack of ductility and toughness at low temperatures. The mechanical properties can be improved significantly due to microstructure refinement obtained by thermomechanical treatments at relatively low temperatures [1–5]. The martensitic initial structure was found to be very promising to receive a fine-grained structure by warm deformation [5]. For instance, according to Reference [6], an as-quenched Fe-1.2wt.%C steel had better workability in comparison with that in an initially spheroidized condition. One of the attractive strategies of warm deformation to receive an excellent balance of strength, ductility and toughness in a wide range of temperatures is the formation of an elongated fine-grained (EFG) structure with nanoscale carbide particles and a strong deformation texture [7], that has recently been applied in industry [8]. This microstructure can be obtained using warm deformation of as-quenched MCLA steels with the martensitic structure [7–10]—this working is sometimes defined as tempforming [11]. During warm deformation, various processes, including precipitation and/or spheroidization of carbide particles, dynamic recovery, dynamic recrystallization, elongation of ferritic grans along the deformation direction and texture formation, can develop in MCLA steels with the martensitic structure [11–13]. The completeness and order of these processes substantially depend on warm deformation parameters, such as temperature, strain rate and strain magnitude [14].

In comparison with ferrite-perlite steels, production of ultrafine grain structure in as-quenched low- or high-alloyed steels usually requires relatively low strain (up to 50%) due to early development of dynamic recrystallization (DRX) [12,15–17]. Besides, further grain refinement during DRX can be obtained at increased strain rate and/or decreased deformation temperature [18]. For example, warm deformation of an as-quenched Fe-0.45wt.%C steel at 550–600 °C resulted in the formation of an ultra-fine-grained ferritic structure with nanoscale carbides due to the development of DRX [15,19,20]. The nanoscale carbides can impede boundary motion through the pinning effect [10,21,22], thereby stabilizing the structure.

On the other hand, warm deformation of an as-quenched Fe-0.4wt.%C-2wt.%Si-1wt.%Cr-1wt.%Mo steel during rolling at 500 °C resulted in the formation of a fibrous fine-grained structure with a strong deformation texture that provided excellent mechanical properties [14]. Furthermore, a <110>//rolling direction (RD) deformation texture in ferrite became weaker while ferritic grains became coarser and more equiaxed with an increase in deformation temperature. In addition, an increase in deformation temperature resulted in the weakening of a <110>//RD texture component during swaging of an as-quenched Fe-0.43wt.%C-0.63wt.%Cr steel in the temperature range of 500 to 600 °C [23].

Analysis of the available literature suggested that the microstructure evolution during warm deformation of as-quenched MCLA steels was mainly studied for quite a narrow temperature interval, while low temperatures remain almost unexplored. Meanwhile, evolution of the martensite microstructure during warm deformation also deserves additional attention, particularly due to the high density of lattice defects in such materials. The relationship between texture formation and DRX during warm deformation should also be studied more comprehensively. Thus, the aim of this work was to examine the martensite microstructure evolution and mechanical behavior of an as-quenched MCLA steel during warm deformation at 400–700 °C.

In the present article, mechanical behavior during compression tests is studied, and activation energy and threshold stress are calculated at different temperatures (Section 3.1 Mechanical Behavior). Additionally, the texture and structure evolution are investigated (Section 3.2 Microstructure Evolution). Finally, a processing map is constructed and discussed in association with the observed structure evolution (Section 3.3 Processing Map Analysis).

2. Materials and Methods

A medium-carbon low-alloy steel with a nominal chemical composition of (in wt.%) Fe-0.36C-0.91Cr-0.62Mn-0.34Si-0.21Ni-0.15Cu-0.017P-0.022S was used as the program material. Compression specimens measuring 8 × 8 × 10 mm^3 were cut from an as-received hot rolled bar using an electric-discharge machine (EDM), Sodick AQ300L (Sodick Co, Fukui, Japan). To produce a fully martensite structure (which is defined hereafter as the initial one), the specimens were oil-quenched from 860 °C. Compression tests were carried out at 400, 450, 500, 550, 600, 650 or 700 °C, with an initial strain rate of 10^{-2}, $5 \cdot 10^{-2}$, $1.3 \cdot 10^{-3}$, $5 \cdot 10^{-3}$ or 10^{-4} s^{-1} using an Instron 300XL test machine equipped with a radial heating furnace (Instron, Norwood, MA, USA). At least two samples were tested for each condition. The samples were heated up with a rate of 4°/min and held for 10 min at deformation temperatures before the tests. The temperature during compression was controlled by a thermocouple attached to the specimen side.

Microstructure was studied in specimens compressed to a true strain (ε) of 0, 0.3, 0.77 or 1.15 at a strain rate of $1.3 \cdot 10^{-3}$ s^{-1} and temperatures of 400, 500, 600 or 700 °C. At 400 °C, the maximum attained true strain was 0.97. Samples for electron back scattering diffraction (EBSD) and transmission electron microscopy (TEM) analysis were cut from the central part of the deformed specimens by the EDM, then mechanically thinned to 0.1 mm and polished at room temperature using a twin-jet electro-polisher TenuPol-5 (Struers ApS, Ballerup, Denmark) in a mixture of 10% perchloric acid and 90% glacial acetic acid at a voltage of 29 V. The observation plane was normal to the compression direction. EBSD analysis was carried out using a FEI Nova NanoSEM 450 scanning electron microscope (SEM) (FEI Company, Hillsboro, Oregon, USA) equipped with an EDAX Hikari EBSD camera (EDAX, Mahwah, NJ, USA). EBSD maps measuring 20 × 20 µm² or 45 × 45 µm² were obtained with a step size of 50 nm. TEM observations were performed using a Jeol JEM 2100 microscope (JEOL, Tokyo, Japan) operating at an accelerating voltage of 200 kV.

3. Results and Discussion

3.1. Mechanical Behavior

Compression curves of the as-quenched Fe-0.36wt.%C-1wt.%Cr steel obtained at different conditions expectedly show a decrease in flow stress with an increase in deformation temperatures and/or decrease in strain rates (Figure 1). Each curve for temperatures of 400–550 °C shows a peak soon after yielding, and the magnitude of the peak decreased with an increase in temperature (Figure 1a–e). At higher temperatures, the peaks were observed at high strain rates only, i.e. $1.3 \cdot 10^{-3}$–10^{-2} s^{-1} (Figure 1e–g).

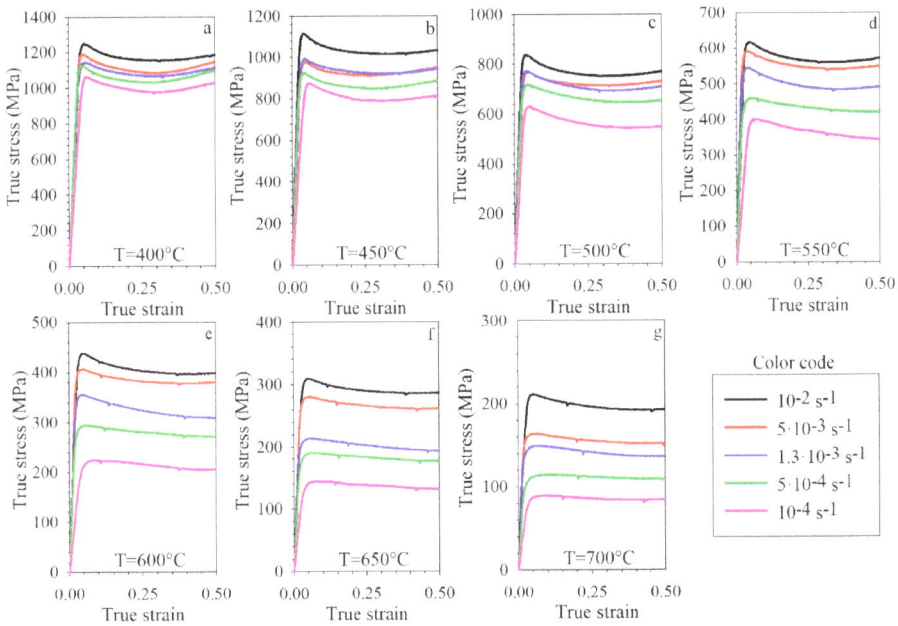

Figure 1. Compression curves of as-quenched Fe-0.36wt.%C-1wt.%Cr steel at temperatures of 400 (a), 450 (b), 500 (c), 550 (d), 600 (e), 650 (f) and 700 °C (g), and different strain rates.

After reaching the peak stress, a work-softening stage was observed (Figure 1a–e). The work-softening became less pronounced with an increase in temperature or with a decrease in strain rate. At 500 °C and low strain rates (10^{-4}–$5 \cdot 10^{-3}$ s^{-1}), or at higher temperatures of 550–700 °C and any examined strain rates, a steady-stage flow stage was observed after softening. Some increase in flow stress at $\varepsilon \sim 0.35$ at relatively low temperatures (400–500 °C) can be associated with friction between the die and the specimen surfaces (Figure 1a,b).

The observed mechanical behavior is typical of as-quenched steels during deformation in the temperature interval of tempering [12,15,20]. Specifically, the stress peak and subsequent work-softening stage can be ascribed to the development of dynamic recrystallization [24,25]. However, the most pronounced work-softening was observed for specimens strained at low temperatures and high strain rates, where dynamic recrystallization was not confirmed by microstructure analysis (see Section 3.2). Another factor which can influence the mechanical behavior can be associated with the martensite decomposition during deformation. Heating to rather low deformation temperatures does not necessarily result in a complete transformation of martensite into a mixture of ferrite and carbides. Therefore, this process proceeds during warm working, increasing both the peak flow stress and the work-softening rate [15].

The results of compression tests showed a strong dependence of mechanical behavior on temperature and strain rate. To establish mechanisms of plastic deformation operated at different conditions, the activation energy of deformation was determined using the Arrhenius equation [26,27]:

$$\dot{\varepsilon} = A\left(\frac{\sigma}{G}\right)^n \exp\left(-\frac{Q}{RT}\right), \qquad (1)$$

where $\dot{\varepsilon}$ is the strain rate (s^{-1}), σ is the flow stress (MPa), G denotes the shear modulus (MPa), R is the universal gas constant (8.314 J·mol^{-1}·K^{-1}), T is the absolute temperature (K), Q is the activation energy of deformation (kJ·mol^{-1}) and A and n are constants.

Since the steady-stage flow stage cannot be recognized at some compression curves, the value of stress corresponding to $\varepsilon = 0.25$ was used for activation energy analysis (Figure 1). Dependences of flow stress on deformation temperature for different strain rates revealed an inflection point at 550–600 °C (Figure 2) that may suggest different controlling mechanisms of deformation in two temperature intervals: 400–550 °C and 600–700 °C.

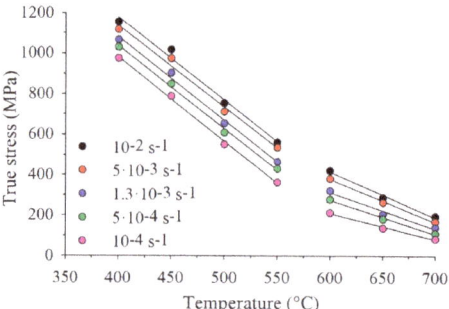

Figure 2. Relationship between true stress ($\varepsilon \sim 0.25$) and compression temperature.

The shear modulus value was estimated using the following equation [26]:

$$G = 6.4 * 10^4 \left(1 - 0.81 \frac{(T-300)}{1810}\right) - \begin{cases} 3.2 * 10^{-2}(T-573)^2, \text{if } T > 573K \\ 2.4 * 10^{-2}(T-923)^2, \text{if } T > 923K \end{cases} \qquad (2)$$

The activation energy of plastic deformation in the temperature range 600–700 °C was determined using the following relations:

$$n = \frac{\partial \ln(\dot{\varepsilon})}{\partial \ln(\sigma/G)}\bigg|_T, \qquad (3)$$

and:

$$Q = R \cdot n \cdot \frac{\partial \ln(\sigma/G)}{\partial (1/T)}\bigg|_{\dot{\varepsilon}}. \qquad (4)$$

The values of n and Q can be found as the slope of the $\ln(\dot{\varepsilon}) - \ln(\sigma/G)$ and $\ln(\sigma/G) - 10^3/T$ plots (Figure 3), respectively. The value of n changed from 6.8 to 5.5 for 600 and 700 °C, respectively.

The activation energy was found to be 316 ± 35 kJ/mol using the mean value of n = 6.2 and by averaging different values of Q at five strain rates.

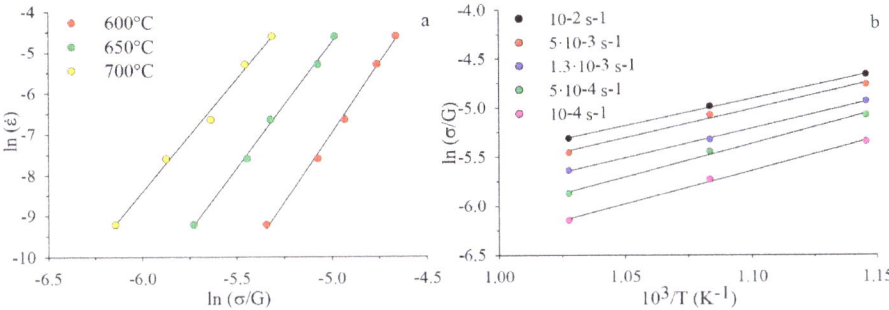

Figure 3. Relationships between $\ln(\dot{\varepsilon}) - \ln(\sigma/G)$ (**a**) and $\ln(\sigma/G) - 10^3/T$ (**b**) in the higher temperature (600–700 °C) interval.

In the lower temperature interval (400–550 °C), the n values were unusually high: 27 at 400 °C and 10 at 550 °C. The activation energy Q = 470 kJ/mol was much higher than that for the interval 600–700 °C; generally, very high values of n and Q are associated with the presence of the threshold stresses σ_{th}. The values of σ_{th} were determined as the stress level below which no strain rate can be detected [28]. When the stress level exceeds σ_{th}, the strain rate is greater than zero. Therefore, the values of σ_{th} can be estimated by extrapolating the $\dot{\varepsilon}^{1/n} - \sigma$ linear relations (plotting for different n) to the zero strain rate [28,29] (Figure 4).

The obtained values of σ_{th} and n for different temperatures are presented in Table 1. At low temperatures, σ_{th} was very high; for example, at 400–450 °C, the threshold stress was a half of the flow stress. However, with an increase in temperature, σ_{th} quickly decreased, reaching σ_{th} = 43 MPa at 600 °C and disappearing at 700 °C. The n values also decreased from 14 at 400 °C to 6 at 600–650 °C.

Equation (1) with the threshold correction can be expressed as follows:

$$\dot{\varepsilon} = A\left(\frac{\sigma - \sigma_{th}}{G}\right)^n \exp\left(-\frac{Q}{RT}\right). \tag{5}$$

The corrected activation energy for temperatures 400–550 °C was found to be Q = 140 ± 35 kJ/mol (n = 9.5 was accepted for the lower temperature domain; Table 1). The obtained value of Q agrees reasonably with the activation energy for pipe self-diffusion in α-iron (174 kJ/mol) [26] or with data obtained in Reference [30] for a Fe-0.6wt.%O steel by creep testing at temperatures of 550–650 °C (185 ± 15 kJ/mol). In the latter case, the main deformation mechanism was non-conservative movement of dislocations also controlled by self-diffusion along the dislocation cores. The higher activation energy reported in Reference [30] can be associated with greater stability of iron oxides against coagulation, in comparison to that of carbides that results in more dispersed oxide particles and a smaller interparticle space.

In the higher temperature interval of deformation (600–700 °C), the obtained value of Q = 243 ± 15 kJ/mol at n = 6.0 is quite close to the activation energy for volume self-diffusion in α-iron (251 kJ/mol) [26]. A little bit higher activation energy in our case can be associated with greater stability of iron oxides against coagulation in comparison to that of carbides that results in more dispersed oxide particles and a smaller interparticle space. For example, it was reported in Reference [30] that an increase in creep testing temperature to 700 °C and corresponding coagulation of oxides resulted in an increase in the activation energy to 259 ± 15 kJ/mol in the Fe-0.6O steel.

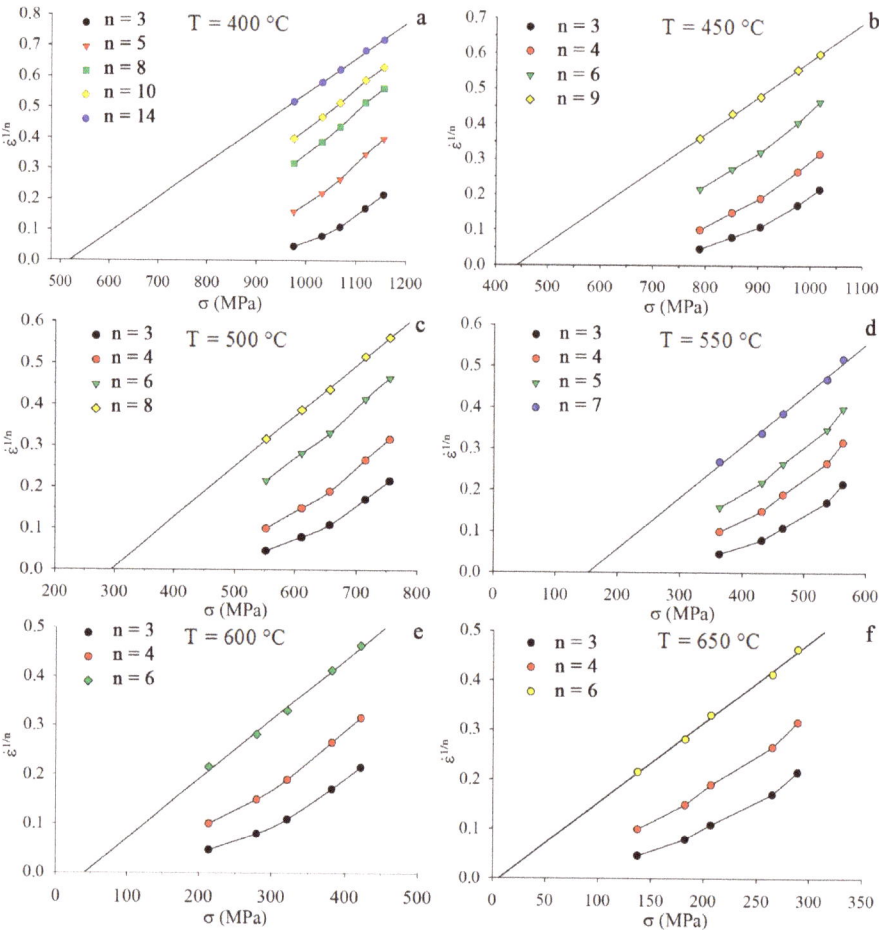

Figure 4. Relationships between $\dot{\varepsilon}^{1/n}$ and σ for temperatures of 400 (**a**), 450 (**b**), 500 (**c**), 550 (**d**), 600 (**e**), and 650 °C (**f**).

Table 1. The values of threshold stress (σ_{th}), n and corrected activation energy of deformation (Q).

Temperature, °C	σ_{th}, MPa	n	Q, kJ·mol^{-1}
400	522	12	
450	443	9	140 ± 35
500	296	8	(at n = 9.5)
550	154	7	
600	43	6	
650	7	6	243 ± 15
700	-	-	(at n = 6.0)

Even higher activation energy of 372.8 kJ·mol^{-1} at temperatures of 550–700 °C and strain rates of 0.001–1.0 s^{-1} was reported in Reference [15] for the as-quenched Fe-0.45wt.%C steel during warm deformation. This difference can be caused by the higher carbon content and a shorter exposure time before deformation (10 min in our case versus 3 min in Reference [15]), resulting in a higher amount of finer carbides which provides an effective obstacle for dislocation and boundary motion during deformation. However, the activation energy value at 550–700 °C in a 1.2%C steel was found to be higher in the as-quenched steel (Q = 331.56 kJ/mol) due to a more pronounced pinning effect by fine cementite particles in comparison to that in the spheroidized condition (Q = 297.94 kJ/mol) [6].

3.2. Microstructure Evolution

Microstructure of the as-quenched Fe-0.36wt.%C-1wt.%Cr steel (Figure 5a) consisted of initial austenitic grains divided by martensite packets, blocks or laths [31]. Subsequent soaking of the as-quenched steel at 400, 500, 600 or 700 °C for 10 min resulted both in the martensite decomposition and in the onset of recovery and recrystallization development [32,33] (Figure 5b–e). Inverse pole figures (IPF) maps show that the finest structure formed at 400 °C (Figure 5b). With an increase in the temperature to 500 or 600 °C (Figure 5c,d), martensite laths gradually transform into relatively large areas of tempered martensite (the microstructure is still martensite; however, the concentration of carbon is already close to that in ferrite [32]). Heating to 700 °C promoted the formation of large ferrite grains with irregular shape primarily surrounded by martensitic packets (Figure 5e). The incomplete martensite decomposition caused by the short exposure at the elevated temperatures can be terminated during further warm deformation [11].

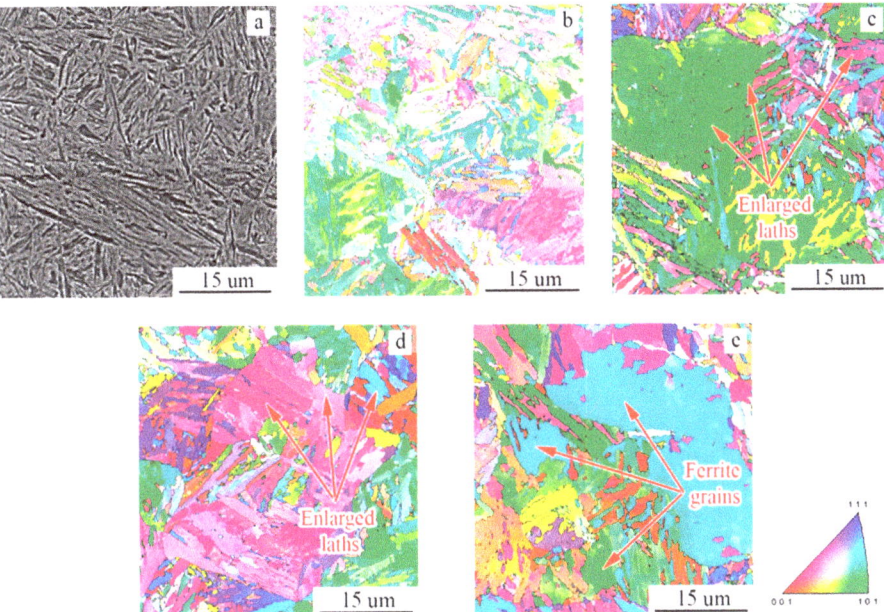

Figure 5. The initial microstructure (**a**) and inverse pole figures (IPF) maps (**b–e**) of Fe-0.36wt.%C-1wt.%Cr steel in as-quenched condition (**a**) and after soaking at 400 (**b**), 500 (**c**), 600 (**d**) or 700 °C (**e**) for 10 min.

Texture component maps showed gradual formation of a two-component (<001>//compression direction (CD) and <111>//CD) deformation texture during compression of the program steel to a true stain of 1.15 at 400, 500, 600 or 700 °C and a strain rate of $1.3 \cdot 10^{-3}$ s^{-1} (Figure 6). However, the texture and microstructure evolution during warm deformation strongly depended on deformation temperature.

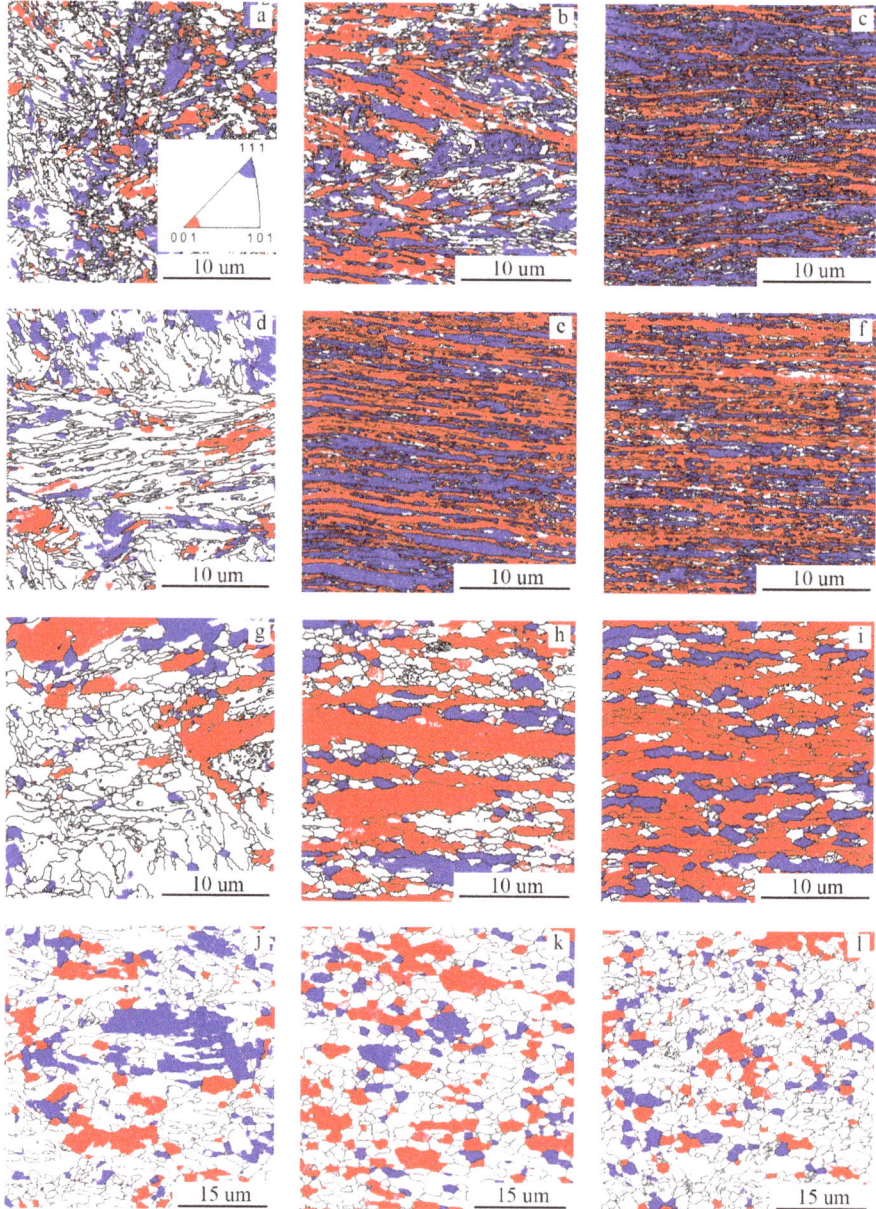

Figure 6. Texture component maps after deformation with a strain rate of $1.3 \cdot 10^{-3}$ s^{-1} at 400 (**a–c**), 500 (**d–f**), 600 (**g–i**) or 700 °C (**j–l**) to ε = 0.3, 0.77 or 1.15. Notation: high-angle boundaries (with misorientations ≥ 15°) are black, low-angle boundaries (with misorientations 2–15°) are white. Color code is shown in Figure 6a. The compression axis is vertical in all cases.

Compression at 400 °C resulted in the formation of a "pancaked" structure comprising highly elongated ferrite grains with a strong <001>//CD texture, and the grains were mainly separated by high-angle boundaries (HABs) (Figure 6a–c). This type of microstructure was obtained by flattening of the initial grains (Figure 5a) due to the geometrical effect of plastic deformation. The grains with the <111>//CD orientation were usually wider in comparison with those with the <001>//CD orientation. The development of substructure in a form of low-angle boundaries (LAB) networks was often observed in the <111>//CD grains. After deformation to $\varepsilon = 0.97$, the volume fractions of the <001>//CD and <111>//CD components reached 32% and 52%, respectively (Table 2).

Table 2. Volume fraction of areas with the <001>//compression direction (CD) (numerator, %) and <111>//CD (denominator, %) texture components in the Fe-0.36wt.%C-1wt.%Cr steel after different compression strain.

Temperature, °C	True Strain (ε)			
	0	0.3	0.77	1.15
400	4/3	9/16	29/28	32/52 *
500	2/1	6/14	49/41	49/31
600	4/6	16/8	40/13	57/19
700	5/3	11/17	12/9	19/10

* for temperature of 400 °C, maximum strain is equal to 0.97.

An increase in the compression temperature to 500 °C did not result in significant changes in the microstructure evolution. The microstructure was mostly composed of areas with both the <001>//CD and <111>//CD orientations—the latter were somewhat thicker after compression to $\varepsilon = 0.77$ (Figure 6e). After $\varepsilon = 1.15$, both structural constituents had a similar morphology of the elongated grains with predominantly high-angle misorientations (Figure 6f). Also, when the strain increased from $\varepsilon = 0.77$ to $\varepsilon = 1.15$, the fraction of the <001>//CD component remained at the same level of 49%, but the <111>//CD component fraction decreased from 41% to 31% (Table 2).

A further increase in the deformation temperature to 600 °C resulted in the formation of a much coarser, partially recrystallized structure (Figure 6g–i). The <001>//CD orientation became the dominant component, and its fraction gradually increased with strain reaching 57% at $\varepsilon = 1.15$. The development of a substructure was evident in the <001>//CD oriented regions. Meanwhile, the <111>//CD orientation fraction was considerably lower, only 19% at $\varepsilon = 1.15$ (Table 2). The microstructure became coarser and more equiaxed after deformation at 700 °C, most probably due to the development of the recrystallization processes (Figure 6g–l). Both the <001>//CD and <111>//CD orientation fractions were reduced dramatically. The morphology of grains with different texture components was quite similar after deformation at the high temperatures. The total fraction of both components did not exceed 30%. It is also worth noting that an increase in ε from 0.3 to 0.77 at 700 °C resulted in a considerable decrease in the fraction of the <111>//CD orientation, from 17% to 9% (Table 2). A similar phenomenon was observed during deformation at 500 °C to $\varepsilon = 0.77$ and $\varepsilon = 1.15$ (Table 2). However, a decrease in the volume fraction of areas with the <111>//CD texture at 600 °C was not detected.

It is also worth noting that the two-component <001>//CD and <111>//CD texture is not always observed after warm deformation. For example, in a Fe-0.4wt.%C-2wt.%Si-1wt.%Cr-1wt.%Mo steel with the initial martensite structure, a single-component <110>//RD deformation texture was observed after warm rolling [11]. The variations in the texture can most probably be attributed to the different character of the material flow depending on the processing scheme.

The dependence of the average distance between HABs (measured along the compression axis) on the deformation temperature and compression strain (Table 3) shows several trends. First, the initial ($\varepsilon = 0$) HABs spacing increased pronouncedly with an increase in the temperature. This finding agrees well with the observed changes in the microstructure (Figure 2). Second, deformation resulted in a continuous decrease in the HABs spacing at each temperature, though with different rates of the microstructure refinement. For instance, at 400 °C, the HABs spacing decreased from 0.78 to 0.19 µm as a result of compression to $\varepsilon = 0.97$. Meanwhile, deformation to $\varepsilon = 1.15$ at 700 °C decreased the boundary spacing from 1.52 to 1.07 µm only. In other words, the microstructure refinement was more evident at lower temperatures, most likely because of sluggish dynamic recrystallization (in contrast to higher temperatures).

Table 3. Average transverse high-angle boundaries (HAB) spacing (µm).

Temperature, °C	True Strain (ε)			
	0	0.3	0.77	1.15
400	0.77	0.44	0.29	0.19 *
500	0.76	0.49	0.30	0.22
600	1.07	0.62	0.57	0.47
700	1.52	1.25	1.28	1.07

* for temperature of 400 °C, maximum strain is equal to 0.97.

TEM analysis (Figures 7–9) showed that after heating to 400 °C, the structure of the steel was comprised of tempered martensite laths with dislocation arrays inside them (Figure 7a) and two types of carbides: fine (<20 nm), almost equiaxed, and coarser rod-shaped particles (length 120 nm and thickness 20 nm). The rod-shaped carbides were predominantly located at the lath boundaries, but some of them were found inside the laths. After compression to ε = 0.3 (Figure 7b), due to the development of dynamic recovery, the dislocation density decreased, and transversal low-angle boundaries were observed in some laths. The fraction of the rod-shaped carbides located inside the laths increased, and this can suggest that boundaries broke from the obstacles during motion, leaving the carbides behind in grain volume. Further straining at 400 °C resulted in a considerable thinning of the laths/grains (Figure 7c). Fine, nearly equiaxed carbides (10 nm) were also observed inside elongated grains. Their formation can be associated with either coalescence of the initial fine carbides or with the fragmentation of the coarser rod-shaped particles.

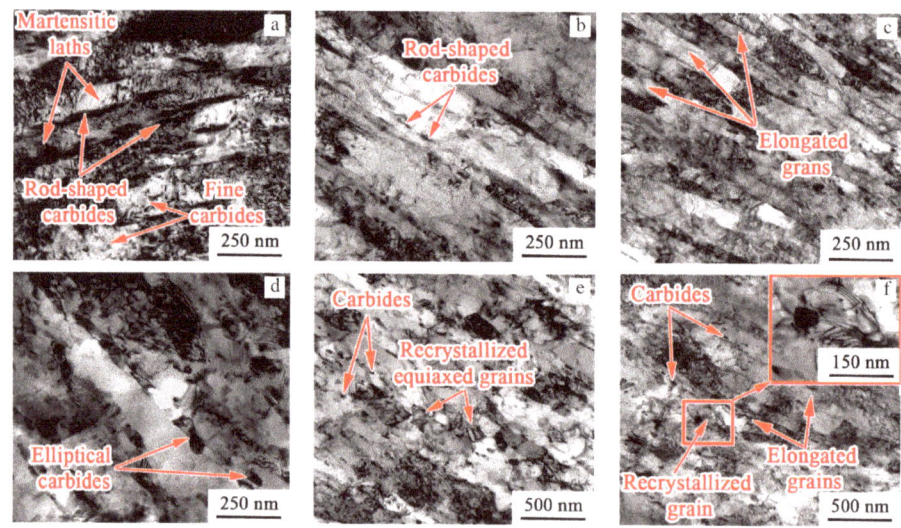

Figure 7. Transmission electron microscopy (TEM) photographs of fine structure after compression at 400 °C (a–c) and 500 °C (d–f) with strain rate of $1.3 \cdot 10^{-3}$ s^{-1} to different strain degrees (ε): 0 (a,d), 0.3 (b), 0.77 (e), 0.97 (c) and 1.15 (f).

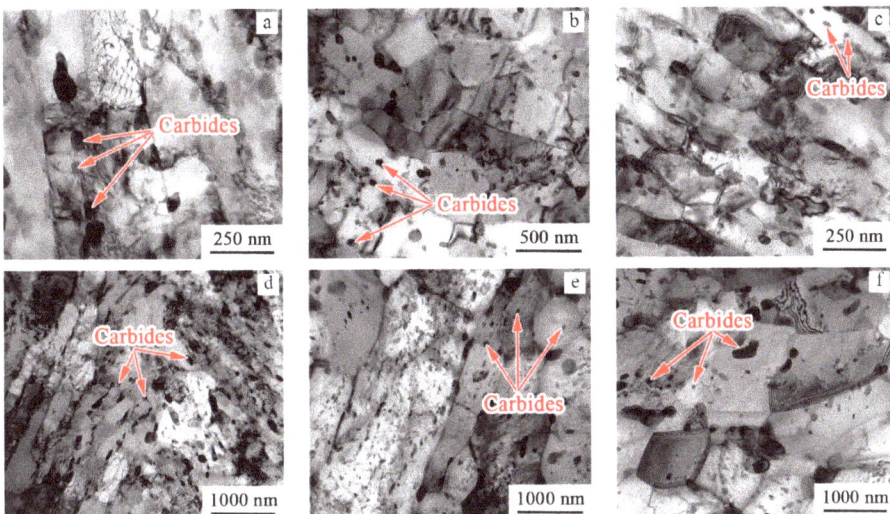

Figure 8. TEM photographs of fine structure after compression at 600 °C (**a–c**) and 700 °C (**d–f**) with strain rate of $1.3 \cdot 10^{-3}$ s^{-1} to different strain degrees (ε): 0 (**a,d**), 0.3 (**b,c**) and 1.15 (**e,f**).

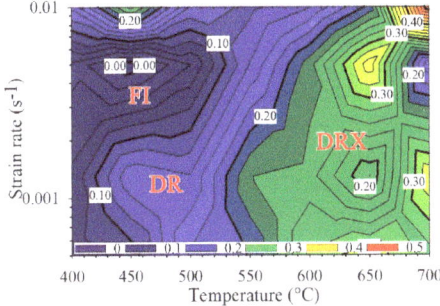

Figure 9. Processing map for warm deformation of the as-quenched Fe-0.36wt.%C-1wt.%Cr steel at $\varepsilon = 0.35$. Notation: FI—flow instability; DR—dynamic recovery; DRX—dynamic recrystallization.

After heating to 500 °C (Figure 7d), the microstructure became somewhat coarser in comparison with that at 400 °C. Some of the laths were almost free of dislocations while some of them contained dislocation arrays and pile-ups. Carbide particles (length 50 nm and thickness 20 nm) of elliptical shape were mainly located at lath boundaries. After deformation, fine elongated grains were observed (Figure 7e,f). Inside the grains, dislocation pile-ups and transversal sub-boundaries formed. In addition, some very fine equiaxed grains (diameter 200 nm) were observed—an example is shown in the insert in Figure 7f at a higher magnification. Note that these grains were often found close to the carbide particles. Some carbides (length 55 nm and thickness 20 nm), arranged in chains, were also found inside the laths/grains (Figure 7e), thereby suggesting boundaries' motion.

With an increase in the heating temperature to 600 °C (Figure 8a) or 700 °C (Figure 8d), a microstructure consisted of rather coarse and elongated ferrite grains/sub-grains instead of martensitic laths. The overall dislocation density was quite low, yet dislocation pile-ups and arrays were found inside some grains. The carbides were coarse and nearly equiaxed (55 nm at 600 °C and 70 nm at 700 °C), however some of them still had an elliptical shape. The carbide particles were found both inside grains/sub-grains and at their boundaries. Compression at these high temperatures resulted in a decrease in dislocation density, obviously due to the development of DRX (Figure 8b,c,e,f). Some fine, dislocation-free equiaxed recrystallized grains with Moir fringes at the boundaries were observed. Carbides were often found at the boundaries of these recrystallized grains. In addition, the carbides became even more equiaxed after deformation at 700 °C, and some increase in size of the carbides from 70 to 110 nm was observed, with an increase in strain from ε = 0 to ε = 1.15.

The activation energy analysis and examination of the microstructure evolution of the as-quenched steel during warm deformation suggests two different controlling mechanisms: dynamic recovery in the interval of 400–550 °C and continuous DRX at 600–700 °C. Specifically, elongated granular structure with fine carbide nanoparticles and high dislocation density inside was found after deformation in the lower-temperature interval. At large strains, some fine equiaxed dislocation-free grains appeared (Figure 5e,f), most probably via discontinuous DRX [25]. However, the pinning effect of carbide nanoparticles and low temperatures made the discontinuous DRX effect insignificant [22]. Moreover, the nanoscale carbide particles in the structure served as obstacles for dislocation motion and influenced on the threshold stresses. An increase in the size of carbides from <20 nm to 55 nm, with a temperature rise from 400 to 600 °C respectively, decreased the threshold stress value from 522 to 43 MPa (Table 1). Specifically, there was not any threshold stress revealed at 700 °C, where the average size of carbide particles achieved 75 nm.

In the current work, the formation of a close to equiaxed structure with low dislocation density and coarser carbides suggested the development at higher temperatures (600–700 °C) of continuous DRX. Similar processes were also observed earlier during deformation at the same temperatures of steels with the initial martensite [12,15,16] or ferrite-perlite [17,22] structures. Meanwhile, the development of discontinuous DRX after large strain at low temperatures (about 500 °C) requires an additional study.

Moreover, the texture development also depended strongly on the temperature and mechanisms of the structure formation. The development of discontinuous DRX inhibited the formation of the <111>//CD orientation fraction in comparison with <001>//CD at a temperature of 500 °C. The formation of different form <111>//CD and <001>//CD orientations (white areas in Figure 6f) and simultaneous drop of the <111>//CD orientation volume fraction (Table 2) can indicate predominant development of discontinuous DRX inside of the <111>//CD oriented volumes. In the lower-temperature interval, the fractions of these components were comparable. Besides, an increase in the warm deformation temperature expectedly led to a less pronounced deformation texture [11].

The carbides precipitation/evolution was also linked to the microstructure evolution. On the one hand, just before the deformation, the solid solution was supersaturated with carbon [32]. Subsequent warm deformation led to intensive formation of the carbide particles which could inhibit dynamic recrystallization [34]. On the other hand, strain-induced decomposition of carbides can be expected [35], however the volume fraction of the carbides did not decrease during deformation. In fact, strain-induced particles were found even in interstitial-free steel after warm deformation [3]. The obtained results suggest dividing of coarser rod-shaped carbides into smaller particles in the lower-temperature interval (Figure 7a,b), and a similar observation was earlier described in Reference [6].

3.3. Processing Map Analysis

The results obtained show a significant dependence of mechanical behavior and microstructure evolution of the as-quenched steel during compression on process parameters (i.e., temperature, strain and strain rate). Generally, elongated microstructure with a developed two-component texture was formed at low temperatures (400 °C and 500 °C), but at high temperatures (600 °C and 700 °C), the formation of equiaxed grains with a weak texture occurred due to the DRX development.

The features of processes in the program steel, occurring during warm deformation, determined the efficiency of power dissipation, which was evaluated in a number of studies using a dynamic material model [24,36–38]. The efficiency of power dissipation (η) during deformation processes can be calculated as follows:

$$\eta = \frac{2m}{m+1}, \quad (6)$$

where m is the strain rate sensitivity:

$$m = \left.\frac{\partial \ln(\sigma)}{\partial \ln(\dot{\varepsilon})}\right|_{\varepsilon,T}. \tag{7}$$

A processing map, representing the efficiency of power dissipation (η) for warm deformation at $\varepsilon = 0.35$, was constructed using the results of the calculation (Figure 9). This value of strain (0.35) was chosen to avoid the influence of friction between the specimen surface and the die at higher strains, especially at low temperatures. The border between the dynamic recovery/dynamic recrystallization areas corresponded to η ~ 0.2 [36], that, depending on a strain rate, fell in the interval 550–600 °C. During deformation with strain rates close to $1.3 \cdot 10^{-3}$ s^{-1}, a rise in deformation temperature led to an increase in the efficiency of power dissipation to η = 0.3 and higher (at 700 °C), which was a consequence of a change in the mechanism of power dissipation from dynamic recovery to dynamic recrystallization. These results were in agreement with the result of microstructure evolution studies in the current work. However, deformation at 400 °C with almost all strain rates and in the temperature range of 400–500 °C with a strain rate of $5 \cdot 10^{-2}$ s^{-1} can result in flow instability with the value of η ~ 0.1. Moreover, the efficiency of power dissipation approaches values, typical of dynamic recovery, at 700 °C. An explanation of this behavior can be related with martensite decomposition during deformation and development of dynamic recovery/recrystallization; however, this point requires additional clarification.

From the viewpoint of practical application, an elongated fine-grained structure with nanoscale carbide particles and strong deformation texture was observed after deformation in the field of 0.1 < η < 0.2, where the energy dissipation occurred due to dynamic recovery. On the other hand, better workability can be obtained during warm deformation in the field, with η > 0.2, where DRX develops. Deformation instability in the field with η < 0.1 can be accompanied by cracking and failure. It should be noted that the as-quenched Fe-1.2wt.%C steel had lower efficiency of power dissipation in the field of deformation instability and, as a result, better processability in comparison with the initially spheroidized steel [6].

4. Conclusions

The mechanisms of the structure formation during warm deformation of the as-quenched Fe-0.36wt.%C-1wt.%Cr steel in a temperature range of 400–700 °C with strain rates of 10^{-2}, $5 \cdot 10^{-2}$, $1.3 \cdot 10^{-3}$, $5 \cdot 10^{-3}$ and 10^{-4} s^{-1} were investigated. The following results were obtained:

1. Two temperature domains can be clearly distinguished, depending on the structure formation mechanisms: low temperatures at 400–550 °C and high temperatures at 600–700 °C. At the low temperatures, deformation was accompanied by dynamic recovery/discontinuous dynamic recrystallization, causing both the rearrangement of dislocations and the formation of a developed substructure. Continuous dynamic recrystallization occurred during deformation at the higher temperatures, resulting in the formation of a coarse, nearly equiaxed grain structure. The activation energy of deformation was 140 ± 35 kJ/mol for the low-temperature domain, and 243 ± 15 kJ/mol for the high-temperature domain.

2. The presence of nanoscale carbide particles in the structure at temperatures of 400–600 °C resulted in the appearance of threshold stresses. The value of threshold decreased from 522 to 43 MPa with an increase in the size of carbides due to a temperature rise from 400 to 600 °C, respectively.

3. A two-component <001>//CD and <111>//CD deformation texture was formed during deformation. Deformation at the low temperatures resulted in the formation of elongated ferritic grains separated mainly by HABs with a strong <001>//CD texture. The grains with the <111>//CD orientation were wider in comparison with those with the <001>//CD orientation. The development of substructure in the form of LABs networks was also observed in the <111>//CD grains. The development of dynamic recrystallization restricted the texture formation.

4. The processing map of warm deformation was constructed. Three main fields of the power dissipation efficiency (η) were observed: dynamic recovery, dynamic recrystallization and flow instability. The border between the areas of dynamic recovery and dynamic recrystallization was at 550–650 °C at all studied strain rates. However, some flow instability was observed at 400 °C and strain rates of 10^{-4}–$5 \cdot 10^{-2}$ s^{-1}, and in the temperature range of 400–500 °C with a strain rate of $5 \cdot 10^{-2}$ s^{-1}.

Author Contributions: D.P., Formal analysis, Writing—Review and Editing, Visualization, Investigation; O.D., Investigation, Formal analysis, Visualization, Validation, Methodology, Writing—Review and Editing; D.S., Methodology, Investigation, Validation; N.S. and S.Z., Writing—Review and Editing; G.S., Conceptualization, Supervision, Writing—Review and Editing. All authors have read and agreed to the published version of the manuscript.

Funding: The authors gratefully acknowledge the financial support from the Russian Science Foundation Grant no. 19-79-30066. The authors are grateful to the personnel of the Joint Research Center, "Technology and Materials", Belgorod National Research University, for their assistance.

Conflicts of Interest: The authors declare no conflict of interest.

References

1. Song, R.; Ponge, D.; Raabe, D.; Speer, J.G.; Matlock, D.K. Overview of processing, microstructure and mechanical properties of ultrafine grained bcc steels. *Mater. Sci. Eng. A* **2006**, *441*, 1–17. [CrossRef]
2. Zhao, J.; Jiang, Z. Thermomechanical processing of advanced high strength steels. *Prog. Mater. Sci.* **2018**, *94*, 174–242. [CrossRef]
3. Akbari, G.H.; Sellars, C.M.; Whiteman, J.A. Microstructural development during warm rolling of an if steel. *Acta Mater.* **1997**, *45*, 5047–5058. [CrossRef]
4. Niechajowicz, A.; Tobota, A. Warm deformation of carbon steel. *J. Mater. Process. Technol.* **2000**, *106*, 123–130. [CrossRef]
5. Miller, R.L. Ultrafine- grained microstructures and mechanical properties of alloy steels. *Met. Trans.* **1972**, *3*, 905–912. [CrossRef]
6. Wu, T.; Gao, Y.W.; Wang, M.Z.; Li, X.P.; Zhao, Y.C.; Zou, Q. Influence of Initial Microstructure on Warm Deformation Processability and Microstructure of an Ultrahigh Carbon Steel. *J. Iron Steel Res. Int.* **2014**, *21*, 52–59. [CrossRef]
7. Kimura, Y.; Inoue, T.; Yin, F.; Tsuzaki, K. Inverse temperature dependence of toughness in an ultrafine grain-structure steel. *Science* **2008**, *320*, 1057–1059. [CrossRef]
8. Kimura, Y.; Inoue, T. Mechanical property of ultrafine elongated grain structure steel processed by warm tempforming and its application to ultra-high-strength bolt. *Tetsu-To-Hagane/J. Iron Steel Inst. Jpn.* **2019**, *105*, 5–23. [CrossRef]
9. Inoue, T.; Kimura, Y.; Ochiai, S. Shape effect of ultrafine-grained structure on static fracture toughness in low-alloy steel. *Sci. Technol. Adv. Mater.* **2012**, *13*. [CrossRef]
10. Kimura, Y.; Inoue, T. Influence of warm tempforming on microstructure and mechanical properties in an ultrahigh-strength medium-carbon low-alloy steel. *Metall. Mater. Trans. A Phys. Metall. Mater. Sci.* **2013**, *44*, 560–576. [CrossRef]
11. Kimura, Y.; Inoue, T.; Tsuzaki, K. Tempforming in medium-carbon low-alloy steel. *J. Alloys Compd.* **2013**, *577*, S538. [CrossRef]
12. Bao, Y.Z.; Adachi, Y.; Toomine, Y.; Xu, P.G.; Suzuki, T.; Tomota, Y. Dynamic recrystallization by rapid heating followed by compression for a 17Ni-0.2C martensite steel. *Scr. Mater.* **2005**, *53*, 1471–1476. [CrossRef]
13. Humphreys, F.J.; Hatherly, M. *Recrystallization and Related Annealing Phenomena*, 2nd ed.; Humphreys, F.J., Hatherly, M., Eds.; Pergamon: Oxford, UK, 2004.
14. Kimura, Y.; Inoue, T.; Fuxing, Y.I.N.; Tsuzaki, K. Delamination toughening of ultrafine grain structure steels processed through tempforming at elevated temperatures. *ISIJ Int.* **2010**, *50*, 152–161. [CrossRef]
15. Li, Q.; Wang, T.S.; Jing, T.F.; Gao, Y.W.; Zhou, J.F.; Yu, J.K.; Li, H.B. Warm deformation behavior of quenched medium carbon steel and its effect on microstructure and mechanical properties. *Mater. Sci. Eng. A* **2009**, *515*, 38–42. [CrossRef]
16. Poorganji, B.; Miyamoto, G.; Maki, T.; Furuhara, T. Formation of ultrafine grained ferrite by warm deformation of lath martensite in low-alloy steels with different carbon content. *Scr. Mater.* **2008**, *59*, 279–281. [CrossRef]

17. Ohmori, A.; Torizuka, S.; Nagai, K.; Koseki, N.; Kogo, Y. Effect of deformation temperature and strain rate on evolution of ultrafine grained structure through single-pass large-strain warm deformation in a low carbon steel. *Mater. Trans.* **2004**, *45*, 2224–2231. [CrossRef]
18. Murty, S.V.S.N.; Torizuka, S.; Nagai, K.; Kitai, T.; Kogo, Y. Dynamic recrystallization of ferrite during warm deformation of ultrafine grained ultra-low carbon steel. *Scr. Mater.* **2005**, *53*, 763–768. [CrossRef]
19. Zhao, X. Microstructural evolution during warm compression of medium carbon steel quenched. *Adv. Mater. Res.* **2008**, *47–50*, 853–856. [CrossRef]
20. Zhao, X.; Yang, X.L.; Jing, T.F. Effect of Initial Microstructure on Warm Deformation Behavior of 45 Steel. *J. Iron Steel Res. Int.* **2012**, *19*, 75–78. [CrossRef]
21. Li, J.; Xu, P.; Tomota, Y.; Adachi, Y. Dynamic recrystallization behavior in a low-carbon martensite steel by warm compression. *ISIJ Int.* **2008**, *48*, 1008–1013. [CrossRef]
22. Song, R.; Ponge, D.; Raabe, D.; Kaspar, R. Microstructure and crystallographic texture of an ultrafine grained C-Mn steel and their evolution during warm deformation and annealing. *Acta Mater.* **2005**, *53*, 845–858. [CrossRef]
23. Dedyulina, O.K.; Salishchev, G.A.; Pertsev, A.S. A Study of the Microstructure and Mechanical Properties of Medium-Carbon Steel 40KhGNM after Warm Swaging. *Met. Sci. Heat Treat.* **2016**, *58*, 132–137. [CrossRef]
24. Al Omar, A.; Cabrera, J.M.; Prado, J.M. Characterization of the hot deformation in a microalloyed medium carbon steel using processing maps. *Scr. Mater.* **1996**, *34*, 1303–1308. [CrossRef]
25. Sakai, T.; Belyakov, A.; Kaibyshev, R.; Miura, H.; Jonas, J.J. Dynamic and post-dynamic recrystallization under hot, cold and severe plastic deformation conditions. *Prog. Mater. Sci.* **2014**, *60*, 130–207. [CrossRef]
26. Frost, H.J.; Ashby, M.F. *Deformation-Mechanism Maps: The Plasticity and Creep of Metals and Ceramics*, 1st ed.; Pergamon Press: Oxford, UK, 1982.
27. Čadek, J. *Creep in Metallic Materials*, 1st ed.; Elsevier: Amsterdam, The Netherlands; Oxford, UK; New York, NY, USA; Tokyo, Japan, 1988.
28. Gibeling, J.C.; Nix, W.D. Description of Elevated Temperature Deformation in Terms of Threshold Stresses and Back Stresses—A Review. *Mater. Sci. Eng.* **1980**, *45*, 123–135. [CrossRef]
29. Li, Y.; Langdon, T.G. A simple procedure for estimating threshold stresses in the creep of metal matrix composites. *Scr. Mater.* **1997**, *36*, 1457–1460. [CrossRef]
30. Dudko, V.A.; Kaibyshev, R.O.; Belyakov, A.N.; Sakai, Y.; Tsuzaki, K. Plastic flow of the mechanically alloyed Fe-0.6%O at temperatures of 550–700 °C. *Phys. Met. Metallogr.* **2009**, *107*, 516–521. [CrossRef]
31. Morito, S.; Huang, X.; Furuhara, T.; Maki, T.; Hansen, N. The morphology and crystallography of lath martensite in alloy steels. *Acta Mater.* **2006**, *54*, 5323–5331. [CrossRef]
32. Krauss, G. Tempering of Lath Martensite in Low and Medium Carbon Steels: Assessment and Challenges. *Steel Res. Int.* **2017**, *88*, 1–18. [CrossRef]
33. Chen, K.; Jiang, Z.; Liu, F.; Yu, J.; Li, Y.; Gong, W.; Chen, C. Effect of quenching and tempering temperature on microstructure and tensile properties of microalloyed ultra-high strength suspension spring steel. *Mater. Sci. Eng. A* **2019**, *766*, 138272. [CrossRef]
34. NajafiZadeh, A.; Jonas, J.J.; Yue, S. Grain refinement by dynamic the simulated warm-rolling of recrystallization during interstitial free steels. *Metall. Mater. Trans. A* **1992**, *23*, 2607–2617. [CrossRef]
35. Shabashov, V.A.; Korshunov, L.G.; Zamatovskii, A.E.; Litvinov, A.V.; Sagaradze, V.V.; Kositsyna, I.I. Deformation-induced dissolution of carbides of the Me(V, Mo)-C type in high-manganese steels upon the friction effect. *Phys. Met. Metallogr.* **2012**, *113*, 914–921. [CrossRef]
36. Prasad, Y.V.R.K.; Sasidhara, S. *Hot Working Guide: A Compendium of Processing Maps*, 1st ed.; ASM International: Materials Park, OH, USA, 1997.
37. Prasad, Y.V.R.K.; Gegel, H.L.; Doraivelu, S.M.; Malas, J.C.; Morgan, J.T.; Lark, K.A.; Barker, D.R. Modeling of dynamic material behavior in hot deformation: Forging of ti-6242. *Metall. Trans. A* **1984**, *15A*, 1883–1892. [CrossRef]
38. Vafaeenezhad, H.; Seyedein, S.H.; Aboutalebi, M.R.; Eivani, A.R. An investigation of workability and flow instability of Sn-5Sb lead free solder alloy during hot deformation. *Mater. Sci. Eng. A* **2018**, *718*, 87–95. [CrossRef]

© 2020 by the authors. Licensee MDPI, Basel, Switzerland. This article is an open access article distributed under the terms and conditions of the Creative Commons Attribution (CC BY) license (http://creativecommons.org/licenses/by/4.0/).

Article

Study on Texture and Grain Orientation Evolution in Cold-Rolled BCC Steel by Reaction Stress Model

Ning Zhang [1], Li Meng [1,*], Wenkang Zhang [2] and Weimin Mao [3]

[1] Metallurgical Technology Institute, Central Iron and Steel Research Institute, Beijing 100081, China; zhangning@cisri.com.cn
[2] Technology Centre of Shanxi Taigang Stainless Steel Co. Ltd., Taiyuan 030003, China; zhangwk@tisco.com.cn
[3] School of Materials Science and Engineering, University of Science and Technology Beijing, Beijing 100083, China; wmmao@ustb.edu.cn
* Correspondence: mengl@cisri.com.cn or li_meng@126.com

Received: 29 June 2020; Accepted: 4 August 2020; Published: 6 August 2020

Abstract: The evolution of texture and grain orientations in a cold-rolled steel of BCC structure was simulated by a reaction stress (RS) model. The results show that cold-rolled texture could be assessed based on a RS model because the stress and strain are considered to remain consistent in the deformation process. The strain consistency is actualized by the cooperation of two plastic strains and an elastic strain. The accumulation range of each reaction stress and different activation abilities of {110}<111> and {112}<111> slip systems strongly affect the calculated deformation textures. The values of reaction stress are influenced by elastic anisotropy; however, the effects are greatly reduced because its corresponding reaction stress accumulation is limited. Typical α-fiber and γ-fiber textures are achieved when the reaction stress accumulation coefficients α_{ij}s are chosen suitably. Furthermore, the α_{ij} values that are selected based on statistically calculated textures can also be used to simulate the orientation change of multiple orientations. The existence of reaction stress is able to stabilize crystallographically symmetrical orientations under rolling deformation, in which the Schmid factors of several slip systems are identical.

Keywords: BCC structure; reaction stress (RS) model; rolling; texture; orientation

1. Introduction

Rolling technology is widely used in manufacturing metals and alloys. In general, deformation texture forms during rolling, and the resultant texture will affect the following processes and determine properties of final products [1,2]. Understanding the formation and evolution of deformation texture can provide theoretical insight for texture control, thus improving material properties [3]. Other than experimental observation, the evolution of texture and crystallographic orientation can be evaluated based on simulation, which requires a reasonable and suitable model dedicated to polycrystalline plastic deformation.

Multiple models for polycrystalline plastic deformation have been developed from the original Sachs model and Taylor model. In recent years, grain interaction (GIA) model, viscoplastic self-consistent (VPSC) model, advanced lamel (ALAMEL) models and others have been proposed. With development of these models, the simulated deformation textures have been continuously improved and nicely match experimental observation [4–10]. Regarding the stress and strain consistency during simulation, it is actualized inside grain groups by GIA and ALAMEL models, in which 8 grains and 2 grains are set as a group respectively, and different relaxation is implemented inside the groups [4–7,11]. The VPSC model considers the stress and strain continuity between every deformed grain and its surrounding matrix statistically [9–11].

Compared to the models developed based on the Taylor principle, the reaction stress (RS) model evaluates the deformation process based on the stress condition. At the beginning, only external stress is considered, and the slip system with the highest Schmid factor is activated. As soon as the active slip induces plastic strain, reaction stress by neighboring grains is taken into consideration. The reaction stress is calculated based on statistical matrix. These external and intergranular reaction stresses are combined to determine the following processes [12,13]. In simulation, the intergranular continuity of strain and stress is achieved in a natural way, and the calculation method does not have a complicated hypothesis [12–15]. RS model has been increasingly applied to metals and alloys with different crystal structures, including BCC, FCC and HCP structures. For BCC structured metals and alloys, this model has been applied to tungsten, which is elastically isotropic [12–15].

Ferritic steel with BCC structure has been widely applied in industry. For example, interstitial-free (IF) steel, low carbon steel and silicon steel are being used in automatic, architectural, and electronic fields. In general, the rolling deformation texture of BCC steel is characterized as α-fiber (rolling direction (RD) of the rolled plate//<110>) and γ-fiber (normal direction (ND) of the rolled plate//<111>). The deformation textures directly affect recrystallization behaviors and anisotropic properties of the final products [1,16–18]. Therefore, it is important to understand the evolution of deformation texture. Meanwhile, the evolution of specific deformation orientation of typical texture components needs to be studied as well. Moreover, BCC structured steel has demonstrated an elastic anisotropy, in which Young's modulus in <111> crystal direction is 6 times of that in <100> direction [19]. The elastic anisotropy and its effect on the deformation process have to be taken into consideration for simulation. Therefore, as a further theoretical study based on RS model, this study helps the development of simulation work by RS model, making it more systematically and comprehensively understood.

In this paper, a RS model is used to investigate the rolling deformation behaviors of a BCC structured steel and the simulated results are compared to the experimental results and results reported in the literature. The effects of reaction stress values and ranges in different directions, strain values, the ratio of critical resolved shear stress (CRSS) of {110}<111> slip system to CRSS of {112}<111> slip system are discussed, and the influence of elastic anisotropy is also considered. The study could help to clarify the deformation process of BCC structured steel.

2. Simulation and Experimental Methods

This study focuses on the rolling deformation behavior of BCC structured steel via both theoretical simulation and experimental characterization. RS model is used to mimic the texture evolution and orientation change of typical texture components. The RS model states that, the stress of a deformed grain undergoing in a polycrystal can be depicted as Equation (1) [12–15]:

$$[\sigma_{ij}] = \sigma_y \begin{bmatrix} 0.5 & 0 & 0 \\ 0 & 0 & 0 \\ 0 & 0 & -0.5 \end{bmatrix} + \sigma_y \begin{bmatrix} 0 & \sigma'_{12}/\sigma_y & \sigma'_{13}/\sigma_y \\ \sigma'_{21}/\sigma_y & \sigma'_{22}/\sigma_y & \sigma'_{23}/\sigma_y \\ \sigma'_{31}/\sigma_y & \sigma'_{32}/\sigma_y & 0 \end{bmatrix} \quad (1)$$

σ_y is the yield strength. The index i and j = 1, 2, 3, and 1, 2, 3 represents rolling direction (RD), transverse direction (TD) and normal direction (ND), respectively. σ'_{ij} represents the reaction stress, and it can be calculated based on intergranular stress and strain balance:

$$\sigma'_{ij} = -\left(\frac{G(g) \times G(m)}{G(g) + G(m)}\right) \times \varepsilon^p_{ij}(g) \quad (2)$$

$G(g)$ and $G(m)$ correspond to the shear modulus of the deformed grain and matrix, respectively, which are dependent on the orientations of the deformed grain and matrix. $\varepsilon^p_{ij}(g)$ is the shear strain of an active slip in the deformed grain.

σ_y can be illustrated by Frank-Read theory:

$$\sigma_y = \frac{\tau_c}{\mu} = \frac{Gb}{\mu d} \tag{3}$$

G is G(g), b and μ are the burgers vector and the Schmid factor corresponding to the active slip system, d is the effective dislocation distance, which decreases with the increasing strain [15]. The connection between $[\sigma_{ij}]$ and the parameters, including b, μ, d, $\varepsilon_{ij}^p(g)$ as well as crystal orientation, can be obtained, and more details can be found in the literature [12–15].

What needs to be considered is that reaction stress varies in a certain range. Therefore, the accumulation coefficient α_{ij} for reaction stress is introduced to impose the upper-limit for reaction stress σ_{ij}', as shown in Equation (4). α_{ijs} varies in the range of 0–1, and the value imposes significant effect on the texture evolution [12,15]:

$$[\sigma_{ij}]_{lim} = \begin{bmatrix} \sigma_{11} \equiv -\sigma_{33} - \sigma_{22}' & \sigma_{12}' \leq \alpha_{12}\sigma_y/2 & \sigma_{13}' \leq \alpha_{13}\sigma_y/2 \\ \sigma_{21}' \leq \alpha_{12}\sigma_y/2 & \sigma_{22}' \leq \alpha_{22}\sigma_y/2 & \sigma_{23}' \leq \alpha_{23}\sigma_y/2 \\ \sigma_{31}' \leq \alpha_{13}\sigma_y/2 & \sigma_{32}' \leq \alpha_{23}\sigma_y/2 & \sigma_{33} \equiv -\sigma_y/2 \end{bmatrix} \tag{4}$$

As the initial orientations in this study, 1000 random orientations are chosen. The calculation step is $\Delta\varepsilon_{33} = -0.0005$. The calculated results at different strains are compared to experimental results and results reported in the literature. The effects of reaction stress accumulation coefficient α_{ij}, the activation ability of {110}<111> and {112}<111> slip systems as well as strain values ε_{33} on the deformation texture are discussed systematically. In particular, the value of α_{ij} has been proven to have significant effect on the texture evolution in BCC structured tungsten [15], and thus its effect on the calculated texture is evaluated in detail. Based on the analysis, simulation parameters are optimized and are used to discuss the orientation evolution of typical texture components in BCC steel. Surface shear stress has limited effect on texture evolution. One of the reasons is that rolling is usually performed at room temperature and thus there is no considerable shear stress by surface friction. Our earlier study on tungsten shows that small surface shear stress may lead to slight a deviation of the texture component or a decrease of texture intensity, while the major texture distribution is not affected [15]. Therefore, this study does not discuss the effect of surface shear stress.

Two types of experiments were carried out in the study. First, a forged Fe-3 wt%.Si steel was annealed at 800 °C for 2 h to obtain a microstructure with an average grain size of ~200 μm, and then a 10 mm × 10 mm × 10 mm sample was cut from it. After being polished at the lateral plane, the sample was compressed to 5% reduction. Afterwards, the initially polished lateral plane was measured with a Zeiss GeminiSEM500 field emission scanning electron microscope (SEM) (Zeiss, Braunschweig, Germany.) to observe the slip traces, by which the strain distribution could be evaluated. The goal is to prove that the strain consistency is due to both penetrating plastic strain and local plastic strain. This hypothesis is proposed when studying the deformation behavior of BCC structured tungsten but is not verified experimentally because of difficulty to obtain experimental data. With reliable experimental results, the distribution of penetrating and local strain is discussed. Second, a Fe-0.46 wt%Si steel was cold rolled to different reductions. Its textures before and after rolling were measured by X-ray diffraction (XRD) (Bruker, Billerica, MA, USA) using Bruker D8 advance X-ray diffractometer. The texture evolution was compared to the simulation results so that the RS model dedicated to steels with BCC structure can be comprehensively studied and its validity will be eventually confirmed.

3. Results and Discussion

The microstructure of 5% compressed silicon steel sample and the compression direction are shown in Figure 1a. Examples of typical slip traces inside grains are given in Figure 1b, and multiple slip traces along different directions in deformed grains are highlighted and deformed grains are labeled as A, B, C and D. Each slip trace corresponds to activation of a specific slip system. At the early stage of rolling deformation, both penetrating slip and local slip can be observed in many grains. Various local slips can be generated simultaneously in a single deformed grain, such as in grain A. The same phenomenon has been reported in [15]. The penetrating plastic strains in adjacent grains do not always coordinate with each other during the deformation, and the strain continuity is achieved via the coordination of penetrating plastic slip and local plastic slip. Elastic strain is not shown in the figure, however, its role in reducing strain inconsistency cannot be ignored, and moreover its corresponding elastic reaction stress contributes to stress continuity.

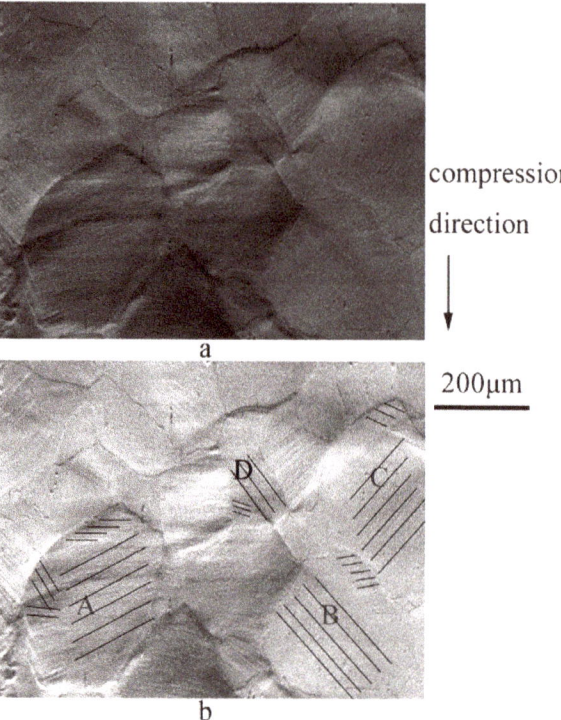

Figure 1. Lateral microstructure of silicon steel compressed to 5% reduction. (**a**) SEM micrograph; (**b**) schematic penetrating and local slip traces in some grains.

The texture evolution at different strains was calculated using various reaction stress accumulation coefficient α_{ij}, and the results when only considering {110}<111> slip are shown in Figure 2(a1–d5). The effect of elastic anisotropy is not considered here. Taylor texture is obtained at low strain when α_{ijs} are set to be 0.2 in Figure 2(a1,b1), and the values of reaction stress do not exceed $0.1\sigma_y$. Under this condition, some texture components near γ-fiber, especially near {111}<110> component, are weakened gradually with strain, while at the same time Taylor texture is further strengthened. When ε_{33} reaches −1.1, Taylor texture becomes a major texture component and α-fiber texture starts to appear in Figure 2(c1). When the α_{ij} value increases, the calculated texture starts to deviate from α-fiber and the intensity of Taylor texture decreases. To sum up, the calculation results are similar to that of tungsten, and strong Taylor texture—which is a typical simulated rolling texture in BCC polycrystals under Taylor principle—is achieved with {110}<111> slip [15].

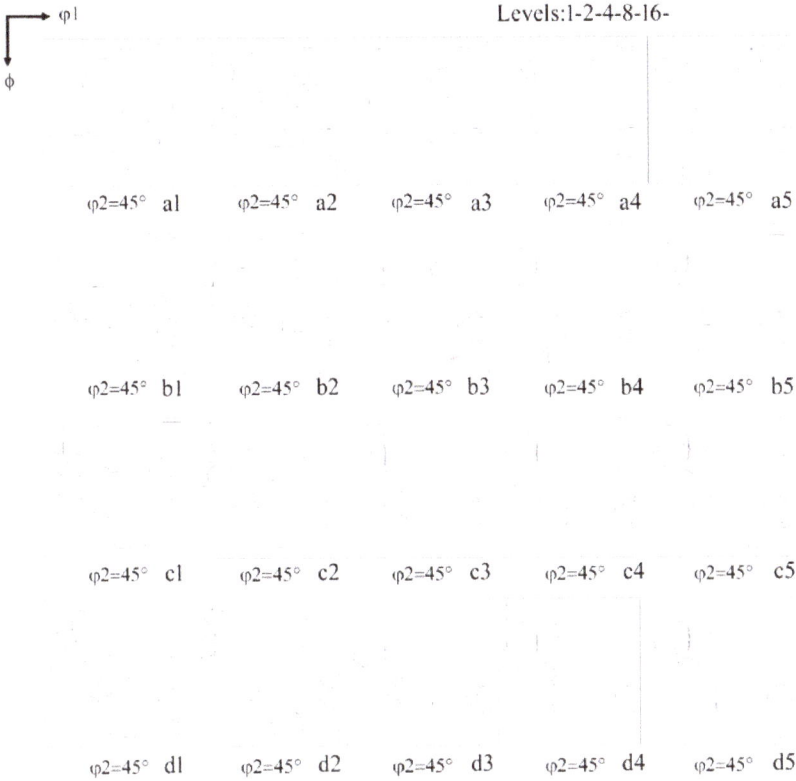

Figure 2. Calculated rolling deformation textures using reaction stress model under different ε_{33} and $\alpha_{ij}s$, {110}<111> slip. $\varepsilon_{33} = -0.36$: (**a1**) $\alpha_{ij}s = 0.2$; (**a2**) $\alpha_{ij}s = 0.4$; (**a3**) $\alpha_{ij}s = 0.6$; (**a4**) $\alpha_{ij}s = 0.8$; (**a5**) $\alpha_{ij}s = 1$; $\varepsilon_{33} = -0.69$: (**b1**) $\alpha_{ij}s = 0.2$; (**b2**) $\alpha_{ij}s = 0.4$; (**b3**) $\alpha_{ij}s = 0.6$; (**b4**) $\alpha_{ij}s = 0.8$; (**b5**) $\alpha_{ij}s = 1$; $\varepsilon_{33} = -1.1$: (**c1**) $\alpha_{ij}s = 0.2$; (**c2**) $\alpha_{ij}s = 0.4$; (**c3**) $\alpha_{ij}s = 0.6$; (**c4**) $\alpha_{ij}s = 0.8$; (**c5**) $\alpha_{ij}s = 1$; $\varepsilon_{33} = -1.6$: (**d1**) $\alpha_{ij}s = 0.2$; (**d2**) $\alpha_{ij}s = 0.4$; (**d3**) $\alpha_{ij}s = 0.6$; (**d4**) $\alpha_{ij}s = 0.8$; (**d5**) $\alpha_{ij}s = 1$.

The elastic anisotropy of Fe should be taken into consideration in the simulation because of its effect on multiple parameters, including reaction stress, slip system, orientation change and texture evolution. The textures at $\varepsilon_{33} = -1.1$ are calculated with consideration of elastic anisotropy. The results are shown in Figure 3. By comparing Figure 3(a1–a5) and Figure 2(c1–c5), it is obvious that the effect of elastic anisotropy is weak when other parameters remain unchanged. The orientation dependent G_{ij} affects the value of σ'_{ij}, while it could not accumulate with no limits owing to the existence of reaction stress accumulation upper-limits. Therefore, the influence of elastic anisotropy on the calculated texture is restricted under current conditions. However, it may affect grains with certain orientations, and thus it should still be included in the simulation presented in the following sections.

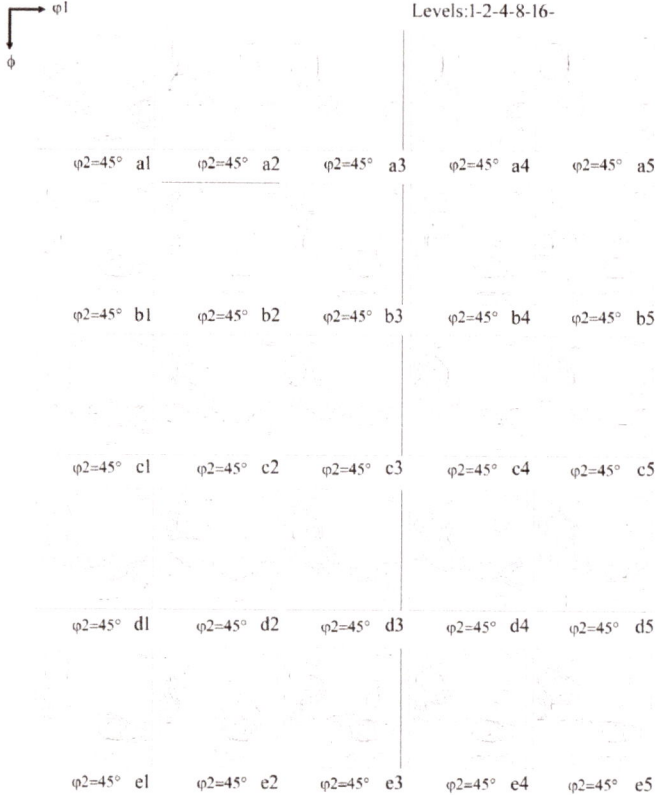

Figure 3. Effects of $\alpha_{ij}s$ on calculated rolling deformation textures at $\varepsilon_{33} = -1.1$ when considering different activation abilities of {110}<111> and {112}<111> slip systems. {110}<111> slip: (**a1**) $\alpha_{ij}s = 0.2$; (**a2**) $\alpha_{ij}s = 0.4$; (**a3**) $\alpha_{ij}s = 0.6$; (**a4**) $\alpha_{ij}s = 0.8$; (**a5**) $\alpha_{ij}s = 1$; {112}<111> slip: (**b1**) $\alpha_{ij}s = 0.2$; (**b2**) $\alpha_{ij}s = 0.4$; (**b3**) $\alpha_{ij}s = 0.6$; (**b4**) $\alpha_{ij}s = 0.8$; (**b5**) $\alpha_{ij}s = 1$; $CRSS_{\{110\}<111>}/CRSS_{\{112\}<111>} = 0.95$: (**c1**) $\alpha_{ij}s = 0.2$; (**c2**) $\alpha_{ij}s = 0.4$; (**c3**) $\alpha_{ij}s = 0.6$;(**c4**) $\alpha_{ij}s = 0.8$; (**c5**) $\alpha_{ij}s = 1$; $CRSS_{\{110\}<111>}/CRSS_{\{112\}<111>} = 1$: (**d1**) $\alpha_{ij}s = 0.2$; (**d2**) $\alpha_{ij}s = 0.4$; (**d3**) $\alpha_{ij}s = 0.6$;(**d4**) $\alpha_{ij}s = 0.8$; (**d5**) $\alpha_{ij}s = 1$; $CRSS_{\{110\}<111>}/CRSS_{\{112\}<111>} = 1.05$: (**e1**) $\alpha_{ij}s = 0.2$; (**e2**) $\alpha_{ij}s = 0.4$; (**e3**) $\alpha_{ij}s = 0.6$;(**e4**) $\alpha_{ij}s = 0.8$; (**e5**) $\alpha_{ij}s = 1$.

The activation abilities of {110}<111> and {112}<111> slip systems are known to greatly influence the calculated texture. With different critical resolved shear stress ratios of {110}<111> vs {112}<111> slip systems (CRSS $_{\{110\}<111>}$/CRSS$_{\{112\}<111>}$), the change of calculated texture with varied α_{isj} is also illustrated in Figure 3. Compared to the result under {110}<111> slip, {112}<111> slip promotes the formation of α-fiber texture in Figure 3(b1–b5). When α_{ijs} is low, {112}<111> slip results in higher percentage of {001}<110> texture. Stronger {001}<110> texture has been reported in silicon steel with higher Si content, and it is suggested that solute Si in the steel makes the activation of {112}<111> slip systems easier, resulting in the formation of {001}<110> texture [20]. This assumption can also be applied to the calculated result shown in Figure 3. Besides {001}<110> texture, α-fiber component between {112}<110> and {111}<110> is also present. When CRSS $_{\{110\}<111>}$/CRSS$_{\{112\}<111>}$ = 1.05, the calculated texture is similar to that when only {112}<111> slip considered, as shown in Figure 3(e1–e5). The slight decrease of {001}<110> texture in Figure 3 proves the enhanced effect of {112}<111> slip on {001}<110> texture. The results in Figure 3(c1–c5,d1–d5) are due to both {110}<111> and {112}<111> slips. In other words, when CRSS $_{\{110\}<111>}$/CRSS$_{\{112\}<111>}$ values are either 0.95 or 1, alternative {110}<111> and {112}<111> slips are activated during rolling deformation. When the activation of {110}<111> slip systems becomes easier, as shown in Figure 3(c1), the distribution of near γ-fiber texture is relatively more continuous.

According to the results in Figures 2 and 3, a set of low α_{ijs} contribute to formation of typical α-fiber and γ-fiber textures in BCC steel but also cause the γ-fiber texture to deviate to Taylor component. Taylor texture is obtained due to reaction stress, and thus it is assumed that lower α_{ijs} may lead to more exact γ-fiber texture. The effect of α_{13} and α_{23} on the deformation texture is shown in Figure 4 when α_{12} and α_{22} are set as 0.2 and as 0 or 0.1, respectively. α_{ijs} are selected based on the methods reported in our previous works [12–15]. In general, compared to ε_{12}, ε_{13} and ε_{23} tend to occur easier and show distinct characteristic under rolling deformation [12,21]. The value of $CRSS_{\{110\}<111>}/CRSS_{\{112\}<111>}$ is chosen to be 0.95, suggesting that {110}<111> slip is presumed to be more likely to be activated [20]. As shown in Figure 4, when α_{13} and α_{23} vary between 0 and 0.08, the calculated texture is a mixture of α-fiber and γ-fiber textures [2,22–24]. Therefore, it can be deduced that for BCC steel, the maximum accumulation of σ'_{12} is about $0.1\sigma_y$, and the existence of σ'_{12} is critical for obtaining γ-fiber texture. In comparison, the accumulation of σ'_{13} and σ'_{23} is restricted in a smaller range. Comparing Figure 4A(a1–c3) to Figure 4B(a1–c3), the {111}<110> texture is attributed to σ'_{22}, while the value of α_{22} still needs to remain small to prevent forming uneven γ-fiber. When α_{13} and α_{23} are discussed in a small range, respectively, their effects on texture evolution interacts with each other and shows a relationship with α_{22}. In detail, at the same α_{23}, higher α_{13} increases α-fiber texture, while different α-fiber texture components occur due to varied α_{22}, and {111}<110> texture is achieved when α_{22} is 0.1. At the same α_{13}, higher α_{23} promotes {111}<112> texture, however, the {111}<112> texture deviates towards Taylor texture when α_{13} increases to 0.08. It concludes that the calculation process is similar to that when tungsten is used, whereas the specific α_{ij} values corresponding to the reasonable simulated textures are different [15].

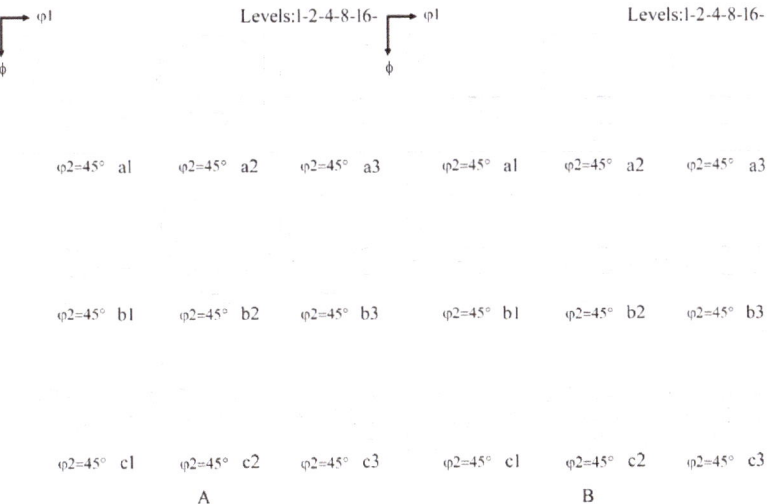

Figure 4. The effect of α_{13}, α_{23} and α_{22} not higher than 0.08 on calculated texture when using α_{12} = 0.2 and $CRSS_{\{110\}<111>}/CRSS_{\{112\}<111>}$ = 0.95. (A) α_{22} = 0: (a1) α_{13} = α_{23} = 0; (a2) α_{13} = 0.04, α_{23} = 0; (a3) α_{13} = 0.08, α_{23} = 0; (b1) α_{13} = 0, α_{23} = 0.04; (b2) α_{13} = α_{23} = 0.04; (b3) α_{13} = 0.08, α_{23} = 0.04; (c1) α_{13} = 0, α_{23} = 0.08; (c2) α_{13} = 0.04, α_{23} = 0.08; (c3) α_{13} = α_{23} = 0.08. (B) α_{22} = 0.1: (a1) α_{13} = α_{23} = 0; (a2) α_{13} = 0.04, α_{23} = 0; (a3) α_{13} = 0.08, α_{23} = 0; (b1) α_{13} = 0, α_{23} = 0.04; (b2) α_{13} = α_{23} = 0.04; (b3) α_{13} = 0.08, α_{23} = 0.04; (c1) α_{13} = 0, α_{23} = 0.08; (c2) α_{13} = 0.04, α_{23} = 0.08; (c3) α_{13} = α_{23} = 0.08.

Based on previous results and discussion, the simulation process and selected parameters can be further optimized so that the calculated textures match more closely to the real deformation process [2,23,24]. As shown in Figure 5, typical α-fiber and γ-fiber textures are obtained at 50%

reduction, and both textures are enhanced with increasing strain. {111}<110> texture as well as other α-fiber texture components becomes stronger while {111}<112> texture is weakened at higher strain. When the accumulation of σ'_{23} is permitted, compared to the situation of relaxing σ'_{23} in Figure 5(a1–a3), stronger {111}<112> texture is shown in Figure 5(c1–c3). When comparing Figure 5(a1–a3) and Figure 5(b1–b3), the preferred {110}<111> slip leads to stronger {112}<110> texture than {001}<110> texture. The stronger {111}<110> texture is also observed at higher strain.

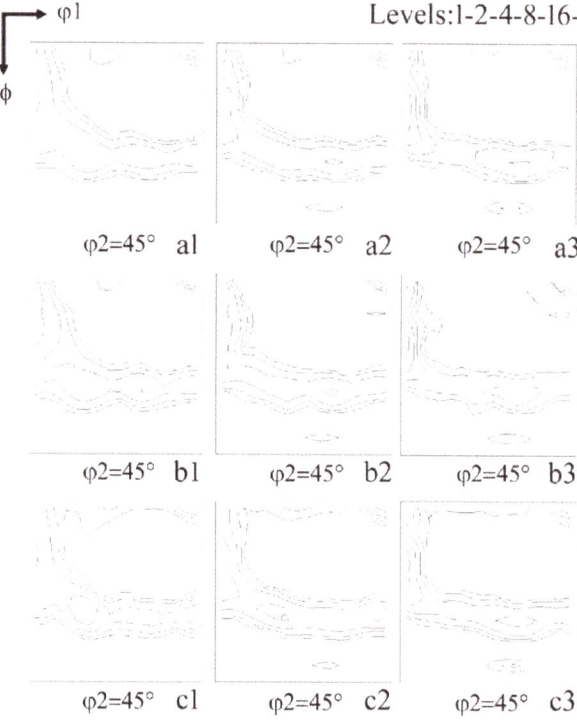

Figure 5. The calculated rolling deformation textures under different strains when using $a_{12} = 0.2$, $a_{13} = 0.04$, $a_{22} = 0.04$ while different a_{23} and CRSS$_{\{110\}<111>}$/CRSS$_{\{112\}<111>}$ values. $a_{23} = 0$, CRSS$_{\{110\}<111>}$/CRSS$_{\{112\}<111>}$ = 0.95: **(a1)** $\varepsilon_{33} = -0.69$, **(a2)** $\varepsilon_{33} = -1.1$, **(a3)** $\varepsilon_{33} = -1.6$; $a_{23} = 0$, CRSS$_{\{110\}<111>}$/CRSS$_{\{112\}<111>}$ = 1: **(b1)** $\varepsilon_{33} = -0.69$, **(b2)** $\varepsilon_{33} = -1.1$, **(b3)** $\varepsilon_{33} = -1.6$; $a_{23} = 0.04$, CRSS$_{\{110\}<111>}$/CRSS$_{\{112\}<111>}$ = 0.95: **(c1)** $\varepsilon_{33} = -0.69$, **(c2)** $\varepsilon_{33} = -1.1$, **(c3)** $\varepsilon_{33} = -1.6$.

Since the simulated textures in Figure 5 show characteristics of a typical rolling deformation texture of BCC structured steel, the selection of corresponding calculation parameters is proved to be reasonable. The experimental measurement shown in Figure 6 illustrates the texture evolution of a cold rolled Fe-Si steel. A weak texture with an intensity less than 1.4 is displayed in the sample prior to rolling, which is similar to that used in the simulation. The cold-rolling process generated α-fiber and γ-fiber textures in 52% ($\varepsilon_{33} = -0.73$) steel, as shown in Figure 6b. When the rolling reduction increases to 80% ($\varepsilon_{33} = -1.6$), the peak of γ-fiber changes from {111}<112> to {111}<110> in Figure 6c. At the same time, α-fiber texture is also enhanced. The texture demonstrates a similar evolution process as the simulated results in Figure 5. Therefore, RS model is considered to be applicable for texture simulation for BCC steel, and the parameters in Figure 5 can be used for simulating this rolling process. Confirmed by both theoretical and experimental analysis, for this Si-Fe material, the {110}<111> slip system is easier to be activated to achieve a strong {112}<110> texture at high strain, and the accumulation of ε_{23} should be permitted while in lower range than Figure 5(c1–c3).

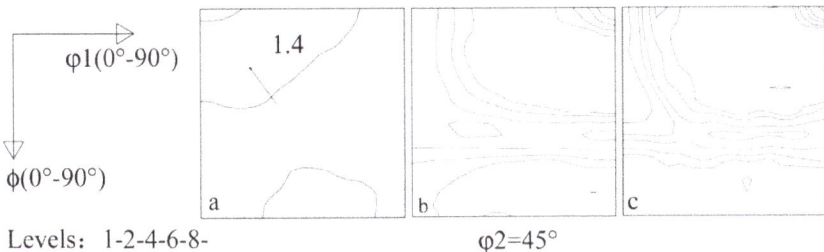

Figure 6. Measured textures of a Fe-0.46wt%Si steel. (a) Prior to rolling; (b) 52% reduction; (c) 80% reduction.

To sum up, when calculating the texture evolution in BCC steel, the applied assumptions include a moderate accumulation of σ'_{12} and much narrower range of the rest of σ'_{ijs}. Similar to relaxed constraint (RC) model, the strain consistency of ε_{13} and ε_{23} is easy to be fulfilled or be compensated by local slip. Through this compensation strategy, the strain continuity in some directions is relaxed and the simulated result is more reasonable than that by Taylor model [25,26]. Although modification of simulation parameters will cause minor changes in the simulated results, major texture distribution and evolution trend remains stable.

It is worth noting that RS model considers the deformation process in a statistical view and the orientation change of major grains contributes to the statistical texture evolution. RS model can also be used to analyze the orientation evolution of many texture components. Based on previous discussions, the calculation parameters used in Figure 5(a1–a3) are selected for theoretically analyzing the evolutions of {001}, {110}, α-fiber and γ-fiber orientations under rolling deformation. These orientations are commonly observed in BCC steels: {001} orientation corresponds to columnar grains; {110} orientations including Goss and brass components are usually obtained after hot rolling silicon steel; α-fiber and γ-fiber texture can be achieved after both hot and cold rolling.

In Figure 7(a1,a3), {001}<100>((45° 0° 45°)) and {001}<110>((0° 0° 45°)) orientations are stable under rolling deformation, and no changes in orientation are observed. Cube orientation shows meta-stability during rolling in silicon steel [27], and stable end {001}<110> orientation is widely reported [28,29]. {001}<120>((20° 0° 45°)) orientation tends to rotate towards α-fiber and deviate from {001}-fiber simultaneously in Figure 7(a2). A similar crystal rotation route is observed in other researchers' work [27,30] as well as in our previous report when investigating the occurrence of {001}<112>-{113}<361> texture in silicon steel. Stability in {112}<110>((0° 35° 45°)) orientation is also confirmed, and the α-fiber components between it and {001}<110> components rotate to them alternatively in Figure 7(b1–b3), depending on which component it is initially closer to. In consequence, {001}<110> and {112}<110> textures on α-fiber become stronger with strain, as shown in Figure 5. As far as the γ-fiber components are concerned, as shown in Figure 7(c1–c3), {111}<110>((0° 55° 45°)) orientation rotates towards {112}<110> orientation, whereas {111}<112>((30° 55° 45°)) orientations show strong stability and barely shift to other orientations. The (15° 55° 45°) orientation firstly rotates to {111}<110> orientation and then transits to {112}<110> orientation. Therefore, it can be deduced that strengthening {111}<110> texture with increasing strain may be not only due to its stability, but also affected by rotation from other orientations. For {110} shear texture components in Figure 7(d1–d3), Goss((90° 90° 45°)) orientation is stable at the early stage but rotates to {111}<112> orientation at higher strain; {110}<229>((75° 90° 45°)) orientation rotates to γ-fiber closer to {111}<112> orientation in the beginning of rolling and then turns to near {111}<110> orientation; brass orientation((60° 90° 45°)) shows a transition route from γ-fiber towards α-fiber in the end. These transition routes for {110} orientations are consistent with the experimental data reported in our early work [31]. Overall, for most of the typical orientations in Figure 7(a1–d3), the rotation preferentially revolving towards α-fiber or γ-fiber is predicted.

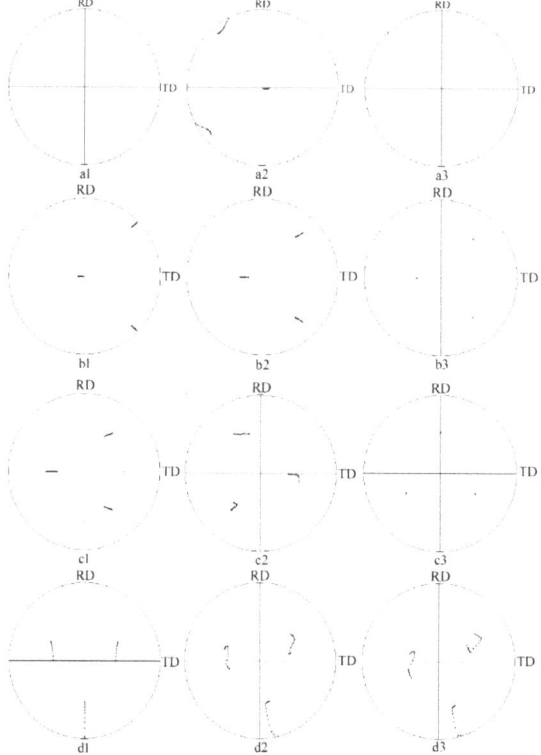

Figure 7. Calculated orientation change of typical texture components of BCC steel during rolling deformation to $\varepsilon_{33} = -1.1$, {200} pole figures. (**a1**) (45° 0° 45°); (**a2**) (20° 0° 45°); (**a3**) (0° 0° 45°); (**b1**) (0° 10° 45°); (**b2**) (0° 20° 45°); (**b3**) (0° 35° 45°); (**c1**) (0° 55° 45°); (**c2**) (15° 55° 45°); (**c3**) (90° 55° 45°); (**d1**) (90° 90° 45°); (**d2**) (75° 90° 45°); (**d3**) (60° 90° 45°).

4. Discussion

The influence of the reaction stress accumulation coefficients α_{ijs} on the stress tensor is evident by restricting the changing range of reaction stress. When the value of reaction stress σ'_{ij} exceeds the upper-limit, it is set to be equal to the upper-limit value. According to the calculation results and discussion in this paper, the reaction stress should not be accumulated to a high level, however, moderate or even minor upper-limits are allowed. During the calculation, small reaction stress in the whole stress tensor may cause occurrence of a new slip, thus the accumulation of pre-existing reaction stress can be eliminated. Meanwhile, change in reaction stress value also affects the stress condition of neighboring grains, and the plastic slip of neighboring grains can also terminate accumulation of reaction stress. Under the combined effect of external and reaction stress with reasonable upper-limits, multiple slips are generated, and the simulated evolution tends to closely match the real situation when the calculation step is sufficiently refined.

On the other hand, moderate or even smaller α_{ij} values are the intrinsic requirement of RS model. For doing so the strain that corresponds to the elastic reaction stress will remain as elastic strain during the process. As shown in Figure 1, besides penetrating the plastic slip and elastic slip, the intergranular strain consistency requires the participation of a local plastic slip. In the case that reaction stress is restricted to the upper-limits, considering the low α_{ij} values, the corresponding elastic strain may be much lower than the strain value used in Equation (2), thus the effect of local slip on strain consistency cannot be ignored.

During the deformation, an active slip usually causes crystal orientation to change, and the strain is accumulated if no other slip systems are activated. For polycrystals, the strain caused by active slip leads to discontinuity between a deformed grain and its environment, and this discontinuity shall be reduced by spontaneous reaction stress [15]. In this study, some crystal symmetrical orientations show stability to a certain extent under rolling deformation. Figure 7 illustrates that, {001}<100> and {110}<001> orientations are stable at the early rolling stage. For these symmetrical orientations, the Schmid factors of two or more slip systems are identical under external stress. For example, after a slip occurs in a grain, cube oriented single crystal shows one-directional crystal rotation around <001> axis [28,32], and then an opposite slip system with the highest Schmid factor will be activated by the accumulated reaction stress in polycrystals. The alternative activation of two slip systems leads to the orientation stability. Meanwhile the strains caused by alternative slips are compensated by each other, and so the total strain is reduced.

5. Summary

A simulation for cold rolling deformation texture in a BCC steel is well established based on a RS model. The model considers stress and strain consistency in a natural way, in which the strain consistency is achieved by cooperation of two plastic strains and an elastic strain. The reaction stress poses significant effect on texture evolution, and varied accumulation upper-limits in different directions are indicated. With different values of reaction stress accumulation coefficient α_{ijs}, {110}<111> slip leads to Taylor or γ-fiber texture, and {112}<111> slip strengthens α-fiber texture, including {001}<110> texture component. The elastic anisotropy influences the values of reaction stress, while the upper-limit of reaction stress noticeably reduces the effect of elastic anisotropy. As the α_{12} is set to be 0.2, typical α-fiber and γ-fiber texture are achieved even when low values of other α_{ijs} are applied in simulation. The accumulation of σ'_{12} is not high, however, its effect on texture formation is not trivial. In the meantime, the change of other reaction stress σ'_{ijs} affects the distribution of γ-fiber texture. The selected α_{ij} values based on calculated texture results can also be used to simulate the orientation change of many specific orientations, and reaction stress is shown to help crystallographically symmetrical orientations, of which the Schmid factors of several slip systems are identical, remaining stable under rolling deformation.

Author Contributions: Conceptualization, N.Z. and L.M.; software, N.Z.; formal analysis, N.Z.; investigation, N.Z.; resources, L.M. and W.Z.; data curation, N.Z.; writing—original draft preparation, N.Z.; writing—review and editing, L.M.; funding acquisition, L.M., W.Z. and W.M. All authors have read and agreed to the published version of the manuscript.

Funding: This research was funded by the National Key Research and Development Program of China (grant number 2017YFB0903901), the National Natural Science Foundation of China (grant number 51571024) and the Shanxi Provincial Science and Technology Major Special Project (grant number 20191102004). And the APC was funded by the National Key Research and Development Program of China (grant number 2017YFB0903901).

Acknowledgments: Thank you to Sheng Zhong for his English editing help.

Conflicts of Interest: The authors declare no conflict of interest.

References

1. Park, J.T.; Szpunar, J.A. Evolution of recrystallization texture in nonoriented electrical steels. *Acta Mater.* **2003**, *51*, 3037–3051. [CrossRef]
2. Kestens, L.A.I.; Pirgazi, H. Texture Formation in Metal Alloys with Cubic Crystal Structures. *Mater. Sci. Technol.* **2016**, *32*, 1303–1315. [CrossRef]
3. Das, A.J. Calculation of crystallographic texture of BCC steels during cold rolling. *Mater. Eng. Perform.* **2017**, *26*, 2708–2720. [CrossRef]

4. Engler, O.; Crumbach, M.; Li, S. Alloy-dependent rolling texture simulation of aluminium alloys with a grain-interaction (GIA) model. *Acta Mater.* **2005**, *53*, 2241–2257. [CrossRef]
5. Mu, S.J.; Tang, F.; Gottstein, G. A cluster-type grain interaction deformation texture model accounting for twinning-induced texture and strain-hardening evolution: Application to magnesium alloys. *Acta Mater.* **2014**, *68*, 310–324. [CrossRef]
6. Van Houtte, P.S.; Li, M.; Seefeldt, M.; Delannay, L. Deformation texture prediction: From the Taylor model to the advanced Lamel model. *Int. J. Plast.* **2005**, *21*, 589–624. [CrossRef]
7. Molinari, A.; Canova, G.R.; Ahzi, S. A self consistent approach of the large deformation polycrystal viscoplasticity. *Acta Metall.* **1987**, *35*, 2983–2994. [CrossRef]
8. Lebensohn, R.A.; Tomé, C.N. A self-consistent anisotropic approach for the simulation of plastic deformation and texture development of polycrystals-Application to zirconium alloys. *Acta Metall. Mater.* **1993**, *41*, 2611–2624. [CrossRef]
9. Lebensohn, R.A.; Tomé, C.A.; Castañeda, P.P. Self-consistent modelling of the mechanical behaviour of viscoplastic polycrystals incorporating intragranular field fluctuations. *Philos. Mag.* **2007**, *87*, 4287–4322. [CrossRef]
10. Lebensohn, R.A.; Turner, P.A.; Signorelli, J.W.; Canova, G.R.; Tomé, C.A. Calculation of intergranular stresses based on a large strain visco-plastic self-consistent model. *Mod. Sim. Mats. Sc. Eng.* **1998**, *6*, 447–465. [CrossRef]
11. Xie, Q.G.; Van Bael, A.; Sidor, J.; Moerman, J.; Van Houtte, P. A new cluster-type model for the simulation of textures of polycrystalline metals. *Acta Mater.* **2014**, *69*, 175–186. [CrossRef]
12. Mao, W.M. Intergranular mechanical equilibrium during rolling deformation of polycrystalline metals based on Taylor principles. *Mater. Sci. Eng. A* **2016**, *672*, 129–134. [CrossRef]
13. Mao, W.M.; Yu, Y.N. Effect of elastic reaction stress on plastic behaviors of grains in polycrystalline aggregate during tensile deformation. *Mater. Sci. Eng. A* **2004**, *367*, 277–281. [CrossRef]
14. Mao, W.M. Influence of Intergranular Mechanical Interactions on Orientation Stabilities during Rolling of Pure Aluminum. *Front. Mater. Sci.* **2018**, *12*, 322–326. [CrossRef]
15. Zhang, N.; Mao, W.M. Study on the cold rolling deformation behavior of polycrystalline tungsten. *Int. J. Refract. Met. Hard Mater.* **2019**, *80*, 210–215. [CrossRef]
16. Ray, R.K.; Jonas, J.J.; Hook, R.K. Cold rolling and annealing textures in low carbon and extra low carbon steels. *Int. Mater. Rev.* **1994**, *39*, 129–172. [CrossRef]
17. Guo, Y.H.; Wang, Z.D.; Li, S.W.; Zou, W.W. Comparison of texture and properties of IF steel with ELC Steel. *J. Northeast. Univ. (Nat. Sci.)* **2007**, *28*, 1713–1716.
18. Shen, Y.F.; Xue, W.Y.; Guo, Y.H. Effect of cold rolling and annealing on texture evolution and mechanical properties of IF steel sheet. *Steel Res. Int.* **2010**, *81*, 146–149.
19. Sung, J.K.; Lee, D.N.; Wang, D.H.; Koo, Y.M. Efficient generation of cube-on-face crystallographic Texture in iron and its alloys. *ISIJ Int.* **2011**, *51*, 284–290. [CrossRef]
20. Mao, W.M.; Yang, P. *Material Science Principles on Electrical Steels*; Higher Education Press: Beijing, China, 2013; pp. 156–157.
21. Hirsch, J.; Lücke, K. Mechanism of deformation and development of rolling texture in polycrystalline Fcc metals-II. *Acta Metall.* **1988**, *36*, 2883–2904. [CrossRef]
22. Raabe, D.; Lücke, K. Rolling and annealing textures of BCC metals. *Mater. Sci. Forum* **1994**, *157*, 597–610. [CrossRef]
23. Kestens, L.A.I.; Jacobs, S. Texture control during the manufacturing of nonoriented electrical steels. *Text. Stress Microstruct.* **2008**, *2008*, 173083. [CrossRef] [PubMed]
24. Bate, P.S.; da Fonseca, J.Q. Texture development in the cold rolling of IF steel. *Mater. Sci. Eng. A* **2004**, *380*, 365–377. [CrossRef]
25. Van Houtte, P. A comprehensive mathematical formulation of an extended Taylor-Bishop-Hill model featuring relaxed constraints, the Renouard-Wintenberger theory and a strain rate sensitivity model. *Text. Microstruct.* **1988**, *8*, 313–350. [CrossRef]
26. Leffers, T.; Jensen, D.J. Evaluation of the effect of initial texture on the development of deformation texture. *Text. Microstruct.* **1986**, *16*, 231–254. [CrossRef]
27. Sha, Y.H.; Sun, C.; Zhang, F. Strong cube recrystallization texture in silicon steel by twin-roll casting process. *Acta Mater.* **2014**, *76*, 106–117. [CrossRef]

28. Walter, J.L.; Hibbard, W.R. Texture of cold-rolled and recrystallized crystal of Silicon-Iron. *Trans. Metall. Soc. AIME* **1958**, *212*, 731–737.
29. Tsuji, N.; Tsuzaki, K.; Maki, T. Effect of initial orientation on the cold rolling behavior of solidified columnar crystals in a 19% Cr ferritic stainless steel. *ISIJ Int.* **1992**, *32*, 1319–1328. [CrossRef]
30. Zhang, N.; Yang, P.; Mao, W.M. {001}<120>-{113}<361> recrystallization textures induced by initial {001} grains and related microstructure evolution in heavily rolled electrical steel. *Mater. Character.* **2016**, *119*, 225–232. [CrossRef]
31. Zhang, N.; He, C.X.; Yang, P. Effect of {110}<229> and {110}<112> grains on texture evolution during cold rolling and annealing of electrical steel. *ISIJ Int.* **2016**, *56*, 1462–1469. [CrossRef]
32. Abe, H.; Matsuo, M.; Ito, K. Cold rolling and recrystallization textures of Silicon-Iron crystals rolled in (100)[001] orientation. *Trans. JIM* **1963**, *4*, 28–32. [CrossRef]

© 2020 by the authors. Licensee MDPI, Basel, Switzerland. This article is an open access article distributed under the terms and conditions of the Creative Commons Attribution (CC BY) license (http://creativecommons.org/licenses/by/4.0/).

Article

Kinetics of Capability Aging in Ti-13Nb-13Zr Alloy

Myoungjae Lee [1], In-Su Kim [1,2], Young Hoon Moon [1], Hyun Sik Yoon [3], Chan Hee Park [2] and Taekyung Lee [1,*]

[1] School of Mechanical Engineering, Pusan National University, Busan 46241, Korea; leo1991@pnu.ac.kr (M.L.); sionsy@kims.re.kr (I.-S.K.); yhmoon@pusan.ac.kr (Y.H.M.)
[2] Advanced Metals Division, Korea Institute of Materials Science, Changwon 51508, Korea; chpark@kims.re.kr
[3] Department of Naval Architecture and Ocean Engineering, Pusan National University, Busan 46241, Korea; lesmodel@pusan.ac.kr
* Correspondence: taeklee@pnu.ac.kr

Received: 27 July 2020; Accepted: 10 August 2020; Published: 11 August 2020

Abstract: Metals for biomedical implant applications require a simultaneous achievement of high strength and low Young's modulus from the viewpoints of mechanical properties. The American Society for Testing and Materials (ASTM) standards suggest two types of processing methods to confer such a mechanical performance to Ti-13Nb-13Zr alloy: solution treatment (ST) and capability aging (CA). This study elucidated the kinetics of CA process in Ti-13Nb-13Zr alloy. Microstructural evolution and mechanical change were investigated depending on the CA duration from 10 min to 6 h. The initial ST alloy possessed the full α'-martensitic structure, leading to a low strength, low Young's modulus, and high ductility. Increasing CA duration increased mechanical strength and Young's modulus in exchange for the reduction of ductility. Such a tendency is attributed to the decomposition of α' martensite into ($\alpha+\beta$) structure, particularly hard α precipitates. Mechanical compatibility (i.e., Young's modulus compensated with a mechanical strength) of Ti-13Nb-13Zr alloy rarely increased by changing CA duration, suggestive of the intrinsic limit of static heat treatment.

Keywords: titanium; Ti-13Nb-13Zr; aging; microstructure; tensile properties; Young's modulus

1. Introduction

Ti-6Al-4V alloy is one of the most used titanium alloys due to its good mechanical properties, lightweight characteristics, and high corrosive resistance. The alloy has been used in various industries including aerospace, automotive, and biomedical applications. For the last section, however, researchers have claimed that toxic Al and V ions are released in a human body by wear and corrosion of the Ti-6Al-4V alloy [1]. This led to various endeavors to replace those alloying elements with safer ones, such as Nb, Ta, and Zr [2]. Ti-13Nb-13Zr alloy is one of accomplishments from such endeavors, which is registered in the American Society for Testing and Materials (ASTM) standards for surgical implant applications [3].

Besides the cytotoxicity, implant materials have two requirements in terms of mechanical properties: high strength and low Young's modulus. The latter is the main interest, because it is much more difficult to control in comparison to the former. Young's modulus of human bone tissue is reported to be 10–30 GPa [4]. Although Ti and its alloys have been used for biomedical applications, their elastic moduli are still higher than that of bone tissue [5]. Excessive Young's moduli of traditional biomedical metals (e.g., Co-Cr-Mo, Grade 4 Ti, Ti-6Al-4V, and stainless steels) lead to a stress-shielding effect; they decrease an amount of stress applied to bone tissue near an implant, resulting in the weakening of human bone [6]. Compared with those materials, Ti-13Nb-13Zr is also beneficial to biomedical applications due to its moderately low Young's modulus.

Mechanical properties of Ti alloys, including Young's modulus, are tailored by either plastic deformation or heat treatment. Both approaches change a microstructure of alloy towards the way beneficial to mechanical improvement. For example, Park et al. [7] employed a cross-rolling of water-quenched Ti-13Nb-13Zr to exploit a beneficial effect of dynamic globularization. Lee et al. [6] recently suggested a cold caliber rolling to confer high strength and low Young's modulus to Ti-13Nb-13Zr alloy. According to their study, the improvement stemmed from the effective grain refinement and formation of metastable martensitic phases. In contrast, relatively less attention has been paid to the other type of approach (i.e., heat treatment). Although Geetha et al. [8] and Majumdar et al. [9] carried out an extensive study of heat treatment, they focused on heat-treatment temperature and cooling rate rather than heat-treatment duration. The objective of this study is to clarify the kinetics of heat treatment in Ti-13Nb-13Zr alloy. Specifically, this study elucidates the relationship between the heat-treatment duration and mechanical compatibility (i.e., Young's modulus compensated with a mechanical strength) of Ti-13Nb-13Zr alloy.

2. Materials and Methods

The Ti-13Nb-13Zr alloy used in this study possessed a specific chemical composition of 13.8 mass% Nb, 14.0 mass% Zr, 0.06 mass% Fe, 0.07 mass% O, 0.01 mass% C, 0.007 mass% N, and balance Ti. Samples were prepared in the dimension of 10 × 10 × 5 mm^3. Geetha et al. [10] reported β-transus temperature (T_β) of 1,008 K for Ti-13Nb-13Zr alloy. Accordingly, Ti-13Nb-13Zr specimens were solution treated (ST) for 1073 K for 1 h and then soaked in a water bath to induce a rapid cooling [6,7]. Sections of these specimens were subsequently aged at 773 K for 10 min, 60 min, 180 min, and 360 min, respectively, followed by air cooling. ASTM standard for the Ti-13Nb-13Zr alloy [3] refers to this aging process as a "capability aging" (CA). Hereafter, the aged specimens are denoted as CA10, CA60, CA180, and CA360 where the numbers indicate their CA duration in minutes.

The samples were water-abraded with #400, #800, #1200, #2400 emery paper, and then treated in different ways for characterization methods. No evidence for oxidation has been found after the mechanical polishing. Samples for X-ray diffraction (XRD, D8 Advance, Bruker, Billerica, MA, USA) analysis were mirror-polished using 3-μm and 1-μm ethanol-based diamond suspensions, respectively. Samples for scanning electron microscopy (SEM, SU8020, Hitachi, Tokyo, Japan) were prepared in the same manner followed by the final polishing with 0.05-μm colloidal silica. Afterwards, SEM specimens were chemically etched in an aqueous solution containing 2% nitric acid and 2% fluoric acid. Samples for electron backscatter diffraction (EBSD, FEI QUANTA 3D FEG, Hillsboro, OR, USA) analysis were electropolished at 35 V in a solution of 5 mL perchloric acid (6%) and 95 mL methanol. XRD measurement was performed in 2θ range from 30° to 40° at a scan speed of 0.75°·min^{-1}. EBSD data were analyzed using the TSL OIM software ver. 8 with a confidence index higher than 0.1.

Young's modulus of heat-treated Ti-13Nb-13Zr alloys was determined in two ways. First, the alloys were machined into a cylindrical tensile specimen with gauge length in 10 mm and gauge diameter in 2.5 mm. Tensile tests were repeated three times at a strain rate of 10^{-3} s^{-1} to secure a data reproducibility. Second, Young's modulus (E) was measured using the velocities of an ultrasonic longitudinal wave (V_L) and transverse wave (V_T), expressed as follows [11]:

$$E = [\rho V_T^2(3V_L^2 - 4V_T^2)]/(V_L^2 - V_T^2), \tag{1}$$

where ρ is the density of the investigated alloys.

3. Results

Figure 1 demonstrates microstructures of the investigated Ti-13Nb-13Zr alloys. ST alloy was investigated through the EBSD analysis as the SEM observation rarely confirmed its fine laths. The average width of its laths was determined to be 0.51 ± 0.24 μm (numbers including the standard deviation). The EBSD micrograph also confirmed a high frequency of submicron-sized twins inside relatively thick laths. These results are in good accordance with microstructural characteristics observed by transmission electron microscopy (TEM) [12]. The EBSD analysis yielded a volume fraction of β phase of 0.6%; however, this phase is highly likely to be absent in the ST alloy considering the thermomechanical processing route as well as an experimental error of measurement. This leaves the deduction that the ST alloy was fully composed of α-variant phases (i.e., α, α′, and α″). The detailed phase constituents in the investigated Ti-13Nb-13Zr alloys are further discussed in Section 4.

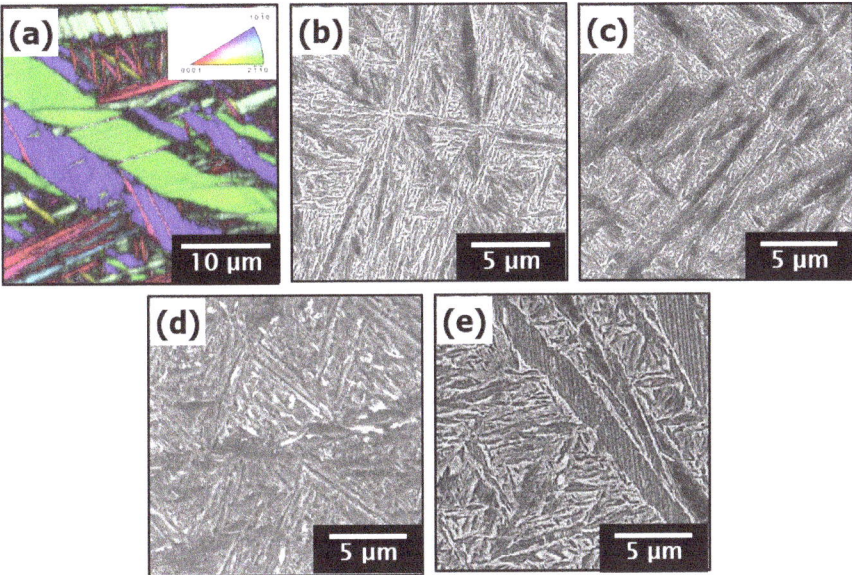

Figure 1. Microstructure of the investigated Ti-13Nb-13Zr alloys: (**a**) electron backscatter diffraction (EBSD) inverse pole figure map of ST alloy; (**b**) SEM micrograph of CA10, (**c**) CA60, (**d**) CA180, and (**e**) CA360 alloys.

Although CA samples presented the similar lath structure, the average width increased with an increase in CA duration: 1.64 ± 0.5 µm for CA10, 2.42 ± 1.3 µm for CA60, 2.73 ± 1.1 µm for CA180, and 3.01 ± 1.5 µm for CA360. This is not surprising because the longer CA process induced more active atomic diffusion and grain growth. The length of α laths extends more rapidly when compared with the thickness due to the diffusional ledge mechanism [13]. It is also noted that the CA360 alloy was fabricated on the basis of the ASTM standard [3]. This standardized procedure rapidly increased the lath width six-fold in comparison to that of the ST sample. The rate of grain growth became retarded as the alloy was aged for a longer duration. For example, the lath size increased by 0.78 µm after CA treatment from 10 min to 60 min, whereas it increased by 0.28 µm from 180 min to 360 min despite 3.6-time longer duration of the heat treatment.

Figure 2 presents the XRD line profile of the investigated Ti-13Nb-13Zr alloys. All samples show peaks at 2θ angle of 34.5°, 38°, and 39.5°. These peaks correspond to the α-variant phases, which are hardly distinguishable via XRD analysis due to a negligible difference in 2θ angles among the phases [6]. Nevertheless, the peaks for ST alloy are considered as the indication of α′ martensite, as discussed in Section 4. It is of particular note that CA180 and CA360 alloys exhibited a "shoulder" peak in the vicinity of the peak generated at 2θ angle of 38°, as marked by the arrows in Figure 2. Such a peak corresponds to (110)β as confirmed in previous studies [14–16].

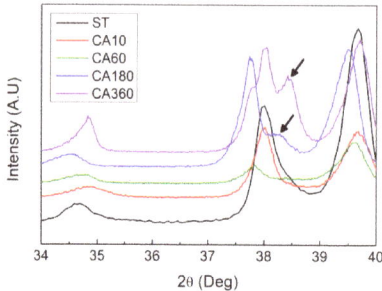

Figure 2. XRD line profiles of the investigated Ti-13Nb-13Zr alloys. The arrows indicate the "shoulder" peak corresponding to (110)β phase.

Figure 3 shows a tensile deformation behavior of the investigated Ti-13Nb-13Zr alloys. Obviously, the flow curves of CA alloys are distinguished from that of the ST alloy in the aspects of an increase in yield strength (YS), ultimate tensile strength (UTS), and Young's modulus (E) as well as a decrease in elongation to failure (EL). Even the short-time (10 min) CA treatment considerably increased the mechanical strength in exchange for the EL reduction. Such trends (i.e., increasing strengths and decreasing EL) were intensified with an increase in CA duration. Meanwhile, ST alloy exhibited a remarkable hardening (i.e., UTS—YS) of 247 MPa, which was suppressed after applying the CA process, as shown in Figure 3b. The hardening decreased with increasing CA duration: 82 MPa for CA10, 76 MPa for CA60, 66 MPa for CA180, and 54 MPa for CA360.

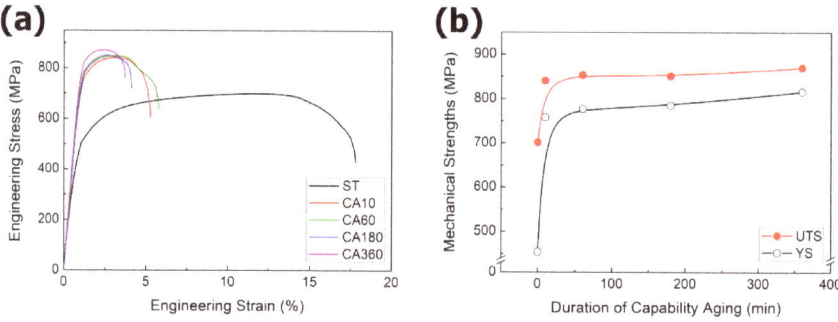

Figure 3. (a) Engineering stress-strain curves and (b) variation in mechanical strengths of the investigated Ti-13Nb-13Zr alloys as a function of capability aging (CA) duration. The data at the zero duration in Figure 3b indicate YS and UTS of ST alloy.

Figure 4 demonstrates increasing elastic moduli with increasing duration of the CA process, which is similar to the tendency for YS and UTS shown in Figure 3b. The trends were consistent in both measurements (i.e., tensile test and ultrasonic method). The data obtained from the slope of flow curve in the elastic regime were consistently higher than those measured via the ultrasonic method. Such a difference is acceptable in light of the intrinsic nature of the ultrasonic method, consistent with previous studies [11,17]. The mechanical properties of ST and CA360, fabricated on the basis of the ASTM standard, are in good accordance with those reported earlier [3,6,7], ensuring the data reliability of the present measurements. Table 1 summarizes the data of mechanical properties for the investigated materials.

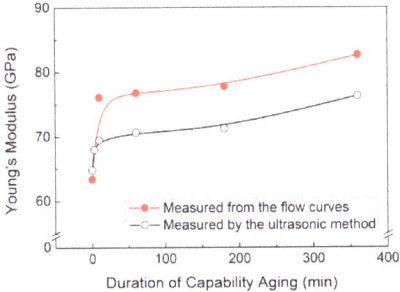

Figure 4. Variation in elastic moduli of the investigated Ti-13Nb-13Zr alloys as a function of CA duration. The data at the zero duration indicate Young's moduli of ST alloy.

Table 1. Mechanical properties of the investigated ST and CA alloys.

Sample	YS (MPa)	UTS (MPa)	E (GPa) *	E (GPa) †	EL (%)
ST	454	701	63.4	64.8	17.8
CA10	758	840	76.1	69.5	5.3
CA60	777	853	76.8	70.7	5.8
CA180	785	851	77.8	71.3	4.1
CA360	816	870	82.6	76.3	3.7

* Measured from the slope of flow curves in the elastic regime; † measured via the ultrasonic method.

4. Discussion

The Ti-13Nb-13Zr alloy can have five types of phases depending on a thermomechanical processing route: hexagonal close-packed (hcp) α, body-centered cubic β, hcp ω, hcp α' martensite, and orthorhombic α'' martensite. The α and β phases are stable phases, while the others are metastable. Ti-13Nb-13Zr alloy has a significantly lower T_β as compared with Ti-6Al-4V alloy [10]. Davidson et al. [18] suggested a martensite starting and finishing temperatures of 823 K and 758 K, respectively. Accordingly, the water quenching from the temperature above T_β induced the formation of full α'-martensitic structure in ST alloy. Kobayashi et al. [19] conducted a TEM study related to phase transformation in Ti-13Nb-13r alloy during a heat treatment at 873 K. A rapid cooling from the β domain causes a phase transformation into the full α'-martensitic phase, as consistent with the literature above. The subsequent heat treatment at 873 K leads to the formation of α'', ω, and β phase in order with increasing duration from 5 min to 24 h. The size and volume fraction of α precipitates are proportional to the heat-treatment temperature [20,21], of which the present condition (773 K) was lower than that adopted by the previous work (873 K) [19]. This resulted in significantly small phases in the present CA alloys. In other words, α'' and ω phases, as well as β in the short-time aged samples, were too small to be observed through SEM and EBSD methods. Meanwhile, α phase precipitated at the phase interfaces and twin boundaries of the initial α' martensite, resulting in the decomposition into stable α and β phases [22]. Their conclusions are consistent with the present results of (i) the full α'-martensitic structure of ST alloy, (ii) the presence of β peaks for CA180 and CA360 samples, and (iii) the decomposition of α' martensite into α and β phases during the CA process.

Young's modulus is an intrinsic material property determined by the atomic bonding forces. Such forces are dependent not only on a crystal structure but also on the distance between atoms, and thus affected by alloying elements, heat treatment, and plastic deformation [23,24]. In the present study, the primary factor for Young's modulus is the presence of various phases. The lowest Young's modulus of ST alloy arose from the full α'-martensitic structure [9]. The subsequent CA process induced the formation of hard α precipitates possessing higher strength and Young's modulus [25]. Increasing fraction of this phase with increasing CA duration gave rise to an increment in strength and Young's modulus of the investigated Ti-13Nb-13Zr alloys, as presented in Figures 3 and 4. The reduction of hardening with increasing CA duration is also understood in terms of increasing the fraction of α precipitates.

Recalling Figure 4, Young's modulus rapidly increased in the early stage of CA process, followed by the retarded increment in the late stage. An additional sample was fabricated by applying a 3-min CA process to confirm this tendency. It exhibited Young's modulus of 68 GPa based on the ultrasonic method, which was close to that of CA10 rather than ST alloy. These results imply that the α' decomposition into (α+β) structure occurred as soon as the alloy was subjected to the CA process.

The results obtained in this work have proven the proportional relationship between mechanical strength and Young's modulus. Such a relation is unfavorable for a biomedical application that requires high strength and a low Young's modulus. This is further supported in terms of mechanical compatibility defined as the ratio of YS to Young's modulus. Figure 5 presents the mechanical compatibilities of the investigated Ti-13Nb-13Zr alloys as well as reported values [6,7,9,12,18,26,27]. Neither the ST nor CA process gave rise to a meaningful improvement in mechanical compatibility, suggestive of an intrinsic limit of static heat treatment. It is thus required to find an alternative thermomechanical process inhibiting the precipitation of hard α phase, such as the combination of martensitic transformation and grain refinement suggested recently [6].

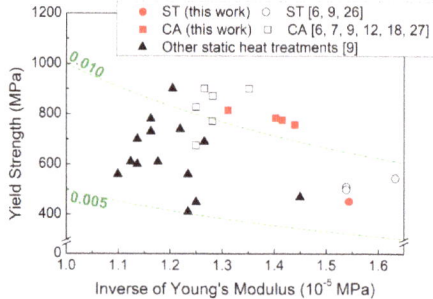

Figure 5. Mechanical compatibilities of the investigated and reported Ti-13Nb-13Zr alloys [6,7,9,12,18,26,27]. The green lines indicate the corresponding value of mechanical compatibility (i.e., YS/E).

5. Conclusions

This study has clarified the kinetics of heat treatment (i.e., CA treatment suggested by the ASTM standard) in Ti-13Nb-13Zr alloy in terms of microstructural evolution and changes in mechanical compatibility. The initial ST alloy was composed of the full α'-martensitic structure, which was decomposed into stable α and β phases during the subsequent CA process. The precipitation of the hard α phase with increasing CA duration resulted in the simultaneous increase in mechanical strength and Young's modulus. Such a tendency was unfavorable for biomedical application that demands high strength and low Young's modulus. Indeed, the CA process rarely enhanced the mechanical compatibility of Ti-13Nb-13Zr alloys regardless of the treatment duration. It is thus required to employ an alternative processing route, rather than ST or CA treatments, that enables mechanical strengthening without the precipitation of hard α phase.

Author Contributions: Conceptualization, T.L, M.L., and C.H.P.; methodology, T.L. and C.H.P.; investigation, M.L. and I.-S.K.; resources, T.L., C.H.P., Y.H.M., and H.S.Y.; writing—original draft preparation, M.L. and T.L.; writing—review and editing, M.L. and T.L.; supervision, T.L.; project administration, T.L. and H.S.Y.; funding acquisition, T.L. All authors have read and agreed to the published version of the manuscript.

Funding: This work was supported by the National Research Foundation of Korea (NRF) grant funded by the Korean government (MSIT) (No. 2018R1C1B6002068).

Conflicts of Interest: The authors declare no conflict of interest.

References

1. Michalska, J.; Sowa, M.; Piotrowska, M.; Widziołek, M.; Tylko, G.; Dercz, G.; Socha, R.P.; Osyczka, A.M.; Simka, W. Incorporation of Ca ions into anodic oxide coatings on the Ti-13Nb-13Zr alloy by plasma electrolytic oxidation. *Mater. Sci. Eng. C* **2019**, *104*, 109957. [CrossRef] [PubMed]
2. Kolli, R.; Devaraj, A. A Review of Metastable Beta Titanium Alloys. *Metals* **2018**, *8*, 506. [CrossRef]
3. ASTM F1713-08(2013). *Standard Specification for Wrought Titanium-13Niobium-13Zirconium Alloy for Surgical Implant Applications (UNS R58130)*; ASTM International: West Conshohocken, PA, USA, 2013.
4. Niinomi, M. Mechanical biocompatibilities of titanium alloys for biomedical applications. *J. Mech. Behav. Biomed. Mater.* **2008**, *1*, 30–42. [CrossRef] [PubMed]
5. Niinomi, M.; Kuroda, D.; Fukunaga, K.I.; Morinaga, M.; Kato, Y.; Yashiro, T.; Suzuki, A. Corrosion wear fracture of new β type biomedical titanium alloys. *Mater. Sci. Eng. A* **1999**, *263*, 193–199. [CrossRef]
6. Lee, T.; Lee, S.; Kim, I.-S.; Moon, Y.H.; Kim, H.S.; Park, C.H. Breaking the limit of Young's modulus in low-cost Ti-Nb-Zr alloy for biomedical implant applications. *J. Alloys Compd.* **2020**, *828*, 154401. [CrossRef]
7. Park, C.H.; Park, J.-W.W.; Yeom, J.-T.T.; Chun, Y.S.; Lee, C.S. Enhanced mechanical compatibility of submicrocrystalline Ti-13Nb-13Zr alloy. *Mater. Sci. Eng. A* **2010**, *527*, 4914–4919. [CrossRef]
8. Geetha, M.; Singh, A.K.; Muraleedharan, K.; Gogia, A.K.; Asokamani, R. Effect of thermomechanical processing on microstructure of a Ti-13Nb-13Zr alloy. *J. Alloys Compd.* **2001**, *329*, 264–271. [CrossRef]
9. Majumdar, P.; Singh, S.B.; Chakraborty, M. The role of heat treatment on microstructure and mechanical properties of Ti-13Zr-13Nb alloy for biomedical load bearing applications. *J. Mech. Behav. Biomed. Mater.* **2011**, *4*, 1132–1144. [CrossRef]
10. Geetha, M.; Kamachi Mudali, U.; Gogia, A.K.; Asokamani, R.; Raj, B. Influence of microstructure and alloying elements on corrosion behavior of Ti-13Nb-13Zr alloy. *Corros. Sci.* **2004**, *46*, 877–892. [CrossRef]
11. Majumdar, P.; Singh, S.B.; Chakraborty, M. Elastic modulus of biomedical titanium alloys by nano-indentation and ultrasonic techniques-A comparative study. *Mater. Sci. Eng. A* **2008**, *489*, 419–425. [CrossRef]
12. Lee, T.; Heo, Y.-U.U.; Lee, C.S. Microstructure tailoring to enhance strength and ductility in Ti-13Nb-13Zr for biomedical applications. *Scr. Mater.* **2013**, *69*, 785–788. [CrossRef]
13. Wang, J.N.; Yang, J.; Xia, Q.; Wang, Y. On the grain size refinement of TiAl alloys by cyclic heat treatment. *Mater. Sci. Eng. A* **2002**, *329–331*, 118–123. [CrossRef]
14. Saji, V.S.; Choe, H.C. Electrochemical corrosion behaviour of nanotubular Ti-13Nb-13Zr alloy in Ringer's solution. *Corros. Sci.* **2009**, *51*, 1658–1663. [CrossRef]
15. Zhou, L.; Yuan, T.; Li, R.; Tang, J.; Wang, M.; Li, L.; Chen, C. Microstructure and mechanical performance tailoring of Ti-13Nb-13Zr alloy fabricated by selective laser melting after post heat treatment. *J. Alloys Compd.* **2019**, *775*, 1164–1176. [CrossRef]
16. Lee, T.; Park, K.-T.T.; Lee, D.J.; Jeong, J.; Oh, S.H.; Kim, H.S.; Park, C.H.; Lee, C.S. Microstructural evolution and strain-hardening behavior of multi-pass caliber-rolled Ti-13Nb-13Zr. *Mater. Sci. Eng. A* **2015**, *648*, 359–366. [CrossRef]
17. Kumar, P.; Mahobia, G.S.; Singh, V.; Chattopadhyay, K. Lowering of elastic modulus in the near-beta Ti-13Nb-13Zr alloy through heat treatment. *Mater. Sci. Technol.* **2020**, *36*, 717–725. [CrossRef]
18. Davidson, J.A.; Mishra, A.K.; Kovacs, P.; Poggie, R.A. New surface-hardened, low-modulus, corrosion-resistant Ti13Nb13Zr alloy for total hip arthroplasty. *Biomed. Mater. Eng.* **1994**, *4*, 231–243.
19. Kobayashi, S.; Nakagawa, S.; Nakai, K.; Ohmori, Y. Phase Decomposition in a Ti-13Nb-13Zr Alloy during Aging at 600 °C. *Mater. Trans.* **2002**, *43*, 2956–2963. [CrossRef]
20. Moffat, D.L.; Larbalestier, D.C. The competition between martensite and omega in quenched Ti-Nb alloys. *Metall. Mater. Trans. A* **1988**, *19*, 1677–1686. [CrossRef]
21. Hussein, M.A.; Azeem, M.; Kumar, A.M.; Al-Aqeeli, N.; Ankah, N.K.; Sorour, A.A. Influence of Thermal Treatment on the Microstructure, Mechanical Properties, and Corrosion Resistance of Newly Developed Ti20Nb13Zr Biomedical Alloy in a Simulated Body Environment. *J. Mater. Eng. Perform.* **2019**, *28*, 1337–1349. [CrossRef]
22. Gil Mur, F.X.X.; Rodríguez, D.; Planell, J.A. Influence of tempering temperature and time on the α'-Ti-6Al-4V martensite. *J. Alloys Compd.* **1996**, *234*, 287–289. [CrossRef]

23. Hao, Y.; Yang, R.; Niinomi, M.; Kuroda, D.; Zhou, Y.; Fukunaga, K.; Suzuki, A. Young's modulus and mechanical properties of Ti-29Nb-13Ta-4.6Zr in relation to α″ martensite. *Metall. Mater. Trans. A* **2002**, *33*, 3137–3144. [CrossRef]
24. Zhou, Y.L.; Niinomi, M.; Akahori, T. Effects of Ta content on Young's modulus and tensile properties of binary Ti-Ta alloys for biomedical applications. *Mater. Sci. Eng. A* **2004**, *371*, 283–290. [CrossRef]
25. Matsumoto, H.; Watanabe, S.; Hanada, S. Microstructures and mechanical properties of metastable [beta] TiNbSn alloys cold rolled and heat treated. *J. Alloys Compd.* **2007**, *439*, 146–155. [CrossRef]
26. Baptista, C.A.R.P.; Schneider, S.G.; Taddei, E.B.; da Silva, H.M. Fatigue behavior of arc melted Ti–13Nb–13Zr alloy. *Int. J. Fatigue* **2004**, *26*, 967–973. [CrossRef]
27. Zhentao, Y.; Lian, Z. Influence of martensitic transformation on mechanical compatibility of biomedical β type titanium alloy TLM. *Mater. Sci. Eng. A* **2006**, *438–440*, 391–394. [CrossRef]

© 2020 by the authors. Licensee MDPI, Basel, Switzerland. This article is an open access article distributed under the terms and conditions of the Creative Commons Attribution (CC BY) license (http://creativecommons.org/licenses/by/4.0/).

Article

On the Applicability of Stereological Methods for the Modelling of a Local Plastic Deformation in Grained Structure: Mathematical Principles

Stanislav Minárik [1,*] and Maroš Martinkovič [2]

[1] Advanced Technologies Research Institute, Faculty of Materials Science and Technology in Trnava, Slovak University of Technology, 917 24 Trnava, Slovakia
[2] Institute of Production Technologies, Faculty of Materials Science and Technology in Trnava, Slovak University of Technology, 917 24 Trnava, Slovakia; maros.martinkovic@stuba.sk
* Correspondence: stanislav.minarik@stuba.sk

Received: 1 July 2020; Accepted: 5 August 2020; Published: 12 August 2020

Abstract: Analysis of systems and structures from their cross-sectional images finds applications in many branches. Therefore, the question of content, quantity, and accuracy of information obtained from various techniques based on cross-sectional views of structures is particularly important. Application of conventional techniques for two-dimensional imaging on the analysis of structure from a cross-sectional image is limited. The reason for this limitation is the fact that these techniques use a fixed cross-sectional plane and therefore cannot check the 3D structural changes caused by deformation. Geometric orientation of a grained structure must be considered when data, scanned from a cross section, is processed in order to obtain information about local deformation in this structure. The so-called degree of structure orientation in 3D can be estimated experimentally from the cross-sectional image of the structure by the statistical (Saltykov) method of oriented testing lines. Subsequently if the correlation between orientation and deformation were to be known a detailed map of local deformation in the structure could be revealed. Unfortunately, exact theoretical works dealing with the assessment of local deformation by means of change of structure orientation in 3D are still missing. Our work seeks to partially remove this shortcoming. In our work we are interested in how the transformation of the image of a grained structure in a cross-sectional plane reflects structure deformation. An initial shape of grains is assumed which is transformed into a deformed shape by analytic calculation. We present brief mathematical derivations aimed at the problem of single grain-surface area deformation. The main goal of this work led to the design of a computationally low consuming procedure for quantification of local deformation in a grained structure based on the distortion of the image of this structure in a cross-sectional view.

Keywords: local deformation; plastic deformation; grained structure; grain shape; grain surface area; degree of grain orientation; deformation tensor

1. Introduction

Final properties of components and systems prepared by means of forming technology are affected by mechanical working of the material. Essentially the forming process is a general method based on the deformation effects of external forces. These effects are macroscopic and do not reflect microscopic changes in the structure of material in all their diversity. The macroscopic strain cannot reflect all microstructural changes in the volume of deformed areas of the structure mainly in the case if only the surface layers of the material are deformed. To obtain exact information about changes in microstructure of materials in the whole volume of the bulk sample it is necessary to use suitable methods. Methods based on acoustic wave emission, analysis of dislocation, microhardness measurement, slip band observation, micromeshes method or macroscopic screw method can be used for this purpose and they are well developed. However, none of the mentioned available methods give analytical formulas relating local values of structure parameters and strain in any position of structure. The mentioned formulas are essential for a quantitative description of local changes induced by plastic deformation in the microstructure.

Grain boundary (interface between grains) is the main microstructural parameter in polycrystalline material. Grains have an isometric dimension when the grained structure is isotropic. The mean size of grains (grains size distribution) and the specific surface area of the grain boundaries are sufficient structural characteristics in this case. Grains have anisometric dimension when the grained structure is anisotropic and it is necessary to describe their orientation in space in this case.

In this context the question is how the spatial orientation of the grains themselves can be defined generally and exactly. It is possible to suspect intuitively that grain surface orientation in this sense can be evaluated by the prevailing orientation of the grain surface area elements. The next question is how can local deformation be estimated from the mentioned orientation of the grain surface.

Experimental observation and quantitative evaluation of shape and orientation of grains, domains or cells in grained, foam and cellular structures have been at the centre of attention of several authors already for some time [1,2]. Experimental observation has shown that the process of deformation is closely associated with the character of the crystallographic structure. Analysis of experimental data reveals that the grain boundaries and grain interactions during deformation play an important role in the process of deformation [3–6]. Various experimental procedures to characterize size, deformation, orientation, and position of single grain within the polycrystalline structure have been developed [7–11]. Most of these methods are already considered to be standard experimental techniques.

Calculation of grain boundary distortion during plastic straining has proven to be important in the theory of crystal plasticity and stereology [12–16]. Modelling of single crystal plasticity by numerical methods allows discussion of deviations of experimental results from ideal crystal behavior [17–19]. Information about the change of the grain surface area per unit volume as a function of deformation parameters is important in the production of steels [20]. The amount of grain surface and grain edge per unit volume are parameters important in kinetic theory since both of them are heterogeneous nucleation sites. Recently published works demonstrate that problems of geometrical orientation of structure and distortion of structural units induced by deformation play an important role in different areas. We can mention for example biology where it is essential to clarify the organlevel tissue deformation dynamics to understand the morphogenetic mechanisms of organ development and regeneration. Construction and geometrical analyses of quantitative whole-organ tissue deformation maps must be realized for the evaluation of major morphogen and anisotropic tissue deformation along an axis [21].

2. Motivation and Aims

In recent years modern techniques and methods to observe crystal structure orientation have become available. For example, the image correlation method [22,23] is widely applied now. There is also a precise technique based on the triple points of grain boundaries [24,25] which can be followed by an electron backscatter diffraction (EBSD) method [26]. However, these methods have significant limitations (difficult sample preparation, requirement of conductive coatings, stability of electron beam for best data acquisition, etc.). In addition, all these methods are based on 2D observations. 3D measurement can only be done by stereology and high-resolution computed tomography (CT with nanofocus) could also handle 3D image reconstruction [27–29]. Stereological techniques provide quantitative information about a three-dimensional material from data scanned from two- dimensional planar sections of the structure. These techniques and methods are widely used today especially in the investigation of the character of coarse-grained and foamed structures.

Previous work has shown that there are several ways to evaluate grain orientation experimentally. However, the most acceptable way is the estimation of anisotropy based on stereological principle enabling determination of the "degree of grain orientation". An anisotropic microstructure consists of (planar or linearly) oriented components which can be easily evaluated using stereological methods. We believe that results of such evaluation (degree of grain orientation in the volume of the plastically deformed structure) can be converted to values of local parameters of deformation. Therefore, it is sufficient to determine the degree of grain orientation to obtain a value of local strain.

We focus on the mathematical treatment of grain surface-deformation in our work. The problem of a grain-surface distortion induced by plastic deformation has been solved. We start with mathematics of grain behavior during linear deformation while we search for a formula enabling calculation of grain surface area change caused by deformation of grain using deformation tensor. Next, we propose a quantity for evaluation of degree of grain surface area orientation with respect to a certain vector in 3D space and discuss the correlation of this quantity with deformation parameters. Our contribution finishes with a brief commentary on the suggested technique of quantitative analysis, and outlines some key areas where further study is recommended.

We emphasize that strictly geometric orientation of the grain in 3D space is discussed in our work. The crystal structure within the grain is not analyzed (even though the concept of grain orientation is usually related with crystallographic orientation). A quantity κ is defined to evaluate the degree of orientation of the grain surface in relation to any direction. The presented work is focused on finding a correlation between the change in the degree of geometrical orientation of a grain and the deformation experienced by it. The case of a grain subjected to plastic deformation is studied in detail. The obtained results were applied for the particular case of a randomly oriented grained structure.

We present only a mathematical model for describing the deformation of a fictional, simplified grained structure—no real material is specified, nor is the actual deformation process described. We contribute thereby to the theoretical works on this topic published in the past [30–34]. A purely theoretical approach is presented. However, it has been shown that the discussed deformation mechanism, i.e., elongation (or the shortening) of grains under applied load, exists in a real-world material [35,36].

Stereological Method Based on Oriented Testing Lines

Direction of grain boundary orientation is the same as the direction of deformation. If the deformation scheme is known, grain boundaries can be decomposed into isotropic, planar, and linear oriented components using stereology method oriented test lines according to Saltykov [1] on propeller oriented metallographic prepared cutting planes. Grain boundaries are surfaces in volume and relative surface area, the average surface area to unit volume is measured—the isotropic, planar, and linear oriented parts of relative surface area. Intersections of cutting plane and the grain boundary surfaces are lines in the cutting plane. The average number of intersections per unit length of oriented test lines with grain boundaries P_L in the cutting plain is measured.

(a) Linear orientation of grain surface in sample volume

In the case of linear orientation of surfaces, the volume cutting plane, on which analysis is realized, must be parallel with the direction of orientation. In the perpendicular plane no orientation is observed. Test lines oriented parallel with the direction of orientation (1) intersect only the isotropic part of the surfaces in volume, test lines oriented perpendicular with direction of orientation (2) intersect isotropic and oriented part of surfaces in volume (see Figure 1). For the isotropic part of the relative surface area is as follows:

$$(S_V)_{IS} = 2(P_L)_P \tag{1}$$

where subscript IS means isotropic-randomly oriented and P means parallel orientation of test lines with direction of orientation. For the linear oriented part of the relative surface area it is as follows:

$$(S_V)_{OR} = \frac{\pi}{2}[(P_L)_O - (P_L)_P] \tag{2}$$

where OR means oriented and subscript O perpendicular (orthogonal) orientation of test lines to direction of orientation. Total relative surface area is carried out by the sum of both parts:

$$(S_V)_{TOT} = \frac{\pi}{2}(P_L)_O + \left(2 - \frac{\pi}{2}\right)(P_L)_P, \qquad (3)$$

where subscript *TOT* means total.

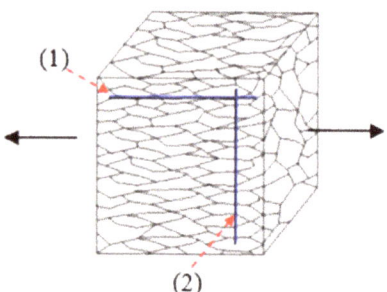

Figure 1. Test lines in the cutting plane at the linear orientation of the surfaces in the volume.

(b) Planar orientation of grain surface in sample volume

In the case of planar orientation of surfaces in the volume cutting plane on which analysis is realized it must be perpendicular to the orientation plane. In the parallel plane no orientation is observed. Test lines oriented parallel to the plane of orientation (1) intersect only the isotropic part of the surfaces in volume, test lines oriented perpendicular to plane of orientation (2) intersect isotropic and oriented part of surfaces in volume (see Figure 2). For the isotropic part of the relative surface area it is as follows:

$$(S_V)_{IS} = 2(P_L)_P \qquad (4)$$

where subscript *IS* means isotropic-randomly oriented and *P* parallel orientation of test lines with plane of orientation. For the planar oriented part of the relative surface area it is as follows:

$$(S_V)_{OR} = (P_L)_O - (P_L)_P \qquad (5)$$

where *OR* means oriented and subscript *O* means perpendicular (orthogonal) orientation of test lines to the plane of orientation. Total relative surface area is carried out by the sum of both parts:

$$(S_V)_{TOT} = (P_L)_O + (P_L)_P \qquad (6)$$

where subscript *TOT* means total.

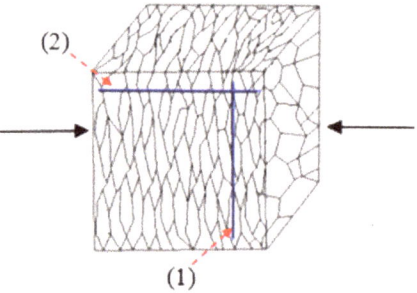

Figure 2. Test lines in the cutting plane at the planar orientation of the surfaces in the volume.

(c) Linear–planar orientation of grain surface in sample volume

In the case of linear–planar orientation of surfaces in volume the orientation is observed in all three orthogonal cutting planes, because the axis of linear orientation is perpendicular to the direction of the plane of orientation (direction of plane is normal to this)—see Figure 3. Each orientation component—planar and linear must be analyzed independently. Measurement must be realized on minimally two cutting planes. There are three possible cutting planes: cutting plane parallel with direction of linear orientation and parallel with plane of orientation, cutting plane perpendicular to direction of linear orientation and perpendicular to plane of orientation and cutting plane parallel with direction of linear orientation and perpendicular to plane of orientation.

The cutting plane parallel with the direction of linear orientation parallel with the plane of orientation test line, parallel with the direction of linear orientation (1) (see Figure 3) intersects only the isotropic part of the surfaces in volume (parameter $(P_L)_{PP}$), test lines oriented perpendicular with the direction of orientation (2) intersect the isotropic and linear oriented part of the surfaces in volume (parameter $(P_L)_{OP}$). For the isotropic and linear oriented part of the relative surface area it is:

$$(S_V)_{IS} = 2(P_L)_{PP} \tag{7}$$

$$(S_V)_{ORL} = \frac{\pi}{2}[(P_L)_{OP} - (P_L)_{PP}] \tag{8}$$

where subscript ORL means linear oriented. The cutting plane perpendicular to the direction of linear orientation and perpendicular to the plane of the orientation test line parallel with the plane of orientation (6) intersects the isotropic and linear oriented parts of the surfaces in volume (parameter $(P_L)_{OP}$) (identical to test line 2), test lines oriented perpendicular to the plane of orientation (5) intersect the isotropic, linear and planar oriented parts of the surfaces in volume (parameter $(P_L)_{OO}$). For the planar oriented part of the relative surface area it is as follows:

$$(S_V)_{ORP} = (P_L)_{OO} - (P_L)_{OP} \tag{9}$$

where subscript ORP means planar oriented.

On the cutting plane parallel with the direction of linear orientation and perpendicular to the plane of orientation test line parallel to the direction of linear orientation (the same as parallel to the plane orientation) (3) intersects only the isotropic part of the surfaces in volume (parameter $(P_L)_{PP}$) (identical to test line 1), the test line oriented perpendicular to the plane of orientation (4) intersects the isotropic, linear, and planar oriented parts of the surfaces in volume (parameter $(P_L)_{OO}$) (identical to test line 5). It is necessary for the determination of the quantitative parameters of the surfaces in volume with linear–planar orientation to use the system of three tested lines on two cutting planes and it is possible to form different combinations of tested lines (according to the identification in Figure 3)—(2 or 6) and (4 or 5) and (1 or 3).

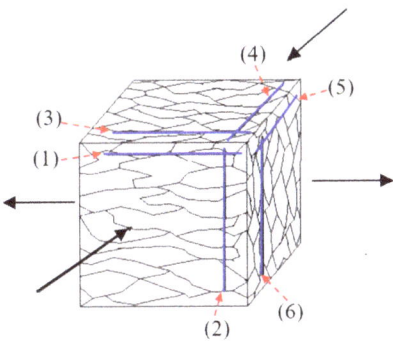

Figure 3. Test lines in the cutting planes at the linear-planar orientation of the surfaces in the volume.

(d) Experimental Quantification of degree of grain orientation

The degree of grain boundary orientation O can be estimated experimentally by means of formula:

$$O = \frac{(S_V)_{OR}}{(S_V)_{TOT}} \tag{10}$$

The degree of linear orientation surface (2D) in volume (3D) O_L of the linear orientated system of the surfaces in volume and degree of planar orientation O_P of planar orientated system of the surfaces in the volume are as follows:

$$O_L = 1 - 2\left\{\frac{(P_L)_P}{\frac{\pi}{2}(P_L)_O + \left(2 - \frac{\pi}{2}\right)(P_L)_P}\right\} \tag{11}$$

$$O_P = 1 - 2\left\{\frac{(P_L)_P}{(P_L)_O + (P_L)_P}\right\}$$

For the linear–planar oriented system of surfaces the degree of linear orientation O_{LP}, degree of planar orientation O_{PL}, and degree of total orientation O_{TLP} can be calculated by the following formulas:

$$O_{LP} = 1 - 2\left\{\frac{2(P_L)_{PP} + (P_L)_{OO} - (P_L)_{OP}}{4(P_L)_{PP} + \pi\left[(P_L)_{OP} - \frac{\pi}{2}(P_L)_{PP}\right] + 2[(P_L)_{OO} - (P_L)_{OP}]}\right\}$$

$$O_{PL} = 1 - 2\left\{\frac{(P_L)_{PP} + \frac{\pi}{4}\left[(P_L)_{OP} - \frac{\pi}{2}(P_L)_{PP}\right]}{2(P_L)_{PP} + \frac{\pi}{2}\left[(P_L)_{OP} - \frac{\pi}{2}(P_L)_{PP}\right] + (P_L)_{OO} - (P_L)_{OP}}\right\} \tag{12}$$

$$O_{TLP} = 1 - 2\left\{\frac{(P_L)_{PP}}{2(P_L)_{PP} + \frac{\pi}{2}\left[(P_L)_{OP} - \frac{\pi}{2}(P_L)_{PP}\right] + (P_L)_{OO} - (P_L)_{OP}}\right\}$$

In the following text we deal with the theoretical analysis of the change of spatial orientation of grain during deformation and the possibilities of generalization of the mentioned relations (11) to (12).

3. Transformation of Grain-Surface Area at Linear Deformation

When a mechanical loading is applied most of the grains in a structure are submitted to a deformation. The deformation occurs according to the different orientations and influences from neighboring grains. If the grain is deformed a change in orientation and size of grain surface area are induced which leads to local heterogeneous plasticity. This is evident in the case where the typical grain prolongation in the direction of the loading force is visible (see Figures 4 and 5).

This section provides an improved mathematical analysis of grain surface area transformation during linear grain deformation. Here we offer mathematical expressions usable in some models requiring the quantification of the deformed grain boundary surface. Relationships between coefficients of deformation tensor and change of grain surface are discussed. Results show how the specific area of the grain boundaries changes during local straining.

Let us discuss isotropic, linear elastic material with a grained structure. The parametrization of the surface of any single grain in 3D lies in the determination of the coordinates of each point on the surface:

$$X_i = X_i(u, v) \tag{13}$$

where i = 1, 2, 3 and u, v are parameters that changes in appropriate intervals. Then the position vectors of each point on the surface of the grain with any possible shape in undeformed configurations are as follows:

$$\vec{R}(u, v) = [X_1(u, v), X_2(u, v), X_3(u, v)], u \in \langle u_1, u_2 \rangle v \in \langle v_1, v_2 \rangle \tag{14}$$

Position vectors of each point on the surface of the same grain in the deformed state are as follows:

$$\vec{r}(u, v) = [x_1(u, v), x_2(u, v), x_3(u, v)], u \in \langle u'_1, u'_2 \rangle v \in \langle v'_1, v'_2 \rangle \tag{15}$$

Figure 4. Scheme illustrating a view of an undeformed grained structure cross-section (for example on a metallographic cut). Scheme shows the cross-section of some individual grains.

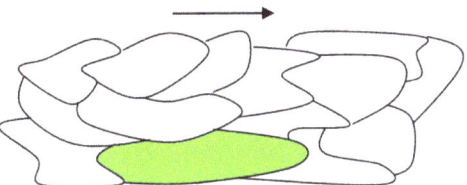

Figure 5. Scheme illustrating a view of the same grained structure cross-section in the deformed configuration. Scheme shows grain surface distortion and typical prolongation of grains.

In the general case, any 3D deformation can be described by a 3 × 3 deformation tensor:

$$\vec{\vec{\varepsilon}} = \begin{pmatrix} \varepsilon_{11}, & \varepsilon_{12}, & \varepsilon_{13} \\ \varepsilon_{21}, & \varepsilon_{22}, & \varepsilon_{23} \\ \varepsilon_{31}, & \varepsilon_{32}, & \varepsilon_{33} \end{pmatrix}. \tag{16}$$

The deformation of grain for a small strain is linear. Therefore, using the notation mentioned above the deformation of the grain can be described by the following anisomorphic linear transformation:

$$\vec{r} = \vec{\vec{\varepsilon}} \cdot \vec{R} \text{ or } x_i = \varepsilon_{ij} X_j \tag{17}$$

where $i, j = 1, 2, 3$ and ε_{ij} are coefficients of the deformation tensor (16). It is no problem to verify that $\varepsilon_{ij} = \delta_{ij}$ in the undeformed case (where δ_{ij} is the Kronecker symbol). We used the typical Einstein's summation convention in the second formula in (17) whereby when an index variable appears twice in a single term it implies summation of that term over all the values of the index.

For simplicity this convention is used in all the following text.

The surface element vectors for undeformed grain are determined as follows:

$$d^2 \vec{S}_{undefor} = \left(\frac{\partial X_2}{\partial u} \frac{\partial X_3}{\partial v} - \frac{\partial X_3}{\partial u} \frac{\partial X_2}{\partial v} \right) \vec{i} - \left(\frac{\partial X_1}{\partial u} \frac{\partial X_3}{\partial v} - \frac{\partial X_3}{\partial u} \frac{\partial X_1}{\partial v} \right) \vec{j} + \left(\frac{\partial X_1}{\partial u} \frac{\partial X_2}{\partial v} - \frac{\partial X_2}{\partial u} \frac{\partial X_1}{\partial v} \right) \vec{k} \tag{18}$$

One of these vectors is shown in Figure 6. Using transformation (17) the vector of any surface element of the grain in the deformed state (see Figure 7) can be determined:

$$d^2 \vec{S}_{defor} = \varepsilon_{2j} \varepsilon_{3k} \left(\frac{\partial X_j}{\partial u} \frac{\partial X_k}{\partial v} - \frac{\partial X_j}{\partial v} \frac{\partial X_k}{\partial u} \right) \vec{i} - \varepsilon_{1j} \varepsilon_{3k} \left(\frac{\partial X_j}{\partial u} \frac{\partial X_k}{\partial v} - \frac{\partial X_j}{\partial v} \frac{\partial X_k}{\partial u} \right) \vec{j} + \varepsilon_{1j} \varepsilon_{2k} \left(\frac{\partial X_j}{\partial u} \frac{\partial X_k}{\partial v} - \frac{\partial X_j}{\partial v} \frac{\partial X_k}{\partial u} \right) \vec{k}, \tag{19}$$

where $i, j, k = 1, 2, 3$.

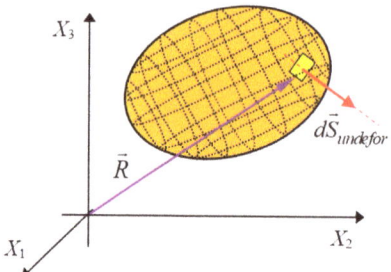

Figure 6. Undeformed grain.

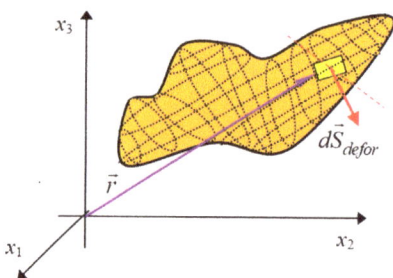

Figure 7. Deformed grain surface.

After suitable re-indexing of the sums in Formula (19) the change of the mentioned vector of the surface element during deformation can be written as follows (please see details in Appendix A):

$$d^2\vec{S}_{defor} = \vec{\sigma} \cdot d^2\vec{S}_{undefor} \qquad (20)$$

where $\vec{\sigma}$ can be introduced as a tensor of grain surface deformation. Coefficients of this tensor are related to the coefficients of the deformation tensor.

$$\sigma_{wm} = (-1)^{w+m} \det\left(\vec{\varepsilon}_{mw}\right) \qquad (21)$$

where $\det\left(\vec{\varepsilon}_{mw}\right)$ is the subdeterminant of the matrix of the deformation tensor $\vec{\varepsilon}$. This subdeterminant is determined from the matrix of tensor $\vec{\varepsilon}$ by deleting the m-th row and w-th column. So, it is easily shows that coefficients of the tensor of grain surface deformation can be found as follows:

$$\sigma_{wm} = \varepsilon_{ik}\varepsilon_{jn} - \varepsilon_{in}\varepsilon_{jk'} \qquad (22)$$

where indexes:

$$i = \frac{9 - 2w - (-1)^w}{4}, \quad j = \frac{15 - 2w + (-1)^w}{4} \qquad (23)$$

$$k = \frac{12 - 2m - \{3 + (-1)^m\}(-1)^{m+w}}{4}, \quad n = \frac{12 - 2m + \{3 + (-1)^m\}(-1)^{m+w}}{4}, \quad w, m = 1, 2, 3. \qquad (24)$$

Then the size of the vector (20) of the deformed surface element is given as follows (please see details in Appendix B):

$$\left(d^2 S_{defor}\right)^2 = \Gamma_{jnks} \frac{\partial X_j}{\partial u} \frac{\partial X_n}{\partial u} \frac{\partial X_k}{\partial v} \frac{\partial X_s}{\partial v}, \tag{25}$$

where coefficients:

$$\Gamma_{jnks} = \sum_{\substack{V([p,q]), p \neq q \\ p,q \in \{1,2,3\}}} \varepsilon_{pj} \varepsilon_{qk} \left(\varepsilon_{pn}\varepsilon_{qs} - \varepsilon_{qn}\varepsilon_{ps}\right) \tag{26}$$

Notation under the summation symbol in (26) means the sum over all variations of indexes p, q while $p \neq q$. The entire surface area of the deformed grains can be determined by formula:

$$S_{defor} = \int_{v_1}^{v_2} \int_{u_1}^{u_2} \sqrt{\Gamma_{jklm} \frac{\partial X_j}{\partial u} \frac{\partial X_k}{\partial v} \frac{\partial X_l}{\partial v} \frac{\partial X_m}{\partial u}} \, du \, dv. \tag{27}$$

where $j, k, l, m = 1, 2, 3$. We note, that we permanently use Einstein's summation convention and the sum of 81 members is under the square root in Formula (27).

4. Quantitative Analysis of Grain Orientation: Projection Factor

In this section we define a novel scalar variable κ for quantification of the degree of grain-surface area orientation. First, there is briefly described motivation for such a way of introducing this variable. We believe that the variable κ helps to evaluate the grain orientation in a topological sense that is recognizably determined by the grain-surface area shape.

Orientations of surface elements of the grain in 3D are determined by the vectors $\vec{dS}$. These vectors are perpendicular to the grain-surface everywhere. Topologically the grain-surface area is considered to be a closed 2D manifold. Therefore the following holds:

$$\int_{(surface)} \vec{dS} = \vec{0} \tag{28}$$

Grain boundaries are very important for energy balance of the grained structure. A special type of perturbance of this balance is induced by the loading force and for this reason a certain direction of orientation of the grain-surface elements becomes preferred under load. However, Formula (28) remains satisfied under load even in the unloaded condition. At the center of our attention is the question how could the correlation between the preferred orientation of the grain surface elements and the local structure deformation parameters be quantified.

Due to the planar character of the infinitesimal surface element dS there is no problem to define the orientation of this element in 3D just by means of the vector $\vec{dS}$. Then quantitative evaluation of the degree of orientation of this single surface element to the direction determined by the unit vector $\vec{\tau}$ can be performed using absolute value of scalar product:

$$\left|\vec{\tau} \cdot \vec{dS}\right| = dS|\cos(\alpha)| = dS_p \tag{29}$$

where α is the angle between vectors $\vec{\tau}$ and $\vec{dS}$. The size of projection of the surface element dS_p to the plane perpendicular to the vector $\vec{\tau}$ is determined by quantity (29). Let S be the surface area of the whole grain. Then the quantity (29) calculated per unit surface area of the grain, i.e.:

$$\frac{dS|\cos(\alpha)|}{S} \tag{30}$$

can determine a contribution of the element dS to the degree of whole grain-surface orientation to the direction determined by the unit vector $\vec{\tau}$. We need to introduce a variable κ enabling evaluation of the degree of whole grain-surface orientation to the direction determined by the unit vector $\vec{\tau}$ which would be also appropriate for quantitative analysis of grain surface distortion. The contribution rate of the surface element dS to the variable κ can be intuitively proposed by generalization of the quantity (31) as follows:

$$d\kappa = \frac{dS}{S}\xi(\alpha), \qquad (31)$$

where $\xi(\alpha)$ is a value depending only on the angle α. Function $\xi(\alpha)$ must be designed in accordance with the requirement that the value κ corresponds to the quantities obtained from stereological measurements. In our work this function has been chosen in such a way that the limit of the values of variable κ correspond to values obtained in procedures realized on analysis of particle orientation using the Saltykov method (see Section 2 for an understanding of the principles). Consistent with this method we require:

$$\lim_{\alpha \to \pi/2} \kappa = 1, \quad \lim_{\alpha \to 0} \kappa = -1 \qquad (32)$$

So the following function can be used in this case:

$$\zeta(\alpha) = 1 - 2|\cos(\alpha)| \qquad (33)$$

and contribution (31) can be written in the following form:

$$d\kappa = \{1 - 2|\cos(\alpha)|\}\frac{dS}{S} = \frac{dS}{S} - 2\frac{\left|\vec{\tau}\cdot d\vec{S}\right|}{S}. \qquad (34)$$

Then the degree of whole grain-surface orientation to the direction determined by the unit vector $\vec{\tau}$ can be evaluated by the integration of (34) over the entire grain surface:

$$\kappa = 1 - 2\frac{1}{S}\int\limits_{(Surface)} \left|\vec{\tau}\cdot d\vec{S}\right| \qquad (35)$$

In the following sections we examine how the quantity κ defined by (35) behaves during deformation and whether it allows analysis of the grain-surface area distortion induced by deformation. Namely if there is a vector $\vec{\gamma}$ satisfying:

$$\kappa(\vec{\gamma}) = \kappa_{max} \qquad (36)$$

i.e., if the value of κ is maximal for vector $\vec{\gamma}$, this vector determines the preferred whole grain-surface area orientation (or grain orientation in a topological sense).

The degree of whole grain-surface orientation to the direction determined by the unit vector $\vec{\tau}$ in both deformed and undeformed configurations can be evaluated by Formula (35):

$$\kappa_{undefor} = 1 - 2\frac{\int\limits_{(Surface)}\left|\vec{\tau}\cdot d\vec{S}_{undefor}\right|}{\int\limits_{(Surface)}\left|d\vec{S}_{undefor}\right|}, \quad \kappa_{defor} = 1 - 2\frac{\int\limits_{(surface)}\left|\vec{\tau}\cdot d\vec{S}_{defor}\right|}{\int\limits_{(surface)}\left|d\vec{S}_{defor}\right|}. \qquad (37)$$

Let $\vec{\rho}$ be the unit vector parallel to the vector of the grain-surface element of undeformed grain, $d\vec{S}_{undefor} = \vec{\rho}dS_{undefor}$ i.e.,: where $|\vec{\rho}| = 1$.

Taking into account the Formula (38) and considering the fact:

$$\vec{\tau} \cdot d^2\vec{S}_{defor} = \vec{\tau} \cdot \left(\vec{\sigma} \cdot d^2\vec{S}_{undefor}\right) = \left(\vec{\sigma}^T \cdot \vec{\tau}\right) \cdot d^2\vec{S}_{undefor}, \qquad (38)$$

where $\vec{\sigma}^T$ is the transposed tensor to the grain surface deformation tensor $\vec{\sigma}$ determined by (22), Formulas (37) can be rewritten in the following form:

$$\kappa_{undefor} = 1 - 2\frac{\int\limits_{(Surface)} \left|\vec{\tau} \cdot d\vec{S}_{undefor}\right|}{\int\limits_{(Surface)} \left|\vec{\rho}\right| dS_{undefor}}, \quad \kappa_{defor} = 1 - 2\frac{\int\limits_{(Surface)} \left|\left(\vec{\sigma}^T \cdot \vec{\tau}\right) \cdot d\vec{S}_{undefor}\right|}{\int\limits_{(Surface)} \left|\vec{\sigma} \cdot \vec{\rho}\right| dS_{undefor}} \qquad (39)$$

As can be concluded from comparison of the above Equation (39), transformation of the value κ during grain deformation can be realized as follows:

$$\vec{\tau} \rightarrow \vec{\sigma}^T \cdot \vec{\tau} \wedge \vec{\rho} \rightarrow \vec{\sigma} \cdot \vec{\rho} \qquad (40)$$

Equation (39) correspond to Formulas (11) to (12). They can be considered as a generalized formula for the degree of grain orientation measured by the Saltykov method. As can be seen from (39) the transformation of κ during grain deformation is relatively complicated. We will not examine the properties of this transformation in detail. However, we want to point out the fact that that it is sufficient to integrate over the surface of the undeformed grain in both of the Formula (39). So, to determine the value κ in deformed configuration we need to know only the shape of the undeformed grain and the coefficients of the deformation tensor ε_{ij}. We believe, that κ is a scalar variable acceptable for the quantitative evaluation of grain-surface distortion during local deformation of the grained structure.

5. The Case of Plastic Deformation

The fundamental assumption used to establish the theory of plasticity is that plastic deformation is isochoric or volume preserving. Next, we apply the results obtained in previous sections to the case of plastic grain deformation, i.e., deformation in which the change of volume of grain is zero.

Conservation of grain volume during the plastic deformation process can be described by the formula:

$$V_{undefor} = V_{defor} \qquad (41)$$

The Gauss formula can be used for determination of grain volume in both undeformed ($V_{undefor}$) and deformed (V_{defor}) configurations:

$$\int\limits_{(S_{undefor})} \vec{R} \cdot d\vec{S}_{undefor} = \int\limits_{(V_{undefor})} div(\vec{R}) dV = 3V_{undefor}, \quad \int\limits_{(S_{defor})} \vec{r} \cdot d\vec{S}_{defor} = \int\limits_{(V_{defor})} div(\vec{r}) dV' = 3V_{defor}. \qquad (42)$$

A comparison of the Equation (42) shows:

$$\int\limits_{(S_{undefor})} \vec{R} \cdot d\vec{S}_{undefor} = \int\limits_{(S_{defor})} \vec{r} \cdot d\vec{S}_{defor} = \int\limits_{(S_{undefor})} \left\{\vec{\sigma}^T \cdot \left(\vec{\varepsilon} \cdot \vec{R}\right)\right\} \cdot d\vec{S}_{undefor} \qquad (43)$$

where Formulas (17) and (20) were used. It is necessary to integrate over the same (but any) surface area on both sides of the previous Equation (43). Therefore, for plastic deformations the following applies:

$$\vec{R} = \vec{\sigma}^T \cdot \left(\vec{\varepsilon} \cdot \vec{R}\right), \vec{\sigma}^T = \vec{\varepsilon}^{-1} \text{ respectively.} \tag{44}$$

As follows from Formula (44) the transposed tensor of the grain surface deformation tensor is inverse to the deformation tensor in the case of plastic deformation. If we consider Formula (43), the following equation can be written in the case of plastic deformation (please see details in Appendix C):

$$\{(\varepsilon_{1n}\varepsilon_{2j}\varepsilon_{3k} - \varepsilon_{2n}\varepsilon_{1j}\varepsilon_{3k} + \varepsilon_{3n}\varepsilon_{1j}\varepsilon_{2k}) - (\delta_{1n}\delta_{2j}\delta_{3k} - \delta_{2n}\delta_{1j}\delta_{3k} + \delta_{3n}\delta_{1j}\delta_{2k})\}\left(\frac{\partial X_j}{\partial u}\frac{\partial X_k}{\partial v} - \frac{\partial X_j}{\partial v}\frac{\partial X_k}{\partial u}\right)X_n = 0. \tag{45}$$

Equation (45) must be satisfied for any function $X_i = X_i(u, v)$ in the case of grain volume conservation during plastic deformation. If (45) is expanded without further assumptions, this leads to the following 27 equations:

$$\varepsilon_{1n}(\varepsilon_{2j}\varepsilon_{3k} - \varepsilon_{2k}\varepsilon_{3j}) - \varepsilon_{2n}(\varepsilon_{1j}\varepsilon_{3k} - \varepsilon_{1k}\varepsilon_{3j}) + \varepsilon_{3n}(\varepsilon_{1j}\varepsilon_{2k} - \varepsilon_{1k}\varepsilon_{2j}) -$$
$$-\delta_{1n}(\delta_{2j}\delta_{3k} - \delta_{2k}\delta_{3j}) + \delta_{2n}(\delta_{1j}\delta_{3k} - \delta_{1k}\delta_{3j}) - \delta_{3n}(\delta_{1j}\delta_{2k} - \delta_{1k}\delta_{2j}) = 0 \ldots \forall\, n, j, k \tag{46}$$

i.e., the following equation must be satisfied in the case of plastic deformation (please see details in Appendix D):

$$\det\left(\vec{\varepsilon}\right) = 1 \tag{47}$$

Next, the conditions for the deformation tensor can be considered in the case of plastic deformation of a homogeneous and isotropic grained structure:

$$\varepsilon_{ji} = 0 \text{ for all } i \neq j \neq k \wedge \det\left(\vec{\varepsilon}\right) = \varepsilon_{ii}\varepsilon_{jj}\varepsilon_{kk} = 1. \tag{48}$$

Elements of the deformation tensor are directly related to grain prolongation (contraction) δ ($-1 < \delta < \infty$) in this case. If the loading force is applied in the direction of the x-axis the following holds:

$$\varepsilon_{11} = \delta + 1, \quad \varepsilon_{22} = \varepsilon_{33} = \frac{1}{\sqrt{\delta + 1}} \tag{49}$$

and due to Formula (44) the degree of grain surface area orientation (39) can be modified to the following form:

$$K_{defor} = 1 - 2\frac{\int_{(Surface)} \left|\left(\vec{\varepsilon}^{-1} \cdot \vec{\tau}\right) \cdot d\vec{S}_{undefor}\right|}{\int_{(Surface)} \left|\vec{\varepsilon}^{-1} \cdot \vec{p}\right| dS_{undefor}} \quad 9. \tag{50}$$

Detailed information about the shape of the grain is necessary for the practical application of the result (39) and (50) because integrals in these formulas are integrated over the entire surface of the grain. However, there is no possibility to determine the real shape of each grain in the material structure exactly. Therefore, deformation of the model of the grained structure is usually investigated. Models of the grained structure consist of grains with various idealized shapes. For instance, deformation of spheres can be considered since spheres are not space filling and have no edges. Deformation of cubes can be also investigated, and mathematical analysis can be simplified, but cubes are unfortunately clearly poor for approximations to the shapes of real grains. The tetrakaidecahedronal shape of the grain is very suitable for modeling because it is space filling and approximately identical with the shape of real grains observed metallographically in the undeformed state [37,38].

Models of grained structures based on the specific idealized shape of grains allow examination of the relationship between grain orientation and local deformation but the problem of non-zero grain orientation in the case of zero value of initial deformation of the structure often arises from such models [39,40].

Our method presented in the next section does not evaluate the orientation of one idealized grain placed in the structure but investigates the orientation of all grain boundaries localized in a certain area of the grained structure. Integrals in the Formulas (39) and (50) must be integrated over the surface of all grain boundaries in this area. The relationship between grain orientation in this local area and local deformation can be determined from these formulas in this case.

6. Random Matrix Approximation

In this section we explain the applicability of results presented above in the modelling of local deformations in grained structures. Integrals in (39) should be integrated over the entire surface of the grain boundaries when the degree of orientation of the whole grained structure κ is evaluated.

For computational purposes, integrals in the Formula (39) can be written in the form of summation over the infinitely small areas placed around the wholegrain boundary surface (see Figures 8 and 9). A cross-sectional image of such a model of a granular structure is shown in Figure 10. So, we obtain the following:

$$\kappa_{defor} \approx 1 - 2 \frac{\sum_{i=1}^{N} \left| \left(\vec{\sigma}^T \cdot \vec{\tau} \right) \cdot \vec{S}_{undefor}^{(i)} \right|}{\sum_{i=1}^{N} \left| \vec{\sigma} \cdot \vec{\rho}^{(i)} \right| S^{(i)}} \tag{51}$$

where $\vec{\tau} = [\tau_1, \tau_2, \tau_3]$ is any unit vector (i.e., $\tau_1^2 + \tau_2^2 + \tau_3^2 = 1$) and $\vec{\rho}^{(i)}$ is unit vector perpendicular to the i-th grain-surface element of undeformed grain:

$$\vec{S}_{undefor}^{(i)} = \vec{\rho}^{(i)} S^{(i)}. \tag{52}$$

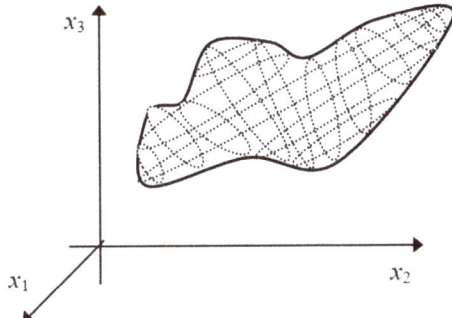

Figure 8. A view of the single grain.

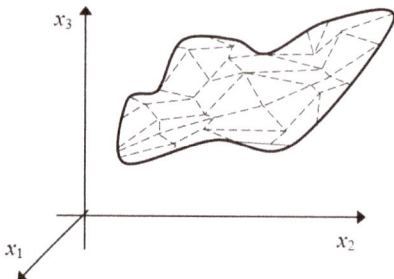

Figure 9. Scheme illustrating a view of the model of the single grain.

Vector $\vec{\rho}^{(i)}$ can be written in the form:

$$\vec{\rho}^{(i)} = [\sin(\eta^{(i)})\cos(\chi^{(i)}),\ \sin(\eta^{(i)})\sin(\chi^{(i)}),\ \cos(\eta^{(i)})]$$

while $\eta^{(i)}$ and $\chi^{(i)}$ are angles which determine the orientation of the i-th grain-surface element. Then the following applies:

$$\vec{S}^{(i)}_{undefor} = \left[S_1^{(i)},\ S_2^{(i)},\ S_3^{(i)}\right] = \left[S^{(i)}\sin(\eta^{(i)})\cos(\chi^{(i)}),\ S^{(i)}\sin(\eta^{(i)})\sin(\chi^{(i)}),\ S^{(i)}\cos(\eta^{(i)})\right] \qquad (53)$$

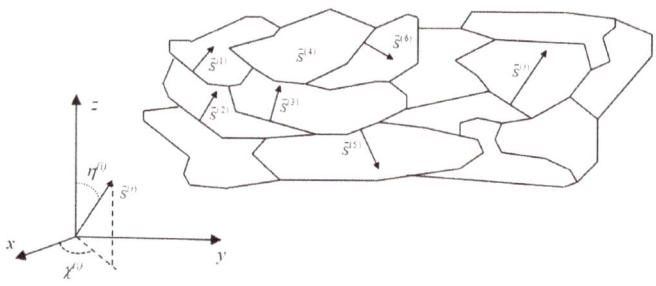

Figure 10. A view of a cross-section of the model of the grained structure. The scheme demonstrates cross-sectional images of small flat areas on the grain surface.

If we use result (22) and notation (53), the sums in the Formula (51) can be determined as (please see details in Appendix E):

$$\sum_{i=1}^{N}\left|\left(\vec{\sigma}^T\cdot\vec{\tau}\right)\cdot\vec{\rho}^{(i)}S^{(i)}\right|=$$

$$=\sum_{i=1}^{N}\Big|\{[(\varepsilon_{22}\varepsilon_{33}-\varepsilon_{23}\varepsilon_{32})\sin(\eta^{(i)})\cos(\chi^{(i)})+(\varepsilon_{23}\varepsilon_{31}-\varepsilon_{21}\varepsilon_{33})\sin(\eta^{(i)})\sin(\chi^{(i)})+(\varepsilon_{21}\varepsilon_{32}-\varepsilon_{22}\varepsilon_{31})\cos(\eta^{(i)})]\tau_1+ \qquad (54)$$
$$+[(\varepsilon_{13}\varepsilon_{32}-\varepsilon_{12}\varepsilon_{33})\sin(\eta^{(i)})\cos(\chi^{(i)})+(\varepsilon_{11}\varepsilon_{33}-\varepsilon_{13}\varepsilon_{31})\sin(\eta^{(i)})\sin(\chi^{(i)})+(\varepsilon_{12}\varepsilon_{31}-\varepsilon_{11}\varepsilon_{32})\cos(\eta^{(i)})]\tau_2+$$
$$+[(\varepsilon_{12}\varepsilon_{23}-\varepsilon_{13}\varepsilon_{22})\sin(\eta^{(i)})\cos(\chi^{(i)})+(\varepsilon_{13}\varepsilon_{21}-\varepsilon_{11}\varepsilon_{23})\sin(\eta^{(i)})\sin(\chi^{(i)})+(\varepsilon_{11}\varepsilon_{22}-\varepsilon_{12}\varepsilon_{21})\cos(\eta^{(i)})]\tau_3\}S^{(i)}\Big|$$

$$\sum_{i=1}^{N}\left|\vec{\sigma}\cdot\vec{p}^{(i)}\right|S^{(i)}=$$

$$=\sum_{i=1}^{N}\Big[\{(\varepsilon_{22}\varepsilon_{33}-\varepsilon_{23}\varepsilon_{32})\cos(\chi^{(i)})\sin(\eta^{(i)})-(\varepsilon_{21}\varepsilon_{33}-\varepsilon_{23}\varepsilon_{31})\sin(\chi^{(i)})\sin(\eta^{(i)})+(\varepsilon_{21}\varepsilon_{32}-\varepsilon_{22}\varepsilon_{31})\cos(\eta^{(i)})\}^{2}+$$

$$+\{(\varepsilon_{13}\varepsilon_{32}-\varepsilon_{12}\varepsilon_{33})\cos(\chi^{(i)})\sin(\eta^{(i)})-(\varepsilon_{13}\varepsilon_{31}-\varepsilon_{11}\varepsilon_{33})\sin(\chi^{(i)})\sin(\eta^{(i)})+(\varepsilon_{12}\varepsilon_{31}-\varepsilon_{11}\varepsilon_{32})\cos(\eta^{(i)})\}^{2}+$$

$$+\{(\varepsilon_{12}\varepsilon_{23}-\varepsilon_{13}\varepsilon_{22})\cos(\chi^{(i)})\sin(\eta^{(i)})-(\varepsilon_{11}\varepsilon_{23}-\varepsilon_{13}\varepsilon_{21})\sin(\chi^{(i)})\sin(\eta^{(i)})+(\varepsilon_{11}\varepsilon_{22}-\varepsilon_{12}\varepsilon_{21})\cos(\eta^{(i)})\}^{2}\Big]^{\frac{1}{2}}S^{(i)} \quad (55)$$

The degree of whole grained structure orientation κ to the direction determined by the unit vector $\vec{\tau}$ was calculated by means of the Formula (51) while the sums in the numerator and denominator in the second member on the right side of this formula were determined by (54) and (55).

Any randomly oriented surface element on the grain boundary $S^{(i)}$ in the given bulk grained structure can be considered in the calculation of κ if the corresponding parameters $\eta^{(i)}$ and $\chi^{(i)}$ are applied. Therefore, random values of angles $\eta^{(i)}$ and $\chi^{(i)}$ should be substituted step by step into Formulas (54) and (55) during the calculation progress. This procedure of determination of the parameter κ can be formulated using the following random matrices directly pertaining to the deformation tensor only:

$$\widetilde{A}_{1}^{(i)} = \begin{pmatrix} \sin(\eta^{(i)})\cos(\chi^{(i)}), & \sin(\eta^{(i)})\sin(\chi^{(i)}), & \cos(\eta^{(i)}) \\ \varepsilon_{21}, & \varepsilon_{22}, & \varepsilon_{23} \\ \varepsilon_{31}, & \varepsilon_{32}, & \varepsilon_{33} \end{pmatrix} \quad (56)$$

$$\widetilde{A}_{2}^{(i)} = \begin{pmatrix} \varepsilon_{11}, & \varepsilon_{12}, & \varepsilon_{13} \\ \sin(\eta^{(i)})\cos(\chi^{(i)}), & \sin(\eta^{(i)})\sin(\chi^{(i)}), & \cos(\eta^{(i)}) \\ \varepsilon_{31}, & \varepsilon_{32}, & \varepsilon_{33} \end{pmatrix} \quad (57)$$

$$\widetilde{A}_{3}^{(i)} = \begin{pmatrix} \varepsilon_{11}, & \varepsilon_{12}, & \varepsilon_{13} \\ \varepsilon_{21}, & \varepsilon_{22}, & \varepsilon_{23} \\ \sin(\eta^{(i)})\cos(\chi^{(i)}), & \sin(\eta^{(i)})\sin(\chi^{(i)}), & \cos(\eta^{(i)}) \end{pmatrix} \quad (58)$$

Matrix $\widetilde{A}_{n}^{(i)}$ is a matrix-valued random variable, i.e., it is a matrix whose elements can be random variables. So, the problem of grained structure orientation can be formulated as a random matrix problem for calculation purposes. In this sense, the Formula (51) can be rewritten in the following form:

$$\kappa_{defor} \approx 1 - 2\frac{\sum_{i=1}^{N}\left(S^{(i)}\left|\sum_{n=1}^{3}\tau_{n}\det\widetilde{A}_{n}^{(i)}\right|\right)}{\sum_{i=1}^{N}\left(S^{(i)}\sqrt{\sum_{n=1}^{3}\left(\det\widetilde{A}_{n}^{(i)}\right)^{2}}\right)} \quad \ldots n = 1, 2, 3 \quad (59)$$

where $\det\widetilde{A}_{n}^{(i)}$ is a determinant of the random matrix $\widetilde{A}_{n}^{(i)}$. Eventually a random $3 \times 3 \times 3$ tensor with the following elements:

$$\widetilde{\Lambda}_{nmk}^{(i)} = \varepsilon_{mk}(1-\delta_{m,n}) + \tfrac{1}{2}\delta_{m,n}\{[1+(-1)^{\alpha_{k}}]\sin(\eta^{(i)})\cos(\chi^{(i)}) + [1+(-1)^{k}]\sin(\eta^{(i)})\sin(\chi^{(i)}) +$$
$$+ [1+(-1)^{\beta_{k}}]\cos(\eta^{(i)})\}$$

$$\text{where}: \alpha_{k} = \sum_{\nu=1}^{k}\frac{1+(-1)^{\nu}}{2}, \beta_{k} = \sum_{\nu=1}^{k}\frac{1-(-1)^{\nu}}{2}, \ldots n, m, k = \{1, 2, 3\} \quad (60)$$

can be imposed and smartly used in the formal mathematical notation as well as in the numerical implementation of the mentioned approximation approach.

7. Algorithmic Issues for the Numerical Treatment of Orientation κ as a Function of Deformation δ

In the previous section the random matrix approximation was presented where grain boundaries are considered to be a large number of planar surface elements. Therefore, all derivatives with respect to parameters u and v in Formulas (19) and (27) are simply constants. The mentioned constants are different for each planar surface and the integrals can be replaced by sums. Random variables $\eta^{(i)}$, $\chi^{(i)}$ and $S^{(i)}$ actually represent the orientation and size of i–th planar surface in the structure.

If the random homogeneous grained structure is considered, the single grain may be of any shape and can have also any orientation. The size and orientation of the mentioned small flat surfaces in the structure are random variables in the approximation. Random matrix approximation is not limited by a particular shape of the grain.

When Formulas (48) and (49) are considered in the homogeneous and isotropic case then results (54) and (55) can be simplified to the following form:

$$\sum_{i=1}^{N} \left| \left(\vec{\sigma} \cdot \vec{\tau} \right)^T \cdot \vec{p}^{(i)} \right| S^{(i)} = \sum_{i=1}^{N} S^{(i)} \left| \frac{1}{(\delta+1)} \sin(\eta^{(i)}) \cos(\chi^{(i)}) \tau_1 + \sqrt{\delta+1} \sin(\eta^{(i)}) \sin(\chi^{(i)}) \tau_2 + \sqrt{\delta+1} \cos(\eta^{(i)}) \tau_3 \right|, \qquad (61)$$

$$\sum_{i=1}^{N} \left| \vec{\sigma} \cdot \vec{p}^{(i)} \right| S^{(i)} = \sum_{i=1}^{N} S^{(i)} \sqrt{ \left\{ \frac{1}{(\delta+1)} \cos(\chi^{(i)}) \sin(\eta^{(i)}) \right\}^2 + \left\{ \sqrt{\delta+1} \sin(\chi^{(i)}) \sin(\eta^{(i)}) \right\}^2 + \left\{ \sqrt{\delta+1} \cos(\eta^{(i)}) \right\}^2 }, \qquad (62)$$

where N is the number of random variables (i.e., number of planar grain boundaries in the random matrix model of the considered grained structure). If the vector $\vec{\tau} = [1, 0, 0]$ (out of plane deformation), we obtain the following formula by means of substituting (61) and (62) into (51):

$$\kappa_{defor} = 1 - \frac{2 \sum_{i=1}^{N} S^{(i)} \left| \sin(\eta^{(i)}) \cos(\chi^{(i)}) \right|}{\sum_{i=1}^{N} S^{(i)} \sqrt{\cos^2(\chi^{(i)}) \sin^2(\eta^{(i)}) + (\delta+1)^3 \{\sin^2(\chi^{(i)}) \sin^2(\eta^{(i)}) + \cos^2(\eta^{(i)})\}}} \qquad (63)$$

Orientation to direction determined by unit vector $\vec{\tau}$ which is parallel to the direction of deformation force is evaluated in this simple case by means of Formula (63). Parameters $\eta^{(i)}$, $\chi^{(i)}$ and $S^{(i)}$ should be randomly generated and substituted in Formula (63) during the numerical simulation of the function $\kappa_{defor}(\delta)$. In addition, orientation of the grained structure in the undeformed configuration must be taken into account during the simulation. Prolongation $\delta = 0$ in the undeformed case and then the following holds:

$$\kappa_{undefor} = 1 - \frac{2 \sum_{i=1}^{N} S^{(i)} \left| \sin(\eta^{(i)}) \cos(\chi^{(i)}) \right|}{\sum_{i=1}^{N} S^{(i)}} \qquad (64)$$

So random values $\eta^{(i)}$, $\chi^{(i)}$ and $S^{(i)}$ must be generated in a such way as to meet the initial condition (64). Considering (63) and (64) we get:

$$\kappa_{defor} = 1 - \frac{\left(1 - \kappa_{undefor}\right) \sum_{i=1}^{N} S^{(i)}}{\sum_{i=1}^{N} S^{(i)} \sqrt{\cos^2(\chi^{(i)}) \sin^2(\eta^{(i)}) + (\delta+1)^3 \{\sin^2(\chi^{(i)}) \sin^2(\eta^{(i)}) + \cos^2(\eta^{(i)})\}}} \qquad (65)$$

It is advantageous to evaluate the orientation of the grained structure by means of the Formula (55) in cases when an extension of this structure is observed caused by the deformation force, i.e., if $0 < \delta$.

In the case where there is contraction of the structure under the deformation force, i.e., $-1 < \delta < 0$, we consider it appropriate to evaluate orientation κ in relation to any vector oriented perpendicular to the deformation force. Vector

$$\vec{\tau} = [0, \cos(\theta), \sin(\theta)] \tag{66}$$

where $0 < \theta < 2\pi$ must be considered in this case of squeezing the structure (in plane deformation) and Formulas (54) and (55) take the form:

$$\sum_{i=1}^{N} \left| \left(\vec{\sigma}^T \cdot \vec{\tau} \right) \cdot \vec{\rho}^{(i)} \right| S^{(i)} = \sum_{i=1}^{N} S^{(i)} \left| \sqrt{\delta+1} \sin(\eta^{(i)}) \sin(\chi^{(i)}) \cos(\theta) + \sqrt{\delta+1} \cos(\eta^{(i)}) \sin(\theta) \right| \tag{67}$$

$$\sum_{i=1}^{N} \left| \vec{\sigma} \cdot \vec{\rho}^{(i)} \right| S^{(i)} = \sum_{i=1}^{N} S^{(i)} \sqrt{\left\{ \frac{1}{(\delta+1)} \cos(\chi^{(i)}) \sin(\eta^{(i)}) \right\}^2 + \left\{ \sqrt{\delta+1} \sin(\chi^{(i)}) \sin(\eta^{(i)}) \right\}^2 + \left\{ \sqrt{\delta+1} \cos(\eta^{(i)}) \right\}^2}. \tag{68}$$

Applying (66), (67), and (68) in Formula (51) we get:

$$\kappa'_{defor} = 1 - 2 \frac{\sum_{i=1}^{N} S^{(i)} \left| \sqrt{\delta+1} \sin(\eta^{(i)}) \sin(\chi^{(i)}) \cos(\theta^{(i)}) + \sqrt{\delta+1} \cos(\eta^{(i)}) \sin(\theta^{(i)}) \right|}{\sum_{i=1}^{N} S^{(i)} \sqrt{\left\{ \frac{1}{(\delta+1)} \cos(\chi^{(i)}) \sin(\eta^{(i)}) \right\}^2 + \left\{ \sqrt{\delta+1} \sin(\chi^{(i)}) \sin(\eta^{(i)}) \right\}^2 + \left\{ \sqrt{\delta+1} \cos(\eta^{(i)}) \right\}^2}} . \tag{69}$$

Therefore, in undeformed configuration it holds that:

$$\kappa'_{undefor} = 1 - \frac{2 \sum_{i=1}^{N} S^{(i)} \left| \sin(\eta^{(i)}) \sin(\chi^{(i)}) \cos(\theta) + \cos(\eta^{(i)}) \sin(\theta) \right|}{\sum_{i=1}^{N} S^{(i)}} \tag{70}$$

and we obtain the following formula for the calculation of the degree of orientation κ:

$$\kappa'_{defor} = 1 - \frac{(\delta+1) \sqrt{\delta+1} \left(1 - \kappa'_{undefor} \right) \sum_{i=1}^{N} S^{(i)}}{\sum_{i=1}^{N} S^{(i)} \sqrt{\cos^2(\chi^{(i)}) \sin^2(\eta^{(i)}) + (\delta+1)^3 \left\{ \sin^2(\chi^{(i)}) \sin^2(\eta^{(i)}) + \cos^2(\eta^{(i)}) \right\}}} . \tag{71}$$

The Monte Carlo method was used to find functions $\kappa_{defor}(\delta)$ and $\kappa'_{defor}(\delta)$. Results of the applied stochastic calculations are shown in graphs in Figures 11 and 12. Individual points of graphs were found by successive application of $N = 10^7$ randomly chosen triples of variables $\{\eta^{(i)}, \chi^{(i)}, S^{(i)}\}$ in Formulas (65) and (71). It was shown that with increasing number of applied random variables N the calculated values κ_{defor} and κ'_{defor} converge to some expected value of the degree of the whole grain- surface orientation for every value δ. This means that the computational stability of the Monte Carlo algorithm has been confirmed in the simulation of the relationship between orientation and deformation of the structure.

Results of the numerical calculation confirmed that Formulas (65) and (71) can be successfully applied for the conversion of the orientation represented by the quantity κ into local plastic deformation characterized by δ in the case of plastic deformation of the grained structure. In addition graphs $\kappa_{defor}(\delta)$ and $\kappa'_{defor}(\delta)$ obtained from these formulas using the Monte Carlo simulation demonstrate the applicability of stereological methods for the modelling of local plastic deformation in grained structures.

However, Formulas (65) and (71) do not represent a general solution. These results can only be applied to a particular case for which there is no rigid solid rotation of grains in the structure and the axes chosen are principal axes of strain (therefore, there are no shear components present). In addition, another limitation is imposed, which is that plastic deformation must be isotropic. Actually, it does not seem that a grain with anisotropic behavior would give complications in the practical implementation of our results.

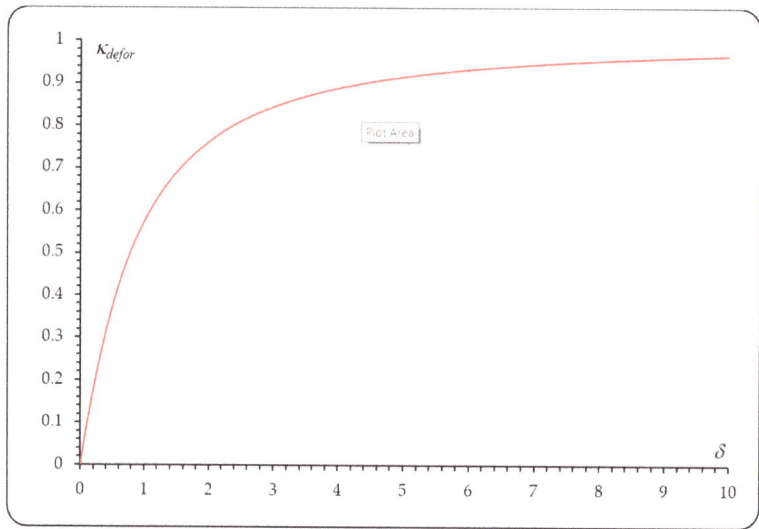

Figure 11. Graph illustrating the result of Monte Carlo simulation of degree of orientation κ as a function of prolongation δ for a plastically deformed random grained structure. Vector $\vec{\tau}$ is parallel to the deformation force.

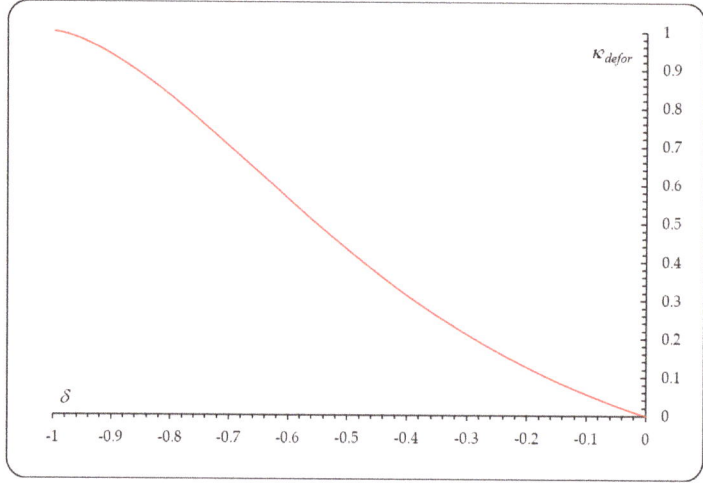

Figure 12. Graph illustrating the result of Monte Carlo simulation of degree of orientation κ as a function of contraction δ for plastically deformed random grained structure. Vector $\vec{\tau}$ is perpendicular to the deformation force.

8. Discussion and Conclusions

In our work we briefly pointed out the possible application of quantities obtained from standard stereological measurements. Our findings show that these quantities can be successfully applied for the modelling of local plastic deformation in grained structures. Our approach is based on the assumption that the grain, as it appears in a two-dimensional cross section, will change its shape and geometric orientation in 3D during plastic deformation. We applied and discussed our results for a homogeneous and isotropic grained structure.

The application is quite simple in the case of plastic deformation. The local plastic deformation in an isotropic and homogeneous grained structure can be evaluated by means of the value of grain elongation (or contraction) δ. However, the grain elongation (contraction) in the real structure cannot be measured directly. Conversely the measurement of the degree of grain orientation κ is available using stereological methods. One of the main benefits of our work is that we found the correlation between the degree of grain orientation κ and grain elongation (contraction) δ (see Formula (65) for elongation or (71) for contraction). A model of random grained structure was established and the mentioned correlation was investigated.

We note that by applying the Formulas (65) and (71), the correlations $\delta(\kappa)$ can be easily determined for various initial grained structure orientations represented by values $\kappa_{undefor}$. The initial orientation of a randomly selected grain in an undeformed random grained structure is arbitrary and $\kappa_{undefor} = 0$ in isotropic cases. However a relatively broad database of graphs $\{\delta(\kappa, \kappa_{undefor})\}$ for the various $\kappa_{undefor}$ could be calculated and recorded when using a computer technique. Subsequently, the value of grain elongation δ for corresponding κ can be found from the relevant graph. However, before this the graph for the quantification of δ must be selected from the database using the value $\kappa_{undefor}$ measured for the undeformed configuration of the grained structure.

Therefore, if a deformation map of the plastically deformed homogeneous and isotropic grained structure is to be investigated in the future, we propose using the results obtained above by the procedure consisting of the following steps:

1. The degree of grain orientation κ in some local area in the given i-th position in the grained structure must be measured in both undeformed ($\kappa^i_{undefor}$) and deformed (κ^i_{defor}) configurations. For example, the Saltykov stereological method is applicable for this purpose.

2. Graph $\delta(\kappa)$ for a given model of grained structure must be identified so as the following is applied: $\delta(\kappa^i_{undefor}) = 0$. We found the graph for the random model of homogeneous and isotropic grained structure by applying the random matrix approximation (see Formulas (65) or (71)) and using Monte Carlo simulation (see Figures 11 and 12). $\kappa^i_{undefor} = 0$ was applied for the homogeneous isotropic case.

3. Value δ_i corresponding to the measured value κ^i_{defor} must be found only from the graph $\delta(\kappa)$. The search procedure is easy: $\delta_i = \delta(\kappa^i_{defor})$. Finally, this value δ_i represents the grain elongation in the i-th position in the structure and allows the local strain in the given i-th position to be evaluated.

However if we have to discuss how the obtained results can be applied to the study of plastic deformation we must also note that the limitations imposed on Formulas (65) or (71) are often too strong to make the presented method useful to study real deformation processes. Namely, also rigid solid rotations of grains must be taken into account in general. Moreover, in order to be applicable in the field of crystal plasticity, the case of anisotropic grain behavior should be discussed. A more general method must be designed which does not lead to ambiguities in the interpretation of the change in the degree of grain orientation. In the general inhomogeneous and anisotropic case, the Formula (9) resulting from random matrix approximation should be used. This formula allows us to find the relationship between the measurable quantities κ^i_{defor} and the coefficients of the deformation tensor ε_{qm} in the local area of the grained structure. However the degree of the grain orientation κ in the given local area of the grained structure will have to be measured in relation to more different vectors $\vec{\tau}^k$ (oriented in various directions) so that we get the appropriate number of equations corresponding to the number of unknown coefficients ε_{qm}. This means the next system of equations must be solved in the inhomogeneous and anisotropic case:

$$\kappa^k_{defor} \approx 1 - 2 \frac{\sum_{i=1}^{N}\left(S^{(i)}\left|\sum_{n=1}^{3}\tau^k_n det\left[\widetilde{A}^{(i)}_n\left(\{\varepsilon_{qm}\}\right)\right]\right|\right)}{\sum_{i=1}^{N}\left(S^{(i)}\sqrt{\sum_{n=1}^{3}\left(det\left[\widetilde{A}^{(i)}_n\left(\{\varepsilon_{qm}\}\right)\right]\right)^2}\right)}, \quad m = 1, 2, 3, \ k = 1, 2, 9 \quad (72)$$

Local values of the coefficients of the deformation tensor ε_{qm} can be determined from the system (73). In the case of plastic deformation one of the equations in the system () may be replaced by the Equation (47). Generally, the verification of the practical applicability of the suggested idea to determine local plastic deformation in a structure requires development of a suitable technique for experimental determination of κ_{defor}. Besides that, it is necessary to develop some numerical methods suitable for software processing that allow coefficients ε_{qm} to be found satisfying a system of Equation (73). System (73) quantified for $\varepsilon_{qm} = \delta_{qm}$ characterizes the initial orientation of the grained structure in relation to the direction given by vectors $\vec{\tau}^k$ in the undeformed configuration and represents the initial conditions:

$$\kappa^k_{undefor} \approx 1 - 2 \frac{\sum_{i=1}^{N}\left(S^{(i)}\left|\sum_{n=1}^{3}\tau^k_n det\left[\widetilde{A}^{(i)}_n\left(\{\varepsilon_{qm}=\delta_{qm}\}\right)\right]\right|\right)}{\sum_{i=1}^{N}\left(S^{(i)}\sqrt{\sum_{n=1}^{3}\left(det\left[\widetilde{A}^{(i)}_n\left(\{\varepsilon_{qm}=\delta_{qm}\}\right)\right]\right)^2}\right)}, \quad m = 1, 2, 3, \ k = 1, 2, 9 \quad (73)$$

This initial condition must be considered in Equation (73) in the case of the deformed configuration and the deformation characteristics represented by tensor ε_{qm} in the local area of the grained structure can be searched by solving this system.

In conclusion, we dealt with the problem of applicability of stereological methods to determine local plastic deformation in a grained structure. One can easily understand that there is a correlation between the change in the grain orientation and grain deformation. This correlation was mathematically demonstrated in our work. Indeed, the expression (50) allows one to investigate this correlation. If the quantity κ clearly corresponds to the degree of grain orientation measured by the Saltykov method, the analysis of local deformation could be realized experimentally on the basis of our results. As an illustration we showed the applicability of our results for a random model of a grained structure. This specific model was used here because of its simplicity but we believe that our method can be applied also for other models.

During the loading process the κ changes and reflects deformation phenomena in the structure. On-line scanning of grain orientation could be of great help for in-situ investigation of the deformation dynamics. In this respect new methods for 3D object imaging may be helpful. Procedures for the practical realization of our method suggested above will have to be modified in this case and this modification will be the subject of a future work. However, we think that the mathematical principles of the method presented in our work will remain the same.

Author Contributions: M.M. is the author of the main idea of the measuring local plastic deformation using stereological methods. He was focused mainly on the experimental measurement of the grains orientation in the material structure using statistical method of oriented testing lines and the technique for local strain analysis based on grain shape. His activities aims to establish the effect of local grain deformation measured on a cross-section to overall mechanical properties and their evolution. S.M. was primarily oriented on mathematical analysis of correlation between orientation and deformation. He was interest in how the image of the structure in a cross-sectional plane changes when the structure is linearly deformed. An initial shape was assumed which was transformed into a deformed shape by analytic calculation. He was also design a computationally low consuming procedure for quantification of local deformation in structure based on distortion of image of this structure in a cross-sectional view. All authors have read and agreed to the published version of the manuscript.

Funding: This publication was funded by the Slovak Research and Development Agency under the contract No. APVV-15-0319.

Acknowledgments: This work was supported by the Operational Programme Research and Innovation for the project: Scientific and Research Centre of Excellence SlovakION for Material and Interdisciplinary Research, code of the project ITMS2014+: 313011W085 co-financed by the European Regional Development Fund.

Conflicts of Interest: The authors declare no conflict of interest.

Appendix A. Transformation of a Grain Surface Element Vector During Deformation

Using the notation (14) and (15) the vector of deformed grain surface element can be written as:

$$d^2 \vec{S}_{defor} = \left(\frac{\partial \vec{r}}{\partial u} \times \frac{\partial \vec{r}}{\partial v}\right) du dv \text{ i.e., in coordinate form :}$$

$$d^2 \vec{S}_{defor} = d^2 S_x^{def} \vec{i} + d^2 S_y^{def} \vec{j} + d^2 S_z^{def} \vec{k} = \tag{A1}$$

$$= \left\{ \left(\frac{\partial x_2}{\partial u}\frac{\partial x_3}{\partial v} - \frac{\partial x_3}{\partial u}\frac{\partial x_2}{\partial v}\right)\vec{i} + \left(\frac{\partial x_1}{\partial v}\frac{\partial x_3}{\partial u} - \frac{\partial x_1}{\partial u}\frac{\partial x_3}{\partial v}\right)\vec{j} + \left(\frac{\partial x_1}{\partial u}\frac{\partial x_2}{\partial v} - \frac{\partial x_2}{\partial u}\frac{\partial x_1}{\partial v}\right)\vec{k} \right\} du dv.$$

Transformation (17) can be applied in (A1) and we get:

$$d^2 \vec{S}_{defor} = \left\{\left(\varepsilon_{2j}\varepsilon_{3k}\frac{\partial X_j}{\partial u}\frac{\partial X_k}{\partial v} - \varepsilon_{2l}\varepsilon_{3m}\frac{\partial X_l}{\partial v}\frac{\partial X_m}{\partial u}\right)\vec{i} + \left(\varepsilon_{1j}\varepsilon_{3k}\frac{\partial X_j}{\partial v}\frac{\partial X_k}{\partial u} - \varepsilon_{1l}\varepsilon_{3m}\frac{\partial X_l}{\partial u}\frac{\partial X_m}{\partial v}\right)\vec{j} + \right.$$

$$\left. + \left(\varepsilon_{1j}\varepsilon_{2k}\frac{\partial X_j}{\partial u}\frac{\partial X_k}{\partial v} - \varepsilon_{1l}\varepsilon_{2m}\frac{\partial X_l}{\partial v}\frac{\partial X_m}{\partial u}\right)\vec{k}\right\} du dv. \tag{A2}$$

After formal unification of index symbols in summations (A2) (indexes take the same values and therefore we can change symbols $m \to k, l \to j$), the coefficients of deformation tensor can be excluded before the brackets:

$$\left(\varepsilon_{2j}\varepsilon_{3k}\frac{\partial X_j}{\partial u}\frac{\partial X_k}{\partial v} - \varepsilon_{2l}\varepsilon_{3m}\frac{\partial X_l}{\partial v}\frac{\partial X_m}{\partial u}\right) = \varepsilon_{2j}\varepsilon_{3k}\left(\frac{\partial X_j}{\partial u}\frac{\partial X_k}{\partial v} - \frac{\partial X_j}{\partial v}\frac{\partial X_k}{\partial u}\right), \tag{A3}$$

$$\left(\varepsilon_{1j}\varepsilon_{3k}\frac{\partial X_j}{\partial v}\frac{\partial X_k}{\partial u} - \varepsilon_{1l}\varepsilon_{3m}\frac{\partial X_l}{\partial u}\frac{\partial X_m}{\partial v}\right) = -\varepsilon_{1j}\varepsilon_{3k}\left(\frac{\partial X_j}{\partial u}\frac{\partial X_k}{\partial v} - \frac{\partial X_j}{\partial v}\frac{\partial X_k}{\partial u}\right), \tag{A4}$$

$$\left(\varepsilon_{1j}\varepsilon_{2k}\frac{\partial X_j}{\partial u}\frac{\partial X_k}{\partial v} - \varepsilon_{1l}\varepsilon_{2m}\frac{\partial X_l}{\partial v}\frac{\partial X_m}{\partial u}\right) = \varepsilon_{1j}\varepsilon_{2k}\left(\frac{\partial X_j}{\partial u}\frac{\partial X_k}{\partial v} - \frac{\partial X_j}{\partial v}\frac{\partial X_k}{\partial u}\right). \tag{A5}$$

It is easy to show that sum (A3) can be written as:

$$\varepsilon_{2j}\varepsilon_{3k}\left(\frac{\partial X_j}{\partial u}\frac{\partial X_k}{\partial v} - \frac{\partial X_j}{\partial v}\frac{\partial X_k}{\partial u}\right) = \varepsilon_{21}\varepsilon_{32}\left(\frac{\partial X_1}{\partial u}\frac{\partial X_2}{\partial v} - \frac{\partial X_1}{\partial v}\frac{\partial X_2}{\partial u}\right) + \varepsilon_{22}\varepsilon_{31}\left(\frac{\partial X_2}{\partial u}\frac{\partial X_1}{\partial v} - \frac{\partial X_2}{\partial v}\frac{\partial X_1}{\partial u}\right) +$$

$$+\varepsilon_{21}\varepsilon_{33}\left(\frac{\partial X_1}{\partial u}\frac{\partial X_3}{\partial v} - \frac{\partial X_1}{\partial v}\frac{\partial X_3}{\partial u}\right) + \varepsilon_{23}\varepsilon_{31}\left(\frac{\partial X_3}{\partial u}\frac{\partial X_1}{\partial v} - \frac{\partial X_3}{\partial v}\frac{\partial X_1}{\partial u}\right) + \varepsilon_{22}\varepsilon_{23}\left(\frac{\partial X_2}{\partial u}\frac{\partial X_3}{\partial v} - \frac{\partial X_2}{\partial v}\frac{\partial X_3}{\partial u}\right) + \tag{A6}$$

$$+\varepsilon_{23}\varepsilon_{32}\left(\frac{\partial X_3}{\partial u}\frac{\partial X_2}{\partial v} - \frac{\partial X_3}{\partial v}\frac{\partial X_2}{\partial u}\right)$$

where we have taken into account that contributions of members characterized by $j = k$ are zero. Formulas (A4) and (A5) can be re-written by the same way and coordinates of the vector of the deformed grain surface element (A2) can be subsequently transformed to the following form:

$$d^2 S_x^{def} = \varepsilon_{2j}\varepsilon_{3k}\left(\frac{\partial X_j}{\partial u}\frac{\partial X_k}{\partial v} - \frac{\partial X_j}{\partial v}\frac{\partial X_k}{\partial u}\right) du dv = (\varepsilon_{21}\varepsilon_{32} - \varepsilon_{22}\varepsilon_{31})\left(\frac{\partial X_1}{\partial u}\frac{\partial X_2}{\partial v} - \frac{\partial X_1}{\partial v}\frac{\partial X_2}{\partial u}\right) du dv +$$

$$+ (\varepsilon_{21}\varepsilon_{33} - \varepsilon_{23}\varepsilon_{31})\left(\frac{\partial X_1}{\partial u}\frac{\partial X_3}{\partial v} - \frac{\partial X_1}{\partial v}\frac{\partial X_3}{\partial u}\right) du dv + (\varepsilon_{22}\varepsilon_{33} - \varepsilon_{23}\varepsilon_{32})\left(\frac{\partial X_2}{\partial u}\frac{\partial X_3}{\partial v} - \frac{\partial X_2}{\partial v}\frac{\partial X_3}{\partial u}\right) du dv \tag{A7}$$

$$d^2 S_y^{def} = -\varepsilon_{1j}\varepsilon_{3k}\left(\frac{\partial X_j}{\partial u}\frac{\partial X_k}{\partial v} - \frac{\partial X_j}{\partial v}\frac{\partial X_k}{\partial u}\right) du dv = -(\varepsilon_{11}\varepsilon_{32} - \varepsilon_{12}\varepsilon_{31})\left(\frac{\partial X_1}{\partial u}\frac{\partial X_2}{\partial v} - \frac{\partial X_1}{\partial v}\frac{\partial X_2}{\partial u}\right) du dv -$$

$$- (\varepsilon_{11}\varepsilon_{33} - \varepsilon_{13}\varepsilon_{31})\left(\frac{\partial X_1}{\partial u}\frac{\partial X_3}{\partial v} - \frac{\partial X_1}{\partial v}\frac{\partial X_3}{\partial u}\right) du dv - (\varepsilon_{12}\varepsilon_{33} - \varepsilon_{13}\varepsilon_{32})\left(\frac{\partial X_2}{\partial u}\frac{\partial X_3}{\partial v} - \frac{\partial X_2}{\partial v}\frac{\partial X_3}{\partial u}\right) du dv, \tag{A8}$$

$$d^2S_z^{def} = \varepsilon_{1j}\varepsilon_{2k}\left(\frac{\partial X_j}{\partial u}\frac{\partial X_k}{\partial v} - \frac{\partial X_j}{\partial v}\frac{\partial X_k}{\partial u}\right)dudv = (\varepsilon_{11}\varepsilon_{22} - \varepsilon_{12}\varepsilon_{21})\left(\frac{\partial X_1}{\partial u}\frac{\partial X_2}{\partial v} - \frac{\partial X_1}{\partial v}\frac{\partial X_2}{\partial u}\right)dudv +$$
$$+(\varepsilon_{11}\varepsilon_{23} - \varepsilon_{13}\varepsilon_{21})\left(\frac{\partial X_1}{\partial u}\frac{\partial X_3}{\partial v} - \frac{\partial X_1}{\partial v}\frac{\partial X_3}{\partial u}\right)dudv + (\varepsilon_{12}\varepsilon_{23} - \varepsilon_{13}\varepsilon_{22})\left(\frac{\partial X_2}{\partial u}\frac{\partial X_3}{\partial v} - \frac{\partial X_2}{\partial v}\frac{\partial X_3}{\partial u}\right)dudv.$$
(A9)

Note that the vector of the same grain surface element in the undeformed configuration is:

$$d^2\vec{S}_{undefor} = d^2S_x^{undef}\vec{i} + d^2S_y^{undef}\vec{j} + d^2S_z^{undef}\vec{k} =$$

$$= \left(\frac{\partial X_2}{\partial u}\frac{\partial X_3}{\partial v} - \frac{\partial X_2}{\partial v}\frac{\partial X_3}{\partial u}\right)dudv\,\vec{i} + \left(\frac{\partial X_1}{\partial v}\frac{\partial X_3}{\partial u} - \frac{\partial X_1}{\partial u}\frac{\partial X_3}{\partial v}\right)dudv\,\vec{j} + \left(\frac{\partial X_1}{\partial u}\frac{\partial X_2}{\partial v} - \frac{\partial X_1}{\partial v}\frac{\partial X_2}{\partial u}\right)dudv\,\vec{k}.$$
(A10)

Comparing the corresponding coordinates of vectors (A2) and (A10) we get:

$$d^2S_x^{def} = (\varepsilon_{21}\varepsilon_{32} - \varepsilon_{22}\varepsilon_{31})d^2S_z^{undef} - (\varepsilon_{21}\varepsilon_{33} - \varepsilon_{23}\varepsilon_{31})d^2S_y^{undef} + (\varepsilon_{22}\varepsilon_{33} - \varepsilon_{23}\varepsilon_{32})d^2S_x^{undef},$$
$$d^2S_y^{def} = -(\varepsilon_{11}\varepsilon_{32} - \varepsilon_{12}\varepsilon_{31})d^2S_z^{undef} + (\varepsilon_{11}\varepsilon_{33} - \varepsilon_{13}\varepsilon_{31})d^2S_y^{undef} - (\varepsilon_{12}\varepsilon_{33} - \varepsilon_{13}\varepsilon_{32})d^2S_x^{undef},$$
$$d^2S_z^{def} = (\varepsilon_{11}\varepsilon_{22} - \varepsilon_{12}\varepsilon_{21})d^2S_z^{undef} - (\varepsilon_{11}\varepsilon_{23} - \varepsilon_{13}\varepsilon_{21})d^2S_y^{undef} + (\varepsilon_{12}\varepsilon_{23} - \varepsilon_{13}\varepsilon_{22})d^2S_x^{undef}$$
(A11)

and vector of the deformed grain surface element can be found as:

$$d^2\vec{S}_{defor} = \{(\varepsilon_{21}\varepsilon_{32} - \varepsilon_{22}\varepsilon_{31})d^2S_z^{undef} - (\varepsilon_{21}\varepsilon_{33} - \varepsilon_{23}\varepsilon_{31})d^2S_y^{undef} + (\varepsilon_{22}\varepsilon_{33} - \varepsilon_{23}\varepsilon_{32})d^2S_x^{undef}\}\vec{i} +$$
$$+ \{-(\varepsilon_{11}\varepsilon_{32} - \varepsilon_{12}\varepsilon_{31})d^2S_z^{undef} + (\varepsilon_{11}\varepsilon_{33} - \varepsilon_{13}\varepsilon_{31})d^2S_y^{undef} - (\varepsilon_{12}\varepsilon_{33} - \varepsilon_{13}\varepsilon_{32})d^2S_x^{undef}\}\vec{j} +$$
$$+ \{(\varepsilon_{11}\varepsilon_{22} - \varepsilon_{12}\varepsilon_{21})d^2S_z^{undef} - (\varepsilon_{11}\varepsilon_{23} - \varepsilon_{13}\varepsilon_{21})d^2S_y^{undef} + (\varepsilon_{12}\varepsilon_{23} - \varepsilon_{13}\varepsilon_{22})d^2S_x^{undef}\}\vec{k}.$$
(A12)

Next it is not difficult to show that:

$$d^2\vec{S}_{defor} = \begin{pmatrix} (\varepsilon_{22}\varepsilon_{33} - \varepsilon_{23}\varepsilon_{32}) & -(\varepsilon_{21}\varepsilon_{33} - \varepsilon_{23}\varepsilon_{31}) & (\varepsilon_{21}\varepsilon_{32} - \varepsilon_{22}\varepsilon_{31}) \\ (\varepsilon_{13}\varepsilon_{32} - \varepsilon_{12}\varepsilon_{33}) & -(\varepsilon_{13}\varepsilon_{31} - \varepsilon_{11}\varepsilon_{33}) & (\varepsilon_{12}\varepsilon_{31} - \varepsilon_{11}\varepsilon_{32}) \\ (\varepsilon_{12}\varepsilon_{23} - \varepsilon_{13}\varepsilon_{22}) & -(\varepsilon_{11}\varepsilon_{23} - \varepsilon_{13}\varepsilon_{21}) & (\varepsilon_{11}\varepsilon_{22} - \varepsilon_{12}\varepsilon_{21}) \end{pmatrix} \cdot \begin{pmatrix} d^2S_x^{und} \\ d^2S_y^{und} \\ d^2S_z^{und} \end{pmatrix}.$$
(A13)

So the grain surface element is transformed during the deformation according to the Formula (20).

Appendix B. Deformed Grain Surface Area

If we use the Formula (A2), the size of surface element area of the deformed grain can be found as follows:

$$\left(d^2S_{defor}\right)^2 = \left(d^2S_x^{def}\right)^2 + \left(d^2S_y^{def}\right)^2 + \left(d^2S_z^{def}\right)^2 = \left\{\left(\varepsilon_{2j}\varepsilon_{3k}\frac{\partial X_j}{\partial u}\frac{\partial X_k}{\partial v} - \varepsilon_{2l}\varepsilon_{3m}\frac{\partial X_l}{\partial v}\frac{\partial X_m}{\partial u}\right)^2 + \right.$$
$$\left. + \left(\varepsilon_{1j}\varepsilon_{3k}\frac{\partial X_j}{\partial v}\frac{\partial X_k}{\partial u} - \varepsilon_{1l}\varepsilon_{3m}\frac{\partial X_l}{\partial u}\frac{\partial X_m}{\partial v}\right)^2 + \left(\varepsilon_{1j}\varepsilon_{2k}\frac{\partial X_j}{\partial u}\frac{\partial X_k}{\partial v} - \varepsilon_{1l}\varepsilon_{2m}\frac{\partial X_l}{\partial v}\frac{\partial X_m}{\partial u}\right)^2\right\}(dudv)^2$$
(A14)

Squaring of first bracket in (A14) we get:

$$\left(\varepsilon_{2j}\varepsilon_{3k}\frac{\partial X_j}{\partial u}\frac{\partial X_k}{\partial v} - \varepsilon_{2l}\varepsilon_{3m}\frac{\partial X_l}{\partial v}\frac{\partial X_m}{\partial u}\right)^2 = \varepsilon_{2j}\varepsilon_{2n}\varepsilon_{3k}\varepsilon_{3s}\frac{\partial X_j}{\partial u}\frac{\partial X_n}{\partial u}\frac{\partial X_k}{\partial v}\frac{\partial X_s}{\partial v} +$$
$$+ \varepsilon_{3m}\varepsilon_{3s}\varepsilon_{2l}\varepsilon_{2n}\frac{\partial X_m}{\partial u}\frac{\partial X_s}{\partial u}\frac{\partial X_l}{\partial v}\frac{\partial X_n}{\partial v} - 2\varepsilon_{2j}\varepsilon_{3m}\varepsilon_{3k}\varepsilon_{2l}\frac{\partial X_j}{\partial u}\frac{\partial X_m}{\partial u}\frac{\partial X_k}{\partial v}\frac{\partial X_l}{\partial v}.$$
(A15)

In the (A15) we can change the symbols of the indexes in the second expression as follows $m \to j$, $s \to n$, $l \to k$, $n \to s$, i.e.,:

$$\varepsilon_{3m}\varepsilon_{3s}\varepsilon_{2l}\varepsilon_{2n}\frac{\partial X_m}{\partial u}\frac{\partial X_s}{\partial u}\frac{\partial X_l}{\partial v}\frac{\partial X_n}{\partial v} = \varepsilon_{3j}\varepsilon_{3n}\varepsilon_{2k}\varepsilon_{2s}\frac{\partial X_j}{\partial u}\frac{\partial X_n}{\partial u}\frac{\partial X_k}{\partial v}\frac{\partial X_s}{\partial v}$$
(A16)

and in the third expression as follows $m \to n, l \to s$:

$$\varepsilon_{2j}\varepsilon_{3m}\varepsilon_{3k}\varepsilon_{2l}\frac{\partial X_j}{\partial u}\frac{\partial X_m}{\partial u}\frac{\partial X_k}{\partial v}\frac{\partial X_l}{\partial v} = \varepsilon_{2j}\varepsilon_{3n}\varepsilon_{3k}\varepsilon_{2s}\frac{\partial X_j}{\partial u}\frac{\partial X_n}{\partial u}\frac{\partial X_k}{\partial v}\frac{\partial X_s}{\partial v}. \tag{A17}$$

Then the square of the first bracket is:

$$\left(\varepsilon_{2j}\varepsilon_{3k}\frac{\partial X_j}{\partial u}\frac{\partial X_k}{\partial v} - \varepsilon_{2l}\varepsilon_{3m}\frac{\partial X_l}{\partial v}\frac{\partial X_m}{\partial u}\right)^2 = \varepsilon_{2j}\varepsilon_{2n}\varepsilon_{3k}\varepsilon_{3s}\frac{\partial X_j}{\partial u}\frac{\partial X_n}{\partial u}\frac{\partial X_k}{\partial v}\frac{\partial X_s}{\partial v} + \\ +\varepsilon_{3j}\varepsilon_{3n}\varepsilon_{2k}\varepsilon_{2s}\frac{\partial X_j}{\partial u}\frac{\partial X_n}{\partial u}\frac{\partial X_k}{\partial v}\frac{\partial X_s}{\partial v} - 2\varepsilon_{2j}\varepsilon_{3n}\varepsilon_{3k}\varepsilon_{2s}\frac{\partial X_j}{\partial u}\frac{\partial X_n}{\partial u}\frac{\partial X_k}{\partial v}\frac{\partial X_s}{\partial v}. \tag{A18}$$

All squared brackets in (A14) can be re-written in the same a way, i.e.,:

$$\left(d^2 S_x^{def}\right)^2 = \left(\varepsilon_{2j}\varepsilon_{2n}\varepsilon_{3k}\varepsilon_{3s} + \varepsilon_{3j}\varepsilon_{3n}\varepsilon_{2k}\varepsilon_{2s} - 2\varepsilon_{2j}\varepsilon_{3n}\varepsilon_{3k}\varepsilon_{2s}\right)\frac{\partial X_j}{\partial u}\frac{\partial X_n}{\partial u}\frac{\partial X_k}{\partial v}\frac{\partial X_s}{\partial v}(dudv)^2, \tag{A19}$$

$$\left(d^2 S_y^{def}\right)^2 = \left(\varepsilon_{1j}\varepsilon_{1n}\varepsilon_{3k}\varepsilon_{3s} + \varepsilon_{3j}\varepsilon_{3n}\varepsilon_{1k}\varepsilon_{1s} - 2\varepsilon_{1j}\varepsilon_{3n}\varepsilon_{3k}\varepsilon_{1s}\right)\frac{\partial X_j}{\partial u}\frac{\partial X_n}{\partial u}\frac{\partial X_k}{\partial v}\frac{\partial X_s}{\partial v}(dudv)^2, \tag{A20}$$

$$\left(d^2 S_z^{def}\right)^2 = \left(\varepsilon_{1j}\varepsilon_{1n}\varepsilon_{2k}\varepsilon_{2s} + \varepsilon_{2j}\varepsilon_{2n}\varepsilon_{1k}\varepsilon_{1s} - 2\varepsilon_{1j}\varepsilon_{2n}\varepsilon_{2k}\varepsilon_{1s}\right)\frac{\partial X_j}{\partial u}\frac{\partial X_n}{\partial u}\frac{\partial X_k}{\partial v}\frac{\partial X_s}{\partial v}(dudv)^2. \tag{A21}$$

Substituting (A19), (A20) and (A21) into the Formula (A14) the size of the deformed grain surface area element can be written as:

$$\left(d^2 S_{defor}\right)^2 = \Gamma_{jnks}\frac{\partial X_j}{\partial u}\frac{\partial X_n}{\partial u}\frac{\partial X_k}{\partial v}\frac{\partial X_s}{\partial v}(dudv)^2, \tag{A22}$$

where:

$$\Gamma_{jnks} = \left\{\varepsilon_{2j}\varepsilon_{2n}\varepsilon_{3k}\varepsilon_{3s} + \varepsilon_{3j}\varepsilon_{3n}\varepsilon_{2k}\varepsilon_{2s} + \varepsilon_{1j}\varepsilon_{1n}\varepsilon_{3k}\varepsilon_{3s} + \varepsilon_{3j}\varepsilon_{3n}\varepsilon_{1k}\varepsilon_{1s} + \varepsilon_{1j}\varepsilon_{1n}\varepsilon_{2k}\varepsilon_{2s} + \varepsilon_{2j}\varepsilon_{2n}\varepsilon_{1k}\varepsilon_{1s} - \\ -2\left(\varepsilon_{2j}\varepsilon_{3n}\varepsilon_{3k}\varepsilon_{2s} + \varepsilon_{1j}\varepsilon_{3n}\varepsilon_{3k}\varepsilon_{1s} + \varepsilon_{1j}\varepsilon_{2n}\varepsilon_{2k}\varepsilon_{1s}\right)\right\} \tag{A23}$$

We note that:

$$\varepsilon_{2j}\varepsilon_{2n}\varepsilon_{3k}\varepsilon_{3s} + \varepsilon_{3j}\varepsilon_{3n}\varepsilon_{2k}\varepsilon_{2s} + \varepsilon_{1j}\varepsilon_{1n}\varepsilon_{3k}\varepsilon_{3s} + \varepsilon_{3j}\varepsilon_{3n}\varepsilon_{1k}\varepsilon_{1s} + \varepsilon_{1j}\varepsilon_{1n}\varepsilon_{2k}\varepsilon_{2s} + \varepsilon_{2j}\varepsilon_{2n}\varepsilon_{1k}\varepsilon_{1s} = \sum_{\substack{V([p,q]), p \neq q \\ p,q \in \{1,2,3\}}} \varepsilon_{pj}\varepsilon_{pn}\varepsilon_{qk}\varepsilon_{qs} \tag{A24}$$

where notation under the summation symbol means the sum over all variations of indexes p, q while $p \neq q$. In addition we use the appropriate changes of the index symbols in some sums in the bracket ($j \leftrightarrow n, k \leftrightarrow s$) and consider that:

$$2\left(\varepsilon_{2j}\varepsilon_{3n}\varepsilon_{3k}\varepsilon_{2s} + \varepsilon_{1j}\varepsilon_{3n}\varepsilon_{3k}\varepsilon_{1s} + \varepsilon_{1j}\varepsilon_{2n}\varepsilon_{2k}\varepsilon_{1s}\right) = \varepsilon_{2j}\varepsilon_{3n}\varepsilon_{3k}\varepsilon_{2s} + \varepsilon_{3j}\varepsilon_{2n}\varepsilon_{2k}\varepsilon_{3s} + \varepsilon_{1j}\varepsilon_{3n}\varepsilon_{3k}\varepsilon_{1s} + \\ +\varepsilon_{3j}\varepsilon_{1n}\varepsilon_{1k}\varepsilon_{3s} + \varepsilon_{1j}\varepsilon_{2n}\varepsilon_{2k}\varepsilon_{1s} + \varepsilon_{2j}\varepsilon_{1n}\varepsilon_{1k}\varepsilon_{2s} = \sum_{\substack{V([p,q]), p \neq q \\ p,q \in \{1,2,3\}}} \varepsilon_{pj}\varepsilon_{qn}\varepsilon_{qk}\varepsilon_{ps}. \tag{A25}$$

Therefore, the size of deformed grain surface area can be written as:

$$S_{defor} = \int_{v_1}^{v_2}\int_{u_1}^{u_2}\sqrt{\Gamma_{jnks}\frac{\partial X_j}{\partial u}\frac{\partial X_n}{\partial u}\frac{\partial X_k}{\partial v}\frac{\partial X_s}{\partial v}}dudv, \tag{A26}$$

where:

$$\Gamma_{jnks} = \sum_{\substack{V([p,q]), p \neq q \\ p,q \in \{1,2,3\}}} \left(\varepsilon_{pj}\varepsilon_{pn}\varepsilon_{qk}\varepsilon_{qs} - \varepsilon_{pj}\varepsilon_{qn}\varepsilon_{qk}\varepsilon_{ps}\right) \tag{A27}$$

which is consistent with Formulas (25) and (26).

Appendix C. Volume Preserving during Plastic Deformation

Integrals in (42) can be written in coordinate form:

$$\int_{(S_{undefor})} \vec{R} \cdot d\vec{S}_{undefor} = \int_{v_1}^{v_2}\int_{u_1}^{u_2} \left(X_1\left(\frac{\partial X_2}{\partial u}\frac{\partial X_3}{\partial v} - \frac{\partial X_2}{\partial v}\frac{\partial X_3}{\partial u}\right) + X_2\left(\frac{\partial X_1}{\partial v}\frac{\partial X_3}{\partial u} - \frac{\partial X_1}{\partial u}\frac{\partial X_3}{\partial v}\right) + X_3\left(\frac{\partial X_1}{\partial u}\frac{\partial X_2}{\partial v} - \frac{\partial X_1}{\partial v}\frac{\partial X_2}{\partial u}\right)\right) dudv \tag{A28}$$

$$\int_{(S_{defor})} \vec{r} \cdot d\vec{S}_{defor} = \int_{v_1}^{v_2}\int_{u_1}^{u_2} \left(x_1\left(\frac{\partial x_2}{\partial u}\frac{\partial x_3}{\partial v} - \frac{\partial x_2}{\partial v}\frac{\partial x_3}{\partial u}\right) + x_2\left(\frac{\partial x_1}{\partial v}\frac{\partial x_3}{\partial u} - \frac{\partial x_1}{\partial u}\frac{\partial x_3}{\partial v}\right) + x_3\left(\frac{\partial x_1}{\partial u}\frac{\partial x_2}{\partial v} - \frac{\partial x_1}{\partial v}\frac{\partial x_2}{\partial u}\right)\right) dudv \tag{A29}$$

Using the transformation (17) the integral in (A29) takes the form:

$$\int_{(S_{defor})} \vec{r} \cdot d\vec{S}_{defor} = \int_{v_1}^{v_2}\int_{u_1}^{u_2} \left\{ \varepsilon_{1n}X_n\left(\varepsilon_{2j}\varepsilon_{3k}\frac{\partial X_j}{\partial u}\frac{\partial X_k}{\partial v} - \varepsilon_{2l}\varepsilon_{3m}\frac{\partial X_l}{\partial v}\frac{\partial X_m}{\partial u}\right) + \right.$$
$$\left. + \varepsilon_{2n}X_n\left(\varepsilon_{1j}\varepsilon_{3k}\frac{\partial X_j}{\partial v}\frac{\partial X_k}{\partial u} - \varepsilon_{1l}\varepsilon_{3m}\frac{\partial X_l}{\partial u}\frac{\partial X_m}{\partial v}\right) + \varepsilon_{3n}X_n\left(\varepsilon_{1j}\varepsilon_{2k}\frac{\partial X_j}{\partial u}\frac{\partial X_k}{\partial v} - \varepsilon_l^1\varepsilon_m^2\frac{\partial X_l}{\partial v}\frac{\partial X_m}{\partial u}\right) \right\} dudv \tag{A30}$$

If we change the index labeling $l \to j$, $m \to k$ in the sums (A30) the following formula results from Equation (43):

$$\varepsilon_{1n}\varepsilon_{2j}\varepsilon_{3k}X_n\left(\frac{\partial X_j}{\partial u}\frac{\partial X_k}{\partial v} - \frac{\partial X_j}{\partial v}\frac{\partial X_k}{\partial u}\right) + \varepsilon_{2n}\varepsilon_{1j}\varepsilon_{3k}X_n\left(\frac{\partial X_j}{\partial v}\frac{\partial X_k}{\partial u} - \frac{\partial X_j}{\partial u}\frac{\partial X_k}{\partial v}\right) + \varepsilon_{3n}\varepsilon_{1j}\varepsilon_{2k}X_n\left(\frac{\partial X_j}{\partial u}\frac{\partial X_k}{\partial v} - \frac{\partial X_j}{\partial v}\frac{\partial X_k}{\partial u}\right) =$$
$$= X_1\left(\frac{\partial X_2}{\partial u}\frac{\partial X_3}{\partial v} - \frac{\partial X_2}{\partial v}\frac{\partial X_3}{\partial u}\right) + X_2\left(\frac{\partial X_1}{\partial v}\frac{\partial X_3}{\partial u} - \frac{\partial X_1}{\partial u}\frac{\partial X_3}{\partial v}\right) + X_3\left(\frac{\partial X_1}{\partial u}\frac{\partial X_2}{\partial v} - \frac{\partial X_1}{\partial v}\frac{\partial X_2}{\partial u}\right), \tag{A31}$$

i.e., in simplified notation:

$$\left(\varepsilon_{1n}\varepsilon_{2j}\varepsilon_{3k} - \varepsilon_{2n}\varepsilon_{1j}\varepsilon_{3k} + \varepsilon_{3n}\varepsilon_{1j}\varepsilon_{2k}\right)X_n\left(\frac{\partial X_j}{\partial u}\frac{\partial X_k}{\partial v} - \frac{\partial X_j}{\partial v}\frac{\partial X_k}{\partial u}\right) =$$
$$= X_1\left(\frac{\partial X_2}{\partial u}\frac{\partial X_3}{\partial v} - \frac{\partial X_2}{\partial v}\frac{\partial X_3}{\partial u}\right) + X_2\left(\frac{\partial X_1}{\partial v}\frac{\partial X_3}{\partial u} - \frac{\partial X_1}{\partial u}\frac{\partial X_3}{\partial v}\right) + X_3\left(\frac{\partial X_1}{\partial u}\frac{\partial X_2}{\partial v} - \frac{\partial X_1}{\partial v}\frac{\partial X_2}{\partial u}\right). \tag{A32}$$

Finally, if we consider the fact that:

$$X_1\left(\frac{\partial X_2}{\partial u}\frac{\partial X_3}{\partial v} - \frac{\partial X_2}{\partial v}\frac{\partial X_3}{\partial u}\right) = \delta_{1n}\delta_{2j}\delta_{3k}\left(\frac{\partial X_j}{\partial u}\frac{\partial X_k}{\partial v} - \frac{\partial X_j}{\partial v}\frac{\partial X_k}{\partial u}\right)X_n, \tag{A33}$$

$$X_2\left(\frac{\partial X_1}{\partial v}\frac{\partial X_3}{\partial u} - \frac{\partial X_1}{\partial u}\frac{\partial X_3}{\partial v}\right) = -\delta_{2n}\delta_{1j}\delta_{3k}\left(\frac{\partial X_k}{\partial u}\frac{\partial X_j}{\partial v} - \frac{\partial X_k}{\partial v}\frac{\partial X_j}{\partial u}\right)X_n, \tag{A34}$$

$$X_3\left(\frac{\partial X_1}{\partial u}\frac{\partial X_2}{\partial v} - \frac{\partial X_1}{\partial v}\frac{\partial X_2}{\partial u}\right) = \delta_{3n}\delta_{1j}\delta_{2k}\left(\frac{\partial X_j}{\partial u}\frac{\partial X_k}{\partial v} - \frac{\partial X_j}{\partial v}\frac{\partial X_k}{\partial u}\right)X_n. \tag{A35}$$

volume preserving can be represented by equation:

$$\left\{\varepsilon_{1n}\varepsilon_{2j}\varepsilon_{3k} - \varepsilon_{2n}\varepsilon_{1j}\varepsilon_{3k} + \varepsilon_{3n}\varepsilon_{1j}\varepsilon_{2k} - \delta_{1n}\delta_{2j}\delta_{3k} + \delta_{2n}\delta_{1j}\delta_{3k} - \delta_{3n}\delta_{1j}\delta_{2k}\right\}\left(\frac{\partial X_j}{\partial u}\frac{\partial X_k}{\partial v} - \frac{\partial X_j}{\partial v}\frac{\partial X_k}{\partial u}\right)X_n = 0 \tag{A36}$$

leading to (46).

Appendix D. Tensor of Plastic Deformation

Formula (46) represents the system of 27 equations for 9 unknown parameters (coefficients of deformation tensor): i.e.,:

$$\begin{aligned}
&\varepsilon_{11}(\varepsilon_{22}\varepsilon_{33}-\varepsilon_{23}\varepsilon_{32})-\varepsilon_{21}(\varepsilon_{12}\varepsilon_{33}-\varepsilon_{13}\varepsilon_{32})+\varepsilon_{31}(\varepsilon_{12}\varepsilon_{23}-\varepsilon_{13}\varepsilon_{22})=1 \ldots \text{ for } \{n,j,k\}=\{1,2,3\}\vee\{1,3,2\}\\
&\varepsilon_{12}(\varepsilon_{23}\varepsilon_{31}-\varepsilon_{21}\varepsilon_{33})-\varepsilon_{22}(\varepsilon_{13}\varepsilon_{31}-\varepsilon_{11}\varepsilon_{33})+\varepsilon_{32}(\varepsilon_{13}\varepsilon_{21}-\varepsilon_{11}\varepsilon_{23})=1 \ldots \text{ for } \{n,j,k\}=\{2,1,3\}\vee\{2,3,1\}\\
&\varepsilon_{13}(\varepsilon_{21}\varepsilon_{32}-\varepsilon_{22}\varepsilon_{31})-\varepsilon_{23}(\varepsilon_{11}\varepsilon_{32}-\varepsilon_{12}\varepsilon_{31})+\varepsilon_{33}(\varepsilon_{11}\varepsilon_{22}-\varepsilon_{12}\varepsilon_{21})=1 \ldots \text{ for } \{n,j,k\}=\{3,1,2\}\vee\{3,2,1\}\\
&\varepsilon_{1n}(\varepsilon_{2j}\varepsilon_{3k}-\varepsilon_{2k}\varepsilon_{3j})-\varepsilon_{2n}(\varepsilon_{1j}\varepsilon_{3k}-\varepsilon_{1k}\varepsilon_{3j})+\varepsilon_{3n}(\varepsilon_{1j}\varepsilon_{2k}-\varepsilon_{1k}\varepsilon_{2j})=0 \ldots \text{ for all other } \{n,j,k\},
\end{aligned} \quad (A37)$$

where $\{i,j,k\}$ is the ordered set of coefficients. Of these 27 equations, only the following 9 are independent:

$$\varepsilon_{11}(\varepsilon_{22}\varepsilon_{33}-\varepsilon_{23}\varepsilon_{32})-\varepsilon_{21}(\varepsilon_{12}\varepsilon_{33}-\varepsilon_{13}\varepsilon_{32})+\varepsilon_{31}(\varepsilon_{12}\varepsilon_{23}-\varepsilon_{13}\varepsilon_{22})=1 \quad (A38)$$

$$\varepsilon_{12}(\varepsilon_{23}\varepsilon_{31}-\varepsilon_{21}\varepsilon_{33})-\varepsilon_{22}(\varepsilon_{13}\varepsilon_{31}-\varepsilon_{11}\varepsilon_{33})+\varepsilon_{32}(\varepsilon_{13}\varepsilon_{21}-\varepsilon_{11}\varepsilon_{23})=1 \quad (A39)$$

$$\varepsilon_{13}(\varepsilon_{21}\varepsilon_{32}-\varepsilon_{22}\varepsilon_{31})-\varepsilon_{23}(\varepsilon_{11}\varepsilon_{32}-\varepsilon_{12}\varepsilon_{31})+\varepsilon_{33}(\varepsilon_{11}\varepsilon_{22}-\varepsilon_{12}\varepsilon_{21})=1 \quad (A40)$$

$$\varepsilon_{11}(\varepsilon_{21}\varepsilon_{32}-\varepsilon_{22}\varepsilon_{31})-\varepsilon_{21}(\varepsilon_{11}\varepsilon_{32}-\varepsilon_{12}\varepsilon_{31})+\varepsilon_{31}(\varepsilon_{11}\varepsilon_{22}-\varepsilon_{12}\varepsilon_{21})=0 \quad (A41)$$

$$\varepsilon_{11}(\varepsilon_{21}\varepsilon_{33}-\varepsilon_{23}\varepsilon_{31})-\varepsilon_{21}(\varepsilon_{11}\varepsilon_{33}-\varepsilon_{13}\varepsilon_{31})+\varepsilon_{31}(\varepsilon_{11}\varepsilon_{23}-\varepsilon_{13}\varepsilon_{21})=0 \quad (A42)$$

$$\varepsilon_{12}(\varepsilon_{21}\varepsilon_{32}-\varepsilon_{22}\varepsilon_{31})-\varepsilon_{22}(\varepsilon_{11}\varepsilon_{32}-\varepsilon_{12}\varepsilon_{31})+\varepsilon_{32}(\varepsilon_{11}\varepsilon_{22}-\varepsilon_{12}\varepsilon_{21})=0 \quad (A43)$$

$$\varepsilon_{12}(\varepsilon_{22}\varepsilon_{33}-\varepsilon_{23}\varepsilon_{32})-\varepsilon_{22}(\varepsilon_{12}\varepsilon_{33}-\varepsilon_{13}\varepsilon_{32})+\varepsilon_{32}(\varepsilon_{12}\varepsilon_{23}-\varepsilon_{13}\varepsilon_{22})=0 \quad (A44)$$

$$\varepsilon_{13}(\varepsilon_{21}\varepsilon_{33}-\varepsilon_{23}\varepsilon_{31})-\varepsilon_{23}(\varepsilon_{11}\varepsilon_{33}-\varepsilon_{13}\varepsilon_{31})+\varepsilon_{33}(\varepsilon_{11}\varepsilon_{23}-\varepsilon_{13}\varepsilon_{21})=0 \quad (A45)$$

$$\varepsilon_{13}(\varepsilon_{22}\varepsilon_{33}-\varepsilon_{23}\varepsilon_{32})-\varepsilon_{23}(\varepsilon_{12}\varepsilon_{33}-\varepsilon_{13}\varepsilon_{32})+\varepsilon_{33}(\varepsilon_{12}\varepsilon_{23}-\varepsilon_{13}\varepsilon_{22})=0. \quad (A46)$$

The mentioned equations can be rewritten as follows:

$$\begin{vmatrix}\varepsilon_{11}&\varepsilon_{21}&\varepsilon_{31}\\\varepsilon_{12}&\varepsilon_{22}&\varepsilon_{32}\\\varepsilon_{13}&\varepsilon_{23}&\varepsilon_{33}\end{vmatrix}=1 \quad \begin{vmatrix}\varepsilon_{12}&\varepsilon_{22}&\varepsilon_{32}\\\varepsilon_{13}&\varepsilon_{23}&\varepsilon_{33}\\\varepsilon_{11}&\varepsilon_{21}&\varepsilon_{31}\end{vmatrix}=1 \quad \begin{vmatrix}\varepsilon_{13}&\varepsilon_{23}&\varepsilon_{33}\\\varepsilon_{11}&\varepsilon_{21}&\varepsilon_{31}\\\varepsilon_{12}&\varepsilon_{22}&\varepsilon_{32}\end{vmatrix}=1 \quad \begin{vmatrix}\varepsilon_{11}&\varepsilon_{21}&\varepsilon_{31}\\\varepsilon_{11}&\varepsilon_{21}&\varepsilon_{31}\\\varepsilon_{12}&\varepsilon_{22}&\varepsilon_{32}\end{vmatrix}=0$$

$$\begin{vmatrix}\varepsilon_{11}&\varepsilon_{21}&\varepsilon_{31}\\\varepsilon_{11}&\varepsilon_{21}&\varepsilon_{31}\\\varepsilon_{13}&\varepsilon_{23}&\varepsilon_{33}\end{vmatrix}=0 \quad (A47)$$

$$\begin{vmatrix}\varepsilon_{12}&\varepsilon_{22}&\varepsilon_{32}\\\varepsilon_{11}&\varepsilon_{21}&\varepsilon_{31}\\\varepsilon_{12}&\varepsilon_{22}&\varepsilon_{32}\end{vmatrix}=0 \quad \begin{vmatrix}\varepsilon_{12}&\varepsilon_{22}&\varepsilon_{32}\\\varepsilon_{12}&\varepsilon_{22}&\varepsilon_{32}\\\varepsilon_{13}&\varepsilon_{23}&\varepsilon_{33}\end{vmatrix}=0 \quad \begin{vmatrix}\varepsilon_{13}&\varepsilon_{23}&\varepsilon_{33}\\\varepsilon_{11}&\varepsilon_{21}&\varepsilon_{31}\\\varepsilon_{13}&\varepsilon_{23}&\varepsilon_{33}\end{vmatrix}=0 \quad \begin{vmatrix}\varepsilon_{13}&\varepsilon_{23}&\varepsilon_{33}\\\varepsilon_{11}&\varepsilon_{21}&\varepsilon_{31}\\\varepsilon_{13}&\varepsilon_{23}&\varepsilon_{33}\end{vmatrix}=0$$

As it can be easily seen, Formula (35) results from these mathematical treatments.

Appendix E. Derivation of Formulas (54) and (55)

No complicated calculations are necessary to determine Formulas (54) and (55). If we consider (21) the tensor of the grain surface deformation can be written as:

$$\vec{\vec{\sigma}}=\begin{pmatrix}(\varepsilon_{22}\varepsilon_{33}-\varepsilon_{23}\varepsilon_{32}) & -(\varepsilon_{21}\varepsilon_{33}-\varepsilon_{23}\varepsilon_{31}) & (\varepsilon_{21}\varepsilon_{32}-\varepsilon_{22}\varepsilon_{31}) \\ (\varepsilon_{13}\varepsilon_{32}-\varepsilon_{12}\varepsilon_{33}) & -(\varepsilon_{13}\varepsilon_{31}-\varepsilon_{11}\varepsilon_{33}) & (\varepsilon_{12}\varepsilon_{31}-\varepsilon_{11}\varepsilon_{32}) \\ (\varepsilon_{12}\varepsilon_{23}-\varepsilon_{13}\varepsilon_{22}) & -(\varepsilon_{11}\varepsilon_{23}-\varepsilon_{13}\varepsilon_{21}) & (\varepsilon_{11}\varepsilon_{22}-\varepsilon_{12}\varepsilon_{21})\end{pmatrix} \quad (A48)$$

and the transposed tensor to the tensor (A48) is:

$$\vec{\sigma}^T = \begin{pmatrix} (\varepsilon_{22}\varepsilon_{33} - \varepsilon_{23}\varepsilon_{32}) & (\varepsilon_{13}\varepsilon_{32} - \varepsilon_{12}\varepsilon_{33}) & (\varepsilon_{12}\varepsilon_{23} - \varepsilon_{13}\varepsilon_{22}) \\ -(\varepsilon_{21}\varepsilon_{33} - \varepsilon_{23}\varepsilon_{31}) & -(\varepsilon_{13}\varepsilon_{31} - \varepsilon_{11}\varepsilon_{33}) & -(\varepsilon_{11}\varepsilon_{23} - \varepsilon_{13}\varepsilon_{21}) \\ (\varepsilon_{21}\varepsilon_{32} - \varepsilon_{22}\varepsilon_{31}) & (\varepsilon_{12}\varepsilon_{31} - \varepsilon_{11}\varepsilon_{32}) & (\varepsilon_{11}\varepsilon_{22} - \varepsilon_{12}\varepsilon_{21}) \end{pmatrix} \quad (A49)$$

is the vector with the following coordinates:

$$\vec{\sigma}^T \cdot \vec{\tau} = \begin{pmatrix} (\varepsilon_{22}\varepsilon_{33} - \varepsilon_{23}\varepsilon_{32})\tau_1 + (\varepsilon_{13}\varepsilon_{32} - \varepsilon_{12}\varepsilon_{33})\tau_2 + (\varepsilon_{12}\varepsilon_{23} - \varepsilon_{13}\varepsilon_{22})\tau_3 \\ (\varepsilon_{23}\varepsilon_{31} - \varepsilon_{21}\varepsilon_{33})\tau_1 + (\varepsilon_{11}\varepsilon_{33} - \varepsilon_{13}\varepsilon_{31})\tau_2 + (\varepsilon_{13}\varepsilon_{21} - \varepsilon_{11}\varepsilon_{23})\tau_3 \\ (\varepsilon_{21}\varepsilon_{32} - \varepsilon_{22}\varepsilon_{31})\tau_1 + (\varepsilon_{12}\varepsilon_{31} - \varepsilon_{11}\varepsilon_{32})\tau_2 + (\varepsilon_{11}\varepsilon_{22} - \varepsilon_{12}\varepsilon_{21})\tau_3 \end{pmatrix} \quad (A50)$$

Subsequent scalar product of vector (A50) with the vector:

$$\vec{\rho}^{(i)} = [\sin(\eta^{(i)})\cos(\chi^{(i)}),\ \sin(\eta^{(i)})\sin(\chi^{(i)}),\ \cos(\eta^{(i)})] \quad (A51)$$

mentioned in Section 6, i.e.,:

$$\left(\vec{\sigma}^T \cdot \vec{\tau}\right) \cdot \vec{\rho}^{(i)} = \begin{pmatrix} (\varepsilon_{22}\varepsilon_{33} - \varepsilon_{23}\varepsilon_{32})\tau_1 + (\varepsilon_{13}\varepsilon_{32} - \varepsilon_{12}\varepsilon_{33})\tau_2 + (\varepsilon_{12}\varepsilon_{23} - \varepsilon_{13}\varepsilon_{22})\tau_3 \\ (\varepsilon_{23}\varepsilon_{31} - \varepsilon_{21}\varepsilon_{33})\tau_1 + (\varepsilon_{11}\varepsilon_{33} - \varepsilon_{13}\varepsilon_{31})\tau_2 + (\varepsilon_{13}\varepsilon_{21} - \varepsilon_{11}\varepsilon_{23})\tau_3 \\ (\varepsilon_{21}\varepsilon_{32} - \varepsilon_{22}\varepsilon_{31})\tau_1 + (\varepsilon_{12}\varepsilon_{31} - \varepsilon_{11}\varepsilon_{32})\tau_2 + (\varepsilon_{11}\varepsilon_{22} - \varepsilon_{12}\varepsilon_{21})\tau_3 \end{pmatrix} \cdot \begin{pmatrix} \cos(\chi^{(i)})\sin(\eta^{(i)}) \\ \sin(\chi^{(i)})\sin(\eta^{(i)}) \\ \cos(\eta^{(i)}) \end{pmatrix} \quad (A52)$$

gives the result:

$$\left(\vec{\sigma}^T \cdot \vec{\tau}\right) \cdot \vec{\rho}^{(i)} =$$
$$= (\varepsilon_{22}\varepsilon_{33} - \varepsilon_{23}\varepsilon_{32})\cos(\chi^{(i)})\sin(\eta^{(i)})\tau_1 + (\varepsilon_{13}\varepsilon_{32} - \varepsilon_{12}\varepsilon_{33})\cos(\chi^{(i)})\sin(\eta^{(i)})\tau_2 + (\varepsilon_{12}\varepsilon_{23} - \varepsilon_{13}\varepsilon_{22})\cos(\chi^{(i)})\sin(\eta^{(i)})\tau_3 +$$
$$+ (\varepsilon_{23}\varepsilon_{31} - \varepsilon_{21}\varepsilon_{33})\sin(\chi^{(i)})\sin(\eta^{(i)})\tau_1 + (\varepsilon_{11}\varepsilon_{33} - \varepsilon_{13}\varepsilon_{31})\sin(\chi^{(i)})\sin(\eta^{(i)})\tau_2 + (\varepsilon_{13}\varepsilon_{21} - \varepsilon_{11}\varepsilon_{23})\sin(\chi^{(i)})\sin(\eta^{(i)})\tau_3 +$$
$$+ (\varepsilon_{21}\varepsilon_{32} - \varepsilon_{22}\varepsilon_{31})\cos(\eta^{(i)})\tau_1 + (\varepsilon_{12}\varepsilon_{31} - \varepsilon_{11}\varepsilon_{32})\cos(\eta^{(i)})\tau_2 + (\varepsilon_{11}\varepsilon_{22} - \varepsilon_{12}\varepsilon_{21})\cos(\eta^{(i)})\tau_3 \quad (A53)$$

that can be written in the form (54).

It is also possible to calculate easily the next scalar product:

$$\vec{\sigma} \cdot \vec{\rho}^{(i)} = \begin{pmatrix} (\varepsilon_{22}\varepsilon_{33} - \varepsilon_{23}\varepsilon_{32}) & -(\varepsilon_{21}\varepsilon_{33} - \varepsilon_{23}\varepsilon_{31}) & (\varepsilon_{21}\varepsilon_{32} - \varepsilon_{22}\varepsilon_{31}) \\ (\varepsilon_{13}\varepsilon_{32} - \varepsilon_{12}\varepsilon_{33}) & -(\varepsilon_{13}\varepsilon_{31} - \varepsilon_{11}\varepsilon_{33}) & (\varepsilon_{12}\varepsilon_{31} - \varepsilon_{11}\varepsilon_{32}) \\ (\varepsilon_{12}\varepsilon_{23} - \varepsilon_{13}\varepsilon_{22}) & -(\varepsilon_{11}\varepsilon_{23} - \varepsilon_{13}\varepsilon_{21}) & (\varepsilon_{11}\varepsilon_{22} - \varepsilon_{12}\varepsilon_{21}) \end{pmatrix} \cdot \begin{pmatrix} \cos(\chi^{(i)})\sin(\eta^{(i)}) \\ \sin(\chi^{(i)})\sin(\eta^{(i)}) \\ \cos(\eta^{(i)}) \end{pmatrix} \quad (A54)$$

which gives the vector with coordinates:

$$\vec{\sigma} \cdot \vec{\rho}^{(i)} = \begin{pmatrix} (\varepsilon_{22}\varepsilon_{33} - \varepsilon_{23}\varepsilon_{32})\cos(\chi^{(i)})\sin(\eta^{(i)}) - (\varepsilon_{21}\varepsilon_{33} - \varepsilon_{23}\varepsilon_{31})\sin(\chi^{(i)})\sin(\eta^{(i)}) + (\varepsilon_{21}\varepsilon_{32} - \varepsilon_{22}\varepsilon_{31})\cos(\eta^{(i)}) \\ (\varepsilon_{13}\varepsilon_{32} - \varepsilon_{12}\varepsilon_{33})\cos(\chi^{(i)})\sin(\eta^{(i)}) - (\varepsilon_{13}\varepsilon_{31} - \varepsilon_{11}\varepsilon_{33})\sin(\chi^{(i)})\sin(\eta^{(i)}) + (\varepsilon_{12}\varepsilon_{31} - \varepsilon_{11}\varepsilon_{32})\cos(\eta^{(i)}) \\ (\varepsilon_{12}\varepsilon_{23} - \varepsilon_{13}\varepsilon_{22})\cos(\chi^{(i)})\sin(\eta^{(i)}) - (\varepsilon_{11}\varepsilon_{23} - \varepsilon_{13}\varepsilon_{21})\sin(\chi^{(i)})\sin(\eta^{(i)}) + (\varepsilon_{11}\varepsilon_{22} - \varepsilon_{12}\varepsilon_{21})\cos(\eta^{(i)}) \end{pmatrix} \quad (A55)$$

and leads to the result (55).

References

1. Saltykov, S.A. *Stereometric Metallography*; Metallurgia: Moscow, Russia, 1970.
2. Vander Voort, G.F. Measuring the grain size of specimens with non-equiaxed grains. *Pract. Metallogr.* **2013**, *50*, 239–251. [CrossRef]
3. Maier, B.; Mihailova, B.; Paulmann, C.; Ihringer, J.; Gospodinov, M.; Stosch, R.; Güttler, B.; Bismayer, U. Effect of local elastic strain on the structure of Pb-based relaxors: A comparative study of pure and Ba- and Bi-doped PbSc0.5Nb0.5O3. *Phys. Rev.* **2009**, *79*, 224108. [CrossRef]

4. Stoudt, M.R.; Levine, L.E.; Creuziger, A.; Hubbard, J.B. The fundamental relationships between grain orientation, deformation-induced surface roughness and strain localization in an aluminum alloy. *Mater. Sci. Eng. A* **2011**, *530*, 107–116. [CrossRef]
5. Wang, X.G.; Witz, J.F.; El Bartali, A.; Dufrénoy, P.; Charkaluk, E. Investigation of grain-scale surface deformation of a pure aluminium polycrystal through kinematic-thermal full-field coupling measurement. In Proceedings of the 13th International Conference on Fracture, Beijing, China, 16–21 June 2013.
6. Luong, M.P. Fatigue limit evaluation of metals using an infrared thermographic technique. *I Mec. Mater.* **1998**, *28*, 155–163. [CrossRef]
7. Moscicki, M.; Kenesei, P.; Wright, J.; Pinto, H.; Lippmann, T.; Borbély, A.; Pyzalla, A.R. Friedels-pair based indexing method for characterization of single grains with hard X-rays. *Mater. Sci. Eng. A* **2009**, *524*, 64–68. [CrossRef]
8. Deng, J.; Rokkam, S. A phase field model. of surface-energy-driven abnormal grain growth in thin films. *Mater. Trans.* **2011**, *52*, 2126–2130. [CrossRef]
9. Gerth, D.; Schwarzer, R.A. Graphical Representation of grain and Hillock orientations in annealed Al-l%Si films. *Textures Microstruct.* **1993**, *21*, 77–93. [CrossRef]
10. Sack, R.A. Indirect evaluation of orientation in polycrystalline materials. *J. Polym. Sci.* **1961**, *51*, 543–560. [CrossRef]
11. Zhang, N.; Tong, W. An experimental study on grain deformation and interactions in an Al-0.5%Mg multicrystal. *Int. J. Plast.* **2004**, *20*, 523–542. [CrossRef]
12. Lewis, H.D.; Walters, K.L.; Johnson, K.A. Particle size distribution by area analysis: Modifications and extensions of the Saltykov method. *Metallography* **1973**, *6*, 93–101. [CrossRef]
13. Xu, Y.H.; Pitot, H.C. An improved stereologic method for three-dimensional estimation of particle size distribution from observations in two dimensions and its application. *Comput. Methods Progr. Biomed.* **2003**, *72*, 1–20. [CrossRef]
14. Rayaprolu, D.B.; Jaffrey, D. Comparison of discrete particle sectioning correction methods based on section diameter and area. *Metallography* **1982**, *15*, 193–202. [CrossRef]
15. Zhao, Z.; Ramesh, M.; Raabe, D.; Cuitiño, A.M.; Radovitzky, R. Investigation of three-dimensional aspects of grain-scale plastic surface deformation of an aluminum oligocrystal. *Int. J. Plast.* **2008**, *24*, 2278–2297. [CrossRef]
16. Patdar, V. A Note on geometry of grain boundaries. *Scr. Metal.* **1986**, *20*, 1227–1230. [CrossRef]
17. Bate, P.S.; Hutchinson, W.B. Grain boundary area and deformation. *Scr. Mater.* **2005**, *52*, 199–203. [CrossRef]
18. Song, K.J.; Wei, Y.H.; Dong, Z.B.; Ma, R.; Zhan, X.H.; Zheng, W.J.; Fang, K. Constitutive model coupled with mechanical effect of volume change and transformation induced plasticity during solid phase transformation for TA15 alloy welding. *Appl. Math. Modell.* **2015**, *39*, 2064–2080. [CrossRef]
19. Sharen, J.; Cummins, P.; Cleary, W. Using distributed contacts in DEM. *Appl. Math. Modell.* **2011**, *35*, 1904–1914.
20. Singh, S.B.; Bhadeshia, H.K.D.H. Topology of grain deformation. *Mater. Sci. Technol.* **1998**, *14*, 832–834. [CrossRef]
21. Yoshihiro, M.; Takayuki, S. Bayesian inference of whole-organ deformation dynamics from limited space-time point data. *J. Theor. Biol.* **2014**, *357*, 74–85.
22. Sutton, M.A.; Orteu, J.J.; Schreier, H.W. *Image Correlation for Shape, Motion and Deformation Measurements*; Hardcover: Boston, MA, USA, 2009; ISBN 978-0-387-78746-6.
23. Keating, T.J.; Wolf, P.R.; Scarpace, F.L. An improved method of digital image correlation. *Photogramm. Eng. Remote Sens.* **1975**, *41*, 993–1002.
24. Gallucci Masteghin, M.; Ornaghi Orlandi, M. Grain-boundary resistance and nonlinear coefficient correlation for SnO_2-based varistors. *Mat. Res.* **2016**, *19*, 1286–1291. [CrossRef]
25. Sindre, B.; Knut, M.; Nes, E. Subgrain Structures Characterized by Electron. Backscatter Diffraction (EBSD). *Mater. Sci. Forum* **2014**, *794*, 3–8.
26. Zou, H.F.; Zhang, Z.F. Application of electron backscatter diffraction to the study on orientation distribution of intermetallic compounds at heterogeneous interfaces (Sn/Ag and Sn/Cu). *J. Appl. Phys.* **2010**, *108*, 103518. [CrossRef]

27. Zacher, G.; Paul, T. *Research Applications. Qualitative and Quantitative Investigation of Soils and Porous Rocks by Using Very High Resolution X-Ray CT Imaging. Unsaturated Soils*; Khalili, N., Russell, A., Khoshghalb, A., Eds.; Taylor & Francis Group: London, UK, 2014; ISBN 978-1-138-00150-3.
28. Khalid, A.; Alshibli, A.; Reed, H. *Advances in Computed Tomography for Geomaterials: GeoX 2010*; John Wiley & Sons: Hoboken, NJ, USA, 2012.
29. Bay, B.K.; Smith, T.S.; Fyhrie, D.P.; Saad, M. Digital volume correlation: Three-dimensional strain mapping using X-ray tomography. *Exp. Mech.* **1999**, *39*, 217–226. [CrossRef]
30. Konrad, N.; Wenxiong, H. A study of localized deformation pattern in granular media. *Comput. Methods Appl. Mech. Eng.* **2004**, *193*, 2719–2743.
31. Ronaldo, I.; Borja, A.A. Computational modeling of deformation bands in granular media. I. Geological and mathematical framework. *Comput. Methods Appl. Mech. Eng.* **2004**, *193*, 2667–2698.
32. Wellmann, C.; Wriggers, P. A two-scale model of granular materials. *Comput. Methods Appl. Mech. Eng.* **2012**, *205*, 46–58. [CrossRef]
33. Benedetti, I.; Aliabadi, M.H. A three-dimensional cohesive-frictional grain-boundary micromechanical model for intergranular degradation and failure in polycrystalline materials. *Comput. Methods Appl. Mech. Eng.* **2013**, *265*, 36–62. [CrossRef]
34. Martinkovic, M.; Pokorny, P. Estimation of local plastic deformation in cutting zone during turning. *Key Eng. Mater.* **2015**, *662*, 173–176. [CrossRef]
35. Martinkovic, M.; Minarik, S. Short notes on the grains modification by plastic deformation. In *Plastic Deformation*; Hubbard, D., Ed.; NOVA: New York, NY, USA, 2016; pp. 1–44.
36. Takahashi, J.; Suito, H. Evaluation of the Accuracy of the Three-Dimensional Size Distribution Estimated from the Schwartz–Saltykov Method. *Metall. Mater. Trans. A* **2003**, *34*, 171–181. [CrossRef]
37. Jensen, D. Estimation of the size distribution of spherical, disc-like or ellipsoidal particles in thin foils. *J. Phys. D Appl. Phys.* **1995**, *28*, 549. [CrossRef]
38. Riosa, P.R.; Glicksmanb, M.E. Modeling polycrystals with regular polyhedra. *Mater. Res.* **2006**, *9*, 231–236. [CrossRef]
39. Jae-Yong, C.; Qin, R.; Bhaeshia, H.K.D.H. Topology of the Deformation of a Non-uniform Grain Structure. *ISIJ Int.* **2009**, *49*, 115–118.
40. Zhu, Q.; Sellars, C.M.; Bhadeshia, H.K.D.H. Quantitative metallography of deformed grains. *Mater. Sci. Technol.* **2007**, *23*, 757. [CrossRef]

© 2020 by the authors. Licensee MDPI, Basel, Switzerland. This article is an open access article distributed under the terms and conditions of the Creative Commons Attribution (CC BY) license (http://creativecommons.org/licenses/by/4.0/).

Article

Use of the Correlation between Grain Size and Crystallographic Orientation in Crystal Plasticity Simulations: Application to AISI 420 Stainless Steel

Jesús Galán-López * and Javier Hidalgo *

Department of Materials Science and Engineering, Delft University of Technology, Mekelweg 2, 2628 CD Delft, The Netherlands
* Correspondence: J.GalanLopez@tudelft.nl (J.G.-L.); J.HidalgoGarcia@tudelft.nl (J.H.)

Received: 3 August 2020; Accepted: 14 September 2020; Published: 16 September 2020

Abstract: Crystal plasticity models attempt to reproduce the complex deformation processes of polycrystalline metals based on a virtual representation of the real microstructure. When choosing this representation, a compromise must be made between level of detail at the local level and statistical significance of the aggregate properties, also taking into account the computational cost of each solution. In this work, the correlation between crystallographic orientation and grain size is considered in the definition of virtual microstructures for the simulation of the mechanical behavior of AISI 420 stainless steel (consisting of a ferrite matrix with large carbide precipitates), in order to improve the accuracy of the solution without increasing model complexity or computation time. Both full-field (DAMASK) and mean-field models (Visco Plastic Self Consistent (VPSC)) are used together in combination with experimental results to study the validity of the assumptions done in each of the models.

Keywords: AISI 420; crystal plasticity; representative volume element; VPSC; DAMASK

1. Introduction

The good high temperature properties achievable by martensitic/ferritic steels containing 9–12 wt.% chromium [1–3] will enable an increase in the operating temperatures and pressures of future power plants, resulting in better efficiency and reduction of greenhouse gas emissions. The role of typical M23C6 and MX nitride precipitates on arresting the grain growth and drop in dislocation density during high temperature creep is not fully understood. In particular, coarse M23C6 carbides might cause heterogeneous strains in the ferritic matrix during deformation, creating preferential sites for low angle subgrain development that is believed to impair creep resistance [4]. In order to implement these materials in power plants and other applications, accurate models to describe their plastic behavior are required.

Several physically-based modelling approaches exist, which consider to a greater or lesser extent different microstructural features as the origin of the mechanical response. In particular, crystal plasticity (CP) models take into account the inherent anisotropy of crystals and crystalline aggregates, which is crucial for the prediction of the mechanical behavior of polycrystalline materials. CP models can be categorized in two main groups. Mean-field CP models are deformation models that consider that each grain (defined as a region of material with a distinct crystallographic orientation) is subjected to a homogeneous mechanical state. They include the classical Taylor [5] and Sachs [6] theories—that correspond, respectively, to the upper and lower bounds of the aggregate response—and more complex relaxed constraints models such as ALAMEL [7] and Visco Plastic Self Consistent (VPSC) [8,9] which, at least in principle, better approaches the real behavior. In contrast, in full-field models, a single grain is spatially resolved by a large number of material points with different mechanical states. Depending on the solver used, current full field models are based on finite element formulations, and are referred to as the Crystal Plasticity Finite Element Method (CPFEM) [10,11], or use fast Fourier transforms in the more recent Crystal Plastic Fast Fourier Transform (CPFFT) models [12,13]. Although full-field models offer a more detailed representation of deformation processes, due to computational limitations they can only simulate a relatively small region of the material, while mean-field models allow the study of the material with greater statistical significance. Mean-field and full-field approaches have been found to be capable of reproducing the mechanical behavior of materials and the evolution of textures upon deformation. However, the prediction of local strain development in chromium steels containing large precipitates and other materials is restricted to the use of full-field models.

In all CP models, information about the material microstructure and the deformation mechanisms at the crystal level needs to be supplied. A synthetic microstructure matching (or partially matching) the stereology and texture of the real material is typically referred to as a Representative Volume Element (RVE). Being able to virtually reconstruct microstructures can be challenging and is the first step to an accurate prediction of the material mechanical behavior. The material microstructure is commonly represented by a discrete set of material grains defined by their crystallographic orientation, cell structure, volume fraction and morphology. The set of grains must be large enough to be representative of the real material but, at the same time, keep a reasonable trade-off with the high computational demands of excessively large synthetic microstructures [14]. Virtual microstructures can be produced based on exact reconstruction methods from an experimentally measured region of the material large enough to ensure statistical homogeneity [15–17]. However, measuring large material microstructure regions in three dimensions—by, for example, electron back-scatter diffraction (EBSD) tomography—is demanding and requires dedicated equipment. Alternatively, a statistically representative volume element can be generated based on the material microstructure statistics [18,19]. Reproducing stereology and texture in the synthetic microstructure is not a trivial task, but different approaches can be found in the literature. The usage of Voronoi tessellations is a popular method for the generation of polycrystalline microstructures, since the nucleation and growth of grains is imitated to some extent by the cell growth process [20,21]. Although the application of multilevel Voronoi tessellations makes it possible to reproduce polycrystals that adhere to complex the distribution of grain size and shape, including the generation of bi-modal microstructures [22,23], a precise synthetic reproduction of the microstructural features of the material is still very challenging [24,25]. For the purposes of this work, the algorithm developed by Groeber et al. [26] proved a reasonable statistical equivalence for most of the relevant morphological and crystallographic parameters of the microstructure. Alternatively, Neper software [22,23] can be used for the generation of complicated microstructures.

Additionally, material properties need to be defined for each phase, including elastic stiffness constants and parameters specific for each deformation mode, such as hardening parameters and strain rate sensitivity. Using appropriate material parameters is as relevant as using a representative set of grains. An optimization algorithm is typically used to find the set of material parameters that best reproduce experimental data. This is a challenging task, since the fitting procedure is also conditioned by the selection of the initial microstructure. The large synthetic microstructures required for high statistical significance often result in simulations too computationally expensive when full-field CP models are employed. Mean-field CP models are computationally more efficient, and therefore less restrictive with the number of grains considered. Although mean-field models cannot reproduce the material behavior at a local level, their averaging capacity makes them ideal to calculate macroscopic behavior when the material is homogenous enough at larger scales. Hence, mean-field CP represents an interesting tool for performing the calibration of material parameters in an efficient manner. The fitted hardening model can then be applied in full-field models to obtain a complete and faithful picture of the material mechanical behavior at different scales.

This work aims to determine an optimal RVE for a chromium steel microstructure consisting of a ferrite matrix with large carbide precipitates. The effect of the size and definition of different virtual microstructures on the simulation of the mechanical behavior of this steel is studied and discussed. A novel method consisting of the correlation between crystallographic orientation and grain size is considered in the definition of virtual microstructures, in order to improve the accuracy of the solution without increasing model complexity or computation time. Both full-field (DAMASK) and mean-field (VPSC) models are used in combination with experimental results to study the validity of the assumptions in each of the models.

2. Materials and Methods

A fully annealed AISI 420 steel rolled sheet with a thickness of 0.45 ± 0.1 mm was used for this study. AISI 420 steel contains 0.32 wt.% C, 0.2 wt.% Si, 0.3 wt.% Mn and 13.7 wt.% Cr as the main alloying elements.

A FEI Quanta 450 scanning electron microscope equipped with a Field Emission Gun (FEG-SEM) and EDAX-TSL, OIM Data Collection software were used to obtain EBSD patterns. The set-up conditions are detailed as: acceleration voltage of 20 kV, spot size #5 corresponding to a beam current of 2.4 nA, working distance of 16 mm, tilt angle of 70°, and a step size of 50 nm in a hexagonal scan grid. The specimen preparation for EBSD consisted of grinding and polishing the specimens with a final polishing step to 0.03 µm colloidal silica solution for 60 min to mitigate the plastic strains introduced at the surface during the process. TSL OIM®® Analyses 6.0 software was used for post-processing and analysis of the orientation data. A grain confidence index (CI) standardization was applied to the raw data, with a minimum tolerance angle and grain size of 5° and 6 pixels, respectively. It was considered that grains are formed by multiple pixel rows. Thereafter, neighbor-orientation correlation with a tolerance angle of 5° and a minimum confidence index of 0.1 was implemented.

An Instron 5500R electromechanical tensile testing machine was used to generate experimental curves. The experiments were carried out at room temperature, in extension control mode and using a load cell of 50 kN. ASTM E8/E8M−13a standard [27] was followed using a sub-size tensile test specimen geometry. Tensile specimens were machined with the long axis (gauge section) oriented along the sheet rolling direction. The elongation during the tensile test was recorded by a clip-on extensometer with knife-edges, a gauge length of 7.8 mm and a maximum extension of ±2.5 mm. Uniaxial tension experiments were carried out at strain rates of $0.01\ s^{-1}$, $0.001\ s^{-1}$ and $0.0001\ s^{-1}$.

An additional experiment was performed using a miniature flat dog-bone type specimen of 10 mm × 2 mm × 0.45 mm gauge. The specimen was subjected to tensile elongation of 1.5 mm with a cross-head displacement of 0.005 mm/s in a Deben micromechanical tester. The change in length between the center of two indents placed along the gauge and separated by 1 mm distance was measured to calculate the real plastic strain attained. A square area of approximately $60 \times 50\ \mu m^2$ was delimited by extra indents at which EBSD analysis was conducted in the unstrained condition and after approximately 15% deformation. The same procedure as for microstructure characterization by EBSD was followed before the interrupted test and after making the indents.

3. Modelling

3.1. CPFFT by DAMASK

Full-field crystal plasticity simulations were performed by the spectral solver based on FFT (Fast Fourier Transform) provided by the DAMASK software. Some of the constitutive equations used for the elastic and plastic deformation of synthetic microstructures are broadly presented here. A complete description of the simulation procedure can be found in [12].

The deformation in the continuum theory of crystal plasticity is described as a multiplicative decomposition into reversible and irreversible parts of the deformation gradient F, where the reversible part F_e accounts for the elastic lattice distortion and rigid rotation, and the irreversible part F_p for the plastic distortion that arises due to slip:

$$F = F_e F_p \tag{1}$$

The stress in the elastic strain regime is expressed in form of the second Piola–Kirchhoff stress tensor S, and depends only on the elastic strain expressed as the Green–Lagrange strain tensor E and the material specific stiffness C, according to:

$$S = C : E \qquad (2)$$

$$E = 1/2\left(F_e F_e^T - 1\right) \qquad (3)$$

For cubic crystals in this study, the elastic stiffness matrix is composed of three independent terms, C_{11}, C_{12} and C_{44}. It is worth to note that reversible dislocation glide, i.e., dislocation anelasticity, is not considered in the model. The evolution of plastic strain is given by:

$$\dot{F}_p = L_p F_p \qquad (4)$$

where L_p is the plastic velocity gradient, which can be determined using the shear rates of all slip systems:

$$L_p = \sum_{\beta=1}^{N} \dot{\gamma}^\eta m_\beta \otimes n_\beta \qquad (5)$$

N denotes the number of slip systems. In the present work, twelve slip systems {110}bcc are considered, based on [18,19], while the {211}bcc systems are not taken into account. For the $M_{23}C_6$ carbides, twelve {111}fcc slips systems are considered. m is the slip direction of the slip plane, and n its normal. According with the visco-plastic formulation of Asaro and Needleman [28], the shear rate $\dot{\gamma}^\eta$ on the system η depends on the resolved shear stress, τ^η, and the critical resolved shear stress (CRSS), τ_C^η:

$$\dot{\gamma}^\eta = \dot{\gamma}_0 \left|\tau^\eta / \tau_C^\eta\right|^n sign(\tau^\eta) \qquad (6)$$

In the above expression, the exponent n is related to strain rate sensitivity (as further discussed in Section 5), and $\dot{\gamma}_0$ is the reference shear rate. Both parameters are material dependent.

3.2. VPSC90

VPSC90 [8,9] is an implementation of the VPSC model originally developed by Lebensohn and Tomé [8,9]. In the VPSC model, each grain in a polycrystal is considered as an ellipsoidal inclusion in a homogeneous medium with the aggregate properties of the polycrystal. This formalism is solved in VPSC90 using an optimization algorithm based on the gradient descent method. Moreover, VPSC90 takes into account the elastic behavior of the polycrystal and allows the usage of the same pheno-power hardening law used in DAMASK (and further explained in Section 3.3).

If the stress and plastic strain rate in each grain are defined, respectively, by the tensors S_g and $\dot{E}_g$, then the corresponding polycrystal magnitudes are calculated using a weighted average with respect to the volumetric fraction of each grain (denoted here by w_g). The total strain E in the time increment Δt is obtained adding the elastic strain, calculated using the stiffness tensor C (see the elastic self-consistent method in [8,9]), and the contribution of the plastic strain rate:

$$S = \sum_{g=1}^{Ng} w_g S_g \qquad (7)$$

$$E = C^{-1} : S + \Delta t \sum_{g=1}^{Ng} w_g \dot{E}_g \qquad (8)$$

Polycrystal and individual grain magnitudes are also correlated by the interaction equation:

$$\dot{E}_p - \dot{E}_g = -\widetilde{M}_g : (S - S_g) \tag{9}$$

where $\dot{E}_p$ is the plastic strain rate of the polycrystal (the multiplier of Δt in Equation (8)) and $\widetilde{M}_g$ is the interaction tensor, which depends on the solution of the viscoplastic Eshelby problem in which grain g is considered an ellipsoidal inclusion in a homogeneous matrix [8,9].

The plastic strain rate of a grain, $\dot{E}_g$, can be easily calculated as the symmetric part of the symmetric and skew-symmetric decomposition of the velocity gradient, and then the velocity gradient is correlated with the shear induced slip using Equations (5) and (6). The anti-skew symmetric part of the velocity gradient tensor corresponds to a solid rigid rotation which is used to update the orientation of the grain.

3.3. Hardening Law

The widely adopted phenomenological power law is used for the description of slip hardening, both in DAMASK and VPSC simulations. According to this law, the evolution of critical shear stress in the slip plane β, is given by its time derivative $\dot{\tau}_C^\beta$:

$$\dot{\tau}_C^\beta = \sum_\eta h_{\beta\eta} \dot{\gamma}^\eta \tag{10}$$

The instantaneous slip-system hardening moduli $h_{\beta\eta}$, in general, depends on the history of slip and provides information about additional hardening caused by interactions of fixed slip systems β and active slip systems η, such that $h_{\beta\eta}$ is determined by:

$$h_{\beta\eta} = q^{\beta\eta} \left[h_0 \left(1 - \tau_C^\eta / \tau_{sat} \right)^a \right] \tag{11}$$

where h_0, τ_C^η and τ_{sat} are, respectively, the reference hardening, the critical shear stress of plane η, and the saturation shear stress. These parameters depend on the crystal structure and the slip system. The parameter a has not a direct physical meaning but has a direct influence on the development of hardening (typically, $a \geq 1$). The latent hardening parameter, $q^{\beta\eta}$, defines the interaction between system β and η. It is set to one if β and η are coplanar, and to 1.4 otherwise [10].

3.4. Fitting of Parameters

In order to fit the hardening parameters of the model, two procedures were implemented as Python scripts using the VPSC and DAMASK models. Essentially, the methods use experimental data and different optimization algorithms in which successive simulations are run to find the set of parameters that best replicates the material response. The goal in every case is to find the set of hardening parameters for which the difference between the simulated and measured tensile curves is minimised.

A modified Nedler–Mead (NM) simplex algorithm was used for the fitting with DAMASK, following a procedure similar to the one described in [29]. The NM simplex algorithm [30] stands out for its simplicity and easy implementation. Its deterministic character and independence of gradient information make it suitable for the relatively low dimensional optimization inverse problem. The algorithm iteratively adjusts the hardening parameters until the error meets a given tolerance. The bounds of calibrated parameters were defined based on typical values for ferrite [31–34]. The $M_{23}C_6$ carbides are considered to behave as rigid elastic particles, of which elastic constants were obtained by ab-initio calculations in [35].

In the case of the fitting using VPSC, an estimation of the error gradient with respect to the parameters is found and, using this gradient, a new guess of the optimal parameters is made. This process is repeated until the solution converges to an optimal value. More specifically, the minimisation is performed using the Levenberg–Marquardt Algorithm (LMA), that interpolates between the Gauss–Newton algorithm and the method of gradient descent [36].

3.5. Virtual Microstructures

For the DAMASK simulations, different RVEs are generated. Using the Dream 3D software [26,37], two virtual 3D microstructures of different sizes are created, both of them with a grain size distribution and morphology matching the experimental results, as well as the same fraction and size of carbides.

In order to reproduce the experimental ferrite texture and its correlation with grain size with maximum accuracy, crystallographic orientations are assigned independently to the ferrite grains in different size ranges, according to the following procedure:

1. A grain list is generated with the average orientation and area of each ferrite grain in the EBSD experiment (grains are defined with an orientation tolerance of maximum 5°).
2. The list is divided in five bins corresponding to different size ranges, such that each bin contains the same number of grains (approximately 500 grains in the available experimental data). The number of bins was decided such that the number of grains in each bin is enough to reproduce the experimental texture with a low error.
3. For each of these bins, an orientation distribution function (ODF) is calculated using the generalized harmonics series expansion method [38], assuming orthorhombic symmetry.
4. Orientations are randomly assigned to the grains in the RVE according to the ODF calculated for the corresponding grain size range. The orientations are sampled from the discrete texture obtained evaluating the ODF in a regular grid in Euler space, with a spacing of 5°.

Additionally, a third RVE with equal topology to the smaller one is generated, in which crystallographic orientations are directly assigned by the StatsGenerator filter in Dream 3D, based on the EBSD data.

For the VPSC simulations, six different "phases" are considered. Five of them correspond to each of the grain size bins of the ferrite phase calculated as previously explained (step 2), and the sixth one to the carbides. The volumetric fraction of each phase is directly obtained from the grain size distribution, while discrete textures are defined from the list of grains in each size bin. Moreover, an average grain shape is calculated for each of the size bins, and the effect of grain size in hardening behavior is taken into account by modifying the hardening parameters for each ferrite phase according to the Hall–Petch relationship, such that τ_C^η in Equation (11) for the phase with average grain size d takes the value:

$$\tau_C^\eta(d) = \tau_C^\eta(0) + \frac{k_y}{\sqrt{d}} = \tau_C^\eta(d_{avg}) + k_y \left(\frac{1}{\sqrt{d}} - \frac{1}{\sqrt{d_{avg}}} \right) \tag{12}$$

where k_y is the strengthening coefficient, which is fitted together with the hardening parameters, and $\tau_C^\eta(d_{avg})$ is the CRSS value for the average diameter d_{avg}.

4. Results

4.1. Experimental and Virtual Microstructures

The initial microstructure is characterized using two EBSD experiments along different planes obtained through the method explained in Section 2. Inverse pole figure maps, showing averaged grain orientations, are presented in Figure 1. Considering two scans on perpendicular sections allows the derivation of certain aspects of the 3D shape of the grains in a stereological manner without requiring demanding 3D experiments.

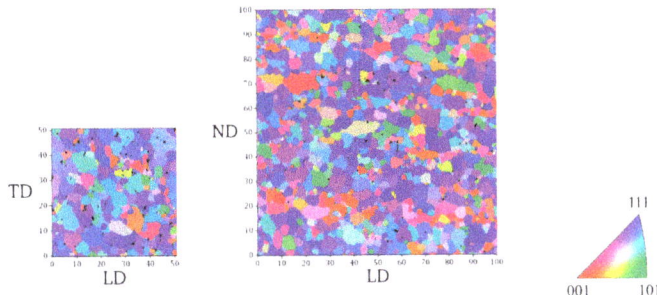

Figure 1. EBSD inverse pole figure (IPF) maps (in the ND direction) along two perpendicular planes: LD–TD on the left, and LD–ND on the right. Carbides are represented in black. Dimensions are given in micrometres.

Equivalent grain diameter, calculated from grain area, and average length of the axis of the equivalent ellipsoid in the rolling direction are calculated combining the results from both measurements. The scan on the LD–TD section is used to calculate the average length of the equivalent ellipsoid axis in the transversal direction, and the experiment along the LD–ND plane is used to calculate the average length of the equivalent ellipsoid axis in the normal direction. In total, 2641 grains and 363 carbides are considered. The diameter and axis length average quantities are always calculated using the geometric average, under the assumption that these magnitudes follow lognormal distributions. Histograms and fitted distributions are shown in Figure 2. It can be seen in the figure that grain diameter and axis lengths indeed follow a distribution which is close to a lognormal function.

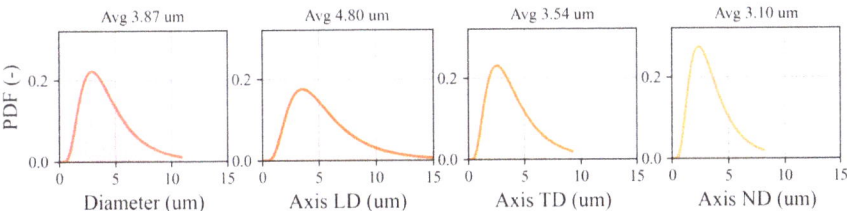

Figure 2. Histograms of grain sizes and equivalent ellipsoid axes and fitted lognormal distributions (continuous lines). On top of each graph, the average value is displayed.

The obtained ODFs for different grain sizes, calculated through the method described in Section 3.5, are shown in Figure 3. The texture of each size bin is displayed in random units with an average value of one so, in order to obtain the probability density for a given orientation and size, it is necessary to multiply by the probability density for the corresponding grain size, which is given in Figure 2. The smallest grains, which represent a small volume fraction, present an almost random texture, with a very weak gamma fibre. The next two bins, which also correspond to relatively small grains and therefore a small volume fraction, show a sharper texture, dominated by the {554}225 component. As the grain size increases, the texture approaches a stronger and more homogeneous gamma fibre. Since the volume fraction corresponding to this size bin is much higher, this is the predominant texture of the ferrite phase (discussed later).

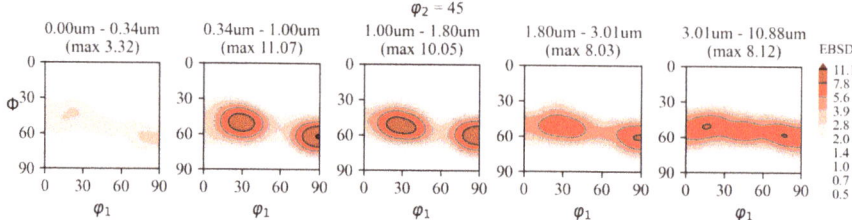

Figure 3. Grain size dependent crystallographic textures. Each $\varphi_2 = 45°$ section corresponds to the ODF calculated for the grains in a different size range (the range for each bin is indicated above the graph, as well as the maximum value of the ODF).

Figure 4 shows three different representative volume elements (RVEs) of the AISI 420 microstructure used in the DAMASK simulations. As explained in Section 3.5, these RVEs are created using Dream 3D, based on the grain size statistics and crystallographic texture collected from the microstructure characterization. RVE$_{GSODF20}$ and RVE$_{ODF20}$ have a size of $20 \times 20 \times 20$ voxels, with a total of 676 ferrite grains and 178 carbides. Crystallographic orientations are assigned to RVE$_{GSODF20}$ extracting discrete textures from the grain size dependent ODFs presented in Figure 3. In the case of RVE$_{ODF20}$, the orientations are assigned by the StatsGenerator filter in Dream 3D, using as input the ferrite texture calculated independently of grain sizes. RVE$_{GSODF50}$ is generated analogously to RVE$_{GSODF20}$, but it has a size of $50 \times 50 \times 50$ voxels, with a total of 3086 ferrite grains and 542 carbides.

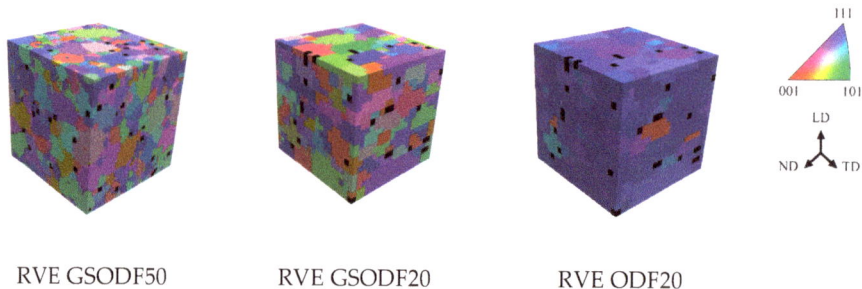

RVE GSODF50   RVE GSODF20   RVE ODF20

Figure 4. IPF maps in the ND direction of the three generated Representative Volume Elements (RVEs).

The grain size distributions obtained from the generated RVEs are compared with the experimental ones in Figure 5. Since the DAMASK simulations are not dependent on a length scale, it is considered in grain size calculations that the average diameter in the RVE matches the experimental one, so all the RVEs have exactly the same average value as that which was measured. Moreover, the grain sizes in RVE50 result in almost the same lognormal distribution as for the EBSD data. For RVE20, there is a larger difference in the standard deviation. A smooth grain size histogram is not reproduced with the reduced number of grains in this RVE.

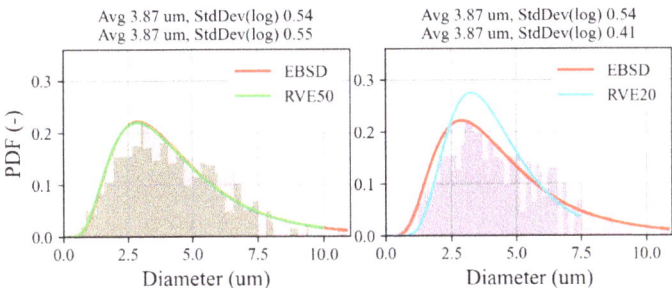

Figure 5. Grain size distribution calculated the RVEs shown in Figure 4 compared with the distribution obtained from the EBSD measurment. The corresponding fitted lognormal distribution is represented with a continuous line.

Crystallographic textures are shown in Figure 6. The experimental ODF, calculated from EBSD data, corresponds to a typical recrystallization texture after cold rolling, with a strong gamma fibre. The difference between the ODF calculated from the different RVEs and the ODF calculated from experimental data. is relatively small (lower than 3.24% and 3.25% for the two GSODF RVEs and 4.17% for RVE$_{ODF20}$) when compared with the experimental error of the ODF calculated from the EBSD data (3.12%). However, while the textures of the two GSODF RVEs are slightly weaker than the experimental one, the one generated by Dream 3D is much sharper (maximum value of 24.42, compared with 7.89 in the experimental ODF).

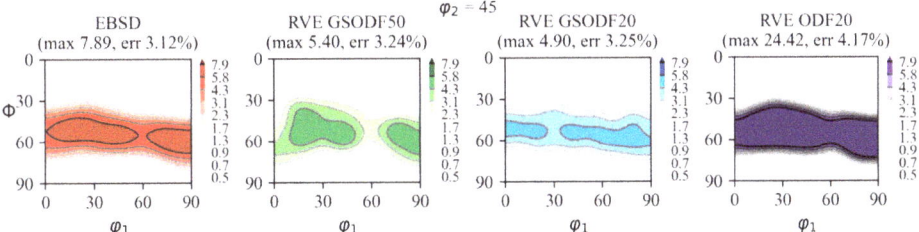

Figure 6. Orientation distribution function ($\varphi_2 = 45°$ section) calculated for the EBSD experiment and the three RVEs shown in Figure 4. The error shown on top indicates the difference between the corresponding discrete texture and the ODF calculated from the EBSD data, or between the calculated ODF and the raw data And it is calculated as the percentage of variance in intensity for all the points in a discrete grid of 5 degrees in the convenient region of Euler space ($90° \times 90° \times 90°$ for cubic crystal symmetry and orthorhombic sample symmetry).

In Figure 7, grain size dependent crystallographic textures are displayed. These ODFs are obtained applying the same procedure used to calculate the grain size dependent textures of the EBSD measurements (presented in Figure 3), described in Section 3.5. Notice that, due to the different grain size distribution and the binning method used, the size ranges vary between the different microstructures. Both RVE$_{GSODF50}$ and RVE$_{GSODF20}$ present a good correspondence with the experimental textures, with a sharp {554}225 component in smaller grains and a more intense and somewhat more homogeneous gamma fibre for the largest ones. However, RVE$_{ODF20}$ shows a very sharp texture for all grain sizes, always with an intense gamma fibre.

It must be noted that, when the GSODF method is used to assign crystallographic orientations in RVE$_{GSODF50}$ and RVE$_{GSODF20}$, the textures are reproduced with similar accuracy, although the number of grains in RVE$_{GSODF20}$ is much lower.

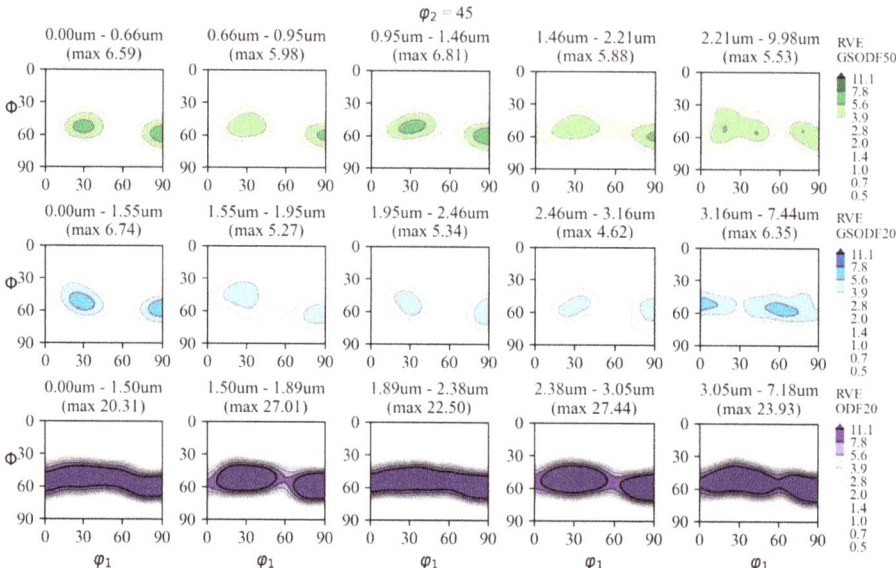

Figure 7. Grain size dependent ODF, calculated for the three RVEs shown in Figure 4. Each $\varphi_2 = 45°$ section corresponds to the ODF calculated for the grains in a different size range.

For the VPSC simulations, the volumetric fraction of each of the five size ranges of ferrite grains is calculated from the grain size distribution in Figure 2. Discrete textures are directly obtained from the list of grains in each size bin of Figure 3 (the original number of ferrite grains is 2641, so approximately 528 grains are used in each ferrite phase). The grain shape is defined differently for each of the ferrite phases, using the axis lengths calculated for each bin presented in Figure 8. The grain size (equivalent diameter) of each ferrite phase is also used in Equation (12) to calculate hardening parameters. The carbide phase is defined with the average shape calculated in the EBSD experiments (ellipsoid axis lengths of 0.78, 0.56 and 0.46 um).

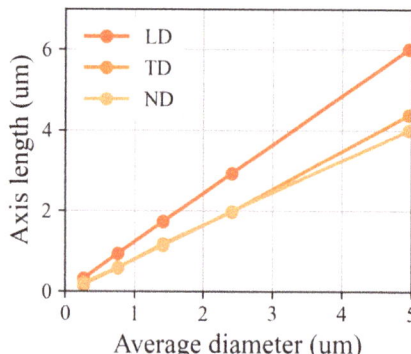

Figure 8. Average length of the equivalent ellipsoid axis for each of the grain size bins into which the EBSD data is divided (see also Figure 2).

4.2. Fitting of Hardening Parameters

Table 1 shows the ferrite and $M_{23}C_6$ carbides materials parameters obtained following the two fitting procedures described in Section 3.4. Relatively similar values of the hardening parameters are obtained using the DAMASK and VPSC models. The most noticeable difference is in the exponent a in expression (11). Simulations showed that the stress partitioned to carbides did not grow to reach values high enough to induce plastic flow, according to the $\tau_{C,0}$ value obtained from the hardness of the $M_{23}C_6$ carbide [39].

Table 1. Adopted and calibrated material parameters for ferrite and $M_{23}C_6$ carbides. The parameters fitted for Visco Plastic Self Consistent (VPSC) correspond to the average grain size ($\tau_c^s(d_{avg})$ in Equation (12)).

Parameter	Unit	Ferrite VPSC GSODF	Ferrite DAMASK RVE$_{ODF20}$	$M_{23}C_6$ Adopted
C_{11}, C_{12}, C_{44}	GPa	233, 135, 128 [32,33]		472, 216, 135 [35]
$\dot{\gamma}_0$	s^{-1}	1×10^{-6} [32,33]		1×10^{-6}
n_{slip}	-	62.4	65	200
$\tau_{C,0}$	MPa	72.5	77	1200
τ_{sat}	MPa	241.0	226	2000
h_0	MPa	2659.9	2534	20
a	-	2.83	1.75	1.1
k_y	MPa m$^{1/2}$	0.738	-	-

Figure 9 (left) shows the simulated curves obtained using the calibrated parameters with DAMASK and VPSC. Both methods led to a reasonable, good approximation of the observed material behavior in the range of studied strain rates. However, as shown in Figure 9 (right), lower errors with respect to the experimental curve are reached using the VPSC model. At strains above 0.02, the error in the predicted stress always lays below 1%, which is the estimated variance of experimental curves. The divergences with the experimental curve increase at lower strains and none of the two models are able to accurately reproduce the material behavior in the yielding region. This can be consequence, as discussed in [40], of low initial hardening in the AISI 420 steel, due to thermal residual stresses and different population of dislocation defects at grain boundaries, which is not accounted for by any of the models.

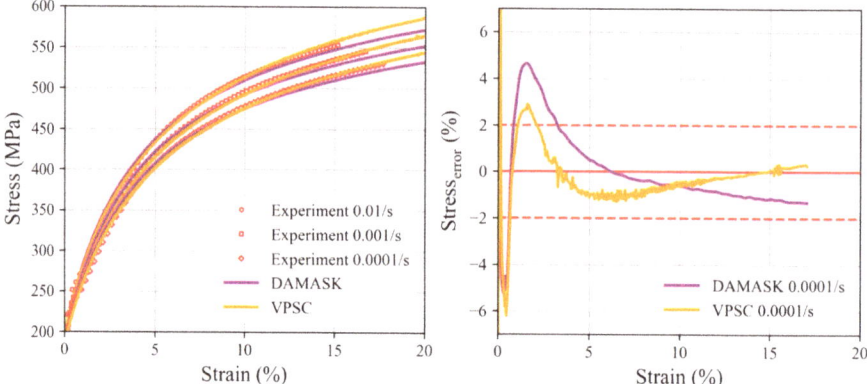

Figure 9. (**Left**) Experimental tensile test curves at different strain rates compared to simulated curves resulting during the calibration process using VPSC with GSODF full texture and Crystal Plasticity Finite Element Method (CPFEM) with RVE$_{ODF20}$. (**Right**) Strain dependence of the error of the predicted stress relative for the 0.0001/s tensile curve (first, cubic spline interpolation was applied for 0 to 0.2 strain and 500 points).

4.3. Deformed Microstructure

After a tensile deformation of 15% is applied to the material as explained in Section 2, new EBSD experiments are performed. In this case, four different scans are performed in different locations of the sample, but on the same LD–ND plane. The obtained IPF maps (in the ND direction) are shown in Figure 10.

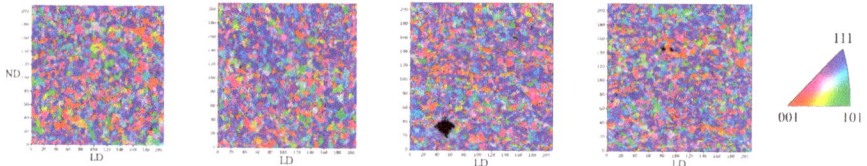

Figure 10. IPF color maps (ND direction) corresponding to the four EBSD measurements performed on the deformed microstructure (uniaxial tension, 15% deformation).

Figure 11 shows the deformed RVEs resulting from DAMASK simulations. The grain size distributions and crystallographic textures corresponding to these RVEs are shown in Figure 12 and are compared with those calculated from the EBSD experiments. The DAMASK simulations are able to capture the change of grain size distribution that takes place during the deformation of the material, as well as reproduce the deformed crystallographic texture. Although in all cases the error with respect to the experimental ODF is lower than 4%, better results are obtained from the GSODF RVEs. It is also observed that the usage of a larger number of points in RVE$_{GSODF50}$ allows a more accurate reproduction of the experimental grain size histogram and crystallographic texture.

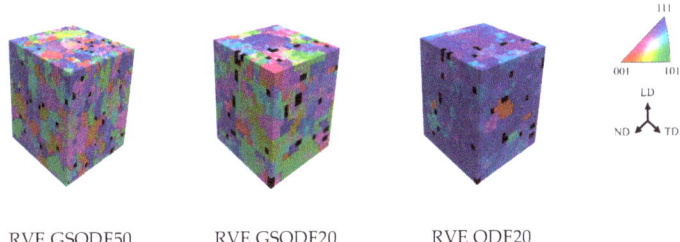

Figure 11. Deformed RVEs after DAMASK simulation under uniaxial tension (deformation: 15%). IPF colors (ND direction), carbides in black.

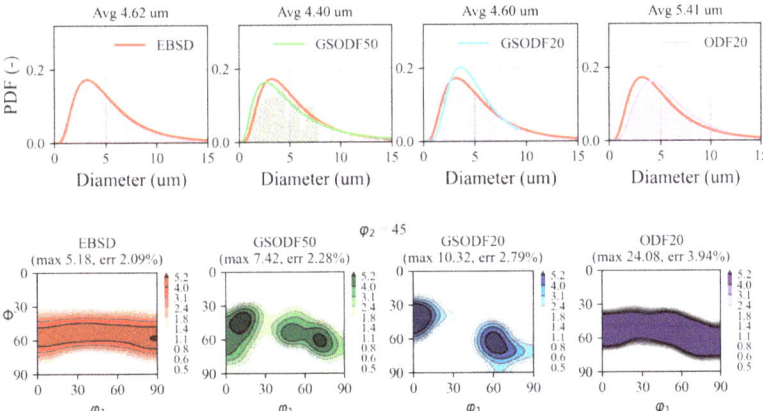

Figure 12. Obtained grain size distributions and crystallographic textures ($\varphi_2 = 45°$ sections) in EBSD experiments and DAMASK simulations. For each grain size distribution, the average grain size is shown on top. On top of the textures, it is shown which is the maximum value and the error with respect to the ODF calculated from the EBSD data, or between the calculated ODF and the raw data (see caption of Figure 6).

The crystallographic textures resulting from the VPSC simulations for each strain rate are shown in Figure 13. In the three simulations, a very similar texture is obtained. Although slightly sharper, these textures are very close to the one calculated from the EBSD measurements (also displayed in the figure). Indeed, the difference between the resulting VPSC textures and the deformed one is almost the same as between the discrete deformed texture and the calculated ODF (2.20% instead of 2.09%).

It is also possible to perform an analysis of the correlation between crystallographic texture and grain size in the deformed microstructure, as the one realized with the initial microstructure in Section 4.1. Figure 14 shows the obtained textures for five different size bins for the EBSD data in Figure 10, the three deformed RVEs in Figure 13, and the VPSC simulation at 0.0001/s in Figure 13. Regarding the EBSD experiments, it is observed that, in contrast with the initial microstructure, the deformed one does not present a strong correlation of crystallographic texture with grain size. Instead, a sharp gamma fibre texture is observed for all size bins, only a bit weaker for the smallest sizes. However, in the simulations, the crystallographic texture shows a clear dependence on grain size. Although the full-field approach used in DAMASK allows the simulation of the grain fragmentation and merging phenomena that take place during the deformation of the material (compare the grain size distributions in Figure 12 with the initial ones presented in Figure 5), grain size dependent textures are not simulated with the same accuracy as the total texture (in Figure 12). In all simulations, a sharper texture is observed for the largest grains, presumably as a consequence of the low number of very large grains in the generated RVEs. In the case of the VPSC simulation, the sharpness of the textures at different grain sizes after the simulation resembles the initial one (c.f. Figure 3). In the used VPSC model (at difference of, for instance, the model presented in [41]), grains are not fragmented or merged, they can only be reoriented (and deformed), and therefore grain size evolution cannot be simulated, making it impossible to accurately reproduce the grain size dependent textures. Nevertheless, the reorientation of the existing grains produces a noticeable change in the size dependent textures, approaching the strong gamma fibre texture observed in the experiments.

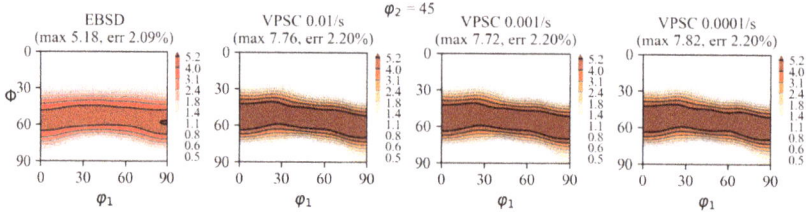

Figure 13. Comparison between deformed texture after uniaxial tension of dog bone specimen deformation 15% at 0.005/s) and DAMASK simulation (deformation: 15% and different strain rates), using parameters fitted with RVE 20 ODF.

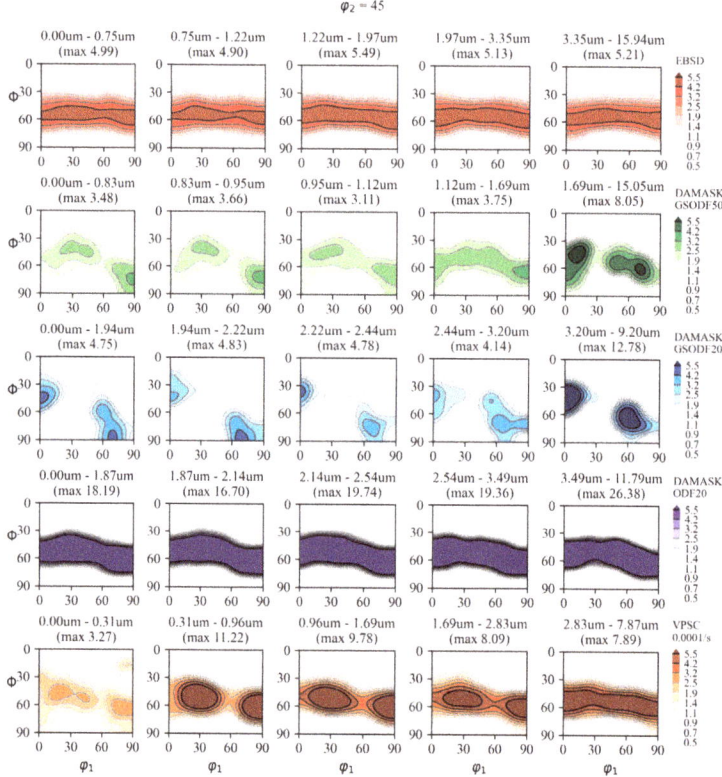

Figure 14. Grain size dependent ODFs (deformation: 15%) obtained from the EBSD measurements, the DAMASK simulations with the GSODF50, GSODF20 and ODF20 RVEs, and the VPSC simulation at a strain rate of 0.0001/s.

5. Discussion

Both DAMASK and VPSC approaches arrive at a relatively similar set of calibrated parameters for ferrite, as shown in Table 1. According to the adopted hardening law, plastic flow begins only when favourable oriented slip systems are activated. This is controlled in a single crystal by $\tau_{C,0}$, which is considered to account for all the strengthening mechanisms, such as the presence of grain boundaries, dislocations, or small precipitates, which affect the material yielding. An estimation of $\tau_{C,0}$ in the polycrystal can be made by connecting this parameter with the yield stress (σ_y) and the arithmetic mean of the Taylor factor, M, using the expression:

$$\sigma_y = M\tau_{C,0} \tag{13}$$

A yield stress of 222 MPa was measured by the 0.2% offset method at 0.0001 s^{-1} strain rate. Considering that the Taylor factor M for the ferrite in present steel is 2.9, and applying Equation (13), a value of 76.5 MPa is obtained for $\tau_{C,0}$, which is close to the respective parameters calibrated using DAMASK and VPSC (c.f. Table 1). It is worth noting that former calculation translates the local CRSS to the yield stress using a simple "averaged" Taylor factor, which is equivalent to applying the full constraint Taylor model. This is a very basic model, but it can provide an accurate initial guess of the boundaries of $\tau_{C,0}$.

The strain rate sensitivity parameter (exponent n in Equation (11)) can be deduced directly from the comparison of the tensile curves at different strains rates [42]. Considering that the virtual-work principle $\tau d\gamma = \sigma d\varepsilon$ applies [43], then:

$$\dot{\varepsilon} = \dot{\varepsilon}_0 \left(\frac{\sigma(\dot{\varepsilon})}{\sigma(\dot{\varepsilon}_0)} \right)^n \Rightarrow \frac{\sigma(\dot{\varepsilon})}{\sigma(\dot{\varepsilon}_0)} = \left(\frac{\dot{\varepsilon}}{\dot{\varepsilon}_0} \right)^{1/n} \tag{14}$$

Applying Equation (14) to the experimental tensile data, under the assumption that quasi-static conditions are fulfilled at a strain rate of $\dot{\varepsilon}_0 = 0.0001/s$, the value $n = 65$ is obtained. Figure 15 shows how the experimental results compare with this value, as well as the curves corresponding to different n values. The value $n = 65$ is used for the DAMASK simulations at different strain rates. The value fitted using VPSC, 62.4, is very close to the experimental result and, as Figure 15 shows, will produce a similar strain rate dependence of tensile behavior.

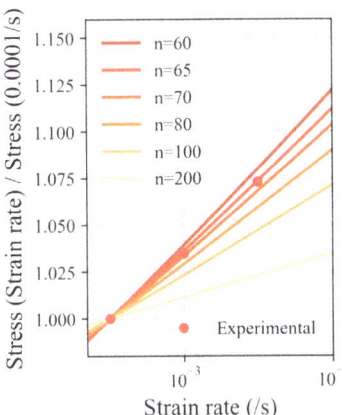

Figure 15. Strain rate dependence of stress for different values of the strain rate sensitivity exponent. The squares represent the strain rate dependence of the material extracted from the three tensile curves at different strain rates in Figure 9, assuming that $\dot{\varepsilon}_0 = 0.0001/s$.

Although both DAMASK and VPSC can be successfully used for the fitting of the hardening parameters, the method that uses the VPSC model has the advantage of requiring much less computational resources. Indeed, while the fitting process using the DAMASK model needed more than 12 h, using eight processors, to find the optimal values for a single strain rate, the method using VPSC allows the performance of the optimization in approximately 2 h for the three strain rates considered (only one processor is used for each simulation, so the three simulations at different strain rates can be run in parallel). These differences in running times are even more significant when it is taken into account that, since the used optimization algorithms can only find local minima, it is necessary to repeat the process several times with different initial parameters. It is therefore desirable to be able to fit the hardening parameters always using the faster VPSC method. In order to evaluate this possibility, the DAMASK simulations are repeated using the parameters fitted with the VPSC model. For completeness, the VPSC simulations are also repeated using the parameters fitted with DAMASK.

Figure 16 shows the tensile curves simulated with the DAMASK and VPSC models using the parameters obtained with the two calibration methods, as well as the simulated transversal and normal strains. As shown in Figure 9, both fitting methods accurately represent the experimental results. However, clear differences are observed when the fitted parameters are applied to simulations with different models and RVEs. It is remarkable that, when the parameters fitted using VPSC are applied to the RVEs used with DAMASK, a better approximation is obtained with more detailed RVEs. A relatively good match is obtained with $RVE_{GSODF50}$, followed by $RVE_{GSODF20}$ and, finally, RVE_{ODF20}. However, when the parameters fitted using DAMASK and RVE_{ODF20} are employed, the next RVE with the best match is $RVE_{GSODF20}$, although the original microstructure is better reproduced by $RVE_{GSODF50}$, with five times more grains. A similar trend is observed in the graphs of strain in the transversal and normal directions (also shown in Figure 16). $RVE_{GSODF50}$ and VPSC yield similar results, while $RVE_{GSODF20}$ is slightly different, and RVE_{ODF20} shows a completely different behavior.

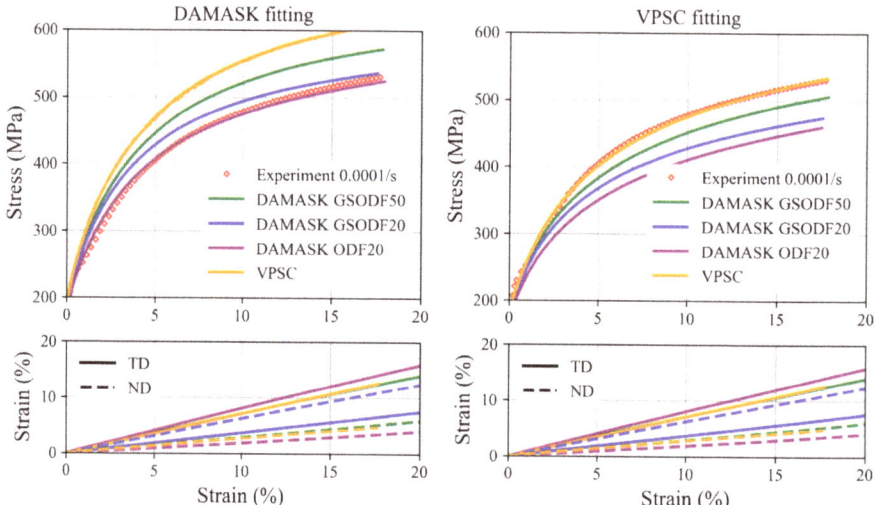

Figure 16. Experimental and simulated tensile test curves and simulated transversal and normal strain resulting from: application of material parameters set obtained after calibration with VPSC and GSODF full texture (at left); and application of material parameter set obtained after calibration with CPFEM and RVE_{ODF20} (at right).

To sum up, it appears that the use of the VPSC model to fit hardening parameters for DAMASK does not present any inconvenience that is not present when a small RVE is directly used in DAMASK.

On the contrary, it approaches the response obtained from highly detailed RVEs, supposedly closer to the real behavior of the material.

Similar conclusions can be extracted from the grain size distributions and crystallographic textures, presented in Figure 17 (c.f. Figure 12). The grain size distributions and crystallographic textures obtained with RVE$_{GSODF50}$ and RVE$_{GSODF20}$ and the parameters fitted using VPSC are practically the same as when using the parameters fitted using RVE$_{ODF20}$. Surprisingly, the experimental grain size distribution and crystallographic texture are better reproduced in the RVE$_{ODF20}$ simulations that use the parameters fitted with VPSC than in those performed with the parameters obtained with DAMASK using the same RVE. Therefore, it is clear that VPSC in combination with the grain size dependent textures approach is an effective method for obtaining an adequate set of materials parameters: mechanical behavior is better correlated with the quality of the RVE used in the simulation, experimental microstructural evolution is reproduced with higher accuracy, and the time and computational resources required to perform the fitting are much lower.

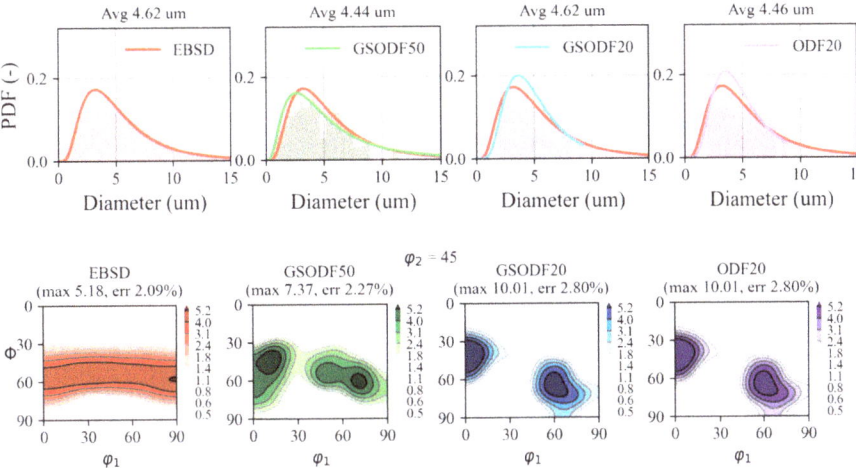

Figure 17. Obtained grain size distributions and crystallographic textures ($\varphi_2 = 45°$ sections) in EBSD experiments and DAMASK simulations using the model parameters fitted with VPSC. For each grain size distribution, the average grain size is shown on top. On top of the textures, it is show which is the maximum value and the error with respect to the ODF calculated from the EBSD data, or between the calculated ODF and the raw data (see caption of Figure 6).

6. Conclusions

Both the DAMASK and VPSC crystal plasticity models can be used to perform simulations of the mechanical behavior and microstructural evolution of AISI420 steel. However, how the real microstructure of the material is represented in the model and plays a fundamental role in the outcome of the simulations. In this work, the correlation between grain size and crystallographic texture has been applied in mean-field and full-field crystal plasticity simulations, after the analysis of EBSD data showed that there is a sharpening of the crystallographic texture for grains of larger size. The use of this correlation enables a more truthful description of the real microstructure, and the additional data makes it possible to perform a deeper analysis of the results.

In the VPSC simulations, grain shape, crystallographic texture and hardening law are defined dependent on grain size. Using a homogenization model like VPSC allows the consideration of a virtual microstructure representative enough from a statistical point of view and with enough efficiency to be used for the fitting of material parameters. The obtained model is further applied in full-field DAMASK simulations, where the correlation between grain size and crystallographic texture is introduced by assigning the orientations of the RVE grains depending on their size. The simulations are capable of reasonably reproducing the correlations observed in the deformed microstructure.

Author Contributions: J.G.-L. and J.H. equally contributed to conceptualisation, design of methodology, formal analysis and in the writing—original draft preparation, review and editing. J.G.-L. focused on the VPSC90 simulations and in the implementation of GSODF concept into representative volume elements. J.H. focused on the simulations with DAMASK software and performed and analyzed the experimental results. All authors have read and agreed to the published version of the manuscript.

Funding: This research has received funding from the European Union's Research Fund for Coal and Steel (RFCS) research program under grant agreement RFCS-2015-709418 (MuSTMeF) and from the VMAP—ITEA 16010 project via the ITEA 3 cluster of the European research initiative EUREKA.

Acknowledgments: The authors want to acknowledge L. Kestens and J. Sietsma for their wise comments and R. Petrov and F. Vercruysse for performing the EBSD experiments.

Conflicts of Interest: The authors declare no conflict of interest.

References

1. Masuyama, F. History of power plants and progress in heat resistant steels. *ISIJ Int.* **2001**, *41*, 612–625. [CrossRef]
2. Rojas, D.; Garcia, J.; Prat, O.; Sauthoff, G.; Kaysser-Pyzalla, A.R. 9%Cr heat resistant steels: Alloy design, microstructure evolution and creep response at 650 °C. *Mater. Sci. Eng. A* **2011**, *528*, 5164–5176. [CrossRef]
3. Vivas, J.; Capdevila, C.; Altstadt, E.; Houska, M.; San-Martín, D. Importance of austenitization temperature and ausforming on creep strength in 9Cr ferritic/martensitic steel. *Scr. Mater.* **2018**, *153*, 14–18. [CrossRef]
4. Vivas, J.; Capdevila, C.; Altstadt, E.; Houska, M.; Sabirov, I.; San-Martín, D. Microstructural Degradation and Creep Fracture Behavior of Conventionally and Thermomechanically Treated 9% Chromium Heat Resistant Steel. *Met. Mater. Int.* **2019**, *25*, 343–352. [CrossRef]
5. Taylor, G.I. Plastic strain in metals. *Plast. Strain Met.* **1938**, *62*, 307–324.
6. Sachs, G. Plasticity problems in metals. *Trans. Faraday Soc.* **1928**, *24*, 84–92. [CrossRef]
7. Houtte, P.V.; Li, S.; Seefeldt, M.; Delannay, L. Deformation texture prediction: From the Taylor model to the advanced Lamel model. *Int. J. Plast.* **2005**, *21*, 589–624.
8. Lebensohn, R.A.; Tomé, C.N. A self-consistent anisotropic approach for the simulation of plastic deformation and texture development of polycrystals: Application to zirconium alloys. *Acta Metall. Mater.* **1993**, *41*, 2611–2624. [CrossRef]
9. Galán, J.; Verleysen, P.; Lebensohn, R.A. An improved algorithm for the polycrystal viscoplastic self-consistent model and its integration with implicit finite element schemes. *Model. Simul. Mater. Sci. Eng.* **2014**, *22*, 55023. [CrossRef]
10. Roters, F.; Eisenlohr, P.; Hantcherli, L.; Tjahjanto, D.D.; Bieler, T.R.; Raabe, D. Overview of constitutive laws, kinematics, homogenization and multiscale methods in crystal plasticity finite-element modeling: Theory, experiments, applications. *Acta Mater.* **2010**, *58*, 1152–1211. [CrossRef]
11. Roters, F.; Eisenlohr, P.; Bieler, T.; Raabe, D. *Crystal Plasticity Finite Element Methods: In Materials Science and Engineering*; John Wiley & Sons: Hoboken, NJ, USA, 2010. [CrossRef]
12. Roters, F.; Diehl, M.; Shanthraj, P.; Eisenlohr, P.; Reuber, C.; Wong, S.L.; Maiti, T.; Ebrahimi, A.; Hochrainer, T.; Fabritius, H.O.; et al. DAMASK—The Düsseldorf Advanced Material Simulation Kit for modeling multi-physics crystal plasticity, thermal, and damage phenomena from the single crystal up to the component scale. *Comput. Mater. Sci.* **2019**, *158*, 420–478. [CrossRef]
13. Eisenlohr, P.; Diehl, M.; Lebensohn, R.A.; Roters, F. A spectral method solution to crystal elasto-viscoplasticity at finite strains. *Int. J. Plast.* **2013**, *46*, 37–53. [CrossRef]
14. Chatterjee, K.; Echlin, M.P.; Kasemer, M.; Callahan, P.G.; Pollock, T.M.; Dawson, P. Prediction of tensile stiffness and strength of Ti-6Al-4V using instantiated volume elements and crystal plasticity. *Acta Mater.* **2018**, *157*, 21–32. [CrossRef]
15. Zaefferer, S.; Wright, S.I.; Raabe, D. Three-Dimensional Orientation Microscopy in a Focused Ion Beam–Scanning Electron Microscope: A New Dimension of Microstructure Characterization. *Metall. Mater. Trans. A* **2008**, *39*, 374–389. [CrossRef]

16. Balzani, D.; Scheunemann, L.; Brands, D.; Schröder, J. Construction of two- and three-dimensional statistically similar RVEs for coupled micro-macro simulations. *Comput. Mech.* **2014**, *54*, 1269–1284. [CrossRef]
17. Pirgazi, H. On the alignment of 3D EBSD data collected by serial sectioning technique. *Mater. Charact.* **2019**, *152*, 223–229. [CrossRef]
18. Swaminathan, S.; Ghosh, S.; Pagano, N.J. Statistically equivalent representative volume elements for unidirectional composite microstructures: Part I—Without damage. *J. Compos. Mater.* **2006**, *40*, 583–604. [CrossRef]
19. Brands, D.; Balzani, D.; Scheunemann, L.; Schröder, J.; Richter, H.; Raabe, D. Computational modeling of dual-phase steels based on representative three-dimensional microstructures obtained from EBSD data. *Arch. Appl. Mech.* **2016**, *86*, 575–598. [CrossRef]
20. Zhang, P.; Balint, D.; Lin, J. Controlled Poisson Voronoi tessellation for virtual grain structure generation: A statistical evaluation. *Philos. Mag.* **2011**, *91*, 4555–4573. [CrossRef]
21. Bargmann, S.; Klusemann, B.; Markmann, J.; Schnabel, J.E.; Schneider, K.; Soyarslan, C.; Wilmers, J. Generation of 3D representative volume elements for heterogeneous materials: A review. *Prog. Mater. Sci.* **2018**, *96*, 322–384. [CrossRef]
22. Quey, R.; Dawson, P.R.; Barbe, F. Large-scale 3D random polycrystals for the finite element method: Generation, meshing and remeshing. *Comput. Methods Appl. Mech. Eng.* **2011**, *200*, 1729–1745. [CrossRef]
23. Kasemer, M.; Quey, R.; Dawson, P. The influence of mechanical constraints introduced by β annealed microstructures on the yield strength and ductility of Ti-6Al-4V. *J. Mech. Phys. Solids* **2017**, *103*, 179–198. [CrossRef]
24. Vittorietti, M.; Kok, P.J.J.; Sietsma, J.; Jongbloed, G. Accurate representation of the distributions of the 3D Poisson-Voronoi typical cell geometrical features. *Comput. Mater. Sci.* **2019**, *166*, 111–118. [CrossRef]
25. Vittorietti, M.; Kok, P.J.J.; Sietsma, J.; Li, W.; Jongbloed, G. General framework for testing Poisson-Voronoi assumption for real microstructures. *Appl. Stoch. Models Bus. Ind.* **2020**. [CrossRef]
26. Groeber, M.; Ghosh, S.; Uchic, M.D.; Dimiduk, D.M. A framework for automated analysis and simulation of 3D polycrystalline microstructures. Part 2: Synthetic structure generation. *Acta Mater.* **2008**, *56*, 1274–1287. [CrossRef]
27. ASTM-E8-E8M-13a. *Standard Test Methods for Tension Testing of Metallic Materials*; ASTM International: West Conshohocken, PA, USA, 2013; Volume E8 E8M-13a, p. 28.
28. Asaro, R.J.; Needleman, A. Overview no. 42 Texture development and strain hardening in rate dependent polycrystals. *Acta Metall.* **1985**, *33*, 923–953. [CrossRef]
29. Chakraborty, A.; Eisenlohr, P. Evaluation of an inverse methodology for estimating constitutive parameters in face-centered cubic materials from single crystal indentations. *Eur. J. Mech. A/Solids* **2017**, *66*, 114–124. [CrossRef]
30. Nelder, J.A.; Mead, R. A Simplex Method for Function Minimization. *Comput. J.* **1965**, *7*, 308–313. [CrossRef]
31. Yalcinkaya, T.; Brekelmans, W.A.M.; Geers, M.G.D. BCC single crystal plasticity modeling and its experimental identification. *Model. Simul. Mater. Sci. Eng.* **2008**, *16*, 85007. [CrossRef]
32. Tasan, C.C.; Hoefnagels, J.P.M.; Diehl, M.; Yan, D.; Roters, F.; Raabe, D. Strain localization and damage in dual phase steels investigated by coupled in-situ deformation experiments and crystal plasticity simulations. *Int. J. Plast.* **2014**, *63*, 198–210. [CrossRef]
33. Maresca, F.; Kouznetsova, V.G.; Geers, M.G.D. On the role of interlath retained austenite in the deformation of lath martensite. *Model. Simul. Mater. Sci. Eng.* **2014**, *22*, 45011. [CrossRef]
34. Maresca, F.; Kouznetsova, V.G.; Geers, M.G.D. Reduced crystal plasticity for materials with constrained slip activity. *Mech. Mater.* **2016**, *92*, 198–210. [CrossRef]
35. Liu, Y.; Jiang, Y.; Xing, J.; Zhou, R.; Feng, J. Mechanical properties and electronic structures of M23C6 (M = Fe, Cr, Mn)-type multicomponent carbides. *J. Alloy. Compd.* **2015**, *648*, 874–880. [CrossRef]
36. Press, W.H.; Flannery, B.P.; Teukolsky, S.A.; Vetterling, W.T. *Numerical Recipes in C: The Art of Scientific Computing*; Cambridge University Press: Cambridge, UK, 1988.
37. Groeber, M.A.; Jackson, M.A. DREAM.3D: A Digital Representation Environment for the Analysis of Microstructure in 3D. *Integr. Mater. Manuf. Innov.* **2014**, *3*, 56–72. [CrossRef]
38. Van Houtte, P. A method for the generation of various ghost correction algorithms—The example of the positivity method and the exponential method. *Texture Stress Microstruct.* **1991**, *13*, 199–212. [CrossRef]

39. Inoue, A.; Arakawa, S.; Masumoto, T. Effect of Alloying Elements on Defect Structure and Hardness of M23C6 Type Carbides. *Trans. Jpn. Inst. Met.* **1979**, *20*, 585–592. [CrossRef]
40. Hidalgo, J.; Vittorietti, M.; Farahani, H.; Vercruysse, F.; Petrov, R.; Sietsma, J. Influence of M23C6 Carbides on the Heterogeneous Strain Development in Annealed 420 Stainless Steel. Available online: https://ssrn.com/abstract=3558254 (accessed on 15 September 2020).
41. Zecevic, M.; Lebensohn, R.A.; McCabe, R.J.; Knezevic, M. Modeling of intragranular misorientation and grain fragmentation in polycrystalline materials using the viscoplastic self-consistent formulation. *Int. J. Plast.* **2018**, *109*, 193–211. [CrossRef]
42. Galán-López, J.; Verleysen, P. Simulation of the plastic response of Ti–6Al–4V thin sheet under different loading conditions using the viscoplastic self-consistent model. *Mater. Sci. Eng. A* **2018**, *712*, 1–11. [CrossRef]
43. Przybyla, C.; Adams, B.; Miles, M. Methodology for Determining the Variance of the Taylor Factor: Application in Fe-3%Si. *J. Eng. Mater. Technol.* **2007**, *129*. [CrossRef]

© 2020 by the authors. Licensee MDPI, Basel, Switzerland. This article is an open access article distributed under the terms and conditions of the Creative Commons Attribution (CC BY) license (http://creativecommons.org/licenses/by/4.0/).

Article

Statistical Crystal Plasticity Model Advanced for Grain Boundary Sliding Description

Alexey Shveykin *, Peter Trusov and Elvira Sharifullina

Department of Mathematical Modeling of Systems and Processes, Perm National Research Polytechnic University, 614990 Perm, Russia; tpv@matmod.pstu.ac.ru (P.T.); elvira16_90@mail.ru (E.S.)
* Correspondence: alexey.shveykin@gmail.com

Received: 19 August 2020; Accepted: 14 September 2020; Published: 16 September 2020

Abstract: Grain boundary sliding is an important deformation mechanism, and therefore its description is essential for modeling different technological processes of thermomechanical treatment, in particular the superplasticity forming of metallic materials. For this purpose, we have developed a three-level statistical crystal plasticity constitutive model of polycrystalline metals and alloys, which takes into account intragranular dislocation sliding, crystallite lattice rotation and grain boundary sliding. A key advantage of our model over the classical Taylor-type models is that it also includes a consideration of grain boundaries and possible changes in their mutual arrangement. The constitutive relations are defined in rate form and in current configuration, which makes it possible to use additive contributions of intragranular sliding and grain boundary sliding to the strain rate at the macrolevel. In describing grain boundary sliding, displacements along the grain boundaries are considered explicitly, and changes in the neighboring grains are taken into account. In addition, the transition from displacements to deformation (shear) characteristics is done for the macrolevel representative volume via averaging, and the grain boundary sliding submodel is attributed to a separate structural level. We have also analyzed the interaction between grain boundary sliding and intragranular inelastic deformation. The influx of intragranular dislocations into the boundary increases the number of defects in it and the boundary energy, and promotes grain boundary sliding. The constitutive equation for grain boundary sliding describes boundary smoothing caused by diffusion effects. The results of the numerical experiments are in good agreement with the known experimental data. The numerical simulation demonstrates that analysis of grain boundary sliding has a significant impact on the results, and the multilevel constitutive model proposed in this study can be used to describe different inelastic deformation regimes, including superplasticity and transitions between conventional plasticity and superplasticity.

Keywords: crystal plasticity; multilevel statistical constitutive model; grain boundary sliding; superplasticity; 1420 alloy

1. Introduction

Superplastic (SP) deformation is promising for manufacturing complex-shaped parts with the required properties from metals and alloys. Using this approach, one can produce lightweight large-size structures without welded seams or with a small number of welded seams and joints. Its application makes it possible to reduce the number of technological operations, to carry out the forming process with low stresses and to decrease the consumption of materials. The structures produced by this method have smooth surfaces and exhibit only slight surface deviations from the desired geometry, which is of particular importance for thickness variation minimization and for high-precision filling of dies [1–10].

In technological processes, the SP deformation regime develops at relatively moderate (about 0.5–0.6) homologous temperatures so that the size and equiaxed shape of individual grains remain unchanged [11–13]. At a higher temperature, the SP deformation can be realized under continuous dynamic recrystallization (DRX) conditions. However, in this case, expensive heat-resistant equipment is required, and it is almost impossible to guarantee the necessary continuity (absence of excessive porosity) of the complex-shaped parts during their manufacture. Besides, the residual stresses arising after high-temperature processing may be high, and the shape of an object may distort. Our study addresses the 1st type SP regime, called "structural SP with steady grain's state". This term has a more precise meaning compared to the pioneer works on superplasticity (e.g., [14]), in which only the average grain structure parameters were kept unchanged, and the material was redistributed among grains during the continuous DRX (the term "steady state structural SP" is used in the literature [15] in this broad sense).

In most works, the properties of the materials in the SP regime are determined on the basis of the results of uniaxial tensile tests. For many alloys, including those widely used in industry, within some temperature and strain rate ranges, there occurs a stress-strain dependence of staged nature, which can be attributed to the effects of different deformation mechanisms, their interaction with each other and a change in their roles during deformation. Similar dependences (bell-shaped curves) are given in the literature for aluminum alloys close to single-phase alloys [16–20], magnesium alloys [21,22] and titanium alloys [23,24]. These analyses reported the absence of a pronounced deformation localization zone (a neck), which allowed researchers to make an assumption about the uniformity of the stress–strain state.

In [25,26], basing on the analysis of information from most of the reviewed sources, the authors set forth a scenario that has been realized in deformation testing alloys close to single-phase alloys with a transition to the SP regime under prescribed temperature and strain rate conditions (the test temperature does not exceed 0.7 of the homologous temperature).

At the initial deformation stage (an ascending branch of the tension curve), the effects of intragranular dislocation sliding (IDS) and grain rotation mechanisms become stronger, which causes a crystallographic texture to occur, whereas the role of grain boundary sliding (GBS) remains insignificant.

At the transient stage (the tension curve gradually bends), the role of GBS increases. The grain boundaries are more prepared for this sliding because of the influx of lattice dislocations that increase the boundary internal energy and improve the smoothing of the grain surfaces by grain boundary diffusion. Depending on the initial structure of the material, test temperature and strain rate conditions, recrystallized grains begin to grow at the expense of the non-recrystallized phase and an integral decrease in the density of intragranular dislocations. During the deformation process, a grain rotation occurs. The GBS causes a shear of adjacent grains against one another.

The specific features of the structural SP regime (part of the curve that corresponds to a steady-state flow stress or its gradual decrease) are as follows. The GBS, accompanied by the accommodative processes of IDS and grain boundary diffusion, grain rotations and shear of adjacent grains against one another, is prevailed. The stability of the structure (all grains remain unchanged, and the structure becomes an equiaxial ultrafine-grained structure) is achieved via the simultaneous realization of accommodative IDS and grain rotation. GBS (as well as dislocation sliding at an transient stage) leads to weakening or disappearance of the texture formed after the first stage of testing.

Therefore, the available complex deformation scenario for the implementation of even the simplest (uniaxial at the macrolevel) SP tests includes several interacting mechanisms with changing roles and significant structure evolution of the material. A similar situation is typical for the technological processes involving SP. Temperature and strain rate conditions may change significantly in different parts of the structure, and thus these processes will proceed in different ways in these objects. This causes a need for the development of mathematical models for describing the peculiar features of existing metal-forming technologies.

Much research has been published regarding the multilevel models developed within the framework of the crystal plasticity for describing IDS, twinning and rotation of crystallite lattices [27–36]. An important direction towards obtaining this class of constitutive models and an accurate description of traditional mechanisms (dislocation motion and interaction) in the context of solid-state physics includes a consideration of other significant deformation mechanisms. In order to describe the structural SP with steady grain's state and preparatory stage, it is necessary to constitute an improved single-crystal plasticity model capable of accounting for GBS, grain growth and some other deformation mechanisms mentioned above. GBS is the leading mechanism in the SP deformation and in the transition to it, and therefore its description is an issue of primary importance. However, to the best of our knowledge, this problem has not received due attention in the literature.

In [37], the activity of GBS was estimated in terms of the direct crystal plasticity model by analyzing tangential stresses on grain boundaries. In [38,39], where the deformation of grains was described with allowance for crystal plasticity, the grain boundary was modelled as a separate deformable medium (phenomenological relations were applied), but the grain sliding was not taken into account explicitly. In [40], almost the same approach is proposed, the mixture model with a smooth transition between phases is considered (the grain is described by the crystal plasticity model, the boundary is modeled by the J_2—plasticity theory).

As noted above, the grain growth can play a significant role at the preparatory stage before the onset of SP regime. Recently, direct (based on solving the boundary volume problem at mesolevel via FEM) [41–43] and self-consistent crystal plasticity models [44,45] have been developed for describing the discontinuous DRX. However, it should be noted that these models include rather complicated procedures needed in grained restructuring and to redefine the stress field so that they satisfy a balance equation.

A self-consistent crystal plasticity model that is able to account for GBS and some other SP mechanisms has been introduced in a recently published article [46], the authors of which set forth an idea to describe "superplastic mechanisms (GBS, diffusion etc.)" in a complex manner by making use of an additional inelastic component for each grain. Although this approach shows some promise, it can be assumed that it hardly be used to formulate the kinetic equations needed to determine a given component via physical analysis (taking into account the interaction of GBS with another deformation mechanisms), also, it will lead to a significant complication of numerical procedures; besides, it is difficult to interpret the grain interaction effect using this approach. Direct crystal plasticity models are difficult to develop for GBS describing because of the need to describe the loss of continuity. For this reason, there occurs a necessity to apply special methods for solving boundary value problems. In recent years, some attempts have been made towards developing such mathematical apparatus (for example, in [47]), yet its successful practical application requires serious additions, including those promoting an increase in its computational efficiency. The issue of resource intensity is of great importance in solving boundary value problems applied to modeling SP technologies. Due to the totality of the arguments given, we have chosen a statistical model.

The structure of the model, in which part of inelastic strain rate due to GBS is separated at the macrolevel, is proposed in [48]. The models of similar structure are also used in [49,50], where the authors analyzed GBS in nanomaterials without considering physical processes that govern it (for instance, the equations for critical stresses were dropped from the consideration).

In [26], the first version of a multilevel statistical constitutive model for describing SP regime and transitions to it was developed. The model was formulated in the rate form in the current configuration, which made it possible to apply the hypothesis of additive contributions of IDS and GBS to strain rate at the macroscale level. In a series of works [51–54], the choice of this approach for formulating geometrically nonlinear equations was justified. It turned out that, in the absence of GBS, the model offers the results, which agree well with the data obtained using other mesoscale models, including those with the most popular formulation of constitutive equations in terms of the unloaded configuration [28,55].

In this paper, we describe in detail the modifications necessary to adapt the statistical crystal plasticity model to the consideration of GBS. In contrast to the model presented in [26], the structure of the model proposed here, as well as the kinetic equations used to describe SP and GBS effects, are changed. This modified model is a basis for developing a constitutive model advanced by an explicit account of discontinuous DRX (grain growth) and its effect on other deformation mechanisms, which are needed to describe both SP and transitions to it (an appropriate publication is scheduled for the near future). Since many deformation mechanisms operate and interact during the SP regime, we use here the simplest phenomenological relations (at the corresponding scales). Probable modifications of these relations are planned to perform in our further study on this topic. On the other hand, the presented constitutive model and numerical computations have revealed very important and interesting aspects regarding the crystal plasticity development for the simulation of deformation with active GBS.

Section 2 includes a description of a three-level constitutive model and presents mesolevel equations for IDS and crystallite lattice rotations, as well as relations for GBS. Section 3 discusses the simulation results obtained in this study and gives their analysis.

2. Three-Level Model with Description of GBS, IDS and Grain Lattice Rotations

In contrast to the classical Taylor-type statistical models, the advanced model proposed in this study is able to account for the interfaces between crystallites. Intergranular boundaries are considered in the model as separate objects, connected at each moment of time with boundaries (grain facets). For these objects, internal variables are introduced, and kinetic equations for their evolution are formulated. A set of intergranular boundaries forms a separate structural level (level for description of GBS), and the model is modified into a three-level model: macrolevel (representative macrovolume), structural level for description of GBS that can be considered as a macrolevel submodel, and mesolevel (grains with boundaries). The schematic diagram of the model structure is given in Figure 1.

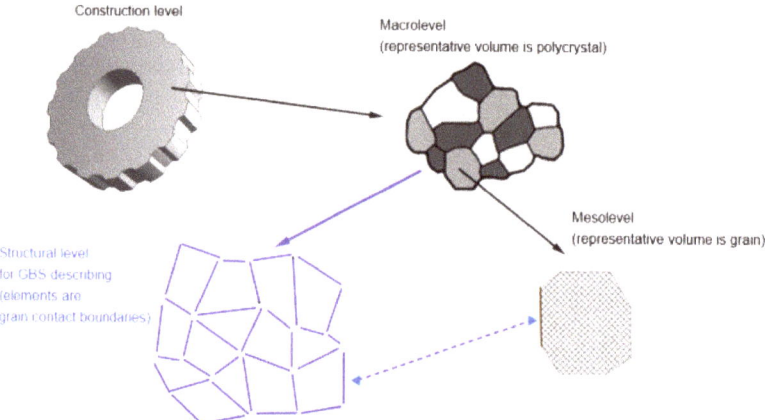

Figure 1. Schematic diagram of the model structure. Additions to a typical statistical crystal plasticity model structure are marked by blue color.

At the initial moment, the aggregates of grains and intergranular boundaries are constructed, and the relations between them are assigned (for the intergranular boundary, the numbers of the grains adjacent to it are set). Due to GBS, the mutual arrangement of grains in the statistical model will change and this will gradually change the numbers of grains that form intergranular boundaries. The initial structure (samples of grains and intergranular boundaries) can be formed both coincident with spatial packing (e.g., the grain structure can be approximated by using Voronoi tessellation) and in a random way (due to the statistical equality of the model elements).

The calculation results confirm the closeness of the integral parameters obtained at the macrolevel, including the change in the GBS proportion in the total strain rate and the change in the distribution function of grain lattice orientations, for different generations of the sample relations. This approach can be called a modification of the "statistical model which takes into account intergranular boundaries, mutual grain arrangement and its change". As regards the possibility of redefining the relations between the elements, it is similar to the movable cellular automata method [56].

For simplicity of the presentation, the formulations are given for isothermal deformations because a maintained constant temperature is observed in most tests and technological modes of SP forming. The changes necessary to consider the non-isothermal case are easy to introduce into the constitutive model.

We suppose that the macroscale stresses integrally characterize the elastic bonds in the material, and therefore the macroscopic stresses are equal to average mesoscale stresses

$$\mathbf{K} = \langle \mathbf{k} \rangle \tag{1}$$

where $\mathbf{K}, \mathbf{k}$ is the weighted Kirchhoff stress tensor at macro- and mesolevels (the grain index is omitted), $\mathbf{k} = \overset{\circ}{\rho}/\rho\, \sigma$, σ is the Cauchy stress tensor, $\overset{\circ}{\rho}$ is the crystallite density in the reference configuration, and ρ is the crystallite density in the current configuration. The continuity conditions are assumed to be fulfilled along the grain contact boundaries (no material discontinuities are present along the normal to the boundary).

The strain rate due to GBS $\mathbf{Z}_{gb}^{in}$ at the macrolevel is determined by averaging the GBS shear rates along the grain contact boundaries. Instead of standard (for the statistical Taylor type model) transmission of the velocity gradient ($\mathbf{l} = \mathbf{L}$) to the lower level, the following relation is used

$$\mathbf{l} = \mathbf{L} - \mathbf{Z}_{gb}^{in} \tag{2}$$

where $\mathbf{l}, \mathbf{L}$ is the transposed velocity gradient at meso and macrolevels, respectively. It can be said that some of the applied kinematic impacts are realized due to GBS at the macrolevel without contributing to the deformation of grains. Theoretically, there may be a situation where, after reaching a certain state of boundaries, the entire instantaneous deformation is realized due to GBS (among the grain layers), i.e., $\mathbf{L} = \mathbf{Z}_{gb}^{in}$, $\mathbf{l} = 0$. This situation may occur in the bicrystal subjected to simple shear loading in the boundary plane. Relations describing GBS and an example that illustrates (2) are given below. First, let us consider the relations at the mesolevel.

To describe the deformation of each crystallite (the crystallite index is omitted), the following system of elastoviscoplastic equations is used

$$\begin{cases} \mathbf{k}^{cor} \equiv \dot{\mathbf{k}} + \mathbf{k} \cdot \boldsymbol{\omega} - \boldsymbol{\omega} \cdot \mathbf{k} = \mathbf{\Pi} : (\mathbf{l} - \boldsymbol{\omega} - \mathbf{z}^{in}), \\ \mathbf{l} = \mathbf{L} - \mathbf{Z}_{gb}^{in}, \\ \mathbf{z}^{in} = \sum_{i=1}^{K} \dot{\gamma}^{(i)} \mathbf{b}^{(i)} \otimes \mathbf{n}^{(i)}, \\ \dot{\gamma}^{(k)} = \dot{\gamma}_0 \left(\tau^{(k)} / \tau_c^{(k)} \right)^m H(\tau^{(k)} - \tau_c^{(k)}), k = 1, \ldots, K, \\ \tau^{(k)} = (\mathbf{b}^{(k)} \otimes \mathbf{n}^{(k)}) : \mathbf{k}, k = 1, \ldots, K, \\ \dot{\tau}_c^{(k)} = \left[\text{relations for } \dot{\tau}_c^{(k)} \right], k = 1, \ldots, K, \\ \boldsymbol{\omega} = [\text{relations for } \boldsymbol{\omega}], \mathbf{o} \cdot \mathbf{o}^T = \boldsymbol{\omega}. \end{cases} \tag{3}$$

In relations (3), summation over one and two contracted index is denoted by "·" and ":", the outer tensor product is denoted by "⊗", $\mathbf{k}^{cor} \equiv \dot{\mathbf{k}} + \mathbf{k} \cdot \boldsymbol{\omega} - \boldsymbol{\omega} \cdot \mathbf{k}$ is the rate of change (corotational derivative) of the Kirchhoff stress tensor, in which the principle of frame-indifference is satisfied, $\boldsymbol{\omega}$ is the spin tensor of the rigid movable coordinate system (RMCS) related to the crystallite lattice, $\mathbf{o}$ is the RMCS orientation tensor, $\mathbf{\Pi}$ is the fourth order tensor of crystallite elastic properties (its components are constant in the crystallite RMCS basis), $\mathbf{z}^{in}$ is the inelastic strain rate component at the mesolevel; $\gamma^{(k)}, \dot{\gamma}^{(k)}, \tau^{(k)}, \tau_c^{(k)}$ are the cumulative shear, shear rate, effective tangential and critical tangential stress of the slip system k, $\mathbf{b}^{(k)}, \mathbf{n}^{(k)}$ is the direction vector and normal to the plane of the slip system k, H(·) is the Heaviside function (hereinafter), and m is a dimensionless parameter. A doubled number of slip systems is used (only positive shears along them are considered). Note that in the constitutive equation, the transposed relative velocity gradient (l–ω) is used as a measure of the total strain rate minus the RMCS rotation [54]. Parameter $\dot{\gamma}_0$ is assumed to be related to the strain rate intensity $\dot{\gamma}_0 = l_u = \sqrt{(l-\omega):(l-\omega)^T}$ (a similar relationship of the nominal shear rate due to IDS with macroscopic strain rate modulus is introduced in [57]).

It was shown in [53,54] that for small elastic deformations typical of metals and alloys and in the absence of GBS, the mesolevel model (3), describing the elastoviscoplastic deformation, yields results close to the data obtained by other alternative mesolevel models, including those with the most commonly used formulation of constitutive relations in the unloaded configuration [28,55]. However, use of the rate formulation (3) in the current configuration is preferable for the model development in the case of an increase in the number of considered mechanisms (in our case, we take into account GBS), since it is possible to accept the strain rate component's additive hypothesis.

The RMCS, in which the crystallite properties are assumed to be constant, is coupled with the symmetry elements of the crystallites, which remain unchanged during the IDS; the definition of the spin in the absence of GBS is discussed in detail in [51,54]. It was shown in [58] that the use of a full constrained Taylor spin $\bar{\omega}_T = \frac{1}{2}(l-l^T) + \sum_{i=1}^{K} \frac{1}{2}\dot{\gamma}^i(\mathbf{n}^i \otimes \mathbf{b}^i - \mathbf{b}^i \otimes \mathbf{n}^i)$ gives similar results. The GBS effect on grain rotation is demonstrated, e.g., in [59]. When GBS is activated along the grain boundaries, it becomes possible to rotate it additionally due to the action of shear stresses along the boundaries; this effect is described by the component added to the specified spin [26]

$$\bar{\omega}_{gb} = \chi \dot{\gamma}_0 \mu_\tau H(\|\langle \mathbf{z}_{gb}^{in}\rangle_{grain}\|), \quad (4)$$

here χ is the model parameter characterizing the viscous boundary resistance to rotations, μ_τ is the moment due to the tangential forces acted on all grain boundary facets, $\langle \mathbf{z}_{gb}^{in}\rangle_{grain}$ is the mean strain rate due to the GBS ($\langle \rangle_{grain}$ denotes the averaging over the total number of boundary facets of the current grain), and the norm of the 2nd rank tensor $\mathbf{T}$ is defined as $\|\mathbf{T}\| = \sqrt{\mathbf{T}:\mathbf{T}^T}$.

An important characteristic of the mesoscale model is that it includes kinetic equations for critical shear stresses $\tau_c^{(k)}$ along the slip systems. Relations for describing the change $\tau_c^{(k)}$ have been proposed in many works, for instance, in [57,60,61]. The purpose of our research is to describe GBS and, as an example illustrating the application of this model, we consider the deformation in the SP regime. According to the above scenario, the grain structure in the SP regime can be considered as a recrystallized structure. In this case, the increase in $\tau_c^{(k)}$ due to the intragranular barriers can be neglected (in the general case, one can use the equations given in the articles already mentioned, or in [26], or in any other publications). On the other hand, since consideration is given here to a fine-grained structure, then it is reasonable that there occurs an increase in the resistance of dislocations to motion due to the boundary defects (orientation mismatch dislocations) and generation of near-boundary dislocation pile-ups during the misorientation between the neighboring grains [62,63].

In order to take into account the boundary effect on IDS, the shear rate relation used in direct crystal plasticity models is usually modified so that it can be applied to the near boundary region. In [64–66], the well-known relation [60] was modified to describe the evolution of the dislocation density over the slip systems, taking into account the pile-ups of dislocations near the boundary. In [67], for the near boundary region, the application of Arrhenius-type relation for IDS shear rates made it possible to analyze the additional activation energy proportional to the energy spent on orientation mismatch dislocations' generation; the shear rates in the near boundary area obtained in [68] turned out to be dependent on the mutual orientation of the slip systems and the shear stresses acting on them.

In accordance with the basic principles of constructing statistical models, the model proposed in this work effectively takes into account the rising of resistance to IDS due to orientation mismatch dislocations and dislocation piles-up near the boundary by increasing of the critical shear stresses $\tau_c^{(k)}$ for the grain (as a homogeneous element in a statistical sample). In the absence of GBS at the grain boundary, we apply the following approximate relation

$$\dot{\tau}_{c[withoutGBS]}^{(k)} = \psi(R_{(N)} - R_{(N)neib})H(R_{(N)} - R_{(N)neib}),$$
$$R_{(N)} = \sum_{i=1}^{K} \dot{\gamma}^{(i)} \mathbf{N} \cdot \mathbf{b}^{(i)} H(\mathbf{N} \cdot \mathbf{b}^{(i)}), \quad (5)$$
$$R_{(N)neib} = \sum_{i=1}^{K} \dot{\gamma}^{(i)}_{neib} \mathbf{N} \cdot \mathbf{b}^{(i)}_{neib} H(\mathbf{N} \cdot \mathbf{b}^{(i)}_{neib}).$$

where $\mathbf{N}$ is the (outer) external normal to the current grain boundary, $R_{(N)}$ is the estimate (of density) of the rate of dislocation inflow into the outer boundary from it, $R_{(N)neib}$ is the estimate (of density) of the rate of passage of dislocations through the boundary into the neighboring grain. The index "neib" denotes the neighboring grain and ψ is the model parameter (Pa).

After the specified GBS occurs on the current grain boundary, a linear combination of the characteristics of the neighboring grains on this boundary in (5) is used, which is similar to the determination of shear stresses for the intergranular boundary carried out by (9). According to (5), in the case of the incompatibility of IDS in the neighboring grains, part of the lattice dislocations forms orientation mismatch dislocations in the boundary, which is taken into account through an increase in the resistance to IDS (in sight, the model can be expanded through the explicit introduction of variables for the densities of various types of defects).

With account for the possible activity of GBS, (5) takes the form

$$\dot{\tau}_c^{(k)} = \begin{cases} \left[(R_{(N)} - R_{GBS(N)})/R_{(N)}\right]\dot{\tau}_{c[without\ GBS]}^{(k)}, & \text{if } R_{(N)} \geq R_{GBS(N)}, \\ -\alpha\left[(R_{GBS(N)} - R_{(N)})\right](\tau_c^{(k)} - \tau_{c0})H(\tau_c^{(k)} - \tau_{c0}), & \text{if } R_{(N)} < R_{GBS(N)}, \end{cases} \quad (6)$$

$$R_{GBS(N)} = \beta \sum_{i=1}^{4} (\dot{\gamma}^{(i)}_{gb})_{(N)},$$

where $R_{GBS(N)}$ is the estimate of the rate of increase in the grain boundary dislocation density, on which orientation mismatch dislocations are dissociated during GBS, β is the dimensionless parameter, $0 < \beta < 1$, $\sum_{i=1}^{4}(\dot{\gamma}^{(i)}_{gb})_{(N)}$ is the total value of the shear rates along the grain facet with the normal $\mathbf{N}$ against its size (four possible directions of positive shear rates are considered), α the dimensionless parameter, and τ_{c0} is the minimum possible critical stress for a grain with a low defect density.

The quantity $(\dot{\gamma}^{(i)}_{gb})_{(N)}$ is defined as $(\dot{\gamma}^{(i)}_{gb})_{(N)} = \frac{S^{(i)}_{gb}}{v} v_{gb0} \left(\tau^{(i)}_{gb}/\tau^{(i)}_{cgb}\right)^n H(\tau^{(i)}_{gb} - \tau^{(i)}_{cgb})$, where $S^{(i)}_{gb}$ is the area of the boundary facet, v is the grain volume, v_{gb0} is the characteristic rate of grain boundary dislocation displacement along the facet, $\tau^{(i)}_{gb}, \tau^{(i)}_{cgb}$ are the tangential and critical stresses for the contact boundaries, n is a dimensionless parameter; the relations for determining the above quantities are considered below. Use of (6) makes it possible to explicitly take into account the fact that, during the GBS, some part of orientation mismatch dislocations dissociates in grain boundary dislocations, which results in a weakening of the grain boundary hardening or causes its softening (depending on the ratio of IDS and GBS rates).

According to the above relations, GBS does not occur for a fine-grained material with a structure prepared for SP at a low homological temperature, and therefore a significant grain boundary hardening may happen due to IDS, which is mainly caused by the boundary effects. At a temperature required for SP, GBS will be active and, according to (6), the critical IDS stresses will decrease. Thus, the model is evidently capable of taking into account the effect of GBS on IDS.

During the GBS, relative displacements of crystallites (grains, subgrains) occur along their common boundary due to the motion of grain boundary dislocations. Note that, in the absence of an explicit dislocation description of (as in the mesoscale model, describing intragranular deformation), the averaged characteristics (shears) are used. To determine the kinematic characteristics of GBS for each grain boundary, a basis of vectors $\mathbf{n}_{gb}$, $\mathbf{b}_{gb}^{(1)}$ и $\mathbf{b}_{gb}^{(2)}$ is introduced into a three-dimensional space: $\mathbf{n}_{gb}$ is a normal to the facet plane, $\mathbf{b}_{gb}^{(1)}$ and $\mathbf{b}_{gb}^{(2)}$ are the mutually orthogonal unit vectors of displacement directions in the facet plane. The dyad of sliding direction and normal determines the GBS slip system.

When constructing the model under study, a transition from the local displacements along the contact boundaries to the deformation characteristics is carried out for the macrolevel representative volume. The movement of grain boundary dislocations in the slip plane is considered, and the viscoplastic relation is used to set their average displacement rate. Thus, we write [69]

$$\dot{\gamma}_{gb}^{(i)} = \frac{S_{gb}^{(i)}}{V} v_{gb0} \left(\tau_{gb}^{(i)} / \tau_{cgb}^{(i)} \right)^n H(\tau_{gb}^{(i)} - \tau_{cgb}^{(i)}) = \\ = \dot{\gamma}_{gb0}^{(i)} \left(\tau_{gb}^{(i)} / \tau_{cgb}^{(i)} \right)^n H(\tau_{gb}^{(i)} - \tau_{cgb}^{(i)}), \quad i = 1, 2, \ldots, 4K_{gb}, \sum, \quad (7)$$

where $v_{gb0} = b_{gb}^{(i)} \rho_{gb}^{(i)} v_0 \Delta l_{gb}$ is the characteristic displacement rate of a set of grain boundary dislocations along the entire boundary facet, $b_{gb}^{(i)}$ is the modulus of the analog of Burgers vector for GBS (for GBS slip system i), $\rho_{gb}^{(i)}$ is the density of grain boundary dislocations (dimensionality of 1/m), which is introduced per unit of the boundary facet area as a ratio of the general length of grain boundary dislocations in the boundary facet and the facet area, $v_0 \Delta l_{gb}$ is some characteristic velocity of grain boundary dislocations at the boundary facet segment), $\dot{\gamma}_{gb0}^{(i)} \stackrel{\text{ob}}{=} \frac{S_{gb}^{(i)}}{V} v_{gb0}$, $S_{gb}^{(i)}$ is the examined facet boundary area, V is the macrolevel representative volume size, n is the parameter, and $4K_{gb}$ is the general number of base displacement directions on all boundary facets in the representative volume at the macrolevel. Generally speaking, rigorous determination of v_{gb0} requires the introduction of another scale-structural level (microscale) in the model so that one can get an opportunity to explicitly consider the characteristics of the varying defect structure (change $\rho_{gb}^{(i)}$ with account for the interaction with lattice dislocations). Such an in-depth description may be useful for further development of the model. Analysis of the results obtained indicates that at this stage of research we shall confine ourselves to considering a three-level model, which makes it possible to achieve the objectives formulated. Therefore, in order to describe the characteristic displacement rate of grain-boundary dislocations moved along the boundary facet v_{gb0}, we have taken an approximate relationship between the strain rate intensity $l_u = \sqrt{(1-\omega):(1-\omega)^T}$ and the grain size r: $v_{gb0} = l_u r$ (similar relationship between the nominal shear rate due to IDS and the macroscopic strain rate modulus was introduced in [57]). In (7), use is made of the variables which characterize the tangential stresses $\tau_{gb}^{(i)}$ in the boundary and the critical stresses $\tau_{cgb}^{(i)}$ needed to activate GBS. The accounting of temperature and rate sensitivity is carried out using the kinetic relations for $\tau_{cgb}^{(i)}$.

The inelastic component of the relative velocity gradient due to GBS $\mathbf{Z}_{gb}^{in}$ is determined by the shear rates (7) as

$$\mathbf{Z}_{gb}^{in} = \sum_{i=1}^{4K_{gb}} \dot{\gamma}_{gb}^{(i)} \mathbf{b}_{gb}^{(i)} \otimes \mathbf{n}_{gb}^{(i)}, \quad (8)$$

where K_{gb} is the number of intergranular boundaries considered in the model, for which internal variables are specified, and the corresponding grains are assigned (during GBS, these neighboring grains change by others).

For illustration, we consider here a simplest test case, assuming that cubical grains and boundaries are identical. Suppose that, at the current time, deformation develops only at the expanse of GBS, i.e., due to shears along the boundaries indicated in blue in Figure 2. This situation may take place when the tangential stresses on the corresponding GBS slip systems are equal to the critical stresses, i.e., shearing occurs in the grain layers.

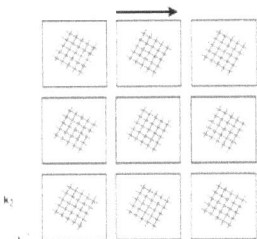

Figure 2. Test case scheme. In statistical sampling the upper grains are considered to be bordering the lower grains.

According to (7), the shear rate in a representative volume generated by the shear along one boundary is calculated as

$$\dot{\gamma}_{gb}^{(i)} = \frac{S_{gb}^{(i)}}{V} v_{gb0} \left(\tau_{gb}^{(i)}/\tau_{cgb}^{(i)}\right)^n H(\tau_{gb}^{(i)} - \tau_{cgb}^{(i)}) = \frac{S_{gb}^{(i)}}{V} l_u r \left(\tau_{gb}^{(i)}/\tau_{cgb}^{(i)}\right)^n H(\tau_{gb}^{(i)} - \tau_{cgb}^{(i)}), \quad i = 1, 2, \ldots, 4K_{gb}, \sum_i,$$

where r is the grain size. We get the boundary area $S_{gb}^{(i)} = r^2$, the grain volume $v = r^3$ and the size of a representative macrovolume $V = Nr^3$ and write

$$\dot{\gamma}_{gb}^{(i)} = \frac{r^2}{Nr^3} l_u r \left(\tau_{gb}^{(i)}/\tau_{cgb}^{(i)}\right)^n H(\tau_{gb}^{(i)} - \tau_{cgb}^{(i)}) = \frac{1}{N} l_u \left(\tau_{gb}^{(i)}/\tau_{cgb}^{(i)}\right)^n H(\tau_{gb}^{(i)} - \tau_{cgb}^{(i)}), \quad i = 1, 2, \ldots, 4K_{gb}, \sum_i.$$

In the case under consideration, a set of N GBS slip systems develops actively, and for each of these modes the shear indicated above will develop as well. Thus, $Z_{gb}^{in} = \sum_{i=1}^{4K_{gb}} \dot{\gamma}_{gb}^{(i)} b_{gb}^{(i)} \otimes n_{gb}^{(i)} = l_u \left(\tau_{gb}^{(i)}/\tau_{cgb}^{(i)}\right)^n k_1 \otimes k_2$ and, if we assume that for active boundaries $\tau_{gb}^{(i)} = \tau_{cgb}^{(i)}$, and the spin of RMCS for grains $\omega = 0$, then we have $Z_{gb}^{in} = l_u k_1 \otimes k_2$. This indicates that the described kinematic effect associated with a simple shear will develop at the expanse of GBS.

The tangential stresses for GBS slip systems at the intercrystallite boundary, which, in the context of the accepted model representation, is assumed to be associated with one of the adjacent grains (the neighboring grains are replaced by others in this case), are determined as

$$\tau_{gb}^{(i)} = (b_{gb}^{(i)} \otimes n_{gb}^{(i)}) : \sigma_{gb'}$$
$$\sigma_{gb} = \frac{\sigma}{2} + \frac{1}{2}((\gamma_{max} - \gamma_{gb})/\gamma_{max} \sigma_A + \gamma_{gb}/\gamma_{max} \sigma_B), \tag{9}$$

where σ_{gb} is the characteristic stress for the boundary under study (index is omitted), γ_{max} is the shear value at which the grains constituting the boundary are changed (loss of contact with the original neighboring grain A), σ, σ_A, σ_B are the stresses in the (base) grain coupled with the boundary, grain A and grain B which replaces grain A on the boundary, γ_{gb} is the shear which occurred till the current moment of time along all the four base directions in the boundary, $\gamma_{gb} = \sqrt{\left(\gamma_{gb}^{(1)} - \gamma_{gb}^{(3)}\right)^2 + \left(\gamma_{gb}^{(2)} - \gamma_{gb}^{(4)}\right)^2}$,

$\gamma_{gb}^{(i)}$ are the cumulative shears that occurred in the boundary till the current moment of time (in four different directions, $\mathbf{b}_{gb}^{(3)} = -\mathbf{b}_{gb}^{(1)}, \mathbf{b}_{gb}^{(4)} = -\mathbf{b}_{gb}^{(2)}$).

At $\gamma_{gb} = 0$, the characteristic stress is defined by the average stress for the initially adjacent base grain and grain A, and at $\gamma_{gb} = \gamma_{max}$ by averaging stresses for the base grain and grain B. In determining the cumulative shear, the shear rate direction along the boundary (vector $\sum_{i=1}^{4} \dot{\gamma}_{gb}^{(i)} \mathbf{b}_{gb}^{(i)}$) is prescribed, and it corresponds to the onset of the GBS along the facet under study; the spatial arrangement of crystallites is taken into account. The value γ_{max}, characterizing the moment at which the grains are changed for current contact boundary, can be obtained as follows

$$\gamma_{max} = l_{gb} S_{gb} / V \qquad (10)$$

where l_{gb}, S_{gb} denote the boundary size in the shear direction and the boundary area, and V is the representative macrovolume size. After the total shear reaches this limiting value, the shear $\gamma_{gb}^{(i)}$ for the intercrystallite boundary examined here is reset to 0. Formula (10) is simplified for the homogeneous grain structure: $\gamma_{max} = l_{gb} S_{gb} / V \approx v/V \approx 1/N$, where N is the number of grains in a representative volume.

The important element of the constitutive model is that it involves kinetic equations for the grain boundary critical shear stresses $\tau_{cgb}^{(i)}$, $i = 1, 2, \ldots, 4K_{gb}$ (to shorten the writing of formulas, the index of the number of the GBS slip system is omitted). In the general case, these relations should take into account an increase (decrease) in the grain structure resistance to GBS, the emergence of hard-phase particles to the boundaries during DRX and the mechanical smoothing of the boundaries; a version of these relationships is given in [26]. In this work, attention has been focused on the SP regime at the final stage of the SP deformation test, when the indicated factors can be neglected (it is assumed that the corresponding processes have been completed). Therefore, the GBS hardening is expressed as

$$\dot{\tau}_{cgb} = (F + G)H(\tau_{cgb} - \tau_{cgb}*) \qquad (11)$$

The constituent F characterizes a decrease in the critical stresses caused by an increase in the boundary energy due to the influx of intragranular dislocations. The term G characterizes the processes which are associated with grain-boundary diffusion and which lead to a decrease in critical stresses (diffusion-induced adjustment of the boundaries). The critical stresses cannot be smaller than a certain minimum level $\tau_{cgb}*$.

The rate of decrease in critical stresses caused by an increase in the boundary energy due to the influx of intragranular dislocations (this effect is considered, for instance, in [70,71]) is described by the following equation

$$F = -\sum_{j=1}^{2} \sum_{k=1}^{K_j} \mu_j g \dot{\gamma}^{(k)(j)} \left(\mathbf{N}^{(k)(j)} \cdot \mathbf{b}^{(k)(j)} \right) H\left(\mathbf{N}^{(k)(j)} \cdot \mathbf{b}^{(k)(j)} \right) + v \sum_{i=1}^{4} \dot{\gamma}_{gb}^{(i)} \qquad (12)$$

where j is the index denoting the neighboring crystallite adjacent to the boundary; as in (7), consideration is given to the initial neighboring grain A and grain B that replaces it, μ_j is the weight coefficient equal to $(\gamma_{max} - \gamma_{gb})/(2\gamma_{max})$ for grain A and $\gamma_{gb}/(2\gamma_{max})$ for grain B (thus, in (12), the same linear combination as in (9) is used); $\dot{\gamma}^{(k)(j)}$ is the shear rate along the slip system k in the crystallite j adjacent to the facet under study, K_j is the number of slip systems in the grain j, $\mathbf{b}^{(k)(j)}$ is the unit vector of the slip direction k in the crystallite j, $\mathbf{N}^{(k)(j)}$ is the outer normal to boundary facet plane for the crystallite j and g is the model parameter characterizing the relationship between the boundary energy and the critical shear stress of GBS. The last term describes a decrease in the boundary energy due to GBS, which causes the softening value to decrease, v is the model parameter characterizing the corresponding effect, four possible directions of positive shear rates are considered; the value of F cannot be positive.

The constituent for the rate of change in critical stresses caused by the processes associated with grain boundary diffusion is defined as [26]

$$G = -c \exp\left(-\frac{U_d}{\kappa\theta}\right)|\sigma_{bb}/\bar{\sigma}|^q \qquad (13)$$

where $\sigma_{bb} = \left|\mathbf{b}_{gb}^{(i)} \cdot \sigma_{gb} \cdot \mathbf{b}_{gb}^{(i)}\right|$ is the normal stress in the shear direction along the boundary $\mathbf{b}_{gb}^{(i)}$ (in the area orthogonal to $\mathbf{b}_{gb}^{(i)}$), calculated by analyzing the stresses σ_{gb} characteristic of the boundary under study (7), $\bar{\sigma}$, c are the model parameters, U_d is the activation energy for grain boundary diffusion, θ is the temperature, κ is the Boltzmann constant. Term (13) characterizes a decrease in the critical shear stress of GBS due to diffusion-induced boundary smoothing, facilitating the dissociation processes of the orientational mismatch dislocations into grain boundary dislocations and lattice dislocations, dislocation climb.

3. Results and Discussion

The model described in this paper was applied to simulate the uniaxial tensile tests in which a representative volume of polycrystalline aluminum alloy 1420 (Al-5.5%Mg-2.2%Li-0.12%Zr) was investigated.

Since in most experiments the rate of crosshead displacement is assumed to be constant [25], in our numerical experiment the decreasing strain rate corresponding to this case is set in the tension direction (along the OX_3-axis of the laboratory coordinate system), i.e., $D_{33} = \dot{H}_{33} = D_0/(1+D_0 t)$, where D_{33} is the strain rate tensor component, H_{33} is the logarithmic Hencky strain measure component, D_0 is the initial strain rate, and t is the time. The rest of the components of the (transposed) velocity gradient, which determines the kinematic effect used in the model, are found from the condition of uniaxial stress state [72]. The experiments were performed at the constant temperature of 523 K, which corresponds to a homological temperature of 0.56 (it is assumed that the equilibrium exchange rate with the environment having this temperature is significantly higher than the rate of possible dissipation-related temperature change during inelastic deformation). Deformation was considered only at the third stage of the SP deformation test [16] in the deformation interval from $H_{33} = 1.1$ to $H_{33} = 1.45$. Therefore, in modeling the final stage of this experiment, we set $D_0 = 3.33 \cdot 10^{-4} s^{-1}$ (for experiment with initial strain rate $0.001 s^{-1}$, these data were used for identification procedure) and $D_0 = 3.33 \cdot 10^{-3} s^{-1}$ (for experiment with initial strain rate $0.01 s^{-1}$, these data were used for verification). It is worth mentioning here that at this stage of research we intend to create the structure of a crystal plasticity model capable of describing GBS. For simplicity of the structure representation, the model does not take into account grain growth; only SP regime (changes in grain sizes are ignored) is modeled. Thus, a comparison with the experimental curve is performed at the third stage of the experiment, where grain growth is neglected in contrast to the early stages.

Within the framework of the proposed statistical model, the representative volume of the alloy 1420 was described through sampling from 343 f.c.c. crystallites, each of which had 26 neighboring boundary facets. The model parameters used in the numerical experiments performed were the same: constants for an observer in a rigid moving coordinate system associated with a lattice, elastic crystallite modulus $p_{1111} = 106.8$ GPa, $p_{1122} = 60.4$ GPa, $p_{1212} = 28.3$ GPa; viscoelastic law parameters for IDS m = 10; spin parameter $\chi = 1.5(MN \cdot m)^{-1}$; parameters of the hardening law due to the boundary effect $\psi = 1.7N$, $\beta = 1$, $\alpha = 0.1$; parameters of the submodel describing GBS n = 5, g = 40MPa, $c \exp\left(-\frac{U_d}{\kappa\theta}\right) = 0.03$ MPa/s, $\bar{\sigma} = 35$ MPa, $q = 3$.

The state of the material is determined by the values of the internal variables, which depend on the history of impacts. In our model, such variables are the critical shear stresses for IDS and GBS and the crystallite orientation tensors. In the numerical experiments, the deformation is considered only at the final stage of the SP deformation test. To take into account deformation at previous stages, we set different values for different strain rates: $\tau_c^{(k)} = 20$ MPa, $\tau_{cgb}^{(i)} = 35$ MPa (grain size $r = 2$ μm) for $D_0 = 3.33 \cdot 10^{-4} \text{s}^{-1}$ and $\tau_c^{(k)} = 32$ MPa, $\tau_{cgb}^{(i)} = 50$ MPa (grain size $r = 1.5$ μm) for $D_0 = 3.33 \cdot 10^{-3} \text{s}^{-1}$. In both cases, a blurred stretch texture was taken as the initial texture.

Figure 3 gives the dependences of the Cauchy stress tensor component Σ_{33} at the macroscale on the component H_{33} of the logarithmic strain measure and the corresponding experimental data [16] (the dependence of stresses on the linear strain measure component from [16] is transformed into the dependence on H_{33}) for two different initial strain rates.

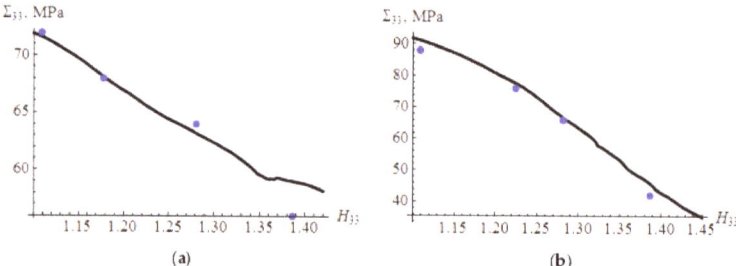

Figure 3. Dependences of stress intensity $\Sigma_u = \Sigma_{33}$ on H_{33} obtained through modeling and during the experimental tensile tests (dots) of 1420 alloy [16] for the initial strain rate: (**a**) $D_0 = 3.33 \cdot 10^{-4} \text{s}^{-1}$ (identification), (**b**) $D_0 = 3.33 \cdot 10^{-3} \text{s}^{-1}$ (verification).

It is seen that the simulation results are in a good agreement with the experimental results. Some deviations in stresses can be attributed to the fact that the grains are intensively replaced due to GBS (Figure 4).

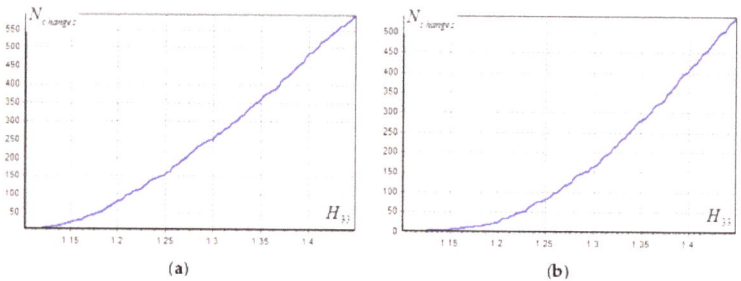

Figure 4. Number of contacting grain changes versus strain H_{33} for the initial strain rate: (**a**) $D_0 = 3.33 \cdot 10^{-4} \text{s}^{-1}$, (**b**) $D_0 = 3.33 \cdot 10^{-3} \text{s}^{-1}$.

It follows from Figure 4 that, as the deformation increases, the number of replaced neighboring grains increases in a monotone way. Figure 5 shows the inelastic strain rate values obtained due to GBS $(Z_{gb}^{in})_{33}$. These values correspond to the current strain rate D_{33} and are determined at equal intervals of the strain H_{33}.

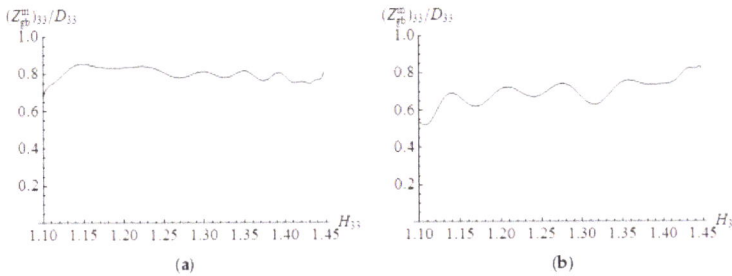

Figure 5. Inelastic strain rate values obtained at the expanse of the GBS $(Z_{gb}^{in})_{33}$, corresponding to the current strain rate D_{33} and determined at equal intervals of the strain H_{33} for the initial strain rate: (a) $D_0 = 3.33 \cdot 10^{-4} \text{s}^{-1}$, (b) $D_0 = 3.33 \cdot 10^{-3} \text{s}^{-1}$.

Note that the lines in Figure 5 are somewhat arbitrary: Z_{gb}^{in} changes at each step, the diagrams contain the values $(Z_{gb}^{in})_{33}$ averaged on some sequent steps made at equal intervals of the strain H_{33}. These diagrams characterize a contribution of GBS to the strain rate at the corresponding moments. This contribution turns out to be great, which provides evidence that GBS plays an essential role in superplastic deformation. At lower strain rates (the time taken for achieving the prescribed deformation increases), the shear rates vary gradually, and a slightly larger contribution of GBS takes place (Figures 4 and 5). The latter can be explained by the more significant effects of diffusion mechanisms, and the model developed is able to take this fact into account.

It is known that in the SP regime the GBS effect results in blurring the texture formed at the initial stage of the SP deformation test [73,74]. Figure 6 presents the pole figures constructed for the initial state ($H_{33} = 1.1$) and the state occurred after reaching the strain $H_{33} = 1.45$ (the figures illustrate the simulations at the initial strain rate $D_0 = 3.33 \cdot 10^{-4} \text{s}^{-1}$; the results obtained for the higher initial strain rate are in good qualitative agreement with the abovementioned data).

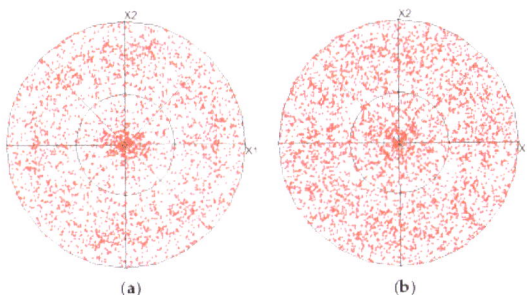

Figure 6. Pole figures for directions <111> (projection from the OX$_3$-axis): (a) for the initial moment ($H_{33} = 1.1$), (b) after tension at $H_{33} = 1.45$.

The results obtained demonstrate that the model can be used to implement the deformation scenario described above at the third stage in the SP regime with a gradual decrease in the flow stress (Figure 3). This regime is characterized by the predominance of the GBS mechanism (Figures 4 and 5); the equiaxial fine-grained structure is stable due to the simultaneous implementation of accommodative IDS and grain rotations, when the shape change of grains occurs at each moment of time, but, because of unceasing rotations, the resulting grain deformation is close to zero. GBS leads to blurring the texture formed after the first stage of testing (Figure 6).

At this stage of our investigation, the most important aim was to develop a general structure of the model, which should include all the key physical mechanisms of deformation and take into account their interactions. Within the framework of this structure, specific relationships can be modified on the basis of more detailed information from solid-state physics and through the use of the experimental data. The findings about the action and interaction of mechanisms, changes in the material structure, along with the data on the evolution of the stress–strain state at the macroscale level, are useful for parameter identification. Actually, a simple uniaxial loading at the macrolevel is a three-dimensional loading at the mesoscale, in which all the deformation modes are manifested themselves, and corresponding information makes it possible to use the results of simple tests to develop a model suitable for studying complex three-dimensional loads that can be observed in technological processes.

Note that the issue about the end of grain growth remains debatable. The scenario given above implies that the process of DRX is completed before the onset of the SP regime. This permits describing the SP regime (the final stage of the SP test) with the proposed constitutive model, which does not take into account grain growth. At the same time, the structure of the model, which involves the relations describing an increase in the initially (in the state before SP deformation testing) recrystallized grains due to the absorption of the material of nonrecrystallized grains, will allow one to consider other options, in particular, those including the continuation of the DRX until the end of the deformation process, to describe changes in the stress–strain state of the material during the entire SP test. The advanced model holds promise for simulating technological processes, in which the history of impacts (in particular, grain growth) may differ significantly in different parts of the products.

4. Conclusions

Even in the case of a simple uniaxial tensile test with access to the SP regime, a complex deformation process scenario takes place. According to this scenario, different physical mechanisms act and interact with each other, their significance changes and the material structure evolves greatly. Practically the same situation may arise in the technological processes which are based on the use of SP. This generates a need to develop mathematical models for describing changes in the material structure and the physical and mechanical properties of the material depending on its state. Recent studies show that a multilevel approach involving internal variables, physical theories of viscoplasticity, and an explicit description of the material structure and physical deformation mechanisms, holds the most promise. One of the main directions in this class model development is to expand the number of significant mechanisms to be taken into account.

In the investigation, we have modified the statistical crystal plasticity model for GBS describing. This model serves as the basis for an improved constitutive model that accounts for SP deformation and transitions to it. In contrast to classical Taylor-type statistical models, this model provides a consideration of the boundaries between crystallites as separate objects, but connected at each moment of time with boundaries (grain facets). For these objects, parameters characterizing their properties are introduced and kinetic equations are formulated. A set of grain boundaries forms a separate structural level (the level of GBS description), and the model is modified into a three-level one. The kinetic equations proposed for GBS include accounting for a decrease in the critical stresses due to an increase in the boundary energy caused by the influx of intragranular dislocations and due to the processes associated with grain-boundary diffusion (diffusion-induced adjustment of the boundaries). The model accounts for the fact that, due to GBS, some orientational boundary mismatch dislocations dissociate into grain-boundary dislocations, which leads, depending on the ratio of IDS and GBS rates, to the weakening of hardening or to softening for IDS. Besides, when GBS is active on the grain boundaries, the model includes an additional spin term due the action of shear stresses on the boundaries. The results of the numerical experiments are found to be in satisfactory agreement with the experimental data, the model reproduces the scenario of the deformation process at the final stage of the SP deformation uniaxial tensile tests.

The authors believe that the constitutive model proposed in this study and analysis of the numerical results has revealed some important and interesting aspects related to the development of crystal plasticity for modeling deformation with active GBS. On the basis of the proposed structure of a multilevel statistical constitutive model, an improved model is currently being developed by means of an explicit account of discontinuous DRX (grain growth) and its effect on other deformation mechanisms. The use of this advanced model will allow an accurate description of the SP regime, a transition to this regime and exit from it, which ensures correct modeling of technological processes (the history of impacts may differ significantly in different parts of the articles manufactured during these processes).

Author Contributions: Conceptualization, P.T. and A.S.; methodology, P.T. and A.S.; software, A.S. and E.S.; validation, A.S., P.T. and E.S.; formal analysis, A.S. and E.S.; investigation, A.S., P.T. and E.S.; data curation, A.S. and E.S.; writing—original draft preparation, A.S.; writing—review and editing, P.T. and A.S.; visualization, A.S. and E.S.; supervision, P.T.; funding acquisition, P.T. All authors have read and agreed to the published version of the manuscript.

Funding: The authors gratefully acknowledge the financial support from the Russian Science Foundation Grant no. 17-19-01292.

Conflicts of Interest: The authors declare no conflict of interest.

References

1. Sherby, O.D.; Wadsworth, J. Superplasticity and superplastic forming processes. *Mater. Sci. Technol.* **1985**, *1*, 925–936. [CrossRef]
2. Henshall, C.A.; Wadsworth, J.; Reynolds, M.J.; Barnes, A.J. Design and manufacture of a superplastic-formed aluminum-lithium component. *Mater. Des.* **1987**, *8*, 324–330. [CrossRef]
3. Baudelet, B. Industrial aspects of superplasticity. *Mater. Sci. Eng. A* **1991**, *137*, 41–55. [CrossRef]
4. Somani, M.C.; Sundaresan, R.; Kaibyshev, O.A.; Ermatchenko, A.G. Deformation processing in superplasticity regime-production of aircraft engine compressor discs out of titanium alloys. *Mater. Sci. Eng. A* **1998**, *243*, 134–139. [CrossRef]
5. Barnes, A.J. Industrial applications of superplastic forming: Trends and prospects. *Mater. Sci. Forum* **2001**, *357–359*, 3–16. [CrossRef]
6. Xing, H. Recent development in the mechanics of superplasticity and its applications. *J. Mater. Process. Technol.* **2004**, *151*, 196–202. [CrossRef]
7. Zeng, Z.P.; Zhang, Y.S.; Zhou, Y.; Jin, Q.L. Superplastic forming of aluminum alloy car body panels. *MSF* **2005**, *475–479*, 3025–3028. [CrossRef]
8. Barnes, A.J.; Raman, H.; Lowerson, A.; Edwards, D. Recent application of superformed 5083 aluminum alloy in the aerospace industry. *Mater. Sci. Forum* **2012**, *735*, 361–371. [CrossRef]
9. Wang, G.F.; Jia, H.H.; Gu, Y.B.; Liu, Q. Research on quick superplastic forming technology of industrial aluminum alloys for rail traffic. *Defect. Diffus. Forum* **2018**, *385*, 468–473. [CrossRef]
10. Bhatta, L.; Pesin, A.; Zhilyaev, A.; Tandon, P.; Kong, C.; Yu, H. Recent development of superplasticity in aluminum alloys: A review. *Metals* **2020**, *10*, 77. [CrossRef]
11. Dupuy, L.; Blandin, J.-J. Damage sensitivity in a commercial Al alloy processed by equal channel angular extrusion. *Acta Mater.* **2002**, *50*, 3253–3266. [CrossRef]
12. Kulas, M.-A.; Green, W.P.; Taleff, E.M.; Krajewski, P.E.; McNelley, T.R. Deformation mechanisms in superplastic AA5083 materials. *Metall. Mater. Trans. A* **2005**, *36*, 1249–1261. [CrossRef]
13. Liu, S.F.; Cai, Y.; Wu, S.L. Low temperature superplasticity of 5083 aluminum alloy. *Adv. Mater. Res.* **2014**, *941–944*, 116–119. [CrossRef]
14. Bochvar, A.A.; Sviderskaya, Z.A. The phenomenon of superplasticity in zinc-aluminum alloys. *Acad. Sci. Bull.* **1945**, *9*, 821–824. (In Russian)
15. Padmanabhan, K.A.; Basariya, M.R. A theory of steady state structural superplasticity in different classes of materials: A materials-agnostic analysis. *Mater. Sci. Eng. A* **2019**, *744*, 704–715. [CrossRef]
16. Berbon, P.B.; Tsenev, N.K.; Valiev, R.Z.; Furukawa, M.; Horita, Z.; Nemoto, M.; Langdon, T.G. Fabrication of bulk ultrafine-grained materials through intense plastic straining. *Metall. Mater. Trans. A* **1998**, *29*, 2237–2243. [CrossRef]
17. Furukawa, M.; Utsunomiya, A.; Matsubara, K.; Horita, Z.; Langdon, T.G. Influence of magnesium on grain refinement and ductility in a dilute Al–Sc alloy. *Acta Mater.* **2001**, *49*, 3829–3838. [CrossRef]

18. Kaibyshev, R.; Goloborodko, A.; Musin, F.; Nikulin, I.; Sakai, T. The role of grain boundary sliding in microstructural evolution during superplastic deformation of a 7055 aluminum alloy. *Mater. Trans.* **2002**, *43*, 2408–2414. [CrossRef]
19. Sakai, G.; Horita, Z.; Langdon, T.G. Grain refinement and superplasticity in an aluminum alloy processed by high-pressure torsion. *Mater. Sci. Eng. A* **2005**, *393*, 344–351. [CrossRef]
20. Mazilkin, A.; Myshlyaev, M. Microstructure and thermal stability of superplastic aluminium—Lithium alloy after severe plastic deformation. *J. Mater. Sci.* **2006**, *41*, 3767–3772. [CrossRef]
21. Watanabe, H.; Mukai, T.; Higashi, K. Superplasticity in a ZK60 magnesium alloy at low temperatures. *Scr. Mater.* **1999**, *40*, 477–484. [CrossRef]
22. Figueiredo, R.B.; Langdon, T.G. Achieving superplastic properties in a ZK10 magnesium alloy processed by equal-channel angular pressing. *J. Mater. Res. Technol.* **2017**, *6*, 129–135. [CrossRef]
23. Sergueeva, A.V.; Stolyarov, V.V.; Valiev, R.Z.; Mukherjee, A.K. Superplastic behaviour of ultrafine-grained Ti–6A1–4V alloys. *Mater. Sci. Eng. A* **2002**, *323*, 318–325. [CrossRef]
24. Ko, Y.G.; Lee, C.S.; Shin, D.H.; Semiatin, S.L. Low-temperature superplasticity of ultra-fine-grained Ti-6Al-4V processed by equal-channel angular pressing. *Metall. Mater. Trans. A* **2006**, *37*, 381. [CrossRef]
25. Sharifullina, E.R.; Shveykin, A.I.; Trusov, P.V. Review of experimental studies on structural superplasticity: Internal structure evolution of material and deformation mechanisms. *PNRPU Mech. Bull.* **2018**, 103–127. [CrossRef]
26. Trusov, P.V.; Sharifullina, E.R.; Shveykin, A.I. Multilevel model for the description of plastic and superplastic deformation of polycrystalline materials. *Phys. Mesomech.* **2019**, *22*, 402–419. [CrossRef]
27. McDowell, D.L. A perspective on trends in multiscale plasticity. *Int. J. Plast.* **2010**, *9*, 1280–1309. [CrossRef]
28. Roters, F.; Eisenlohr, P.; Hantcherli, L.; Tjahjanto, D.D.; Bieler, T.R.; Raabe, D. Overview of constitutive laws, kinematics, homogenization and multiscale methods in crystal plasticity finite-element modeling: Theory, experiments, applications. *Acta Mater.* **2010**, *58*, 1152–1211. [CrossRef]
29. Diehl, M. Review and outlook: Mechanical, thermodynamic, and kinetic continuum modeling of metallic materials at the grain scale. *MRS Commun.* **2017**, *7*, 735–746. [CrossRef]
30. Beyerlein, I.J.; Knezevic, M. Review of microstructure and micromechanism-based constitutive modeling of polycrystals with a low-symmetry crystal structure. *J. Mater. Res.* **2018**, *33*, 3711–3738. [CrossRef]
31. Knezevic, M.; Beyerlein, I.J. Multiscale modeling of microstructure-property relationships of polycrystalline metals during thermo-mechanical deformation. *Adv. Eng. Mater.* **2018**, *20*, 1700956. [CrossRef]
32. Trusov, P.V.; Shveykin, A.I. *Multilevel Models of Mono- and Polycrystalline Materials: Theory, Algorithms, Application Examples*; SB RAS: Novosibirsk, Russia, 2019. (In Russian)
33. Romanova, V.A.; Balokhonov, R.R.; Schmauder, S. Numerical study of mesoscale surface roughening in aluminum polycrystals under tension. *Mater. Sci. Eng. A* **2013**, *564*, 255–263. [CrossRef]
34. Ardeljan, M.; Beyerlein, I.J.; McWilliams, B.A.; Knezevic, M. Strain rate and temperature sensitive multi-level crystal plasticity model for large plastic deformation behavior: Application to AZ31 magnesium alloy. *Int. J. Plast.* **2016**, *83*, 90–109. [CrossRef]
35. Yaghoobi, M.; Ganesan, S.; Sundar, S.; Lakshmanan, A.; Rudraraju, S.; Allison, J.E.; Sundararaghavan, V. PRISMS-Plasticity: An open-source crystal plasticity finite element software. *Comput. Mater. Sci.* **2019**, *169*, 109078. [CrossRef]
36. Yaghoobi, M.; Allison, J.E.; Sundararaghavan, V. Multiscale modeling of twinning and detwinning behavior of HCP polycrystals. *Int. J. Plast.* **2020**, *127*, 102653. [CrossRef]
37. Doquet, V.; Barkia, B. Combined AFM, SEM and crystal plasticity analysis of grain boundary sliding in titanium at room temperature. *Mech. Mater.* **2016**, *103*, 18–27. [CrossRef]
38. Wei, Y.J.; Anand, L. Grain-boundary sliding and separation in polycrystalline metals: Application to nanocrystalline fcc metals. *J. Mech. Phys. Solids* **2004**, *52*, 2587–2616. [CrossRef]
39. Wei, Y.; Bower, A.F.; Gao, H. Enhanced strain-rate sensitivity in fcc nanocrystals due to grain-boundary diffusion and sliding. *Acta Mater.* **2008**, *56*, 1741–1752. [CrossRef]
40. Cheng, T.-L.; Wen, Y.-H.; Hawk, J.A. Diffuse interface approach to modeling crystal plasticity with accommodation of grain boundary sliding. *Int. J. Plast.* **2019**, *114*, 106–125. [CrossRef]
41. Mellbin, Y.; Hallberg, H.; Ristinmaa, M. Recrystallization and texture evolution during hot rolling of copper, studied by a multiscale model combining crystal plasticity and vertex models. *Model. Simul. Mater. Sci. Eng.* **2016**, *24*, 075004. [CrossRef]

42. Zhao, P.; Wang, Y.; Niezgoda, S.R. Microstructural and micromechanical evolution during dynamic recrystallization. *Int. J. Plast.* **2018**, *100*, 52–68. [CrossRef]
43. Ruiz Sarrazola, D.A.; Pino Muñoz, D.; Bernacki, M. A new numerical framework for the full field modeling of dynamic recrystallization in a CPFEM context. *Comput. Mater. Sci.* **2020**, *179*, 109645. [CrossRef]
44. Zhou, G.; Li, Z.; Li, D.; Peng, Y.; Zurob, H.S.; Wu, P. A polycrystal plasticity based discontinuous dynamic recrystallization simulation method and its application to copper. *Int. J. Plast.* **2017**, *91*, 48–76. [CrossRef]
45. Tang, T.; Zhou, G.; Li, Z.; Li, D.; Peng, L.; Peng, Y.; Wu, P.; Wang, H.; Lee, M.-G. A polycrystal plasticity based thermo-mechanical-dynamic recrystallization coupled modeling method and its application to light weight alloys. *Int. J. Plast.* **2019**, *116*, 159–191. [CrossRef]
46. Zecevic, M.; Knezevic, M.; McWilliams, B.; Lebensohn, R.A. Modeling of the thermo-mechanical response and texture evolution of WE43 Mg alloy in the dynamic recrystallization regime using a viscoplastic self-consistent formulation. *Int. J. Plast.* **2020**, *130*, 102705. [CrossRef]
47. Rezaei, S.; Rezaei Mianroodi, J.; Khaledi, K.; Reese, S. A nonlocal method for modeling interfaces: Numerical simulation of decohesion and sliding at grain boundaries. *Comput. Methods Appl. Mech. Eng.* **2020**, *362*, 112836. [CrossRef]
48. Shveykin, A.I.; Sharifullina, E.R. Development of multilevel models based on crystal plasticity: Description of grain boundary sliding and evolution of grain structure. *NST* **2015**, *6*. [CrossRef]
49. Zhao, Y.; Toth, L.; Massion, R.; Skrotzki, W. Role of grain boundary sliding in texture evolution for nanoplasticity. *Adv. Eng. Mater.* **2017**, *20*. [CrossRef]
50. Toth, L.S.; Skrotzki, W.; Zhao, Y.; Pukenas, A.; Braun, C.; Birringer, R. Revealing grain boundary sliding from textures of a deformed nanocrystalline Pd–Au alloy. *Materials* **2018**, *11*, 190. [CrossRef]
51. Trusov, P.V.; Shveykin, A.I.; Yanz, A.Y. Motion decomposition, frame-indifferent derivatives, and constitutive relations at large displacement gradients from the viewpoint of multilevel modeling. *Phys. Mesomech.* **2017**, *20*, 357–376. [CrossRef]
52. Trusov, P.V.; Shveykin, A.I. On motion decomposition and constitutive relations in geometrically nonlinear elastoviscoplasticity of crystallites. *Phys. Mesomech.* **2017**, *20*, 377–391. [CrossRef]
53. Shveikin, A.I.; Trusov, P.V. Correlation between geometrically nonlinear elastoviscoplastic constitutive relations formulated in terms of the actual and unloaded configurations for crystallites. *Phys. Mesomech.* **2018**, *21*, 193–202. [CrossRef]
54. Trusov, P.V.; Shveykin, A.I.; Kondratev, N.S. Multilevel metal models: Formulation for large displacement gradients. *NST* **2017**, *8*. [CrossRef]
55. Anand, L. Single-crystal elasto-viscoplasticity: Application to texture evolution in polycrystalline metals at large strains. *Comput. Methods Appl. Mech. Eng.* **2004**, *193*, 5359–5383. [CrossRef]
56. Psakhie, S.G.; Horie, Y.; Ostermeyer, G.P.; Korostelev, S.Y.; Smolin, A.Y.; Shilko, E.V.; Dmitriev, A.I.; Blatnik, S.; Špegel, M.; Zavšek, S. Movable cellular automata method for simulating materials with mesostructure. *Theor. Appl. Fract. Mech.* **2001**, *37*, 311–334. [CrossRef]
57. Beyerlein, I.J.; Tomé, C.N. A dislocation-based constitutive law for pure Zr including temperature effects. *Int. J. Plast.* **2008**, *24*, 867–895. [CrossRef]
58. Shveikin, A.I.; Trusov, P.V. Multilevel models of polycrystalline metals: Comparison of relations describing the crystallite lattice rotations. *NST* **2019**, *10*. [CrossRef]
59. Watanabe, H.; Kurimoto, K.; Uesugi, T.; Takigawa, Y.; Higashi, K. Accommodation mechanisms for grain boundary sliding as inferred from texture evolution during superplastic deformation. *Philos. Mag.* **2013**, *93*, 2913–2931. [CrossRef]
60. Kocks, F.; Mecking, H. Physics and phenomenology of strain hardening: The FCC case. *Prog. Mater. Sci.* **2003**, *48*, 171–273. [CrossRef]
61. Gérard, C.; Cailletaud, G.; Bacroix, B. Modeling of latent hardening produced by complex loading paths in FCC alloys. *Int. J. Plast.* **2013**, *42*, 194–212. [CrossRef]
62. Bieler, T.R.; Eisenlohr, P.; Zhang, C.; Phukan, H.J.; Crimp, M.A. Grain boundaries and interfaces in slip transfer. *Curr. Opin. Solid State Mater. Sci.* **2014**, *18*, 212–226. [CrossRef]
63. Kalidindi, S.R.; Vachhani, S.J. Mechanical characterization of grain boundaries using nanoindentation. *Curr. Opin. Solid State Mater. Sci.* **2014**, *18*, 196–204. [CrossRef]

64. Lefebvre, S. Etude Expérimentale et Simulation Numérique du Comportement Mécanique de Structures Sub-Micrométriques de Cuivre: Application aux Interconnexions dans les Circuits Intégrés. Ph.D. Thesis, Ecole centrale de Paris, Châtenay-Malabry, France, 2006.
65. Haouala, S.; Segurado, J.; LLorca, J. An analysis of the influence of grain size on the strength of FCC polycrystals by means of computational homogenization. *Acta Mater.* **2018**, *148*, 72–85. [CrossRef]
66. Rubio, R.A.; Haouala, S.; LLorca, J. Grain boundary strengthening of FCC polycrystals. *J. Mater. Res.* **2019**, *34*, 2263–2274. [CrossRef]
67. Ma, A.; Roters, F.; Raabe, D. A dislocation density based constitutive law for BCC materials in crystal plasticity FEM. *Comput. Mater. Sci.* **2007**, *39*, 91–95. [CrossRef]
68. Mayeur, J.R.; Beyerlein, I.J.; Bronkhorst, C.A.; Mourad, H.M.; Hansen, B.L. A crystal plasticity study of heterophase interface character stability of Cu/Nb bicrystals. *Int. J. Plast.* **2013**, *48*, 72–91. [CrossRef]
69. Sharifullina, E.; Shveykin, A.; Trusov, P. Multilevel model of polycrystalline materials: Grain boundary sliding description. *IOP Conf. Ser. Mater. Sci. Eng.* **2017**, *286*, 012026. [CrossRef]
70. Pshenichnyuk, A.I.; Astanin, V.V.; Kaibyshev, O.A. The model of grain-boundary sliding stimulated by intragranular slip. *Philos. Mag. A* **1998**, *77*, 1093–1106. [CrossRef]
71. Lapera, M.; Spader, D.; Ghonem, H. A coupled, physics-based matrix-grain boundary model for creep of carbide strengthened nickel-based superalloys—I. Concepts and formulation. *Mater. Sci. Eng. A* **2020**, *769*, 138421. [CrossRef]
72. Trusov, P.V.; Shveykin, A.I.; Nechaeva, E.S.; Volegov, P.S. Multilevel models of inelastic deformation of materials and their application for description of internal structure evolution. *Phys. Mesomech.* **2012**, *15*, 155–175. [CrossRef]
73. Bricknell, R.H.; Edington, J.W. Textures in a superplastic Al-6Cu-0.3Zr alloy. *Acta Metall.* **1979**, *27*, 1303–1311. [CrossRef]
74. Pérez-Prado, M.T.; González-Doncel, G. Texture Changes During Deformation of a 7475 Superplastic Aluminum Sheet Alloy. *Textures Microstruct.* **2000**, *34*, 33–42. [CrossRef]

© 2020 by the authors. Licensee MDPI, Basel, Switzerland. This article is an open access article distributed under the terms and conditions of the Creative Commons Attribution (CC BY) license (http://creativecommons.org/licenses/by/4.0/).

Article

Accurate Estimation of Brittle Fracture Toughness Deterioration in Steel Structures Subjected to Large Complicated Prestrains

Hiroaki Kosuge [1,*], Tomoya Kawabata [1], Taira Okita [1] and Hidenori Nako [2]

1. Department of Systems Innovation, Graduate School of Engineering, The University of Tokyo, 7-3-1, Hongo, Bunkyo, Tokyo 113-8656, Japan; kawabata@fract.t.u-tokyo.ac.jp (T.K.); okita@race.t.u-tokyo.ac.jp (T.O.)
2. Kobe Steel Ltd., 1-5-5, Takatsukadai, Nishi, Kobe, Hyogo 651-2271, Japan; nako.hidenori@kobelco.com
* Correspondence: kosuge@fract.t.u-tokyo.ac.jp; Tel.: +81-3-5841-6509

Received: 26 August 2020; Accepted: 23 September 2020; Published: 25 September 2020

Abstract: Studies have suggested that brittle fractures occur in steel because microcracks in the brittle layer at grain boundaries propagate as a result of the increase in piled-up dislocations. Therefore, prestraining can approach the limits of a material, which could lead to a decrease in fracture toughness. However, strains are tensors comprising multiple components, so the effect of prestrain on fracture toughness is not simple. Additionally, the mechanism of change in critical stress due to prestrain has not been thoroughly investigated. For the lifetime evaluation of steel structures with a complicated load history, it is important to generalize the effect of complicated prestrain on the decrease in fracture toughness. In this paper, a single prestrain was applied in a direction different from the crack opening direction. A general three-point bending test was employed for fracture evaluation. Numerical analyses using the strain gradient plasticity (SGP) theory, which is a method based on the finite element method (FEM) are carried out; conventional macroscopic material damage rules are considered as well. Using these FEM analyses, the critical stress is calculated. Finally, the change in critical stress can be expressed by the yield point increase and dislocation density and formulated based on the identified micromechanisms.

Keywords: brittle fracture; prestrain direction; strain gradient plasticity; critical stress; dislocation density

1. Introduction

Iron is the most abundant element on earth and is used in many applications, including high-rise buildings, large structures, infrastructure (e.g., pipelines for transporting natural gas and storage tanks for natural gas), and ships. As the size of these structures continues to increase, the risk of brittle fractures steadily increases. However, brittle fractures are intrinsically inevitable in carbon steel material due to its lattice structure. To ensure the safety of steel structures by preventing brittle fractures, a deeper understanding of brittle fractures is required.

To begin this study, the historical aspects of the mechanisms of brittle fracture in steel are introduced. Brittle fractures are considered to originate from a microcrack in a heterogeneous microstructure. The most famous governing formula that combines microcracks and fractures is Griffith's equation, which is shown in Equation (1) [1]:

$$\sigma_f = \sqrt{\frac{2E\gamma}{\pi c}} \tag{1}$$

where σ_f is the critical stress, E is the Young's modulus, γ is the surface energy per unit area, and c is the crack length. Equation (1) is derived based on the second law of thermodynamics and Inglis's solution [2] and indicates that a cleavage fracture occurs due to crack propagation when the stress exceeds σ_f. However, this critical stress observed in practice is much lower than the theoretical fracture strength. Therefore, different influential mechanisms on the actual fracture in steel have been investigated. Stroh [3] suggested that piled-up dislocations from plastic deformation affected the fracture. Later, this theory was applied to Smith's [4] and Petch's equation [5,6]. Smith [4] suggested that the critical stress is greatly affected by the thickness of carbide particles at the grain boundaries and formulated the critical condition as follows:

$$\sigma_f = \sqrt{\frac{4E\gamma_p}{\pi t} - \left\{\tau_{eff}\sqrt{\frac{d}{t}} + \frac{4\tau_i}{\pi}\right\}^2} \tag{2}$$

where γ_p is the plastic work required to create a unit area of fracture surface, τ_{eff} is the effective shear stress, d is the grain diameter, t is the thickness of the carbide layer, and τ_i is the frictional stress. This equation is based on experimental results, which showed that brittle cracks always initiated by the occurrence of microcracks in the brittle layer, such as carbide particles located in grain boundaries, as shown in Figure 1.

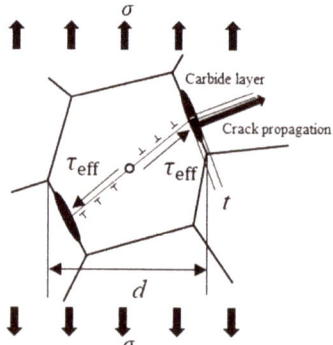

Figure 1. Brittle fracture micromechanism.

As shown in Figure 1, the dislocation motion is generated by plastic deformation, and the dislocations pile up to the carbide layer as a result of shear stress when tensile stress is applied at a distance. The piled-up dislocations create a stress concentration at the carbide layer, which induces the nucleation of a microcrack. The mechanism of brittle fracture is the nucleation and propagation of the brittle crack, which initiates in the carbide layer.

Petch [5,6] improved this concept using major microstructural parameters, yield stress and grain diameter and proposed a practical formula that is dominated by the nucleation of microcracks. On the basis of dislocation theory, Cottrell [7] suggested that the critical conditions for cleavage fracture are determined not only by the nucleation of microcracks but also by the growth of the microcracks.

Using these models, a number of application studies of critical stress have been carried out over many years. Based on the knowledge that brittle fractures occur according to the weakest link model [8], the Weibull stress model was proposed by Beremin [9]. This theory states that the critical condition can be accurately estimated by using the probability of microcrack existence, which is determined with Griffith's equation. A number of studies have been carried out based on the Beremin model. Bordet [10,11] postulated that the nucleation and growth of microcracks can be treated independently. In recent studies, Lei [12,13] suggested that an improvement in accuracy can be made by reconsidering the Weibull stress model based on the experience that brittle cracks never occur without yielding. However, there are some mandatory parameters that are difficult to determine in this improved model, which makes it difficult to apply this model in actual engineering applications.

In addition to these approaches purely regarding the critical stress condition of the material itself, a number of studies of critical stress after prestraining have been investigated. Based on Smith's model, plastic deformation leads to an increase in piled-up dislocations and directly results in an increase in the risk of brittle fracture. Actually, microcrack nucleation induced by plastic deformation was recognized in experimental observations [14,15]. Therefore, subjecting a material to plastic prestrain means getting close to the limit of the material and leading to a decrease in fracture toughness. Some previous works have focused on the effect of prestrain on fracture toughness. Sukedai et al. [16] showed that the fracture toughness of Charpy impact test samples decreased when the samples were subjected to a tensile prestrain. Miki et al. [17] showed that critical crack tip opening displacement (CTOD) decreased when a sample was subjected to either tensile or compressive prestrains. From their experimental results, they verified that a single prestrain led to a decrease in fracture toughness. However, strain is a tensor comprising multiple components, which makes it difficult to understand the effects of prestrain. The generalized effects of prestrain, including cyclic or directional prestrain, have not been quantitatively elaborated. Although many structural members are subjected to complicated strain cycles, such as simple uniaxial tensile-compressive strain cycles and multidirectional strain cycles, where the crack opening direction is different, a sufficient mechanism of toughness deterioration has not been discovered. Recently, Kosuge et al. [18] revealed the effects of random tensile-compressive uniaxial prestrain on the ductile–brittle transition temperature (DBTT) through experiments and finite element method (FEM) analyses, which included the effects of piled-up dislocations. They showed that the relationship between the number of piled-up dislocations and the amount of shift in the DBTT of a fracture toughness test sample has a one-to-one correspondence. Hence, it is clear that, regarding toughness degradation under plastic prestrains, the number of piled-up dislocations is the main factor that controls brittle fracture performance. However, this relationship is effective only for one fracture toughness specimen configuration, so generalization to actual steel structures is necessary.

Different practical studies combining experiments and numerical analyses reported that the critical stress increased [19], decreased [20], or remained constant [21] after prestraining. Therefore, it is reasonable to assume that the critical stress is affected by prestrain in some way, but the mechanism by which the critical stress is affected under different prestraining conditions has not been clarified because sufficient knowledge has not been obtained. Critical stress is very important for determining the occurrence of brittle fractures because the stress distribution at the crack tip can be easily calculated with commercial FEM software. To predict and prevent catastrophic brittle fractures, the effect of prestrain on the change in critical stress must be thoroughly understood because most structures are subjected to complicated strain histories (e.g., earthquakes for on-land structures and heavy wave impacts for offshore structures).

In this paper, based on the abovementioned background, the change in critical stress under various prestraining conditions was investigated by combining experiments and numerical analyses. The effect of the prestrain direction was investigated by applying the prestrain in a different direction than the crack opening direction. The critical stress was determined using FEM analysis, and the results were compared according to the prestrain direction. Additionally, FEM analysis based on crystal plasticity was conducted to examine the mechanism in more detail and to formulate the generalized effect of prestrain on critical stress.

2. Experiment

In this section, the methods used in this study for prestraining and fracture testing are presented and the results from these tests are explained. The effect of the prestrain direction was evaluated by implementing four different prestrain directions.

2.1. Preparation of Testing and Prestraining

In this research, a thermomechanical control process (TMCP) steel plate [22] with a thickness of 32 mm was used. TMCP steel is applied in various fields, such as natural gas pipelines [23], offshore structures [24], tanks [25], ships [26], buildings [27], and bridges [28]. The yield strength and tensile strength of the TMCP steel in this study are about 500 MPa and 600 MPa, respectively. The test specimens subjected to prestraining were machined from the mid-thickness position so that the longitudinal direction was vertical to the rolling direction, as shown in Figure 2a. The selected amounts of prestrain were 1%, 3%, and 7%, which were measured as the change in the gage length distance. To investigate the effect of the prestrain on the yield stress or uniform elongation, a tensile test was carried out using round bar test specimens after applying the prestrain depicted in Figure 2b. Additionally, the main mode of evaluation in this study, the fracture toughness test, was performed using three-point bending specimens, as shown in Figure 2c. There were no fatigue precracks at the tip of the notch, and only a sharp notch shape remained. Regarding both specimens, the important parameter, θ, is defined as the angle between the prestrain direction and the longitudinal direction of the specimen, and θ was changed to 0°, 45°, 60° and 90°.

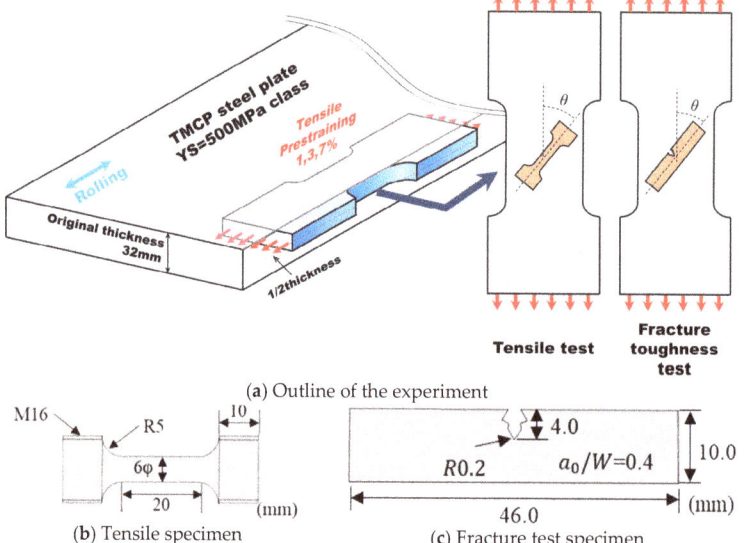

Figure 2. Schematic illustration of the experiment and configuration of the tensile and fracture test specimens. TMCP: thermomechanical control process. (**a**) Outline of the experiment, (**b**) Tensile specimen, (**c**) Fracture test specimen.

2.2. Tensile Tests

True stress-true strain curves calculated from the experiment are shown in Figure 3. After applying the prestrain, the curves shifted to the right according to the amount of prestrain. As shown in Figure 3, strain hardening by prestraining was observed at all angles, wherein the yield stress increased and uniform elongation decreased. Additionally, the shape of the curve did not change significantly, indicating that the aging had little effect on the samples.

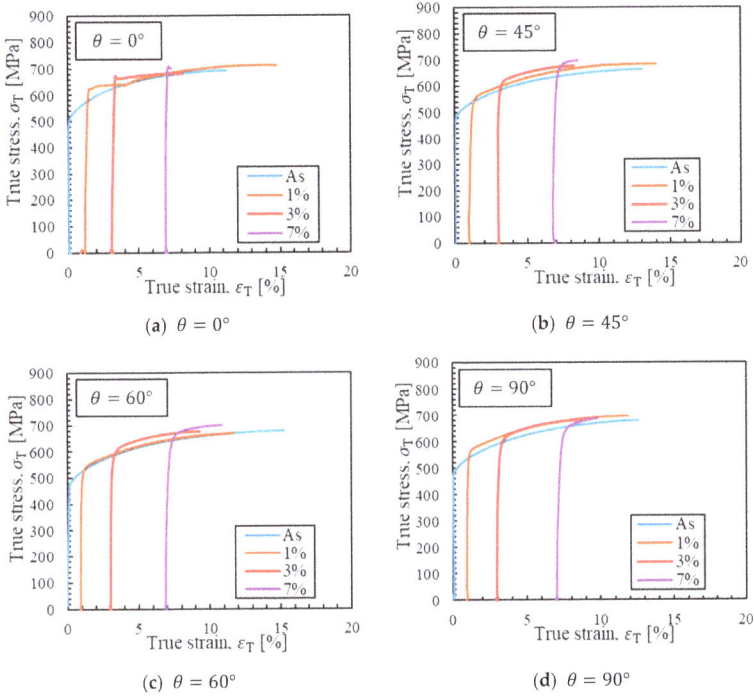

Figure 3. True stress-true strain curves before/after prestraining. (a) $\theta = 0°$, (b) $\theta = 45°$, (c) $\theta = 60°$, (d) $\theta = 90°$.

2.3. Fracture Test

Three specimens were prepared for each prestrain condition, and a general three-point bending test was carried out at −160 °C. To observe the effect of prestrain on fracture toughness, fracture test specimens with no prestrain were also prepared. Additionally, the TMCP steel plate generally has strong anisotropy due to the growth of texture created during the unrecrystallization temperature range, so the fracture toughness is considered to be different in each direction. Thus, specimens without prestrain were prepared in four directions ($\theta = 0°$, 45°, 60° and 90°, and especially $\theta = 0°$ is vertical and $\theta = 90°$ is parallel to rolling direction). As mentioned in Section 2.1, this specimen does not have fatigue cracks, so this test cannot be regarded as a CTOD test, which is specified in the International Organization for Standardization (ISO) standard [29]. Therefore, the CTOD obtained from this experiment is denoted as "quasi-CTOD" hereinafter and is distinguished from the standardized CTOD. However, quasi-CTOD was calculated by Equations (3)–(9) [30,31].

$$\delta_{quasi.cr} = \frac{K^2}{m\sigma_y E} + f_p \frac{r_p(W-a_0)}{r_p(W-a_0)+a_0+z} V_p \quad (3)$$

$$m = 4.9 - 3.5 YR \quad (4)$$

$$r_p = 0.43 \quad (5)$$

$$f_p = F(B) \cdot f_{p@B=25\ mm}(YR) \quad (6)$$

$$f\left(\frac{a_0}{W}\right) = \frac{3\left(\frac{a_0}{W}\right)^{\frac{1}{2}}\left[1.99 - \left(\frac{a_0}{W}\right)\left(1-\frac{a_0}{W}\right)\left(2.15 - 3.93\frac{a_0}{W} + 2.7\left(\frac{a_0}{W}\right)^2\right)\right]}{2\left(1+2\frac{a_0}{W}\right)\left(1-\frac{a_0}{W}\right)^{\frac{3}{2}}} \quad (7)$$

$$F(B) = 0.8 + 0.2\exp\{-0.019(B-25)\} \quad (8)$$

$$f_{p@B=25\,mm}(YR) = -1.4(YR)^2 + 2.8(YR) - 0.35 \quad (9)$$

where m is the constraint factor, K is the stress intensity factor, f_p is the plasticity correction factor, a_0 is the initial crack length, W is the width, B is the thickness, σ_y is the yield stress, YR is the yield ratio, V_p is the plastic displacement of the clip gage, and $\delta_{quasi.cr}$ is the "quasi-CTOD". In this test, material constants were given as shown in Table 1.

Table 1. Value of each parameter in the equations used for crack tip opening displacement (CTOD) calculations.

Parameter	Value	Units
σ_y	500	MPa
E	206	GPa
YR	0.806	-
B	10	mm
W	10	mm
a_0	4	mm
S	80	mm
z	−1.5	mm

According to the experimental data, $\delta_{quasi.cr}$ was calculated using these equations and was compared for each prestrain condition, as shown in Figure 4. The vertical axis of Figure 4 shows the nondimensional toughness variation index $\delta_{quasi.cr}$ (after prestraining) divided by $\delta_{quasi.cr\_As}$ (with no prestrain) for each direction. As shown in Figure 4, the samples exhibit remarkable embrittlement in all prestrain conditions, although there was a difference depending on the angle.

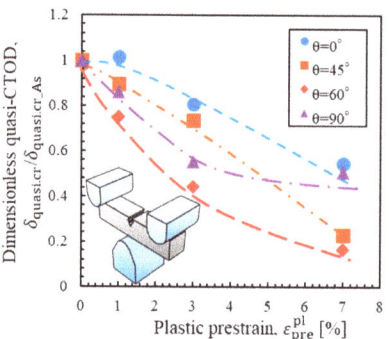

Figure 4. Toughness degradation in critical CTOD due to prestrain in different directions.

3. Calculation of Critical Stress

In this section, the critical stress was calculated based on the experimental results shown in the previous section. FEM analysis was carried out as a continuous simulation covering all processes, including prestraining and fracture tests. Finally, the critical stress was obtained for each experiment, and the results were compared among the different prestrain conditions.

3.1. Simulation of the Prestraining Process

In this analysis, a combined hardening rule, which combines isotropic and kinematic hardening rules, was used for the simulation. The steel material undergoes strain hardening with plastic deformation and yield surface changes. The isotropic hardening rule defines the magnitude of the yield surface, and the kinematic hardening rule represents the movement of the yield surface. In this study, because the prestrain direction was different from the crack opening direction, the Bauschinger effect was considered to be undergoing load reversal. In the case where load reversal occurs, the yield surface moves while changing its magnitude, so the combined hardening rule is considered to be appropriate. The combined hardening rule is defined in Equations (10) and (11).

$$\sigma^0 = \sigma|_0 + Q_\infty\left(1 - e^{-\beta \bar{\varepsilon}^{pl}}\right) \tag{10}$$

$$\dot{\alpha}_{ij} = C\frac{1}{\sigma^0}(\sigma_{ij} - \alpha_{ij})\dot{\bar{\varepsilon}}^{pl} - \gamma\alpha_{ij}\dot{\bar{\varepsilon}}^{pl} \tag{11}$$

Isotropic hardening is expressed by Equation (10) [32], whereas kinematic hardening is expressed by Equation (11) [33]. In these equations, σ^0 is the equivalent stress that decides the magnitude of the yield surface, $\sigma|_0$ is the yield stress with no plastic strain, Q_∞ and β are intrinsic material parameters, C is the initial kinematic hardening coefficient, and γ is the parameter that expresses the damping factor of the kinematic hardening coefficient. In this analysis, each parameter was determined by trial-and-error, as shown in Table 2.

Table 2. Determination of the parameters in the combined hardening rule.

Parameter	Value	Units
$\sigma\|_0$	500	MPa
Q_∞	200	MPa
β	4.5	-
C	10,000	MPa
γ	81	-

3.2. Simulation of the Fracture Test

A three-point bending test for evaluating fracture toughness was simulated with an FEM model. Similar to the simulation of the prestraining process, the combined hardening rule was applied in this step. In this analysis, the "Map Solution" technique [34] was used to transfer all analysis results, including the back stress tensor and plastic strain from the prestraining simulation, to the initial conditions of the fracture test model, which has completely different mesh divisions. Then, the fracture test can be simulated, including the parameters of plastic strain and back stress from the prestraining simulation. The prestrain direction can be arbitrarily reproduced by rotating the model during mapping from the prestraining process to the fracture test simulation. In the fracture test process, the combined hardening parameters were optimized for each angle condition, as shown in Table 3, to accurately estimate the stress–strain relationship after prestraining. Using these parameters, a full process analysis including the prestraining process and the subsequent fracture test was carried out.

Table 3. Determination of the parameters for the fracture test.

Parameter	$\theta = 0°$	$\theta = 45°$	$\theta = 60°$	$\theta = 90°$	Units
$\sigma\|_0$	750	750	750	750	MPa
Q_∞	500	600	500	550	MPa
β	3	3	3	3	-
C	5000	5000	5000	5000	MPa
γ	81	81	81	81	-

Next, the critical stress was calculated based on the analysis mentioned above. As shown in Figure 5, the representative fracture initiation point was regarded as the point with the largest value of maximum principal stress at the time when the bending test specimen fractured. Under this assumption, the maximum principal stress at the fracture point was regarded as the critical stress. The largest value of maximum principal stress appeared near the notch root on the mid-thickness plane. The critical stress was calculated under all prestrain conditions, including the "As" condition (without prestraining). Figure 5 shows the change in the critical stress under each prestrain condition. The critical stress without prestrain, σ_{cr_As}, was different from the critical stress with prestrain in different directions. Therefore, the critical stress after prestrain, σ_{cr}, was made dimensionless through a comparison with σ_{cr_As}. As shown in Figure 5, the critical stress tended to increase with the increase in prestrain under all prestrain directions, and the ratio of the increase varies with respect to the prestrain direction. Based on these results, the mechanism of the increase in the critical stress was investigated from the viewpoint of both the continuum macroscopic model and the crystal plasticity model in the next section.

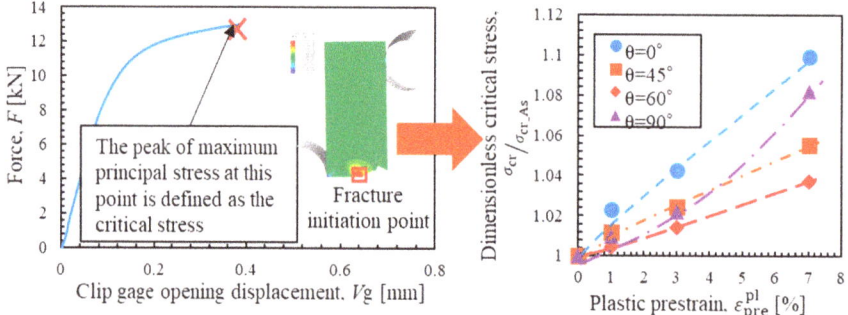

Figure 5. Change in the critical stress due to prestrain in different directions.

4. Mechanism of Change in Critical Stress

In the previous section, it was revealed that the critical stress clearly increased with respect to the amount of prestrain. The increase ratio varied with respect to the prestrain direction. Hence, to evaluate the safety of structures, it is important to understand how the critical conditions change when a complicated prestrain is applied, and the mechanism that can explain the experimental results should be investigated in detail. In this section, the critical stress change mechanism is clarified based on both the continuum macroscopic model and the crystal plasticity model.

4.1. Analysis of the Macroscopic Model

In this subsection, the candidate mechanism is investigated based on a macroscopic model. It is reasonable that material damage progresses by plastic deformation and does not occur during elastic deformation [12,13]. Thus, the critical stress was divided into elastic and plastic components, as shown in Equation (12).

$$\sigma_{cr} = \sigma_{el} + \sigma_{pl} \qquad (12)$$

where σ_{el} is the elastic component and σ_{pl} is the plastic component of critical stress. The critical stress is considered to change by changing these values through different mechanisms.

First, the mechanism of the elastic component is discussed. The elastic component is considered to change through yield surface evolution. The yield condition with the combined hardening rule is defined in Equation (13).

$$\bar{\sigma} = \sigma^0 + \omega(\theta)\bar{\alpha} \qquad (13)$$

where $\bar{\sigma}$ is the von Mises stress, $\bar{\alpha}$ is the equivalent back stress, and σ^0 is the magnitude of the yield surface calculated in Equation (10). Additionally, $\omega(\theta)$ is a function for θ that has the largest value at $\theta = 0°$ and the smallest value at $\theta = 90°$. This is because the Bauschinger effect occurs during load reversal. In this study, while the largest value of the maximum principal stress at the fracture point is defined as the critical stress, the yield stress component is not explicitly defined inside of the maximum principal stress, so the "quasi-yield" stress is defined for convenience, as shown in Figure 6a.

The maximum principal stress exhibits a step increase at a constant value of equivalent plastic strain, after which it increases monotonically with respect to the equivalent plastic strain. The quasi-yield stress ($\sigma_{y,cr}$) is defined as the maximum principal stress at this inflection point. Figure 6b shows the change in $\sigma_{y,cr}$ due to prestrain in the different directions. Similar to Figure 5, the quasi-yield stress after prestrain was made dimensionless through comparison with σ_{y,cr_As}. As shown in Figure 6b, the change in the dimensionless quasi-yield stress, $\sigma_{y,cr}/\sigma_{y,cr\_As}$, exhibited large variations depending on the prestrain direction. This may be because of the Bauschinger effect. The extent of the Bauschinger effect should vary with respect to the prestrain direction, and its effects are strongest at $\theta = 90°$ and weakest at $\theta = 0°$. Thus, the ratio of decrease in the dimensionless quasi-yield stress due to the Bauschinger effect is considered to be largest at $\theta = 90°$.

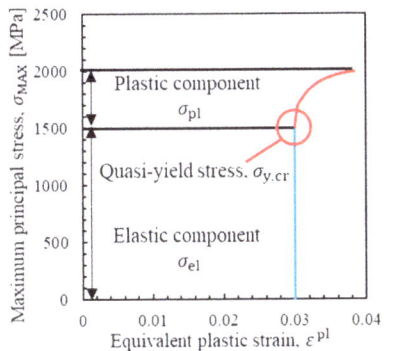 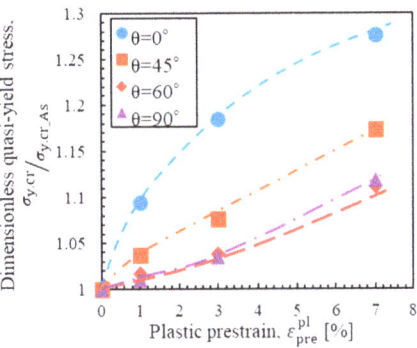

Figure 6. Change in quasi-yield stress due to prestrain in different directions. (a) Definition of $\sigma_{y,cr}$, (b) Change in the dimensionless quasi-yield stress.

However, the ratio of increase in $\sigma_{y,cr}$ at $\theta = 90°$ is larger than that at $\theta = 60°$, as shown in Figure 6b. This is because Q_∞ is larger at $\theta = 90°$ than at $\theta = 60°$, as shown in Table 3. This may be related to the strong texture due to the strong TMCP.

At this point, the plastic component of the critical stress must be discussed. Lei [12,13] showed that elastic deformation did not contribute to brittle fracture, i.e., only plastic deformation contributed to brittle fracture. Lei [12,13] defined plastic stress as the cut-off stress ($\sigma_{cut-off}$), and in this study, the critical stress change mechanism is discussed based on the assumption that only $\sigma_{cut-off}$ contributes to brittle fracture. Figure 7 shows the change in $\sigma_{cut-off}$ for each prestrain direction. Similar to Figures 5 and 6, $\sigma_{cut-off}$ after prestrain was made dimensionless through comparison with $\sigma_{cut-off\_As}$. As shown in Figure 7, the dimensionless cut-off stress, $\sigma_{cut-off}/\sigma_{cut-off\_As}$, decreases with increasing prestrain at $\theta = 0°$, 45° and 60°. This indicates that the plastic deformability before brittle crack initiation was reduced by the prestrain, which made the material brittle. However, the ratio of decrease in the dimensionless cut-off stress exhibits substantial variations with respect to the prestrain direction; the dimensionless cut-off stress even increases at $\theta = 90°$.

As described above, the discussion was advanced by dividing the critical fracture stress into the yield stress term and the stress term that occurs during plastic deformation. However, the latter term cannot be understood by using continuum mechanics alone. In particular, the mechanism of decrease in $\sigma_{cut-off}$ should be discussed in detail to accurately estimate the critical stress. In the next subsection, the detailed mechanism of the change in $\sigma_{cut-off}$ is discussed using an FEM analysis based on the crystal plasticity model.

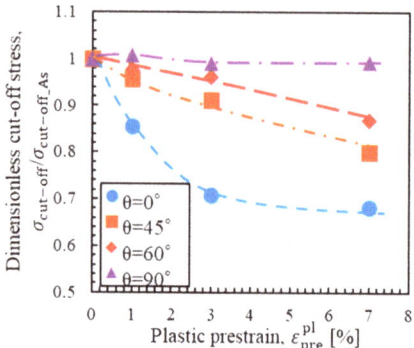

Figure 7. Change in the dimensionless cut-off stress due to prestrain in different directions.

4.2. Analysis of the Crystal Plasticity Model

In this subsection, the strain gradient plasticity (SGP) theory was used based on the crystal plasticity model. First, the theory of SGP is introduced. The SGP theory was proposed in 1970 by Ashby [35], and to date, many studies pertaining to SGP formulation have been carried out [36–39]. The SGP theory is based on the assumption that not only the plastic strain but also the plastic strain gradient should contribute to the work of deformation in a body. Additionally, one of the advantages of this approach is that the events that occur during plastic deformation at the grain boundary can be explained using the polycrystalline model and SGP theory. The steel material is polycrystalline and has stress and strain gradients in each grain even under macroscopically homogeneous strain conditions. However, in the conventional continuum macroscopic model, the stress and strain cannot be calculated for every grain; only the average value is calculated. To understand the phenomenon of polycrystallinity, it is important to apply a polycrystalline model. Additionally, the computational cost associated with the use of the SGP theory is much lower than that of the conventional crystal plasticity finite element method (CPFEM) because the SGP theory simplifies the average crystal orientation of each grain. Additionally, the SGP theory can calculate the dislocation density expressed by the plastic strain gradient and equivalent plastic strain. As mentioned in Section 1, brittle fracture of steel occurs when a microcrack is generated due to piled-up dislocations in the brittle phase, so it is important to deepen the understanding of dislocation distributions to discuss the embrittlement mechanism. There are two types of dislocations: geometrically necessary dislocations (GNDs) and statistically stored dislocations (SSDs). GNDs depend on the plastic strain gradient, whereas SSDs depend on the equivalent plastic strain. Martínez et al. [40] showed that the densities of these dislocations can be calculated with Equations (14) and (15).

$$\rho_{GND} = \bar{r}\frac{\eta^P}{b} \tag{14}$$

$$\rho_{SSD} = [\sigma_{ref} f(\varepsilon^P)/(M\alpha\mu b)]^2 \tag{15}$$

where b is the magnitude of the Burgers vector, $\bar{r}$, M and α are intrinsic material parameters, μ is the shear modulus, and η^P is the equivalent plastic strain gradient. Note that η^P and $\sigma_{ref} f(\varepsilon^P)$ can be calculated with Equations (16)–(18).

$$\eta^P = \sqrt{\frac{1}{4}\sum_{i,j,k}\varepsilon^P_{ij,k}\varepsilon^P_{ij,k}} \tag{16}$$

$$\varepsilon^P_{ij,k} = \frac{\partial \varepsilon^P_{ij}}{\partial x_k} \tag{17}$$

$$\sigma_{ref} f(\varepsilon^P) = \sigma_{yield} \left(\frac{E}{\sigma_{yield}}\right)^N (\varepsilon^P + \sigma_{yield}/E)^2 \tag{18}$$

where N is the strain hardening coefficient and σ_{yield} is the yield stress. Additionally, Gudmundson [39] showed that the principle of virtual work can be calculated with Equation (19).

$$\int_\Omega \left[\sigma_{eij}\delta\varepsilon_{ij} + q_{ij}\delta\varepsilon^P_{ij} + m_{ijk}\delta\varepsilon^P_{ij,k}\right]dV + \int_{S\Gamma}\left[\check{M}_{ij}\delta\check{\varepsilon}_{ij} + \hat{M}_{ij}\delta\hat{\varepsilon}_{ij}\right]dS = \int_{S^{ext}}\left[\sigma_{ij}n_j\delta u_i + m_{ijk}n_k\delta\varepsilon^P_{ij}\right]dS \tag{19}$$

where σ_{eij} is the elastic stress, q_{ij} is the plastic stress, m_{ijk} is the moment stress, M_{ij} is the moment traction, $\check{\varepsilon}_{ij}$ is the average strain, and $\hat{\varepsilon}_{ij}$ is the difference in strain between two grains. In Equation (19), the first term on the left side represents the work in the grain, the second term on the left side represents the work in the grain boundary, and the term on the right side represents the work by external force.

In this analysis, a polycrystalline microstructure was simulated so that the average grain diameter was approximately 30 μm, and a prestrain test was simulated. It is well known that TMCP steel plates for actual applications have ausformed bainite microstructures, which contain many secondary phases inside prior austenite grains [22]. Kitade et al. [41] carried out a series of sophisticated experiments and reported that the brittle fracture initiation site for this type of steel is located at martensite–austenite (MA) constituents formed at high angle grain boundaries and that there are scarcely high value areas of kernel average misorientation (KAM). The KAM is an index obtained from electron backscatter diffraction (EBSD) analysis corresponding to the dislocation density inside of prior austenite grains just before brittle crack initiation. Through these experimental findings, the authors believe that this assumption of grain size is appropriate.

First, the model was loaded with a 1% prestrain in the vertical direction, and then the fracture test was simulated so that the model was loaded with tensile strain in the direction of θ rotation. As mentioned above, SGP theory can calculate the individual dislocation densities, ρ_{GND} and ρ_{SSD}. The total dislocation density (ρ_{TD}), which is a sum of ρ_{GND} and ρ_{SSD}, was defined as the evaluation value of material damage. Figure 8 shows the history of $\rho_{TD,max}$ for each condition. Here, $\rho_{TD,max}$ is denoted as the value at the center of the element in which ρ_{TD} exhibited the maximum value. Similarly, $\rho_{GND,max}$ and $\rho_{SSD,max}$ are the values of ρ_{GND} and ρ_{SSD} from the element that exhibited $\rho_{TD,max}$. In all cases, the elements that exhibited the maximum ρ_{TD} were always located close to the grain boundary. Additionally, equivalent plastic strain in the horizontal axis denoted the global value and calculated as the average value of all element. As shown in Figure 8, the history of $\rho_{TD,max}$ varies greatly depending on the prestrain direction. When the changes in $\rho_{GND,max}$ and $\rho_{SSD,max}$ are compared, it is clear that $\rho_{GND,max}$ is largely changed for each prestrain direction, whereas $\rho_{SSD,max}$ shows a constant change that is approximately independent of the prestrain direction. This is because $\rho_{SSD,max}$ depends on the equivalent plastic strain, as shown in Equation (15). Focusing on the change in the history of $\rho_{GND,max}$ evolution, $\rho_{TD,max}$ and $\rho_{GND,max}$ continuously increase only when $\theta = 0°$, whereas when $\theta \neq 0°$, these values decrease and then increase again. This may reflect the fact that piled-up dislocations move in the opposite direction after load reversal. When $\theta = 45°$ and $60°$, it is understood that load reversal did not occur in the completely opposite direction, but partial load reversal occurred; hence, $\rho_{TD,max}$ slightly decreased just after the beginning of the secondary load. It can be regarded that material damage does not progress while $\rho_{TD,max}$ decreases. Therefore, during some range just after load reversal when $\theta = 45°, 60°$, and $90°$, it is supposed that the damage did not progress. According to Figure 8, the range was the largest in the condition of $\theta = 90°$, which can be the reason why the amount of decrease in $\sigma_{cut-off}$ is quite small in the case of $\theta = 90°$, as shown in Figure 7. In this range where the material damage did not progress, the amount of equivalent plastic strain increased, so the equivalent stress increased, as shown in Equation (10). This mechanism made $\sigma_{cut-off}$ larger at $\theta = 90°$.

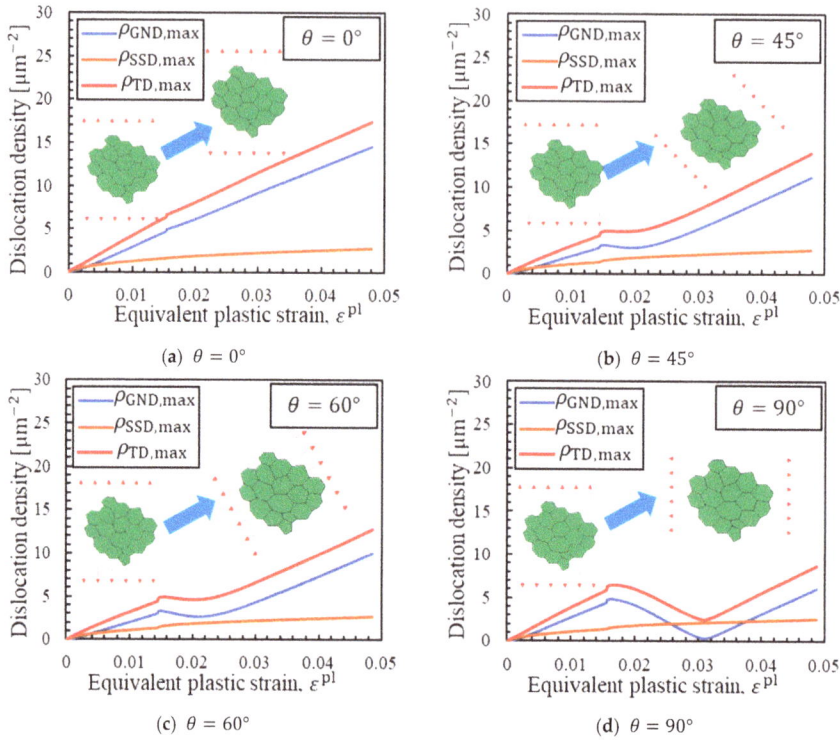

Figure 8. Change in peak dislocation density for each prestrain condition. (a) $\theta = 0°$, (b) $\theta = 45°$, (c) $\theta = 60°$, (d) $\theta = 90°$.

Based on the abovementioned assumption, a quantitative discussion was carried out. As shown in Figure 9a, the plastic strain required to reach a certain $\rho_{TD,cr}$ is considered. In this analysis, the plastic strain required to reach $\rho_{TD,cr}$, 17.5 μm^{-2}, after applying the prestrain was denoted as $\varepsilon^{pl}_{fract}(\theta)$. Material damage progresses continuously in the condition of $\theta = 0°$; therefore, $\varepsilon^{pl}_{fract}(\theta)$ is smallest in the condition of $\theta = 0°$. In contrast, the range where the damage did not progress was largest at $\theta = 90°$; therefore, $\varepsilon^{pl}_{fract}(\theta)$ was largest at $\theta = 90°$. Brittle fracture occurred in the originating microcrack due to the density of piled-up dislocations; thus, the criterion of the brittle fracture can be simplified as the critical value of $\rho_{TD,max}$. Here, the assumption that brittle fracture occurred when ρ_{TD} reached 17.5 μm^{-2} was tentatively made. Based on this assumption, $\varepsilon^{pl}_{fract}(\theta)$ describes the amount of plastic strain applied after finishing the prestrain until fracture. As mentioned above, because the equivalent stress is determined by the equivalent plastic strain, the increase in $\varepsilon^{pl}_{fract}(\theta)$ directly corresponds to the increase in $\sigma_{cut-off}$. Figure 9b shows the relationship between $\varepsilon^{pl}_{fract}(\theta)$ and dimensionless $\sigma_{cut-off}$ when 1% prestrain was applied. As shown in Figure 9b, there is a strong correlation between $\varepsilon^{pl}_{fract}(\theta)$ and $\sigma_{cut-off}$; hence, $\sigma_{cut-off}$ increases with increasing $\varepsilon^{pl}_{fract}(\theta)$. Furthermore, when the amount of prestrain increased, $\rho_{TD,max}$ increased and $\varepsilon^{pl}_{fract}(\theta)$ decreased. As a result, $\sigma_{cut-off}$ is considered to decrease, and this result matches Figure 7.

Thus, it is understood that the change in $\sigma_{cut-off}$ can be explained by the change in $\varepsilon^{pl}_{fract}(\theta)$, i.e., the way material damage progresses due to piled-up dislocations. In the next section, based on these mechanisms, a formulation is presented for the change in critical stress after prestrain application.

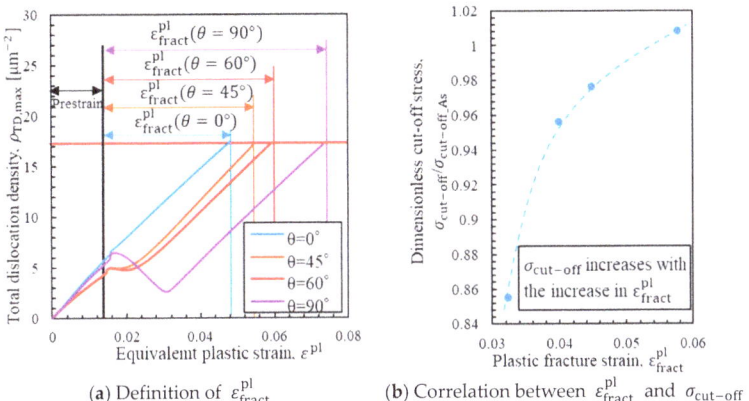

Figure 9. Mechanism of change in $\sigma_{\text{cut-off}}$ during loading with different prestrain directions. (a) Definition of $\varepsilon_{\text{fract}}^{\text{pl}}$, (b) Correlation between $\varepsilon_{\text{fract}}^{\text{pl}}$ and $\sigma_{\text{cut-off}}$.

5. Formulation of Critical Stress Change from Various Prestrains

In the previous section, the change in critical stress can be expressed by the increase in yield stress and the decrease in $\sigma_{\text{cut-off}}$. Based on this theory, the formulation of critical stress is carried out. First, the general form for the elastic component of the critical stress in the fracture test is proposed in Equation (20).

$$\sigma_{\text{el}} = \sigma_{\text{el\_As}} + \varphi(\eta)\left[Q_{\infty}\left(1 - e^{-\beta\varepsilon_{\text{pre}}^{\text{pl}}}\right) + \omega(\theta)\bar{a}\right] \tag{20}$$

where $\sigma_{\text{el\_As}}$ is the elastic component of critical stress without prestraining, $\varphi(\eta)$ is a function of η that is set equal to one for uniaxial stress conditions, $\varepsilon_{\text{pre}}^{\text{pl}}$ is the plastic prestrain, $\omega(\theta)$ is a function of θ, and $Q_{\infty}\left(1 - e^{-\beta\varepsilon_{\text{pre}}^{\text{pl}}}\right) + \omega(\theta)\bar{a}$ represents the combined hardening rule. As shown in Figure 10a, the amount of increase in yield stress is larger with high stress triaxiality. Therefore, because the sharp-notched bending specimen used in this study has high triaxiality, the amount of increase in yield stress increases. Here, it was assumed that the increase in yield stress rises linearly with respect to triaxiality, and $\varphi(\eta)$ was set to 1.60 for this specimen in this paper. It should be noted, however, that in order to determine this value accurately, it must be determined experimentally using specimens with different multiaxiality. Additionally, $\omega(\theta)$ is the function that expresses the degree of the Bauschinger effect, which is a function of the period. As shown in Figure 10b, $\omega(\theta)$ is largest at $\theta = 0°$ and smallest at $\theta = 90°$. In this study, $\omega(\theta)$ was set to 1.0 at $\theta = 0°$ and −0.20 at $\theta = 90°$.

Hereafter, $\sigma_{\text{cut-off}}$ is discussed. As mentioned above, $\sigma_{\text{cut-off}}$ is considered to be greatly affected by the difference in piled-up dislocations. When load reversal occurs, dislocations pile up in the opposite direction, and there is some range where material damage does not progress, which restrains the decrease in $\sigma_{\text{cut-off}}$. However, when load reversal occurs, dislocations not only pile up like GNDs but are also simply stored like SSDs. Therefore, the effect of material damage due to both piled-up dislocations and stored dislocations should be considered.

First, the effect of piled-up dislocations is considered. This effect strongly depends on the prestrain direction. Here, the damage due to piled-up dislocations is defined in Equation (21).

$$f\left(\varepsilon_{\text{pre}}^{\text{pl}}, \theta\right) = \lambda\omega(\theta)\left\{1 - \exp\left(-\zeta\varepsilon_{\text{pre}}^{\text{pl}}\right)\right\} \tag{21}$$

Equation (21) is a monotonically increasing function with respect to $\varepsilon_{\text{pre}}^{\text{pl}}$ and expresses the material damage due to piled-up dislocations. Additionally, this function is designed to converge to $\lambda\omega(\theta)$ in the case of a sufficiently high value of $\varepsilon_{\text{pre}}^{\text{pl}}$, which corresponds to the phenomenon that the parameter of brittle fracture toughness, such as the critical CTOD, converges to a certain low value even if there is a sufficiently high amount of prestrain. $\omega(\theta)$ is the same function used in Equation (20), which corresponds to how dislocations are moved in the opposite direction. The range where material damage does not progress increases with increasing θ. Therefore, Equation (21) shows that material damage decreases with a high value of θ.

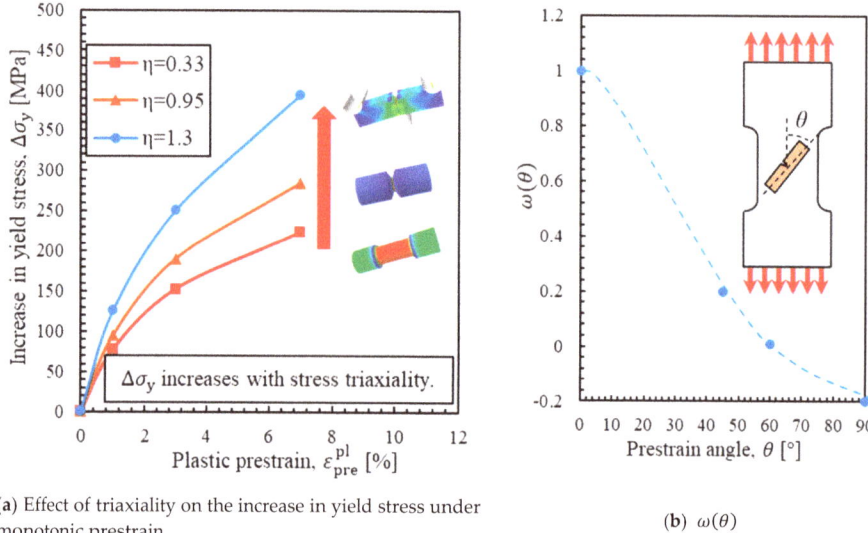

(a) Effect of triaxiality on the increase in yield stress under monotonic prestrain

(b) $\omega(\theta)$

Figure 10. Introduction of coefficients of the proposed elastic component. (a) Effect of triaxiality on the increase in yield stress under monotonic prestrain, (b) $\omega(\theta)$.

Hereafter, the effect of stored dislocations is considered. This damage is considered to be independent of the prestrain direction and depends only on the amount of prestrain because the stored dislocations are considered to increase monotonically with respect to the equivalent plastic strain, whereas piled-up dislocations strongly depend on the prestrain direction. The damage due to stored dislocations is defined in Equation (22).

$$g\left(\varepsilon_{pre}^{pl}\right) = \kappa\left\{1 - \exp\left(-\xi \varepsilon_{pre}^{pl}\right)\right\} \tag{22}$$

Equation (22) is a monotonically increasing function with respect to ε_{pre}^{pl} and converges to κ in the case of a sufficiently high value of ε_{pre}^{pl}, similar to Equation (21). Combining Equation (21) and Equation (22), the decrease in $\sigma_{cut-off}$ is defined in Equation (23).

$$\Delta\sigma_{cut-off} = -\sigma_{cut-off\_As}\left[\lambda\omega(\theta)\left\{1 - \exp\left(-\zeta\varepsilon_{pre}^{pl}\right)\right\} + \kappa\left\{1 - \exp\left(-\xi\varepsilon_{pre}^{pl}\right)\right\}\right] \tag{23}$$

Finally, the change in critical stress is calculated by combining Equation (20) and Equation (23). The critical stress after prestraining can be determined with Equation (24).

$$\sigma_{cr,est} = \sigma_{cr\_As} + \varphi(\eta)\left[Q_\infty\left(1 - e^{-\beta\varepsilon_{pre}^{pl}}\right) + \omega(\theta)\bar{\alpha}\right] - \sigma_{cut-off\_As}\left(f\left(\varepsilon_{pre}^{pl}, \theta\right) + g\left(\varepsilon_{pre}^{pl}\right)\right) \tag{24}$$

Figure 11 shows a comparison between the critical stress calculated from Equation (24) and that calculated from FEM analysis. As shown in Figure 11, the critical stress can be estimated with high precision using Equation (24). The parameters shown in Figure 11 were determined based on SGP theory. Additionally, except $\varphi(\eta)$, these are considered to depend on microstructure, so these values can be used in different test specimen if the same material is used. Thus, the critical stress can be estimated for specimens with various levels of prestrain in different directions subjected to fracture tests with different levels of triaxiality.

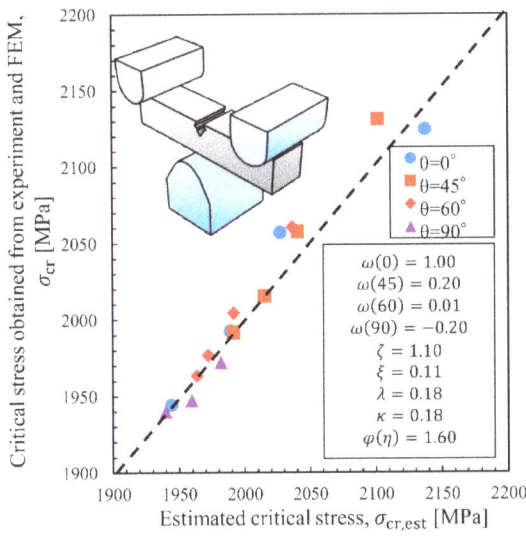

Figure 11. Accuracy of the proposed formula established to estimate critical stress.

6. Conclusions

In this study, the effect of the prestraining direction and its amount on the critical stress of brittle fracture was investigated. Additionally, a new formula for accurately estimating critical stress was proposed based on the micromechanism obtained from the SGP theory, from which the following findings were obtained:

- It was verified that the critical stress increased due to the application of prestrain at any angle, and the ratio of this increase varied strongly with respect to the prestrain direction. This finding is different from Griffith's equation or Smith's model, in which the critical stress is determined only by the length of the microcrack or the thickness of the carbide particles and the remote stress.
- It was shown that the increase in critical stress can be separately explained by the increase in yield stress and the decrease in $\sigma_{\text{cut-off}}$. Additionally, the decrease in $\sigma_{\text{cut-off}}$ represented the effect of embrittlement and strongly depended on the way dislocations were piled up. Using analysis based on the SGP theory, the change in $\sigma_{\text{cut-off}}$ was shown to be affected by piled-up dislocations that moved in the opposite direction during load reversal.
- The change in critical stress can be formulated based on the micromechanisms. The critical stress was calculated by the triaxiality of the fracture test and the amount and direction of prestrain. It was shown that critical stress can be estimated with high accuracy by giving appropriate parameters.

In this study, many new findings on the critical stress that have not been previously discussed were presented. However, there are still many problems to be solved in this study. First, the present prestrain was introduced in a uniaxial state, whereas the strain in an actual structure is always loaded in a multiaxial state. In addition, in the critical stress formulation, there are many material-dependent parameters, and it is desirable to construct a universal model that can be applied to any material. Additionally, this model has been developed on the assumption that dislocations dominate the damage, but the mechanism may differ in a material in which the brittle crack initiation position is located at large and hard particles, such as MA or cementite, inside of grains. The effect of the microstructure on the critical stress after prestrain should be investigated in future studies.

Author Contributions: H.K. carried out the experiment and numerical simulation. He also wrote the manuscript with support from all other members. T.K. designed the whole project and the experimental plan. He also improved the manuscript suitable for publication. T.O. continually conducted important debates on both continuum mechanics and crystal plasticity, and gave important inspirations to the principal author. He also gave valid advice as an author on the final manuscript. H.N. gave useful advice at any time during the research in order to clarify the engineering purpose of the project, He also made important suggestions on the interpretation of the experimental results. All authors have read and agreed to the published version of the manuscript.

Funding: This research was funded by JSPS KAKENHI Grant Number JP19H00802.

Acknowledgments: The authors would like to thank Akiyasu Morita and Yuta Yaguchi for offering high performance in many complicated experiments.

Conflicts of Interest: The authors declare no conflict of interest.

References

1. Griffith, A.A. The phenomena of rupture and flow in solid. *Philos. Trans. Ser. A* **1920**, *221*, 163–198.
2. Inglis, C.E. Stresses in a Plate Due to the Presence of Cracks and Sharp Corners. *Trans. Inst. Nav. Archit.* **1913**, *55*, 219–241.
3. Stroh, A. A theory of the fracture of metals. *Adv. Phys.* **1957**, *6*, 418–465. [CrossRef]
4. Smith, E. The nucleation and growth of cleavage microcracks in mild steel. *Phys. Basis Yield Fract. Conf. Proc.* **1966**, *1966*, 36–46.
5. Hall, E.O. The deformation and aging of mild steel. *Proc. Phys. Soc. Sect. B* **1951**, *64*, 747–753. [CrossRef]
6. Petch, N.J. The Cleavage Strength of Polycrystals. *J. Iron Steel Inst.* **1953**, *174*, 25–28.
7. Cottrell, A.H. Theory of brittle fracture in steel and similar metals. *Trans. Metall. Soc. AIME* **1959**, *212*, 192–201.
8. Epstein, B. Statistical approach to brittle fracture. *J. Appl. Phys.* **1948**, *19*, 140–147. [CrossRef]
9. Beremin, F.M. A local criterion for cleavage fracture of a nuclear pressure vessel steel. *Metall. Trans. A* **1983**, *14*, 2277–2287. [CrossRef]
10. Bordet, S.R.; Karstensen, A.D.; Knowles, D.M.; Wiesner, C.S. A new statistical local criterion for cleavage fracture in steel. Part I: Model presentation. *Eng. Fract. Mech.* **2005**, *72*, 435–452. [CrossRef]
11. Bordet, S.; Karstensen, A.; Knowles, D.; Wiesner, C. A new statistical local criterion for cleavage fracture in steel. Part II: Application to an offshore structural steel. *Eng. Fract. Mech.* **2005**, *72*, 453–474. [CrossRef]
12. Lei, W.-S. A cumulative failure probability model for cleavage fracture in ferritic steels. *Mech. Mater.* **2016**, *93*, 184–198.
13. Lei, W.-S. A discussion of "An engineering methodology for constraint corrections of elastic–plastic fracture toughness—Part II: Effects of specimen geometry and plastic strain on cleavage fracture predictions" by C. Ruggieri, R.G. Savioli, R.H. Dodds [Eng. Fract. Mech. 146 (2015) 185–209]. *Eng. Fract. Mech.* **2017**, *178*, 527–534. [CrossRef]
14. McMahon, C.; Cohen, M. Initiation of cleavage in polycrystalline iron. *Acta Met.* **1965**, *13*, 591–604. [CrossRef]
15. Gurland, J. Observations on the fracture of cementite particles in a spheroidized 1.05% c steel deformed at room temperature. *Acta Met.* **1972**, *20*, 735–741.
16. Sukedai, E.; Hid, M. Effect of Tensile Pestrain on Ductile-Brittle Transition Temperture of Low Carbon Steel. *Mater. Sci. Monogr.* **1982**, *15*, 112–118.
17. Miki, C.; Sasaki, E.; Kyuba, H.; Takenoi, I. Deterioration of Fracture Toughness of Steel by Effect of Tensile and Compressive Prestrain. *J. JSCE* **2000**, *640*, 165–175.
18. Kosuge, H.; Kawabata, T.; Okita, T.; Murayama, H.; Takagi, S. Establishment of damage estimation rules for brittle fracture after cyclic plastic prestrain in steel. *Mater. Des.* **2020**, *185*, 108222. [CrossRef]
19. Yoshinari, H.; Enami, K.; Koseki, T.; Shimanuki, H.; Aihara, S. Ductile and brittle fracture initiation behavior for compressively prestrained steel. *J. Soc. Nav. Arch. Jpn.* **2001**, *2001*, 559–567. [CrossRef]
20. Bordet, S.R.; Tanguy, B.; Bugat, S.; Moinereau, D.; Pineau, A. Cleavage Fracture Micromechanisms Related to WPS Effect in RPV Steel. *Fract. Nano Eng. Mater. Struct.* **2008**, *16*, 835–836.

21. Tagawa, T.; Itoh, A.; Miyata, T. Ouantitative prediction of embrittlement due to pre-strain for low carbon steels. *Q. J. Jpn. Weld. Soc.* **1996**, *2*, 429–434. [CrossRef]
22. Nishioka, K.; Ichikawa, K. Progress in thermomechanical control of steel plates and their commercialization. *Sci. Technol. Adv. Mater.* **2012**, *13*, 023001. [CrossRef] [PubMed]
23. International Organization for Standardization. *Petroleum and Natural Gas Industries—Steel Pipe for Pipeline Transportation Systems*; ISO 3183:2019; ISO: Geneva, Switzerland, 2019.
24. The American Petroleum Institute (API). Steel Plates Produced by Thermo Mechanically Controlled Processing for Offshore Structures. In *API Specification 2W*, 6th ed.; API: Washington, DC, USA, 2019.
25. ASTM International. *Standard Specification for Steel Plates for Pressure Vessels, Produced by Thermo-Mechanical Control Process (TMCP)*; ASTM A841/A841M-17; ASTM International: West Conshohocken, PA, USA, 2017.
26. International Association of Classification Societies. *Normal and Higher Strength Hull Structural Steels*; IACS W11, Rev.9; IACS: London, UK, 2017.
27. The Japanese Iron and Steel Federation (JISF). *TMCP Steel for Building (TMCP325, TMCP355)*; MDCR 0016-2016; JISF: Tokyo, Japan, 2016.
28. The Japanese Iron and Steel Federation. *Rolled Steel with 500N/mm2 Yield Strangth and 700N/mm2 Yield Strength for Welded Structure*; MDCR 0014-2004; JISF: Tokyo, Japan, 2005.
29. International Organization for Standardization. *Metallic Materials—Unified Method of Test for the Determination of Quasistatic Fracture Toughness*; ISO12135:2016; ISO: Geneva, Switzerland, 2016.
30. Kawabata, T.; Tagawa, T.; Sakimoto, T.; Kayamori, Y.; Ohata, M.; Yamashita, Y.; Tamura, E.-I.; Yoshinari, H.; Aihara, S.; Minami, F.; et al. Proposal for a new CTOD calculation formula. *Eng. Fract. Mech.* **2016**, *159*, 16–34. [CrossRef]
31. International Organization for Standardization. *Metallic Materials—Method of Test for the Determination of Quasistatic Fracture Toughness of Welds*; ISO15653:2018; ISO: Geneva, Switzerland, 2018.
32. Chaboche, J.L. Constitutive equations for cyclic plasticity and cyclic viscoplasticity. *Int. J. Plast.* **1989**, *5*, 247–302. [CrossRef]
33. Lemaitre, J.; Chaboche, J.L. *Mechanics of Solid Materials*; Cambridge University Press: Cambridge, UK, 1994; 584p, ISBN 0521477581/9780521477581.
34. *Abaqus*, version 2018; Dassault Systèmes®: Vélizy-Villacoublay, France, 2018.
35. Ashby, M.F. The deformation of plastically non-homogeneous materials. *Philos. Mag.* **1970**, *21*, 399–424. [CrossRef]
36. Fleck, N.; Hutchinson, J. A phenomenological theory for strain gradient effects in plasticity. *J. Mech. Phys. Solids* **1993**, *41*, 1825–1857. [CrossRef]
37. Fleck, N.; Muller, G.; Ashby, M.; Hutchinson, J. Strain gradient plasticity: Theory and experiment. *Acta Met. Mater.* **1994**, *42*, 475–487. [CrossRef]
38. Fleck, N.A.; Hutchinson, J.W. A reformulation of strain gradient plasticity. *J. Mech. Phys. Solids* **2001**, *49*, 2245–2271. [CrossRef]
39. Gudmundson, P. A unified treatment of strain gradient plasticity. *J. Mech. Phys. Solids* **2004**, *52*, 1379–1406. [CrossRef]
40. Martínez-Pañeda, E.; Betegón, C. Modeling damage and fracture within strain-gradient plasticity. *Int. J. Solids Struct.* **2015**, *59*, 208–215. [CrossRef]
41. Kitade, A.; Kawabata, T.; Kimura, S.; Takatani, H.; Kagehira, K.; Mitsuzumi, T. Clarification of micromechanism on Brittle Fracture Initiation Condition of TMCP Steel with MA as the trigger point. *Procedia Struct. Integr.* **2018**, *13*, 1845–1854. [CrossRef]

© 2020 by the authors. Licensee MDPI, Basel, Switzerland. This article is an open access article distributed under the terms and conditions of the Creative Commons Attribution (CC BY) license (http://creativecommons.org/licenses/by/4.0/).

Article

Fatigue Crack Initiation of Metals Fabricated by Additive Manufacturing—A Crystal Plasticity Energy-Based Approach to IN718 Life Prediction

Chun-Yu Ou [1,*], Rohit Voothaluru [2,*] and C. Richard Liu [1,3,*]

1. School of Industrial Engineering, Purdue University, West Lafayette, IN 47907, USA
2. The Timken Company, North Canton, OH 44720, USA
3. Birck Nanotechnology Center, Purdue University, West Lafayette, IN 47907, USA
* Correspondence: ou1@purdue.edu (C.-Y.O.); rohit.voothaluru@timken.com (R.V.); liuch@purdue.edu (C.R.L.)

Received: 31 August 2020; Accepted: 4 October 2020; Published: 6 October 2020

Abstract: There has been a long-standing need in the marketplace for the economic production of small lots of components that have complex geometry. A potential solution is additive manufacturing (AM). AM is a manufacturing process that adds material from the bottom up. It has the distinct advantages of low preparation costs and a high geometric creation capability. However, the wide range of industrial processing conditions results in large variations in the fatigue lives of metal components fabricated using AM. One of the main reasons for this variation of fatigue lives is differences in microstructure. Our methodology incorporated a crystal plasticity finite element model (CPFEM) that was able to simulate a stress–strain response based on a set of randomly generated representative volume elements. The main advantage of this approach was that the model determined the elastic constants (C_{11}, C_{12}, and C_{44}), the critical resolved shear stress (g_0), and the strain hardening modulus (h_0) as a function of microstructure. These coefficients were determined based on the stress–strain relationships derived from the tensile test results. By incorporating the effect of microstructure on the elastic constants (C), the shear stress amplitude ($\frac{\Delta \tau}{2}$) can be computed more accurately. In addition, by considering the effect of microstructure on the critical resolved shear stress (g_0) and the strain hardening modulus (h_0), the localized dislocation slip and plastic slip per cycle ($\frac{\Delta \gamma_p}{2}$) can be precisely calculated by CPFEM. This study represents a major advance in fatigue research by modeling the crack initiation life of materials fabricated by AM with different microstructures. It is also a tool for designing laser AM processes that can fabricate components that meet the fatigue requirements of specific applications.

Keywords: additive manufacturing; fatigue crack initiation; crystal plasticity; finite element model; metal

1. Introduction

Additive manufacturing (AM) is a manufacturing process that adds material from the bottom up. Components fabricated via AM are now being used in motor vehicles, consumer products, medical products, aerospace devices, and even some military projects. General Electric employs AM technologies for the production of fuel nozzles, brackets, and sensor housings for jet engine turbines, and has recently planned to produce more than 100,000 parts this way [1]. NASA uses AM to produce components for rocket engine propulsion systems [2]. According to Wohler's report 2017 [3], approximately 49% of the materials used in AM are metals. The ability to predict the fatigue resistance of components fabricated with AM has become ever more critical with the increasing number of vital components that require high strength.

In the literature that addresses the fatigue data of AM [4], metal parts are shown to have a fatigue performance that displays a great deal of variation; some parts displayed a performance similar to parts that are wrought and cast, whereas others performed far worse than wrought and cast parts, as indicated by Edwards [5] and Kobryn [6]. One of the main reasons for this variation is flexibility in processing conditions, so it is of interest to develop a prediction model that allows one to design a component fabrication process that can meet fatigue resistance requirements.

Recent studies have emphasized experiments on fatigue crack propagation for AM materials. Walker et al. [7] observed that fatigue life has a high variation range from 10^4 to 10^7 cycles. Zhai et al. [8] investigated fatigue crack growth mechanisms in Ti-6Al-4V and concluded that the components fabricated with AM had a lower fatigue crack growth threshold and higher fracture toughness than wrought Ti-6Al-4V. The fatigue crack growth rate resistance was higher for materials made by AM in [9,10]. However, research into a model for predicting fatigue crack initiation in AM materials is still in its early stages.

Fatigue crack initiation is a critical property of components in industrial applications since it accounts for 50% to 90% of fatigue life, especially for low-stress and high-cycle fatigue conditions [11–13]. We often specify a lower stress condition to obtain a higher safety factor. Under these circumstances, fatigue crack initiation becomes more important. However, this major part of fatigue life prediction is mostly ignored by mainstream researchers working on fatigue modeling. As a result, a fundamental study of crack initiation is essential.

In this study, we present a fatigue crack initiation model that can predict the life of metals fabricated by AM. A crystal plasticity finite element model (CPFEM) was developed to compute the stress and stress–strain response. The model was validated using published fatigue testing results. This approach represents a major advance in fatigue research on AM materials.

2. Materials and Methods

The primary objective of this study is to present a methodology to estimate fatigue crack initiation life by means of a crystal plasticity finite element model. Our main focus is on presenting a simulation approach that can include the microstructure variations for metals fabricated by AM. The fatigue initiation model is based on the assumption that cracks are initiated when energy reaches a critical value, which is discussed in Section 2.1. In Section 2.2, mechanical energy is computed with a CPFEM in which the coefficients were determined based on the microstructure of the potential crack initiation zone. We used the experimental results conducted by Yang et al. [14] as a case study to validate the model by nickel-based superalloy samples that were fabricated by laser AM. The experimental setup is summarized in Section 2.3.

2.1. Fatigue Crack Initiation Model

During fatigue tests, most of the energy is absorbed by the material as mechanical energy, which causes elastic deformation ($E_{elastic}$) and plastic deformation ($E_{plastic}$). The remaining energy is transformed into heat ($E_{thermal}$) and diffused. The energy equation can be formulated as in Equation (1):

$$\Delta E(Ni) = -E_{elastic} - E_{plastic} + E_{release} - E_{thermal} \quad (1)$$

The elastic energy can be modeled as in Equation (2), assuming a virtual crack is defined as being penny-shaped [15].

$$E_{elastic} = \frac{4(1-v^2)}{3E}\sigma^3 a^3 \quad (2)$$

where a is the crack radius, v is Poisson's ratio, σ is the normal load applied, and E is the elastic modulus. The cracks tend to be initiated from stress-concentrated areas such as gas pores [14], slip bands [16], and areas lacking diffusion [14,17,18] for AM-fabricated materials. The plastic energy stored at the potential crack-initiated areas can be modeled based on Fine [19] as in Equation (3):

$$E_{plastic} = \frac{\pi t^2 N \delta}{2} \tag{3}$$

where t is the virtual defect size, N is the cycle number, and δ is the energy increase per cycle. The energy released ($E_{release}$) during crack initiation is formulated in Equation (4) based on Bhat and Fine [20]:

$$E_{release} = \pi a^2 \gamma_s \tag{4}$$

where γ_s is the surface energy. According to the assumptions of Tanaka and Mura [21], the total energy absorbed can be cumulative. Cracks are initiated when the energy reaches a critical value. The Bhat and Fine [14,20] model assumes that the total energy is released when the energy change ΔE reaches its maximum value and a crack is spontaneously initiated. As the crack begins to form, the internal energy is reduced due to the release of energy ($E_{release}$) as the crack develops. To find the maximum ΔE for crack initiation, Equation (5) is derived:

$$\frac{\partial \Delta E}{\partial a} = 0 \tag{5}$$

We plug in the elastic energy change and plastic energy change per cycle, as shown in Equation (6), with the assumption that the temperature change is low enough during the fatigue test to neglect the thermal energy.

$$\left(\frac{\partial \Delta E}{\partial a}\right)_N = \frac{\partial}{\partial a}\left(-\frac{4(1-v^2)}{3E}\sigma^2 a^3 - \frac{\pi t^2 N \delta}{2} + \pi a^2 \gamma_s\right) = 0 \tag{6}$$

Since the plastic portion of the energy change is assumed to be caused purely by mechanical loading, the plastic dissipation energy is the product of the shear stress and plastic slip accumulation amplitudes. The crack initiation life is then derived as Equation (7):

$$N_i = \frac{\pi E \gamma_s - 4\sigma^2 a^2 (1-v^2)}{\pi E \rho t \left(\frac{\Delta \tau}{2}\right)\left(\frac{\Delta \gamma_p}{2}\right)} \tag{7}$$

where ρ is the energy efficiency coefficient [22,23], $\left(\frac{\Delta \tau}{2}\right)$ is the shear stress amplitude, and $\left(\frac{\Delta \gamma_p}{2}\right)$ is the plastic slip rate per cycle. The elastic part $4\sigma^2 a^2(1-v^2)$ can be neglected since it is on the order of A^2. The final fatigue crack initiation can be summarized as shown in Equation (8):

$$N_i = \frac{\gamma_s}{\rho t \left(\frac{\Delta \tau}{2}\right)\left(\frac{\Delta \gamma_p}{2}\right)} \tag{8}$$

2.2. Crystal Plasticity Finite Element Model

The CPFEM model was developed based on the computational framework presented by Voothaluru and Liu [22,24,25]. The model applies small deformation kinematical theory [26,27] to the computation of the shear stress amplitude and plastic slip per cycle. The shear stress (τ^s) is calculated by the Cauchy stress tensor (σ), slip direction (s), and Schmid factor (m) as shown in Equation (9).

$$\tau^s = \sigma : (s^s \otimes m^s) \tag{9}$$

The plastic slip rate is calculated with Equation (10) as presented by Hutchinson [28].

$$\dot{\gamma}^s = \dot{\gamma}^0 \left|\frac{\tau^s - x^s}{g^s}\right|^n sgn(\tau^s) \tag{10}$$

where g^s is the slip resistance and x^s is the back stress. We assumed that the slip resistance and back stress obey the hardening law presented by Brown [29] as shown in Equations (11) and (12):

$$g^s = \sum_{s'} H_{dir} \left|\dot{\gamma}^{s'}\right| - q^{ss'} \sum_{s'} H_{dyn} \left|\dot{\gamma}^{s'}\right| \quad (11)$$

$$x^s = A_{dir} \sum_{s'} \dot{\gamma}^s - x^s A_{dyn} \left|\dot{\gamma}^s\right| \quad (12)$$

where H_{dir} and A_{dir} are the isotropic hardening coefficients and H_{dyn} and A_{dyn} are dynamic recovery coefficients. These constitutive equations were coded in ABAQUS software [30,31] to compute the energy during reversible fatigue.

The representative volume element (RVE) was defined as the volume of heterogeneous material that is sufficiently large to be statistically representative of the real component's microstructure. A set of randomly generated RVEs was developed for the microstructure at the potential crack initiation area. The RVE was generated within ABAQUS to have 64,000 C3D8 elements with a random orientation assigned to each element. This allows the development of a 40 × 40 × 40 model which has been generated with 50 different instantiations for orientation inputs that were randomly generated from a Gaussian normal distribution. This approach has been validated for other polycrystalline materials owing to the size independence of the models [32]. The boundary conditions were those followed by Smit [33], Kumar et al. [34], and Zhang [35]. A displacement boundary condition (U_x) was applied to the top face (1—2—3—4). A fixed boundary condition ($U_x = 0$) was applied to the bottom face (5—6—7—8). Periodic boundary conditions were applied to the faces on the sides, (1—2—5—6), (1—4—5—8), (3—4—7—8), and (2—3—6—7), as shown in Figure 1.

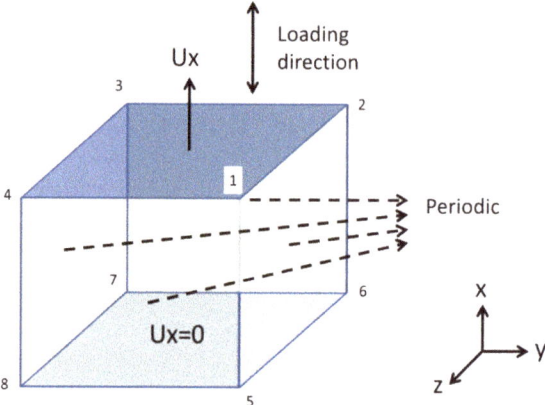

Figure 1. Boundary conditions for the representative volume elements (RVEs) [23].

To include the variation of microstructure, the coefficients were determined as a function of microstructure. The elastic constants (C_{11}, C_{12}, and C_{44}), critical resolved shear stress (g_0), and strain hardening modulus (h_0) were determined from the stress–strain relationships, based on the tensile test results, as in Verma and Biswas's method [36]. This approach estimated the coefficients by trial and error until it achieved an appreciable fit with the experimental data. The shear stress amplitude ($\frac{\Delta \tau}{2}$) and plastic slip per cycle ($\frac{\Delta \gamma_p}{2}$) in Equation (8) can be computed more accurately by the crystal plasticity finite element model if we have precise values for the elastic constants (C_{11}, C_{12}, and C_{44}), the critical resolved shear stress (g_0), and the strain hardening modulus (h_0) as a function of microstructure.

2.3. Experiments

In this study, we took the results from the fatigue crack initiation experiments carried out by Yang et al. [14] to validate the fatigue model's life estimate. Nickel-based superalloy IN718 samples were fabricated by selective laser melting. The processing conditions were adjusted so that the density was around 99.9%. The round bar specimens were 53.6 mm long with a smooth dog-bone shape of 3 mm in diameter. The gauge regions were machined and polished to remove surface defects for the as-built components. Then, the samples were tested with an ultrasonic fatigue testing machine (Shimadzu 2000). Symmetrical push-pull loading was applied using the stress ratio $R = -1$ and a frequency of about 20 kHz.

During fatigue loading, the samples were periodically observed by a scanning electron microscope (SEM). The crack initiation was defined as the smallest cracks that could be observed by the SEM, with a maximum micro-crack size of approximately 30 μm. In this experiment, Yang et al. found that the surface cracks began before the interior cracks. This led us to focus on surface crack initiations for validation of the fatigue model. The formulas for surface crack initiation life based on the experiments were incorporated in Equation (13).

$$\sigma_a = 1280.9 \times N_i^{-0.0618} \tag{13}$$

where σ_a is the applied stress amplitude and N_i is the fatigue crack initiation life.

The tensile test specimen and SLM processing parameters were designed to be identical to the samples of fatigue. The dog-bone design forced the weakest region to be at the gauge area in the middle of the specimen. Thus, the elastic constants (C_{11}, C_{12}, and C_{44}), the critical resolved shear stress (g_0), and the strain hardening modulus (h_0) could be determined based on the microstructure of the weakest crack initiation area.

3. Results

3.1. Fatigue Parameter Estimation

The elastic constants (C_{11}, C_{12}, and C_{44}) and hardening coefficients (g_0, h_0) were determined based on the stress–strain relationship derived from the tensile test results. The elastic constant C_{ijkl} correlated the elastic behaviors with the applied stress, as shown in Equation (14).

$$\sigma_{ij} = \sum C_{ijkl} \varepsilon_{kl} \tag{14}$$

where σ_{ij} is the stress in the i direction in the plane perpendicular to the j direction, and ε_{ij} is the elastic strain response. The model assumes symmetric crystal properties, so only C_{11}, C_{12}, and C_{44} were considered. The critical resolved shear stress (g_0) is the stress required to initiate slip. This parameter determines the stress condition for which the material starts yielding. The strain hardening modulus (h_0) is the parameter that determines the slope of hardening after the point of yield of the material.

As in Verma and Biswas's method [35], the elastic constants, critical resolved shear stress, and strain hardening modulus were determined by the best fit of a small deformation region of the tensile testing measurements, as shown in Table 1. The model's simulation results were a reasonable enough fit to be calibrated by the empirical data, as shown in Figure 2.

Table 1. The elastic constants (C_{11}, C_{12}, and C_{44}), critical resolved shear stress (g_0), and strain hardening modulus (h_0) estimates.

C_{11}	C_{12}	C_{44}	g_0	h_0
230 GPa	170 GPa	100 GPa	0.42 GPa	6.4 GPa

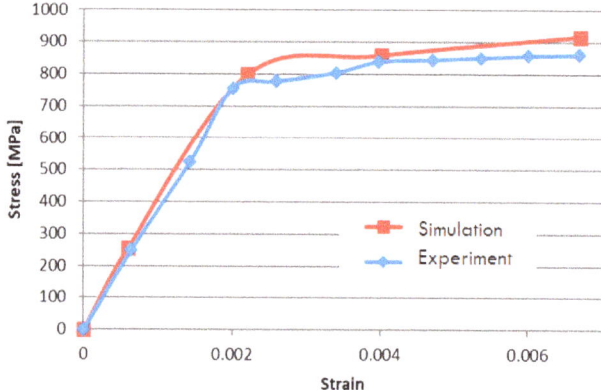

Figure 2. Calibration of coefficients by comparison of experimental data and simulation.

Since the specimens were designed in a round dog-bone shape, we were able to ensure that the stress-concentrated area during the tensile test and the potential fatigue crack initiation region were identical, at the gauge area in the middle of specimen. The comparison of experimental data and the simulation demonstrated that this approach was able to calibrate the coefficients for the microstructure of the weakest region of the samples. The accuracy of this simulation was around 95%, which can improve if needed.

3.2. Fatigue Crack Initiation Life Estimation

After the elastic constants, critical resolved shear stress, and strain hardening modulus were calibrated for the microstructure of the weakest region, we were able to use these coefficients as the inputs for the CPFEM.

In this study, the finite element model was used to compute the shear stress amplitude ($\frac{\Delta \tau}{2}$) and plastic slip per cycle ($\frac{\Delta \gamma_p}{2}$) during fatigue testing with applied stress values (σ_a) of 500, 600, and 700 MPa, as shown in Table 2. The fatigue crack initiation life can be computed by Equation (8). We used the 700 MPa case as a benchmark and estimated the fatigue crack initiation life with the energy ratio, as shown in Equation (15). We assumed that the surface energy (γ_s), energy efficiency coefficient (ρ), and defect size (t) were constant for the different samples, a reasonable assumption since the samples were fabricated under identical processing conditions and their microstructures were similar.

$$\frac{(N_i)_{compute}}{(N_i)_{700}} = \frac{\left(\frac{\Delta \tau}{2}\right)_{700} \left(\frac{\Delta \gamma_p}{2}\right)_{700}}{\left(\frac{\Delta \tau}{2}\right)_{compute} \left(\frac{\Delta \gamma_p}{2}\right)_{compute}} \tag{15}$$

Table 2. Computation results of shear stress amplitude ($\frac{\Delta \tau}{2}$) and plastic slip per cycle ($\frac{\Delta \gamma_p}{2}$).

σ_a	$\frac{\Delta \tau}{2}$	$\frac{\Delta \gamma_p}{2}$
500 MPa	412.4 MPa	1.2×10^{-5} MPa
600 MPa	491.7 MPa	1.4×10^{-5} MPa
700 MPa	578.6 MPa	2.1×10^{-4} MPa

The fatigue crack initiation life prediction results were compared with the empirical data, as shown in Table 3. We assumed that the crack initiation life could be treated as a normal distribution.

The empirical average crack initiation life cycle was estimated by Equation (13) from Yang et al. [14]. The standard deviation (SD) was estimated from the actual empirical data. The computation results were within the range of one standard deviation (SD) from the average crack initiation life. As indicated in Figure 3, our CPFEM model was able to reasonably predict the fatigue crack initiation life of materials fabricated by AM.

Table 3. Comparing fatigue crack initiation life estimation with experimental data.

σ_a	Experiments				Simulations
	Average (μ)	Standard Deviation (SD)	Upper Bound ($\mu + D$)	Lower Bound ($\mu - SD$)	
500 MPa	4.08×10^6	3.08×10^6	6.56×10^6	3.92×10^5	4.32×10^5
600 MPa	2.14×10^5	1.92×10^5	5.33×10^5	1.49×10^5	3.11×10^5
700 MPa	1.76×10^4	–	–	–	Benchmark

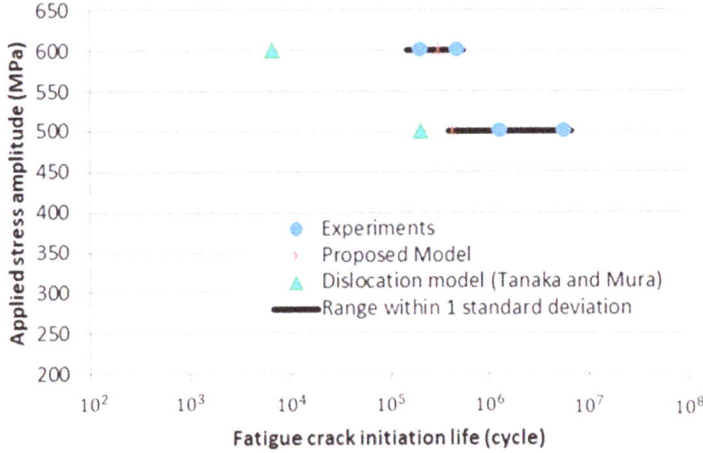

Figure 3. Comparison of fatigue crack initiation life predicted by the proposed model and the Tanaka and Mura model [21].

The proposed model can predict the crack initiation life of materials fabricated by AM with reasonable accuracy within the range of one standard deviation (SD). On the contrary, if we applied the dislocation model [21] to AM materials, the predicted initiation life was outside of one standard deviation, as shown in Figure 3. The main reason for this improvement in accuracy was that the proposed model considered the microstructure effect at the weakest region of the components. The stress and strain response at the potential crack initiation area can be accurately computed by the crystal plasticity finite element model.

In contrast, the dislocation model [21] did not consider the anisotropic nature of the microstructure, which led to poor accuracy. The dislocation model is as shown in Equation (16):

$$N_i = \frac{AW_c}{(\Delta \tau - 2\tau_k)^2} \tag{16}$$

where A is a function of crack mechanism, W_c is the fracture energy per unit area, $\Delta \tau$ is the shear stress, and τ_k is the friction stress. These coefficients are related to the microstructure and slip system at the localized potential crack initiation area. However, Tanaka and Mura neglected the effect of microstructure and assumed that these coefficients were material constants.

The presented fatigue crack initiation model determined the elastic constants (C_{11}, C_{12}, and C_{44}), the critical resolved shear stress (g_0), and the strain hardening modulus (h_0) as a function of microstructure. These coefficients were determined based on the stress–strain relationships derived from tensile test results. By incorporating the effect of microstructure on the elastic constants (C), the shear stress amplitude ($\frac{\Delta \tau}{2}$), as shown in Equation (16), can be computed more accurately. In addition, by considering the effect of microstructure on the critical resolved shear stress (g_0) and the strain hardening modulus (h_0), the localized dislocation slip and plastic slip per cycle ($\frac{\Delta \gamma_p}{2}$), as shown in Equation (17), can be precisely calculated. As discussed in previous papers [16,22–25], the energy efficiency coefficient (ρ), and the maximum persistent slip band width (t_m) are also coefficients related to microstructure. However, due to limited experimental data, we assumed they were a constant value in this study. These coefficients, as a function of microstructure, can be further studied in future work. This approach can improve the estimation of crack initiation life by fundamentally studying the way in which each coefficient is affected by the microstructure at the potential crack initiation area.

Another advantage of the fatigue crack initiation model presented here is that only a few empirical data are needed to get a reasonable prediction of initiation life. The material properties can be determined from the stress–strain relationship, based on the tensile test results. A set of randomly generated RVEs can reasonably simulate the material's microstructure at the weakest region. Although the fatigue crack initiation life variation is large for materials fabricated by AM, the presented model was able to predict the life to within one standard deviation in this case study. For future work, the model can be applied to different crack mechanisms initiated from gas pores, lack of diffusion, and slip band. In addition, actual components' potential crack initiation zones can be scanned using electron backscatter diffraction to develop more representative RVEs for CPFEM models.

In summary, a methodology for predicting the fatigue crack initiation life of metals fabricated by AM was presented. A set of RVEs was randomly generated for the potential crack initiation zone. The elastic constants (C_{11}, C_{12}, and C_{44}), critical resolved shear stress (g_0), and strain hardening modulus (h_0) were determined from the stress–strain relationships based on the tensile test results. The shear stress amplitude ($\frac{\Delta \tau}{2}$) and plastic slip per cycle ($\frac{\Delta \gamma_p}{2}$), as shown in Equation (8), can be computed more accurately by the crystal plasticity finite element model if we have precise values of elastic constants (C_{11}, C_{12}, and C_{44}), critical resolved shear stress (g_0), and strain hardening modulus (h_0) as a function of microstructure. The energy efficiency coefficient (ρ), and maximum persistent slip band width (t_m) are also coefficients related to microstructure. However, due to limited experimental data, we assumed that they were a constant value in this study. These coefficients, as a function of microstructure, can be further studied in future works to improve the model's accuracy.

4. Conclusions

A methodology for predicting the fatigue crack initiation life of metals fabricated by AM was presented. The fatigue crack initiation model determined the elastic constants (C_{11}, C_{12}, and C_{44}), critical resolved shear stress (g_0), and strain hardening modulus (h_0) as a function of microstructure. These coefficients were determined based on the stress–strain relationships, derived from the tensile test results. By incorporating the effect of microstructure on the elastic constants (C), the shear stress amplitude ($\frac{\Delta \tau}{2}$) can be computed more accurately. In addition, by considering the effect of microstructure on the critical resolved shear stress (g_0) and the strain hardening modulus (h_0), the localized dislocation slip and plastic slip per cycle ($\frac{\Delta \gamma_p}{2}$) can be precisely calculated. This study has enabled a major advance in fatigue research by modeling the crack initiation life of materials fabricated by AM through considering these coefficients as a function of microstructure. Using this model, the large variations in the fatigue lives of metal components fabricated using AM can be incorporated into prediction models.

Author Contributions: Conceptualization, C.R.L. and C.-Y.O.; methodology, C.-Y.O., C.R.L., and R.V.; software, R.V.; validation, C.-Y.O.; writing—original draft preparation, C.-Y.O.; writing—review and editing, R.V. and C.R.L.; supervision, C.R.L.; project initiation and administration, C.R.L.; funding acquisition, C.R.L. All authors have read and agreed to the published version of the manuscript.

Funding: This work was supported by the National Science Foundation (CMMI), through grant No. 1562960. Some ideas and materials are patent pending, owned by Purdue University.

Conflicts of Interest: The authors declare no conflict of interest.

References

1. Seifi, M.; Salem, A.A.; Beuth, J.; Harrysson, O.; Lewandowski, J.J. Overview of Materials Qualification Needs for Metal Additive Manufacturing. *JOM* **2016**, *68*, 747–764. [CrossRef]
2. Draper, S.; Locci, I.; Lerch, B.; Ellis, D.; Senick, P.; Meyer, M.; Free, J.; Cooper, K.; Jones, Z. Materials Characterization of Additively Manufactured Components for Rocket Propulsion. In Proceedings of the 66th International Astronautical Congress, Jerusalem, Israel, 12–16 October 2015; pp. 1–9.
3. Wohler's Report, Additive Manufacturing and 3D Printing State of the Industry Annual Worldwide Progress Report. 2017. Available online: https://wohlersassociates.com/press72.html (accessed on 3 April 2017).
4. Li, P.; Warner, D.; Fatemi, A.; Phan, N. Critical assessment of the fatigue performance of additively manufactured Ti–6Al–4V and perspective for future research. *Int. J. Fatigue* **2016**, *85*, 130–143. [CrossRef]
5. Edwards, P.; Ramulu, M. Fatigue performance evaluation of selective laser melted Ti–6Al–4V. *Mater. Sci. Eng. A* **2014**, *598*, 327–337. [CrossRef]
6. Kobryn, P.A.; Semiatin, S.L. Mechanical properties of laser-deposited Ti–6Al–4V. In Proceedings of the Solid Freeform Fabrication Proceedings, Austin, TX, USA, 6–8 August 2001.
7. Walker, K.; Liu, Q.; Brandt, M. Evaluation of fatigue crack propagation behaviour in Ti-6Al-4V manufactured by selective laser melting. *Int. J. Fatigue* **2017**, *104*, 302–308. [CrossRef]
8. Zhai, Y.; Galarraga, H.; Lados, D.A. Microstructure, static properties, and fatigue crack growth mechanisms in Ti-6Al-4V fabricated by additive manufacturing: LENS and EBM. *Eng. Fail. Anal.* **2016**, *69*, 3–14. [CrossRef]
9. Gordon, J.; Haden, C.; Nied, H.; Vinci, R.P.; Harlow, D. Fatigue crack growth anisotropy, texture and residual stress in austenitic steel made by wire and arc additive manufacturing. *Mater. Sci. Eng. A* **2018**, *724*, 431–438. [CrossRef]
10. Yadollahi, A.; Shamsaei, N. Additive manufacturing of fatigue resistant materials: Challenges and opportunities. *Int. J. Fatigue* **2017**, *98*, 14–31. [CrossRef]
11. Choi, Y.; Liu, C.R. Rolling contact fatigue life of finish hard machined surfaces. *Wear* **2006**, *261*, 485–491. [CrossRef]
12. Mughrabi, H. Microstructural mechanisms of cyclic deformation, fatigue crack initiation and early crack growth. *Philos. Trans. R. Soc. A Math. Phys. Eng. Sci.* **2015**, *373*, 20140132. [CrossRef]
13. Kazymyrovych, V.; Bergström, J.; Burman, C. The Significance of Crack Initiation Stage in Very High Cycle Fatigue of Steels. *Steel Res. Int.* **2010**, *81*, 308–314. [CrossRef]
14. Yang, K.; Huang, Q.; Wang, Q.; Chen, Q. Competing crack initiation behaviors of a laser additively manufactured nickel-based superalloy in high and very high cycle fatigue regimes. *Int. J. Fatigue* **2020**, *136*, 105580. [CrossRef]
15. Fine, M.E.; Bhat, S.P. A model of fatigue crack nucleation in single crystal iron and copper. *Mater. Sci. Eng. A* **2007**, *468*, 64–69. [CrossRef]
16. Ou, C.-Y.; Liu, C.R. The Effects of Grain Size and Strain Amplitude on Persistent Slip Band Formation and Fatigue Crack Initiation. *Met. Mater. Trans. A* **2019**, *50*, 5056–5065. [CrossRef]
17. Kruth, J.; Mercelis, P.; Van Vaerenbergh, J.; Froyen, L.; Rombouts, M. Binding mechanisms in selective laser sintering and selective laser melting. *Rapid Prototyp. J.* **2005**, *11*, 26–36. [CrossRef]
18. Gu, D.; Shen, Y. Balling phenomena in direct laser sintering of stainless steel powder: Metallurgical mechanisms and control methods. *Mater. Des.* **2009**, *30*, 2903–2910. [CrossRef]
19. Fine, M. Phase transformation theory applied to elevated temperature fatigue. *Scr. Mater.* **2000**, *42*, 1007–1012. [CrossRef]
20. Bhat, S.P.; Fine, M.E. Fatigue crack nucleation in iron and a high strength low alloy steel. *Mater. Sci. Eng. A* **2001**, *314*, 90–96. [CrossRef]

21. Tanaka, K.; Mura, T. A Dislocation Model for Fatigue Crack Initiation. *J. Appl. Mech.* **1981**, *48*, 97–103. [CrossRef]
22. Voothaluru, R.; Liu, C.R. Determination of lattice level energy efficiency for fatigue crack initiation. *Fatigue Fract. Eng. Mater. Struct.* **2013**, *36*, 670–678. [CrossRef]
23. Ou, C.-Y.; Voothaluru, R.; Liu, C.R. A Methodology for Incorporating the Effect of Grain Size on the Energy Efficiency Coefficient for Fatigue Crack Initiation Estimation in Polycrystalline Metal. *Metals* **2020**, *10*, 355. [CrossRef]
24. Voothaluru, R.; Liu, C.R. A crystal plasticity based methodology for fatigue crack initiation life prediction in polycrystalline copper. *Fatigue Fract. Eng. Mater. Struct.* **2014**, *37*, 671–681. [CrossRef]
25. Rohit, V. A Crystal Plasticity Based Methodology for Modeling Fatigue Crack Initiation and Estimating Material Coefficients to Predict Fatigue Crack Initiation Life at Micro, Nano and Macro Scales. Ph.D. Thesis, Purdue University, West Lafayette, IN, USA, 2014.
26. Taylor, G.I. Plastic Strain in Metals. *Inst. Met.* **1938**, *62*, 307.
27. Asaro, R.J. Crystal Plasticity. *J. Appl. Mech.* **1983**, *50*, 921–934. [CrossRef]
28. Hutchinson, J.W. Bounds and self-consistent estimates for creep of polycrystalline materials. *Proc. R. Soc. Lond. Ser. A Math. Phys. Sci.* **1976**, *348*, 101–127. [CrossRef]
29. Brown, S.; Kim, K.; Anand, L. An internal variable constitutive model for hot working of metals. *Int. J. Plast.* **1989**, *5*, 95–130. [CrossRef]
30. Huang, Y. *A User-Material Subroutine Incorporating Single Crystal Plasticity in the ABAQUS Finite Element Program*; Harvard University Research Report; Harvard University: Cambridge, MA, USA, 1991.
31. Liu, B.; Raabe, D.; Roters, F.; Eisenlohr, P.; Lebensohn, R.A. Comparison of Finite Element and Fast Fourier Transform Crystal Plasticity Solvers for Texture Prediction Model. *Simul. Mater. Sci. Eng.* **2010**, *18*, 085005. [CrossRef]
32. Bedekar, V.; Voothaluru, R.; Yu, D.; Wong, A.; Galindo-Nava, E.; Gorti, S.B.; An, K.; Hyde, R.S. Effect of nickel on the kinematic stability of retained austenite in carburized bearing steels—In-situ neutron diffraction and crystal plasticity modeling of uniaxial tension tests in AISI 8620, 4320 and 3310 steels. *Int. J. Plast.* **2020**, *131*, 102748. [CrossRef]
33. Smit, R.; Brekelmans, W.; Meijer, H. Prediction of the mechanical behavior of nonlinear heterogeneous systems by multi-level finite element modeling. *Comput. Methods Appl. Mech. Eng.* **1998**, *155*, 181–192. [CrossRef]
34. Kumar, R.S.; Wang, A.-J.; McDowell, D.L. Effects of Microstructure Variability on Intrinsic Fatigue Resistance of Nickel-base Superalloys—A Computational Micromechanics Approach. *Int. J. Fract.* **2006**, *137*, 173–210. [CrossRef]
35. Zhang, J.; Prasannavenkatesan, R.; Shenoy, M.M.; McDowell, D.L. Modeling fatigue crack nucleation at primary inclusions in carburized and shot-peened martensitic steel. *Eng. Fract. Mech.* **2009**, *76*, 315–334. [CrossRef]
36. Verma, R.K.; Biswas, P. Crystal plasticity-based modelling of grain size effects in dual phase steel. *Mater. Sci. Technol.* **2016**, *32*, 1553–1558. [CrossRef]

© 2020 by the authors. Licensee MDPI, Basel, Switzerland. This article is an open access article distributed under the terms and conditions of the Creative Commons Attribution (CC BY) license (http://creativecommons.org/licenses/by/4.0/).

Article

The Effect of Equal-Channel Angular Pressing on Microstructure, Mechanical Properties, and Biodegradation Behavior of Magnesium Alloyed with Silver and Gadolinium

Boris Straumal [1,2,3,11,*], Natalia Martynenko [1,4], Diana Temralieva [1,4], Vladimir Serebryany [4], Natalia Tabachkova [1,5], Igor Shchetinin [1], Natalia Anisimova [1,6], Mikhail Kiselevskiy [1,6], Alexandra Kolyanova [4], Georgy Raab [7], Regine Willumeit-Römer [8], Sergey Dobatkin [1,4] and Yuri Estrin [9,10]

1. National University of Science and Technology "MISIS", Leninsky Avenue 4, 119991 Moscow, Russia; nata_roug@mail.ru (N.M.); diana4-64@mail.ru (D.T.); ntabachkova@gmail.com (N.T.); ingvvar@gmail.com (I.S.); n_anisimova@list.ru (N.A.); kisele@inbox.ru (M.K.); dobatkin.sergey@gmail.com (S.D.)
2. Institute of Solid State Physics and Chernogolovka Scientific Center of the Russian Academy of Sciences, Chernogolovka, Leninskiy Prospekt, 14, 119991 Moscow, Russia
3. Institute of Nanotechnology, Karlsruhe Institute of Technology, Eggenstein-Leopoldshafen, 76131 Karlsruhe, Germany
4. A.A. Baikov Institute of Metallurgy and Materials Science of the RAS, Leninskiy Prospekt, 49, 119334 Moscow, Russia; vns@imet.ac.ru (V.S.); sasha-kolianova@yandex.ru (A.K.)
5. A.M. Prokhorov General Physics Institute of the RAS, 119991 Moscow, Russia
6. "N.N. Blokhin National Medical Research Center of Oncology" of the Ministry of Health of the Russian Federation, 119991 Moscow, Russia
7. Ufa State Aviation Technical University, 450077 Ufa, Russia; giraab@mail.ru
8. Institute of Materials Research, Division Metallic Biomaterials, Helmholtz-Zentrum Geesthacht (HZG), 21502 Geesthacht, Germany; Regine.Willumeit@hzg.de
9. Department of Materials Science and Engineering, Monash University, 3800 Clayton, Australia; yuri.estrin@monash.edu
10. Department of Mechanical Engineering, The University of Western Australia, 6009 Crawley, Australia
11. Institute of Solid State Physics of the Russian Academy of Sciences, Ac. Ossipyan str., 2, 142432 Chernogolovka, Russia
* Correspondence: straumal@issp.ac.ru

Received: 17 September 2020; Accepted: 8 October 2020; Published: 10 October 2020

Abstract: The effect of equal channel angular pressing (ECAP) on the microstructure, texture, mechanical properties, and corrosion resistance of the alloys Mg-6.0%Ag and Mg-10.0%Gd was studied. It was shown that ECAP leads to grain refinement of the alloys down to the average grain size of 2–3 µm and 1–2 µm, respectively. In addition, in both alloys the precipitation of fine particles of phases $Mg_{54}Ag_{17}$ and Mg_5Gd with sizes of ~500–600 and ~400–500 nm and a volume fraction of ~9% and ~8.6%, respectively, was observed. In the case of the alloy Mg-6.0%Ag, despite a significant grain refinement, a drop in the strength characteristics and a nearly twofold increase in ductility (up to ~30%) was found. This behavior is associated with the formation of a sharp inclined basal texture. For alloy Mg-10.0%Gd, both ductility and strength were enhanced, which can be associated with the combined effect of significant grain refinement and an increased probability of prismatic and basal glide. ECAP was also shown to cause a substantial rise of the biodegradation rate of both alloys and an increase in pitting corrosion. The latter effect is attributed to an increase in the dislocation density induced by ECAP and the occurrence of micro-galvanic corrosion at the matrix/particle interfaces.

Keywords: biomedical materials; magnesium alloys; equal-channel angular pressing (ECAP); microstructure; X-ray diffraction (XRD); texture; mechanical properties; biodegradation

1. Introduction

The combination of good biocompatibility and acceptable mechanical properties made magnesium alloys one of the most popular groups of materials for bioresorbable implants [1–7]. In recent years, research has been increasingly directed towards the development of magnesium alloys for medical applications. Along with good biocompatibility, these alloys often have additional functional properties. These properties, in combination with a suitable degradation rate, result from a smart selection of alloying elements. For example, it is possible to create an alloy with antibacterial activity. A medical prosthesis or implant made from such an alloy can reduce a risk of wound infection in the postoperative period due to gradual release of metal ions in combination with temporal pH changes inhibiting the growth of bacteria during the degradation of the implanted device. A good example are alloys of the Mg-Zn-Sn system. Not only do they possess good biocompatibility, but they also inhibit the growth of bacteria due to alkalization of the medium during degradation and the concomitant release of Mg, Sn, and Zn ions [8,9]. Gao et al. [10] showed that alloying magnesium with a small amount of Sr and Ga (up to 0.1 wt.%) effectively suppresses the activity of Gram-positive and Gram-negative bacteria (*Staphylococcus aureus, Staphylococcus epidermidis, Escherichia coli*), mainly due to the release of Ga^{3+} ions. Magnesium alloyed with silver also shows similar properties [11–14].

Another important area of research concerns medical magnesium alloys with an antitumor effect. Currently, development of drugs with systemic cytostatic activity for treatment of cancer patients is a truly topical area of medicine. Despite that, surgery is still one of the most used treatments. In this case, there usually is a need for partial replacement of the affected area with an artificial product. Examples include resection of bone affected by osteosarcoma or excision of a part of the esophagus. Due to their biodegradability, magnesium alloys are promising candidates for such applications. An emerging area of application of Mg alloys is oncology. Indeed, biodegradability potentially enables deployment of implants made from magnesium alloys as a platform for targeted delivery of gradually released cytostatic agents that suppress the growth and vitality of tumor cells in the area of neoplasm formation. Recent studies of medicinal magnesium alloys have been aimed at developing alloy compositions that would allow for inhibiting the vital activity of cancer cells, while not exerting a detrimental effect on normal cells. Thus, Wang et al. [15] showed that a Mg-Zn-Y-Nd alloy has a greater cytotoxicity to esophageal cancer cells than 317L stainless steel, which is frequently used in the production of stents. Studies carried out on giant cell tumors of bone incubated with samples of zoledronic acid-loaded magnesium-strontium alloys also demonstrated the advantages of this approach [16]. It was shown in [17] that the appropriate degradation of even pure magnesium leads to the inhibition of the growth of the MG63 osteosarcoma cell line, while the addition of Ag and Y enhances this effect. Studies carried out on the U2OS cancer cell line showed the effectiveness of the additions of Zn [18] and La [19] as inhibitors of cell growth, the effect being enhanced with an increase in the content of these elements. In addition, the present authors also carried out research aimed at studying the effect of the addition of rare earth elements [20] and Ag [21] on the vitality of tumor cells. These previous studies allow us to conclude that the use of magnesium alloys with an appropriate composition and degradation behavior is a promising avenue for treatment of cancer patients.

However, it should be emphasized that, quite often, the medical magnesium alloys may contain alloying elements which negatively affect their mechanical properties. This may preclude their use in medical devices. Indeed, orthopedic implants must have a sufficiently high level of strength, while stents require good ductility. Potential negative effects of alloying may be compensated for by the microstructure modification of the alloy. This can be achieved, for example, by plastic deformation imparting the desired properties to the material. From this viewpoint, equal-channel angular pressing (ECAP) looks particularly promising [22]. Due to significant grain refinement this technique generally produces, it can improve the mechanical characteristics of metallic materials quite substantially.

However, in contrast to the body-centered cubic and face-centered cubic metals, which deform well at room temperature [23–25], the low number of slip systems in magnesium and its alloys makes it necessary to carry out deformation processing at elevated temperatures. In this case, the accumulation of crystal lattice defects with a density sufficient for the formation of an ultrafine-grained (UFG) structure is hindered by the processes of recovery and recrystallization. Therefore, it is advisable to carry out ECAP with a stepwise decrease in the processing temperature for microstructure refinement [26–29] that would allow accumulation of sufficiently high dislocation density.

In this work, the effect of ECAP on the structure, texture, mechanical properties, and corrosion behavior of magnesium-based alloys Mg-6.0%Ag and Mg-10.0%Gd was studied. These alloys are attractive candidates for treatment of cancer patients. The earlier studies on alloys of the Mg-Ag [21,30] and Mg-Gd [31–33] systems showed that this deformation method is promising for the refinement of the microstructure and the associated increase in strength and ductility of alloys of these systems.

2. Materials and Methods

The alloys investigated, Mg-6.0%Ag and Mg-10.0%Gd, were obtained by smelting in a Nabertherm induction furnace (Nabertherm, Lilienthal, Germany) at a temperature of 720 °C using a mixture of Ar^+ with 3 vol.% SF_6 as a protective gas. Subsequently, cast ingots with a diameter of 60 mm underwent a T4 heat treatment (annealing at 425 °C for 16 h for Mg-6.0%Ag and at 525 °C for 8 h for Mg-10.0%Gd, followed by quenching in water). To obtain billets 10 mm in diameter required for ECAP, ingots of the alloys were extruded with an extrusion ratio of 1:25. The extrusion was conducted at a temperature of 425 °C and a ram speed of 1.1 mm/s for Mg-6.0%Ag alloy and at a temperature of 400 °C and a ram speed of 2.2 mm/s for Mg-10.0%Gd alloy. Before extrusion, the alloys were preheated for 60 min at deformation temperatures. After extrusion, both alloys were annealed again at a homogenization temperature (425 °C for 2 h and 525 °C for 1 h for Mg-6.0%Ag and Mg-10.0%Gd, respectively) and cooled by quenching in water. The resulting condition will be referred to as the initial state of the alloys. Route Bc ECAP was carried out with temperature being dropped in discrete steps after defined strain increments (Figure 1a). The intersection angle between the entry and exit channels of the ECAP die was 120° (Figure 1b), and the total number of passes was 12.

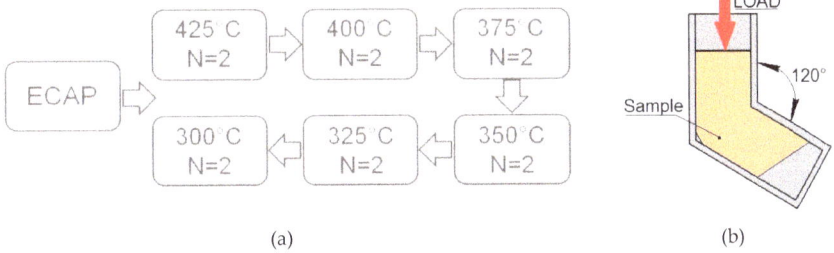

Figure 1. The equal channel angular pressing (ECAP) processing regime employed (**a**) and the scheme of ECAP process (**b**). N denotes the number of passes at a given temperature.

The study of the microstructure of the alloys in the initial state and after ECAP was carried out in the direction parallel to the direction of ECAP and extrusion preceding annealing for the initial state of the alloys. The microstructure of the alloys in the initial state and after ECAP was investigated using an optical microscope Axio Observer D1m (Carl Zeiss, Jena, Germany). The microstructure of alloys after ECAP was also studied by transmission electron microscopy (TEM) using a JEOL JEM 2010 microscope (Jeol, Tokyo, Japan). The operating voltage was 200 kV. The foils for TEM studies were prepared by ion-milling in a precision ion polishing system (Gatan, PIPS II, Gatan Inc. Pleasanton, CA, USA). A quantitative assessment of structural components was carried out by the method of random secants using the software Image Expert Professional 3 (Version 3, Moscow, Russia).

Phase analysis by means of X-ray diffraction (XRD) was performed on a Rigaku Ultima IV diffractometer (Rigaku, Japan) using Co-Kα radiation and a graphite monochromator for a diffracted beam. The spectra were analyzed using the PDXL software (Rigaku, Japan) by the Rietveld method with the PDF-2 powder diffractogram database (ICDD). Texture measurements were carried out using a DRON-7 X-ray diffractometer (SPE "Burevestnik," St. Petersburg, Russia) in CuKα radiation with the aid of the Texx [34] and Texxor [35] software (IMET RAS, Moscow, Russia). Five incomplete direct pole figures {10$\bar{1}$2}, {11$\bar{2}$0}, {10$\bar{1}$3}, {0004}, and {10$\bar{1}$4} were obtained with a maximum inclination angle $\alpha_{max} = 70°$ and a step size of 5° in the radial angle α and the azimuth angle β on a pole figure. The orientation distribution functions (ODFs) were calculated from the measured pole figures presented as a superposition of a large number (1,000) of standard distributions with a small scatter. The centers of standard functions were located on a regular three-dimensional grid in the orientation space [35]. From these ODFs, complete pole figures were also calculated. The volume fractions of the major orientations were estimated using the ODF as described in [36]. Using the Euler angles and the volume fractions of the orientations, the generalized Schmid factors for the existing deformation systems and the inverse orientation factors were calculated, following the procedure reported in [36]. The texture analysis was carried out in the directions parallel and perpendicular to the pressing axis.

The mechanical properties of the alloys were evaluated based on uniaxial tensile tests carried out at room temperature in an Instron 3382 testing machine (Instron, High Wycombe, UK) with an extension rate of 1 mm/min. The tests were carried out on flat samples with a cross-section of 2 mm × 1 mm and a gauge length of 5.75 mm.

The degradation of the alloys in vitro was studied at 37 °C. Before testing, the samples in the form of disks 10 mm in diameter and 1.5 mm in thickness were grinded with abrasive paper (from P800 to P2500), cleaned using an ultrasonic bath, and then sterilized by immersing them for four hours in 70% ethanol and drying under sterile conditions. The specimens (no less than four for each condition) were incubated in a DMEM (Dulbecco's Modified Eagle Medium; Sigma-Aldrich, St. Louis, MO, USA) culture medium for four days. After incubation, the specimens were rinsed in a solution of Cr_2O_3, $AgNO_3$, $Ba(NO_3)_2$, and distilled water for one minute in order to remove degradation products from their surfaces. The mass loss was determined by weighing the specimens with a Sartorius Pro 11 scale (ISO 9001; Sartorius Lab Instruments GmbH & Co, Göttingen, Germany) with an accuracy of three decimal places, or 0.001 g. The degradation rate, DR (in mm/year) was calculated using the following equation (ASTM_G1-03-E):

$$DR = 8.76 \times 10^4 \times \frac{\Delta m}{A \times t \times \rho} \tag{1}$$

where Δm is the mass loss in grams, t is the immersion time in hours, A is the specimen surface area in cm^2, and ρ is the density of the alloy in g/cm^3.

The surface of the samples after degradation tests was inspected using an instrumental microscope MMI-2 (NPZ, Novosibirsk, USSR).

3. Results

Figure 2 shows the microstructure of the alloys Mg-6.0%Ag and Mg-10.0%Gd in the initial state and after ECAP. The microstructure of both alloys after annealing consisted of grains of a supersaturated solid solution of silver or gadolinium in magnesium. No second-phase particle precipitation after quenching in water was detected metallographically and by XRD. The average grain size in the initial state was 70.9 ± 4.0 μm for the alloy Mg-6.0%Ag and 70.8 ± 4.0 μm for the alloy Mg-10.0%Gd (Figure 2a,b).

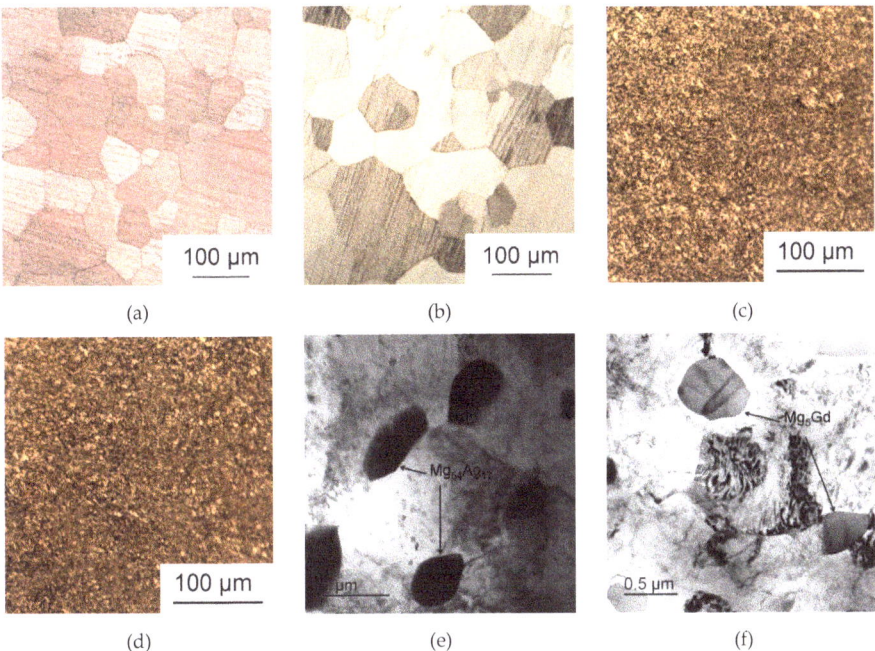

Figure 2. Structure of the alloys Mg-6.0%Ag (a,c,e) and Mg-10.0%Gd (b,d,f) in the initial state (a,b) and after ECAP (c–f).

ECAP led to a substantial refinement of the microstructure of both alloys. It should be noted that this microstructure was generally homogeneous without signs of bimodality (Figure 2c,d). The grain size decreased to ~2–3 μm in the case of the Mg-6.0% Ag alloy and to ~1–2 μm in the case of the Mg-10.0%Gd alloy. ECAP deformation conducted at elevated temperatures with cooling of the billets at room temperature led to the precipitation of second-phase particles in both alloys. In the Mg-6.0%Ag alloy, the precipitation of second-phase particles with a size of ~500–600 nm, located mainly at the grain boundaries, was found. In alloy Mg-10.0%Gd, precipitation of second-phase particles was observed mainly at triple points. The particles were round and ~400–500 nm in size (Figure 2e,f).

To identify the stoichiometric composition of precipitated particles and to determine their volume fraction, the XRD phase analysis of the alloys was carried out before and after deformation (Figure 3). Investigation of the Mg-6.0%Ag alloy in the initial state showed that after quenching, the alloy was entirely in a single-phase condition represented by a supersaturated solid solution of silver in magnesium. By contrast, in Mg-10.0%Gd alloy, about ~1.9% of a second phase, identified as $Mg_{46}Gd_9$, was detected. According to our analysis, this phase had face-centered cubic structure (space group $F\bar{4}3m$). In earlier studies [37] a compound with about the same stoichiometric composition was already identified as a Mg_5Gd. Furthermore, according to literature, the occurrence of this compound (Mg_5Gd) was also identified in Mg-10.0%Gd alloy after aging treatment [38]. It can be surmised that in our case this phase was formed during the extrusion process prior to annealing and did not have enough time to dissolve during the initial heat treatment. After ECAP, the volume fraction of particles of the $Mg_{46}Gd_9$ (Mg_5Gd) phase was found to be about 18.6%.

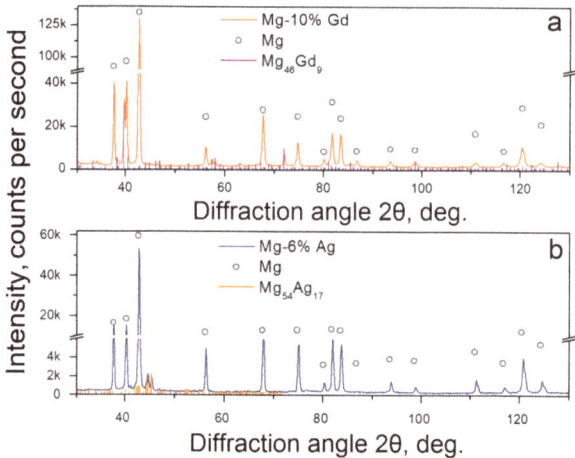

Figure 3. XRD spectra of the alloys Mg-10.0%Gd (**a**) and Mg-6.0%Ag (**b**) after ECAP. (Note that the $Mg_{46}Gd_9$ phase is nearly identical with the Mg_5Gd phase).

Investigation of the Mg-6.0% Ag alloy after ECAP also showed precipitation of second-phase particles with a volume fraction of ~9%, apparently during heating process before and during the ECAP processing. We identified this phase as $Mg_{54}Ag_{17}$ (space group Immm), which is in good agreement with the literature data [11,30,39].

Table 1 and Figure 4 show a summary of the mechanical properties of the alloys at the room temperature before and after ECAP. The yield stress (YS) of the alloy Mg-6.0%Ag in the initial state was 162 ± 3 MPa and its ultimate tensile strength (UTS) was 239 ± 1 MPa. After ECAP, a drop in these characteristics to 44 ± 6 MPa and 224 ± 5 and, respectively, was observed. At the same time, tensile elongation (El) of the alloy after ECAP rose to 30.6 ± 3.0% - to be compared to 16.0 ± 0.3% in the initial state. By contrast, ECAP of Mg-10.0%Gd led to an increase in both strength and ductility characteristics. The respective values of YS, UTS, and El rose from 123 ± 7 MPa, 185 ± 4 MPa, and 13.2 ± 1.7% in the initial state to 211 ± 1 MPa, 258 ± 2 MPa, and 18.0 ± 3.6% in the post-ECAP condition.

Table 1. Mechanical properties of the Mg-6.0%Ag and Mg-10.0%Gd alloys before and after ECAP.

	Processing	YS, MPa	UTS, MPa	El, %	Grain Size, μm
Mg-6.0%Ag	Initial state	162 ± 3	239 ± 1	16.0 ± 0.3	70.9 ± 4.0
	ECAP	44 ± 6	224 ± 5	30.6 ± 3.0	2–3
Mg-10.0%Gd	Initial state	123 ± 7	185 ± 4	13.2 ± 1.7	70.8 ± 4.0
	ECAP	211 ± 1	258 ± 2	18.0 ± 3.6	1–2

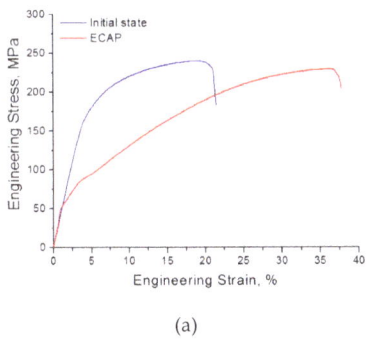 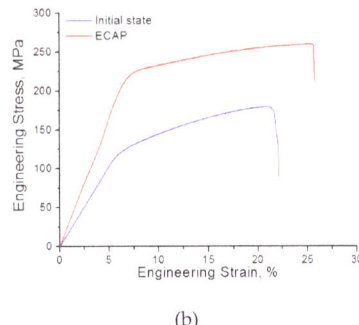

Figure 4. Engineering stress–strain response of the alloys Mg-6.0%Ag (**a**) and Mg-10.0%Gd (**b**) before and after ECAP.

It should also be noted that for the Mg-10.0%Gd alloy processed by ECAP the elastic portion of the engineering stress vs. engineering strain curve has a greater slope than for the alloy in the initial condition (Figure 4b). We believe that the reason for such behavior is an increase in Young's modulus due to texture and structure transformations that occur in the alloy after ECAP deformation. Our calculations (to be presented in forthcoming publications) confirm this hypothesis at a qualitative level. Another interesting issue for discussion is the occurrence of an inflection in the deformation curve for alloy Mg-6.0%Ag after ECAP (Figure 4a). This behavior (also observed for Mg-4.0%Ag deformed in compression [30]) may be associated with the activation of new slip systems during uniaxial tension or deformation twinning. While the latter mechanism cannot be ruled out on theoretical grounds, no twins were detected in the ECAP-processed Mg-6.0%Ag after uniaxial tension. At this stage, the question of the mechanism responsible for the occurrence of the inflection remains open.

In addition to the influence of ECAP on the mechanical characteristics through microstructure changes, the texture variation usually also comes into play. Therefore, we conducted texture measurements in the initial state and after ECAP (both in the longitudinal and the transversal sections of the billets) (Figure 5). The pole figures (PFs) of Mg-6.0%Ag exhibited a transformation from a sharp prismatic texture in the initial state to an inclined basal one. The inclination angle α = 45–50° relative to the longitudinal direction (LD) and a rotation angle β = 25–30° relative to the normal direction (ND) clockwise in the cross-section of the billet were found after ECAP (see pole figure (PF) {0004} in Figure 5a). The texture in the longitudinal section of the billet in the initial state was a rather weak prismatic type. It transformed to a sharp basal texture inclined at α = 85° relative to ND and rotated by an angle β = 45° relative to LD counterclockwise after ECAP (see PF {0004} in Figure 5b). Unlike this texture behavior, the texture of the alloy Mg-10.0%Gd in the initial and the post-ECAP condition was rather dispersed and did not exhibit sharp peaks, neither in the longitudinal nor in the transversal section of the billets (Figure 5c,d).

The results on the degradation rate for both alloys studied in the initial state and after ECAP are presented in Figure 6a. These data show that the microstructure changes and modification of phase composition occurring as a result of ECAP lead to an increase in the degradation rate. Thus, the degradation rate of the alloy Mg-6.0%Ag after ECAP rose to 3.74 ± 0.56 mm/year compared to 0.43 ± 0.10 mm/year in the initial state. In the case of the Mg-10.0%Gd alloy, an increase in the degradation rate from 0.15 ± 0.04 mm/year in the initial state to 2.19 ± 0.78 mm/year after ECAP was observed.

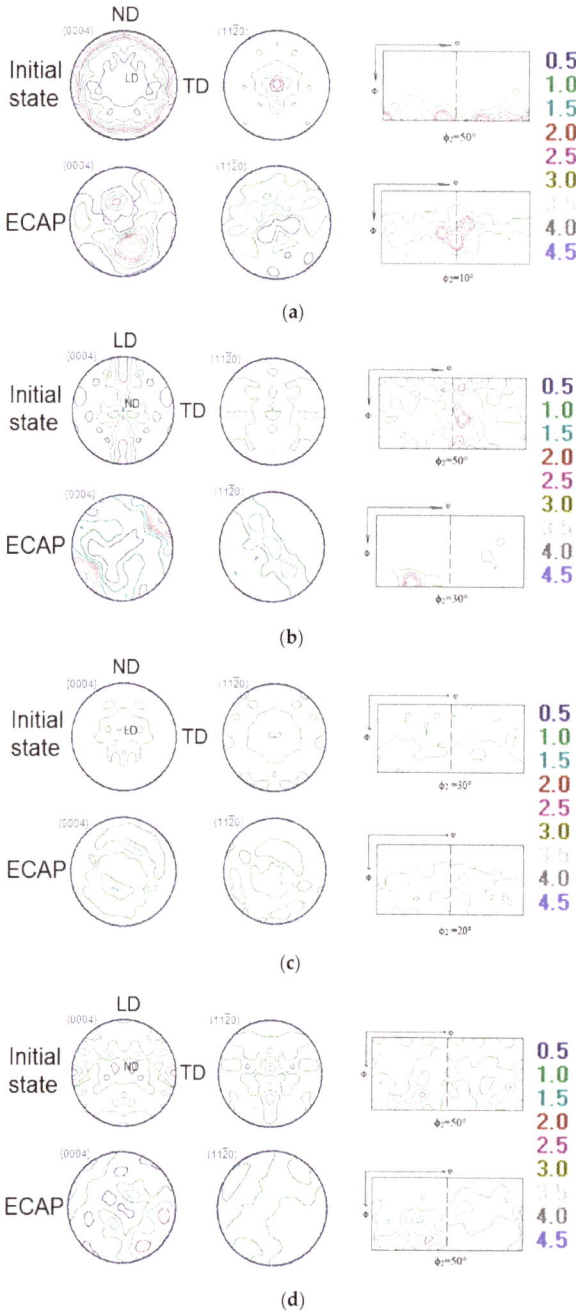

Figure 5. {0004} and {11$\bar{2}$0} pole figures and cross-sections of orientation distribution functions (ODF) of the Mg-6.0%Ag (**a,b**) and Mg-10.0%Gd (**c,d**) alloys in the initial state and after ECAP for transversal (**a,c**) and longitudinal (**b,d**) sections of billets. (The notation LD, TD, and ND corresponds to the longitudinal, transverse, and normal direction, respectively).

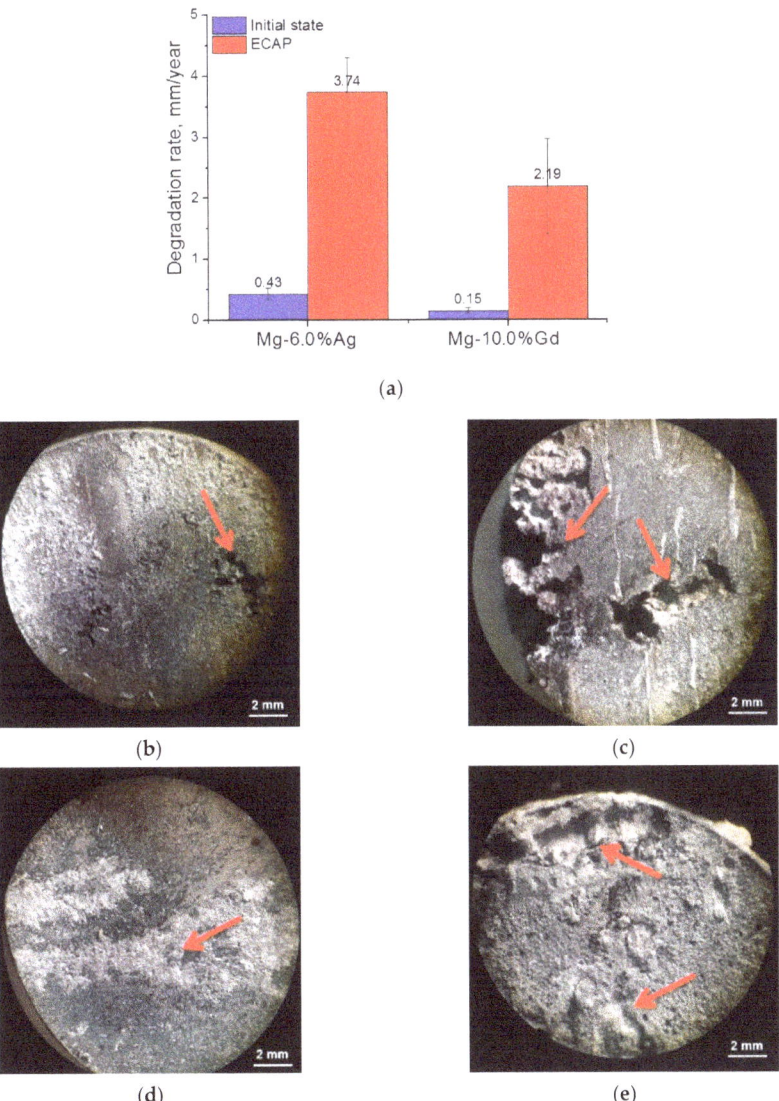

Figure 6. Degradation rate (**a**) and the surface of the samples after testing but before cleaning for alloy Mg-6.0%Ag (**b**,**c**) and alloy Mg-10.0%Gd (**d**,**e**). The images (**b**,**d**) correspond to the initial state and (**c**,**e**) to the post-ECAP state of the respective alloys. The arrows indicate sites of localized degradation.

Inspection of the sample surface after degradation tests showed nonuniformity of degradation of Mg-6.0%Ag in the initial state and especially after ECAP (Figure 6b,c). Samples of annealed Mg-6.0%Ag did show signs of localization of degradation but, unlike in the post-ECAP condition, pitting did not have a through-thickness character. In the case of the Mg-10.0%Gd alloy, nonuniform degradation was observed only in the post-ECAP samples (Figure 6e). Only minor signs of localization of corrosion were detected. To summarize these results, one can state that localization of degradation was most pronounced in the Mg-6.0%Ag alloy processed by ECAP. In this case, a profuse through-thickness pitting of samples was observed. For the ECAP-processed Mg-10.0%Gd alloy, a strong localization of corrosion was also found, but pitting was not as expressed as in the case of the ECAP-processed Mg-6.0%Ag alloy.

The observed acceleration of the biodegradation rate of both alloys due to ECAP would generally be regarded as an undesirable effect, which would put in question the usefulness of this processing for medical implant applications of the alloys considered. However, the situation is entirely different for specifically targeted applications in oncological patients. Here, a bioresorbable implant is used not only as a mechanically strong scaffold, but also as a therapeutic agent inhibiting the growth and proliferation of tumor cells. It should also be stated that the increase in degradation rate is still relatively moderate, which in the long run is tolerated by the healthy tissue [40].

4. Discussion

The studies reported here have shown that ECAP has a strong effect on the mechanical and biodegradation characteristics of magnesium alloys Mg-6.0%Ag and Mg-10.0%Gd. The effect is different for the two alloys, though. In the case of Mg-10.0%Gd, a typical situation is found: a decrease in the grain size and precipitation of second-phase particles lead to an increase in the yield strength and the ultimate tensile strength. A different picture is observed for the Mg-6.0%Ag alloy. In this case, ECAP leads to grain refinement to a scale similar to that observed in the Mg-10.0%Gd alloy. The size of the precipitated particles has a similar magnitude, as well. However, in the case of ECAP-processed Mg-6.0%Ag alloy, a decrease in both strength characteristics (both YS and UTS) is observed. On a positive side, doubling of the tensile elongation that characterizes ductility of the alloy is achieved (Figure 4, Table 1). In the case of Mg-10.0%Gd, an increase in ductility is also observed, but it is not as significant as for Mg-6.0%Ag. The most likely reason for this behavior is a change in texture induced by ECAP. As was shown for Mg-6.0%Ag, a sharp inclined basal texture occurred as a result of ECAP (Figure 5a,b). The formation of an inclined basal texture in magnesium alloys is known to have a negative effect on their strength, while having a beneficial effect on ductility. This is what was observed for the Mg-6.0%Ag alloy. A similar behavior was already found for binary Mg-2.0%Ag and Mg-4.0%Ag alloys [11] and the ternary Mg-1.0%Zn-0.3%Ca alloy [41]. A negative effect of the inclined basal texture on the yield stress was previously shown for the alloys AZ31 [42] and AM60 [43]. In the case of the Mg-10.0%Gd alloy considered here, the texture did not undergo significant changes; a weak texture in the initial state did not change its dispersed character as a result of ECAP (Figure 5c,d). However, the magnitude of the orientation factors calculated for the basal, prismatic, and pyramidal slip systems, as well as one of the main twinning systems in magnesium alloys, $\{10\bar{1}2\}<\bar{1}011>$, can provide an explanation for the observed increase in ductility.

Texture orientation factors for slip systems are determined by the values of the respective Schmid factors and are their reciprocals. Therefore, a decrease in the value of an orientation factor may indicate an increase in the probability of dislocation glide on the corresponding slip system. In our case, the estimation of the orientation factors for the Mg-10.0%Gd alloy before and after ECAP, calculated for the longitudinal section of the billet, indicates the highest probability of dislocation slip on the basal planes. Indeed, the orientation factor does not show big changes, while slip on other planes weakens (Table 2). At the same time, the calculation of the orientation factors for the cross-section of the billet indicates an increase in the probability of prismatic slip, as reflected in a decrease in the value of the respective orientation factor. An increase in the probability of slip on prismatic planes can be the reason for a slight rise of the ductility of the Mg-10.0%Gd alloy. We have already observed a similar effect on a magnesium alloy with Y and Nd treated by ECAP [44]. An improvement of the ductility of magnesium alloys due to the formation of a prismatic texture was also demonstrated earlier for the quaternary Mg-4.4%Al-0.9%Zn-0.4%Mn alloy [45]. In the case of the Mg-6.0%Ag alloy, the calculation of the orientation factors indicates the most probable slip of dislocations mainly on the basal planes, which is also confirmed by the type of the observed texture. However, it should be noted that this trend is most clearly expressed in the transversal cross section of the billet. The calculation of the orientation factors for the twinning system {10$\bar{1}$2}<$\bar{1}$011> for both alloys showed a decrease in the probability of twinning, which was confirmed by the absence of twins in the structure of both alloys after ECAP.

Table 2. Orientation factors for deformation systems of the alloys studied. (The abbreviations CS and LS stand for a (transversal) cross section and a longitudinal section, respectively).

State of the Alloys			Basal {0001}<1120>	Prismatic {10$\bar{1}$0}<1120>	Pyramidal <c+a>	Twinning {10$\bar{1}$2}<$\bar{1}$011>
Mg-6.0%Ag	CS	Initial state	7.5	3.6	4.9	4.7
		ECAP	4.0	5.1	5.3	5.9
	LS	Initial state	4.7	5.1	4.4	4.4
		ECAP	4.3	5.1	5.1	5.5
Mg-10.0%Gd	CS	Initial state	4.7	5.7	5.1	5.3
		ECAP	4.9	5.3	5.9	6.6
	LS	Initial state	4.5	4.7	4.9	5.0
		ECAP	4.5	5.1	5.0	5.2

The homogeneity of the microstructure of both alloys processed by ECAP suggests that the observed changes in their mechanical characteristics are associated with the structure refinement, precipitation of particles, and texture evolution. The calculation of the orientation factors shows that the basal slip prevails in the ECAP-modified alloy Mg-6%Ag. This holds true both for a longitudinal section of a processed billet (parallel to the direction of the tensile tests conducted to determine the mechanical characteristics) and for a transversal cross section of the billet. In the case of Mg-10%Gd, the magnitude of the calculated orientation factors indicates the activation of non-basal sliding. This assertion is valid only for the longitudinal section of the billet. It can be considered as established that for the alloy Mg-6%Ag the mechanical characteristics are primarily affected by texture effects, microstructure refinement and particle precipitation playing a subordinate role. By contrast, for the alloy Mg-10%Gd all three factors appear to influence the mechanical properties in accord. The combined effect of grain refinement and precipitation of particles of the Mg$_5$Gd phase (with the volume fraction of about 19%) is responsible for strengthening, while the activation of prismatic sliding accounts for an enhancement of ductility. However, it is known that the greatest strengthening effect after aging is observed in alloys containing from 10 to 20 wt.% Gd [46]. At the same time, the aging of the quenched alloy Mg-10%Gd has a little effect on its Vickers hardness [47], which in our case indicates that the contribution of grain refinement to strengthening still prevails over the contribution of particles. It is interesting that the equilibrium phase of Mg$_5$Gd precipitates in the form of plates parallel to prismatic planes {10$\bar{1}$0}$_{Mg}$ at the early stages of aging [46,48]. In our case, the main process of nucleation of phase particles is apparently observed at the initial stages of ECAP (sequentially through the nucleation of phases β″ (Mg$_3$Gd) → β′ (Mg$_{3-5}$Gd) → β$_1$ (Mg$_{3-5}$Gd) → β (Mg$_5$Gd)), but the coagulation and particle growth occurs with an increase in the number of passes (and, consequently, the number of heating before and during deformation). In the final state (after 12 ECAP passes), we mostly deal with large, rounded overaged particles, but we do not exclude the possibility of presence of small number of lamellar particles. At the same time, the studies carried out on the Mg–1.5Sn–1.4Zn–0.2Mn alloy with the addition of 0.3Ag indicated the formation of the particles ε′ (Mg$_{54}$Ag$_{17}$) [49]. These particles were rod-shaped and was normal to (0001)$_{Mg}$. Probably, in our case, there can also be a fraction of fine particles ε′, which inhibit dislocation slip, but the most part of the phase in the structure of the alloy Mg-6%Ag after ECAP is a rounded coagulated Mg$_{54}$Ag$_{17}$ particle ~500–600 nm in size (Figure 2e).

The study of the corrosion properties of the alloys showed that ECAP leads to a significant acceleration of the degradation process for both Mg-6.0%Ag and Mg-10.0%Gd alloys. The reason for the recorded increase in the degradation rate is the structural and phase changes occurring in the alloys during severe plastic deformation. On the one hand, grain refinement leads to a significant increase in the grain boundary area and in the density of dislocations. It is known that an increase in the dislocation density in metals and alloys often leads to a deterioration of their corrosion resistance [50,51]. Magnesium and its alloys are no exception to that rule. Thus, it was previously shown that grain refinement in pure magnesium caused by ECAP leads to an increase in its degradation rate and the formation of deeper pitting owing to a greater density of crystal lattice defects [51]. In the case of the ternary Mg-4.0%Zn-2.0%Ni alloy, it was shown that an increase in the extrusion temperature, which promotes a decrease in the dislocation density, led to a reduction of the degradation rate [52]. It was also shown that a reduced density of dislocations in the alloy LAZ832-0.2%Zr after extrusion is also one of the reasons for its good corrosion resistance [53]. On the other hand, there are also a number of reports asserting that the formation of an ultrafine-grained (UFG) structure in magnesium alloys can reduce the degradation rate due to the accelerated formation of a protective oxide-hydroxide layer and an ensuing increase of corrosion resistance [54–56]. However, it should be noted that no UFG structure was formed in the alloys studied. In the case of the Mg-4.0%Zn-2.0%Ni and LAZ832-0.2%Zr alloys, the dislocation density played an important role in the degradation rate. However, it was not the only factor governing the magnitude of the degradation rate. The occurrence of a second phase in the structure and its distribution affected the corrosion resistance substantially. It is known that the presence of more corrosion-resistant phases in magnesium alloys can lead to micro-galvanic corrosion, when a less resistant matrix acts as an anode, and second-phase particles act as a cathode. In our case, an increase in the degradation rate for both alloys was accompanied by the precipitation of particles of a second phase: $Mg_{54}Ag_{17}$ in the Mg-6.0%Ag alloy and Mg_5Gd in the Mg-10.0%Gd alloy. That is, particles of $Mg_{54}Ag_{17}$ and Mg_5Gd phases, acting as a cathode in relation to the less stable magnesium matrix, lead to an increase in the degradation rate of the alloys studied. At the same time, accelerated degradation, predominantly at the matrix/particle interface, can also promote a stronger localization of corrosion, as was observed for ECAP-treated alloys. Such behavior was also found for other magnesium alloys. For example, in the case of binary Mg-Sc alloys, an increase in the Sc content to 0.3 wt% led to an improvement of corrosion resistance, while a further increase in the Sc content led to the precipitation of Mg-Sc particles and an increase in the degradation rate by galvanic corrosion [57]. Similar behavior was demonstrated for Mg-Bi-Al-based alloys [58], Mg-6.0%Gd-2.0%Y-1.0%Zn-0.3%Zr alloy [59], Mg-Li-Ca alloys [60], etc. Not only does the micro-galvanic mechanism of corrosion cause the degradation of the matrix in the immediate vicinity of the particles, but it also causes grain-boundary corrosion if the precipitate particles reside at grain boundaries. For both alloys considered in this paper, the particle sites were the same. In addition, the formation of a protective film during corrosion of magnesium alloys was observed, mainly consisting of degradation products and compounds formed during the reaction with components of the corrosive medium. Although this film protects against degradation, it is rather unevenly distributed over the surface of the sample, often leaving unprotected spots. In such particle locations degradation proceeds at an accelerated rate, first around a particle, and then along the grain boundaries. In this case, separation of a grain (and sometimes of several grains) from the sample surface can occur, which entails the formation of pitting. Pits of this kind were observed especially clearly for alloys processed by ECAP, where the concentration of the second phase was high. The micro-galvanic mechanism of corrosion and grain-boundary corrosion of magnesium alloys are described in several works, for example, in [61–63].

Thus, the increase in the degradation rate of the Mg-6.0%Ag and Mg-10.0%Gd alloys after ECAP can be explained by a deformation-induced increase in the density of crystal lattice defects and the precipitation of second-phase particles, which promoted the occurrence of micro galvanic corrosion. Therefore, a potential way to improve the properties of Mg-based alloys, including their corrosion resistance, should focus on developing regimes of deformation in which precipitation of particles would be suppressed, or their amount would be minimized. This can be realized by lowering the initial deformation temperature, reducing the number and duration of heating cycles before and during ECAP, as well as reducing the number of deformation steps and the associated temperature drops. A suitable method to reduce the dislocation density can be the use of low-temperature annealing after

ECAP (below the temperature corresponding to the onset of active decomposition of a supersaturated solid solution). In trying to optimize the properties of the alloys one should bear in mind their intended application in bioresorbable implants. Hence, processing regimes need to be tuned to ensure the level of the degradation rate desirable for a given implant application.

5. Conclusions

1. Processing by ECAP was shown to cause a decrease in the average grain size down to 2–3 μm for the Mg-6.0%Ag alloy and 1–2 μm for the Mg-10.0%Gd alloy. In addition, ECAP-induced precipitation of particles of the $Mg_{54}Ag_{17}$ and Mg_5Gd phases with a size of 500–600 nm and 400–500 nm and a volume fraction of ~9 and ~18.6%, respectively, was observed.
2. ECAP was also found to lead to an increase in both strength and ductility of the Mg-10.0%Gd alloy. In the case of the Mg-6.0%Ag alloy, a decrease in the UTS and especially in the YS was observed. This was accompanied with an almost twofold increase in tensile ductility as represented by the strain to failure: from 16.0 ± 0.3% in the initial state to 30.6 ± 3.0% after ECAP.
3. An increase in the ductility of the Mg-6.0%Ag alloy combined with a decrease in the strength characteristics is associated with the formation of a sharp inclined basal texture and the predominant slip of dislocations on the basal planes. The improvement in the ductility of the Mg-10.0%Gd alloy achieved by ECAP is associated with an increase in the probability of dislocation slip on prismatic planes, with a retained high probability of slip on the basal planes.
4. For both alloys studied, ECAP led to a significant increase in the biodegradation rate and to a pronounced pitting. In the case of the Mg-6.0%Ag alloy, the degradation rate increased from 0.43 ± 0.10 mm/year in the initial state to 3.74 ± 0.56 mm/year after ECAP. For the Mg-10.0%Gd alloy, the degradation rate after ECAP increased to 2.19 ± 0.78 mm/year compared to 0.15 ± 0.04 mm/year in the initial state. The observed rise of the degradation rate after ECAP is associated with an increase in the dislocation density after deformation and the occurrence of micro galvanic corrosion due to the precipitation of particles of the $Mg_{54}Ag_{17}$ and Mg_5Gd phases.

The pronounced ECAP-induced rise of the biodegradation rate is undesirable for many medical applications. However, this property can be useful for obtaining alloys considered as a platform for local delivery of anticancer agents to reduce the tumor volume or to prevent recurrence of tumor growth in cancer patients. Indeed, the associated accelerated release of alloying elements with potential antitumor activity directly into the area of tumor growth can be considered as an additional remedy in antitumor therapy, complementing the standard course of systemic therapy.

Author Contributions: Conceptualization, B.S., Y.E. and R.W.-R.; Methodology, N.M., N.A., N.T., I.S., G.R. and V.S.; Software, N.M., N.T., G.R., A.K. and I.S.; Validation, M.K. and S.D.; Formal Analysis, B.S., S.D., Y.E. and R.W.-R.; Investigation, N.M., N.A., D.T., N.T., A.K., I.S. and V.S.; Resources, Y.E., S.D. and R.W.-R.; Data Curation, Y.E., B.S. and R.W.-R.; Writing—Original Draft Preparation, N.M.; Writing, Review and Editing, Y.E., S.D. and R.W.-R.; Visualization, N.M. and B.S.; Supervision, Y.E.; Project Administration, B.S. and R.W.-R.; Funding Acquisition, B.S. and R.W.-R. All authors have read and agreed to the published version of the manuscript.

Funding: This research was funded through RSF grant # 18-45-06010 and Helmholtz-RSF Joint Research Groups Grant (Grant # HRSF-0025).

Acknowledgments: Our thanks go to B. Wiese who supplied the Mg-6.0%Ag and Mg-10.0%Gd alloys used in this work.

Conflicts of Interest: The authors declare no conflict of interest.

References

1. Han, H.-S.; Loffredo, S.; Jun, I.; Edwards, J.; Kim, Y.-C.; Seok, H.-K.; Witte, F.; Mantovani, D.; Glyn-Jones, S. Current status and outlook on the clinical translation of biodegradable metals. *Mater. Today* **2019**, *23*, 57–71. [CrossRef]
2. Zheng, Y.; Gu, X.; Witte, F. Biodegradable metals. *Mater. Sci. Eng. R Rep.* **2014**, *77*, 1–34. [CrossRef]

3. Parfenov, E.; Kulyasova, O.; Mukaeva, V.; Mingo, B.; Farrakhov, R.; Cherneikina, Y.; Yerokhin, A.; Zheng, Y.; Valiev, R. Influence of ultra-fine grain structure on corrosion behaviour of biodegradable Mg-1Ca alloy. *Corr. Sci.* **2020**, *163*, 108303. [CrossRef]
4. Kiani, F.; Wen, C.; Li, Y. Prospects and strategies for magnesium alloys as biodegradable implants from crystalline to bulk metallic glasses and composites—A review. *Acta Biomater.* **2020**, *103*, 1–23. [CrossRef]
5. Costantino, M.; Schuster, A.; Helmholz, H.; Meyer-Rachner, A.; Willumeit-Römer, R.; Luthringer-Feyerabend, B.J. Inflammatory response to magnesium-based biodegradable implant materials. *Acta Biomater.* **2020**, *101*, 598–608. [CrossRef]
6. Merson, D.L.; Brilevesky, A.; Myagkikh, P.; Tarkova, A.; Prokhorikhin, A.; Kretov, E.; Frolova, T.S.; Vinogradov, A. The Functional Properties of Mg–Zn–X Biodegradable Magnesium Alloys. *Materials* **2020**, *13*, 544. [CrossRef] [PubMed]
7. Bazhenov, V.E.; Koltygin, A.; Komissarov, A.; Li, A.; Bautin, V.; Khasenova, R.; Anishchenko, A.; Seferyan, A.; Komissarova, J.; Estrin, Y. Gallium-containing magnesium alloy for potential use as temporary implants in osteosynthesis. *J. Magnes. Alloy.* **2020**, *8*, 352–363. [CrossRef]
8. Zhao, W.; Wang, J.; Weiyang, J.; Qiao, B.; Wang, Y.; Li, Y.; Jiang, D.-M. A novel biodegradable Mg-1Zn-0.5Sn alloy: Mechanical properties, corrosion behavior, biocompatibility, and antibacterial activity. *J. Magnes. Alloy.* **2020**, *8*, 374–386. [CrossRef]
9. Jiang, W.; Wang, J.; Liu, Q.; Zhao, W.; Jiang, D.; Guo, S. Low hydrogen release behavior and antibacterial property of Mg-4Zn-xSn alloys. *Mater. Lett.* **2019**, *241*, 88–91. [CrossRef]
10. Gao, Z.; Song, M.; Liu, R.-L.; Shen, Y.; Ward, L.; Cole, I.; Chen, X.; Liu, X. Improving in vitro and in vivo antibacterial functionality of Mg alloys through micro-alloying with Sr and Ga. *Mater. Sci. Eng. C* **2019**, *104*, 109926. [CrossRef]
11. Tie, D.; Feyerabend, F.; Müller, W.D.; Schade, R.; Liefeith, K.; Kainer, K.; Willumeit, R. Antibacterial biodegradable Mg-Ag alloys. *Eur. Cells Mater.* **2013**, *25*, 284–298. [CrossRef] [PubMed]
12. Feng, Y.; Zhu, S.; Wang, L.; Chang, L.; Hou, Y.; Guan, S. Fabrication and characterization of biodegradable Mg-Zn-Y-Nd-Ag alloy: Microstructure, mechanical properties, corrosion behavior and antibacterial activities. *Bioact. Mater.* **2018**, *3*, 225–235. [CrossRef] [PubMed]
13. Shuai, C.; Zhou, Y.; Yang, Y.; Gao, C.; Peng, S.; Wang, G. Ag-Introduced Antibacterial Ability and Corrosion Resistance for Bio-Mg Alloys. *BioMed Res. Int.* **2018**, *2018*, 1–13. [CrossRef] [PubMed]
14. Dai, Y.; Liu, H.; Tang, Y.; Xu, X.; Long, H.; Yan, Y.; Luo, Z.; Zhang, Y.; Yu, K.; Zhu, Y. A Potential Biodegradable Mg-Y-Ag Implant with Strengthened Antimicrobial Properties in Orthopedic Applications. *Metals* **2018**, *8*, 948. [CrossRef]
15. Wang, S.; Zhang, X.; Li, J.; Liu, C.; Guan, S. Investigation of Mg–Zn–Y–Nd alloy for potential application of biodegradable esophageal stent material. *Bioact. Mater.* **2020**, *5*, 1–8. [CrossRef] [PubMed]
16. Li, M.; Wang, W.; Zhu, Y.; Lu, Y.; Wan, P.; Yang, K.; Zhang, Y.; Mao, C. Molecular and cellular mechanisms for zoledronic acid-loaded magnesium-strontium alloys to inhibit giant cell tumors of bone. *Acta Biomater.* **2018**, *77*, 365–379. [CrossRef]
17. Dai, Y.; Tang, Y.; Xu, X.; Luo, Z.; Zhang, Y.; Li, Z.; Lin, Z.; Zhao, S.; Zeng, M.; Sun, B.; et al. Evaluation of the mechanisms and effects of Mg–Ag–Y alloy on the tumor growth and metastasis of the MG63 osteosarcoma cell line. *J. Biomed. Mater. Res. Part B Appl. Biomater.* **2019**, *107*, 2537–2548. [CrossRef]
18. Wu, Y.; He, G.; Zhang, Y.; Liu, Y.; Li, M.; Wang, X.; Li, N.; Li, K.; Zheng, G.; Zheng, Y.; et al. Unique antitumor property of the Mg-Ca-Sr alloys with addition of Zn. *Sci. Rep.* **2016**, *6*, 21736. [CrossRef]
19. Shuai, C.; Liu, L.; Shuai, C.; Gao, C.; Zhao, M.-C.; Yi, L.; Peng, S. Lanthanum-Containing Magnesium Alloy with Antitumor Function Based on Increased Reactive Oxygen Species. *Appl. Sci.* **2018**, *8*, 2109. [CrossRef]
20. Anisimova, N.; Kiselevskiy, M.; Martynenko, N.; Straumal, B.; Willumeit-Römer, R.; Dobatkin, S.; Estrin, Y. Cytotoxicity of biodegradable magnesium alloy WE43 to tumor cells in vitro: Bioresorbable implants with antitumor activity? *J. Biomed. Mater. Res. Part B Appl. Biomater.* **2019**, *108*, 167–173. [CrossRef]
21. Estrin, Y.; Martynenko, N.; Anisimova, N.; Temralieva, D.; Kiselevskiy, M.; Serebryany, V.; Raab, G.; Straumal, B.; Wiese, B.; Willumeit-Römer, R.; et al. The Effect of Equal-Channel Angular Pressing on the Microstructure, the Mechanical and Corrosion Properties and the Anti-Tumor Activity of Magnesium Alloyed with Silver. *Materials* **2019**, *12*, 3832. [CrossRef] [PubMed]
22. Valiev, R.; Langdon, T.G. Principles of equal-channel angular pressing as a processing tool for grain refinement. *Prog. Mater. Sci.* **2006**, *51*, 881–981. [CrossRef]

23. Muñoz, J.A.; Higuera, O.F.; Cabrera-Marrero, J.-M. Microstructural and mechanical study in the plastic zone of ARMCO iron processed by ECAP. *Mater. Sci. Eng. A* **2017**, *697*, 24–36. [CrossRef]
24. Bochvar, N.; Rybalchenko, O.V.; Shangina, D.; Dobatkin, S. Effect of equal-channel angular pressing on the precipitation kinetics in Cu-Cr-Hf alloys. *Mater. Sci. Eng. A* **2019**, *757*, 84–87. [CrossRef]
25. Rybal'Chenko, O.; Dobatkin, S.; Kaputkina, L.; Raab, G.; Krasilnikov, N. Strength of ultrafine-grained corrosion-resistant steels after severe plastic deformation. *Mater. Sci. Eng. A* **2004**, 387–389. [CrossRef]
26. Bryła, K. Microstructure and mechanical characterisation of ECAP-ed ZE41A alloy. *Mater. Sci. Eng. A* **2020**, *772*, 138750. [CrossRef]
27. Li, W.; Shen, Y.; Shen, J.; Shen, D.; Liu, X.; Zheng, Y.; Yeung, K.W.; Guan, S.; Kulyasova, O.B.; Valiev, R.Z. In vitro and in vivo studies on pure Mg, Mg–1Ca and Mg–2Sr alloys processed by equal channel angular pressing. *Nano Mater. Sci.* **2020**, *2*, 96–108. [CrossRef]
28. Martynenko, N.; Lukyanova, E.; Anisimova, N.; Kiselevskiy, M.; Serebryany, V.; Yurchenko, N.; Raab, G.; Birbilis, N.; Salishchev, G.; Dobatkin, S.; et al. Improving the property profile of a bioresorbable Mg-Y-Nd-Zr alloy by deformation treatments. *Materialia* **2020**, *13*, 100841. [CrossRef]
29. Minárik, P.; Veselý, J.; Král, R.; Bohlen, J.; KUBÁSEK, J.; Janeček, M.; Stráská, J. Exceptional mechanical properties of ultra-fine grain Mg-4Y-3RE alloy processed by ECAP. *Mater. Sci. Eng. A* **2017**, *708*, 193–198. [CrossRef]
30. Bryła, K.; Horky, J.; Krystian, M.; Lityńska-Dobrzyńska, L.; Mingler, B. Microstructure, mechanical properties, and degradation of Mg-Ag alloy after equal-channel angular pressing. *Mater. Sci. Eng. C* **2020**, *109*, 110543. [CrossRef]
31. Alizadeh, R.; Ngan, A.H.W.; Pereira, P.H.R.; Huang, Y.; Langdon, T.G.; Mahmudi, R. Microstructure, Texture, and Superplasticity of a Fine-Grained Mg-Gd-Zr Alloy Processed by Equal-Channel Angular Pressing. *Met. Mater. Trans. A* **2016**, *47*, 6056–6069. [CrossRef]
32. Fu, Y.; Sun, J.; Yang, Z.; Xu, B.; Han, J.; Chen, Y.; Jiang, J.; Ma, A. Aging behavior of a fine-grained Mg-10.6Gd-2Ag alloy processed by ECAP. *Mater. Charact.* **2020**, *165*, 110398. [CrossRef]
33. Sun, J.; Xu, B.; Yang, Z.; Zhou, H.; Han, J.; Wu, Y.; Song, D.; Yuan, Y.; Zhuo, X.; Liu, H.; et al. Achieving excellent ductility in high-strength Mg-10.6Gd-2 Ag alloy via equal channel angular pressing. *J. Alloy. Compd.* **2020**, *817*, 152688. [CrossRef]
34. Serebryany, V.N.; Kurtasov, S.F.; Savyolova, T. Pole Figure Measurement Plan Influence on Accuracy ODF Coefficients Determined by Modified Harmonic Method. *Mater. Sci. Forum* **2005**, *495*, 1693. [CrossRef]
35. Savyolova, T.; Kourtasov, S. ODF Restoration by Orientations Grid. *Mater. Sci. Forum* **2005**, *495*, 301–306. [CrossRef]
36. Serebryany, V.N.; Rokhlin, L.L.; Monina, A.N. Texture and anisotropy of mechanical properties of the magnesium alloy of Mg-Y-Gd-Zr system. *Inorg. Mater. Appl. Res.* **2014**, *5*, 116–123. [CrossRef]
37. Fornasini, M.L.; Manfrinetti, P.; Gschneidner, K.A. GdMg5: A complex structure with a large cubic cell. *Acta Crystallogr. Sect. C Cryst. Struct. Commun.* **1986**, *42*, 138–141. [CrossRef]
38. Campos, M.D.R.S.; Blawert, C.; Mendis, C.L.; Mohedano, M.; Zimmermann, T.; Proefrock, D.; Zheludkevich, M.L.; Kainer, K.U. Effect of Heat Treatment on the Corrosion Behavior of Mg-10Gd Alloy in 0.5% NaCl Solution. *Front. Mater.* **2020**, *7*. [CrossRef]
39. Persson, K. *Materials Data on Mg54Ag17 (SG:71) by Materials Project*; Lawrence Berkeley National Lab. (LBNL): Berkeley, CA, USA, 2020. [CrossRef]
40. Kraus, T.; Fischerauer, S.F.; Hänzi, A.C.; Uggowitzer, P.J.; Löffler, J.F.; Weinberg, A. Magnesium alloys for temporary implants in osteosynthesis: In vivo studies of their degradation and interaction with bone. *Acta Biomater.* **2012**, *8*, 1230–1238. [CrossRef]
41. Martynenko, N.; Lukyanova, E.; Serebryany, V.; Prosvirnin, D.; Terentiev, V.; Raab, G.; Dobatkin, S.; Estrin, Y. Effect of equal channel angular pressing on structure, texture, mechanical and in-service properties of a biodegradable magnesium alloy. *Mater. Lett.* **2019**, *238*, 218–221. [CrossRef]
42. Suh, J.; Victoria-Hernández, J.; Letzig, D.; Golle, R.; Volk, W. Effect of processing route on texture and cold formability of AZ31 Mg alloy sheets processed by ECAP. *Mater. Sci. Eng. A* **2016**, *669*, 159–170. [CrossRef]
43. Akbaripanah, F.; Fereshteh-Saniee, F.; Mahmudi, R.; Kim, H. Microstructural homogeneity, texture, tensile and shear behavior of AM60 magnesium alloy produced by extrusion and equal channel angular pressing. *Mater. Des.* **2013**, *43*, 31–39. [CrossRef]

44. Martynenko, N.; Lukyanova, E.A.; Serebryany, V.N.; Gorshenkov, M.; Shchetinin, I.; Raab, G.; Dobatkin, S.V.; Estrin, Y. Increasing strength and ductility of magnesium alloy WE43 by equal-channel angular pressing. *Mater. Sci. Eng. A* **2018**, *712*, 625–629. [CrossRef]
45. Dobatkin, S.; Galkin, S.; Estrin, Y.; Serebryany, V.; Diez, M.; Martynenko, N.; Lukyanova, E.; Perezhogin, V. Grain refinement, texture, and mechanical properties of a magnesium alloy after radial-shear rolling. *J. Alloy. Compd.* **2019**, *774*, 969–979. [CrossRef]
46. Nie, J. Precipitation and Hardening in Magnesium Alloys. *Met. Mater. Trans. A* **2012**, *43*, 3891–3939. [CrossRef]
47. Vostrý, P.; Smola, B.; Stulíková, I.; Von Buch, F.; Mordike, B.L. Microstructure Evolution in Isochronally Heat Treated Mg–Gd Alloys. *Phys. Status Solidi A* **1999**, *175*, 491–500. [CrossRef]
48. Gao, X.; He, S.; Zeng, X.; Fu, P.; Ding, W.; Nie, J. Microstructure evolution in a Mg–15Gd–0.5Zr (wt.%) alloy during isothermal aging at 250 °C. *Mater. Sci. Eng. A* **2006**, *431*, 322–327. [CrossRef]
49. Shi, Z.-Z.; Zhang, W.-Z. Enhanced age-hardening response and microstructure study of an Ag-modified Mg–Sn–Zn based alloy. *Philos. Mag. Lett.* **2013**, *93*, 473–480. [CrossRef]
50. Kim, H.S.; Kim, W.J. Annealing effects on the corrosion resistance of ultrafine-grained pure titanium. *Corr. Sci.* **2014**, *89*, 331–337. [CrossRef]
51. Song, D.; Ma, A.; Jiang, J.; Lin, P.; Yang, D.; Fan, J. Corrosion behavior of equal-channel-angular-pressed pure magnesium in NaCl aqueous solution. *Corr. Sci.* **2010**, *52*, 481–490. [CrossRef]
52. Niu, H.-Y.; Deng, K.-K.; Nie, K.-B.; Wang, C.-J.; Liang, W.; Wu, Y. Degradation behavior of Mg-4Zn-2Ni alloy with high strength and high degradation rate. *Mater. Chem. Phys.* **2020**, *249*, 123131. [CrossRef]
53. Sun, Y.; Wang, R.; Peng, C.; Cai, Z. Microstructure and corrosion behavior of as-extruded Mg-xLi-3Al-2Zn-0.2Zr alloys (x = 5, 8, 11 wt.%). *Corr. Sci.* **2020**, *167*, 108487. [CrossRef]
54. Birbilis, N.; Ralston, K.; Virtanen, S.; Fraser, H.L.; Davies, C.H.J. Grain character influences on corrosion of ECAPed pure magnesium. *Corr. Eng. Sci. Technol.* **2010**, *45*, 224–230. [CrossRef]
55. Ralston, K.; Birbilis, N. Effect of Grain Size on Corrosion: A Review. *Corrosion* **2010**, *66*, 075005. [CrossRef]
56. Dobatkin, S.V.; Lukyanova, E.A.; Martynenko, N.; Anisimova, N.; Kiselevskiy, M.V.; Gorshenkov, M.V.; Yurchenko, N.Y.; I Raab, G.; Yusupov, V.S.; Birbilis, N.; et al. Strength, corrosion resistance, and biocompatibility of ultrafine-grained Mg alloys after different modes of severe plastic deformation. *IOP Conf. Ser. Mater. Sci. Eng.* **2017**, *194*, 12004. [CrossRef]
57. Zhang, C.; Wu, L.; Liu, H.; Huang, G.; Jiang, B.; Atrens, A.; Pan, F. Microstructure and corrosion behavior of Mg-Sc binary alloys in 3.5 wt.% NaCl solution. *Corr. Sci.* **2020**, *174*, 108831. [CrossRef]
58. Liu, Y.; Cheng, W.-L.; Liu, Y.-H.; Niu, X.-F.; Wang, H.-X.; Wang, L.-F.; Cui, Z.-Q. Effect of alloyed Ca on the microstructure and corrosion behavior of extruded Mg-Bi-Al-based alloys. *Mater. Charact.* **2020**, *163*, 110292. [CrossRef]
59. Wang, Y.; Zhang, Y.; Wang, P.; Zhang, D.; Yu, B.; Xu, Z.; Jiang, H. Effect of LPSO phases and aged-precipitations on corrosion behavior of as-forged Mg–6Gd–2Y–1Zn–0.3Zr alloy. *J. Mater. Res. Technol.* **2020**, *9*, 7087–7099. [CrossRef]
60. Ding, Z.-Y.; Cui, L.-Y.; Zeng, R.-C.; Zhao, Y.-B.; Guan, S.-K.; Xu, D.K.; Lin, C.-G. Exfoliation corrosion of extruded Mg-Li-Ca alloy. *J. Mater. Sci. Technol.* **2018**, *34*, 1550–1557. [CrossRef]
61. Esmaily, M.; Svensson, J.; Fajardo, S.; Birbilis, N.; Frankel, G.; Virtanen, S.; Arrabal, R.; Thomas, S.; Johansson, L. Fundamentals and advances in magnesium alloy corrosion. *Prog. Mater. Sci.* **2017**, *89*, 92–193. [CrossRef]
62. Dong, J.-H.; Tan, L.; Ren, Y.; Yang, K. Effect of Microstructure on Corrosion Behavior of Mg–Sr Alloy in Hank's Solution. *Acta Metall. Sin.* **2018**, *32*, 305–320. [CrossRef]
63. Zhang, Y.; Li, J.; Lai, H.; Xu, Y. Effect of Homogenization on Microstructure Characteristics, Corrosion and Biocompatibility of Mg-Zn-Mn-xCa Alloys. *Materials* **2018**, *11*, 227. [CrossRef] [PubMed]

© 2020 by the authors. Licensee MDPI, Basel, Switzerland. This article is an open access article distributed under the terms and conditions of the Creative Commons Attribution (CC BY) license (http://creativecommons.org/licenses/by/4.0/).

Article

Bridging the Gap between Bulk Compression and Indentation Test on Room-Temperature Plasticity in Oxides: Case Study on SrTiO$_3$

Xufei Fang [1,*,†], Lukas Porz [1,†], Kuan Ding [1] and Atsutomo Nakamura [2,3,*]

1. Department of Materials and Earth Sciences, Technical University of Darmstadt, 64287 Darmstadt, Germany; porz@ceramics.tu-darmstadt.de (L.P.); ding@ceramics.tu-darmstadt.de (K.D.)
2. Department of Materials Physics, Nagoya University, Furo-cho, Chikusa-ku, Nagoya 464-8603, Japan
3. PRESTO, Japan Science and Technology Agency (JST), 7, Gobancho, Chiyoda-ku, Tokyo 102-0076, Japan
* Correspondence: fang@ceramics.tu-darmstadt.de (X.F.); anaka@nagoya-u.jp (A.N.)
† Equal Contribution.

Received: 19 September 2020; Accepted: 12 October 2020; Published: 14 October 2020

Abstract: Dislocation-based functionalities in inorganic ceramics and semiconductors are drawing increasing attention, contrasting the conventional belief that the majority of ceramic materials are brittle at room temperature. Understanding the dislocation behavior in ceramics and advanced semiconducting materials is therefore critical for the mechanical reliability of such materials and devices designed for harvesting the dislocation-based functionalities. Here we compare the mechanical testing between indentation at nano-/microscale and bulk uniaxial deformation at macroscale and highlight the dislocation plasticity in single crystal SrTiO$_3$, a model perovskite. The similarities and differences as well as the advantages and limitations of both testing protocols are discussed based on the experimental outcome of the crystal plasticity, with a focus on the pre-existing defect population being probed with different volumes across the length scales ("size effect"). We expect this work to pave the road for studying dislocation-based plasticity in various advanced functional ceramics and semiconductors.

Keywords: dislocation plasticity; ceramics; SrTiO$_3$; nanoindentation; bulk deformation

1. Introduction

Dislocations, one-dimensional line defects, are one of the major carriers of plastic deformation in crystalline materials. There has been increasing attention on dislocation-based functionality in ceramic materials in recent years [1–7], albeit the common belief is that the majority of ceramic materials are brittle. In fact, a substantial number of ceramics and semiconductors (single crystal) can be deformed plastically by dislocations at room temperature [8–14]. Dislocations are, in nature, crystalline imperfections that may impact the mechanical properties of the materials and devices; therefore, understanding the dislocation-based mechanics in such materials is critical for assessing their mechanical reliability. As many applications are processed, used or stored at room temperature, this work will mainly focus on the dislocation behaviors at room temperature.

In general, there are three endeavors being made to introduce dislocations into ceramics. First, dislocations can be produced in a well-arranged manner by interface method via direct bonding like the bi-crystal technique [15–18]. Another approach is via novel processing techniques such as thin film growth method using plasma-assisted molecular beam epitaxy [4], spark plasma sintering [19], and flash sintering [20], with the latter two methods focusing on bulk materials. The third approach is by mechanical deformation, where most of the efforts have been made at high temperatures [2,7,12,21,22], while for room temperature study, "ductile" ceramics [9,10,12,23–27] have been of particular interest due to their capability of dislocation-mediated plastic deformation. To be more specific, mechanical deformation can be further divided into bulk deformation and micro-/nanoscale (e.g., indentation) testing, with the latter rising rapidly in recent years. The wide range of length scales covered by such testing methods is compatible with the application of the materials and devices spanning from macroscale centimeter size to micro-/nanoscale in MEMS and NEMS (micro and nanoelectromechanical systems).

Bulk deformation has been very often used for high-temperature deformation for ceramics with thermal activation of dislocation mobility, while for room-temperature tests, the focus was mainly on "ductile" ceramics such as $SrTiO_3$ [11,23,28,29], $KNbO_3$ [30], MgO [8], and ZnS [12,14]. The large volume that is probed in bulk tests also gives a much higher chance of encountering the effect of the pre-existing defects such as dislocation, micro-cracks, etc. In addition, due to the limited slip systems in ceramic materials, deforming polycrystalline ceramics is only possible at elevated temperatures and is therefore out of the scope of this work. Bulk deformation on single crystals is demanding and less cost-effective due to the standing challenge for fabricating high-purity bulk single crystals of most advanced materials; hence bulk tests find limitation in investigating the dislocation behaviors of the majority of advanced functional ceramics. Yet, it offers attractive perspectives such as generating well-aligned dislocation arrays to serve functionality evaluation [1,21,22].

On the other hand, in compatibility with the ever-decreasing scale of many functional devices and materials, nano-/micromechanical methods [31] are becoming much more applicable and feasible. Among which, (nano-)indentation method is seen as a highly versatile, high-throughput and cost-effective method [32,33], because the probed volume can be varied easily over many orders of magnitude, e.g., in the range of 10^{-3} µm^3 to 10^4 µm^3 when selecting readily available tip radii between 90 nm and 25 µm [27]. The onset of dislocation activities in indentation is indicated by the "pop-in" event (a sudden excursion of the displacement at nominally constant load), which is usually seen as the signal for incipient plasticity, namely, the onset of plastic deformation. In addition, the highly confined volume being probed using such methods very often is defect-free or with very low defect density. Hence the material may behave differently in comparison to the bulk tests. A well-known phenomenon based on such size effect is termed as "smaller is stronger" [34]. Such a small scale method greatly favors the site-specific investigation, yet it may also find its limitation in the study of dislocation-based functionality due to the very local probed region in sub-micrometer regime.

The comparison discussed above clearly shows a highly desired need to bridge the gap between these two methods across the length scales as well as to overcome the shortcomings of each method. The question posed in this work is how to best combine these two methods (bulk and indentation tests) to better understand the dislocation behaviors. To this end, addressing the similarities and differences of the deformation mechanisms obtained in both methods and at both length scales is required.

In this paper, we propose a new experimental protocol for probing the mechanisms of the incipient plasticity in single crystal $SrTiO_3$ at room temperature via combined bulk compression and indentation tests. $SrTiO_3$ is a prototypical perovskite with cubic structure and has been frequently used in various functional devices. It is also a well-known "ductile" ceramic oxide at room temperature as has been discussed before, with the slip system {110}<110> being activated. The stress and strain analyses, the activation of slip planes at macroscale and micro-/nanoscale, and the effect of pre-existing defect density in different tested volumes have been compared. The differences and similarities as well as the advantages and limitations of both methods are compared to shed light on the dislocation-based plasticity to draw a complete deformation picture in $SrTiO_3$.

2. Experimental Method

2.1. Materials

Single crystal SrTiO$_3$ as a model ceramic oxide is used in the current work. The crystals were grown by the Verneuil method from high purity SrTiO$_3$ powder (99.9 wt % and Sr/Ti = 1.04 [13]) and high purity SrCO$_3$ powder (99.99 wt %) and were used for the present study (Shinkosha Co., Ltd., Yokohama, Japan). Samples used for both bulk deformations have a size of 3 mm × 3 mm × 7.5 mm, and for indentation have a size of 2.6 mm × 2.6 mm × 6.5 mm. All sample faces were polished by diamond abrasives and then finished by polishing with colloidal silica. The surface roughness measured by Atomic Force Microscopy (AFM, Veeco, Plainview, NY, USA) is less than 1 nm.

2.2. Mechanical Testing

Uniaxial bulk compression tests were performed on a uniaxial compression machine (Shimadzu Co. Ltd., Kyoto, Japan, AG-10kNX) along the [310] and [001] axis in air at room temperature with a constant strain rate of 1.0×10^{-5} s^{-1} and 1.0×10^{-1} s^{-1}, respectively. A camera (Nikon Co. Ltd., Tokyo, Japan, D7200, 60 frames/sec for high strain rate test, and 24 frames/sec for low strain rate test) is used to capture the real-time deformation information such as slip lines formation on the surface of the samples.

Indentation tests were performed in air at room temperature on the (001) surface, which can be chemically etched to reveal the dislocation structures. The continuous stiffness measurement (CSM) technique is used on G200 (Keysight, Santa Rosa, CA, USA), with a constant indentation strain rate of 0.05 s^{-1}. A harmonic displacement oscillation of 2 nm was applied with a frequency of 45 Hz. A Berkovich diamond indenter and spherical diamond indenters with various tip radii (Synton MDP, Nidau, Switzerland) were used for the indentation tests. The machine compliance was first calibrated on fused silica. The effective tip radius, for simplicity, has been determined by fitting the initial elastic portion of the load-displacement curves using Hertzian elastic contact [35] when the indentation depth (here the pop-in depth) is much smaller than the tip radius. For detailed comparison of different fitting methods for tip radii calibration, it has been described elsewhere by Li et al. [36]. For each test condition, 25 indents were performed to ensure reproducibility.

2.3. Characterization

The specimens were chemically etched for ~20 s in 15 mL 50% HNO$_3$ with 16 drops of 50% HF to reveal the dislocation patterns on the surface before and after deformation. Afterwards, the surface and etch pits were characterized using a scanning electron microscope (SEM, TESCAN MIRA3-XMH, Brno, Czech Republic) with an acceleration voltage of 5 kV.

3. Results

3.1. Bulk Deformation

We begin by presenting the results on samples with a loading axis along the [310] direction. The stress-strain curves in Figure 1 illustrate clearly a linear elastic response of the material deformation prior to the onset of initial yielding (as indicated by the dashed lines), after which the materials deform elasto-plastically. The strain rate effect is evidenced in Figure 1, namely, a higher strain rate leads to a higher yield stress (150 MPa for 1.0×10^{-1} s^{-1} and 118 MPa for 1.0×10^{-5} s^{-1}). Meanwhile, the fracture strain is approximately the same and is independent of the strain rate (6.0% for 1.0×10^{-1} s^{-1} and 6.0% for 1.0×10^{-5} s^{-1}).

The optical images (insets in Figure 1) demonstrate a few slip lines and some random crack-like features (based on the image contrast and post-mortem examination) inside the crystals. Microcracks, however, contribute only a minor fraction to the overall strain, which is mostly carried by dislocations as long as the cracks are confined locally or predominantly propagate vertically. Even though some cracks may form in the SrTiO3 crystal, which is often inevitable, dislocation-based deformation can continue to a large plastic strain, unless the crystal actually shatters. In addition, the loading direction [310] favors only two slip directions during compression in SrTiO3, whose slip systems at room temperature are {110}<110> (six in total) as reported in bulk compression tests [23,28,29].

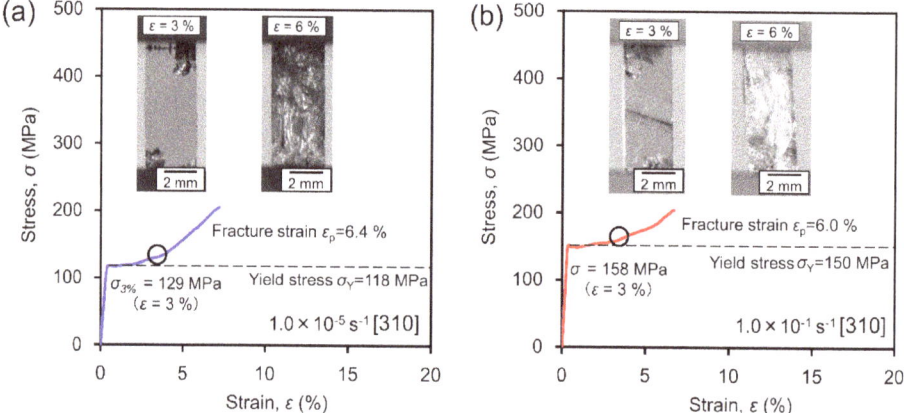

Figure 1. Stress-strain curves and corresponding real-time images captured during deformation for bulk deformation on single crystal SrTiO3 along the [310] direction: (**a**) Strain rate 1.0×10^{-5} s^{-1}; (**b**) Strain rate 1.0×10^{-1} s^{-1}.

In addition, it is evident that in bulk deformation, there is a larger chance of probing pre-existing defects or easily causing local stress concentrations due to the contact issue, which is critical for crack initiation in the majority of ceramic materials with brittle nature, especially for the case of loading in [310] with only one slip direction favored. The crack formation is clearly evidenced by the in situ image captured using a high-speed camera in Figure 1a, where the dark region in the up-right corner indicates localized non-uniform deformation, most likely due to local contact at the sample edge.

The stress-strain curves in Figure 2 correspond to a different loading direction in [001]. Analogous to the results obtained along the [310] direction, a clear linear elastic response and the strain rate effect are observed, namely, a higher strain rate leads to a higher yield stress (135 MPa for 1.0×10^{-1} s^{-1} and 112 MPa for 1.0×10^{-5} s^{-1}). The fracture strain again is independent of the strain rate (13.1% for 1.0×10^{-1} s^{-1} and 13.6% for 1.0×10^{-5} s^{-1}). However, it is much larger in this case than that shown in Figure 1 along the [310] loading direction, which is less than half of the fracture strain obtained in the [001] loading direction. In addition, the slip patterns can be clearly captured during the deformation along [001] direction, as shown by the parallel black strips, which are perpendicular to the loading directions. This is not surprising based on the calculation of the Schmid factor and the room-temperature slip system in SrTiO3. The plastic deformation therefore is obviously mediated by the slip activation via the motion of pre-existing dislocations and multiplication to general new dislocations, as the homogeneous dislocation nucleation in SrTiO3 would require a shear stress of ~17 GPa (identical to the theoretical strength), which is not occurring in bulk deformation but in nanoindentation with a sharp tip [27]. In comparison to the present result (Figure 2) with only multiple parallel slip lines that are perpendicular to the loading axis, it is interesting to note the different slip patterns observed post-mortem in polarized light microscope [11,23,29], where the 45° intersection of the slip planes was also revealed, which was direct proof for the four equivalent slip planes. A detailed discussion on the deformation mechanisms will be presented later in Section 4.

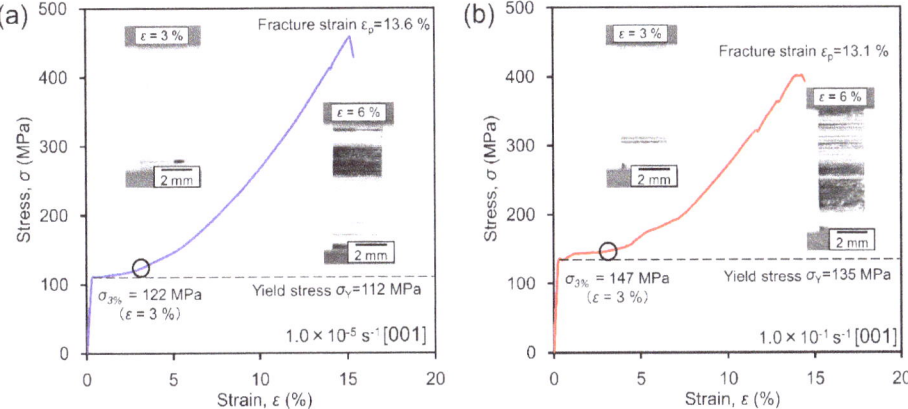

Figure 2. Stress-strain curves and corresponding real-time images captured during deformation for bulk deformation on single crystal SrTiO$_3$ along the [001] direction: (**a**) Strain rate 1.0×10^{-5} s^{-1}; (**b**) Strain rate 1.0×10^{-1} s^{-1}, adopted and modified from [13].

3.2. Indentation Tests

A representative indentation load-displacement curve (red line) is presented in Figure 3a, which clearly depicts the elastic portion (fitted with Hertzian theory, blue line) and the following pop-in event. The pop-in signifies the transition from purely elastic to elasto-plastic deformation, in this case, mediated by the dislocation activation. Here, by dislocation activation it refers to a combination of dislocation nucleation, multiplication and glide motion due to the relatively large tip that was used (Figure 3a, with an effective tip radius of $R_{fit} = 1.4$ µm), as revealed by the post-mortem etch pit study and SEM characterization. Strictly speaking, for indentation with larger tips, it remains a great challenge to discern all these three dislocation activities simply based on the load-displacement curve. An in-depth discussion on this point will be presented in Section 4. Note that the indentation depth at pop-in is about 60 nm, which is very small in comparison to the effective tip radius. The corresponding surface etch pits captured by the SEM image in Figure 3b clearly demonstrate the activated slip planes in both <100> and <110> directions, which belong to the {110}-45° planes and {110}-90° planes, respectively. The degrees 45° and 90° indicate the inclined angles between the slip planes and the indented (001) plane, as schematically illustrated in Figure 4.

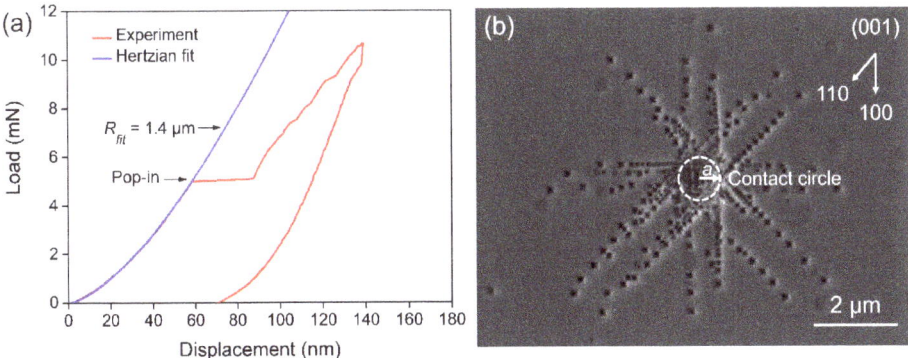

Figure 3. (**a**) Indentation load-displacement curves with pop-in event on (001) surface of single crystal SrTiO$_3$; (**b**) SEM images showing dislocation etch pits.

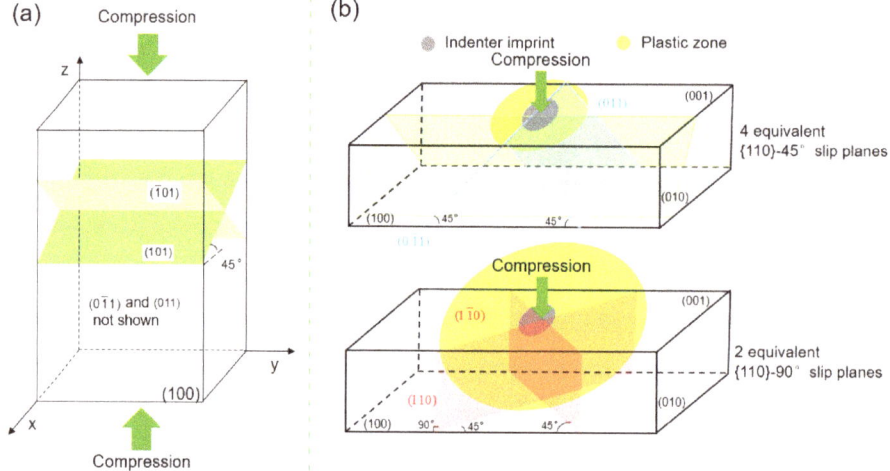

Figure 4. Comparison of the activation of slip systems: (**a**) slip activation in bulk compression along [001] direction; (**b**) slip activation in indentation test on (001) surface, note the angles 45° and 90° indicate the inclined angle between the slip plane and the indented surface.

It is noted that due to the higher confinement of deformation in the indentation test, the crack formation, if there were any, can be better suppressed especially for a small indenter tip, e.g., sharp Berkovich tip at small load [37], under which the maximum shear stress easily reaches the shear strength prior to the fracture strength, was achieved, and thus dislocation nucleation is promoted [27,37].

4. Analyses and Discussions

4.1. Stress Comparison and Activation of the Slip Systems

In order to compare the similarities and differences in the dislocation-mediated plasticity, we first compare the slip systems that can be activated during deformation. For simplicity, we consider in both tests the loading direction along the [001] direction. In bulk deformation illustrated in Figure 4a, only two representative slip planes are presented while there are four equivalent slip planes (all of which are inclined 45° to the [001] loading direction) that can be activated. Contrasting the bulk deformation, two additional slip planes that are 90° inclined to the surface can be activated (two red planes in Figure 4b) during indentation due to the non-zero resolved shear stress on these two planes. A detailed analysis is presented later. These additionally activated slip planes comprise one of the major differences between the bulk uniaxial deformation and the indentation tests.

The onset of the plastic deformation can be correlated to dislocation nucleation, multiplication and glide motion at room temperature. Homogeneous dislocation nucleation in a perfect crystal requires that the maximum shear stress reaches the theoretical shear strength, which is about $G/2\pi$, with G being the shear modulus. Heterogeneous dislocation nucleation (i.e., dislocation nucleation from pre-existing defects), dislocation multiplication and dislocation glide motion can be activated at a much lower stress as in the case of bulk deformation.

The different stress distribution as well as the stress states in bulk and indentation deformation result in different mechanical responses with respect to the slip plane activation shown in Figure 4. In uniaxial bulk compression, the maximum shear stress is expressed with respect to the normal stress by $\tau_{max}^{bulk} = \frac{\sigma_{uniaxial}}{2}$, with $\sigma_{uniaxial}$ being the uniaxial stress along the compression direction. In the case of loading in [001] direction (with a Schmid factor of 0.5), the maximum shear stress in bulk compression, in an ideal case, lies along the 45° planes inclined to the [001] loading direction and is equal to the critically resolved shear stress τ_{CRSS}^{bulk}, as has been validated by Patterson et al. [29].

In indentation, the maximum shear stress is $\tau_{max}^{indent} = 0.31(\frac{6E_r^2}{\pi^3 R^2} P_{pop-in})^{1/3}$ [35]. Here, R is the effective tip radius, and P_{pop-in} corresponds to the load at the onset of the first pop-in (the elastic limit). The reduced modulus E_r is calculated from the elastic constants of the indenter and the specimen by $\frac{1}{E_r} = \frac{1-v_i^2}{E_i} + \frac{1-v_s^2}{E_s}$. With $E_i = 1140$ GPa and $v_i = 0.07$ for the diamond tip, $E_s = 264$ GPa and $v_s = 0.237$ for SrTiO$_3$ [29], it gives $E_r = 224$ GPa. Following Swain et al. [38], the resolved maximum shear stress on the {110}-45° slip planes is of 0.46 p_0 (with p_0 being the mean pressure) at the position of about 0.5 a (a is the contact radius) beneath the indentation surface along the central axis, while the resolved maximum shear stress on the {110}-90° slip planes is 0.33 p_0 at a position approximately 0.5 a directly below the circle of the contact. Hence, the activation of the {110}-45° slip planes occurs prior to the {110}-90° slip planes [39], and the dislocations on the {110}-90° slip planes are all initiated from the edge of the contact circle and travel away from the indenter in <110> direction, as illustrated in Figure 3b.

For comparison, the τ_{CRSS}^{bulk} is the stress for dislocation glide in bulk deformation and should be close to the lattice friction stress, τ_f, at the yield of plastic deformation (incipient plasticity in bulk), where the effect of dislocation-dislocation hardening (in later stage with large strain) does not need to be regarded due to a very low dislocation density at this stage in single crystal [29]. In indentation, however, the friction stress τ_f must not be correlated to the resolved maximum shear stress at pop-in (as it correlates mainly to dislocation nucleation, being either homogeneous or heterogeneous), but rather can be estimated from the dislocation pile-ups [9,40] as revealed by the etch pit patterns (Figure 3b). A nice correlation between τ_{CRSS}^{bulk} and τ_f^{indent} has been found in single crystal SrTiO$_3$ [29,40], which gives a value of about 60–90 MPa at room temperature. Consider $\tau_{CRSS}^{bulk} = \tau_{max}^{bulk} = \frac{\sigma_{uniaxial}}{2}$ and the obtained yield stress from Figure 2, a good agreement is confirmed by our experiment as well.

4.2. Strain Comparison

The strain in bulk deformation can be directly read from the stress-strain curves, for instance, in Figures 1 and 2. The strain analysis in indentation, however, is less straightforward. The elastic strain under spherical indentation can be estimated according to Field et al. [41] via $\varepsilon = 0.2\, a/R$. At the critical condition of pop-in occurrence, there is:

$$\varepsilon_c = 0.2\, a_c / R \tag{1}$$

where R is the effective tip radius, a_c is the contact radius at pop-in and is estimated by [41]:

$$a_c = \sqrt{h_c R} \tag{2}$$

With $h_c = 60$ nm being the indentation depth at pop-in and the effective tip radius $R = 1.4$ μm obtained in Figure 3a, this gives the estimation of the strain at the pop-in:

$$\varepsilon_c = 0.2\, \sqrt{h_c/R} \approx 4\% \tag{3}$$

This strain corresponds to the elastic limit in indentation test and is much larger than the elastic limit in bulk deformation, which is smaller than 0.5% in Figures 1 and 2.

The estimation of plastic strain is more complicated under spherical indentation depending on the deformation stage [41] as well as the tip size with respect to the defect density being probed (Section 4.3). For simplicity, however, we still adopt $\varepsilon = 0.2\, a/R$ as an upper bound for the estimation of the plastic strain beyond the pop-in. In this case, we take the post-mortem SEM image in Figure 3b and determine the ultimate contact radius $a = 500$ nm, with $R = 1.4$ μm it gives $\varepsilon_p \approx 7\%$. It is noteworthy that both {110}-45° and {110}-90° slip planes (Figure 4b) have been activated at this plastic strain during indentation, while only the {110}-45° slip planes were activated during bulk compression tests although with a much higher plastic strain (13.6% and 13.1% in Figure 2).

4.3. Indentation Pop-in Related to Defect Population

In comparison to bulk deformation, the indentation pop-in has been frequently used as a powerful tool for understanding the incipient plasticity at micro-/nanoscale, with a focus on the dislocation nucleation as well as multiplication and motion of pre-existing dislocations, as has been extensively studied in metallic materials [42]. In contrast, the pop-in mechanisms in ceramics have been less addressed. Therefore, a detailed discussion on the indentation pop-in is made here.

Considering single crystal ceramic or semiconductor materials, the most relevant defects for the crystal plasticity are the pre-existing dislocations and point defects prior to the mechanical loading. It remains yet a challenging topic to quantify the impact of the defects individually from the pop-in statistics. Nevertheless, these defects present in ceramics, independent of whether they are pre-existing dislocations or point defects, are very often rather far away from the indenter tip and therefore only need rather low stresses in comparison to the homogeneous dislocation nucleation ($G/2\pi$), which as discussed above, occurs only at nanoscale testing such as using a sharp indenter [27,43]. As a result, these defects can most likely still be activated before the stress for homogeneous dislocation nucleation is reached underneath the indenter. However, regularly the x-axis of a pop-in statistical distribution is specified by the maximum shear stress available beneath the indenter tip even though this is not actually the critical stress level for the relevant defects to be activated [27,44]. Recent models in metallic materials have been suggested, which effectively and accurately convert pop-in statistics into a defect strength and density [42,44,45]. While these approaches are accurate, their up-front time investment makes them less convenient to accompany the development of understanding.

Instead, we suggest to start with a simple consideration with the basic question: How many defects will be in the volume underneath the indenter? Therefore, both volumetric defect density as well as the relevant volume need to be known. Defect density is either directly specified in volumetric units or can be directly converted to it when approximating line defects, such as dislocations, as point defects with Equation (4).

$$\rho_{volumetric} = \rho_{areal}^{2/3} \tag{4}$$

It is tempting to use the areal dislocation density and contrast it to the contact area, which we avoid for two reasons. One, using a volumetric density allows using the approach for all types of defects, which can be helpful later. Two, it is more difficult to reasonably approximate a representative area underneath the indent. In particular, a representative area is *not* the contact area of the indenter. Instead, it is a much larger area where the stress is sufficient to activate defects. The radius of a reasonable area can be approximated from the half sphere discussed below where we believe it is more intuitive to adopt the volumetric perspective.

Calculating the volume, which is stressed above a certain value, is a bit more cumbersome. It will be approximated here as a half sphere with a radius r, which is defined as the distance from the tip where the stress is lower than the stress required to activate a defect. Here, $\tau_{critical}$ is the stress required to activate a defect, τ_{local} is the stress in the distance from the tip r, and a is the contact radius. The maximum stress underneath the indenter is labeled as τ_{max}, where either the experimentally observed maximum stress can be inserted or the theoretical shear stress used, depending on the individual needs.

$$\tau_{critical} = \tau_{local} = \tau_{max} * \left(\frac{a}{r}\right)^2 \tag{5}$$

This equation can be re-arranged as:

$$r = a * \sqrt{\frac{\tau_{max}}{\tau_{critical}}} \tag{6}$$

The contact radius can be expressed in two ways. Firstly, it can be calculated according to $a = \left(\frac{3PR}{4E_r}\right)^{1/3}$ [35], with the tip radius R, reduced modulus E_r and indentation load P obtained from the experimental data. This expression has the clear advantage that it can be used for direct comparison with experimental data. Its disadvantage is, however, that it makes comparison between different tip radii difficult because the value of the load P varies with tip radius. Regarding the load P that is needed to reach a particular stress τ, which depends on the tip radius, it will be replaced with an expression that purely relies on tip radius. By combining $a = \left(\frac{3PR}{4E_r}\right)^{1/3}$ and $\tau_{max} = 0.31\left(\frac{6E_r^2}{\pi^3 R^2}P\right)^{1/3}$, we have:

$$a = R\frac{\pi \tau_{max}}{0.62 E_r} \qquad (7)$$

The contact radius a in dependence on the tip radius is retrieved by relating it to the tip radius R, and the reduced elastic modulus $E_r = 224$ GPa and the maximum shear stress τ_{max} (which is the theoretical shear stress in the case of homogeneous dislocation nucleation) can be inserted.

Combining these equations, the volume can be calculated with an experimental load measured by Equation (8) and for purely theoretical analysis with Equation (9):

$$V = \frac{2}{3}\pi r^3 = \frac{2}{3}\pi \left(\left(\frac{3PR}{4E_r}\right)^{\frac{1}{3}} \sqrt{\frac{\tau_{max}}{\tau_{critical}}}\right)^3 \qquad (8)$$

$$V = \frac{2}{3}\pi r^3 = \frac{2}{3}\pi \left(R\frac{\pi \tau_{max}}{0.62 E_r} \sqrt{\frac{\tau_{max}}{\tau_{critical}}}\right)^3 \qquad (9)$$

When the volume is known, the number n of expected defects can be estimated by multiplying the volume with the defect density, e.g., by $n = \rho V$.

As the probed volume is dependent on the tip radius by the third power, the range of the volume can be nicely tuned between e.g., in the range of 10^{-3} µm^3 to 10^4 µm^3 when selecting readily available tip radii over a wide range [27]. For 25 µm tip radius, the volume equals to 10^{-14}–10^{-13} m^3 while the volume for a 1.4 µm tip radius is only 10^{-17}–10^{-16} m^3. Consider the case of pre-existing dislocation analysis, when contrasting the approximated volumetric density of ~10^{15} m^3 for a dislocation density of ~10^{10} m^2 by multiplying volume and density, it becomes clear that in one case dislocations should be readily detectable while in the other case next to no pre-existing dislocations are found. For a 25 µm tip radius, this approximation suggests to find n = 10–100 dislocations, while for a 1.4 µm tip radius, only n = 0.01–0.1 should be found. Hence, the pop-in behavior at 25 µm tip radius should show severe impact by pre-existing dislocations while the pop-in behavior at 1.4 µm tip radius should show next to no impact by pre-existing dislocations. However, it is worthy to note that in Figure 5, the maximum shear stress for 1.4 µm tip radius is yet much lower than the theoretical value (blue line). This is most probably attributed to the free surface being probed by the indenter, and the surface imperfections can serve as heterogeneous dislocation nucleation sites (e.g., surface steps [46]) to dramatically reduce the stress level. In the extreme case of a small volume (e.g., an effective tip radius R = 90 nm [27]), the chance to find a defect in the probed volume will be nearly zero and hence statistically irrelevant, meaning that predominantly homogeneous dislocation nucleation will be observed.

The overall distribution and transition of the maximum shear stress for different tip radii was experimentally demonstrated in Figure 5 by our indentation experiments in single crystal SrTiO$_3$ on the (001) surface. Similar results were reported on metallic material such as Mo single crystal by Bei et al. [45], where a series of different indenter sizes were tested. With the simple calculation suggested above, it is easy for the experimentalists to get a rough idea of which defect density can be best tested with what tip radius.

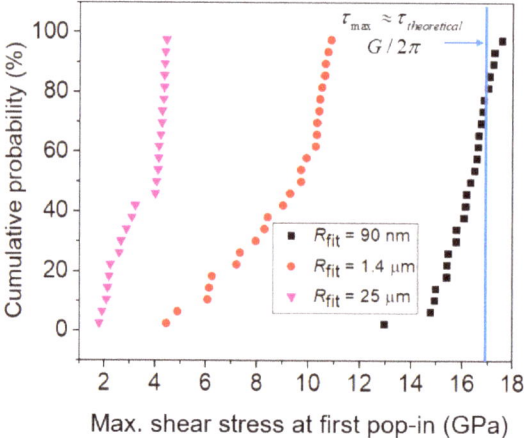

Figure 5. Indentation pop-in statistics on single crystal (001)-SrTiO$_3$ with tip radii of 25 µm, 1.4 µm and 90 nm. The data on 25 µm and 90 nm were adopted from [27].

On the other hand, it should be noted that the maximum shear stress occurs at about 0.5 a below the indenter (Figure 6a). When the tip is sharp and no defect is detected in the probed volume, then the maximum shear stress (τ_{max}) is responsible for nucleating dislocations homogeneously. In this case, the maximum shear stress is the critical shear stress ($\tau_{critical}$) for dislocation nucleation. However, when the number of defects increases as the probed volume becomes large underneath a larger indenter, it also becomes likely to find defects very close to the tip. Such defects would then already respond to smaller stresses (such as lattice friction stress $\tau_{friction}$) in order to become mobile (Figure 6b). Thus, the observed force for the "pop-in" becomes low, as shown in Figure 5 for the 25 µm tip radius. In consequence, the response for indentation test with a large tip radius and high density of defect would become very similar to a bulk compression test.

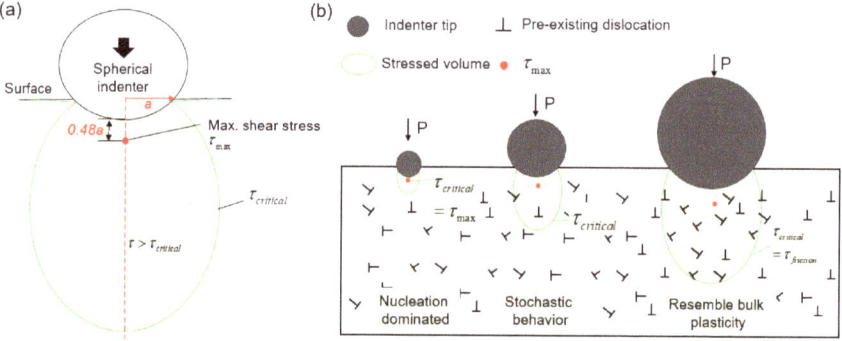

Figure 6. Schematic illustration of (**a**) the maximum shear stress and the critical shear stress under a spherical indenter; (**b**) effect of the defect population with respect to the volume being probed beneath the indenter tip, image modified based on [45]. Note that the critical shear stress is equal to the maximum shear stress when the stress volume is defect-free, so that homogeneous dislocation is activated. The complication of surface defects on heterogeneous dislocation nucleation is not included in this map.

4.4. Pros & Cons between Bulk and Indentation Tests

In order to comprehensively compare the similarities/differences and advantages/limitations between bulk and indentation tests on ceramic materials, we summarize the relevant aspects in Table 1. We expect this table to serve as a primary guideline for designing experiments suiting various purposes at different length scales. We note that in situ indentation tests in SEM and/or TEM are not included, but they can serve as very powerful approaches to directly demonstrate, for instance, dislocation nucleation, dislocation/grain boundary interaction, as well as crack initiation and propagation [47–52] for highly site-specific purposes. Another emerging technique using micro-pillar compression for plastic deformation of oxide ceramics has been recently reported on flash sintered TiO_2 [20] and yttria-stabilized zirconia [53]. Micro-pillar compression assembles the stress state of bulk compression but exhibits strong size effect [34] and is very often more favorable in metallic materials [31,54,55].

The purpose of this work is to pave the road for understanding dislocation mechanics in ceramics across the length scale using the combinatorial approach of bulk deformation and indentation tests. Both methods are common experimental practices now. Together with other techniques such as etch pit method [9,40], SEM, AFM (atomic force microscopy) [56], ECCI (electron channeling contract imaging) [57] and TEM, the dislocation structures and the slip systems can be quantified to shed light on the crystal plasticity in less studied or new classes of ceramics, with a primary goal to avoid crack formation.

Table 1. Comparison of the advantages and limitations between bulk and indentation tests.

	Bulk Deformation (Larger than mm Size)	(Nano)Indentation (Nano-/Microscale)
Pros	<ul><li>Well-established experimental protocol</li><li>Well-alignment of dislocations [1,2,21] in large deformation to facilitate functional properties study</li><li>Easy to handle with larger sized samples</li><li>Relatively easy to conduct tests at high temperature up to 1100 °C and higher [1,22,58,59]</li><li>Relatively easy to conduct tests at cryogenic temperature in liquid nitrogen at 77 K [60]</li></ul>	<ul><li>High-throughput testing on small or low-dimensional sample</li><li>Fast screening of multiple materials</li><li>Easy to induce dislocations without cracks below a critical tip radius</li><li>Site-specific investigation for dislocation interaction with other types of defects such as point defects, pre-existing dislocations and interfaces (grain boundaries, second phase boundary, etc.)</li><li>Low defect population in small volume</li></ul>
Cons	<ul><li>Less cost-effective for large samples</li><li>Large bulk samples are not always available, especially for many advanced functional ceramics and semiconductors</li><li>Imperfect alignment easily causes cracking due to the local stress concentration/pre-existing cracks</li><li>Difficult to activate slip systems (especially secondary slip systems) due to limited stress level</li><li>High defect population in large volume</li></ul>	<ul><li>Limited to local small regions</li><li>Machine dynamics, especially in load-controlled system for most indentation systems</li><li>Challenging for high-temperature tests due to tip reaction, thermal drift, etc. [61–64].</li><li>State-of-the-art temperature is limited to 1100 °C [65]</li><li>Challenging for cryogenic temperature tests, e.g., in liquid nitrogen at 77 K</li></ul>

5. Summary

Understanding the dislocation behaviors in advanced functional ceramic and semiconductor materials is one of the greatest challenges in light of the dislocation-based mechanical properties and emerging dislocation-based functionality. This cannot be achieved alone by either conventional bulk deformation testing, which requires either large samples that are very often unavailable, or by nano-/microscale mechanical testing that focuses on limited volume with limitation to account for the effect of the defect population. A better understanding of the crystal plasticity requires the combinatorial approach using both bulk deformation and indentation tests, as is presented and discussed here.

The similarities and differences of the deformation mechanisms at the different length scales need to be considered to construct a complete understanding of deformation behavior. While bulk deformation is best for testing bulk properties, indentation is highly versatile in the volume probed and can hence be applied for site-specific testing of individual mechanisms.

Author Contributions: X.F. and A.N. conceived the research idea, X.F. and L.P. drafted the first manuscript. X.F., K.D. and A.N. analyzed the experimental data. All authors discussed the data and commented on the manuscript. All authors have read and agreed to the published version of the manuscript.

Funding: This research was funded by Deutsche Forschungsgemeinschaft: 414179371; Deutsche Forschungsgemeinschaft: 398795637; JST PRESTO: JPMJPR199A; JSPS KAKENHI: JP19H05786, JP17H06094 and JP18H03840.

Acknowledgments: X.F. gratefully acknowledges the financial support of Athene Young Investigator Program (TU Darmstadt) and the Deutsche Forschungsgemeinschaft (DFG, No. 414179371). L.P. is indebted to the funding from DFG (No. 398795637). A.N. thanks the financial support of JST PRESTO Grant Number JPMJPR199A and JSPS KAKENHI Grant Numbers JP19H05786, JP17H06094 and JP18H03840, Japan. We acknowledge the support by the DFG and the Open Access Publishing Fund of Technical University of Darmstadt. We thank Kensuke Yasufuku for his assistance on bulk deformation tests and K. Durst for the access to the indentation equipment. We would like to thank the three anonymous reviewers for their valuable comments.

Conflicts of Interest: The authors declare no conflict of interest.

References

1. Nakamura, A.; Matsunaga, K.; Tohma, J.; Yamamoto, T.; Ikuhara, Y. Conducting nanowires in insulating ceramics. *Nat. Mater.* **2003**, *2*, 453–456. [CrossRef]
2. Ikuhara, Y. Nanowire design by dislocation technology. *Prog. Mater. Sci.* **2009**, *54*, 770–791. [CrossRef]
3. Szot, K.; Speier, W.; Bihlmayer, G.; Waser, R. Switching the electrical resistance of individual dislocations in single-crystalline $SrTiO_3$. *Nat. Mater.* **2006**, *5*, 312–320. [CrossRef] [PubMed]
4. Sun, B.; Haunschild, G.; Polanco, C.; Ju, J.Z.; Lindsay, L.; Koblmuller, G.; Koh, Y.K. Dislocation-induced thermal transport anisotropy in single-crystal group-III nitride films. *Nat. Mater.* **2019**, *18*, 136–140. [CrossRef] [PubMed]
5. Khafizov, M.; Pakarinen, J.; He, L.; Hurley, D.H. Impact of irradiation induced dislocation loops on thermal conductivity in ceramics. *J. Am. Ceram. Soc.* **2019**, *102*, 7533–7542. [CrossRef]
6. Kim, S.I.; Lee, K.H.; Mun, H.A.; Kim, H.S.; Hwang, S.W.; Roh, J.W.; Yang, D.J.; Shin, W.H.; Li, X.S.; Lee, Y.H.; et al. Dense dislocation arrays embedded in grain boundaries for high-performance bulk thermoelectrics. *Science* **2015**, *348*, 109–114. [CrossRef]
7. Ren, P.; Höfling, M.; Koruza, J.; Lauterbach, S.; Jiang, X.; Frömling, T.; Khatua, D.K.; Dietz, C.; Porz, L.; Ranjan, R.; et al. High Temperature Creep-Mediated Functionality in Polycrystalline Barium Titanate. *J. Am. Ceram. Soc.* **2019**. [CrossRef]
8. Argon, A.S.; Orowan, E. Plastic deformation in mgo single crystals. *Philos. Mag.* **1964**, *9*, 1003–1021. [CrossRef]
9. Gaillard, Y.; Tromas, C.; Woirgard, J. Quantitative analysis of dislocation pile-ups nucleated during nanoindentation in MgO. *Acta Mater.* **2006**, *54*, 1409–1417. [CrossRef]
10. Johnston, W.G.; Gilman, J.J. Dislocation Multiplication in Lithium Fluoride Crystals. *J. Appl. Phys.* **1960**, *31*, 632–643. [CrossRef]
11. Brunner, D.; Taeri-Baghbadrani, S.; Sigle, W.; Rühle, M. Surprising Results of a Study on the Plasticity in Strontium Titanate. *J. Am. Ceram. Soc.* **2001**, *84*, 1161–1163. [CrossRef]
12. Oshima, Y.; Nakamura, A.; Matsunaga, K. Extraordinary plasticity of an inorganic semiconductor in darkness. *Science* **2018**, *360*, 772–774. [CrossRef] [PubMed]
13. Nakamura, A.; Yasufuku, K.; Furushima, Y.; Toyoura, K.; Lagerlöf, K.; Matsunaga, K. Room-Temperature Plastic Deformation of Strontium Titanate Crystals Grown from Different Chemical Compositions. *Crystals* **2017**, *7*, 351. [CrossRef]
14. Oshima, Y.; Nakamura, A.; Lagerlöf, K.P.D.; Yokoi, T.; Matsunaga, K. Room-temperature creep deformation of cubic ZnS crystals under controlled light conditions. *Acta Mater.* **2020**, *195*, 690–697. [CrossRef]
15. Feng, B.; Ishikawa, R.; Kumamoto, A.; Shibata, N.; Ikuhara, Y. Atomic Scale Origin of Enhanced Ionic Conductivity at Crystal Defects. *Nano Lett.* **2019**, *19*, 2162–2168. [CrossRef]

16. Furushima, Y.; Nakamura, A.; Tochigi, E.; Ikuhara, Y.; Toyoura, K.; Matsunaga, K. Dislocation structures and electrical conduction properties of low angle tilt grain boundaries in LiNbO$_3$. *J. Appl. Phys.* **2016**, *120*. [CrossRef]
17. Lee, H.S.; Mizoguchi, T.; Mistui, J.; Yamamoto, T.; Kang, S.J.L.; Ikuhara, Y. Defect energetics in SrTiO$_3$ symmetric tilt grain boundaries. *Phys. Rev. B* **2011**, *83*. [CrossRef]
18. Ikuhara, Y.; Nishimura, H.; Nakamura, A.; Matsunaga, K.; Yamamoto, T.; Lagerloef, K.P.D. Dislocation Structures of Low-Angle and Near-Sigma 3 Grain Boundaries in Alumina Bicrystals. *J. Am. Ceram. Soc.* **2003**, *86*, 595–602. [CrossRef]
19. Adepalli, K.K.; Kelsch, M.; Merkle, R.; Maier, J. Enhanced ionic conductivity in polycrystalline TiO$_2$ by "one-dimensional doping". *Phys. Chem. Chem. Phys. Pccp* **2014**, *16*, 4942–4951. [CrossRef]
20. Li, J.; Wang, H.; Zhang, X. Nanoscale stacking fault–assisted room temperature plasticity in flash-sintered TiO$_2$. *Sci. Adv.* **2019**, *5*, eaaw5519. [CrossRef]
21. Johanning, M.; Porz, L.; Dong, J.; Nakamura, A.; Li, J.-F.; Rödel, J. Influence of dislocations on thermal conductivity of strontium titanate. *Appl. Phys. Lett.* **2020**, *117*, 021902. [CrossRef]
22. Nakamura, A.; Lagerlöf, K.P.D.; Matsunaga, K.; Tohma, J.; Yamamoto, T.; Ikuhara, Y. Control of dislocation configuration in sapphire. *Acta Mater.* **2005**, *53*, 455–462. [CrossRef]
23. Gumbsch, P.; Taeri-Baghbadrani, S.; Brunner, D.; Sigle, W.; Ruhle, M. Plasticity and an inverse brittle-to-ductile transition in strontium titanate. *Phys. Rev. Lett.* **2001**, *87*, 085505. [CrossRef] [PubMed]
24. Johnston, W.G.; Gilman, J.J. Dislocation Velocities, Dislocation Densities, and Plastic Flow in Lithium Fluoride Crystals. *J. Appl. Phys.* **1959**, *30*, 129–144. [CrossRef]
25. Gaillard, Y.; Tromas, C.; Woirgard, J. Pop-in phenomenon in MgO and LiF: Observation of dislocation structures. *Philos. Mag. Lett.* **2010**, *83*, 553–561. [CrossRef]
26. Tromas, C.; Gaillard, Y.; Woirgard, J. Nucleation of dislocations during nanoindentation in MgO. *Philos. Mag.* **2006**, *86*, 5595–5606. [CrossRef]
27. Fang, X.; Ding, K.; Janocha, S.; Minnert, C.; Rheinheimer, W.; Frömling, T.; Durst, K.; Nakamura, A.; Rödel, J. Nanoscale to microscale reversal in room-temperature plasticity in SrTiO$_3$ by tuning defect concentration. *Scr. Mater.* **2020**, *188*, 228–232. [CrossRef]
28. Taeri, S.; Brunner, D.; Sigle, W.; Rühle, M. Deformation behaviour of strontium titanate between room temperature and 1800 K under ambient pressure. *Z. Met.* **2004**, *95*, 433–446. [CrossRef]
29. Patterson, E.A.; Major, M.; Donner, W.; Durst, K.; Webber, K.G.; Rodel, J. Temperature-Dependent Deformation and Dislocation Density in SrTiO$_3$ (001) Single Crystals. *J. Am. Ceram. Soc.* **2016**, *99*, 3411–3420. [CrossRef]
30. Mark, A.F.; Castillo-Rodriguez, M.; Sigle, W. Unexpected plasticity of potassium niobate during compression between room temperature and 900 °C. *J. Eur. Ceram. Soc.* **2016**, *36*, 2781–2793. [CrossRef]
31. Dehm, G.; Jaya, B.N.; Raghavan, R.; Kirchlechner, C. Overview on micro- and nanomechanical testing: New insights in interface plasticity and fracture at small length scales. *Acta Mater.* **2017**, *142*, 248–282. [CrossRef]
32. Pharr, G.M.; Oliver, W.C. Measurement of Thin Film Mechanical Properties Using Nanoindentation. *Mrs Bull.* **1992**, *17*, 28–33. [CrossRef]
33. Oliver, W.C.; Pharr, G.M. An improved technique for determining hardness and elastic modulus using load and displacement sensing indentation experiments. *J. Mater. Res.* **1992**, *7*, 1564–1583. [CrossRef]
34. Greer, J.R.; De Hosson, J.T.M. Plasticity in small-sized metallic systems: Intrinsic versus extrinsic size effect. *Prog. Mater. Sci.* **2011**, *56*, 654–724. [CrossRef]
35. Johnson, K.L. *Contact Mechanics*; Cambridge University Press: Cambridge/London, UK, 1985.
36. Li, W.; Bei, H.; Qu, J.; Gao, Y. Effects of machine stiffness on the loading–displacement curve during spherical nano-indentation. *J. Mater. Res.* **2013**, *28*, 1903–1911. [CrossRef]
37. Javaid, F.; Bruder, E.; Durst, K. Indentation size effect and dislocation structure evolution in (001) oriented SrTiO$_3$ Berkovich indentations: HR-EBSD and etch-pit analysis. *Acta Mater.* **2017**, *139*, 1–10. [CrossRef]
38. Swain, M.V.; Lawn, B.R. A Study of Dislocation Arrays at Spherical Indentations in LiF as a Function of Indentation Stress and Strain. *Phys. Stat. Sol.* **1969**, *35*, 909–923. [CrossRef]
39. Javaid, F.; Stukowski, A.; Durst, K. 3D Dislocation structure evolution in strontium titanate: Spherical indentation experiments and MD simulations. *J. Am. Ceram. Soc.* **2017**, *100*, 1134–1145. [CrossRef]

40. Javaid, F.; Johanns, K.E.; Patterson, E.A.; Durst, K. Temperature dependence of indentation size effect, dislocation pile-ups, and lattice friction in (001) strontium titanate. *J. Am. Ceram. Soc.* **2018**, *101*, 356–364. [CrossRef]
41. Field, J.S.; Swain, M.V. A simple predictive model for spherical indentation. *J. Mater. Res.* **1993**, *8*, 297–306. [CrossRef]
42. Gao, Y.; Bei, H. Strength statistics of single crystals and metallic glasses under small stressed volumes. *Prog. Mater. Sci.* **2016**, *82*, 118–150. [CrossRef]
43. Morris, J.R.; Bei, H.; Pharr, G.M.; George, E.P. Size Effects and Stochastic Behavior of Nanoindentation Pop In. *Phys. Rev. Lett.* **2011**, *106*, 165502. [CrossRef] [PubMed]
44. Jin, K.; Xia, Y.; Crespillo, M.; Xue, H.; Zhang, Y.; Gao, Y.F.; Bei, H. Quantifying early stage irradiation damage from nanoindentation pop-in tests. *Scr. Mater.* **2018**, *157*, 49–53. [CrossRef]
45. Bei, H.; Xia, Y.Z.; Barabash, R.I.; Gao, Y.F. A tale of two mechanisms: Strain-softening versus strain-hardening in single crystals under small stressed volumes. *Scr. Mater.* **2016**, *110*, 48–52. [CrossRef]
46. Zimmerman, J.A.; Kelchner, C.L.; Klein, P.A.; Hamilton, J.C.; Foiles, S.M. Surface step effects on nanoindentation. *Phys. Rev. Lett.* **2001**, *87*, 165507. [CrossRef]
47. Matsunaga, K.; Ii, S.; Iwamoto, C.; Yamamoto, T.; Ikuhara, Y. In situ observation of crack propagation in magnesium oxide ceramics. *Nanotechnology* **2004**, *15*, S376–S381. [CrossRef]
48. Kondo, S.; Shibata, N.; Mitsuma, T.; Tochigi, E.; Ikuhara, Y. Dynamic observations of dislocation behavior in $SrTiO_3$ by in situ nanoindentation in a transmission electron microscope. *Appl. Phys. Lett.* **2012**, *100*, 181906. [CrossRef]
49. Kondo, S.; Mitsuma, T.; Shibata, N.; Ikuhara, Y. Direct observation of individual dislocation interaction processes with grain boundary. *Sci. Adv.* **2016**, *2*, e1501926. [CrossRef]
50. Kondo, S.; Ishihara, A.; Tochigi, E.; Shibata, N.; Ikuhara, Y. Direct observation of atomic-scale fracture path within ceramic grain boundary core. *Nat. Commun.* **2019**, *10*, 2112. [CrossRef]
51. Nie, A.; Bu, Y.; Huang, J.; Shao, Y.; Zhang, Y.; Hu, W.; Liu, J.; Wang, Y.; Xu, B.; Liu, Z.; et al. Direct Observation of Room-Temperature Dislocation Plasticity in Diamond. *Matter* **2020**, *2*, 1222–1232. [CrossRef]
52. Sernicola, G.; Giovannini, T.; Patel, P.; Kermode, J.R.; Balint, D.S.; Britton, T.B.; Giuliani, F. In situ stable crack growth at the micron scale. *Nat. Commun.* **2017**, *8*, 1–9. [CrossRef] [PubMed]
53. Cho, J.; Li, J.; Wang, H.; Li, Q.; Fan, Z.; Mukherjee, A.K.; Rheinheimer, W.; Wang, H.; Zhang, X. Study of deformation mechanisms in flash-sintered yttria-stabilized zirconia by in-situ micromechanical testing at elevated temperatures. *Mater. Res. Lett.* **2019**, *7*, 194–202. [CrossRef]
54. Malyar, N.V.; Micha, J.S.; Dehm, G.; Kirchlechner, C. Size effect in bi-crystalline micropillars with a penetrable high angle grain boundary. *Acta Mater.* **2017**, *129*, 312–320. [CrossRef]
55. Malyar, N.V.; Micha, J.S.; Dehm, G.; Kirchlechner, C. Dislocation-twin boundary interaction in small scale Cu bi-crystals loaded in different crystallographic directions. *Acta Mater.* **2017**, *129*, 91–97. [CrossRef]
56. Gaillard, Y.; Tromas, C.; Woirgard, J. Study of the dislocation structure involved in a nanoindentation test by atomic force microscopy and controlled chemical etching. *Acta Mater.* **2003**, *51*, 1059–1065. [CrossRef]
57. Zaefferer, S.; Elhami, N.-N. Theory and application of electron channelling contrast imaging under controlled diffraction conditions. *Acta Mater.* **2014**, *75*, 20–50. [CrossRef]
58. Otsuka, K.; Kuwabara, A.; Nakamura, A.; Yamamoto, T.; Matsunaga, K.; Ikuhara, Y. Dislocation-enhanced ionic conductivity of yttria-stabilized zirconia. *Appl. Phys. Lett.* **2003**, *82*, 877–879. [CrossRef]
59. Fang, X.; Jia, J.; Feng, X. Three-point bending test at extremely high temperature enhanced by real-time observation and measurement. *Measurement* **2015**, *59*, 171–176. [CrossRef]
60. Luo, H.; Lu, W.; Fang, X.; Ponge, D.; Li, Z.; Raabe, D. Beating hydrogen with its own weapon: Nano-twin gradients enhance embrittlement resistance of a high-entropy alloy. *Mater. Today* **2018**, *21*, 1003–1009. [CrossRef]
61. Wheeler, J.M.; Armstrong, D.E.J.; Heinz, W.; Schwaiger, R. High temperature nanoindentation: The state of the art and future challenges. *Curr. Opin. Solid State Mater. Sci.* **2015**, *19*, 354–366. [CrossRef]
62. Trenkle, J.C.; Packard, C.E.; Schuh, C.A. Hot nanoindentation in inert environments. *Rev. Sci. Instrum.* **2010**, *81*, 073901. [CrossRef] [PubMed]
63. Li, Y.; Fang, X.; Xia, B.; Feng, X. In situ measurement of oxidation evolution at elevated temperature by nanoindentation. *Scr. Mater.* **2015**, *103*, 61–64. [CrossRef]

64. Li, Y.; Fang, X.; Qu, Z.; Lu, S.; Li, H.; Zhu, T.; Yu, Q.; Feng, X. In situ full-field measurement of surface oxidation on Ni-based alloy using high temperature scanning probe microscopy. *Sci. Rep.* **2018**, *8*, 6684. [CrossRef] [PubMed]
65. Minnert, C.; Oliver, W.C.; Durst, K. New ultra-high temperature nanoindentation system for operating at up to 1100 °C. *Mater. Des.* **2020**, *192*, 108727. [CrossRef]

Publisher's Note: MDPI stays neutral with regard to jurisdictional claims in published maps and institutional affiliations.

© 2020 by the authors. Licensee MDPI, Basel, Switzerland. This article is an open access article distributed under the terms and conditions of the Creative Commons Attribution (CC BY) license (http://creativecommons.org/licenses/by/4.0/).

Article

Effect of 3D Representative Volume Element (RVE) Thickness on Stress and Strain Partitioning in Crystal Plasticity Simulations of Multi-Phase Materials

Faisal Qayyum [1,*,†], **Aqeel Afzal Chaudhry** [2,†], **Sergey Guk** [1], **Matthias Schmidtchen** [1], **Rudolf Kawalla** [1] **and Ulrich Prahl** [1]

1. Institute of Metal Forming, Technische Universität Bergakademie Freiberg, 09599 Freiberg, Germany; sergey.guk@imf.tu-freiberg.de (S.G.); matthias.schmidtchen@imf.tu-freiberg.de (M.S.); rudolf.kawalla@imf.tu-freiberg.de (R.K.); ulrich.prahl@imf.tu-freiberg.de (U.P.)
2. Geotechnical Institute, Technische Universität Bergakademie Freiberg, 09599 Freiberg, Germany; aqeel.chaudhry@ifgt.tu-freiberg.de
* Correspondence: faisal.qayyum@student.tu-freiberg.de
† These authors contributed equally to this work.

Received: 27 August 2020; Accepted: 9 October 2020; Published: 17 October 2020

Abstract: Crystal plasticity simulations help to understand the local deformation behavior of multi-phase materials based on the microstructural attributes. The results of such simulations are mainly dependent on the Representative Volume Element (RVE) size and composition. The effect of RVE thickness on the changing global and local stress and strain is analyzed in this work for a test case of dual-phase steels in order to identify the minimal RVE thickness for obtaining consistent results. 100 × 100 × 100 voxel representative volume elements are constructed by varying grain size and random orientation distribution in DREAM-3D. The constructed RVEs are sliced in depth up to 1, 5, 10, 15, 20, 25, 30, 40, and 50 layers to construct different geometries with increasing thickness. Crystal plasticity model parameters for ferrite and martensite are taken from already published data and assigned to respective phases. Although the global stress/strain behavior of different RVEs is similar (<5% divergence), the local stress/strain partitioning in RVEs with varying thickness and grain size shows a considerable variation when statistically compared. It is concluded that two-dimensional (2D) RVEs can be used for crystal plasticity simulations when global deformation behavior is of interest. Whereas, it is necessary to consider three-dimensional (3D) RVEs, which have a specific thickness and number of grains for determining stabilized and more accurate local deformation behavior. This estimation will help researchers in optimizing the computation time for accurate mesoscale simulations.

Keywords: crystal plasticity; DAMASK; representative volume element; dual-phase steel; local deformation behavior

1. Introduction

The micro-structure of a material plays an important role in defining the mechanical properties [1] and service life of a component [2,3]. Numerical models can help to understand and improve the component's life by providing detailed insight into the local [4,5] and global [6–8] deformation behaviors. A concept of Representative Volume Element (RVE) is used in order to numerically simulate continuous yet locally heterogeneous materials [9,10]. A lot of work in the recent past has been carried out to estimate the local deformation behavior of single and multi-phase materials based on two-dimensional (2D) and three-dimensional (3D) RVEs [11–14]. 2D and 3D RVEs can be constructed using single [15] or multilayer [16] Voronoi tessellation using measured or virtual local grain size, phase, and orientation distribution data [17]. They can also be constructed by processing EBSD map of a material appropriately [18], and specifically 3D RVEs can be constructed by multi-layer EBSD mapping of a local region while using Focused Ion Beam (FIB) milling [19].

Using RVEs in CP simulations has been common practice for quite some time, and most researchers used 2D RVEs, as they are easier to collect and they can help compare the locally observed deformation behavior with simulation results [20–22]. In recent studies [19,23], it was shown that 3D RVEs—compared with 2D RVEs—yield nearly accurate local stress–strain evolution results, yet the effect of RVE thickness on results was not analyzed in these studies. The computational costs in CP simulations are quite high and largely depend on the size of RVE, especially for high-resolution full phase simulations, it can take weeks to yield the desired results [24]. It is important to know the effective RVE thickness relative to the material, average grain size, and develop a simulation model accordingly. Such estimation leads to reduced computational cost while maintaining the accuracy of the results. In the past, researchers analyzed the effects of RVE size and applied boundary conditions on the deformation behavior of heterogeneous materials [4,25,26]. Scale-dependent elastic and elastoplastic deformation behaviors of periodic [27] and random [28] composites were analyzed.

DREAM-3D [29] is an open-source tool that is available to construct RVEs from experimentally or analytically available micro-structural data. Recently, the coupling of DREAM-3D with DAMASK has been made easier by introducing a pipeline object in order to directly export the generated RVE to a readable geometry file [30]. This technique is used in the current study to construct virtual RVEs with different grain sizes. Recently, researchers suggested a methodology for calibrating the DAMASK models using a benchmark 1000 grain RVE [18] by comparing the results with experimental flow curves. In another work, they calibrated the model by comparing it with the in-situ acoustic emission data [31]. In both of these publications, the calibrated model was used to carry out full phase simulations for the TRIP steel matrix and TRIP steel Mg-PSZ composite. EBSD maps were used as 2D geometries for simulations. Although local evolution of stress, strain, dislocation density, transformation, and twinning were analyzed, it was reported that the results are only qualitatively accurate for comparison as there is no third dimension for these attributes to evolve in. It was concluded that 3D RVEs should be considered for accurate full phase simulations. Considering such work to be capital and computationally intensive, selecting appropriate RVE thickness with respect to the material grain size is very important. Recently, Diehl et al. [32] investigated the influence of neighborhood on stress and strain partitioning in DP steel microstructures and showed the relevance of subsurface microstructures in this regard. It was concluded that structural changes farther than three times the average grain size have a negligible effect on the region of interest. However, the conclusions were made based on a fixed grain size range and limited statistical analysis.

Considerable research has been carried out in the development of RVEs from experimental and statistical data [33]. A very detailed review of the generation of 3D representative volume elements for heterogeneous materials was recently published [34]. Several experimental, physical-based, and geometrical methods were discussed, along with the available commercial and open-source tools in this work. Researchers have also used the constructed RVEs in order to run mesoscale simulations and they have analyzed the effect of different RVEs on the outcome of simulations [35–38]. CP simulations are increasingly being used to model and analyze complex problems from the single crystal up to the component scale [39]. It has been reported that the simulation results get better with increasing RVE size, as larger RVEs are better informed and incorporate more microstructural features [34,40]. Zeghadi et al. [41,42] tried to establish such a relationship for single-phase materials. However, it was mentioned while concluding the study to check results for a bigger data set and more grains on the free surface. Harris et al. [43,44] carried out a detailed study for determining the appropriate RVE size for 3D microstructural material characterization that is based on the multi-phase composite gas separation membrane. They showed that the developed statistical model was able to predict the experimental observations reasonably. A study that identifies the optimal finite thickness of an RVE—relative to the material grain size—in multi-phase materials for optimal simulation time without compromising the accuracy of results is still missing. Different researchers in the past have approached this problem differently. The method that is presented in the current work is unique, fast, and effective in establishing the desired output.

In this work, the effect of RVE thickness—with the intention of identifying the amount of finite thickness required for consistent results—on the changing global and local stress and strain is analyzed for a test case of dual-phase steels. RVEs were constructed by varying mean feature ESD ($\mathcal{D}_f$) between 3 to 18 μm using DREAM-3D with grains having randomly assigned orientation distributions. The constructed RVEs were sliced in up to 1, 5, 10, 15, 20, 25, 30, 40, and 50 layers to produce different geometries comprising the same microstructure—with increasing thickness. Crystal plasticity model parameters for ferrite and martensite are taken from already published data and assigned to respective phases. Probability Distribution Functions (PDF) and Cumulative Distribution Functions (CDF) of all simulation results are compared in order to estimate the solution convergence with changing grain size statistically. In the end, a simple function is proposed for calculating the sufficient RVE thickness that is necessary for obtaining a converged solution.

Section 2 provides the details of material data, RVE construction, simulation scheme, and CP material model parameters. Section 3 presents the results that were obtained in this study. In Section 4, the results are discussed in comparison with state of the art, and insight into the outlook is provided. Eventually, the study is concluded in Section 5.

2. Numerical Simulation Model Development

For the numerical simulation modeling, RVEs were virtually constructed while using open source tool DREAM-3D [29]. The case of dual-phase steel is considered as a case study in the current work. Micro-structural attributes were adopted from the literature [45]. 3D RVEs were constructed—with varying grain size—using open source tool DREAM-3D [29]. These constructed RVEs were sliced into varying thickness ranges from one layer to 50 layers and used as input geometries for numerical simulations. Crystal plasticity based open source tool DAMASK was used to model and solve numerical simulations by adopting already published methodology [22].

2.1. RVE Construction and Geometry Files Production

Dual-phase steel is chosen as a case study in the current work, because, on the one hand, it is a simple material that consists of ferrite matrix and embedded martensite islands (consisting of laths with different crystallographic orientations) in it, while, on the other hand, it is an excellent example of multi-phase materials with two phases comprising of drastically different mechanical properties. The microstructural attributes i.e., grain size and martensite percentage, were adopted from the previous work of Jiang et al. [46]. It has been reported that, in DP steels, the grain size of ferrite ranges from 5 to 25 μm, whereas martensite grain size ranges from 3 to 18 μm. Considering this upper and lower grain size limit, DREAM-3D was used for the construction of virtual RVEs with varying grain sizes, while keeping the ferrite and martensite grain size ratio to be 1:1 intentionally to keep the problem at hand simple.

Table 1 provides the total number of grain in each 100^3 RVE for a corresponding ESD given as a dimensionless entity. A pipeline was built with the initialization of virtual data generation while using a stats generator filter. The ratio of 90 vol. % primary phase(ferrite), 10 vol. % secondary phase(martensite) was chosen for all cases to simplify the simulation model results and generalize the comparison of statistical data. As the crystal structure of both Ferrite and Martensite is body-centered cubic. Therefore, they were generated as "cubic equiaxed" phases. Different RVEs were generated by varying the ESD statistical distribution data. The grain size distribution of ferrite and martensite for each RVE taken is given in Table 1.

The synthetic volume size of all RVEs was kept 100 × 100 × 100 voxels (where the size of each voxel is 1 μm³) and specified number of particles could pack in the defined volume, which means that more grains packed in the defined volume when $\mathcal{D}_f$ was small and lesser when $\mathcal{D}_f$ was large as given in last column of Table 1. For both phases, grain shape was stated to be "ellipsoid". The maximum number of iterations (swaps) allowed was 100,000. Using already published methodology [30], the generated RVEs were recorded as .xdmf file, which is readable by Paraview for visualization, and .geom files were saved, which are readable by DAMASK. Figure 1 presents the general flow of information in the pipeline with associated inputs and outputs.

Table 1. Grain size distribution data for ferrite and martensite used for the construction of Representative Volume Elements (RVEs). The grain sizes are in µm and "Total Grains" is a dimensionless entity.

Nomenclature	Ferrite Grains			Martensite Grains			Total Grains
	Min.	Max	Avg.	Min.	Max	Avg.	
A	5.1	7.6	6.35	3.5	5.7	4.6	8400
B	6.6	9.9	8.25	5.2	8.4	6.8	3700
C	8.5	12.7	10.6	5.4	8.8	7.1	1900
D	9.7	14.5	12.1	8.2	13.2	10.7	1200
E	11.2	16.8	14.0	9.8	15.8	12.8	770

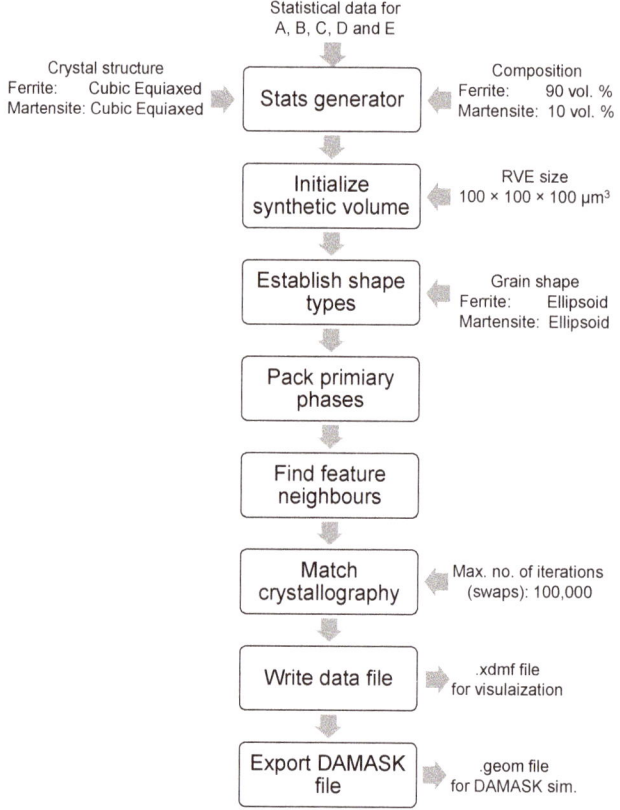

Figure 1. Block diagram of the DREAM-3D pipeline for RVE generation with associated inputs and outputs of various pipeline blocks. Table 1 provides details of statistical input data for A, B, C, D, and E.

Figure 2 shows a schematic diagram representing the generated 3D RVEs and their slicing sequence for the construction of simulation geometries. RVE-A and E are shown for representation, whereas B, C, and D were modeled in a similar fashion. Ferrite (F) and Martensite (M) are shown separately in the sliced images for better visualization of phase distributions. This research's primary objective is to identify the optimal RVE thickness with respect to the material grain size for obtaining converged local results. For this reason, each constructed 3D RVE with varying grain size was sliced into geometries with increasing thickness from one to 50 layers. Slicing sequence is also shown in Figure 2 for each RVE with yellow dotted lines, whereas green dotted lines in the schematic diagram represent more sliced geometries that are not shown in the current figure, but were constructed for simulations. There are 100,000 gaussian mesh elements in one layer, which means that each geometry contains 100,000 gaussian mesh elements multiplied by the number of layers considered in each case. A complete grain comes into account with a lesser number of slices in the case of small $\mathcal{D}_f$, whereas more slices are needed in order to represent full-grain with bigger $\mathcal{D}_f$.

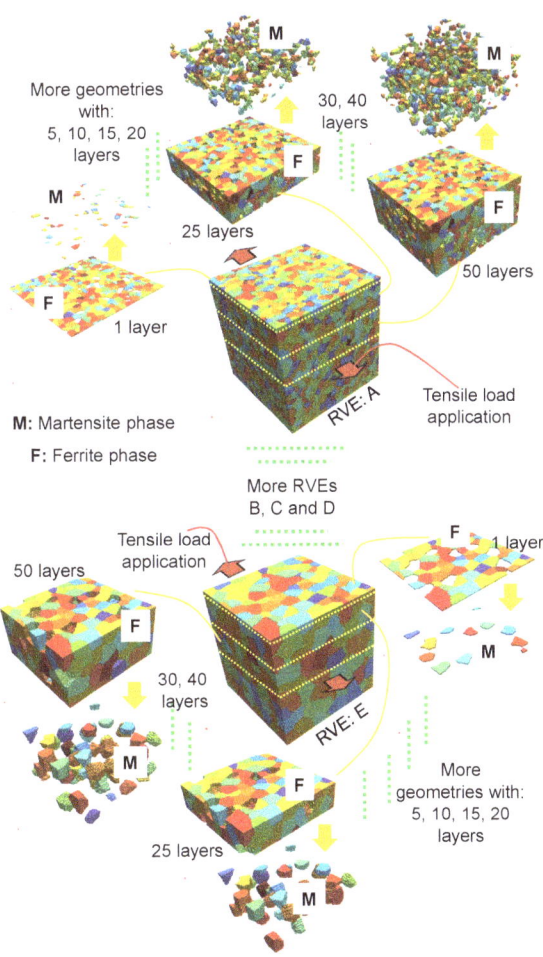

Figure 2. Schematic diagram representing the generated 3D RVEs and their slicing sequence for the construction of simulation geometries. RVE-A and E are shown in this figure, whereas B, C, and D were modeled similarly. The slicing sequence is also shown in the figure with a yellow dotted line, and green dotted lines represent more sliced geometries that are not shown here but were constructed for simulations. Ferrite (F) and Martensite (M) are shown separately in the sliced images for better visualization of phase distribution in the RVEs.

2.2. Material Properties and Loading Conditions

Although there is a large difference in the mechanical properties of ferrite and martensite, in current research, both of the phases were assigned elastic-viscoplastic properties by adopting already developed phenomenological power law available in DAMASK [47]. Details of the hardening model with equations is provided in Appendix A for the readers who are not familiar with the constitutive equations used in DAMASK. The elastic coefficients, variables defining plastic flow behavior and fitting parameters for both phases were adopted from the already published literature [45]. Table 2 presents the adopted parameters. In the spectral method each grid point is considered as a computation point which is assigned a phase and initial crystallographic orientation before the simulations start. In the current work, mixed boundary conditions were applied uni-axially along x-direction as follows, while keeping the out-of-plain surfaces in all geometries stress-free:

$$\dot{F}_{ij} = \begin{bmatrix} 1 & 0 & 0 \\ 0 & * & 0 \\ 0 & 0 & * \end{bmatrix} \times 10^{-3}.s^{-1} \quad (1)$$

$$P_{ij} = \begin{bmatrix} * & * & * \\ * & 0 & * \\ * & * & 0 \end{bmatrix} Pa \quad (2)$$

Periodicity of the solution is inherent to the spectral method due to the FOURIER approximation of the deformation gradient field [47]. Yet, it is known that the immediate neighborhood mainly influences the strain heterogeneity, and the influence of artificial periodicity introduced by the boundary description is confined to a narrow zone [22]. Therefore, in the current study, this effect is ignored. For strongly different phases i.e., ferrite and martensite, the reference stiffness has a strong influence on stability and convergence rate, as was shown earlier by Michel et al. [48]. The readers are encouraged to refer to the work by Diehl et al. [19] for the explanation of the assumptions in the FFT regarding the reference stiffness that are used within the framework of the crystal plasticity provided by DAMASK.

Table 2. Parameters for ferrite and martensite adopted from literature [45] for the numerical simulation modeling.

Parameter Definition	Symbol	Attributes for Ferrite	Attributes for Martensite	Unit
First elastic stiffness constant with normal strain	C_{11}	233.3	417.4	GPa
Second elastic stiffness constant with normal strain	C_{12}	135.5	242.4	GPa
First elastic stiffness constant with shear strain	C_{44}	128.0	211.1	GPa
Shear strain rate	$\dot{\gamma}_0$	1	1	$10^{-3}/s$
Initial Shear resistance on [111]	S_0 [111]	95	406	MPa
Saturation shear resistance on [111]	S_∞ [111]	222	873	MPa
Initial Shear resistance on [112]	S_0 [112]	96	457	MPa
Saturation shear resistance on [112]	S_∞ [112]	412	971	MPa
Slip hardening parameter	h_0	1.0	563	GPa
Interaction hardening parameter	$h_{\alpha,\beta}$	1.0	1.0	-
Stress exponent	n	20	20	-
Curve fitting parameter	w	2.0	2.0	-

The simulation results were post-processed by using already available subroutines in the DAMASK installation module, and data were further statistically analyzed using the Seaborn library in Python. The local stress–strain distributions were visualized using open source tool Paraview [49].

3. Results

If global/averaged stress–strain behavior of all the simulations is plotted and compared for multiple layers of a specific RVE, as shown in Figure 3, similar results with slight variation of slope are observed for varying thickness of geometry (Figure 3a,b) or variation of grain size (Figure 3c). It can be clearly observed in these figures that the trend of stress–strain curve is the same for all grain sizes with slight variation (higher stress response to same strain with increasing grain size).

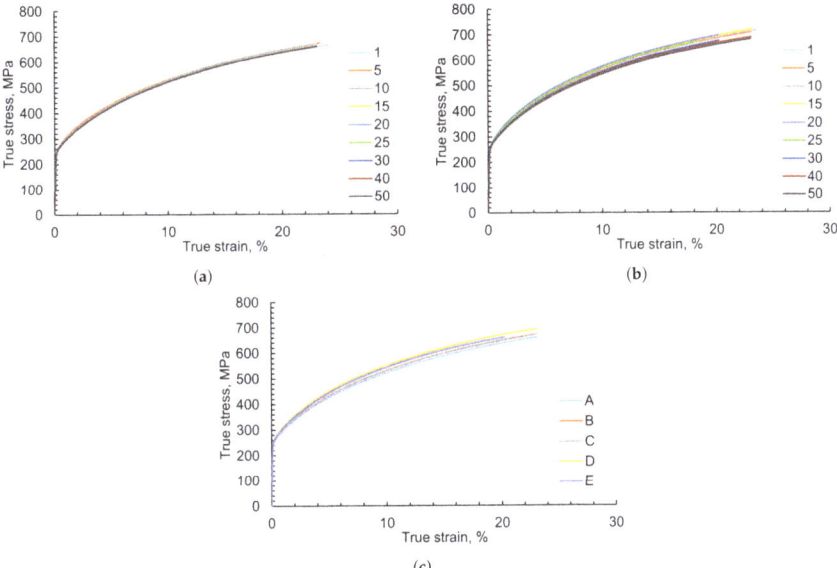

Figure 3. Comparison of global stress–strain curves of different cases (**a**) RVE-A all layers' comparison, (**b**) RVE-E all layers' comparison, (**c**) showing true stress strain curves for RVEs with different grain sizes. It is observed that there is not much difference in the results and almost any geometry yields very similar results. Some numerical simulation models crashed due to the non convergence of global stress/strain equilibrium at the spectral code level, and therefore are presented up to that certain level of strain.

Although stress–strain curves are used by engineers and scientists to understand the overall material behavior, they can be very misleading in the case of such full phase simulations where the local results may vary drastically. Still, the averages' response remains the same. To make this point, during the post-processing of the data, the RVEs were sectioned as schematically represented in Figure 4. This scheme was adopted to expose the top surface and middle section of the RVE-E as an example case. Local stress and strain values are represented with the same scale to show how they change with varying RVE thickness at 25% of true strain.

Figure 5 shows the local von Mises true stress distribution in all geometries of RVE-E at 25% of true strain. Only the ferrite phase is shown here for better visualization by filtering out the martensite phase, which, due to very high stresses (≈2.5 GPa), distorts the scale. It is observed that there is a high contrast of stress distribution in the 01-layer RVE with some areas of very high stresses and others with very low stresses. As the thickness of the RVE is increased from 01 layers to 50 layers, the stresses on the surface diminish and they are relatively more homogeneously distributed within the matrix, and the high contrast for stresses diminishes. There is very less difference in the local stress distribution of 40-layer and 50-layer simulations. In these 3D simulations, it is observed that, although the phase interface is more prone to higher stresses, it is not always the case. In the middle section, it is observed that the local stress distribution becomes consistent with similar areas of high and low stresses. This similarity in obtained results—with increased RVE thickness—represents the convergence of the point to point local solutions.

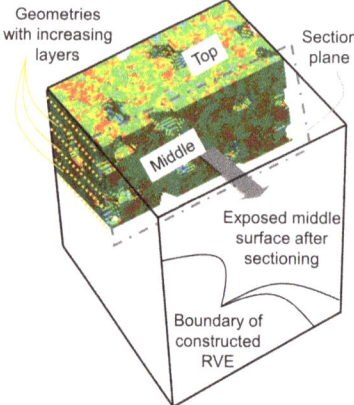

Figure 4. Schematic diagram showing the sectioning scheme to expose the middle and top surfaces for visual comparison of local stress and strain in the upcoming figures. Local von Mises true stress distribution in the ferrite phase has been used here just as an example for color coding of the section of interest.

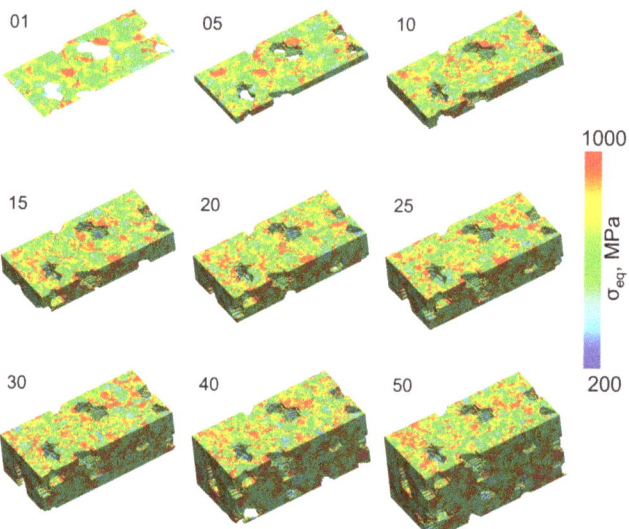

Figure 5. Local von Mises true stress distribution—in ferrite phase—for all geometries of RVE-E at 25% of global true strain.

Figure 6 shows local von Mises true strain distribution in all geometries of RVE-E at 25% of global true strain. In this figure, both—ferrite and martensite—phases are shown. Embedded martensite grains undergo negligible plastic strain during overall deformation and, hence, exhibit almost zero strain in Figure 6 (pointed out by green arrows). In 01-layer simulation results, it is observed that the local strain contrast is quite large with sharp strain channels around martensite grains oriented 45° to the applied load direction. As the RVE thickness increases from 01-layer to 50-layers, it is observed that this sharp strain contrast on the top surface diminishes due to strain distribution in the third dimension. In the middle section of simulated geometry, it is observed that the local strain distribution converges with similar solution output in case of 40-layer and 50-layer geometries, respectively (pointed out by red arrows). The magnitude and position of local strain distribution in these cases are identical.

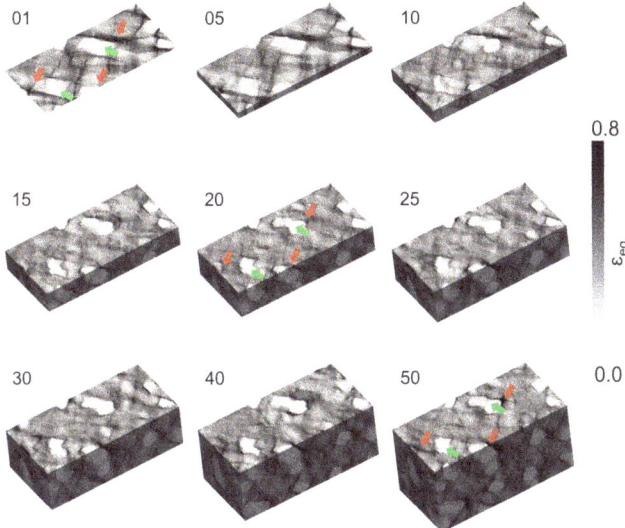

Figure 6. Local von Mises true strain distribution—in ferrite and martensite phases—for all geometries of RVE-E at 25% of global true strain. Red arrows represent areas of high strain in ferrite matrix, and green arrows represent areas of low strain in martensite grains.

The visual comparison of local stress and strain distribution in multiple varying geometries to observe the convergence of results—as shown in Figures 5 and 6—is a very challenging task. Visual inspections are primarily dependent on subjective choices; therefore, statistical data analysis tools are adopted in the current research in order to work out the convergence of the observed results. For statistical analysis, PDFs and CDFs of true local stress and strain distributions in each phase are constructed for each simulated geometry at the maximum global strain. Local stress and strain distributions in each phase of each geometry are compared. This detailed comparison is shown in Figure 7 by intelligently grouping data in different subplots.

It is observed that, with small $\mathcal{D}_f$, i.e., in the case of RVE-A, as shown in Figure 7a, the local stress, and strain distribution in both phases is quite different for 01-layer geometry as compared with thicker geometries. It is observed that, with increasing geometry thickness, the PDFs and CDFs become similar after more than 10-layers.

When $\mathcal{D}_f$ increases, i.e., in case of RVE-C as shown in Figure 7c, similar trend of convergence with increasing geometry thickness is observed. The distribution varies up to 20-layer geometry for the current case and it does not change with a further increase in geometry thickness. It is observed that, with increasing $\mathcal{D}_f$ in all RVEs i.e., in Figure 7a–e, the PDFs and CDFs converge with increasing geometry thicknesses, but more geometry thickness is needed when $\mathcal{D}_f$ is large. It is an expected response because for RVEs with large $\mathcal{D}_f$ more geometry thickness is needed to define a grain completely and hence the flow of stresses and strains around it becomes possible.

The stress and strain distribution behavior of martensite in thin geometries (one-layer to 15-layers) is different from thick geometries (20-layers to 50-layers), as observed in Figure 7(iii,iv). One-layer simulations in Figure 7(iii) represent a large and packed strain distribution profile compared to higher thickness results. Distribution is relatively more dispersed over a broad strain range, independent of $\mathcal{D}_f$. The stress distribution profile of one-layer simulations in Figure 7(iv) for all RVEs shows two peaks and a wider dispersion, whereas the distribution in close to bell shape when the RVE thickness is increased.

In Figure 7a–e, it is observed that at 50-layers geometry—due to very less change in the local stress and strain distribution—a converged solution for all RVEs is obtained. The PDF and CDF plots for both phases and all geometries with varying $\mathcal{D}_f$ are compared in Figure 8. It is observed that the curves accurately match with a slight difference in the peak values. When considering the same composition and material properties in all cases, this comparison confirms the convergence of the obtained results. From these data, one can interpret that with increasing layers in the RVE, the material volume and number of mesh points increase, or more specifically, a total number of randomly oriented grains increase. Therefore, the statistical behavior converges towards an average and, hence, produces a false notion of local convergence. This interpretation is not correct, as it can be verified by comparing local plots in Figures 5 and 6 that the local point-to-point convergence of the results happens, which is captured by the statistical comparison presented in Figure 7.

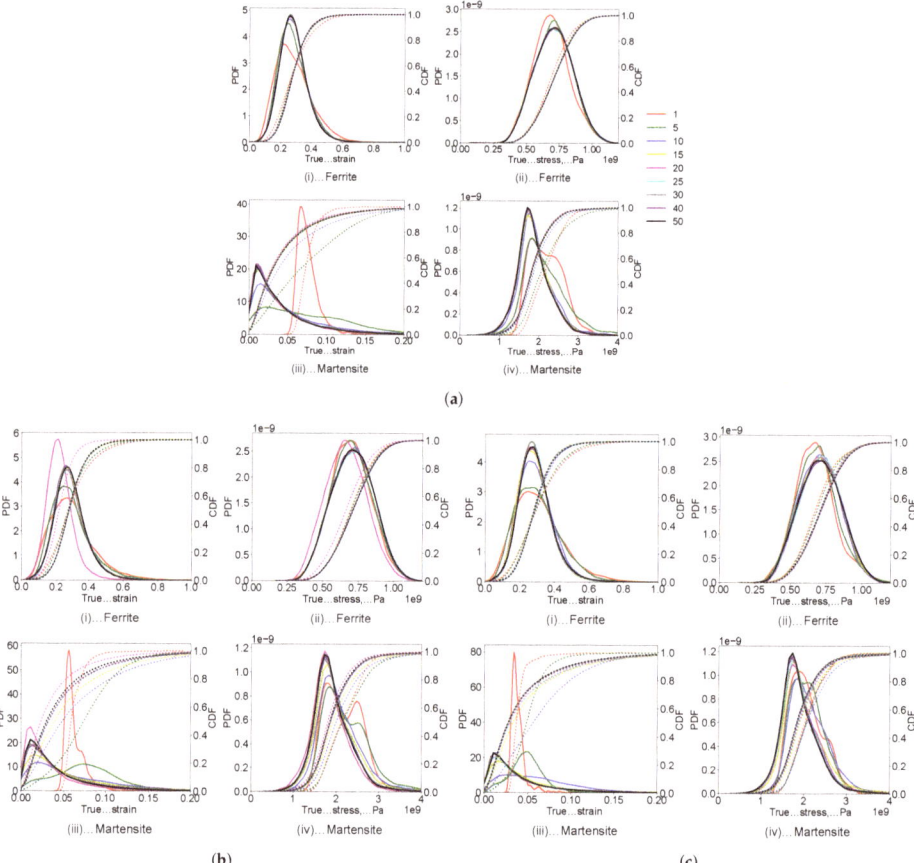

Figure 7. *Cont.*

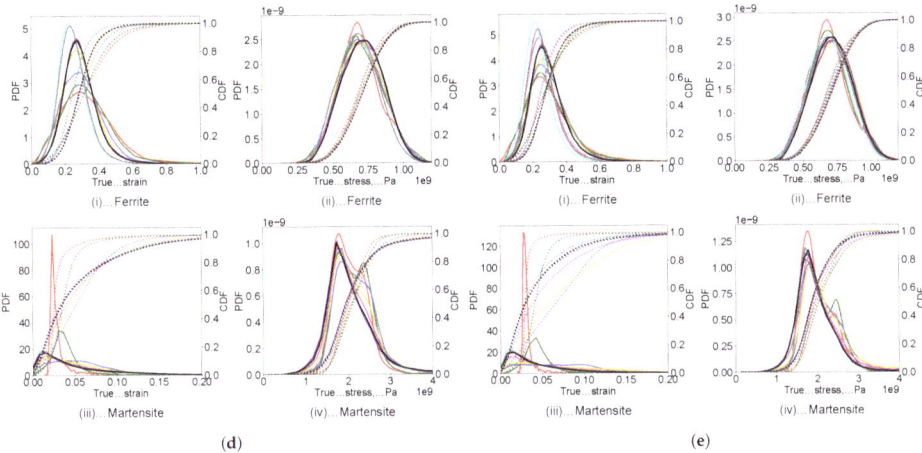

Figure 7. Comparison of (**i**) local strain distribution in ferrite, (**ii**) local stress distribution in ferrite, (**iii**) local strain distribution in martensite, (**iv**) local stress distribution in martensite, using Probability Distribution Functions (PDF) and Cumulative Distribution Functions (CDF) curves for RVEs with varying $\mathcal{D}_f$, (**a**) RVE-A, (**b**) RVE-B (**c**), RVE-C (**d**), RVE-D (**e**), and RVE-E at 25% of global true strain.

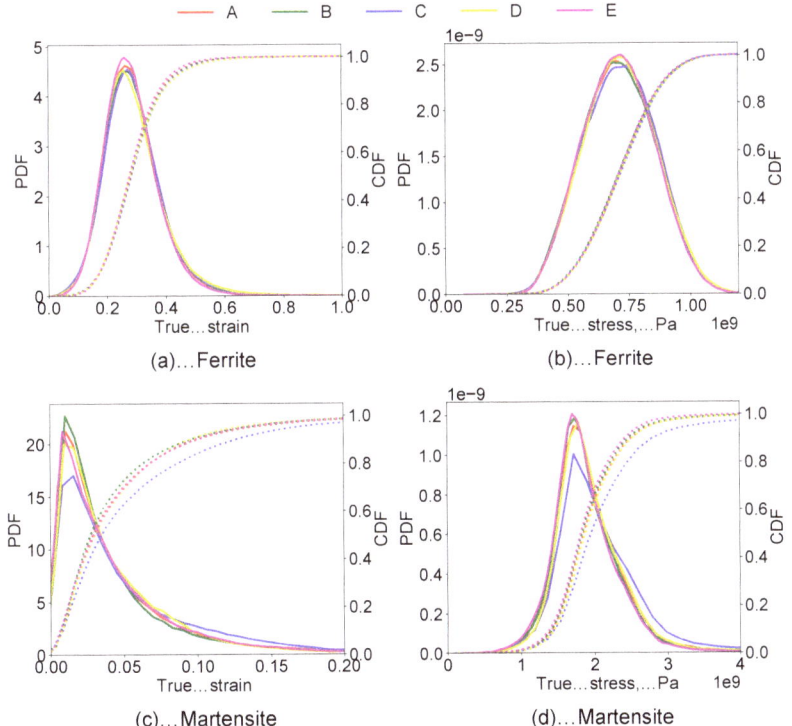

Figure 8. Comparison of (**a**) local strain distribution in ferrite, (**b**) local stress distribution in ferrite, (**c**) local strain distribution in martensite, and (**d**) local stress distribution in martensite, using PDF and CDF curves for RVEs with varying $\mathcal{D}_f$ at 25% of global true strain.

Although with increasing geometry thickness, the stress and strain distributions for both phases in Figure 7 are observed to move in multiple dimensions with varying shape. The maxima were noted and normalized against 50-layer geometries to simplify the convergence criteria. Peak values of PDF normalized against the 50-layer thickness simulation for ferrite strain are shown in Figure 9a and for ferrite stress in Figure 9b. Here, it is important to mention that the convergence of a result was analyzed against 50-layer thickness simulations, assuming them as perfectly converged, which might not be the case.

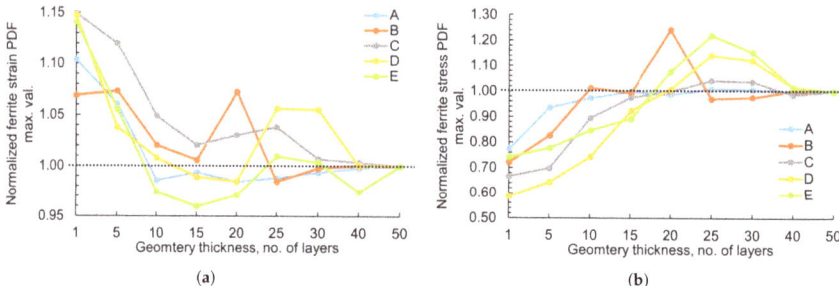

Figure 9. Peak values of PDF normalized against the 50-layer thickness simulation for (**a**) ferrite strain and (**b**) ferrite stress for $\mathcal{D}_f$ between 8.2 μm to 18.2 μm at 25% of the true global strain. Dotted lines represent normalization index of 1. It is observed that for all $\mathcal{D}_f$, increasing the RVE thickness converges the solution.

4. Discussion

Full phase simulation models are extensively used to study microstructural attributes' effects on local and global material deformation behaviors. The validation of such simulation models simultaneously on a global and local scale is a challenge that has not been addressed in the existing literature [32,41,42], but with its limitations. Crystal plasticity-based full-phase simulation models rely upon many constitutive and fitting parameters identified by comparing the experimental evolution of deformation mechanisms with the simulation results. These averaged stress–strain curves are compared to represent the accuracy of simulation results [17,22]. Based on such—global results based calibrated—models, the local material behavior is analyzed and studied. This methodology is useful for the inexpensive employment of such models. From the results of the current work, it is now clear that such methodology can be misleading as a wide variety of microstructures, which might result in very different local results, can yield the same global results.

It is shown in Figures 5 and 6 that the problems arise when the local deformation behaviors are compared. Owing to time and capital-intensive critical task, not many researchers in the past have carried out such analysis and observed that results only match qualitatively [19,31]. It has been reported by earlier researchers [19] that, in 2D DAMASK simulations, there is a high local stress and strain contrast due to the non-availability of the third direction. Experimental analysis of stresses and strains in 3D is almost impossible due to which the 3D simulation results cannot be validated. Additionally, paradoxically, if 3D EBSD data are measured, then there is no sample left to perform experimentation. On the other hand, if in-situ tests are executed on a sample to analyze deformation behavior, 3D EBSD data cannot be collected.

In the past, it has been elaborated that the 3D results are different from the 2D results, and realistic 3D geometries are better than the generalized 3D geometries [19]. How much of the 3D dimension is required for a converged solution was reported by Diehl et al. [32] by running simulations with fixed grain size RVEs. In this study, the effect of varying grain size of ferrite and martensite in DP steel is analyzed in order to study its effect. RVEs with varying mean feature size were synthetically constructed for dual-phase steel case in the current work. The RVEs were sliced to construct geometries with varying thicknesses from one to 50 layers. In 2D simulations with actual EBSD data or synthetically constructed RVEs, non-realistic high contrast stress and strain distribution are observed. It is evident that there is a drastic difference in the local results in 2D geometry when compared with 3D geometry.

The current work has not analyzed the effect of free surface on the convergence of the results, which can be important when comparing the converged local solutions with in-situ test observations. This should be kept in consideration while adopting the current methodology for such analysis. The readers are encouraged to refer to [32,50] for such modelling methodology.

In current work, statistical probability and cumulative distribution curves were compared for each phase in order to comprehensively analyze the local stress and strain distributions. The following relationship can be drawn for the obtained results, which is in accordance with the previous publications [32,42]:

$$\text{converged solution for } \mathcal{D}_f \propto \frac{1}{\mathcal{G}_\mathcal{N}} \propto \mathcal{L}_\mathcal{N} \tag{3}$$

An empirical function can be drawn and it generalizes the convergence results trend by normalizing the peak of stress and strain PDFs in the ferrite phase. Figure 10 shows the constructed empirical convergence criteria from the results of RVE-A, B, and C converged simulations. Only the first three points were considered in the development of the proposed empirical model. This is because the convergence of results here is used as a relative term against 50-layer thickness simulations, which might be misleading in the case of larger grain size RVEs. More simulations are needed with bigger RVEs in the future to be sure about their convergence. It is understood from previous work [18] that a large number of grains is vital for a solution convergence and, therefore, it is not good practice to reduce the total number of grains below 500 in an RVE.

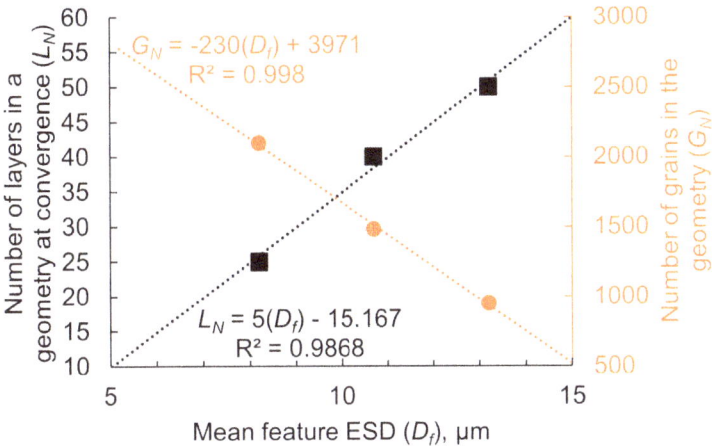

Figure 10. Constructed empirical convergence criteria from results of RVE-A, B and C converged simulations, where black trend line represented the function between mean feature ESD and required number of layers for converged solution. Orange trend line represents the relationship between mean feature ESD and number of grains that can be packed in a 100 × 100 × 100 µm³ geometry.

This criterion helps in setting the upper bound limit of $\mathcal{D}_f$ to 15. The number of layers required for a specific $\mathcal{D}_f$ can be calculated using the following derived empirical function:

$$\mathcal{L}_\mathcal{N} \geq 5\mathcal{D}_f - 15 \tag{4}$$

this equation is very specific and it is only valid for the similar grain sizes of two phases having 100^3 µm³ RVE size. To generalize the conclusions for a more general use, the equation can be modified as:

$$t_{RVE} \geq \left(\frac{500 D_f}{100} - 15\right) \times 2 \text{ µm} \tag{5}$$

This given criterion is the new finding of the current research. It should be adopted and fulfilled while carrying out full phase simulations and discussing the local stress and strain evolution in a given microstructure.

5. Conclusions

In this research, the effect of RVE thickness on the changing global and local results is analyzed for an exemplary case of dual-phase steels. This research's primary objective is to identify the optimal RVE thickness with respect to the defined feature size for obtaining converged local results. RVEs were constructed by varying the mean feature sizes between 3 to 18 µm using DREAM-3D with grains having randomly assigned orientation distributions. The constructed RVEs were sliced into 1, 5, 10, 15, 20, 25, 30, 40, and 50 layers to construct different geometries with increasing thickness. Crystal plasticity model parameters for ferrite and martensite phases were adopted from already published work. The global and local simulation results were directly and statistically compared in order to draw the following conclusions from the study:

1. As long as the orientation distribution and composition of the material is the same—even with changing mean feature size, the total number of grains and geometry thickness—the global deformation and stress–strain behavior of the material does not change drastically (<5% divergence)
2. Although the global stress–strain behavior of different RVEs is similar, a large variation in the local stress—strain of RVEs with varying thickness and the feature size is observed during the visual and statistical comparison.
3. Stress and strain probability distribution in 2D RVEs is drastically different from 3D RVEs. In 2D RVEs, the strain prediction in the soft phase is higher, in the hard phase is lower, whereas the stress prediction in the soft phase is lower and the hard phase is higher.
4. It is necessary to consider 3D RVEs, which are at least five times larger than the average grain size for stabilized and more accurate local deformation behavior determination. This estimation helps in optimizing the computation time for accurate mesoscale simulations.
5. Figure 10 and Equations (4) & (5) provide a criterion for choosing the mean feature ESD and effective RVE thickness. This criterion should be fulfilled for full phase simulations carried out using DAMASK in the future for obtaining converged simulation results for multi-phase materials i.e., DP steel.

Author Contributions: Conceptualization, F.Q. and A.A.C.; methodology, F.Q.; software, A.A.C.; validation, F.Q., A.A.C., and S.G.; formal analysis, A.A.C.; investigation, F.Q.; resources, S.G. and M.S.; data curation, F.Q.; writing–original draft preparation, F.Q. and A.A.C.; writing–review and editing, F.Q., A.A.C.; visualization, F.Q. and M.S.; supervision, S.G., M.S. and U.P.; project administration, R.K. and U.P.; funding acquisition, F.Q. and U.P. All authors have read and agreed to the published version of the manuscript.

Funding: This research was funded by Deutsche Forschungsgemeinschaft (DFG, German Research Foundation) within the framework of collaborative research group "TRIP Matrix Composites" project number 54473466 – SFB 799. The APC is also funded by SFB799.

Acknowledgments: The authors acknowledge the DAAD Faculty Development for Candidates (Balochistan), 2016 (57245990)—HRDI-UESTP's/UET's funding scheme in cooperation with the Higher Education Commission of Pakistan (HEC) for sponsoring the stay of Faisal Qayyum at IMF TU Freiberg. This work is conducted within the DFG funded collaborative research group "TRIP Matrix Composites" (SFB 799). The authors gratefully acknowledge the German Research Foundation (DFG) for the financial support of the SFB 799. The authors also acknowledge the support of Martin Diehl & Franz Roters (MPIE, Düsseldorf) for their help regarding the functionality of DAMASK.

Conflicts of Interest: The authors declare no conflict of interest.

Nomenclature

Acronyms

Symbol	Description	Unit
RVE	Representative Volume Element	–
EBSD	Electron Back Scatter Diffraction	–
CP	Crystal Plasticity	–
TRIP	Transformation Induced Plasticity	–

Greek Symbols

Symbol	Description	Unit
ESD	Estimated Sphere Diameter	μm
Mg-PSZ	Magnesium Partially Stabilized Zirconia	–
PDF	Probability Distribution Function	–
CDF	Cummulative Distribution Function	–
G_N	Number of considered grains	–
L_N	Number of considered layers	–
$\mathcal{D}_f$	Mean feature estimated sphere diameter	μm
t_{RVE}	RVE thickness	μm

Appendix A. Phenomenological Crystal Plasticity Model

The model is adopted for the body-centered cubic (bcc) crystals of the phenomenological crystal plasticity description by Peirce et al. [51]. The microstrcuture is parametrized in terms of a slip resistance $S^\alpha_{\{011\}}$ on each of the 12 $\{011\}\langle 111\rangle$ slip systems, and $S^\alpha_{\{211\}}$ on each of the 12 $\{211\}\langle 111\rangle$ slip systems which are indexed by $\alpha = 1, \ldots, 24$. These resistances increase asymptotically towards S^α_∞ with shear γ according to the relationship

$$\dot{S}^\alpha = h_0(1 - S^\alpha/S^\alpha_\infty)^w h_{\alpha\beta} \dot{\gamma}^\beta \tag{A1}$$

with interaction $(h_{\alpha\beta})$ and fitting (w, h_0) parameters. Given a set of current slip resistances, shear on each system evolves at a rate of

$$\dot{\gamma}^\alpha = \dot{\gamma}_0 \left|\frac{\tau^\alpha}{S^\alpha}\right|^n \mathrm{sgn}(\tau^\alpha) \tag{A2}$$

with $\tau^\alpha = \mathbf{S}.(\mathbf{b}^\alpha \otimes \mathbf{n}^\alpha)$, a reference shear rate $\dot{\gamma}_0$ and a stress exponent n. The superposition of shear on all slip systems in turn determines the plastic velocity gradient:

$$\mathbf{L}_P = \dot{\gamma}^\alpha \, \mathbf{b}^\alpha \otimes \mathbf{n}^\alpha \tag{A3}$$

where $\mathbf{b}^\alpha$ and $\mathbf{n}^\alpha$ are unit vectors along the slip direction and slip plane normal respectively.

Using the quasi-static strain rate of 1×10^{-4} all RVEs were loaded in tension for 2500 s (load direction with reference to the top surface is shown in Figure 2 with red arrows) to reach a total strain of average 25%.

References

1. Clyne, T.; Withers, P. *An Introduction to Metal Matrix Composites*; Cambridge University Press: Cambridge, UK, 1995.
2. Ullah, M.; Wu, C.S.; Qayyum, F. Prediction of crack tip plasticity induced due to variation in solidification rate of weld pool and its effect on fatigue crack propagation rate (FCPR). *J. Mech. Sci. Technol.* **2018**, *32*, 3625–3635. [CrossRef]
3. Mukhtar, F.; Qayyum, F.; Anjum, Z.; Shah, M. Effect of chrome plating and varying hardness on the fretting fatigue life of AISI D2 components. *Wear* **2019**, *418*, 215–225. [CrossRef]
4. Berisha, B.; Hirsiger, S.; Hippke, H.; Hora, P.; Mariaux, A.; Leyvraz, D.; Bezençon, C. Modeling of anisotropic hardening and grain size effects based on advanced numerical methods and crystal plasticity. *Arch. Mech.* **2019**, *71*, 489–505.

5. Hazanov, S. Hill condition and overall properties of composites. *Arch. Appl. Mech.* **1998**, *68*, 385–394. [CrossRef]
6. Ullah, M.; Pasha, R.A.; Chohan, G.Y.; Qayyum, F. Numerical simulation and experimental verification of CMOD in CT specimens of TIG welded AA2219-T87. *Arab. J. Sci. Eng.* **2015**, *40*, 935–944. [CrossRef]
7. Hussain, N.; Qayyum, F.; Pasha, R.A.; Shah, M. Development of multi-physics numerical simulation model to investigate thermo-mechanical fatigue crack propagation in an autofrettaged gun barrel. *Def. Technol.* **2020**, in press. [CrossRef]
8. Mukhtar, F.; Qayyum, F.; Elahi, H.; Shah, M. Studying the Effect of Thermal Fatigue on Multiple Cracks Propagating in an SS316L Thin Flange on a Shaft Specimen Using a Multi-Physics Numerical Simulation Model. *Strojniški Vestnik J. Mech. Eng.* **2019**, *65*, 565–573. [CrossRef]
9. Zaoui, A. Changement d'échelle: Motivation et méthodologie. In *Homogénéisation en Mécanique des Matériaux, Tome 1: Matériaux Aléatoires Élastiques et Milieux Périodiques*; Hermes Science: Paris, France, 2001; pp. 19–39.
10. Besson, J.; Cailletaud, G.; Chaboche, J.L.; Forest, S. *Mécanique Non Linéaire des Matériaux*; Hermès Science Publications: Paris, France, 2001.
11. Lebensohn, R.A.; Kanjarla, A.K.; Eisenlohr, P. An elasto-viscoplastic formulation based on fast Fourier transforms for the prediction of micromechanical fields in polycrystalline materials. *Int. J. Plast.* **2012**, *32*, 59–69. [CrossRef]
12. Qayyum, F.; Guk, S.; Kawalla, R.; Prahl, U. Experimental investigations and multiscale modeling to study the effect of sulfur content on formability of 16MnCr5 alloy steel. *Steel Res. Int.* **2019**, *90*, 1800369. [CrossRef]
13. Anbarlooie, B.; Hosseini-Toudeshky, H.; Hosseini, M.; Kadkhodapour, J. Experimental and 3D Micromechanical Analysis of Stress–Strain Behavior and Damage Initiation in Dual-Phase Steels. *J. Mater. Eng. Perform.* **2019**, *28*, 2903–2918. [CrossRef]
14. Kim, D.K.; Kim, E.Y.; Han, J.; Woo, W.; Choi, S.H. Effect of microstructural factors on void formation by ferrite/martensite interface decohesion in DP980 steel under uniaxial tension. *Int. J. Plast.* **2017**, *94*, 3–23. [CrossRef]
15. Liu, Y.; Wang, W.; Lévy, B.; Sun, F.; Yan, D.M.; Lu, L.; Yang, C. On centroidal voronoi tessellation—energy smoothness and fast computation. *ACM Trans. Graph. (ToG)* **2009**, *28*, 1–17. [CrossRef]
16. Kok, P.; Spanjer, W.; Vegter, H. A microstructure based model for the mechanical behavior of multiphase steels. *Key Eng. Mater.* **2015**, *651*, 975–980. [CrossRef]
17. Diehl, M.; Groeber, M.; Haase, C.; Molodov, D.A.; Roters, F.; Raabe, D. Identifying Structure–Property Relationships Through DREAM.3D Representative Volume Elements and DAMASK Crystal Plasticity Simulations: An Integrated Computational Materials Engineering Approach. *J. Miner.* **2017**, *69*, 848–855. [CrossRef]
18. Qayyum, F.; Guk, S.; Prüger, S.; Schmidtchen, M.; Saenko, I.; Kiefer, B.; Kawalla, R.; Prahl, U. Investigating the local deformation and transformation behavior of sintered X3CrMnNi16-7-6 TRIP steel using a calibrated crystal plasticity-based numerical simulation model. *Int. J. Mat. Res. (Zeitschrift für Metallkunde)* **2020**, *111*, 392–404. [CrossRef]
19. Diehl, M.; An, D.; Shanthraj, P.; Zaefferer, S.; Roters, F.; Raabe, D. Crystal plasticity study on stress and strain partitioning in a measured 3D dual phase steel microstructure. *Phys. Mesomech.* **2017**, *20*, 311–323. [CrossRef]
20. Liu, C.; Shanthraj, P.; Diehl, M.; Roters, F.; Dong, S.; Dong, J.; Ding, W.; Raabe, D. An integrated crystal plasticity–phase field model for spatially resolved twin nucleation, propagation, and growth in hexagonal materials. *Int. J. Plast.* **2018**, *106*, 203–227. [CrossRef]
21. Diehla, M.; Naunheim, Y.; Yan, D.; Morsdorf, L.; An, D.; Tasan, C.C.; Zaefferer, S.; Roters, F.; Raabe, D. Coupled Experimental-Numerical Analysis of Strain Partitioning in Metallic Microstructures: The Importance of Considering the 3D Morphology. In Proceedings of the BSSM 12th International Conference on Advances in Experimental Mechanics, Sheffield, UK, 29–31 August 2017; pp. 1–2.
22. Tasan, C.C.; Hoefnagels, J.P.M.; Diehl, M.; Yan, D.; Roters, F.; Raabe, D. Strain localization and damage in dual phase steels investigated by coupled in-situ deformation experiments and crystal plasticity simulations. *Int. J. Plast.* **2014**, *63*, 198–210. [CrossRef]
23. Ramazani, A.; Mukherjee, K.; Quade, H.; Prahl, U.; Bleck, W. Correlation between 2D and 3D flow curve modelling of DP steels using a microstructure-based RVE approach. *Mater. Sci. Eng. A* **2013**, *560*, 129–139. [CrossRef]

24. Knezevic, M.; Savage, D.J. A high-performance computational framework for fast crystal plasticity simulations. *Comput. Mater. Sci.* **2014**, *83*, 101–106. [CrossRef]
25. He, Q.C. Effects of size and boundary conditions on the yield strength of heterogeneous materials. *J. Mech. Phys. Solids* **2001**, *49*, 2557–2575. [CrossRef]
26. Gitman, I.; Askes, H.; Sluys, L. Representative volume: Existence and size determination. *Eng. Fract. Mech.* **2007**, *74*, 2518–2534. [CrossRef]
27. Jiang, M.; Jasiuk, I.; Ostoja-Starzewski, M. Apparent elastic and elastoplastic behavior of periodic composites. *Int. J. Solids Struct.* **2002**, *39*, 199–212. [CrossRef]
28. Jiang, M.; Ostoja-Starzewski, M.; Jasiuk, I. Scale-dependent bounds on effective elastoplastic response of random composites. *J. Mech. Phys. Solids* **2001**, *49*, 655–673. [CrossRef]
29. Groeber, M.A.; Jackson, M.A. DREAM. 3D: A digital representation environment for the analysis of microstructure in 3D. *Integr. Mater. Manuf. Innov.* **2014**, *3*, 5. [CrossRef]
30. Barrett, T.J.; Savage, D.J.; Ardeljan, M.; Knezevic, M. An automated procedure for geometry creation and finite element mesh generation: Application to explicit grain structure models and machining distortion. *Comput. Mater. Sci.* **2018**, *141*, 269–281. [CrossRef]
31. Qayyum, F.; Guk, S.; Schmidtchen, M.; Kawalla, R.; Prahl, U. Modeling the Local Deformation and Transformation Behavior of Cast X8CrMnNi16-6-6 TRIP Steel and 10% Mg-PSZ Composite Using a Continuum Mechanics-Based Crystal Plasticity Model. *Crystals* **2020**, *10*, 221. [CrossRef]
32. Diehl, M.; Shanthraj, P.; Eisenlohr, P.; Roters, F. Neighborhood influences on stress and strain partitioning in dual-phase microstructures. *Meccanica* **2016**, *51*, 429–441. [CrossRef]
33. Hitti, K.; Laure, P.; Coupez, T.; Silva, L.; Bernacki, M. Precise generation of complex statistical Representative Volume Elements (RVEs) in a finite element context. *Comput. Mater. Sci.* **2012**, *61*, 224–238. [CrossRef]
34. Bargmann, S.; Klusemann, B.; Markmann, J.; Schnabel, J.E.; Schneider, K.; Soyarslan, C.; Wilmers, J. Generation of 3D representative volume elements for heterogeneous materials: A review. *Prog. Mater. Sci.* **2018**, *96*, 322–384. [CrossRef]
35. Lee, S.B.; Lebensohn, R.; Rollett, A.D. Modeling the viscoplastic micromechanical response of two-phase materials using Fast Fourier Transforms. *Int. J. Plast.* **2011**, *27*, 707–727. [CrossRef]
36. Zhang, C.; Li, H.; Eisenlohr, P.; Liu, W.; Boehlert, C.; Crimp, M.; Bieler, T. Effect of realistic 3D microstructure in crystal plasticity finite element analysis of polycrystalline Ti-5Al-2.5 Sn. *Int. J. Plast.* **2015**, *69*, 21–35. [CrossRef]
37. Prüger, S.; Seupel, A.; Kuna, M. A thermomechanically coupled material model for TRIP-steel. *Int. J. Plast.* **2014**, *55*, 182–197. [CrossRef]
38. Madivala, M.; Schwedt, A.; Wong, S.L.; Roters, F.; Prahl, U.; Bleck, W. Temperature dependent strain hardening and fracture behavior of TWIP steel. *Int. J. Plast.* **2018**, *104*, 80–103. [CrossRef]
39. Roters, F.; Diehl, M.; Shanthraj, P.; Eisenlohr, P.; Reuber, C.; Wong, S.L.; Maiti, T.; Ebrahimi, A.; Hochrainer, T.; Fabritius, H.O.; et al. DAMASK–The Düsseldorf Advanced Material Simulation Kit for modeling multi-physics crystal plasticity, thermal, and damage phenomena from the single crystal up to the component scale. *Comput. Mater. Sci.* **2019**, *158*, 420–478. [CrossRef]
40. Shanthraj, P.; Eisenlohr, P.; Diehl, M.; Roters, F. Numerically robust spectral methods for crystal plasticity simulations of heterogeneous materials. *Int. J. Plast.* **2015**, *66*, 31–45. [CrossRef]
41. Zeghadi, A.; N'guyen, F.; Forest, S.; Gourgues, A.F.; Bouaziz, O. Ensemble averaging stress–strain fields in polycrystalline aggregates with a constrained surface microstructure–Part 1: Anisotropic elastic behaviour. *Philos. Mag.* **2007**, *87*, 1401–1424. [CrossRef]
42. Zeghadi, A.; Forest, S.; Gourgues, A.F.; Bouaziz, O. Ensemble averaging stress–strain fields in polycrystalline aggregates with a constrained surface microstructure–Part 2: Crystal plasticity. *Philos. Mag.* **2007**, *87*, 1425–1446. [CrossRef]
43. Harris, W.M.; Chiu, W.K. Determining the representative volume element size for three-dimensional microstructural material characterization. Part 1: Predictive models. *J. Power Sources* **2015**, *282*, 552–561. [CrossRef]
44. Harris, W.M.; Chiu, W.K. Determining the representative volume element size for three-dimensional microstructural material characterization. Part 2: Application to experimental data. *J. Power Sources* **2015**, *282*, 622–629. [CrossRef]

45. Tjahjanto, D.; Turteltaub, S.; Suiker, A. Crystallographically based model for transformation-induced plasticity in multiphase carbon steels. *Contin. Mech. Thermodyn.* **2008**, *9*, 399–422. [CrossRef]
46. Jiang, Z.; Guan, Z.; Lian, J. Effects of microstructural variables on the deformation behaviour of dual-phase steel. *Mater. Sci. Eng. A* **1995**, *190*, 55–64. [CrossRef]
47. Eisenlohr, P.; Diehl, M.; Lebensohn, R.A.; Roters, F. A spectral method solution to crystal elasto-viscoplasticity at finite strains. *Int. J. Plast.* **2013**, *46*, 37–53. [CrossRef]
48. Michel, J.; Moulinec, H.; Suquet, P. A computational scheme for linear and non-linear composites with arbitrary phase contrast. *Int. J. Numer. Methods Eng.* **2001**, *52*, 139–160. [CrossRef]
49. Ahrens, J.; Geveci, B.; Law, C. Paraview: An end-user tool for large data visualization. In *The Visualization Handbook*; Academic Press: New York, NY, USA, 2005; Volume 717.
50. Maiti, T.; Eisenlohr, P. Fourier-based spectral method solution to finite strain crystal plasticity with free surfaces. *Scr. Mater.* **2018**, *145*, 37–40. [CrossRef]
51. Peirce, D.; Asaro, R.; Needleman, A. An analysis of nonuniform and localized deformation in ductile single crystals. *Acta Metall.* **1982**, *30*, 1087–1119. [CrossRef]

© 2020 by the authors. Licensee MDPI, Basel, Switzerland. This article is an open access article distributed under the terms and conditions of the Creative Commons Attribution (CC BY) license (http://creativecommons.org/licenses/by/4.0/).

Article

Microstructural Influence on Stretch Flangeability of Ferrite–Martensite Dual-Phase Steels

Jae Hyung Kim [1], Taekyung Lee [2,*] and Chong Soo Lee [3,*]

[1] Steel Product Research Lab, POSCO, Pohang 37877, Korea; veritas@posco.com
[2] School of Mechanical Engineering, Pusan National University, Busan 46241, Korea
[3] Graduate Institute of Ferrous Technology, Pohang University of Science and Technology (POSTECH), Pohang 37673, Korea
* Correspondence: taeklee@pnu.ac.kr (T.L.); cslee@postech.ac.kr (C.S.L.)

Received: 22 October 2020; Accepted: 6 November 2020; Published: 9 November 2020

Abstract: This work investigated the microstructural effect on stretch flangeability of ferrite–martensite dual-phase (DP) steels. Three types of DP steels with various martensitic structures were prepared for the research: fibrous martensite in water-quenched (WQ) sample, chained martensite in air-quenched (AQ) sample, and coarse martensite in step-quenched (SQ) sample. The WQ specimen exhibited the highest mechanical strength and hole expansion ratio compared to the AQ and SQ samples despite their similar fraction of martensite. Such a result was explained in view of uniform distribution of fine martensite and high density of geometrically necessary dislocations in the WQ specimen. Meanwhile, most cracks initiated at either rolling or transverse direction during the stretch flangeability test regardless of the martensitic morphology. It was attributed to the highest average normal anisotropy in the direction of 45° to rolling direction.

Keywords: dual-phase steel; martensite; hole expansion ratio; stretch flangeability; normal anisotropy

1. Introduction

Dual-phase (DP) steels composed of hard martensitic particles and soft ferritic matrix have received great attention for automotive applications due to their good combination of strength and elongation [1]. The mechanical properties of ferrite–martensite DP steels strongly depend on the volume fraction and morphology (e.g., size, shape, distribution, continuity) of martensitic constituent. Kim and Thomas [2] compared uniaxial tensile properties of DP steels with various martensitic microstructures, such as fine fibrous martensite induced by intermediate quenching, fine globular martensite by intercritical annealing, and coarse martensite by step quenching. In their work, the tensile strength increased with increasing fraction of martensite following the rule of mixture. The coarse martensitic structure exhibited the lowest elongation due to the occurrence of cleavage fracture in the ferritic matrix. Chang and Preban [3] described tensile properties of DP steel in terms of a ferritic grain size and martensitic volume fraction. Mediratta et al. [4] ascribed the fine dispersion of martensite to homogeneous deformation during low-cycle fatigue tests. Molaei and Ekrami [5] observed the higher fatigue strength of DP steels with fibrous martensite (i.e., the structure composed of parallel and narrow laths [6]) in comparison to the alloy with networked martensite (i.e., the structure connected through a phase interface [6]) due to the tortuous crack propagation in the former microstructure. According to the work of Suh et al. [7], microcracks uniformly propagate in the intermediate-quenched microstructure along the longitudinal and transverse directions during uniaxial tensile tests, whereas strain localization gives rise to poor ductility in the step-quenched microstructure. Kim and Lee [8] reported the change of fracture mode based on the martensitic structure under dynamic loading condition. Das and Chattopadhyay [9] confirmed the lowest yield ratio and highest work-hardening rate in the intermediate-quenched microstructure. Park et al. [10], however, reported a low drawability after this quenching process due to easy initiation and propagation of cracks at uniformly distributed ferrite–martensite interfaces.

Various studies have investigated tensile and fracture properties of DP steels, including the literature mentioned above. In contrast, stretch flangeability of DP steels has received less attention in spite of its importance in the automotive industry, especially for a wheel-forming process where punched holes are expanded to fabricate a designed wheel. Hole expansion ratio (HER) of steel plate is measured to evaluate stretch flangeability in general. Researchers have tried to predict the stretch flangeability of a material using other mechanical properties. For example, HER and ultimate tensile strength (UTS) were reported to follow the inversely proportional relation [11,12]. Comstock et al. [13] concluded that HER was proportional to average normal anisotropy of sheet materials. However, these previous works focused on relatively indirect methods (e.g., regression analysis) to study stretch flangeability of DP steel.

The direct investigation of microstructural influence on stretch flangeability has rarely attracted attention. Yoon et al. [14] recently clarified the relationship between stretch flangeability and microstructural features of ultrahigh-strength DP steel. Although their study provided a meaningful insight for the academia, they investigated only one type of martensitic morphology. There still has been no study focusing on the effect of extensive microstructural factors (e.g., the martensitic morphology) on the stretch flangeability of DP steels. Therefore, the present study investigates such an issue using three types of DP steels possessing totally different martensitic morphologies. The results demonstrated a peculiar tendency that was inconsistent with the previous conclusions. Such a discrepancy was explained by adopting a constitutive analysis.

2. Materials and Methods

DP590 steel was hot-rolled to a thickness of 3 mm for the present study. The material was austenitized at 1000 °C for 1 h, and then processed with three different cooling methods in order to change the martensitic structure. First, sections of the austenitized samples were water-quenched, and then soaked in a furnace at 785 °C for 1 h. Second, sections were air-quenched from the austenitizing temperature, followed by the same heat treatment of intercritical annealing. The final sections were cooled from the austenitizing temperature to 750 °C and then held for 1 h. These samples are denoted as water-quenched (WQ), air-quenched (AQ), and step-quenched (SQ) specimens, respectively.

To characterize microstructure, mirror-polished samples were etched in 2% nital solution and then analyzed using optical microscope (OM) and scanning electron microscope (SEM) in three planes: rolling direction–normal direction (RD-ND) plane, rolling direction–transverse direction (RD-TD) plane, and transverse direction–normal direction (TD-ND) plane. Meanwhile, kernel average misorientation map was obtained using an electron backscatter diffraction (EBSD) analysis. The specimens for this analysis were prepared by electropolishing at 30 V in mixed solution of 95% acetic acid and 5% perchloric acid.

Uniaxial tensile tests were conducted with dog-bone-shaped specimens at constant crosshead speed with an initial strain rate of 5×10^{-3} s^{-1} using a hydraulic mechanical testing machine. The samples were deformed along RD. The normal anisotropy was calculated by measuring the displacement fields on the specimen surface using the digital image correlation technique. Two 2448 × 2048 pixels 14-bit charge coupled device cameras were used to measure the stain along RD and TD simultaneously. Hardness of constituent phases was measured using the nanoindentation method with a Berkovich-type diamond tip. The indentation was performed at a tip speed of approximately 10^{-5} mm·s^{-1} to a depth of 2 μm of the etched specimens. The dwell time was 10 s at the peak load.

The hole expansion test was performed using 90 mm × 90 mm × 1.7 mm specimens with a hole 10 mm in diameter machined at the center. An Erichsen machine equipped with 60° conical punch was used to expand the center hole until the observation of the first through-thickness crack at the hole edge. Specimen holding force was 400 kN and punch speed was 0.25 mm·s^{-1}. The tests were repeated five times for each material for data reproducibility. The fracture surface of each specimen after the hole-expansion test was investigated by SEM.

3. Results

Table 1 summarizes the microstructural and mechanical data of the investigated DP steels. Although the volume fraction of martensite was comparable among the specimens, the cooling rate exerted a significant influence on the other microstructural characteristics. This is consistent with the OM micrographs of the investigated steels (Figure 1). The water quenching from the austenitizing temperature (i.e., 1000 °C) gave rise to full martensitic microstructure, which was then decomposed into ferrite and austenite during the intercritical annealing. The austenite was transformed into martensite again in the second water quenching from 785 °C, resulting in the fibrous martensitic structure for the WQ specimen (Figure 1a). Meanwhile, austenite transformed into ferrite and pearlite during the cooling from 1000 °C in the ambient atmosphere; the ferrite was formed in the prior austenite grains, while the pearlite preferentially nucleated at the prior austenite grain boundaries. The pearlite was transformed into austenite and then martensite through the subsequent intercritical annealing and water quenching processes. As a result, the chained martensitic structure was formed in the AQ specimen (Figure 1b). The SQ specimen showed the coarse martensite with the largest size of phase constituents (i.e., 19.6 µm). In addition, the specimen exhibited an inhomogeneous distribution of ferrite and martensite in RD-ND and TD-ND planes (Figure 1c).

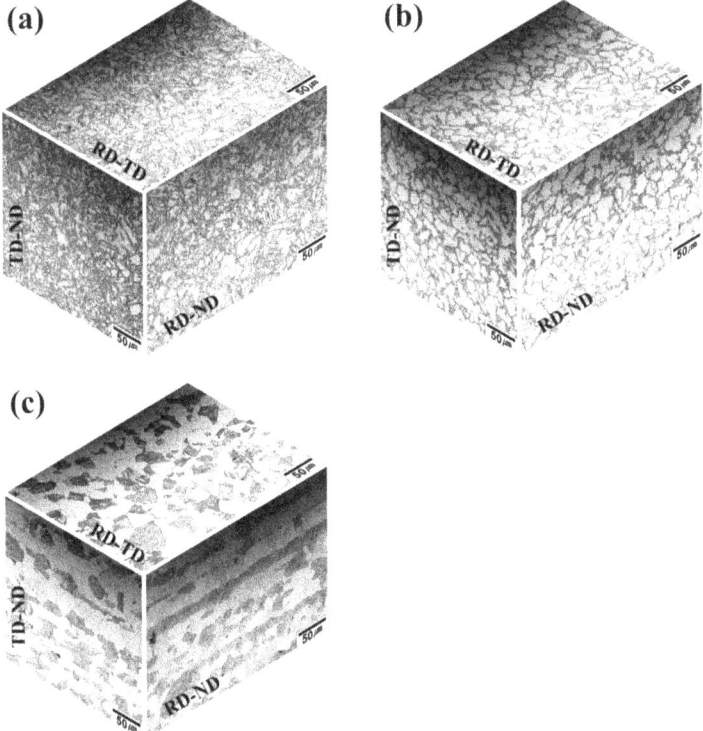

Figure 1. Optical microscope (OM) image of the investigated DP steels in three different planes. (a) WQ specimen, (b) AQ specimen, and (c) SQ specimen.

Table 1. Microstructural parameters and mechanical properties of dual-phase (DP) steel specimens.

Sample	f_M (%)	d_M (μm)	d_f (μm)	YS (MPa)	UTS (MPa)	EL (%)	PUE (%)	λ (%)	r_0	r_{45}	r_{90}	r_m
WQ	34.5	1.8	7.3	391	710	20.7	9.0	130	0.87	0.98	0.88	0.93
AQ	34.7	3.4	12.8	353	668	17.0	6.0	105	0.86	0.99	0.88	0.93
SQ	33.9	19.6	18.1	318	613	18.6	6.3	55	0.80	0.88	0.79	0.84

f_M: volume fraction of martensite, d_M: grain size of martensite, and d_f: grain size of ferrite.

The investigated specimens showed typical mechanical features of DP steels including continuous yielding and high strain-hardening rate (Figure 2). Although the elongation to failure (EL) showed no significant difference, yield strength (YS) and UTS were strongly affected by the cooling methods and resultant microstructures as summarized in Table 1. The WQ specimen exhibited the highest strengths, while the SQ specimen possessed the lowest values in spite of the comparable volume fraction of martensite.

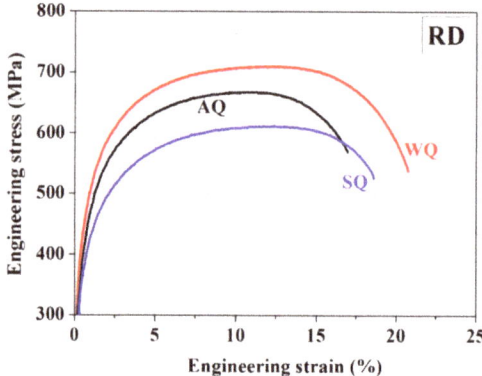

Figure 2. Engineering stress–strain curve for the investigated DP steels measured along RD.

HER values (λ) were determined from five repetitive tests using the following equation:

$$\lambda = (h_1 - h_0)/h_0 \qquad (1)$$

where h_0 of 10 mm is the initial hole diameter and h_1 is the final hole diameter immediately after the initiation of through-thickness crack. Previous works reported the inversely proportional HER–UTS relation, as remarked in Section 1. However, the present results showed the direct proportion between the two properties; the WQ specimen exhibited the highest HER value as well as UTS among the investigated steels, followed by the AQ and SQ specimens (Figure 3). Such a unique tendency will be discussed further in Section 4.

Average normal anisotropy (r_m) was determined as follows:

$$r_m = (r_0 + 2\,r_{45} + r_{90})/4 \qquad (2)$$

where r_0, r_{45}, and r_{90} indicate the normal anisotropy of RD, 45° to RD, and TD, respectively. It should be noted that the WQ and AQ specimens showed similar average normal anisotropy, even though they exhibited the different HER values proportional to their UTS (Table 1). The SQ specimen showed the lowest average normal anisotropy due to the inhomogeneous distribution of ferrite and martensite. With respect to the directions, the highest normal anisotropy was the value measured in the direction of 45° to RD (i.e., r_{45}) regardless of martensitic morphology. The high r_{45} values indicate the resistance to thinning under deformation in the direction of 45° to RD [12], suggesting the retarded necking and suppressed initiation of through-thickness crack during hole expansion. Indeed, 93% of cracks were initiated either at RD or TD direction that possessed the low normal anisotropy (Figure 3).

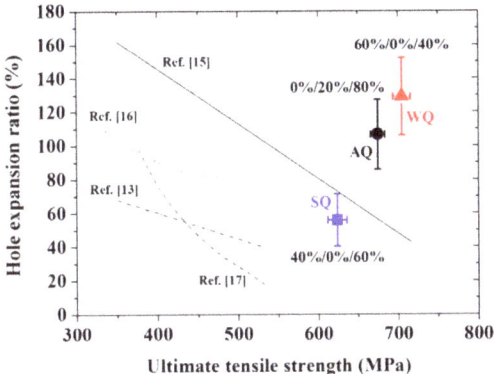

Figure 3. HER–UTS relation of DP steels in the present study as well as the literature. The number in percent indicates the probability of through-thickness crack formed along RD, 45° to RD, and TD, respectively.

4. Discussion

Mechanical strength of DP steel is generally explained in terms of the rule of mixture of ferrite and martensite, because the property increases with increasing fraction of martensitic phase [2]. However, it was shown in this work that the WQ specimen possessed the highest YS and UTS in spite of the similar volume fraction of martensite among the investigated alloys, implying that other factors contributed to the mechanical strength. The yielding of DP steels preferentially occurs in soft ferritic phase rather than hard martensite. Therefore, YS of the investigated DP steels can be analyzed using the following equations with microstructural parameters of ferritic phase [15]:

$$YS = [\sigma_0^2 + \sigma_s^2 + \sigma_{gb}^2 + \sigma_{disl}^2]^{0.5} - \delta_r \quad (3)$$

$$\sigma_0 = [2G/(1-v)] \exp(-2\pi w/b) \quad (4)$$

$$\sigma_s = 4560[wt.\%C] + 83[wt.\%Si] + 37[wt.\%Mn] + 60[wt.\%Cr] \quad (5)$$

$$\sigma_{gb} = k_y d_f^{-0.5} \quad (6)$$

$$\sigma_{disl} = \alpha M G b \rho^{0.5} \quad (7)$$

$$\delta r = 8 f_M^{2}/f_f^{2} \quad (8)$$

where σ_0 (=50 MPa) is the lattice friction stress, σ_s is the solid-solution hardening, σ_{gb} is the grain-boundary hardening, σ_{disl} is the dislocation hardening, and δ_r is the factor for residual-stress effects. G (=80 GPa) is the shear modulus, v is the Poisson's ratio, w is the dislocation width, and b (=2.5 × 10^{-10} m) is the Burger's vector. [wt.%X] indicates the mass percentage of element X, among which [wt.%C] was assumed to be the supersaturated carbon content in ferrite at 770 °C [16]. k_y (=0.55 MPa· m$^{0.5}$) is a Hall–Petch coefficient for high-strength low-alloy steels [16], α (=0.33) is a material constant, d_f is the average ferrite grain size, M (=3) is the Taylor factor, ρ is the dislocation density in ferritic phase, f_M is the volume fraction of martensite, and f_f is the volume fraction of ferrite.

Table 2 summarizes the hardening coefficients of the investigated materials, attributing the difference in YS to two factors: grain-boundary hardening and dislocation hardening. Mechanical properties of DP steel is indeed directly affected by the grain size of ferritic phase as well as the volume fraction of martensite [3]. The finest ferritic grains in the WQ specimen led to the significant Hall–Petch strengthening. In addition, the WQ specimen possessed the highest dislocation hardening; this hardening constituent was determined from the measured YS and the other estimated constituents (i.e., σ_0, σ_s, σ_{gb}, and δ_r). Ferritic grains are plastically deformed due to the volume expansion of martensite during the martensitic transformation [17]. This gives rise to the formation of geometrically necessary dislocations (GNDs) in the ferritic phase to maintain lattice continuity at ferrite–martensite interfaces [18]. As a result, dislocation density in the vicinity of the interface is higher than that in a ferritic grain. Kernel average misorientation maps demonstrate such a tendency obviously (Figure 4). The SQ specimen with the lowest fraction of ferrite–martensite interfaces exhibited the radical change of dislocation density. The WQ specimen, however, showed uniform distribution of high dislocation density due to the finely distributed martensite. Calcagnotto et al. [19] also reported consistent results where a dislocation density was measured to be 2.5×10^{13} m^{-2} in a ferritic grain and 2.5×10^{14} m^{-2} in ferritic area near the ferrite–martensite interface. In conclusion, the uniformly distributed fine martensite resulted in high GNDs and resultant dislocation hardening in the WQ specimen.

Table 2. Hardening constituents in the constitutive model and calculated dislocation density in ferritic phase.

Material	YS (MPa)	σ_0 (MPa)	σ_s (MPa)	σ_{gb} (MPa)	σ_{disl} (MPa)	ρ (m^{-2})
WQ	391			203.6	302.6	2.34×10^{14}
AQ	353	50	139.6	153.7	284.3	2.06×10^{14}
SQ	318			129.3	254.3	1.65×10^{14}

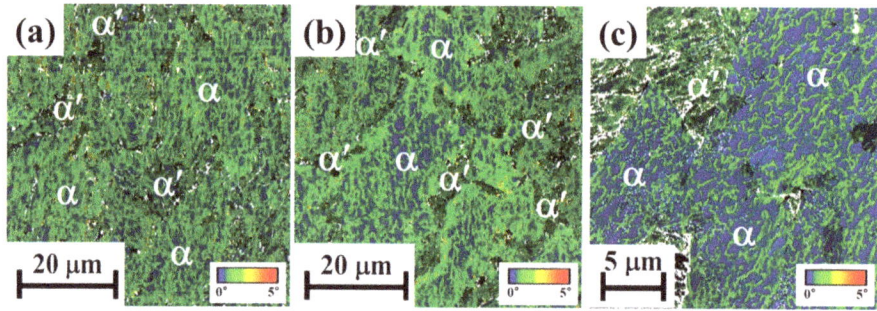

Figure 4. Kernel average misorientation maps for (a) WQ specimen, (b) AQ specimen, and (c) SQ specimen; 0° and 5° are set to be the minimum and maximum misorientation angles, respectively.

A constitutive analysis of strain-hardening behavior provided further insight into the contribution of dislocation hardening to plastic deformation. Three approaches have been commonly used to describe strain hardening of materials: the Hollomon analysis [20], the Crussard–Jaoul (C-J) analysis [21] based on the Ludwik equation [22], and the modified C-J analysis [9,23] on the basis of the Swift equation [24]. This work adopted the modified C-J analysis as this method can clearly explain the hardening behavior of DP steels with various microstructural parameters [23]. The modified C-J analysis is expressed as follows:

$$\ln(d\sigma/d\varepsilon) = (1-m)\ln\sigma - \ln(k_s m) \qquad (9)$$

where m is the strain-hardening exponent and k_s is a material constant.

The investigated DP steels showed the two-stage strain-hardening behavior where the slope changed at a transition strain (Figure 5). In DP steel, plastic deformation is accommodated in soft ferrite first (stage I) and then the hard martensite starts deforming plastically with ferritic phase (stage II) [23,25]. The WQ specimen exhibited the highest strain-hardening exponent at stage I (m_1):

4.2 for WQ, 3.8 for AQ, and 2.8 for SQ. This implies the largest strain hardening in the ferritic phase due to high density of GNDs. This is consistent with the results as aforementioned in Table 2. The SQ specimen possessed the doubled transition strain (i.e., 2.9%) in comparison to the values of the other two materials (1.5% for WQ and 1.4% for AQ). This is attributed to the low density of GNDs in ferritic grains of the SQ specimen that required a higher amount of plastic deformation to reach the transition strain [23]. In contrast to the m_I values, the investigated materials exhibited comparable results with respect to the strain-hardening exponent at stage II (m_{II}): 7.0 for WQ, 6.8 for AQ, and 6.7 for SQ. Tomita and Okabayashi [25] compared experimentally obtained m_{II} values with the calculation based on the rule of mixture. Interestingly, these values showed a good agreement even though the authors did not consider a martensitic morphology. Son et al. [23] reported comparable m_{II} values between coarse- and ultrafine-grained DP steels. Therefore, m_{II} value of DP steel is determined by phase compositions and mechanical properties of each phase constituent, rather than the martensitic morphology and size, which provides a good explanation for the similar m_{II} values confirmed in WQ, AQ, and SQ specimens.

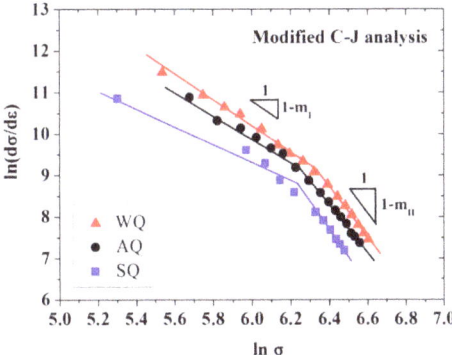

Figure 5. The modified C-J analysis of the investigated DP steels.

It is necessary to consider the influence of sample preparation method prior to discussing the effect of microstructure on HER results. The center hole of HER specimen can be fabricated by either punching or machining process. The interaction between tool and material generates a number of defects on the sheared edge and microcracks in the vicinity of the hole during the punching process, which considerably degrades the stretch flangeability [13,26]. The center hole of HER specimens was fabricated by wire-cut machining in this work to exclude unexpected factors generated from the punching process. Hence, the difference in HER values is considered to arise from microstructural characteristics.

Cracks were initiated at the ferrite–martensite interfaces during the hole-expansion tests in all specimens (Figure 6a–c). This result stemmed from the large difference in hardness between the two phases. There have been many efforts to reduce this gap of hardness. For example, researchers alternated martensitic constituent of DP steel into bainite with lower hardness [27]. In this work, such an effect was achieved by increasing hardness of ferritic phase due to the enhanced dislocation hardening in the WQ specimen, while the hardness of martensite was unchanged (Figure 6d). It is also noted that the deviation of hardness of ferrite was significantly reduced in the WQ specimen in contrast to the SQ specimen. Since GNDs in the matrix cause the local hardening, the hardness could be uniformly increased in the WQ specimen due to its homogeneous distribution of fine martensite as discussed above. In addition to the hardness discrepancy, the fine martensite retarded the crack initiation, as supported by significantly short length of crack in the WQ specimen (Figure 6e).

The AQ specimen exhibited the lower HER in spite of its average normal anisotropy similar to the WQ specimen. This can be explained by the martensitic morphology as well as the aforementioned two factors (i.e., the larger discrepancy in hardness between the two phases and larger grain size).

The rapid increment of microvoids at the joint between martensite reduces the necking formability in the AQ specimen composed of the chained morphology [28]. This gave rise to the lowest post-uniform elongation (PUE). Chen et al. [26] reported the lower HER of twinning-induced plasticity (TWIP) steels than that of interstitial-free steel despite the higher elongation of the TWIP steels. They attributed such results to the absence of PUE resulting from a negative strain-rate sensitivity of TWIP steels. Thus, the low PUE contributed to the low HER of the AQ specimen to some extent.

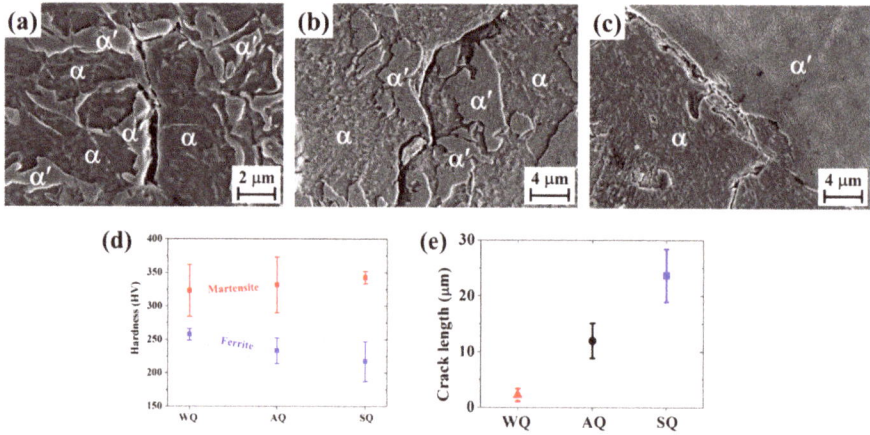

Figure 6. SEM micrograph of (a) WQ specimen, (b) AQ specimen, and (c) SQ specimen near the hole; (d) phase hardness and (e) crack length for the investigated specimens after the hole-expansion test.

5. Conclusions

The present work investigated the effect of microstructure on the stretch flangeability of ferrite–martensite DP steels with various martensitic structures. WQ, AQ, SQ specimens were composed of the fibrous, chained, and coarse martensitic structures, respectively. Although the investigated DP steels possessed the comparable fraction of martensite, the WQ specimen exhibited the highest tensile strength and HER value. The improved mechanical strength is attributed to the Hall–Petch strengthening and dislocation hardening induced by homogeneous distribution of fine martensite as well as high PUE. The enhanced stretch flangeability of the WQ specimen, characterized by its high HER value, resulted from the microstructural homogeneity that retarded the crack initiation and propagation. The high r_{45} values shown in all DP steels indicate the strong resistance to thinning, which well explains the suppressed crack initiation along the 45° direction.

Author Contributions: Conceptualization, J.H.K., T.L., and C.S.L.; investigation, J.H.K. and T.L.; resources, C.S.L.; writing—original draft preparation, J.H.K. and T.L.; writing—review and editing, T.L. and C.S.L.; supervision, C.S.L. All authors have read and agreed to the published version of the manuscript.

Funding: This work was supported by the National Research Foundation of Korea (NRF) grant funded by the Korean government (Ministry of Science and ICT) (No. 2018R1C1B6002068 & 2020R1A4A3079417).

Conflicts of Interest: The authors declare no conflict of interest.

References

1. Jha, G.; Das, S.; Lodh, A.; Haldar, A. Development of hot rolled steel sheet with 600MPa UTS for automotive wheel application. *Mater. Sci. Eng. A* **2012**, *552*, 457–463. [CrossRef]
2. Kim, N.J.; Thomas, G. Effects of morphology on the mechanical behavior of a dual phase Fe/2Si/0.1C steel. *Met. Mater. Trans. A* **1981**, *12*, 483–489. [CrossRef]

3. Peng-Heng, C.; Preban, A. The effect of ferrite grain size and martensite volume fraction on the tensile properties of dual phase steel. *Acta Met.* **1985**, *33*, 897–903. [CrossRef]
4. Mediratta, S.; Ramaswamy, V.; Rao, P. Influence of ferrite-martensite microstructural morphology on the low cycle fatigue of a dual-phase steel. *Int. J. Fatigue* **1985**, *7*, 107–115. [CrossRef]
5. Molaei, M.; Ekrami, A. The effect of dynamic strain aging on fatigue properties of dual phase steels with different martensite morphology. *Mater. Sci. Eng. A* **2009**, *527*, 235–238. [CrossRef]
6. Türkmen, M.; Gündüz, S. Martensite morphology and strain aging behaviours in intercritically treated low carbon steel. *Ironmak. Steelmak.* **2011**, *38*, 346–352. [CrossRef]
7. Suh, D.; Kwon, D.; Lee, S.; Kim, N.J. Orientation dependence of microfracture behavior in a dual-phase high-strength low-alloy steel. *Met. Mater. Trans. A* **1997**, *28*, 504–509. [CrossRef]
8. Kim, S.; Lee, S. Effects of martensite morphology and volume fraction on quasi-static and dynamic deformation behavior of dual-phase steels. *Met. Mater. Trans. A* **2000**, *31*, 1753–1760. [CrossRef]
9. Das, D.; Chattopadhyay, P.P. Influence of martensite morphology on the work-hardening behavior of high strength ferrite–martensite dual-phase steel. *J. Mater. Sci.* **2009**, *44*, 2957–2965. [CrossRef]
10. Park, K.S.; Park, K.-T.; Lee, D.L.; Lee, C.S. Effect of heat treatment path on the cold formability of drawn dual-phase steels. *Mater. Sci. Eng. A* **2007**, *449*, 1135–1138. [CrossRef]
11. Chatterjee, S.; Bhadeshia, H.K.D.H. Stretch-flangeability of strong multiphase steels. *Mater. Sci. Technol.* **2007**, *23*, 606–609. [CrossRef]
12. Narayanasamy, R.; Sathiyanarayanan, C.; Padmanabhan, P.; Venugopalan, T. Effect of mechanical and fractographic properties on hole expandability of various automobile steels during hole expansion test. *Int. J. Adv. Manuf. Technol.* **2009**, *47*, 365–380. [CrossRef]
13. Comstock, R.J.; Scherrer, D.K.; Adamczyk, R.D. Hole Expansion in a Variety of Sheet Steels. *J. Mater. Eng. Perform.* **2006**, *15*, 675–683. [CrossRef]
14. Yoon, J.I.; Jung, J.; Lee, H.H.; Kim, J.Y.; Kim, H.S. Relationships Between Stretch-Flangeability and Microstructure-Mechanical Properties in Ultra-High-Strength Dual-Phase Steels. *Met. Mater. Int.* **2019**, *25*, 1161–1169. [CrossRef]
15. Kang, Y.-L.; Han, Q.-H.; Zhao, X.-M.; Cai, M.-H. Influence of nanoparticle reinforcements on the strengthening mechanisms of an ultrafine-grained dual phase steel containing titanium. *Mater. Des.* **2013**, *44*, 331–339. [CrossRef]
16. Mazaheri, Y.; Kermanpur, A.; Najafizadeh, A. Strengthening Mechanisms of Ultrafine Grained Dual Phase Steels Developed by New Thermomechanical Processing. *ISIJ Int.* **2015**, *55*, 218–226. [CrossRef]
17. Kadkhodapour, J.; Schmauder, S.; Raabe, D.; Ziaei-Rad, S.; Weber, U.; Calcagnotto, M. Experimental and numerical study on geometrically necessary dislocations and non-homogeneous mechanical properties of the ferrite phase in dual phase steels. *Acta Mater.* **2011**, *59*, 4387–4394. [CrossRef]
18. Nye, J. Some geometrical relations in dislocated crystals. *Acta Met.* **1953**, *1*, 153–162. [CrossRef]
19. Calcagnotto, M.; Ponge, D.; Demir, E.; Raabe, D. Orientation gradients and geometrically necessary dislocations in ultrafine grained dual-phase steels studied by 2D and 3D EBSD. *Mater. Sci. Eng. A* **2010**, *527*, 2738–2746. [CrossRef]
20. Movahed, P.; Kolahgar, S.; Marashi, S.; Pouranvari, M.; Parvin, N. The effect of intercritical heat treatment temperature on the tensile properties and work hardening behavior of ferrite–martensite dual phase steel sheets. *Mater. Sci. Eng. A* **2009**, *518*, 1–6. [CrossRef]
21. Jin, J.-E.; Lee, Y.-K. Strain hardening behavior of a Fe–18Mn–0.6C–1.5Al TWIP steel. *Mater. Sci. Eng. A* **2009**, *527*, 157–161. [CrossRef]
22. Markandeya, R.; Nagarjuna, S.; Satyanarayana, D.; Sarma, D. Correlation of structure and flow behaviour of Cu–Ti–Cd alloys. *Mater. Sci. Eng. A* **2006**, *428*, 233–243. [CrossRef]
23. Son, Y.I.; Lee, Y.K.; Park, K.-T.; Lee, C.S.; Shin, D.H. Ultrafine grained ferrite–martensite dual phase steels fabricated via equal channel angular pressing: Microstructure and tensile properties. *Acta Mater.* **2005**, *53*, 3125–3134. [CrossRef]
24. Swift, H. Plastic instability under plane stress. *J. Mech. Phys. Solids* **1952**, *1*, 1–18. [CrossRef]
25. Tomita, Y.; Okabayashi, K. Tensile stress-strain analysis of cold worked metals and steels and dual-phase steels. *Met. Mater. Trans. A* **1985**, *16*, 865–872. [CrossRef]
26. Chen, L.; Kim, J.-K.; Kim, S.-K.; Kim, G.-S.; Chin, K.-G.; De Cooman, B.C. Stretch-Flangeability of High Mn TWIP steel. *Steel Res. Int.* **2010**, *81*, 552–568. [CrossRef]

27. Sudo, M.; Iwai, T. Deformation Behavior and Mechanical Properties of Ferrite-Bainite-Martensite (Triphase) Steel. *Trans. Iron Steel Inst. Jpn.* **1983**, *23*, 294–302. [CrossRef]
28. Park, K.; Nishiyama, M.; Nakada, N.; Tsuchiyama, T.; Takaki, S. Effect of the martensite distribution on the strain hardening and ductile fracture behaviors in dual-phase steel. *Mater. Sci. Eng. A* **2014**, *604*, 135–141. [CrossRef]

Publisher's Note: MDPI stays neutral with regard to jurisdictional claims in published maps and institutional affiliations.

© 2020 by the authors. Licensee MDPI, Basel, Switzerland. This article is an open access article distributed under the terms and conditions of the Creative Commons Attribution (CC BY) license (http://creativecommons.org/licenses/by/4.0/).

Article
Regularized Yield Surfaces for Crystal Plasticity of Metals

Bjørn Holmedal

Department of Material Science and Engineering, Norwegian University of Science and Technology, 7491 Trondheim, Norway; bjorn.holmedal@ntnu.no

Received: 21 October 2020; Accepted: 23 November 2020; Published: 25 November 2020

Abstract: The rate-independent Schmid assumption for a metal crystal results in a yield surface that is faceted with sharp corners. Regularized yield surfaces round off the corners and can be convenient in computational implementations. To assess the error by doing so, the coefficients of regularized yield surfaces are calibrated to exactly interpolate certain points on the facets of the perfect Schmid yield surface, while the different stress predictions in the corners are taken as the error estimate. Calibrations are discussed for slip systems commonly activated for bcc and fcc metals. It is found that the quality of calibrations of the ideal rate-independent behavior requires very large yield-surface exponents. However, the rounding of the corners of the yield surface can be regarded as an improved approximation accounting for the instant, thermal strain-rate sensitivity, which is directly related to the yield-surface exponent. Distortion of the crystal yield surface during latent hardening is also discussed, including Bauschinger behavior or pseudo slip systems for twinning, for which the forward and backward of the slip system are distinguished.

Keywords: crystal plasticity; yield surface; slip systems

1. Introduction

A robust and general formulation with a yield surface for crystal plasticity can conveniently be applied in a framework similar to the continuum-plasticity framework for finite-element calculations, e.g., including elasticity and using a return-mapping algorithm. There are two major reasons for choosing this approach. Firstly, in perfectly rate-independent crystal plasticity finite element (CPFEM) formulations a nonunique stress versus strain-rate relation makes the calculation procedure break down, when it is applied into a finite element method for the solution of boundary value problems for polycrystals [1]. The reason is a nonunique slip-system selection. Uniqueness can be obtained with a positive definite hardening matrix in the hardening [2,3]. However, that kind of a limitation of the hardening law in order to avoid an ill-posed mathematical problem prevents the use of relevant constitutive laws. The rounded crystal yield surface does not suffer from this problem [4–6].

Secondly, the general metal plasticity framework enables crystal elasticity to be part of the model and allows arbitrary models for the critical resolved shear stress, as compared to the viscoplastic or rate-independent asymptotical cases. The activation of different slip systems with different glide planes is of interest, as well as complex models for latent hardening of the slip systems.

The nonuniqueness of the rate-independent approach disappears with the introduction of positive strain-rate sensitivities individually on each slip system. From a physical point of view, this is more realistic since the critical resolved shear stresses of the slip systems will have an instant strain-rate sensitivity. This sensitivity increases with increased temperature. Due to its simplicity, the purely viscous power-law relation [7], has become very popular.

$$\tau^s = \tau_0 \mathrm{sgn}(\dot{\gamma}^s) \left| \frac{\dot{\gamma}^s}{\dot{\gamma}_0} \right|^m \tag{1}$$

Here τ^s is the resolved shear stress and $\dot{\gamma}^s$ is the resolved shear-strain rate for slip system s. The instant strain-rate sensitivity coefficient m, the stress amplitude τ_0 and the strain-rate scale $\dot{\gamma}_0$ are model parameters. A noninstant strain-rate sensitivity, for which an increased strain is required to change the stress, may be included by more advanced models, where τ_0 is allowed to change as a function of strain by a differential equation differently at different strain rates.

The instant strain-rate sensitivity, m, in this model contributes individually to each slip system. After an abrupt strain-rate jump in metal, this contribution typically accounts for about half the instant stress change, whereas the remaining part of the stress changes gradually, adapting to the new strain rate during a short strain transient, see the review [8]. In the power-law, Equation (1), all slip systems will be activated, enabling a simple implementation and with proved mathematical uniqueness [7,9]. The uniqueness holds also in the more general case including an a-thermal threshold value of the critical resolved shear stress [10]. In the limit of vanishing strain-rate sensitivity, the viscoplastic model degenerates to one solution amongst the possible Taylor ambiguity solutions of the rate-independent formulation.

The strain-rate coefficient is commonly set very small to obtain an approximately rate-independent solution or even extrapolated to zero [11]. It was pointed out in [10] that this provides a very similar texture-evolution prediction as to the corresponding rate-independent solution because the same stress corners on the crystal yield surface are selected. However, due to the round-off of the yield surface, the stress predictions are significantly smaller, i.e., the Taylor factor is decreasing with increasing strain-rate sensitivity coefficient m. However, if m is set very small, the rate-dependent, nonlinear equations behave very stiffly, requiring decreasingly small time steps, and the numerical implementations become inefficient or even nonconvergent. The viscoplastic model comes in many variants, e.g., including latent hardening or several slip systems.

At room temperature or below, most metals show weak strain-rate sensitivity. Hence the rate-independent models are attractive, provided the nonuniqueness issue can be dealt with. While the rate-dependent models require finite disturbances, the rate-independent Schmid model can be applied in infinitesimal bifurcation sheet necking analysis [12], allowing forming-limit diagrams to be calculated without calibration of some surface roughness coefficient as input. However, it has been reported that even in the limit of an infinite yield-surface exponent, the application of a regularized yield surface may provide different estimates for certain components the elastoplastic tangent modulus than the perfectly rate independent approach [13]. This is of importance for applications with incremental self-consistent model formulations and probably also for the mentioned stability analysis.

In Taylor-model implementations, many alternative assumptions have been suggested to obtain uniqueness of the slip systems; for more details see the review in [10]. However, the regularized yield surface corresponds exactly to a self-similar iso-value of the rate of plastic work of the flow potential for the viscoplastic power-law formulation. This solution again corresponds to a Taylor ambiguity solution for the selection of active slip systems, namely the limit of vanishing strain-rate sensitivity (see [10]), which showed that this solution corresponds to the one which can approximately, but not exactly, can be obtained applying singular-value decomposition or quadratic programming.

In CPFEM implementations, two strategies have been successfully applied. The first alternative is inversion of nonpositive definite hardening matrices by the singular-value decomposition [14], which enables a fast and stable numerical algorithm; see [10] for a recent discussion. The second approach is a rounded crystal yield surface [4–6], which fits straight into the continuum-plasticity framework. As pointed out by Manik and Holmedal [10], it corresponds to the rate-independent limit of the viscoplastic model. Until recently, return mapping algorithms for yield surfaces with high exponents have not been numerically stable. In a recent publication, Paux et al. [15] avoid this problem by combining the best part of the two available methods, using the regularized yield surface to obtain a smooth, unique tangent modulus, while using the highly efficient Schmid model for the integration of the single crystal constitutive equations. However, progress has recently been reported related to stable, efficient return-mapping algorithms for yield surfaces with high exponents. Thus far, this has been successfully tested for only for continuum plasticity yield surfaces, see [16]. However, the technology is generic and enables stable implementations also of high-exponent regularized yield surfaces for crystals. The details related to adapting this approach and implement return mapping algorithms for crystal yield surfaces will be reported elsewhere.

So far, crystal-plasticity implementations into the finite-element method or into Taylor type of models have mainly assumed either the strain-rate independent plasticity approach or simple visco-plastic behavior as described by power laws. The main reasons for choosing these approaches are the same as they were for choosing continuum-plasticity models some years ago, i.e., model simplicity and numerical convenience. However, the limitation to these specific models for the critical-resolved shear stress, nowadays become a hinder for dealing with more realistic models from a physical-metallurgical point of view, i.e., for dealing with real alloys, containing various types of particles and solid-solution elements applying models that distinguish the instant and the microstructurally induced parts of the strain-rate sensitivity. The framework discussed in this article, enables formulation of models with latent hardening and several sets of active slip systems. This allows yield-surface distortions as predicted by models dealing with strain-path changes related to Bauschinger and cross-hardening effects, e.g., [17–20]. Recently progress on similar distortional-hardening approaches in continuum plasticity models has been reported in [21–26]. The same framework can be applied to pseudo-slip systems applied in deformation-twinning models [27].

In the current work, the quality of the rounded crystal yield surface as an approximation to the rate-independent one will be explored for various set of slip systems activated. The theory is outlined in Section 2, relating the yield surface to iso-surfaces of the potential for the rate-dependent power-law theory and interpreting the regularized yield surface as an approximation accounting for the instant strain-rate sensitivity of the critical resolved shear stresses. Furthermore, calibration strategies are specified. In Section 3, calibrations are made to fcc slip systems, fcc with nonoctahedral slips and bcc with 24 slip systems. Based on the results and theory, the applicability of the crystal yield surfaces is discussed in Section 4 and conclusions are drawn in Section 5.

2. Theory

The strain-rate independent crystal yield surface can mathematically be expressed as a smooth inner convex envelope of the linear facets from each slip system. The ideal Schmid assumption can mathematically be expressed as the following yield criteria:

$$|\sigma : P^s| \leq \tau_c^s \tag{2}$$

Here σ is the stress tensor at yielding, and τ_c^s the critical resolved shear stress for slip on slip system s. Furthermore, $P^s = (m^s \otimes n^s + n^s \otimes m^s)/2$, where m^s is the slip plane unity normal vector, n^s the unity slip direction vector. Single slip system solutions are located on facets and obey the normality rule $D_p = P^s \dot{W}_p / \tau_c^s$, where D_p is the symmetric part of the plastic velocity gradient and $\dot{W}_p = \sigma : D_p$ is the rate of the external plastic work. Only the shape of the yield surface can be derived from Equation (2). The symmetric part of the plastic velocity gradient D_p can be uniquely derived from the normality rule and Equation (2) only for cases of one single active slip system, which defines the facets (and hence indirectly also the corners where adjacent facets intersect). However, for solutions at corners or other facet interactions, the active slip systems are not provided by Equation (2), and then D_p cannot be uniquely determined solely by Equation (2).

2.1. Rate-Independent Regularizations

The inner envelope of the yield surfaces by Equation (2) corresponds to a yield surface with sharp corners. By regularization it can be expressed as one yield criterion, i.e., by a smooth yield surface

$$f(\sigma) = \left(\sum_s \xi^s \left|\frac{\sigma:P^s}{\tau_c^s}\right|^n\right)^{\frac{1}{n}} - 1 = 0$$

$$D_p = \dot{\lambda}\frac{\partial f}{\partial \sigma} = \frac{\dot{\lambda}}{(f+1)^{n-1}} \sum_{a=-N}^{N} \xi^a \left|\frac{\sigma:P^s}{\tau_c^s}\right|^{n-1} \frac{P^s}{\tau_c^a},$$
(3)

which corresponds to the yield criteria in Equation (2) in the limit when $n \to \infty$ and with $\xi^s \equiv 1$. Here f is the yield function. In the absence of corners, the normality rule is now unique everywhere. It follows that the plastic-rate parameter can be found as

$$\dot{\lambda} = \dot{W}_p$$
(4)

Note that the yield surface in Equation (3) is centro-symmetric. For cases where the reverse and forward of a slip system have different critical-resolved shear stresses, a slightly more general yield surface can be adapted, counting both slip directions as two different slip systems:

$$f(\sigma) = \left(\sum_s \xi^s \left\langle\frac{\sigma:P^s}{\tau_c^s}\right\rangle^n\right)^{\frac{1}{n}} - 1$$

$$D_p = \dot{\lambda}\frac{\partial f}{\partial \sigma} = \frac{\dot{\lambda}}{(f+1)^{n-1}} \sum_{a=-N}^{N} \xi^s \left\langle\frac{\sigma:P^s}{\tau_c^s}\right\rangle^{n-1} \frac{P^s}{\tau_c^a}$$
(5)

Here the McCaulay brackets are used, i.e., $\langle x \rangle = \max(x, 0) = \frac{1}{2}(x + |x|)$.

2.2. Relation to Strain-Rate Dependent Formulations

It is interesting to compare Equation (3) to the stress potential for a rate-dependent crystal, obeying the power law [7,9].

$$\tau^s = \tau_0 \text{sgn}(\dot{\gamma}^s)\left|\frac{\dot{\gamma}^s}{\dot{\gamma}_0}\right|^m$$

$$\psi = \frac{m}{m+1}\dot{W}_p = \frac{\tau_0\dot{\gamma}_0 m}{(m+1)}\sum_s \left|\frac{\sigma:P^s}{\tau_0}\right|^{\frac{1+m}{m}}$$
(6)

$$D_p = \frac{\partial \psi}{\partial \sigma} = \dot{\gamma}_0 \sum_s \left|\frac{\sigma:P^s}{\tau_0}\right|^{\frac{1}{m}} P^s$$

In this special case, no true yield surface exists, but shape-invariant flow surfaces for constant values of $\dot{W}_p$ have the same shape and play a similar role as the regularized yield surface, Equation (3). Note the close correspondence between the strain-rate sensitivity exponent m and the yield surface exponent n.

$$m = \frac{1}{n-1}$$
(7)

Note that when ξ^s and τ_c^s are the same for all the slip systems, the regularized yield surface with exponent n will give the same solution as the viscoplastic model with strain-rate sensitivity m at the same plastic work rate. It follows as a corollary that, in this case, the texture evolution predicted by the Taylor model will be the same for these two models.

The instant strain-rate sensitivity influences the yield stress in two distinct ways. Firstly, the magnitude of the critical resolved shear stress changes as a function of the strain rate, i.e., the volume of the yield surface changes. However, for cases with a small strain-rate sensitivity this effect is neglectable, as the critical resolved shear stress then depends only weakly on the logarithm of the strain rate. Secondly, the corners of the yield surface are rounded, locally. Corners are sharper in higher dimensions. Hence the local change of the stress at the rounded yield-surface corners in the five-dimensional stress space becomes significant, even in cases of a very small strain-rate sensitivity. Since most crystal-plasticity solutions are located near these corners, the magnitude of the resulting polycrystal stress tensor will decrease significantly as compared to as predicted by a perfectly rate-independent model with sharp corners. However, the stress itself still will be mildly sensitive to changes of the strain rate, i.e., it is caused by the changed strain-rate sensitivity, not the change of the strain rate. For a real metal at room temperature, the rate-independent model with rounded corners provides a more realistic approximation to the real behavior than a model with sharp corners.

The physical based mechanical threshold strength (MTS) model [28] provides estimates and discussion of the instant strain-rate sensitivity, from which the simplest power-law estimate $m = g_0 \mu b^3 / kT$, clearly reveals its dependence on the temperature T; see also [29] for a discussion of the formulation of the MTS model adequate for crystal plasticity implementations. Here μ is the elastic shear modulus, b the length of the Burgers vector, k the Boltzmann constant and g_0 nondimensional activation energy with magnitude of order unity. According to this estimate a realistic strain-rate sensitivity at room temperature for e.g., aluminum is $m \approx 10^{-2}$, hence a realistic yield surface exponent will be $n \approx 100$, and the rounded corner solution will be 1–2% lower in stress than the ideal corner.

With increasing temperature, the strain-rate sensitivity of the critical resolved shear stress also needs to be accounted for, i.e., by a strain-rate dependent model, where the "volume" of the yield surface depends on the strain rate. Furthermore, with smaller yield-surface exponents the regularization itself will affect not only the corners locally, but also the magnitude of the entire yield surface will shrink. This can be accounted for by adjusting ξ^s, so that the inscribed regularized yield surface touches the centres of the yield surface facets of the nonregularized yield surface. The distribution of the slip rates on the slip systems controls the shape of the yield surface, while the change of the magnitude of the yield surface mainly depends on the total amount of slip activity, i.e., τ_c^s depends on Γ, being the solution of $\dot{\Gamma} = \sum_s |\dot{\gamma}^s|$. Hence, efficient models can be formulated based on the regularized yield surface, also for cases at elevated temperatures.

As discussed in [10], the rate-dependent plastic potential degenerates in its rate-independent limit $m \to 0$ to one particular solution of the Taylor ambiguity, which corresponds exactly to the solution with the regularized yield surface in the limit $n \to \infty$. In practice, this limit can be approached with a stable numerical iterative algorithm that can handle a high exponent of the regularized yield surface or a low strain-rate sensitivity in the rate-dependent formulation. The use of line-search in addition to the Newton-iteration scheme seems very promising; see [16].

2.3. Fitting Stress Points Projected Radially onto the Facets

With decreasing yield-surface exponents, e.g., at elevated temperatures, a calibration of ξ^s becomes increasingly important. As an approximation, using equal values ξ for ξ^s for all the slip systems, Gambin [5] proposed

$$\xi = \frac{N}{\sum_{s=1}^{N} \sum_{r=1}^{N} \left(2\frac{\tau_c^s}{\tau_c^r} |\mathbf{P}^s : \mathbf{P}^r|\right)^n} \tag{8}$$

Here, N is the total number of slip systems. This estimate was obtained as an approximation for fitting one point on each yield surface facet, namely the one in the radial direction of the normal to the facet and being closest to the origin, i.e., $S = 2\mathbf{P}^s \tau_c^s$, where S is the deviatoric part of the stress tensor. The approximately fit gives only one average value ξ for ξ^s that is equal for all involved slip systems. For the case of only one family of slip systems, it can be shown that this corresponds to any of the solutions that exactly go through these points, and Equation (8) can be simplified to

$$\xi = \frac{1}{2^n \sum_{r=1}^{N} |\mathbf{P}^s : \mathbf{P}^r|^n} \tag{9}$$

However, if a fit is made to several families of slip systems with each their constant critical resolved shear stress, there will be different values of ξ^s for each of the involved slip systems. This idea can be extended to cases where three or more families of slip systems are activated and is a more precise solution as compared to the estimate in Equation (8) made by Gambin. The solution, Equation (9), is an important special case of more general cases, for which one equation has to be solved for each involved slip system α.

$$\sum_{\beta=1}^{N} |P^{\alpha} : P^{\beta}|^{n} 2^{n} \left(\frac{\tau_{C}^{\alpha}}{\tau_{C}^{\beta}}\right)^{n} \xi^{\beta} = 1, \alpha = 1, \ldots, N \qquad (10)$$

An important special case is when cases involving two families of centro-symmetric slip systems are considered, and each family has one critical resolved shear stress. In these cases, only two different values of ξ^s are required. These two values can be determined simply by fitting to one arbitrarily chosen facet of each type, i.e., fitted radially to $S = 2P^s \tau_C^s$ for two selected slip systems, s_1 and s_2, one from each family.

$$2^n \xi^{(1)} \sum_{s=1}^{N_1} |P_1^{s_1} : P_1^{s}|^n + 2^n \xi^{(2)} \left(\frac{\tau_C^{s_1}}{\tau_C^{s_2}}\right)^n \sum_{s=1}^{N_2} |P_1^{s_1} : P_2^{s}|^n = 1$$
$$2^n \xi^{(1)} \left(\frac{\tau_C^{s_2}}{\tau_C^{s_1}}\right)^n \sum_{s=1}^{N_1} |P_2^{s_2} : P_1^{s}|^n + 2^n \xi^{(2)} \sum_{s=1}^{N_2} |P_2^{s_2} : P_2^{s}|^n = 1 \qquad (11)$$

It follows that

$$\xi^{(1)} = \frac{\frac{1}{2^n}\left(\sum_{s=1}^{N_2} |P_2^{s_2} : P_2^{s}|^n - \left(\frac{\tau_C^{s_1}}{\tau_C^{s_2}}\right)^n \sum_{s=1}^{N_2} |P_1^{s_1} : P_2^{s}|^n\right)}{\left(\sum_{s=1}^{N_1} |P_1^{s_1} : P_1^{s}|^n\right)\left(\sum_{s=1}^{N_2} |P_2^{s_2} : P_2^{s}|^n\right) - \left(\sum_{s=1}^{N_2} |P_1^{s_1} : P_2^{s}|^n\right)\left(\sum_{s=1}^{N_1} |P_2^{s_2} : P_1^{s}|^n\right)}$$

$$\xi^{(2)} = \frac{\frac{1}{2^n}\left(\sum_{s=1}^{N_1} |P_1^{s_1} : P_1^{s}|^n - \left(\frac{\tau_C^{s_2}}{\tau_C^{s_1}}\right)^n \sum_{s=1}^{N_1} |P_2^{s_2} : P_1^{s}|^n\right)}{\left(\sum_{s=1}^{N_1} |P_1^{s_1} : P_1^{s}|^n\right)\left(\sum_{s=1}^{N_2} |P_2^{s_2} : P_2^{s}|^n\right) - \left(\sum_{s=1}^{N_2} |P_1^{s_1} : P_2^{s}|^n\right)\left(\sum_{s=1}^{N_1} |P_2^{s_2} : P_1^{s}|^n\right)} \qquad (12)$$

For general cases, the set of N linear Equation (10) can be solved to derive the coefficients ξ^{α}. Note that the yield surface is convex only as far as all derived coefficients are positive.

2.4. Fitting the Midpoint of Each Involved Facet of the Crystal Yield Surface

Each facet of the crystal yield surface corresponds to one slip system and can be defined by the combination of one point on the facet and by that P^s points in its normal direction. Equation (3) or Equation (5) are also recognized as a "Facet polynomial" [30]. Following the procedure proposed in that work, the yield surface can be fitted exactly to a certain number of stress points and corresponding symmetric part of plastic velocity gradient tensors. An improved choice, as compared to the approach by Gambin, is to match the center point of each facet, which corresponds to the average of the corner locations of the facet. However, the corner solutions must be found first.

A corner corresponds to a valid deviatoric stress-tensor solution S_{crn} obtained by solving Equation (1) with only five of the slip systems. However, only the corners, for which

$$\left|\frac{S_{crn} : P^s}{\tau_C^s}\right| \leq 1 \qquad (13)$$

for all the slip systems, are on the yield surface. All yield-surface corners belonging to a facet, i.e., to slip system s, can be found as the solutions of all possible combinations of five slip systems that obey Equation (2). A corner solution only involves some of the slip systems and is a common corner of all facets corresponding to these involved slip systems. The midpoint of a facet corresponds to the average of the stress tensors of all its corners, and the normal direction of the facet is parallel to P^s.

These midpoint stress points will be denoted S^k, one for each slip system, i.e., $1 \leq k \leq N$. The coefficients λ^s of this calibration can be found by solving the following linear equation system

$$\sum_{s=1}^{N} \xi^s \left| \frac{S^k : P^s}{\tau_c^s} \right|^n = 1 \qquad (14)$$

Note that the yield surface is convex only if all coefficients $\xi^s \geq 0$, which is always the case with a sufficiently high exponent n. The system from Equation (12) can efficiently be solved iteratively. If the exponent n is too low, but one still wants to proceed, the points corresponding to the negative ξ^s can simply be excluded to obtain a convex surface. When accounting for both slip directions of each slip system, one must include both directions as independent slip systems with each their ξ^s.

For the important cases involving two families of slip systems, only two values of ξ^s are required. Then, it is enough to fit the yield function to two facet center points, one to each type of facets, denoted S^{s1} and S^{s2} for the two families of slip systems, respectively.

$$\xi^{(1)} \sum_{s=1}^{N_1} \left| \frac{S^{s1}:P_1^s}{\tau_c^{(1)}} \right|^n + \xi^{(2)} \sum_{s=1}^{N_2} \left| \frac{S^{s1}:P_2^s}{\tau_c^{(2)}} \right|^n = 1$$

$$\xi^{(1)} \sum_{s=1}^{N_1} \left| \frac{S^{s2}:P_1^s}{\tau_c^{(1)}} \right|^n + \xi^{(2)} \sum_{s=1}^{N_2} \left| \frac{S^{s2}:P_2^s}{\tau_c^{(2)}} \right|^n = 1 \qquad (15)$$

It follows that

$$\xi^{(1)} = \frac{\sum_{s=1}^{N_2} \left|\frac{S^{s2}:P_2^s}{\tau_c^{(2)}}\right|^n - \sum_{s=1}^{N_2} \left|\frac{S^{s1}:P_2^s}{\tau_c^{(2)}}\right|^n}{\left(\sum_{s=1}^{N_1}\left|\frac{S^{s1}:P_1^s}{\tau_c^{(1)}}\right|^n\right)\left(\sum_{s=1}^{N_2}\left|\frac{S^{s2}:P_2^s}{\tau_c^{(2)}}\right|^n\right) - \left(\sum_{s=1}^{N_2}\left|\frac{S^{s1}:P_2^s}{\tau_c^{(2)}}\right|^n\right)\left(\sum_{s=1}^{N_1}\left|\frac{S^{s2}:P_1^s}{\tau_c^{(1)}}\right|^n\right)}$$

$$\xi^{(2)} = \frac{\sum_{s=1}^{N_1} \left|\frac{S^{s1}:P_1^s}{\tau_c^{(1)}}\right|^n - \sum_{s=1}^{N_1} \left|\frac{S^{s2}:P_1^s}{\tau_c^{(1)}}\right|^n}{\left(\sum_{s=1}^{N_1}\left|\frac{S^{s1}:P_1^s}{\tau_c^{(1)}}\right|^n\right)\left(\sum_{s=1}^{N_2}\left|\frac{S^{s2}:P_2^s}{\tau_c^{(2)}}\right|^n\right) - \left(\sum_{s=1}^{N_2}\left|\frac{S^{s1}:P_2^s}{\tau_c^{(2)}}\right|^n\right)\left(\sum_{s=1}^{N_1}\left|\frac{S^{s2}:P_1^s}{\tau_c^{(1)}}\right|^n\right)} \qquad (16)$$

Note that the corner-stress tensor is everywhere divided by the critical resolved shear stress, i.e., without loss of generality the coefficients $\xi^{(1)}$ and $\xi^{(2)}$ can be calculated based on corner solutions, for which one of the resolved shear stresses equals unity and depends only on the ratio $\tau_c^{(2)}/\tau_c^{(1)}$.

3. Regularized Crystal Yield Surfaces as Approximations to the Schmid Criteria

The application of regularized crystal yield surfaces in crystal-plasticity implementations can be convenient, and calibration of proper polynomial expressions is explained in Section 2. The simplest approach is to put all coefficients $\xi^s \equiv 1$ in Equation (3) or Equation (5). This will be a good approximation for sufficiently large exponents n. The question is, however, how large n needs to be for a given set of slip systems. A more refined approach is the simplified solution proposed by Gambin, Equations (8) and (9), or the generalized version for several slip systems in Equations (10)–(12), which will be referred to as "radial" in the discussion below. Finally, the solution that intersects all the midpoints of yield-surface facets is uniquely defined but requires considerably more work to determine the coefficients, as the midpoints of the facets of the yield surface must be derived first. It is interesting and of practical relevance to look at the details for the most important examples for fcc and bcc slip systems.

An ideal regularization should match the facets of the crystal yield surface but slightly round off its sharp corners, i.e., the magnitude of the stress tensor becomes smaller at the corners. An analogy is an inscribed circle in a square, for which the difference is largest in the directions towards the corners. The corners of a cube are further away from an inscribed sphere. This illustrates the fact that corners become sharper from two to three dimensions, and even more in higher dimensions. Hence the corners in the five-dimensional stress space are sharp, and the error is larger close to the corners than elsewhere. By "error" it is here meant the difference between the regularized yield surface and the Schmid surface with sharp corners. Two estimates will be made to describe how well the regularized yield surface fits the Schmid criteria. The first one is the average error at the corners

$$e_{crn} = \frac{1}{N_{crn}} \sum_{n=1}^{N_{crn}} \frac{\|S_{crn,n} - S_{crn,\infty}\|}{\|S_{crn,\infty}\|} \tag{17}$$

Here $S_{crn,n}$ is the stress in the same direction in the stress space like the corner of the Schmid yield surface $S_{crn,\infty}$ (i.e., with infinite exponent n), where N_{crn} is the number of corners of the yield surface. The second measure is the average of the errors at the middle of each of the facets of the yield surface.

$$e_{fac} = \frac{1}{N_s} \sum_{s=1}^{N_s} \frac{S_{fac,s} - S_{fac,\infty}}{S_{fac,\infty}} \tag{18}$$

Here $S_{fac,s}$ is the stress tensor at the exact midpoint of the facet of the Schmid yield surface belonging to slip system s, and N_s is the number of slip systems involved in the yield surface.

3.1. The fcc Case with $\{111\}\langle\bar{1}10\rangle$

The standard fcc case with only one set of slip systems is the simplest to deal with. The solution proposed by Gambin is identically equal to the "radial" solution, where ξ^s takes the same value ξ for all slip systems. This solution is compared to ξ, being estimated by the midpoint fitting as functions of n in Figure 1. The average errors at the corners and at the midpoints of the facets are compared in Figure 2. Note that the midpoint fit has no error in the midpoints, i.e., $e_{fac} = 0$ for all n. For an exponent n larger than about 15, both approaches coincide with the simple approach putting $\xi^s \equiv 1$. As can be seen from Figure 2 this corresponds to about 15% corner error. Cases where $n \geq 15$ correspond to a power-law strain rate sensitivity $m \leq 0.07$, hence in such cases the simple asymptotic solution $\xi^s \equiv 1$ applies.

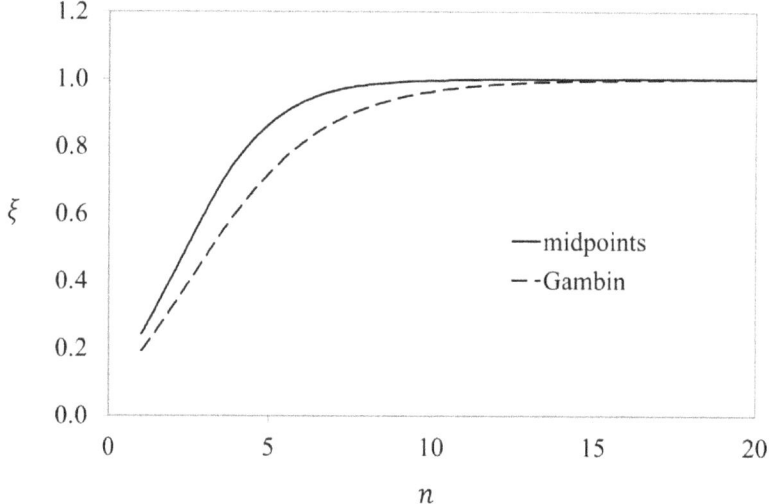

Figure 1. Fitted coefficient of the yield surface as function of the exponent n, for the case of fcc with octahedral slip systems, fitted to facet midpoints and by the regularization proposed by Gambin (1992), [5].

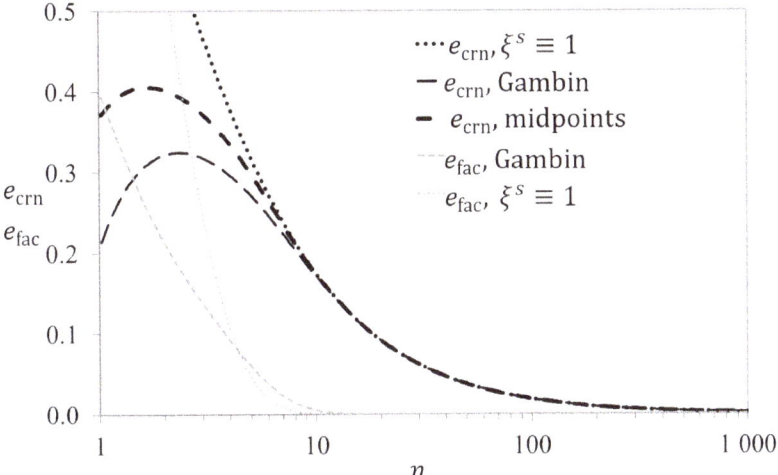

Figure 2. The average error at the corners and at the facet midpoints as functions of n, for yield surface regularizations according to Gambin, fitted to the midpoints or simply by $\xi^s \equiv 1$.

Note that for the most rounded yield surfaces with the smallest exponents, the facet errors increase and only the midpoint fit provides a consistent regularization. The corner errors decrease very slowly, and even at $n = 90$ the error remains about 2%.

3.2. The fcc Case with $\{111\}\langle\bar{1}10\rangle$ and Nonoctahedral $\{001\}\langle\bar{1}10\rangle$

Nonoctahedral $\{001\}\langle\bar{1}10\rangle$ slips can be activated as long as the ratio of their critical resolved shear stress to the one for the $\{111\}\langle\bar{1}10\rangle$ slip systems is less than $\sqrt{3}$. The yield surface has 36 facets corresponding to the 18 slip systems and 186 corners. Figure 3 shows the fitted coefficients for two cases, when the critical resolved shear stress is equal for the two slip systems and when $\tau_c^{\{001\}\langle\bar{1}10\rangle} = 1.5\tau_c^{\{111\}\langle\bar{1}10\rangle}$. The corresponding errors are shown in Figure 4. The errors at a given exponent n are very similar for the two cases, and all the fitting procedures give similar errors for n larger than 10. When the critical resolved shear stresses are equal, the radial fit gives small facet errors and quite similar corner error as the midpoint fit.

When the ratio between the two critical resolved shear stresses equals 1.5, the facets due to the presence of nonoctahedral slips systems contribute to only a small portion of the yield surface, and their midpoints can only be fitted for n larger than about 7. However, their midpoints are still poorly fitted by the Gambin formula and by the radial approach up to values of n close to 30.

3.3. The bcc Case with $\{\bar{2}11\}\langle111\rangle$ and $\{\bar{1}10\}\langle111\rangle$

It is commonly accepted that 48 slip systems consisting with $\{\bar{2}11\}\langle111\rangle$ and $\{\bar{3}11\}\langle111\rangle$ in addition to the basal $\{\bar{1}10\}\langle111\rangle$ are activated during plastic deformation of bcc steel at room temperature. However, for computational efficiency and convenience, calculations are commonly limited to two sets of slip systems, which provides many glide planes, i.e., close to the pencil glide assumption, hence only the 24 slip systems from $\{\bar{2}11\}\langle111\rangle$ and $\{\bar{1}10\}\langle111\rangle$ will be considered in the example here. When $\frac{1}{2}\sqrt{3}\tau_c^{\{\bar{1}10\}\langle111\rangle} < \tau_c^{\{\bar{2}11\}\langle111\rangle} < \frac{2}{3}\sqrt{3}\tau_c^{\{\bar{1}10\}\langle111\rangle}$ all 24 slip bcc systems contribute and the resulting yield surface has 432 corners and 48 facets. Commonly, $\tau_c^{\{\bar{2}11\}\langle111\rangle} = 0.95\tau_c^{\{\bar{1}10\}\langle111\rangle}$ or simply $\tau_c^{\{\bar{2}11\}\langle111\rangle} = \tau_c^{\{\bar{1}10\}\langle111\rangle}$ are used, see e.g., [31]. Figure 5 shows the yield-surface parameter calibrations. With $n > 50$ the parameters are close to unity. Figure 6 shows that the error at the midpoints of the facets vanishes for $n > 20$, whereas the error in the corners is about 5% with $n = 30$, from which it slowly decays proportionally to n^{-1}.

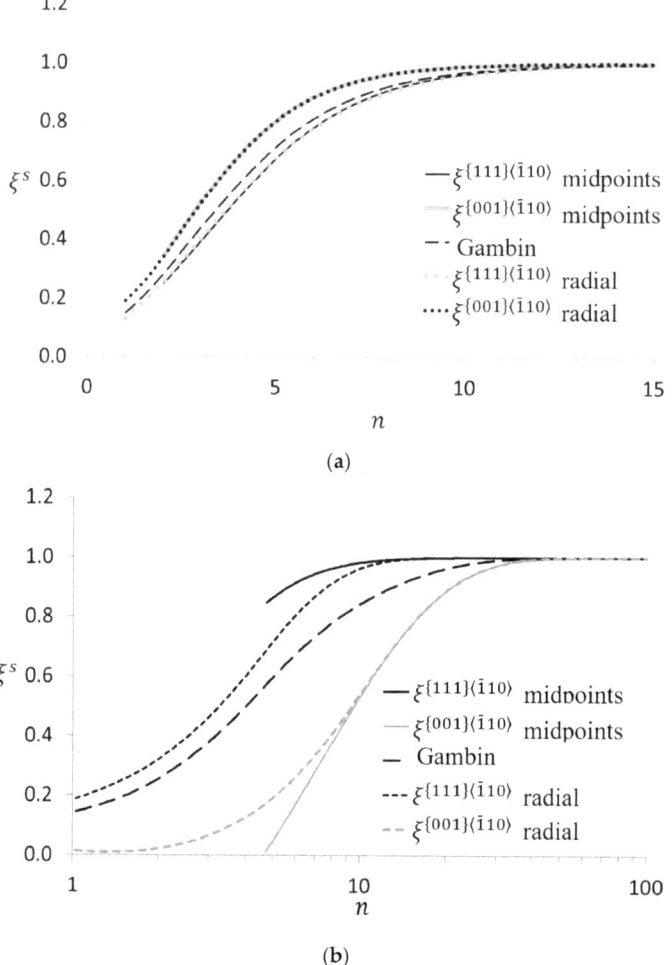

Figure 3. The fitted coefficients of the yield surface as functions of the exponent n, for the case of fcc with (**a**) octahedral $\{111\}\langle\bar{1}10\rangle$ and (**b**) additional nonoctahedral $\{001\}\langle\bar{1}10\rangle$ slip systems, fitted to facet midpoints, by the regularization proposed by Gambin, and fitted radially to the facets. (**a**) $\tau_c^{\{001\}\langle\bar{1}10\rangle} = \tau_c^{\{111\}\langle\bar{1}10\rangle}$ and (**b**) $\tau_c^{\{001\}\langle\bar{1}10\rangle} = 1.5\tau_c^{\{111\}\langle\bar{1}10\rangle}$.

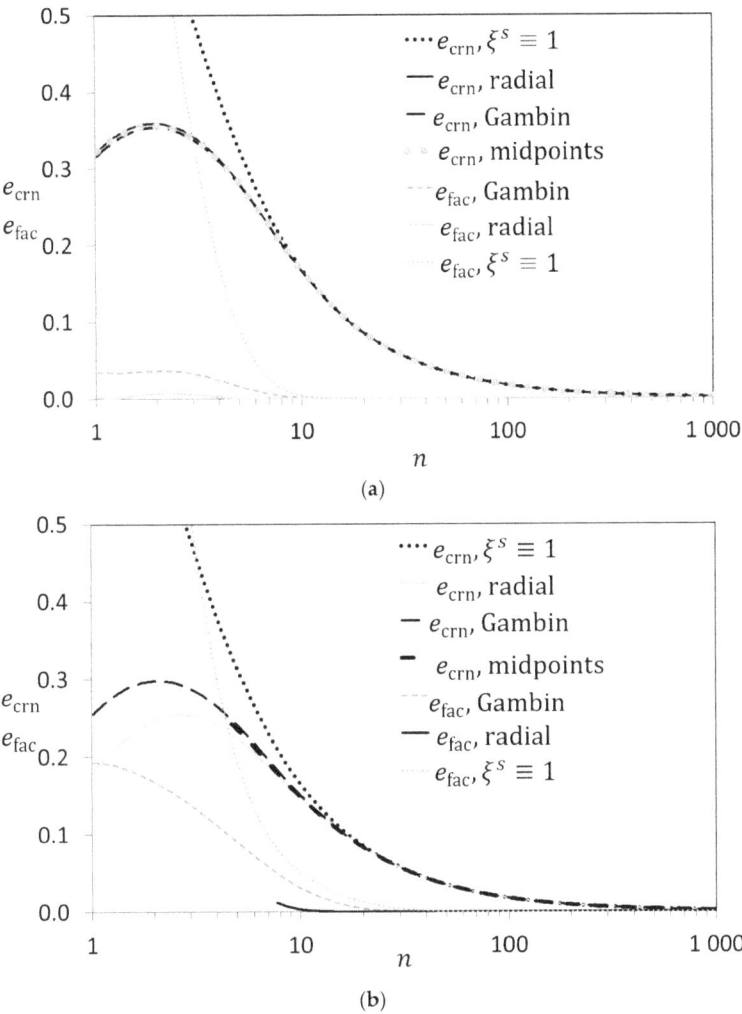

Figure 4. The average error at the corners and at the facet midpoints as a function of n, for the case of fcc with (**a**) octahedral $\{111\}\langle\bar{1}10\rangle$ and (**b**) additional nonoctahedral $\{001\}\langle\bar{1}10\rangle$ slip systems, for yield surface regularizations according to Gambin, fitted to the midpoints, fitted radially or simply by $\xi^s \equiv 1$. (**a**) $\tau_c^{\{001\}\langle\bar{1}10\rangle} = \tau_c^{\{111\}\langle\bar{1}10\rangle}$ and (**b**) $\tau_c^{\{001\}\langle\bar{1}10\rangle} = 1.5\tau_c^{\{111\}\langle\bar{1}10\rangle}$.

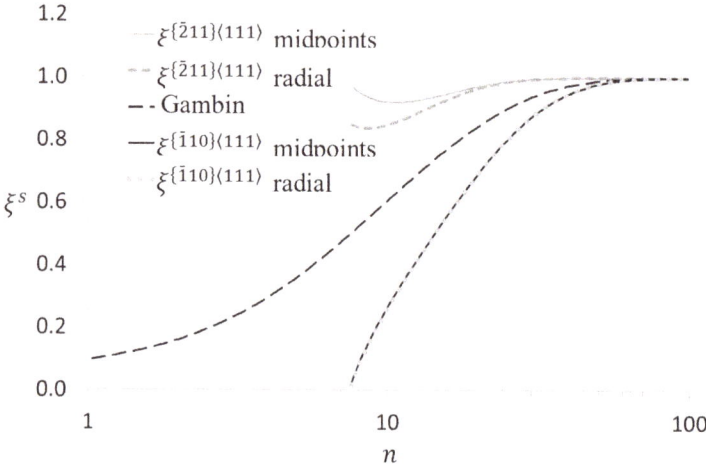

Figure 5. The fitted coefficients of the yield surface as functions of the exponent n, for the case of bcc with $\{\bar{1}10\}\langle111\rangle$ and $\{\bar{2}11\}\langle111\rangle$ slip systems with $\tau_c^{\{\bar{2}11\}\langle111\rangle} = 0.95\tau_c^{\{\bar{1}10\}\langle111\rangle}$, fitted to facet midpoints, by the regularization proposed by Gambin, and radially to the facets.

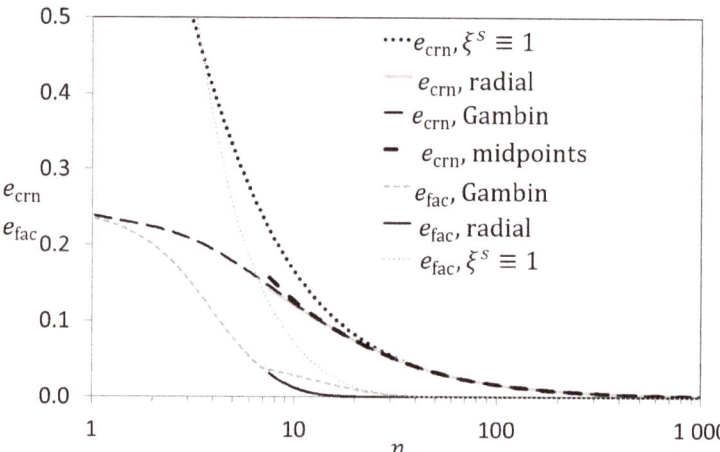

Figure 6. The average error at the corners and at the facet midpoints as a function of n, for the case of bcc with $\{\bar{1}10\}\langle111\rangle$ and $\{\bar{2}11\}\langle111\rangle$ slip systems with $\tau_c^{\{\bar{2}11\}\langle111\rangle} = 0.95\tau_c^{\{\bar{1}10\}\langle111\rangle}$. Yield surface regularizations according to Gambin, fitted to the midpoints, fitted radially or simply by $\xi^s \equiv 1$ are shown.

3.4. Latent Hardening or Twinning

In cases with strain-path changes, the latent hardening of the slip systems may be important. Models for the evolution of individual critical-resolved shear stresses for each slip system are required. Loss of symmetry will split the corners and introduce more corners of the yield surface. By the individual evolutions of the critical resolved shear stresses, the shape of the yield surface is updated at each integration point in time. In order to model complex Bauschinger behavior, as in [18], a distinction is made between forward and backward of the slip system. This requires the use of Equation (5). Similarly, this is important to distinguish twinning from de-twinning behavior described by pseudo slip systems, e.g., [27]. The stress-differential effect will also imply such a distinction, but explanations involving non-Schmid effects will imply a nonassociated yield surface that must be accounted for, see [32].

The simple approach with all coefficients $\xi^s \equiv 1$ is highly efficient when applicable, i.e., at room temperature applying a high exponent, n. Figure 7 shows corner and facet errors for fcc and bcc structures, where the critical resolved shear stress are given a ±5% random variation for cases, where forward and backward slip directions are not distinguished, i.e., with 12 slip systems and correspondingly 12 different ξ^s to be calibrated. Here all ξ^s are individually fitted. The result will be very similar if forward and backward were distinguished, except that twice as many ξ^s would have to be calculated. The calibration quality varies similarly as without this variation. The asymptotic assumption $\xi^s \equiv 1$ applies well for $n > 30$, and the corner errors slowly decay with increasing exponent, n.

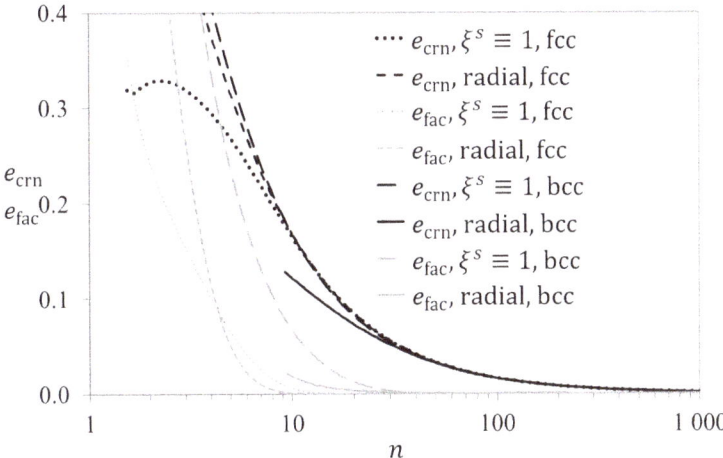

Figure 7. The average error at the corners and at the facet midpoints as a function of n, for the case of bcc with $\{\bar{1}10\}\langle 111\rangle$ and $\{\bar{2}11\}\langle 111\rangle$ slip systems with $\tau_c^{\{\bar{2}11\}\langle 111\rangle} = 0.95 \tau_c^{\{\bar{1}10\}\langle 111\rangle}$ and for the case of fcc with basal slips. In both cases, the critical resolved shear stresses are given an additional ±5% randomly chosen variation of their strengths. Yield surface regularizations fitted radially or simply by $\xi^s \equiv 1$ are shown.

4. Discussion

The results presented here provides an overview of how large the yield-surface exponent n must be for the simplifying assumption $\xi^s = 1$, to be valid. For the case of fcc with primary slip systems, this is $n > 10$, i.e., for instant strain-rate sensitivities smaller than $m \approx 0.1$. Hence, the regularized yield surface can be applied at room temperature and at elevated temperatures up to this limit. In the interpretation of the yield surface exponent in terms of the instant strain-rate sensitivity, the regularized yield surface will provide a more accurate result than the perfectly rate-independent theory, and as a bonus, the Taylor ambiguity vanishes.

However, when nonoctahedral slip systems are included, the number of corners increases, which gives the regularization a greater influence on the resulting yield surface. This must be compensated by decreasing ξ^s, and this must also be done for larger n, in the range between 10 and 50. For aluminum the nonoctahedral slip systems are activated at higher temperatures, for which $n < 10$. The approach taken here is to adjust ξ^s, so that the facet center is touched by the regularized yield surface from the inside. Note, that it is not necessarily the correct choice. Comparison to rate-dependent models should be made, but the results here guides when such an adjustment must be made. Note that adjusting ξ^s is mathematically equivalent to adjusting τ_c^s to $\tau_c^s/(\xi^s)^n$; hence, this adjustment is vital at elevated temperatures.

For the case of bcc, nonbasal slip systems should also be included at room temperature. The results show that the approximation $\xi^s = 1$ holds for n larger than ≈ 50, which is the case for ferritic steels at room temperature and below. Similarly, in cases with latent hardening, either fcc or bcc, the critical resolved shear stresses will vary for each slip system, and there will be many corners; hence n larger than ≈ 50 is also required in these cases for the validity of the approximation $\xi^s = 1$.

The 5-dimensional yield surface corners are sharp, hence the difference between the regularized yield surface and the rate-independent one is largest at the corner solutions of the Schmid yield surface. Furthermore, the number of corners increases faster than the number of facets in higher dimensions, e.g., the fcc $\{\bar{1}10\}\langle 111\rangle$ crystal yield surface has 24 facets and 56 corners. Hence the corner solution will dominate a considerable fraction of the yield surface. Figure 8 shows the first three- and two-dimensional sections of the fcc $\{\bar{1}10\}\langle 111\rangle$ crystal yield surface. Note that what appears as corners and facets in these sections are often projected edges of the surface in the 5-diemensional stress space. The section in Figure 8c is chosen because it contains many true corner solutions, which are marked. The section in Figure 8d contains no corner solutions, but one facet midpoint, as indicated. It can be seen, that at an exponent $n = 100$, which is realistic at room temperature, the regularized solution is not significantly affected by the regularization. At smaller exponents, the entire yield surface shrinks in Figure 8a,c, since these sections are strongly influenced by their many corner solutions. Still, with $\xi^s \equiv 1$, the yield surface approximately goes through the midpoint of the facet in Figure 8b,d, except for the case with $n = 5$ in Figure 8b. At this low exponent, which is realistic at hot deformation, the corners significantly influence and shrink all parts of the yield surface, including the facet midpoints. According to the curve in Figure 1, $\xi^s \equiv 0.86$ would be required to make the facet midpoint touch the Schmid yield surface when $n = 5$.

In the full-constraints Taylor model, for which the velocity-gradient tensor is prescribed, only the corner solutions of the yield surface are realized. Because of stress relaxations due to neighbor-grain interactions that are not accounted for in the Taylor model, some more noncorner solutions may be realized, as seen in recent relaxed-constrains Taylor models e.g., [33,34], similarly as CPFEM with a sufficiently fine mesh [11,35]. However, with any rate-independent model, most of the solutions still will be corner solutions. As pointed out in [10], the corner solutions that are chosen remain the same, as far as the strain rate sensitivity is increased up to less than $m \approx 0.1$ (or corresponding lowering of the exponent as low as $n \approx 10$). Hence the texture change is not sensitive to the choice of m in this range. However, the magnitude of the stress tensor will decrease rapidly with increasing m. In other words, the Taylor factor decreases sensitively with increasing m.

When calibrating the crystal regularized yield surface to the faceted rate-independent one, the error in the corners remains significant even for very large exponents, n. However, this is not really an error. Rather, interpreted in terms of the rate-dependent potential, it is due to the weak strain-rate sensitivity at room temperature, which still has a significant impact and should be included in the models, i.e., by an appropriate choice of the yield-surface exponent. The instant strain-rate sensitivity of a material can be measured by abrupt strain-rate changes; see [36] for a recent discussion. If assuming the viscoplastic power-law behavior, the application of a rounded yield surface is justified when the exponent n matches the instant strain rate sensitivity, i.e., $n = 1 + 1/m$.

In applications, large yield-surface exponents may cause severe numerical challenges, since the associated direction of the symmetric part of the plastic velocity gradient will alternate between the adjacent facets of the considered stress corner during return-mapping iterations. Hence it is unrealistic to apply the regularized yield surface as an approximation to truly rate-independent behavior, without carefully dealing with this numerical challenge first. The approach by Scherzinger [16] shows a way to achieve this by adding a line search to the Newton iterations, ongoing work will be reported soon.

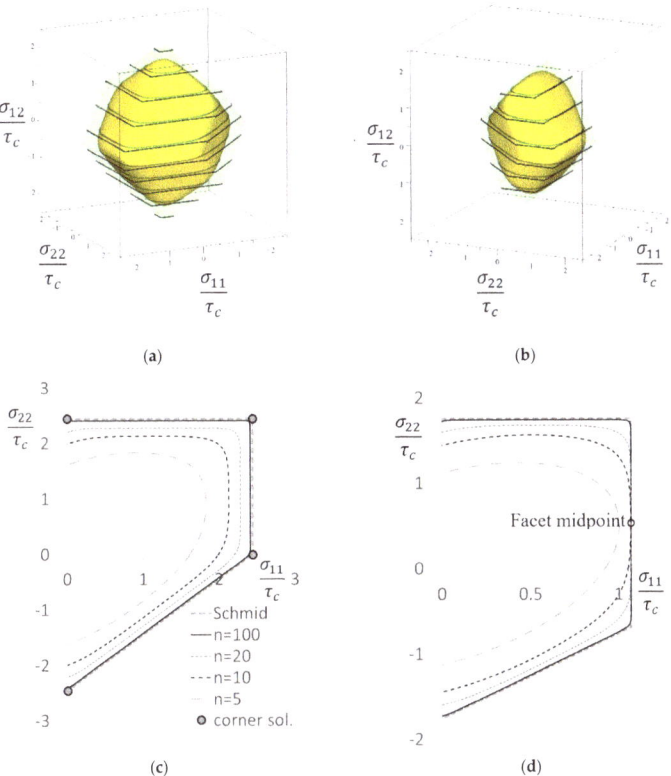

Figure 8. Sections of the fcc $\{\bar{1}10\}\langle111\rangle$ crystal yield surface. (**a**) The σ_{11}-σ_{22}-σ_{12} section with $\sigma_{13} = \sigma_{23} = 0$. (**b**) The σ_{11}-σ_{22}-σ_{12} section with $\sigma_{23} = -0.689\tau_c$ and $\sigma_{13} = 0$. (**c**) The σ_{11}-σ_{22} shear-stress-free section with $\sigma_{33} = \sigma_{12} = \sigma_{13} = \sigma_{23} = 0$. The corner solutions are indicated in this section. (**d**) The section σ_{11}-σ_{22} with $\sigma_{12} = -\sigma_{23} = 0.689\tau_c$ and $\sigma_{13} = 0$. The facet midpoint in this section is indicated. The critical resolved shear stress τ_c is equal for all slip systems and $\xi^s \equiv 1$. In (**a**,**b**), the Schmid surface is compared to $n = 10$. In (**c**,**d**), the indicated yield surface exponents n are compared to the Schmid solution.

Deformation twinning by pseudo slip systems or the stress-differential effect are examples of mechanisms that distort the yield surface. In the continuum-plasticity theory, Bauschinger effects are commonly handled by a back stress, even in complex models accounting for strain-path changes [37], but recently, significant progress has been reported on continuum models involving yield-surface distortions [21–26]. Similarly, crystal-plasticity models with latent hardening predict a shape change of the crystal yield surface during deformation [17,18,20]. The origin of the Bauschinger effect can be a combination of slip-system dependent mechanisms, e.g., dislocation pile-ups, that distorts the yield surface, and composite effects, e.g., strong, large elastic particles, that shifts the yield surface by kinematic hardening. In the proposed framework, both types can be handled simultaneously at the crystal scale, where the yield-surface distortions occur by the critical resolved shear stress changes during simulations. At room temperature, one may apply $\xi^s \equiv 1$, as the exponent n will be larger than 50, i.e., the instant strain-rate sensitivity for most metals at room temperature is $m > 0.02$. The advanced continuum-plasticity models [21–26] can be calibrated based on virtual experiments, performed using models for distortions of the yield surface by virtual experiments calculated by physical-based crystal-plasticity models [17,18,20], which may complement rather demanding mechanical experiments.

5. Conclusions

Mathematical formulations of yield surfaces for metal crystals have been analyzed. Three calibration methods have been applied for fitting regularized yield surfaces to rate-independent faceted ones. Yield-surface calibrations for the fcc crystal, with or without nonoctahedral slip systems, and for the commonly considered 24 slip systems for bcc steels have been analyzed. An exact match of the midpoints of the facets can be obtained, provided the exponent n is sufficiently high. A good match for the Schmid stress corners, can only be obtained by applying a very high yield-surface exponent. However, the deviation at the corners at lower exponents can be justified by the interpretation of the yield surface exponent in terms of the instant strain-rate sensitivity. This makes the use of a regularized yield surface a precise, flexible method at room temperature without having to deal with the Taylor ambiguity. A return-mapping algorithm that can handle large yield-surface exponents is expected to be available soon, which will make the use of a regularized yield surface a flexible, robust and efficient candidate for CPFEM implementations, enabling the use of realistic physical-based models for the critical resolved shear stresses, allowing a distinction between thermal and a-thermal contributions, and optionally accounting for latent hardening and/or twinning.

Funding: This research received no external funding.

Conflicts of Interest: The authors declare no conflict of interest.

References

1. Peirce, D.; Asaro, R.; Needleman, A. An analysis of nonuniform and localized deformation in ductile single crystals. *Acta Met.* **1982**, *30*, 1087–1119. [CrossRef]
2. Hill, R. Generalized constitutive relations for incremental deformation of metal crystals by multislip. *J. Mech. Phys. Solids* **1966**, *14*, 95–102. [CrossRef]
3. Mandel, J. Generalisation de la theorie de plasticite de W. T. Koiter. *Int. J. Solids Struct.* **1965**, *1*, 273–295. [CrossRef]
4. Arminjon, M. A Regular Form of the Schmid Law. Application to the Ambiguity Problem. *Textures Microstruct.* **1991**, *14*, 1121–1128. [CrossRef]
5. Gambin, W. Refined analysis of elastic-plastic crystals. *Int. J. Solids Struct.* **1992**, *29*, 2013–2021. [CrossRef]
6. Gambin, W.; Barlat, F. Modeling of deformation texture development based on rate independent crystal plasticity. *Int. J. Plast.* **1997**, *13*, 75–85. [CrossRef]
7. Hutchinson, J.W. Bounds and self-consistent estimates for creep of polycrystalline materials. *Proc. R. Soc. Lond. Ser. A Math. Phys. Sci.* **1976**, *348*, 101–127. [CrossRef]
8. Diak, B.; Upadhyaya, K.; Saimoto, S. Characterization of thermodynamic response by materials testing. *Prog. Mater. Sci.* **1998**, *43*, 223–363. [CrossRef]
9. Tóth, L.S.; Gilormini, P.; Jonas, J. Effect of rate sensitivity on the stability of torsion textures. *Acta Met.* **1988**, *36*, 3077–3091. [CrossRef]
10. Manik, T.; Holmedal, B. Review of the Taylor ambiguity and the relationship between rate-independent and rate-dependent full-constraints Taylor models. *Int. J. Plast.* **2014**, *55*, 152–181. [CrossRef]
11. Zhang, K.; Holmedal, B.; Mánik, T.; Saai, A. Assessment of advanced Taylor models, the Taylor factor and yield-surface exponent for FCC metals. *Int. J. Plast.* **2019**, *114*, 144–160. [CrossRef]
12. Yoshida, K.; Kuroda, M. Comparison of bifurcation and imperfection analyses of localized necking in rate-independent polycrystalline sheets. *Int. J. Solids Struct.* **2012**, *49*, 2073–2084. [CrossRef]
13. Yoshida, K.; Brenner, R.; Bacroix, B.; Bouvier, S. Effect of regularization of Schmid law on self-consistent estimates for rate-independent plasticity of polycrystals. *Eur. J. Mech. A/Solids* **2009**, *28*, 905–915. [CrossRef]
14. Anand, L.; Kothari, M. A computational procedure for rate-independent crystal plasticity. *J. Mech. Phys. Solids* **1996**, *44*, 525–558. [CrossRef]
15. Paux, J.; Ben Bettaieb, M.; Badreddine, H.; Abed-Meraim, F.; Labergere, C.; Saanouni, K. An elasto-plastic self-consistent model for damaged polycrystalline materials: Theoretical formulation and numerical implementation. *Comput. Methods Appl. Mech. Eng.* **2020**, *368*, 113138. [CrossRef]

16. Scherzinger, W. A return mapping algorithm for isotropic and anisotropic plasticity models using a line search method. *Comput. Methods Appl. Mech. Eng.* **2017**, *317*, 526–553. [CrossRef]
17. Beyerlein, I.J.; Tomé, C.N. Modeling transients in the mechanical response of copper due to strain path changes. *Int. J. Plast.* **2007**, *23*, 640–664. [CrossRef]
18. Holmedal, B.; Van Houtte, P.; An, Y. A crystal plasticity model for strain-path changes in metals. *Int. J. Plast.* **2008**, *24*, 1360–1379. [CrossRef]
19. Kitayama, K.; Tomé, C.; Rauch, E.; Gracio, J.; Barlat, F. A crystallographic dislocation model for describing hardening of polycrystals during strain path changes. Application to low carbon steels. *Int. J. Plast.* **2013**, *46*, 54–69. [CrossRef]
20. Peeters, B.; Kalidindi, S.R.; Van Houtte, P.; Aernoudt, E. A crystal plasticity based work-hardening/softening model for b.c.c. metals under changing strain paths. *Acta Mater.* **2000**, *48*, 2123–2133. [CrossRef]
21. Barlat, F.; Gracio, J.J.; Lee, M.-G.; Rauch, E.F.; Vincze, G. An alternative to kinematic hardening in classical plasticity. *Int. J. Plast.* **2011**, *27*, 1309–1327. [CrossRef]
22. Ha, J.; Kim, J.-H.; Barlat, F.; Lee, M.-G. Continuous strain path change simulations for sheet metal. *Comput. Mater. Sci.* **2014**, *82*, 286–292. [CrossRef]
23. Holmedal, B. Bauschinger effect modelled by yield surface distortions. *Int. J. Plast.* **2019**, *123*, 86–100. [CrossRef]
24. Qin, J.; Holmedal, B.; Hopperstad, O. A combined isotropic, kinematic and distortional hardening model for aluminum and steels under complex strain-path changes. *Int. J. Plast.* **2018**, *101*, 156–169. [CrossRef]
25. Qin, J.; Holmedal, B.; Hopperstad, O. Experimental characterization and modeling of aluminum alloy AA3103 for complex single and double strain-path changes. *Int. J. Plast.* **2019**, *112*, 158–171. [CrossRef]
26. Qin, J.; Holmedal, B.; Zhang, K.; Hopperstad, O.S. Modeling strain-path changes in aluminum and steel. *Int. J. Solids Struct.* **2017**, *117*, 123–136. [CrossRef]
27. Kowalczyk-Gajewska, K. Modelling of texture evolution in metals accounting for lattice reorientation due to twinning. *Eur. J. Mech. A/Solids* **2010**, *29*, 28–41. [CrossRef]
28. Kocks, U.; Mecking, H. Physics and phenomenology of strain hardening: The FCC case. *Prog. Mater. Sci.* **2003**, *48*, 171–273. [CrossRef]
29. Holmedal, B. On the formulation of the mechanical threshold stress model. *Acta Mater.* **2007**, *55*, 2739–2746. [CrossRef]
30. Van Houtte, P.; Yerra, S.K.; Van Bael, A. The Facet method: A hierarchical multilevel modelling scheme for anisotropic convex plastic potentials. *Int. J. Plast.* **2009**, *25*, 332–360. [CrossRef]
31. Van Houtte, P.; Delannay, L.; Kalidindi, S.R. Comparison of two grain interaction models for polycrystal plasticity and deformation texture prediction. *Int. J. Plast.* **2002**, *18*, 359–377. [CrossRef]
32. Kuroda, M.; Kuwabara, T. Shear–band development in polycrystalline metal with strength–differential effect and plastic volume expansion. *Proc. R. Soc. A Math. Phys. Eng. Sci.* **2002**, *458*, 2243–2259. [CrossRef]
33. Manik, T.; Holmedal, B. Additional relaxations in the Alamel texture model. *Mater. Sci. Eng. A* **2013**, *580*, 349–354. [CrossRef]
34. Van Houtte, P. Deformation texture prediction: From the Taylor model to the advanced Lamel model. *Int. J. Plast.* **2005**, *21*, 589–624. [CrossRef]
35. Zhang, K.; Holmedal, B.; Hopperstad, O.; Dumoulin, S.; Gawad, J.; Van Bael, A.; Van Houtte, P. Multi-level modelling of mechanical anisotropy of commercial pure aluminium plate: Crystal plasticity models, advanced yield functions and parameter identification. *Int. J. Plast.* **2015**, *66*, 3–30. [CrossRef]
36. Manik, T.; Holmedal, B. On the criterion for compensation to avoid elastic–plastic transients during strain rate change tests. *Acta Mater.* **2013**, *61*, 653–659. [CrossRef]
37. Mánik, T.; Holmedal, B.; Hopperstad, O.S. Strain-path change induced transients in flow stress, work hardening and r-values in aluminum. *Int. J. Plast.* **2015**, *69*, 1–20. [CrossRef]

Publisher's Note: MDPI stays neutral with regard to jurisdictional claims in published maps and institutional affiliations.

© 2020 by the author. Licensee MDPI, Basel, Switzerland. This article is an open access article distributed under the terms and conditions of the Creative Commons Attribution (CC BY) license (http://creativecommons.org/licenses/by/4.0/).

Article

Mathematical Modeling of Plastic Deformation of a Tube from Dispersion-Hardened Aluminum Alloy in an Inhomogeneous Temperature Field

Oleg Matvienko [1,2,*], Olga Daneyko [1,2] and Tatiana Kovalevskaya [1,2]

1. Tomsk State University of Architecture and Building, 634003 Tomsk, Russia; olya_dan@mail.ru (O.D.); takov47@mail.ru (T.K.)
2. National Research Tomsk State University, 634050 Tomsk, Russia
* Correspondence: matvolegv@mail.ru; Tel.: +7-983-2382-150

Received: 20 October 2020; Accepted: 30 November 2020; Published: 2 December 2020

Abstract: The effect of temperature distribution on a stress–strain state tube made of disperse-hardened aluminum alloy subjected to internal pressure was investigated. The mathematical model is based on equations of physical plasticity theory and principles of mechanics of deformable solids. The results of this investigation demonstrate that varying the outer wall temperature in the range of 200 K at a fixed temperature of the inner wall leads to a significant change in the plastic resistance limit (for the considered tube sizes, this change is approximately 15%). An increase of the tube wall temperature reduces the resistance to plastic deformation. For the same absolute temperature difference between the outer and inner walls, the plastic resistance limit is less for the higher temperature of the inner wall of the tube. A decrease of the distances between the hardening particles at the same volume fraction of second phase leads to a significant increase in the pressure required to achieve plastic deformation of the tube walls. An increase in tube wall temperature reduces the resistance to plastic deformation. For the same absolute temperature difference between the outer and inner walls, the plastic resistance limit is lower for the higher temperature of the inner tube wall. The decrease of the distance between the hardening particles at the same volume fraction of the second phase leads to a significant increase in the pressure required to achieve plastic deformation of the tube walls.

Keywords: deformation; dislocation; dispersion hardening; nanoparticles; stress; strain; plastic deformation; tube; heat exchange; temperature

1. Introduction

A heat exchanger is a device designed to transfer heat from one medium to another in the most efficient way [1]. Heat exchangers, consisting of tube bundles, are widely used in various industries. Heat exchangers are extensively used in power engineering, automotive industry, aviation and space industry and also in household applications (air condition, space heating and refrigeration) [2].

In [3], the mechanical properties of a heat exchange tube made of hardening materials have been comprehensively investigated. In [4,5] von Mises yield criteria and Hencky deformation theory were applied to determine the stress state of the heat exchange tube. In order to simplify the dependence on the stresses from strains, in [5] a bilinear model has been proposed. The simplified bilinear model of the stress–strain curve consists of two straight lines with different slopes. The slope of the first line characterizes the elastic modulus of the material. The slope of the second line reflects the hardening properties of the material.

Numerous studies on the elastic–plastic behavior of pipes subjected to internal pressure have led to development of theories for predicting the tubes' burst failure [6–8].

Comparison of the stress state of the steel pipe under the action of constant pressure applied (separately and simultaneously) to the outer and inner tube walls was performed in [9]. The action of pressure on the outer cylindrical surface of the pipe leads to maximum stress of material. The outer and inner layers of the pipe material are subjected to 85% compression strain when two loads are simultaneously applied.

The authors of [10] presented two stress–strain models based on the theory of shells to describe the stress–strain relationship in the cross-section of a coiled tube. Membrane theory of shells of revolution applied to an elliptic torus has been proven to be a good approach for the axial strain description but inaccurate with regard to circumferential strain. In order to assess the bending effect and correct circumferential strains, a semi-empiric method has been proposed to determine an empiric law relating to the ellipse focal distance with pipe inner pressure.

Zhu and Leis investigated the plastic flow and the elastic–plastic deformation of tubes, which is based on the maximum shear stresses principle [11,12].

In [13], the authors presented evaluation of the stress–strain state and areas of stress concentration accounting for detection of different loads. The finite element model is presented to determine the pipeline sections in the pre-emergency state. Strength analysis of the pipeline showed that buckling or sagging of its sections leads to unallowable stresses. Large pipeline sagging induces plastic deformation.

In [14], plastic deformation induced by unsteady pressure gradients in fluid-filled pipes was investigated. The general expression for waves induced by plastic deformation is derived.

The effect of internal pressure on radial strain of a steel pipe, subjected to monotonic and cyclic loading, was analyzed in [15]. The behavior of circular long tubes subjected to external pressure and axial load under plane strain in conjunction with the constitutive equation taking into account corner formation on the yield surface and the Bauschinger effect are analyzed in [16].

It should also be noted that the main trends of engineering development are use of new materials that significantly increase the specific power of units. Mechanical properties of materials may be improved by dispersion hardening [17,18]. Dispersion-hardened alloys contain fine, submicron and nanoscale particles of another material distributed in the matrix. In such alloys, the matrix assumes most of the load. Due to the large number of insoluble particles in the matrix, a structure resistant to plastic deformation is formed [19].

The dispersed hardening particles of the strengthening phase resist dislocation motion during material loading. Consequently, the strength of material depends on the dislocation structure that forms during plastic deformation. The basic principles of physical theory of plasticity and strain hardening were formulated by Orowan [20], Ashby [21,22], Hirsch [23–25] and Humphreys [26–28]. In [29–31], based on physical plasticity theory, a mathematical model was developed, and investigations of elastoplastic deformation of the tube made from disperse-hardened alloys were carried out.

The effect of internal and external pressure applied to the tube from dispersion-hardened aluminum alloy was investigated in [32–34]. The results of the investigation demonstrate that hardening of the alloy by nanoparticles significantly improves the strength characteristics of the material.

Understanding the heat transfer condition effect on the stress–strain state of the heat exchanger walls is important because the strength properties are dependent on temperature. Thus, it is necessary to evaluate the effect of the temperature field on the stress–strain state of heat exchangers to improve their efficiency and reliability.

The present work is devoted to modeling of plastic deformation of a tube subjected to uniform internal pressure in the inhomogeneous temperature field. It is assumed that the tube is made of an Al-based alloy with incoherent spherical nondeformable nanoparticles.

The mathematical model is based on equations of the balance of the defect structure [35–37] and principles of mechanics of deformable solids [38–40]. In this paper, to study the stress state of the tube made from aluminum alloy strengthened by nanoparticles, numerical simulation was used. This is because the investigated problem is nonlinear as the tensile strength, yield strength and Young's modulus depend on temperature.

2. Methods

2.1. Mathematical Model of Plastic Deformations

The strength and plasticity of composite materials are highly dependent on the structural condition of the materials. Therefore, to predict mechanical properties of materials it is necessary to take into account evolution of the defect structure of materials, namely, generation of defects during plastic deformation, their mutual transformation and annihilation. Kovalevskaya et al. proposed a mathematical model describing plastic deformation of dispersion-strengthened metals [41]. This model is based on concepts of hardening and rest. Physical mechanisms underlying the mathematical model of plastic deformation of disperse-hardened alloys with incoherent particles are described in [42,43].

Presence of a disperse-hardening phase in the material makes the modeling object much more complex than single-phase materials. In the process of plastic deformation, interaction of dislocations with particles leads to formation of new elements of the dislocation structure, which leads to hardening of the material [43,44].

The following types of defects are formed during plastic deformation in FCC alloys with incoherent nanoparticles: shear-forming dislocations with density (ρ_m), prismatic dislocation loops of vacancy (ρ_p^v) and interstitial types (ρ_p^i), dislocation dipoles of vacancy (ρ_d^v) and interstitial types (ρ_d^i), interstitial atoms with concentration (c_i), monovacancies (c_v) and bivacancies (c_{2v}). The start of formation of dipole structures is determined by the achievement of a critical dislocation density [43], the value of which depends on the scale characteristics of the hardening phase [42,43].

The mathematical model includes balance equations for deformation due to line and point defects with regard to generation and annihilation of all types of defects. Generation and annihilation of prismatic dislocation loops occur near incoherent strengthening particles. The model takes into account the deposition of point defects on dislocations, which can lead to both annihilation of dislocations of various types and an increase in their density. The possibility of transition of dislocations in prismatic loops and in dipole configurations to shear dislocations is also taken into account.

The model assumes that the matrix material of the dispersion-hardened alloy is pure aluminum, particles of the hardening phase are nondeformable, and the distance between them does not change during plastic deformation at all temperatures.

The mathematical model uses the following balance equations of the defect dislocation structure [36,45]:

$$\frac{d\rho_m}{da} = (1 - w_s P_{as})\frac{F}{Db} - \frac{2b}{a}(1 - w_s)\rho_m^2 \min(r_a, \rho_m^{-1/2})(c_{2v}Q_{2v} + c_{1v}Q_{1v} + c_i Q_i) + \frac{2b}{a}\alpha\sqrt{\rho}(\rho_p^v(c_{1v}Q_{1v} + c_{2v}Q_{2v}) + \rho_p^i c_i Q_i) + \frac{2b}{ar_a}(\rho_d^i c_i Q_i + \rho_d^v(c_{1v}Q_{1v} + c_{2v}Q_{2v})), \quad (1)$$

$$\frac{d\rho_p^i}{da} = \frac{<\chi>\delta}{2\Lambda_p^2 b} - \frac{2\alpha}{a}\sqrt{\rho}\rho_p^i b(2c_{2v}Q_{2v} + c_i Q_i + 2c_{1v}Q_{1v}), \quad (2)$$

$$\frac{d\rho_p^v}{da} = \frac{<\chi>\delta}{2\Lambda_p^2 b} - \frac{2\alpha}{a}\sqrt{\rho}\rho_p^v b(c_{2v}Q_{2v} + 2c_i Q_i + c_{1v}Q_{1v}), \quad (3)$$

$$\frac{d\rho_d^v}{da} = \frac{1}{\Lambda_p b} - \frac{2b}{ar_a}\rho_d^v(c_{2v}Q_{2v} + c_i Q_i + c_{1v}Q_{1v}) \quad (4)$$

$$\frac{d\rho_d^i}{da} = \frac{1}{\Lambda_p b} - \frac{2b}{ar_a}\rho_d^i(c_{2v}Q_{2v} + c_i Q_i + c_{1v}Q_{1v}), \quad (5)$$

$$\frac{dc_i}{da} = q\frac{\tau_{dyn}}{G} - \frac{c_i}{\dot{a}}[((1 - w_s)\rho_m + \rho_p + \rho_d)b^2 Q_i + Q_{1v}c_{1v} + Q_{2v}c_{2v} + Q_i(c_{1v} + c_{2v})], \quad (6)$$

$$\frac{dc_{1v}}{da} = \frac{q\tau_{dyn}}{6G} - \frac{1}{\dot{a}}[(((1 - w_s)\rho_m + \rho_p + \rho_d)b^2 + c_i + c_{1v})Q_{1v}c_{1v} + Q_i c_i c_{1v} - (Q_{2v} + Q_i)c_i c_{2v}], \quad (7)$$

$$\frac{dc_{2v}}{da} = \frac{5q\tau_{dyn}}{6G} - \frac{1}{\dot{a}}[(((1-w_s)\rho_m + \rho_p + \rho_d)b^2 + c_i)Q_{2v}c_{2v} + Q_ic_ic_{2v} - Q_{1v}c_{1v}^2], \quad (8)$$

$$\dot{a} = \frac{8}{\pi} \frac{\tau^3(((1-\beta_r)\rho_m + \rho_p + \rho_d)(\tau - \tau_a))^{1/3}}{G^{4/3}b^{1/3}(\tau^2 - G^2b^2\xi\beta_r\rho_m)^{1/2}} \times \\ \times \frac{v_D B\beta_r^{1/2}}{\xi^{1/6}F(1-\beta_r)} \exp\left[-\frac{0.2Gb^3 - (\tau - \tau_a)\Lambda b^2}{kT}\right]. \quad (9)$$

Here, a is the shear deformation; $\dot{a}$ is the strain rate; b is the Burgers vector module; F is a dimensionless geometric parameter which characterizes the shape of the shear zone and connects its diameter, perimeter and area; D is the diameter of the shear zone; P_{as} is the probable annihilation of screw dislocations; τ_{dyn} is the excess stress over static resistance to the dislocation movement; $Q_j = Z_j v_D \exp(-U_j^{(m)}/kT)$ is the kinetic coefficient; $U_j^{(m)}$ is the activation energy of the j-th type point defect migration; Z_j^i is the number of places for the j-th type point defect jump ($j = i, v$); v_D is the Debye frequency; k is the Boltzmann constant; T is the temperature of deformation; ω_s is the fraction of screw dislocations; $<\chi>$ is the ratio of the average length of dislocations accumulated on the particles to the particle size; Λ_p is the distance between hardening particles; δ is the particle diameter; q is the intensity of point defect generation; G is the shear modulus; ρ is the dislocations density; α is the coefficient of the interaction between dislocations; ξ is the forest dislocation fraction; $\rho_p = \rho_p^i + \rho_p^v$ is the density of prismatic dislocation loops; $\rho_d = \rho_d^i + \rho_d^v$ is the dislocation density in dipole configurations; β_r is the reacting dislocation fraction; Λ is the average length of free dislocation segment and r_a is the effective capture radius:

$$r_a = \frac{Gb}{4\pi\tau_f} \frac{(2-\nu)}{(1-\nu)}$$

Here, τ_f is the friction stress; ν is the Poisson's ratio.

The athermal resistance (τ_a) to the dislocation movement in the disperse-hardened alloy with incoherent particles is the sum of the friction stress, τ_f, the interaction between forest dislocations, τ_d, and the stress of particle bypass, τ_{Or}, i.e., $\tau_a = \tau_f + \tau_d + \tau_{Or}$.

The intensity of the processes of annihilation of linear defects is determined by the strain rate, which is related to the applied stress and dislocation density by relation, Equation (9). The investigation was conducted for deformation with a constant strain rate. Therefore, Equation (9) is not differential. This is a transcendental equation that allows us to find the value of the applied stress.

It is necessary to set the initial values of the point defect concentrations and dislocation densities for solution of the system of ordinary differential equations that describe the balance of linear and point deformation defects. The former corresponds to a concentration of thermodynamically balanced point defects at a given temperature, whereas the latter corresponds to the unstrained state of the crystal. Under the condition $a = 0$, it is assumed that there are no dislocation prismatic loops and dipole configurations in the crystal, i.e., $c_i^{(0)} = \exp(-U_i^f/kT)$, $c_v^{(0)} = \exp(-U_v^f/kT)$, $c_{2v}^{(0)} = \exp(-U_{2v}^f/kT)$, $\rho_m^{(0)} = 10^{12}$ m^{-2} and $\rho_p^{(0)} = \rho_d^{(0)} = 0$.

The calculations are carried out at the following parameter values for disperse-hardened Al-based alloys [43,45]: $F = 4$, $<\chi> = 4$, $\alpha = 0.5$, $\beta_r = 0.14$, $\xi = 0.5$, $\omega_s = 0.3$, $\tau_f = 10$ MPa, $U_v^f = 1.26$ eV, $U_{2v}^f = 2.16$ eV, $U_i^f = 3.28$ eV, $U_v^m = 0.88$ eV, $U_{2v}^m = 0.69$ eV, $U_i^m = 0.117$ eV and $b = 2.5 \cdot 10^{-10}$ m.

2.2. Elastoplastic Material Properties

Aluminum alloys are known for their good specific strength and corrosion resistance. Dispersion-hardened aluminum alloys consist of a coarse-grained aluminum matrix containing populations of particle Al$_2$O$_3$. Incoherent dispersoids distributed within a metallic matrix provide high strength at ambient and elevated temperatures, as they impede dislocation glide. Use of coarsening-resistant submicron dispersoids such as Al$_2$O$_3$ allows for dispersion-strengthened aluminum with creep resistance at high temperatures (500 °C and above).

Experimental studies [16] show that hardening of aluminum alloys with dispersed particles weakly affects the moduli of elasticity and shear. However, alloy properties depend on temperature. To describe the temperature dependence of the shear modulus, Bell's formula can be used [46–48]:

$$G = \begin{cases} G_0 & \text{at} \quad T < 0.06 T_m \\ G_1\left(1 - \frac{T}{2T_m}\right) & \text{at} \quad 0.06 T_m < T < 0.57 T_m. \end{cases} \quad (10)$$

In Equation (10), $T_m = 933$ K is the melting temperature; $G_0 = 35.017$ GPa and $G_1 = 36.1$ GPa are parameters that characterize the elastic properties of aluminum.

The results of an investigation based on the solution of Equations (1)–(9) show that strain hardening of Al-based materials with an incoherent strengthening phase for the same volume fraction increases with decreasing particle size and the distance between them at all deformation temperatures. An investigation of the effect of the scale characteristics of the strengthening phase on the strain hardening of aluminum-based materials with an incoherent strengthening phase showed that dislocation dipoles are not formed throughout the entire process of plastic deformation in a material with nanosized particles. There is a decrease of shear-forming dislocation density and dislocation density in the prismatic loops (both vacancy and interstitial type) with an increase of deformation temperature (Figure 1).

Figure 2 presents dependences between the flow stresses, τ, and plastic deformation, $a_{pl} = a - \tau_0/G$, of the aluminum-based alloy strengthened with incoherent nanoparticles Al_2O_3.

Plastic deformation starts when the stress intensity in the material is equal to the yield shear stress. Hardening curves are characterized by a monotonic dependence between flow stress and deformation (Figure 2). At low values of plastic deformation, a_{pl}, a significant increase of flow stress values occur. At high a_{pl} values, the stress–strain curve has a horizontal asymptote matching the yield point at $\tau = \tau_\infty$. The simulation results predict that hardening of the material by nanoparticles significantly changes the strength characteristics of the material. A decrease of the distance between particles for the same volume fraction of the hardening phase at the same temperature leads to increasing of the flow stress (see curves 1 and 4, 2 and 5, and 3 and 6), which means hardening of the material. With increasing temperature, the material becomes more plastic, which is accompanied by decreasing of the flow stress (see curves 1–3 and 4–6).

Approximation of the obtained balance between the elements of deformation defects and dependence of the flow stress on the deformation degree allows us to obtain the function of $\tau(a)$ with an error not exceeding 0.1%:

$$\tau = \tau_0 + \tau_1 \frac{a - \tau_0/G}{a_* + a}, \quad (11)$$

where τ_0 is the yield stress, $\tau_1 = \tau_\infty - \tau_0$ is the hardening stress, which characterizes the maximum increase of flow stress during plastic deformation, and a_* is an empirical parameter that determines the rate at which the flow curve reaches the asymptote.

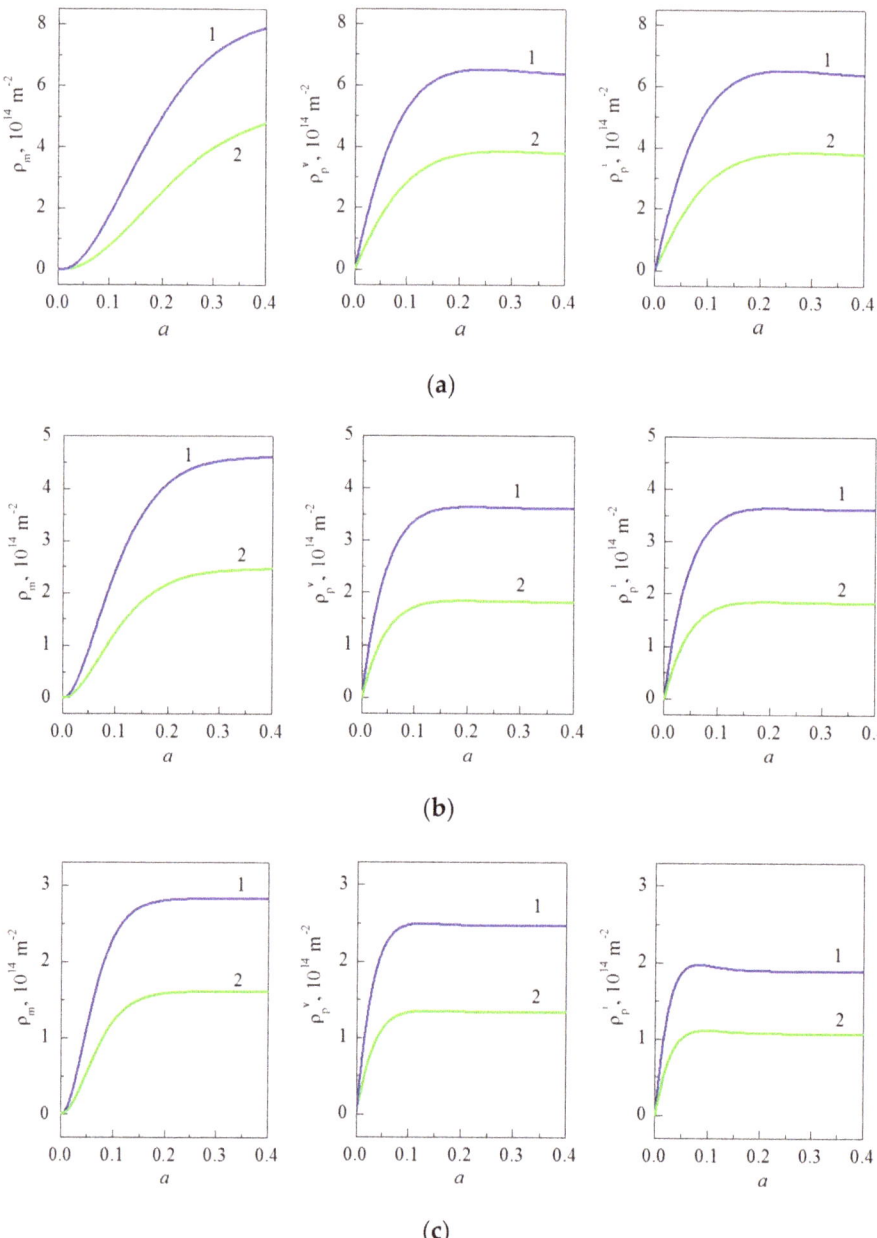

Figure 1. Dependencies of the density of shear-forming dislocations, dislocation density in the prismatic loops of vacancy and interstitial types from deformation. Diameter of particles (δ), nm: 1–10, 2–20; distance between particles (Λ_p), nm: 1–100, 2–200; (**a**) temperature: 293 K, (**b**) temperature: 393 K and (**c**) temperature: 493 K.

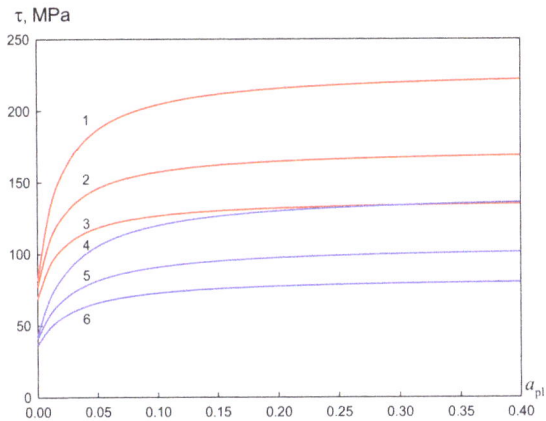

Figure 2. Stress–strain curves for an aluminum-based alloy at various temperatures. Red curves (1–3) correspond to the diameter of strengthening particles, $\delta = 10$ nm, and distances between the particles, $\Lambda_p = 100$ nm; blue curves (4–6) correspond to the diameter of strengthening particles, $\delta = 20$ nm, and distances between the particles, $\Lambda_p = 200$ nm. Temperature of deformation on the curves 1 and 4 is $T = 293$ K, on curves 2 and 5 it is $T = 393$ K and on curves 3 and 6 it is $T = 493$ K.

Figure 3 shows the temperature dependence of the yield stress, τ_0, and hardening stress, τ_1, calculated for various parameters of the strengthening phase. The simulation results show that with increasing temperature the yield stress, τ_0, decreases (see curves 1 and 2). This means that with increasing temperature, the alloy becomes more plastic because plastic deformation occurs at lower stresses. In addition, a decrease of the hardening stress, τ_1, with increasing temperature (see curves 3 and 4) indicates a weakening of the ability of the material to plastic hardening. An increase in the distance between the strengthening particles at the same volume fraction leads to a decrease of the yield stress, τ_0 (see curves 1 and 2), and hardening stress, τ_1 (see curves 3 and 4).

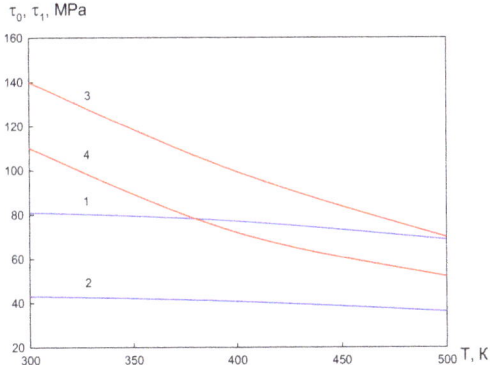

Figure 3. Temperature dependencies of yield stress, τ_0 (blue curves 1 and 2), and hardening stress, τ_1 (red curves 3 and 4). Particles diameter, δ, and distance between particles, Λ_p, on curves 1 and 3 are, respectively, 10 and 100 nm, and on curve 2 they are 20 and 200 nm.

In the considered temperature range ($293K \leq T \leq 493K$), the dependences τ_0 and τ_1 can be approximated by the correlations:

$$\frac{\tau_0}{\tau_{0*}} = C_{00} + \frac{C_{01}}{T_0}T + \frac{C_{02}}{T_0^2}T^2, \qquad (12)$$

$$\frac{\tau_1}{\tau_{1*}} = C_{10} + \frac{C_{11}}{T_0}T + \frac{C_{12}}{T_0^2}T^2 \qquad (13)$$

In Equations (12) and (13) τ_{0*} and τ_{1*} are the yield stress and hardening stress at temperature, $T_0 = 293$ K. The approximation parameters have the following values: $C_{00} = 0.8665$, $C_{01} = 0.3479$, $C_{02} = -0.2144$, $C_{10} = 2.9817$, $C_{11} = -2.7000$ and $C_{12} = 0.7183$.

The values of the material constants: τ_{0*}, τ_{1*} and a_* for various sizes, δ, of hardening particles and the distances between the particles, Λ_p, are presented in Table 1.

Table 1. The material constants τ_{0*}, τ_{1*} and a_* (1).

Hardening Phase Parameters	Material Constants
$\Lambda_p = 100$ nm $\delta = 10$ nm	$\tau_0 = 81.08$ MPa $\tau_1 = 143.1$ MPa $a_* = 0.011$
$\Lambda_p = 200$ nm $\delta = 20$ nm	$\tau_0 = 43.13$ MPa $\tau_1 = 110.13$ MPa $a_* = 0.013$

2.3. Verification of the Results

Verification of the results was carried out by comparison with experimental data presented in previous work [49,50]. Table 2 presents the mechanical properties of the aluminum alloy A356 containing scandium fluoride particles.

Table 2. Properties of A356-based composites containing scandium fluoride particles.

Alloy	τ_0 [MPa]		τ_1 [MPa]	
	Experiment [50]	Results of Modeling	Experiment [50]	Results of Modeling
A356–0.2% ScF3	98 ± 6	102	92 ± 11	87
A356–1% ScF3	109 ± 8	114	141 ± 9	134

Table 2 shows both experimental results, which are presented in [50], and theoretical predictions of the yield stress, τ_0, and the hardening stress, τ_1. By comparing the experimental and theoretical values, one can see that, in general, the results of the predictions are fairly close to the experimental data. The good correlation between the experimental measurements and simulations results proved the correct methods and approaches for the simulation of processes of plastic deformation.

2.4. Temperature Distribution in the Tube Wall

To model the heat transfer, we will assume that the inner wall of the tube ($r = R_{in}$) has a constant temperature, $T = T_{in}$, and the external wall ($r = R_{ex}$), $T = T_{ex}$. Note that if the end walls of the tube are thermally insulated, then the temperature distribution will not depend on the axial coordinate z. In addition, due to the axial symmetry of the problem under consideration, the temperature will not depend on the angular coordinate φ. Thus, the heat transfer equation can be written as:

$$\frac{1}{r}\frac{\partial}{\partial r}\left(r\frac{\partial T}{\partial r}\right) = 0. \qquad (14)$$

The boundary conditions can be formulated as:

$$r = R_{in}: \quad T = T_{in}; \quad r = R_{ex}: \quad T = T_{ex} \qquad (15)$$

Integration of Equation (14) with boundary conditions, Equation (15), allows us to determine the dependence of temperature on the radial coordinate

$$T = T_{in} + (T_{ex} - T_{in}) \frac{\ln r - \ln R_{in}}{\ln R_{ex} - \ln R_{in}} \qquad (16)$$

Figure 4 shows the temperature distribution over the tube wall thickness for various values of T_{in} and T_{ex}. As can be seen from the figures, this distribution is linear and can be approximated by the dependence:

$$T = T_{in} + \frac{\Delta T}{h}(r - R_{in}), \qquad (17)$$

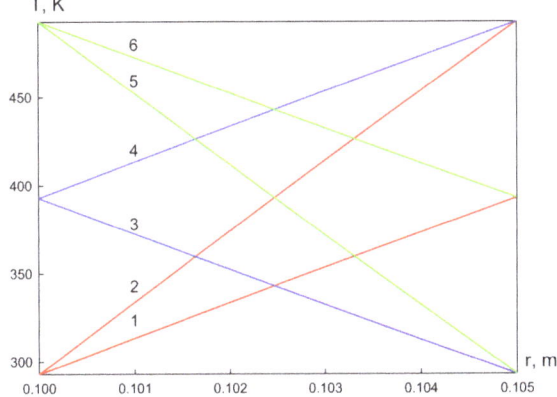

Figure 4. Temperature distribution in the tube wall: radius of the inner wall of the pipe $R_{in} = 0.1$ m, radius of the outer wall of the pipe $R_{ex} = 0.105$ m. The red curves (1 and 2) correspond to an inner wall temperature of 293 K. On curve 1 the outer wall temperature is 393 K; on curve 2 it is 493 K. The blue curves (3 and 4) correspond to an inner wall temperature of 393 K. On curve 3 the outer wall temperature is 293 K; on curve 4 it is 493 K. The green curves (5 and 6) correspond to an inner wall temperature of 493 K. On curve 5 the outer wall temperature is 293 K; on curve 6 it is 493 K.

The analysis showed that the approximation error of Equation (16) increases with an increase in the absolute temperature difference between the external T_{ex} and internal T_{in} walls, and with an increase in the relative wall thickness $(R_{ex} - R_{in})/R_{in}$. However, in the considered range of temperature changes, $(T_{ex} - T_{in})/T_{in} < 0.7$, the linear approximation error does not exceed 1.2%. Thus, to calculate the stress–strain state of the tube walls, we will use dependence, Equation (17).

2.5. Stresses in Tube Walls

Let us consider the stress–strain state of the tube loaded by the internal pressure when the temperature of its inner and outer walls has different values (Figure 5). A mathematical model of the stress–strain state includes the equilibrium equations and relations between deformations and displacements and also between stresses and deformations. According to Timoshenko and Goodier [40], the balance of radial stresses can be described subject to axial symmetry and flat deformation [48,51] by the following equation:

$$\frac{\partial \sigma_{rr}}{\partial r} + \frac{\sigma_{rr} - \sigma_{\varphi\varphi}}{r} = 0 \qquad (18)$$

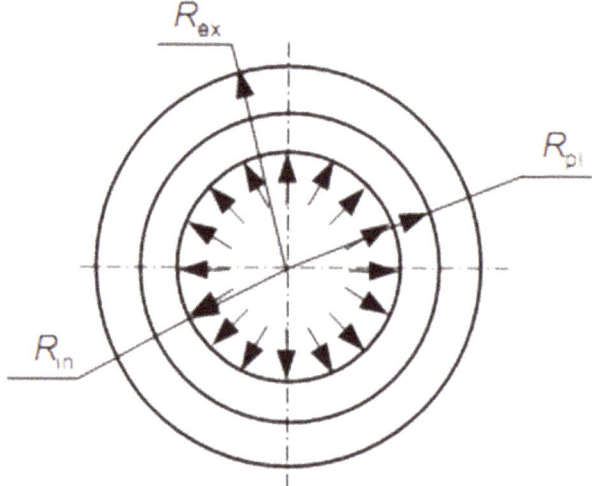

Figure 5. Cross-sectional view of tube deformation.

The boundary conditions for Equation (18) can be written as:

$$r = R_{in}: \sigma_{rr} = -p_{in}; \; r = R_{ex}: \sigma_{rr} = 0 \qquad (19)$$

The deformation of the tube walls is determined by the magnitude of the applied pressure, p_{in}. If p_{in} has small values, then deformation of the tube walls is elastic. As pressure is increased, stresses in the wall of the tube increase. According to the Tresca–Saint-Venant plasticity condition [51,52], a plastic deformation occurs when the maximum tangential stress achieves its ultimate value in the material:

$$|\sigma_{rr} - \sigma_{\varphi\varphi}| = \tau_0 \qquad (20)$$

If the value of the applied pressure becomes equal to the limit of elastic resistance, $p = p_{el}$, then plastic deformation occurs on the inner wall of the tube. At even greater pressure, the plastic state covers an annular layer of radius, R_{pl}, adjacent to the inner surface of the tube. A region will be adjacent to the outer boundary of this layer, in which the elastic state of the material will still be preserved. When the value of the applied pressure reaches the limit of plastic resistance, $p = p_{pl}$, all the material in the thickness of the tube will go into a plastic state.

In this paper we will consider the case when the limit of plastic resistance is reached, that is, the deformation of the entire tube wall is plastic. The analysis of the stress–strain state can be carried out on the basis of equations of the deformational theory of plasticity.

2.6. Displacements and Strains in Tube Walls

The deformation theory of plasticity is based on the theory that volume changes due to plastic deformations do not occur [40,51]. Volumetric deformation occurs only as a result of elastic and temperature stresses. Thus, during plastic deformation in an inhomogeneous temperature field, volumetric deformation is equal to:

$$\varepsilon_V = 3\alpha_T(T - T_{in}) \qquad (21)$$

With small tensile/compression and shear, the volumetric strain is equal to:

$$\varepsilon_V = \varepsilon_{rr} + \varepsilon_{\varphi\varphi} + \varepsilon_{zz} \qquad (22)$$

The components of the strain tensor are determined by the Cauchy relations and, in the presence of axial symmetry and a plane deformed state, they have the form:

$$\varepsilon_{rr} = \frac{du_r}{dr}, \quad \varepsilon_{\varphi\varphi} = \frac{u_r}{r}, \quad \varepsilon_{zz} = 0 \tag{23}$$

In Equation (23), u_r is the radial component of the displacement vector.

Taking into account dependences, Equations (21)–(23), the volumetric deformation can be found from the solution of the equation:

$$\frac{\partial u_r}{\partial r} + \frac{u_r}{r} = 3\alpha_T(T - T_{in}). \tag{24}$$

The solution of this equation, taking into account the temperature dependence, Equation (17), has the form:

$$u_r = 3\alpha_T \frac{\Delta T}{h}\left(\frac{r^2}{3} - \frac{rR_{in}}{2}\right) + \frac{C_*}{r}, \tag{25}$$

where C_* is the integration constant which should be determined.

Equation (25) allows us to determine the components of the strain tensor:

$$\varepsilon_{rr} = \frac{\partial u_r}{\partial r} = 3\alpha_T \frac{\Delta T}{h}\left(\frac{2r}{3} - \frac{R_{in}}{2}\right) - \frac{C_*}{r^2}, \quad \varepsilon_{\varphi\varphi} = \frac{u_r}{r} = 3\alpha_T \frac{\Delta T}{h}\left(\frac{r}{3} - \frac{R_{in}}{2}\right) + \frac{C_*}{r^2}, \tag{26}$$

According to the Duhamel–von Neumann hypothesis, deformation can be represented as the sum of deformations caused by a force load [51,53]

$$\varepsilon_{rr}^\sigma = -\frac{C_*}{r^2}, \quad \varepsilon_{\varphi\varphi}^\sigma = \frac{C_*}{r^2} \tag{27}$$

and deformation caused by thermal expansion:

$$\varepsilon_{rr}^T = 3\alpha_T \frac{\Delta T}{h}\left(\frac{2r}{3} - \frac{R_{in}}{2}\right), \quad \varepsilon_{\varphi\varphi}^T = 3\alpha_T \frac{\Delta T}{h}\left(\frac{r}{3} - \frac{R_{in}}{2}\right) \tag{28}$$

According to [40], strain intensity caused by a force load is equal to:

$$a = \sqrt{2\left((\varepsilon_{rr}^\sigma)^2 + (\varepsilon_{\varphi\varphi}^\sigma)^2\right)} \tag{29}$$

Substitution of dependencies (27) in Equation (29) leads to the expression

$$a = 2\frac{C_*}{r^2} \tag{30}$$

Let us determine the integration constant, C_*, using the condition at the boundary of elastic and plastic deformation areas.

$$\tau_0 = Ga(R_{pl}) = 2G\frac{C_*}{R_{pl}^2} \tag{31}$$

Thus, the strain intensity is equal:

$$a = \frac{\tau_0}{G}\frac{R_{pl}^2}{r^2} \tag{32}$$

2.7. Numerical Method

The balance equations for elements of defect dislocation structure and mechanics of deformable solids were solved numerically by using Runge–Kutta–Merson's fifth order method [54]. Let us write the set of Equations (1)–(8) and (18) in matrix form:

$$\frac{dY}{dX} = F(X, Y) \tag{33}$$

The calculation algorithm for the 5th order Runge–Kutta–Merson method is represented by the equations:

$$\begin{aligned}
Y_{i+1} &= Y_i + \tfrac{1}{6}(k_0 + 4k_3 + k_4), \\
k_0 &= hF(X_i, Y_i), \\
k_1 &= hF\left(X_i + \tfrac{h}{3}, Y_i + \tfrac{k_0}{3}\right), \\
k_2 &= hF\left(X_i + \tfrac{h}{3}, Y_i + \tfrac{k_0 + k_1}{6}\right), \\
k_3 &= hF\left(X_i + \tfrac{h}{2}, Y_i + \tfrac{k_0 + 3k_2}{6}\right) \\
k_4 &= hF\left(X_i + h, Y_i + \tfrac{k_0 - 3k_2}{6} + 2k_3\right)
\end{aligned} \tag{34}$$

The total error of the method is $O(h^5)$.

In the case of long-term calculations that require a large number of calculations, it is possible to reduce the calculation time by using a variable step of the difference grid, h. When using a variable step in the calculations, the difference between adjacent values of the grid function, $\Delta = |Y_{i+1} - Y_i|$, is controlled. If Δ exceeds the specified error, Δ_{max}, the grid step is halved; at small values of Δ, the step is doubled. The conditions for automatic selection of the grid step are represented by Equation (35).

$$h_{i+1} = \begin{cases} \tfrac{1}{2}h_i & \text{if} \quad \Delta_{max} < \Delta \\ h_i & \text{if} \quad \Delta_{max} \leq \Delta \leq 2\Delta_{max}, \\ 2h_i & \text{if} \quad \Delta < 2\Delta_{max} \end{cases} \tag{35}$$

The conditions in Equation (35) make it possible to significantly reduce the computation time of the problem while maintaining the accuracy of the solution.

3. Results and Discussion

Let us proceed to the analysis of the main results of the mathematical modeling. The mathematical model of this tube assumes that its inner and outer radii are, respectively, $R_{in} = 0.1$ m and $R_{ex} = 0.105$ m.

Figure 6 shows the dependences of the plastic resistance limit, p_{pl}, of the tube on the temperature of the outer wall. When the outer wall is heated, the material of the outer layers of the tube becomes more plastic. As a result of this, the pressure required to achieve plastic deformation of the outer wall decreases. In contrast, cooling of the outer wall leads to an increase in the limit of plastic resistance. Note that at a fixed value, T_{in}, varying the temperature of the outer wall in the range of 200 K leads to a significant change in the limit of plastic resistance (for the geometry under consideration, this change is about 15%).

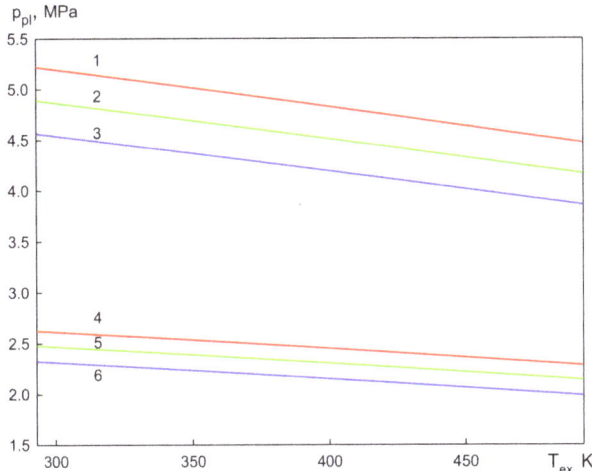

Figure 6. Dependence of the plastic resistance limit on the temperature of the outer wall: $R_{in} = 0.1$ m, $R_{ex} = 0.105$ m. Curves 1–3: $\Lambda_p = 100$ nm, $\delta = 10$ nm; curves 4–6: $\Lambda_p = 200$ nm, $\delta = 20$ nm; curves 1 and 4—temperature of the inner wall of the pipe $T_{in} = 293$ K; curves 2 and 5–$T_{in} = 393$ K and curves 3 and 6–$T_{in} = 493$ K.

Similar dependences characterizing the effect of the temperature of the inner wall on plastic deformation of the tube are illustrated in Figure 7.

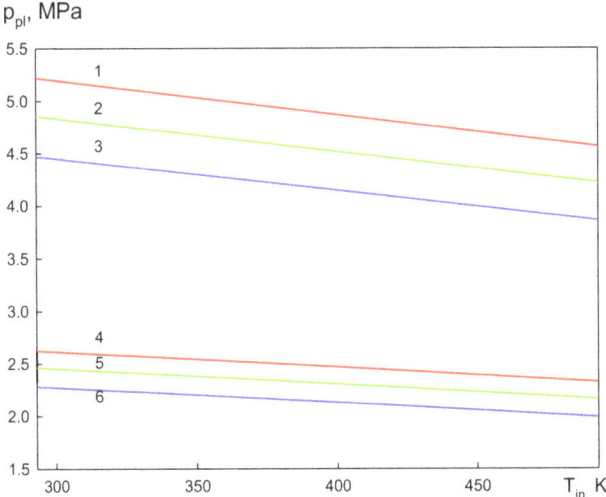

Figure 7. Dependence of the plastic resistance limit on the temperature of the inner wall: $R_{in} = 0.1$ m, $R_{ex} = 0.105$ m. Curves 1–3: $\Lambda_p = 100$ nm, $\delta = 10$ nm; curves 4–6: $\Lambda_p = 200$ nm, $\delta = 20$ nm; curves 1 and 4: temperature of the outer wall of the pipe $T_{ex} = 293$ K; curves 2 and 5: $T_{ex} = 393$ K; curves 3 and 6: $T_{ex} = 493$ K.

An increase of the temperature of the inner wall at a fixed value of the outer wall temperature, T_{ex}, reduces the resistance to plastic deformation of the inner layers of the tube, which causes a decrease, p_{pl}.

When cooling the inner wall of the tube, the limit of plastic resistance increases. The variation, p_{pl}, with variation of the inner wall temperature, T_{in}, in the range of 200 K for the considered tube sizes

($R_{in} = 0.1$ m, $R_{ex} = 0.105$ m) is also 15%. Thus, to calculate the strength characteristics of the heat exchanger, it is necessary to take into account temperature distribution in its walls.

The result of mathematical modeling demonstrates that in alloys with small distances between the hardening particles with the same volume fraction of particles, significantly greater pressure is required to achieve plastic deformation of the tube walls. Thus, a decrease in the distance between particles with the same volume fraction of particles causes hardening of the material, leading to an increase in the limit of plastic resistance. Note that for alloys with different parameters of the hardening phase, the character of the dependences of the limit of plastic resistance on the temperature of the tube wall does not change.

Figures 8 and 9 show the effect of tube dimensions on the value of the limit of plastic resistance. As the wall thickness of the tube increases, its resistance to applied pressure increases. As a result, as can be seen from Figure 8, with an increase of the wall thickness, the limits of the plastic resistance of the tube increase.

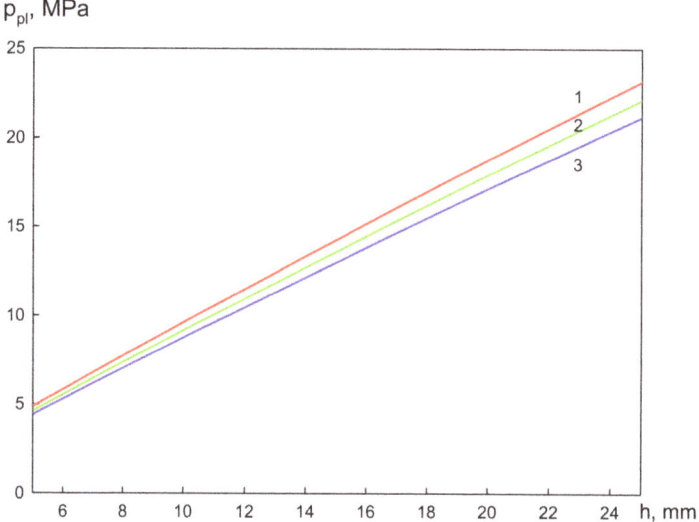

Figure 8. The dependence of the limit of plastic resistance on the wall thickness of the tube $h = R_{ex} - R_{in}$: $T_{in} = 293$ K, $T_{ex} = 393$ K, $\Lambda_p = 100$ nm, $\delta = 10$ nm, curve 1–$R_{in} = 0.1$ m, curve 2–0.105 m and curve 3–0.11 m.

The dependence of the plastic resistance limit of a tube on its dimensions is shown in Figure 9. An increase in the radius of the tube at a fixed wall thickness leads to a decrease of p_{pl}.

The limit of plastic resistance also decreases with increasing temperature of the tube wall. For the same absolute temperature difference between the outer and inner walls, the plastic resistance limit is lower as the temperature of the inner wall of the tube increases.

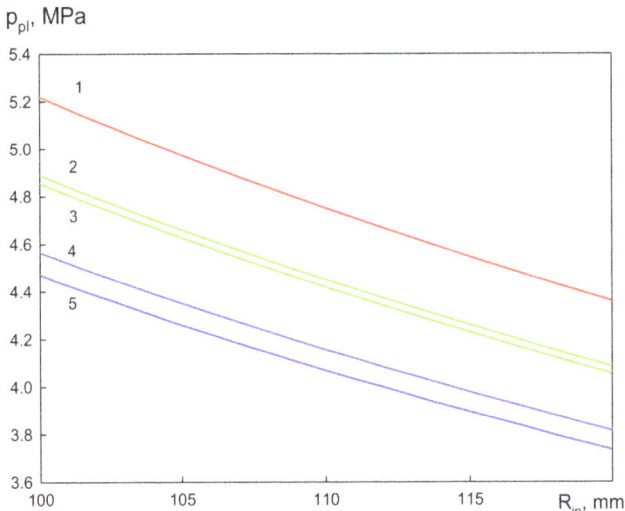

Figure 9. The dependence of the plastic resistance limit on the tube size. $\Lambda_p = 100$ nm, $\delta = 10$ nm, $h = 0.05$ m; curve 1–$T_{in} = 293$ K, $T_{ex} = 293$ K; curve 2–293, 393; curve 3–393, 293; curve 4–293, 493 and curve 5–493, 393.

4. Conclusions

During the investigation, the authors developed the mathematical model and obtained quantitative data to determine the effect of the temperature difference between the outer and inner walls of the tube on its stress–strain state. The results of mathematical modeling showed that with an increase of the thickness of the tube wall, the plastic resistance limit increases. An increase of the tube radius at a fixed wall thickness leads to a decrease of the plastic resistance limit. The material of the outer layer of the tube becomes more plastic when heating of the outer wall occurs, so the pressure required to achieve plastic deformation of the outer wall decreases. The cooling of the outer wall leads to an increase of the plastic resistance limit. An increase of the temperature of the inner wall reduces the resistance to plastic deformation of the inner layers of the tube. When the inner wall of the tube cools, the plastic resistance limit increases. For the same absolute temperature difference between the outer and inner walls, the plastic resistance limit is lower as the temperature of the inner wall of the tube increases. To calculate the strength characteristics of the heat exchanger, it is necessary to take into account the temperature distribution in its walls. The result of mathematical modeling demonstrates that a decrease of the distance between particles with the same volume fraction leads to significantly greater pressure being required to achieve plastic deformation of the tube walls and causes hardening of the material. Investigation of the defect structure formation in the tube walls under conditions of inhomogeneous temperature is planned in following publications.

Author Contributions: Conceptualization, O.M., T.K. and O.D.; Data curation, O.D.; Formal analysis, O.M.; Funding acquisition, T.K. and the Ministry of Science and Higher Education of the Russian Federation (theme No. FEMN-2020-0004); Investigation, O.M. and O.D.; Methodology, O.M., T.K. and O.D.; Project administration, T.K.; Resources, O.M.; Software, O.M.; Supervision, O.D.; Validation, O.M.; Visualization, O.D.; Writing—original draft, O.M.; Writing—review & editing, O.M. and O.D. All authors have read and agreed to the published version of the manuscript.

Funding: The research was carried out within the state assignment of the Ministry of Science and Higher Education of the Russian Federation (theme No. FEMN-2020-0004).

Acknowledgments: The authors acknowledge the Ministry of Science and Higher Education of the Russian Federation (theme No. FEMN-2020-0004).

Conflicts of Interest: The authors declare no conflict of interest.

References

1. Saunders, E.A. *Heat Exchanges: Selection, Design and Construction*; Longman Scientific and Technical: New York, NY, USA, 1988.
2. Northcutt, B.; Mudawar, I. Enhanced design of cross-flow microchannel heat exchanger module for high-performance aircraft gas turbine engines. *J. Heat Transf.* **2012**, *134*, 061801-1–061801-13. [CrossRef]
3. Wang, H.; Sang, Z. Theory solution of Hydraulic expansion tube-to-tube sheet joint residual contact pressure for power-hardening material. *China Pet. Mach.* **2007**, *11*, 24–28.
4. Bouzid, A.H.; Mourad, A.H.I.; Domiaty, E.A. Influence of Bauschinger effect on the residual contact pressure of hydraulically expanded tube-to-tubesheet joints. *Int. J. Press. Vessel. Pip.* **2016**, *146*, 1–10. [CrossRef]
5. Ying, H.; Xuesheng, W.; Qinzhu, C.; Jianfu, W. Bilinear Simplification of Material Model in Theoretical Calculation of Hydraulic Expansion. *Mach. Des. Res.* **2018**, *34*, 199–202.
6. Law, M.; Bowie, G. Failure strain in high yield-to-tensile ratio line pipes. *J. Pipeline Integr.* **2006**, *5*, 25–36.
7. Robertson, A.; Li, H.; Mackenzie, D. Plastic collapse of pipe bends under combined internal pressure and in-plane bending. *Int. J. Press. Vessel. Pip.* **2005**, *82*, 407–416. [CrossRef]
8. Gajdos, L.; Sperl, M. Evaluating the integrity of pressure pipelines by fracture mechanics. *INTECH Appl. Fract. Mech.* **2012**, *10*, 283–310.
9. Chemezov, D.; Bayakina, A.; Goremykin, V.; Prokofiev, A.; Pavlov, E.; Petrenko, A.; Sergeeva, M.; Gusenkov, M. Strain intensity of the steel pipe under the action of external tensile, compressive and combined loads. *ISJ Theor. Appl. Sci.* **2020**, *2*, 30–35. [CrossRef]
10. Bergant, A.; Tijsseling, A.S.; Vítkovský, J.P.; Covas, D.I.; Simpson, A.R.; Lambert, M.F. Parameters affecting water-hammer wave attenuation, shape and timing, part 2: Case studies. *J. Hydraul. Res.* **2008**, *46*, 382–391. [CrossRef]
11. Zhu, X.K.; Leis, B.N. Evaluation of burst pressure prediction models for line pipes. *Int. J. Press. Vessel. Pip.* **2012**, *89*, 85–97. [CrossRef]
12. Zhu, X.K.; Leis, B.N. Theoretical and numerical predictions of burst pressure of pipelines. *J. Press. Vessel. Technol.* **2007**, *129*, 644–652. [CrossRef]
13. Burkov, P.; Yan'nan, V.; Burkova, S. Stress–strain analysis of pipelines laid in permafrost. *IOP Conf. Ser. Earth Environ. Sci.* **2016**, *43*, 012080. [CrossRef]
14. Chohan, R.K. Plastic deformation induced by pressure transients in fluid-filled pipes. *Int. J. Press. Vessel. Pip.* **1988**, *33*, 333–343. [CrossRef]
15. Pleșcan, C.; Stanciu, M.D.; Szasz, M. The Effect of Internal Pressure on Radial Strain of Steel Pipe Subjected to Monotonic and Cyclic Loading. *Materials* **2019**, *12*, 2849. [CrossRef]
16. Tomita, Y.; Shindo, A.; Kim, Y.S.; Michiura, K. Deformation behaviour of elastic-plastic tubes under external pressure and axial load. *Int. J. Mech. Sci.* **1986**, *5*, 263–274. [CrossRef]
17. Chawla, N.; Shen, Y.L. Mechanical Behavior of Particle Reinforced Metal Matrix Composites. *Adv. Eng. Mater.* **2001**, *3*, 357–370. [CrossRef]
18. Beffort, O.; Long, S.; Cayron, C.; Kuebler, J.; Buffat, P.A. Alloying effects on microstructure and mechanical properties of high volume fraction SiC-particle reinforced Al-MMCs made by squeeze casting infiltration. *Compos. Sci. Technol.* **2007**, *67*, 737–745. [CrossRef]
19. Karabasov, Y.S. *Advanced Materials*; MISIS: Moscow, Russia, 2002; 736p.
20. Orowan, E. Discussion on internal stresses. In *Symposium on Internal Stresses in Metals and Alloys*; Institute of Metals: London, UK, 1948; pp. 451–453.
21. Ashby, M.F. Work hardening of dispersion-hardened crystals. *Philos. Mag.* **1966**, *14*, 1157–1178. [CrossRef]
22. Ebeling, R.; Ashby, M.F. Dispersion hardening of copper single crystals. *Philos. Mag.* **1966**, *13*, 805–834. [CrossRef]
23. Hirsch, P.B.; Hymphreys, F.J. Comment on "Dispersion hardening in metals" by E.W. Hart. *Scr. Metall.* **1973**, *7*, 259–265. [CrossRef]
24. Hazzledine, P.M.; Hirsch, P.B. A coplanar Orowan loops model for dispersion hardening. *Philos. Mag.* **1974**, *30*, 1331–1351. [CrossRef]
25. Humphreys, F.J.; Hirsch, P.B. Work-hardening and recovery of dispersion hardened alloys. *Philos. Mag.* **1976**, *34*, 373–390. [CrossRef]

26. Humphreys, F.J.; Stewart, A.T. Dislocation generation at SiO$_2$ particles in an α-brass matrix on plastic deformation. *Surf. Sci.* **1972**, *31*, 389–421. [CrossRef]
27. Hymphreys, F.J.; Martin, J.W. The effect of dispersed phases upon dislocation distributions in plastically deformed copper crystals. *Philos. Mag.* **1967**, *16*, 927–957. [CrossRef]
28. Hymphreys, F.J.; Hirsch, P.B. The deformation of single crystals of copper and copper-zinc alloys containing alumina particles. II. Microstructure and dislocation-particle interactions. *Proc. Phys. Soc.* **1973**, *318*, 73–92.
29. Matvienko, O.V.; Daneiko, O.I.; Kovalevskaya, T.A. Plastic deformation of copper-based alloy reinforced with incoherent nanoparticles. *Russ. Phys. J.* **2017**, *60*, 236–248. [CrossRef]
30. Matvienko, O.V.; Daneyko, O.I.; Kovalevskaya, T.A. Stress-stain state of pipe made of copper-based alloy strengthened with incoherent nanoparticles. *Russ. Phys. J.* **2017**, *60*, 562–569. [CrossRef]
31. Matvienko, O.V.; Daneyko, O.I.; Kovalevskaya, T.A. Dislocation structure of the pipe made of alloy reinforced with incoherent particles under uniform internal pressure. *Russ. Phys. J.* **2017**, *60*, 1233–1242. [CrossRef]
32. Matvienko, O.; Daneyko, O.; Kovalevskaya, T. Mathematical modeling of plastic deformation of a tube from dispersion-hardened aluminum alloy. *MATEC Web Conf.* **2018**, *243*, 00008. [CrossRef]
33. Matvienko, O.V.; Daneyko, O.I.; Kovalevskaya, T.A. Elastoplastic deformation of dispersion-hardened aluminum tube under external pressure. *Russ. Phys. J.* **2018**, *61*, 1520–1528. [CrossRef]
34. Matvienko, O.V.; Daneyko, O.I.; Kovalevskaya, T.A. Elastoplastic deformation of dispersion-hardened aluminum tube under external and internal pressure. *Russ. Phys. J.* **2019**, *62*, 720–728. [CrossRef]
35. Daneyko, O.I.; Kovalevskaya, T.A. Temperature effect on stress–strain properties of dispersion-hardened crystalline materials with incoherent nanoparticles. *Russ. Phys. J.* **2019**, *61*, 1687–1694. [CrossRef]
36. Kovalevskaya, T.A.; Daneyko, O.I. The influence of scale parameters of strengthening phase on plastic shear zone in heterophase alloys with incoherent nanoparticles. *Russ. Phys. J.* **2020**, *62*, 2247–2254. [CrossRef]
37. Daneyko, O.I.; Kulaeva, N.A.; Kovalevskaya, T.A.; Kolupaeva, S.N. Investigation of thermal hardening of the FCC material containing strengthening particles with an L1$_2$ superstructure. *Russ. Phys. J.* **2015**, *58*, 336–342. [CrossRef]
38. Chakrabarty, J. *Theory of Plasticity*; Mc Graw-Hill Book Company: New York, NY, USA; Hamburg, Germany; London, UK; Paris, France; Sydney, Australia; Tokyo, Japan, 1987; Volume VIII, 791p.
39. Matvienko, O.V.; Daneyko, O.I.; Kovalevskaya, T.A. Strengthening particle size effect on residual stresses in dispersion-hardened alloy. *Russ. Phys. J.* **2018**, *61*, 962–973. [CrossRef]
40. Timoshenko, S.P.; Goodier, J.N. *Theory of Elasticity*; McGraw Hill: New York, NY, USA, 2010; 567p.
41. Kovalevskaya, T.A.; Vinogradova, I.V.; Popov, L.E. *Mathematical Modeling of Plastic Deformation in Heterophase Alloys*; TSU: Tomsk, Russia, 1992; 234p.
42. Daneyko, O.I.; Kovalevskaya, T.A.; Matvienko, O.V. The influence of incoherent nanoparticles on thermal stability of aluminum alloys. *Russ. Phys. J.* **2018**, *61*, 1229–1235. [CrossRef]
43. Kovalevskaya, T.; Daneyko, O.; Kulaeva, N.; Kolupaeva, S. Influence of the scale characteristics of the hardening phase with L1$_2$ superstructure on the evolution of deformation point defects. In *AIP Conference Proceedings*; AIP Publishing LLC: Tomsk, Russia, 2016; p. 040003. [CrossRef]
44. Daneyko, O.I.; Kovalevskaya, T.A.; Kulaeva, N.A.; Kolupaeva, S.N.; Shalygina, T.A. Evolution of dislocation subsystem components during plastic deformation depending on parameters of strengthening phase with L1$_2$ superstructure. *Russ. Phys. J.* **2017**, *60*, 821–829. [CrossRef]
45. Daneyko, O.I.; Kovalevskaya, T.A.; Kulaeva, N.A. Modeling of plastic deformation of dispersion-hardened materials with L1$_2$ superstructure particles. *Russ. Phys. J.* **2017**, *60*, 508–514. [CrossRef]
46. Larikov, L.N.; Yurchenko, Y.F. *Thermal Properties of Metals and Alloys*; Naukova Dumka: Kiev, Ukraine, 1985; 312p.
47. Polmear, L.J. *Light Alloys: Metallurgy of Lights Metals*; John Willey and Sons: Melbourne, Australia, 1995; 235p.
48. Gorshkov, A.G.; Starovoitov, E.I.; Tarlakovskii, D.V. *Theory of Elasticity and Plasticity*; Fizmatlit: Moscow, Russia, 2002; 416p.
49. Khrustalev, A.; Vorozhtsov, A.; Kazantseva, L.; Promakhov, V.; Kalashnikov, M.; Eskin, D.; Kurzina, I. Influence of scandium fluoride on the structure and phase composition of Al-Si alloy. *MATEC Web Conf.* **2018**, *243*, 00020. [CrossRef]
50. Matvienko, O.; Daneyko, O.; Kovalevskaya, T.; Khrustalyov, A.; Zhukov, I.; Vorozhtsov, A. Investigation of Stresses Induced Due to the Mismatch of the Coefficients of Thermal Expansion of the Matrix and the Strengthening Particle in Aluminium-Based Composites. *Metals* **2020**, Unpublished.

51. Matvienko, O.V.; Daneyko, O.I.; Kovalevskaya, T.A. Residual stresses induced by elastoplastic unloading in a tube made of dispersion-hardened alloy. *Russ. Phys. J.* **2018**, *61*, 730–742. [CrossRef]
52. Matvienko, O.; Daneyko, O.; Kovalevskaya, T. Mathematical modeling of nanodispersed hardening of FCC materials. *Acta Metall. Sin. Engl. Lett.* **2018**, *31*, 1297–1304. [CrossRef]
53. Matvienko, O.V.; Daneyko, O.I.; Kovalevskaya, T.A. Stress–strain state of dispersion-hardened aluminum tube under external and internal pressure. *Russ. Phys. J.* **2020**, *62*, 1805–1812. [CrossRef]
54. Chapra, S. *Numerical Methods for Engineers*; McGraw: New York, NY, USA, 2015; 518p.

Publisher's Note: MDPI stays neutral with regard to jurisdictional claims in published maps and institutional affiliations.

© 2020 by the authors. Licensee MDPI, Basel, Switzerland. This article is an open access article distributed under the terms and conditions of the Creative Commons Attribution (CC BY) license (http://creativecommons.org/licenses/by/4.0/).

Article

Influence of Trapped Gas on Pore Healing under Hot Isostatic Pressing in Nickel-Base Superalloys

Mahesh R. G. Prasad *, Siwen Gao, Napat Vajragupta and Alexander Hartmaier

ICAMS, Ruhr-Universität Bochum, 44801 Bochum, Germany; siwen.gao@rub.de (S.G.); napat.vajragupta@rub.de (N.V.); alexander.hartmaier@rub.de (A.H.)
* Correspondence: mahesh.prasad@rub.de

Received: 11 November 2020; Accepted: 14 December 2020; Published: 17 December 2020

Abstract: Under the typical hot isostatic pressing (HIP) processing conditions, plastic deformation by dislocation slip is considered the primary mechanism for pore shrinkage, according to experimental observations and deformation mechanism maps. In the present work, a crystal plasticity model has been used to investigate the influence of applied pressure and holding time on porosity reduction in a nickel-base single crystal superalloy. The influence of trapped gas on pore shrinkage is modeled by coupling mechanical deformation with pore–gas interaction. In qualitative agreement with experimental investigations, we observe that increasing the applied pressure or the holding time can effectively reduce porosity. Furthermore, the effect of pore shape on the shrinkage is observed to depend on a combination of elastic anisotropy and the complex distribution of stresses around the pore. Simulation results also reveal that, for pores of the same shape, smaller pores (radius < 0.1 µm) have a higher shrinkage rate in comparison to larger pores (radius $\geq$ 0.1 µm), which is attributed to the increasing pore surface energies with decreasing pore sizes. It is also found that, for smaller initial gas-filled pores (radius < 0.1 µm), HIP can result in very high gas pressures (on the order of GPa). Such high pressures either act as a driving force for argon to diffuse into the surrounding metal during HIP itself, or it can result in pore re-opening during subsequent annealing or mechanical loading. These results demonstrate that the micromechanical model can quantitatively evaluate the individual influences of HIP processing conditions and pore characteristics on pore annihilation, which can help optimize the HIP process parameters in the future.

Keywords: porosity; HIP; superalloys; crystal plasticity; additive manufacturing

1. Introduction

Modern gas turbine blades in aero-engines are usually made of nickel-base superalloys, solidified as single crystals to obtain excellent resistance to extreme working conditions such as temperatures up to 1100 °C as well as static and cyclic loading. In nickel-base single crystal superalloys, the outstanding deformation resistance at elevated temperatures is due to the absence of grain boundaries, precipitate strengthening by coherent cuboidal γ' phases (L1$_2$ crystal structure) homogeneously distributed in the γ (face centered cubic) matrix, and the solid solution strengthening of both γ and γ' phases by adding a high concentration of refractory elements. Nevertheless, a higher degree of dendritic segregation during solidification and void formation in the interdendritic region during casting and extensive homogenization tend to occur, which deteriorates the mechanical properties [1–3]. In order to heal these microvoids, hot isostatic pressing (HIP) is utilized as an advanced thermal treatment [4], which combines plastic deformation, creep, and diffusion bonding [5,6] to reduce the voids through appropriate pressing at high temperatures [7,8]. For nickel-base superalloys, a typical HIP cycle consists of holding the material at $\sim$1280 °C (close to the γ'-solvus temperature) and at a constant pressure of ~200 MPa for 2–5 h [9,10]. In order to optimize the HIP parameters, it is necessary to understand the void reduction in this process by numerical modeling.

Another aspect to be considered during modeling is the presence of gas inside pores and its influence on pore closure. In recent manufacturing techniques such as additive manufacturing (AM) [11], parts are fabricated in an inert atmosphere, and thus the formation of pores filled with gas (argon) is a real possibility [12,13]. As microstructural defects such as porosity can be very detrimental to the effective mechanical property of the fabricated material [14,15], more research has been focused on alleviating these defects [13,16]. Experimental investigations have analyzed the influence of AM process parameters on the resulting porosity and have shown that the porosity can be reduced to a greater extent during the fabrication [13]. Nevertheless, HIP is available to be used as a post-AM process to eliminate porosity and has been in active research in the AM community [17,18]. Although HIP has been shown to eliminate voids in parts fabricated by casting and sintering, its effectiveness in eliminating pores in AM-fabricated parts is uncertain [19]. Some studies claim that HIP can completely eliminate the pores resulting in fully dense AM parts [16–18], while others have shown that HIP has limited effectiveness in removing pores [19,20]. In the works of Tammas-Williams et al. [20], X-ray computed tomography has been used to track individual pores found in AM-fabricated titanium. From their detailed experimental analysis, it has been confirmed that all the pores that appear during the heat treatment stages (subsequent to HIP) can be correlated to the locations of pores in the original build. Evidently, these pores had disappeared during HIP but reappeared during the subsequent annealing. Thus, it is vital to investigate the influence of HIP process parameters on pore shrinkage and also to understand the kinetics and mechanism of pore annihilation by modeling the pore–gas interaction.

At present, there is only one publication [21] focusing on the modeling of void annihilation in a nickel-base single crystal superalloy during HIP. This simulation applies to one void under one HIP condition. Our work aims to develop a crystal plasticity finite element model that is not only able to model the shrinkage of voids but also the influence of trapped gas and its resistance to pore shrinkage. Since void annihilation is mainly attributed to dislocation glide [21], our model is based on the phenomenological crystal plasticity constitutive laws. In addition, the self-diffusion of atoms assists the shrinkage of pores due to thermal disturbance at high temperature [7]. This mechanism is treated as a temperature-dependent coefficient of the strain rate in the current model. The mechanical response of a gas-filled pore is modeled by coupling the deformation of the material and the pressure exerted by the trapped gas on the pore surface. The current work investigates the influence of pore characteristics on pore shirnkage and the effect of HIP parameters such as isostatic pressures and holding times on the pore annihilation. Throughout this manuscript, a distinction between voids and pores is made unless mentioned otherwise; pores that do not contain trapped gas are referred to as voids.

2. Micromechanical Modeling

This section presents the basics of the developed micromechanical model. In the detailed experimental investigation conducted by Epishin et al. [22], it was shown that slip of dislocations on the octahedral planes is the dominant deformation mechanism in a single-crystal superalloy at 1561 K. In addition, the deformation mechanism map [23] also indicates that dislocation glide operates in the considered HIP temperature and stress regimes. Based on these observations, the crystal plasticity (CP) method is used in this work to predict the plastic deformation behavior of the material, and the gas interaction is coupled with the mechanical deformation by using the surface-based fluid cavity model of Abaqus [24]. The CP model is implemented in a user-defined material subroutine (UMAT) to capture the creep response, and the behavior of the gas is modeled by using a user-defined fluid subroutine (UFLUID). A strict insoluble condition is assumed for the gas, hence the mass of gas in the pore does not change. Since the focus of the current work is to systematically analyze the pore closure mechanism during HIP, certain simplifications regarding the pore locations are made. The pores are assumed to be located in the grain bulk and the influence of grain boundaries on the pore closure through vacancy diffusion is not modeled here. In the following subsections, different aspects of the modeling approach, such as the ultra-high-temperature CP model, modeling gas inside pores, and model parameterization, are detailed.

2.1. Crystal Plasticity Model

Starting from the kinematics of deformation [25], the deformation gradient $\mathbf{F}$ on the macroscopic level can be multiplicatively decomposed into an elastic deformation part $\mathbf{F}^e$ and a plastic deformation part $\mathbf{F}^p$ as follows:

$$\mathbf{F} = \mathbf{F}^e \mathbf{F}^p \tag{1}$$

With the help of an average elastic stiffness tensor $\tilde{\mathbb{C}}$ on the macroscopic level, the second Piola–Kirchhoff stress in the intermediate configuration is calculated by

$$\tilde{\mathbf{S}} = \tilde{\mathbb{C}} \left[\frac{1}{2} \left(\mathbf{F}^{eT} \mathbf{F}^e - \mathbf{I} \right) \right] \tag{2}$$

where $\mathbf{I}$ is the second order unit tensor. Based on the dislocation slip controlled plastic deformation mechanism, the common resolved shear stress for each slip system α is

$$\tau_\alpha = \tilde{\mathbf{S}} : \widetilde{\mathbf{M}}_\alpha \tag{3}$$

where $\widetilde{\mathbf{M}}_\alpha$ is the Schmid tensor. Thus, the flow rule is given by

$$\dot{\gamma}_\alpha = \frac{A\,G\,b\,D_0}{k_B\,T} \exp\left(-\frac{Q_{sd}}{RT}\right) \left| \frac{\tau_\alpha}{\hat{\tau}_\alpha^{slip} - \tau_\alpha^{soft}} \right|^{p_1} \text{sign}(\tau_\alpha) \tag{4}$$

where $\dot{\gamma}_\alpha$ is the shear rate, G is the shear modulus, b is the magnitude of Burgers vector of the dislocation, k_B is the Boltzmann constant, A is a material constant, D_0 is the pre-exponential frequency factor of the self-diffusion coefficient, Q_{sd} is the activation energy of self-diffusion, R is the Gas constant, T is the temperature, p_1 is the inverse value of the strain rate sensitivity, and $\hat{\tau}_\alpha^{slip}$ is the slip resistance determined by

$$\dot{\hat{\tau}}_\alpha^{slip} = \sum_{\beta=1}^{N_s} h_0 \chi_{\alpha\beta} \left(1 - \frac{\hat{\tau}_\beta^{slip}}{\hat{\tau}^{sat}}\right)^{p_2} |\dot{\gamma}_\beta| \tag{5}$$

where N_s is the number of slip systems, h_0 is the initial hardening rate, $\chi_{\alpha\beta}$ is the cross hardening matrix, $\hat{\tau}^{sat}$ is the saturation slip resistance due to dislocation density accumulation, and p_2 is a fitting parameter.

The creep behavior of single-crystal CMSX-4 alloy at the HIP treatment temperature (1561 K) has been experimentally investigated by Epishin et al. [21,22]. Please note that this temperature lies above the γ' solvus temperature such that the material consists only of γ phase. It is observed that at this super-solvus temperature, the superalloy is very soft and rapidly deforms under stresses between 4 MPa and 16 MPa. This is due to the dissolution of the strengthening γ'-precipitate at this ultra-high homologous temperature. To account for this time induced softening behavior, τ_α^{soft} is introduced in the flow rule (Equation (4)) to counteract the hardening. Evolution of this high temperature softening stress is given by

$$\tau_\alpha^{soft} = v_1 \tanh(v_2\,\epsilon^\gamma) \tag{6}$$

where v_1 is the softening stress, and v_2 defines the rate of softening. These parameters are computed during the fitting process, described in Section 2.2.

2.2. Model Calibration

The constitutive model described in the previous section is implemented as a user-defined material subroutine. Creep simulations for the CMSX-4 superalloy are performed at a temperature 1561 K and stresses between 4 MPa and 16 MPa. The temperature-dependent elastic constants of CMSX-4 ($c_{11} = 185$ GPa, $c_{12} = 154.9$ GPa, and $c_{44} = 69.4$ GPa) and the experimental creep curves for the various loading test cases are obtained from the works of Epishin et al. [21,22]. These tensile creep curves are used as reference here for comparison and model validation. A simple 3D finite element model is set up to represent a sample of single crystal superalloys. By using the material parameters given in Table 1, a good fit to 10 MPa experimental creep curve is first obtained (see Figure 1). Subsequently, the model is validated using the 8 MPa and 16 MPa tensile creep load cases, as shown in Figure 1. This calibrated creep model can now be used to study the pore shrinkage mechanism during the HIP process.

Table 1. Fitted material parameters for CMSX-4 superalloy. $\hat{\tau}_0^{slip}$ is the initial slip resistance produced by dislocation interaction.

Parameter	Value	Unit
G	69.4	GPa
b	0.254	nm
D_0	3.36×10^{-4} [26]	m^2 s^{-1}
Q_{sd}	292 [26]	kJ mol^{-1}
A	1.2×10^{-10}	-
p_1	3.0	-
p_2	0.05	-
$\hat{\tau}_0^{slip}$	40	MPa
$\hat{\tau}^{sat}$	600	MPa
h_0	60	MPa
v_1	60	MPa
v_2	2.0	-

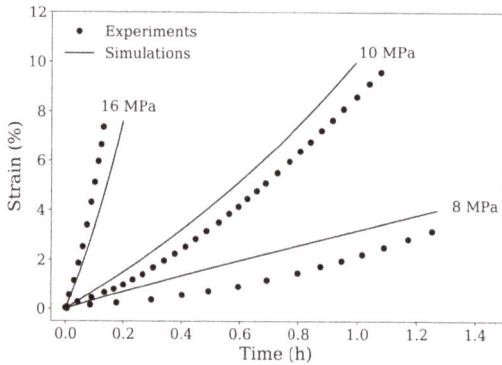

Figure 1. Comparison of simulation results and experimental data for creep under different tensile loads along [100] direction at 1561 K. The experimental creep curves are obtained from the works of Epishin et al. [21,22].

It has been reported in [3,21], by synchrotron tomography investigations, that the shape of the pores in heat-treated CMSX-4 is approximately spherical, and the volume fraction of the voids has been estimated to be about 0.23% [21]. Using this as initial reference porosity, a spherical void with a volume fraction of 0.23% is modeled in the microstructure. Assuming the void is located in the middle of cube-shaped material, only 1/8th dimension of the complete model is used in the calculation, due to symmetrical geometry. The representative 3D setup is shown as a planar view in Figure 2 with loads and boundary conditions. To reach a compromise between the computational cost and the solution accuracy, 1648 hexahedral elements are used. The standard Abaqus self-contact algorithms (frictionless) are set on the free surface of the void to treat the contact interaction during void closure.

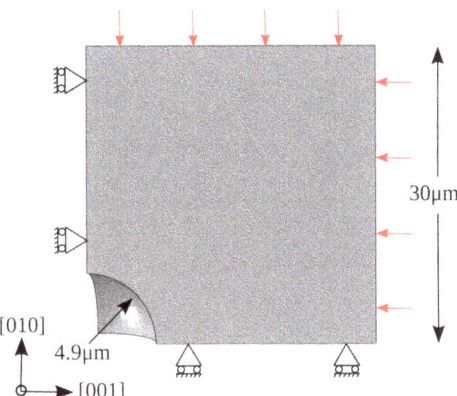

Figure 2. Planar view of a 3D model with 1/8th spherical void with loads and boundary conditions.

Following the experimental HIP conditions reported in [21], in the first step (henceforth referred to as Stage 1), a 103 MPa isostatic pressure is applied on the surrounding borders at 1561 K for 10 min. This is followed by the holding step (henceforth referred to as Stage 2), where the temperature and the pressure are held constant for 6 h. The void volume fraction is plotted against the HIP simulation time in Figure 3. The experimental and the Epishin HIP model volume fraction evolution curves are obtained from [21]. The root mean square error (RMSE) between the experimental and the current model curves is computed to be 0.024. Comparing the curves and considering the RMSE value obtained, reasonably good agreement can be observed between the experimental and the current model simulation results.

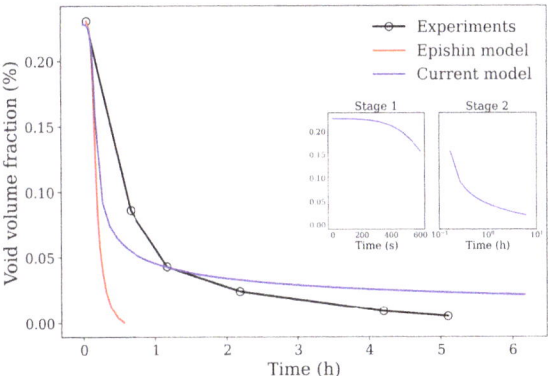

Figure 3. Comparison of void volume fraction evolution curves between experimental and simulation results. The experimental curve is obtained from Epishin et al. [21]. The figure inset indicates the two stages of HIP loading.

The two stages of the HIP process are indicated in the inset of Figure 3 as Stage 1 and Stage 2. It is important to note here that the void closure already begins during Stage 1 and that the closure rate at this stage is very high in comparison to Stage 2. The decrease in the closure rate at the end of Stage 2 is due to the relaxation of the von Mises stresses around the void. The von Mises stress fields as well as the plastic strain contours around the spherical void under the isostatic pressure are depicted in Figure 8. In practice, the material has many voids with various sizes, which will indeed influence the void annihilation behavior. However, our simplified model with its unique round void can capture the main regulation of void reduction and is computationally efficient.

2.3. Modeling Gas inside Pores

The primary difficulty in addressing the mechanical response of a gas-filled pore is the coupling between the deformation of the material and the pressure exerted by the contained gas on the surrounding material. Figure 4 depicts a gas-filled pore subjected to external loads. The response of the material depends not only on the external loads, but also on the pressure exerted by the gas, which, in turn, is affected by the deformation of the material. The surface-based fluid cavity capability in Abaqus [24] provides the coupling needed to analyze such situations in which the cavity (here pore) is filled by fluid (here gas) with uniform properties and state.

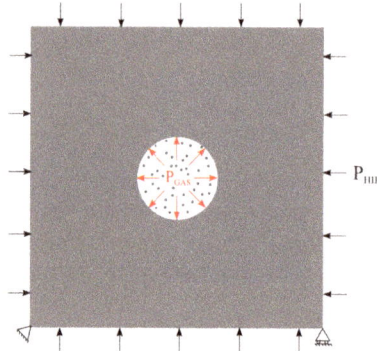

Figure 4. Planar view of a gas-filled spherical pore subjected to external applied load P_{HIP}. The pressure exerted by the trapped gas on the pore surface is represented as P_{GAS}.

The structure is discretized by a finite element mesh and the boundary of the pore is defined by an element-based surface with normals pointing to the inside of the pore. The gas is associated with a node known as the cavity reference node and has a single degree of freedom representing the pressure inside the pore. The cavity reference node is also used in the calculation of the pore volume. The gas inside the pore is assumed compressible, and, using the ideal gas assumption, the density of the gas inside the pore can be calculated as

$$\rho(P,T) = \frac{M_w(P + P_{amb})}{RT} \tag{7}$$

where M_w is the molecular weight of the gas, R is the gas constant, T is the gas temperature, P is the gas pressure, and P_{amb} is the ambient pressure. At the start of the analysis (prior to the first iteration), the reference gas density calculated in the user subroutine UFLUID (for the reference pressure $P = P_{ref}$, and reference temperature $T = T_{ref}$) is used to calculate the mass of gas from the initial pore volume (using Equation (7)). During the analysis, the expected pore volume is calculated from the gas mass and the current density. The parameters used for modeling gas inside pores are listed in Table 2.

Table 2. Parameters used for modeling gas inside pores. The gas considered for simulations is argon.

M_w (kg mol^{-1})	R (JK^{-1}mol^{-1})	P_{ref} (MPa)	P_{amb} (MPa)	T_{ref} (K)
0.039948	8.314	0	0.101	1728.15

During AM, the pressure inside the build chamber is ≈0.1 MPa, and as solidification begins the argon gas gets trapped in the melt pool (from the argon atmosphere). The trapped gas can be assumed to have an internal pressure equal to the AM build chamber pressure. Upon solidification, the temperature decreases to room temperature, but the size of the pore can be assumed to remain constant, as the thermal expansion rate of metals is negligible. During the subsequent HIP loading (before Stage 1), the temperature is raised to $0.7T_m$ (T_m-melting temperature of the metal), and again the pore size is assumed to remain constant. During the HIP loading stages, the pore size decreases due to a combination of externally applied pressure P_{HIP} and the pressure due to surface tension of the pore P_ν. Hence, for pore equilibrium, the gas pressure inside the pore must balance the externally applied pressure and surface tension pressure $P_{GAS} = P_{HIP} + P_\nu$ [27]. The pressure due to surface tension is modeled using the DLOAD subroutine in Abaqus. P_ν is computed as $2\nu/r$, where r is the pore radius and ν is the surface energy of the pore. However, this is not used in the calculation of gas pressure, but rather the equation of state described in Equation (7). Although there may be some limitations to using Equation (7), it is a good approximation to describe the behavior of argon under HIP processing conditions.

3. Numerical Study and Results

In this section, the micromechanical model described above is applied to study the void and pore closure mechanisms and to investigate the relationship between the HIP parameters, pore characteristics, and porosity reduction.

3.1. Influence of HIP Processing Conditions

The isostatic pressure, the holding time, and the temperature play crucial roles in void/pore reduction during the HIP treatment. Compared to the applied pressure and holding time, the void/pore shrinkage is more sensitive to temperature due to the strongly temperature-dependent dissolution of the γ' phase and the self-diffusion effect. As the HIP processing temperature is above the γ'-solvus, we assume that the whole material contains only the soft γ phase. Thus, without the obstruction of γ' precipitates for dislocation glide, the deformation resistance decreases dramatically, so that the void/pore shrinkage turns out to be easier at such high temperatures. The reduction of void and pore volumes under different isostatic pressures and different holding times at 1561 K is shown in Figures 5 and 6, respectively. It is clear that a higher applied pressure and longer holding time lead to better void/pore shrinkage.

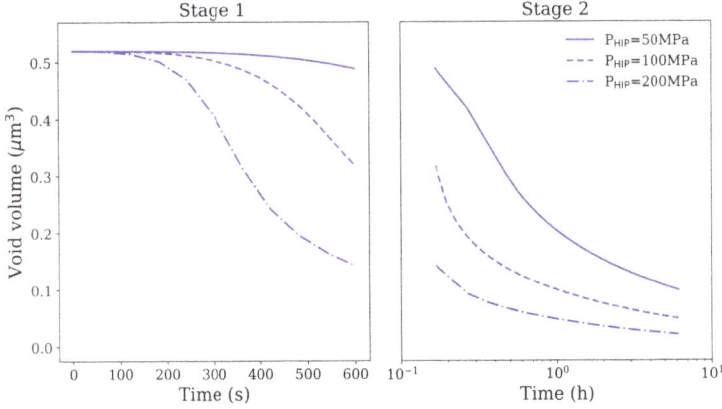

Figure 5. Void volume reduction in Stage 1 and Stage 2 HIP loading for spherical void of radius 1 μm under different magnitudes of the isostatic pressure and different holding times at 1561 K.

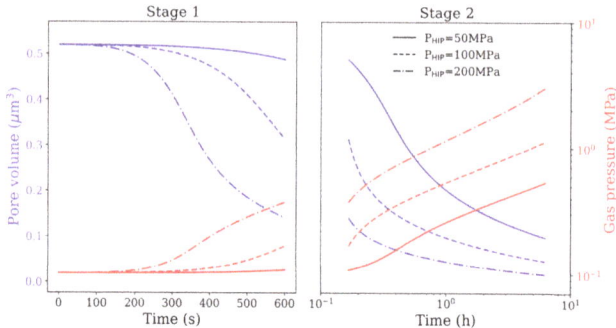

Figure 6. Pore volume reduction and gas pressure evolution in Stage 1 and Stage 2 HIP loading for spherical pore of radius 1 μm under different magnitudes of the isostatic pressure and different holding times at 1561 K.

At constant pressure, the void/pore shrinkage velocity decreases with increasing holding time. This feature is more evident when higher pressure is used. The reason is that the local stress around the void/pore becomes smaller as local plastic strain increases during the process of closure (see Figure 8a,b). The local strain hardening retards further shrinkage. For the gas-filled pores, the evolution of gas pressure is plotted on the secondary y-axis as a function of simulation time for various isostatic pressures (see Figure 6). Since the gas pressure inside the pores does not reach very high values, the resistance to pore shrinkage by the gas is negligible.

The current model addresses the influence of isostatic pressure and holding time on the void/pore closure at temperatures above the γ' solvus. However, it does not simulate void/pore closure below this temperature as the presence and possible evolution of γ' precipitates during the HIP process must be taken into account. In fact, the complicated stress field around the voids/pores could cause the γ' precipitates to raft along different directions, and such rafted precipitates would influence void/pore annihilation and the mechanical properties of the materials after HIP. Future work will address the influence of γ' precipitates on the void/pore healing and the optimal HIP parameters in this context.

3.2. Influence of Pore Shape

For the same HIP conditions, different shapes of voids/pores could give rise to different shrinkage behaviors. Figure 7 depicts the 3D view of model mesh configurations containing idealized spherical and ellipsoidal voids. Due to symmetry, only 1/8th of the complete model is used in simulations. To isolate and study only the shape effect, the spherical and ellipsoidal voids are modeled off the same volume. The radius of the spherical void is 1 μm. The ellipsoid dimensions are calculated by using $4/3\pi R^3 = 4/3\pi abc$ and assuming $b = c$ and $a = 3b$.

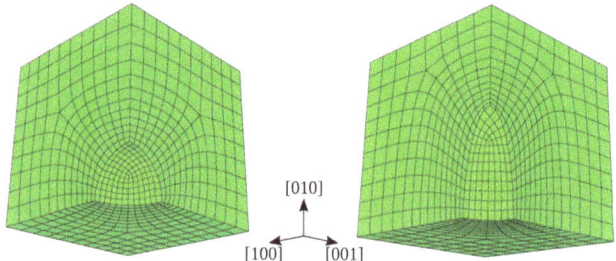

Figure 7. 1/8th spherical (**left**) and 1/8th ellipsoidal (**right**) model mesh configurations used in HIP simulations.

For the spherical void, during Stage 1 loading, stress accumulates around the void's surface as depicted in Figure 8a. Anisotropy in elastic properties contributes to high Mises stress that is observed along the <111> orientation. During the subsequent Stage 2 loading, the material creeps and the stresses around the void relax due to increasing plastic strains. Since the crystal is soft along the axial directions and stiff at the corners, the void deforms along the axial directions to reduce the total mechanical energy of the system. This is depicted in Figure 8b, where high plastic strains are observed along <001> orientations.

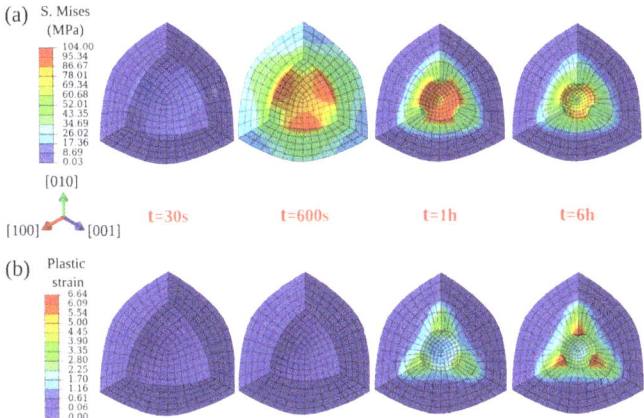

Figure 8. Von Mises stress (**a**) and plastic strain (**b**) evolution around a spherical void (zoomed area) with increasing simulation time under 100 MPa isostatic pressure at 1561 K.

For the ellipsoidal void, although the local von Mises stress is high at the acute angle (see Figure 9a), the closure starts from the waist position and results in a large plastic stain field there (see Figure 9b). This is due to the combined effect of elastic anisotropy and the complex distribution of stresses on the void surface due to its curvature.

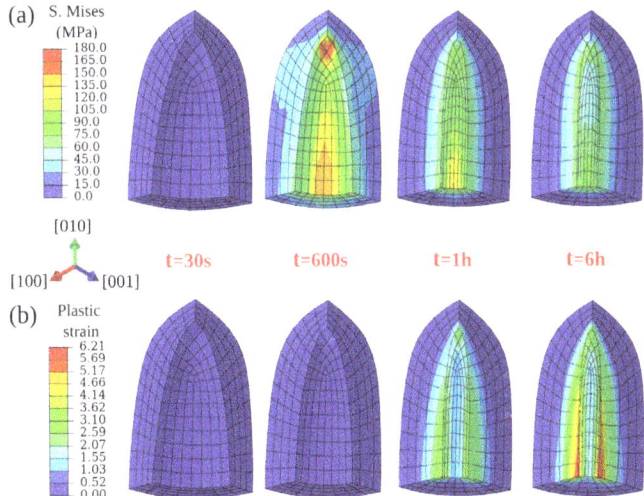

Figure 9. Von Mises stress (**a**) and plastic strain (**b**) evolution around an ellipsoidal void (zoomed area) with increasing simulation time under 100 MPa isostatic pressure at 1561 K.

Figure 10 depicts the corresponding void volume reduction for spherical and ellipsoidal voids. Although the initial void volume is the same for both kinds of voids, the closure rate is higher for the ellipsoidal void. Similar closure behavior is observed between spherical and ellipsoidal pores filled with gas. The evolution of gas pressure as a function of simulation time is plotted in Figure 11 on the secondary y-axis. Owing to the difference in the closure rates between the two-pore shapes, the gas pressure inside the pores also evolves with the difference in magnitudes. Nevertheless, since the pressure due to surface tension (P_v) is very small, the gas pressure inside the pores does not reach very high values.

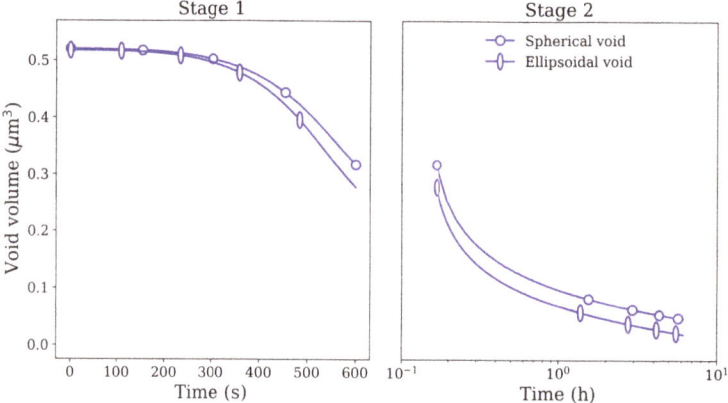

Figure 10. Void volume reduction in Stage 1 and Stage 2 HIP loading for spherical and ellipsoidal voids of identical volume under 100 MPa isostatic pressure at 1561 K.

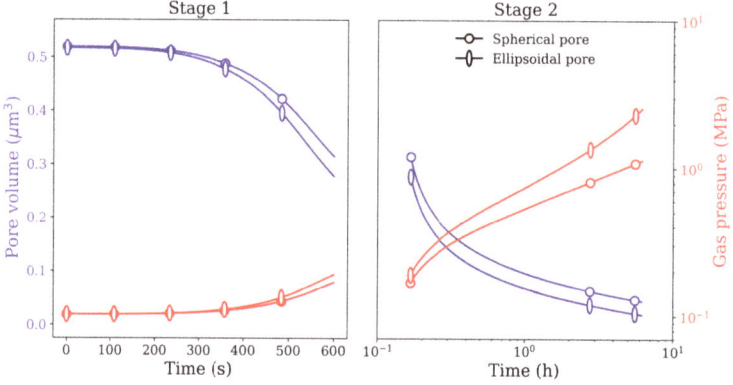

Figure 11. Pore volume reduction in Stage 1 and Stage 2 HIP loading for spherical and ellipsoidal pores of identical volume under 100 MPa isostatic pressure at 1561 K.

Apart from the two idealized void shapes considered in the current study, an additional configuration with an irregular shaped void, consistent with the real case in the scanning electron microscope (SEM) observations [28], is considered for comparison. The void is modeled as a quasi-two-dimensional structure with one layer of element thickness (0.5 μm) in the z-direction, and the whole model cannot be deformed in this direction. Loads and boundary conditions applied are as shown in Figure 12, where the bottom-left corner node is fixed in the x- and y-directions and the bottom-right corner node is fixed in the y-direction only. The other used conditions, element type, and UMAT with the determined material parameters are the same as in the previous models.

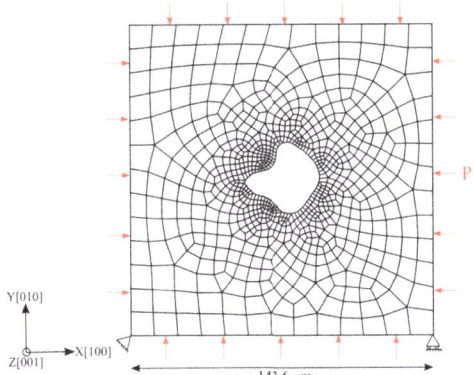

Figure 12. Quasi-2D representative CPFE model with finite element mesh and boundary conditions for the irregular shaped void. 'P' represents the isostatic pressure applied on the outer boundaries.

Figure 13 depicts the evolution of von Mises stress and plastic strain fields around the irregular shaped void during its closure. Void closure starts from the position close to the acute angles where the local von Mises stresses are high. As the void shrinks gradually, the void-centered heterogeneous stress field shrinks, whereas the corresponding plastic strain field expands. By comparing these three void shapes (irregular, spherical, and ellipsoidal), it can be said that the spherical void is the most difficult to annihilate due its geometry and the corresponding stress distribution.

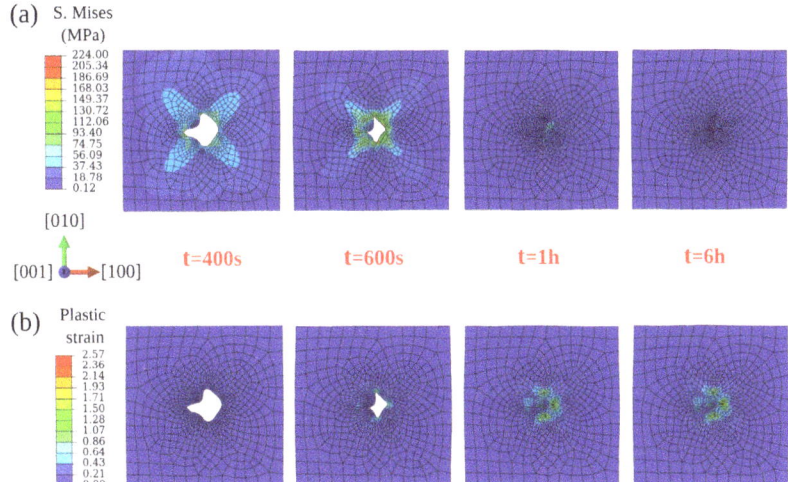

Figure 13. von Mises stress (**a**) and plastic strain (**b**) evolution around irregular-shaped void (Quasi-2D) with increasing simulation time under 100 MPa isostatic pressure at 1561 K.

3.3. Influence of Pore Size

To clarify the effect of void/pore size on the closure behavior, a comparison of three cases with the same spherical void but of different sizes (radii 1 µm, 0.1 µm, and 0.01 µm) is made here. All three void configurations are loaded under 100 MPa isostatic pressure at 1561 K. Figure 14 depicts the evolution of the void volume reduction ratio (defined as the ratio of current void volume to the initial void volume) as a function of simulation time. It can be seen that the void closure rate accelerates with decreasing void size, and the void with 0.01 µm radius closes completely during Stage 1 loading itself.

Here, the voids are considered to be closed when the void surfaces along any two orthogonal axial directions come in contact. After this stage, the solution loses its validity and the computation of void volume becomes significantly difficult.

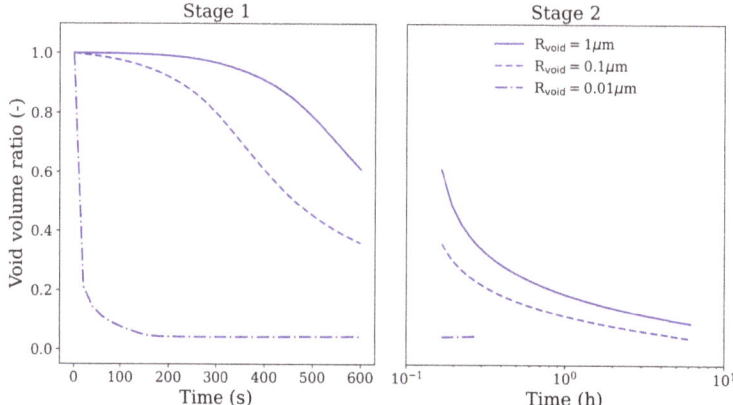

Figure 14. Void volume reduction ratio in Stage 1 and Stage 2 HIP loading for spherical voids of various radii under 100 MPa isostatic pressure at 1561 K.

The increasing closure rates for decreasing void sizes can be attributed to the effect of pressure due to surface tension (P_γ). Since P_γ varies inversely with the void size, its influence on the void surface becomes dominant at smaller void sizes. Similar closure behavior is observed for gas-filled pores of varying sizes. In Figure 15, the final pore volumes (obtained from the HIP simulations) are plotted corresponding to the initial pore volumes on a log-log plot. The gas pressures recorded at the end of each HIP simulation are also plotted corresponding to the initial pore volumes on the secondary y-axis of this figure. It can be seen that the pores with larger initial volume decrease by one or two orders in magnitude, and, as reported earlier, the gas pressure inside these larger pores is negligible and does not offer any resistance to pore closure. However, for smaller initial pore sizes, P_γ becomes more dominant, and it not only accelerates the pore closure but also makes the gas pressures reach very high values (order of GPa).

From the log-log plot, the almost linear relationship between the final and initial pore volumes dictates that these quantities have a power-law relationship. The power-law fitting is done using the nonlinear least squares fitting method [29] as shown in Figure 15, and the exponent value is calculated to be 1.57. This value indicates that the relationship between these two quantities is nonlinear and pore closure accelerates with decreasing initial pore sizes. Figure 15 also depicts the hydrostatic stress distribution in the vicinity of the pores. Snapshots are taken at the end of each simulation for three pores of radii 0.01 μm, 0.1 μm, and 1 μm. The localized concentration of hydrostatic stress around the pores can be observed in all three cases; however, the magnitude and the direction of stress vary significantly. Negative hydrostatic stress states for 1 μm and 0.1 μm pores indicate that the externally applied pressure is dominant in these pores and is thus undergoing closure. In contrast, a very high positive hydrostatic stress state observed around the 0.01 μm pore indicates that there is significant resistance to pore closure offered by the gas trapped within this pore. Although the resistance increases with decreasing pore size, due to large plastic strains, the pore surfaces establish contact and the cavity reference node becomes ill-defined and terminates the simulation.

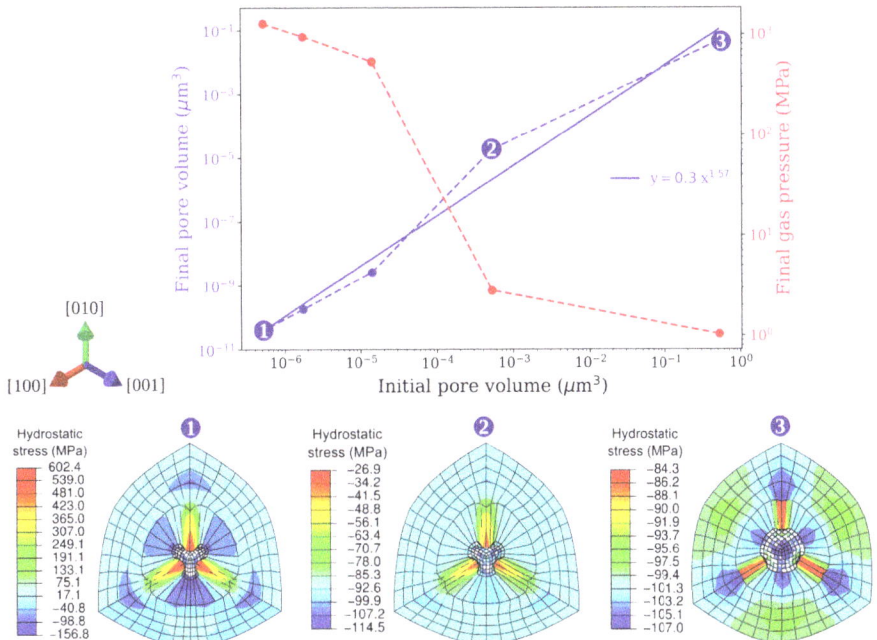

Figure 15. (**Top**) Final pore volume and final gas pressure under HIP loading as a function of initial pore volume. (**Bottom**) Snapshots of hydrostatic stress distribution around pores (zoomed area) taken at the end of simulations. Simulations are done with spherical pores of various radii under 100 MPa isostatic pressure at 1561 K.

During HIP, the complete annihilation of pores is hindered due to argon's low solubility in metal and the non-existing chemical potential difference between the pore and the surrounding environment (HIP chamber) [30]. However, both the solubility of argon in metals and the chemical potential difference generally increase with increasing pressures [31]. The significantly higher pore pressures could favor the dissolution of argon and act as a driving force to expel argon from the pores to the external environment. From the simulations conducted in this study (see Figure 15), it is evident that the pressure inside the pores reaches very high values (GPa scale) only for very small-sized pores (<0.1 µm). Thus, smaller-sized pores with very high pore pressures are more favorable for closure than larger-sized pores (≥0.1 µm radius) as they lack the necessary solubility and the chemical potential difference. Although the current model accounts for the influence of argon inside the pores, it neglects the diffusion of argon into the metal under high pore pressures. As the argon diffusion plays a significant role in pore annihilation, an in-depth investigation in this matter is expected to increase our understanding of pore healing through the HIP. In fact, a micromechanical model can be developed by coupling crystal plasticity with argon diffusion to understand the diffusion characteristics of argon in the material microstructure.

4. Conclusions

Hot isostatic pressing (HIP) simulations using various pressures and holding times were performed on a nickel-base single crystal superalloy at a temperature above the γ'-solvus. A crystal plasticity (CP) finite element model was developed to investigate the relationship between HIP process parameters, pore characteristics, including gas pressure, and pore shrinkage. In this CP-based creep model, pore shrinkage via plastic deformation is controlled by dislocation slip, and the temperature dependency is introduced through an Arrhenius term. Despite using a simple model, the initial simulation results show similar features for the pore closure as the experimental observations. The mechanical response of gas-filled pores is modeled by coupling the deformation of the material and the pressure exerted by the trapped gas on the surrounding material. In line with experimental observations, our simulations also predict significant pore reductions by either increasing the applied pressure or the holding time. It is also found from simulations that, under the same conditions for pores with the same shape, the smaller pores shrink faster than the larger ones due to increasing pore surface energies with decreasing pore sizes. This nonlinear mechanical response gives rise to very high gas pressures (orders of GPa) in smaller gas-filled pores (radius < 0.1 µm) and offers significant resistance to further pore closure. In contrast, the resistance offered by the trapped gas is negligible for larger pores (radius ≥ 0.1 µm) as the gas pressures do not reach very high values due to low pore surface energies. Furthermore, the pore shape strongly affects the rate of shrinkage through a combined effect of elastic anisotropy and the complex distribution of stresses around the pore due to its shape. The current CP model can capture various aspects of pore healing during HIP, and it can be extended to incorporate other aspects, such as diffusion of argon into the metal and the evolution of γ'-precipitates in the vicinity of pore. From a future perspective, an in-depth investigation into the re-growth of pores after HIP (due to the very high pressure in the trapped gas) can increase our understanding of optimizing the HIP process parameters.

Author Contributions: Conceptualization, M.R.G.P., N.V., and A.H.; Data curation, S.G.; Formal analysis, M.R.G.P., S.G., and A.H.; Investigation, S.G. and N.V.; Methodology, M.R.G.P. and S.G.; Project administration, N.V. and A.H.; Resources, A.H.; Supervision, N.V. and A.H.; Writing—original draft, M.R.G.P.; Writing—review and editing, N.V. and A.H. All authors have read and agreed to the published version of the manuscript.

Funding: This research received no external funding.

Acknowledgments: We acknowledge support by the DFG Open Access Publication Funds of the Ruhr-Universität Bochum.

Conflicts of Interest: The authors declare no conflict of interest.

Abbreviations

The following abbreviations are used in this manuscript:

HIP Hot isostatic pressing
AM   Additive manufacturing
CP   Crystal plasticity

References

1. Reed, R.C. *The Superalloys: Fundamentals and Applications*; Cambridge University Press: Cambridge, UK, 2006.
2. Bokstein, B.; Epishin, A.; Link, T.; Esin, V.; Rodin, A.; Svetlov, I. Model for the porosity growth in single-crystal nickel-base superalloys during homogenization. *Scr. Mater.* **2007**, *57*, 801–804. [CrossRef]
3. Link, T.; Zabler, S.; Epishin, A.; Haibel, A.; Bansal, M.; Thibault, X. Synchrotron tomography of porosity in single-crystal nickel-base superalloys. *Mater. Sci. Eng. A* **2006**, *425*, 47–54. [CrossRef]
4. Atkinson, H.; Davies, S. Fundamental aspects of hot isostatic pressing: An overview. *Metall. Mater. Trans. A* **2000**, *31*, 2981–3000. [CrossRef]
5. Mälzer, G.; Hayes, R.; Mack, T.; Eggeler, G. Miniature Specimen Assessment of Creep of the Single-Crystal Superalloy LEK 94 in the 1000 Temperature Range. *Metall. Mater. Trans. A* **2007**, *38*, 314–327. [CrossRef]
6. Wangyao, P.; Lothongkum, G.; Krongtong, V.; Homkrajai, W.; Chuankrerkkul, N. Microstructural Restoration by HIP and Heat Treatment Processes in Cast Nickel Based Superalloy, IN-738. *Chiang Mai J. Sci.* **2009**, *36*, 287–295.
7. Bor, H.; Hsu, C.; Wei, C. Influence of hot isostatic pressing on the fracture transitions in the fine grain MAR-M247 superalloys. *Mater. Chem. Phys.* **2004**, *84*, 284–290. [CrossRef]
8. Kim, M.; Chang, S.; Won, J. Effect of HIP process on the micro-structural evolution of a nickel-based superalloy. *Mater. Sci. Eng. A* **2006**, *441*, 126–134. [CrossRef]
9. Appa Rao, G.; Sankaranarayana, M.; Balasubramaniam, S. Hot Isostatic Pressing Technology for Defence and Space Applications. *Def. Sci. J.* **2012**, *62*, 73–80.

10. Whitesell, H.; Overfelt, R. Influence of solidification variables on the microstructure, macrosegregation, and porosity of directionally solidified Mar-M247. *Mater. Sci. Eng. A* **2001**, *318*, 264–276. [CrossRef]
11. Gu, D.D.; Meiners, W.; Wissenbach, K.; Poprawe, R. Laser additive manufacturing of metallic components: Materials, processes and mechanisms. *Int. Mater. Rev.* **2012**, *57*, 133–164. [CrossRef]
12. Tammas-Williams, S.; Zhao, H.; Léonard, F.; Derguti, F.; Todd, I.; Prangnell, P. XCT analysis of the influence of melt strategies on defect population in Ti–6Al–4V components manufactured by Selective Electron Beam Melting. *Mater. Charact.* **2015**, *102*, 47–61. [CrossRef]
13. Kasperovich, G.; Haubrich, J.; Gussone, J.; Requena, G. Correlation between porosity and processing parameters in TiAl6V4 produced by selective laser melting. *Mater. Des.* **2016**, *105*, 160–170. [CrossRef]
14. Blackwell, P. The mechanical and microstructural characteristics of laser-deposited IN718. *J. Mater. Process. Technol.* **2005**, *170*, 240–246. [CrossRef]
15. Yadollahi, A.; Shamsaei, N.; Thompson, S.M.; Elwany, A.; Bian, L. Effects of building orientation and heat treatment on fatigue behavior of selective laser melted 17-4 PH stainless steel. *Int. J. Fatigue* **2017**, *94*, 218–235. [CrossRef]
16. Frazier, W.E. Metal additive manufacturing: A review. *J. Mater. Eng. Perform.* **2014**, *23*, 1917–1928. [CrossRef]
17. Tammas-Williams, S.; Withers, P.J.; Todd, I.; Prangnell, P.B. The effectiveness of hot isostatic pressing for closing porosity in titanium parts manufactured by selective electron beam melting. *Metall. Mater. Trans. A* **2016**, *47*, 1939–1946. [CrossRef]
18. Beretta, S.; Romano, S. A comparison of fatigue strength sensitivity to defects for materials manufactured by AM or traditional processes. *Int. J. Fatigue* **2017**, *94*, 178–191. [CrossRef]
19. Yadollahi, A.; Shamsaei, N. Additive manufacturing of fatigue resistant materials: Challenges and opportunities. *Int. J. Fatigue* **2017**, *98*, 14–31. [CrossRef]
20. Tammas-Williams, S.; Withers, P.; Todd, I.; Prangnell, P. Porosity regrowth during heat treatment of hot isostatically pressed additively manufactured titanium components. *Scr. Mater.* **2016**, *122*, 72–76. [CrossRef]
21. Epishin, A.; Fedelich, B.; Link, T.; Feldmann, T.; Svetlov, I. Pore annihilation in a single-crystal nickel-base superalloy during hot isostatic pressing: Experiment and modelling. *Mater. Sci. Eng. A* **2013**, *586*, 342–349. [CrossRef]
22. Epishin, A.; Fedelich, B.; Nolze, G.; Schriever, S.; Feldmann, T.; Ijaz, M.F.; Viguier, B.; Poquillon, D.; Le Bouar, Y.; Ruffini, A. Creep of single crystals of nickel-based superalloys at ultra-high homologous temperature. *Metall. Mater. Trans. A* **2018**, *49*, 3973–3987. [CrossRef]
23. Ashby, M.F. Mechanisms of deformation and fracture. *Adv. Appl. Mech.* **1983**, *23*, 117–177.
24. ABAQUS. Analysis User's Manual, Version 6.14. 2017. Available online: http://130.149.89.49:2080/v6.14/books/usb/default.htm (accessed on 16 December 2020).
25. Lee, E. Elastic-plastic deformation at finite strains. *J. Appl. Mech.* **1969**, *36*, 1–6. [CrossRef]
26. MacEwan, J.R.; MacEwan, J.U.; Yaffe, L. Self-diffusion in polycrystalline Nickel. *Can. J. Chem.* **1959**, *37*, 1623–1628. [CrossRef]
27. Stevens, R.A.; Flewitt, P.E.J. Hot isostatic pressing to remove porosity & creep damage. *Mater. Des.* **1982**, *3*, 461–469.
28. Epishin, A.; Svetlov, I. Evolution of pore morphology in single-crystals of nickel-base superalloys. *Inorg. Mater. Appl. Res.* **2016**, *7*, 45–52. [CrossRef]
29. Björck, A.; *Numerical Methods for Least Squares Problems*; Society for Industrial and Applied Mathematics: Philadelphia, PA, USA, 1996.
30. Shao, S.; Mahtabi, M.J.; Shamsaei, N.; Thompson, S.M. Solubility of argon in laser additive manufactured α-titanium under hot isostatic pressing condition. *Comput. Mater. Sci* **2017**, *131*, 209–219. [CrossRef]
31. Boom, R.; Kamperman, A.A.; Dankert, O.; Van Veen, A. Argon solubility in liquid steel. *Metall. Mater. Trans. B* **2000**, *31*, 913–919. [CrossRef]

Publisher's Note: MDPI stays neutral with regard to jurisdictional claims in published maps and institutional affiliations.

© 2020 by the authors. Licensee MDPI, Basel, Switzerland. This article is an open access article distributed under the terms and conditions of the Creative Commons Attribution (CC BY) license (http://creativecommons.org/licenses/by/4.0/).

Article

Heterogeneous Deformation Behavior of Cu-Ni-Si Alloy by Micro-Size Compression Testing

Sari Yanagida [1,2], Takashi Nagoshi [3,*], Akiyoshi Araki [2], Tso-Fu Mark Chang [1,2], Chun-Yi Chen [1,2], Equo Kobayashi [2], Akira Umise [1,2], Hideki Hosoda [1,2], Tatsuo Sato [1,2] and Masato Sone [1,2,*]

1. Institute of Innovative Research, Tokyo Institute of Technology, Kanagawa 226-8503, Japan; yanagida.sari@kobelco.com (S.Y.); chang.m.aa@m.titech.ac.jp (T.-F.M.C.); chen.c.ac@m.titech.ac.jp (C.-Y.C.); umise.a.aa@m.titech.ac.jp (A.U.); hosoda.h.aa@m.titech.ac.jp (H.H.); sato.tatsuo8@gmail.com (T.S.)
2. Department of Material Science and Engineering, Tokyo Institute of Technology, Kanagawa 226-8503, Japan; a.akiyoshi.1111@gmail.com (A.A.); kobayashi.e.ad@m.titech.ac.jp (E.K.)
3. National Institute of Advanced Industrial Science and Technology, Ibaraki 305-8564, Japan
* Correspondence: nagoshi-t@aist.go.jp (T.N.); sone.m.aa@m.titech.ac.jp (M.S.); Tel.: +81-45-924-5043 (M.S.)

Received: 13 November 2020; Accepted: 18 December 2020; Published: 21 December 2020

Abstract: The aim of this study is to investigate a characteristic deformation behavior of a precipitation strengthening-type Cu-Ni-Si alloy (Cu-2.4Ni-0.51Si-9.3Zn-0.15Sn-0.13Mg) by microcompression specimens. Three micropillars with a square cross-section of $20 \times 20 \times 40$ μm^3 were fabricated by focused ion beam (FIB) micromachining apparatus and tested by a machine specially designed for microsized specimens. The three pillars were deformed complicatedly and showed different yield strengths depending on the crystal orientation. The micromechanical tests revealed work hardening by the precipitation clearly. Electron backscattered diffraction analysis of a deformed specimen showed a gradual rotation of grain axis at the grain boundaries after the compression test.

Keywords: alloy; microcompression test; precipitation strengthening; work hardening; deformation

1. Introduction

When size of materials decreases from bulk size to micrometer scale, the dislocations could easily escape to the free surface before having interaction with each other, and the dislocation density would decrease, which leads to a dislocation starved condition and causes increased strength. This is a widely accepted explanation of size-dependent deformation behavior known as the "size effect" observed in crystalline metallic materials [1]. The investigation expands to variety of metallic materials such as noncubic metals [2,3] and nanocrystalline materials [4,5]. The dislocation motion could be affected by several internal defects such as vacancies [6], grain boundaries [5] and precipitations [7].

Uchic et al. [8] reported a microcompression testing method to effectively investigate mechanical property of small materials. Yet, in a compression test, the specimen is collapsed into a flat shape during the testing, and the rupture strength cannot be measured. Furthermore, buckling occurs easily and high aspect ratio specimens cannot be used. In this technique, the constraint between the specimen and indenter is only in one axis along the yield direction, so friction between the specimen and the indenter occurs and leads to heterogeneous stress distribution in the specimen [9]. To resolve this problem, Kiener et al. [10] developed a way to conduct microtensile testing. In this technique, the specimen's gripper part and the indenter are constrained in two directions [11], so homogeneity deformation of the specimen can be gained.

In this study, micromechanical properties of a precipitation-hardened copper alloy, Cu-Ni-Si alloy, are investigated. Cu-Ni-Si alloy is a precipitation strengthening-type alloy including δ-Ni$_2$Si precipitates and receives a lot of attention for its applications for electronic components thanks to its low cost, high strength and good electrical and thermal conductivity [12–14]. These properties are also beneficial for use in microelectromechanical systems (MEMS). However, effects of the precipitates in microscale are not yet thoroughly investigated. In our previous studies, we reported microtensile tests of a pure copper and a Cu-Ni-Si alloy with microtensile specimens fabricated by focused ion beam (FIB) milling system. In the microtensile tests, both of the specimens showed characteristic large serrations during deformation, which were not observed in the bulk samples, and the obvious necking deformation led to a decrease in the flow stress [15]. For deep understanding of micromechanical behavior in precipitate strengthened metals, comparisons of the results obtained from compression and tensile tests would be valuable to understand mechanical properties of the alloys in microscale in order to utilize them to the micro- or nano-electromechanical systems. In this study, the micromechanical property of Cu-2.4Ni-0.51Si-9.3Zn-0.15Sn-0.13Mg alloy under microcompression testing is discussed.

2. Materials and Methods

2.1. Sample Composition and Heat Treatment Process

The material investigated in this study was supplied by Furukawa Electric Inc. in a sheet form (1.2mm thickness) with the chemical composition Cu-2.4Ni-0.51Si-9.3Zn-0.15Sn-0.13Mg (mass%). For the Cu-Ni-Si alloy, a heat treatment process was conducted first. The heat treatment process involved homogenization at 1223 K for 18 ks, cold rolling for 90% in thickness reduction rate and solution treatment process at 1123 K for 600 s. In order to reveal the effect of the precipitation, aging at 723 K for 300 s was performed, where the temperature and time needed for the precipitation hardening were determined in previous studies [13,15]. The grain size of the Cu-Ni-Si alloy was about 30 μm.

2.2. Fabrication Process

The microsized specimens were fabricated by a FIB (FB2100, Hitachi, Tokyo, Japan) from the bulk sample described in former Section 2.1 [16,17]. In the beginning, the sliced sample with the thickness of about 50 m was fabricated from middle part of the bulk sample. For fabrication of the pillar, the specimen was tilted at −45 ± 2 degrees to avoid tapering of the pillar. For the compression test, three micropillars (specimens A, B and C) with different grain geometry as shown in Figure 1, having the identical size of 20 × 20 μm² in cross section and 40 μm in length were fabricated.

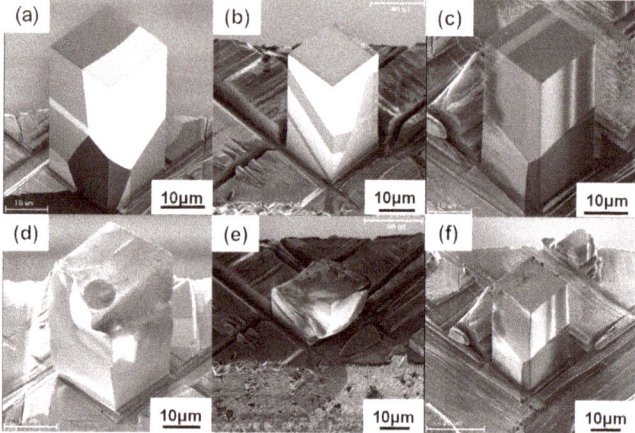

Figure 1. Scanning ion microscopy images of the three microcompression specimen AA: (**a,d**), specimen B (**b,e**), and specimen C: (**c,f**). The images were taken before: (**a–c**) and after; (**d–f**) the compression test.

2.3. Micromechanical Testing

A testing machine specially designed for microsized specimens was developed in our group [18]. The displacement was controlled at 0.1 μm/s by a piezoactuator. The test temperature was 295 ± 2K. The specimen was observed before and after the mechanical tests by a scanning electron microscope (SEM, S-4300 SE, Hitachi, Tokyo, Japan). After the test, a micropillar of the Cu-Ni-Si alloy was milled by the FIB to show cross-section of the deformed part and then observed by a scanning ion microscope (SIM) equipped in the FIB and analyzed by electron backscattered diffraction (EBSD, Bruker AXS GmbH, Karlsruhe, Germany and S-4300 SE, Hitachi, Tokyo, Japan).

3. Results and Discussion

3.1. Mechanical Behavior by Microcompression Test

SIM images of the three Cu-Ni-Si alloy micropillars before the compression test are shown in Figure 1a–c. SIM is the imaging technique using FIB in which crystallographic difference enhanced in their contrast thus can easily reveal grains and twin boundaries on surface of the micropillars. Figure 1d–f shows images of the three pillars after the microcompression test. The results indicated the deformation took place in all of the grains in the pillar. In SIM images of all of the deformed micropillars, thin, curvy and numerous slip lines with some small protrusion were observed. This tendency was also observed in pure Cu specimen which grains were constrained by grain boundaries [15].

Furthermore, after the deformation as shown in Figure 1d–f, the specimens were twisted a little. This shows grains in the microcompression specimen are constrained by grain boundaries causing the specimen to deform heterogeneously. Especially the black circle as shown in Figure 1d could be originated from the rotation deformation of a grain under microdiamond indenter. On the other hand, in a microtensile specimen of Cu-Ni-Si alloy as previous reported, the specimen was constrained in one direction only hence each grains had more freedom to deform, and the microtensile specimen deformed homogeneity [16].

Figure 2 shows the engineering stress-engineering strain (S_E-S_E) curves of the three specimens A, B and C. The yield strengths of specimens A, B and C were 278, 303 and 206 MPa, respectively. The difference in the yield stress was caused by dominantly crystal orientation difference, grain boundary, grain size and specimen size effect. In our previous work, the influence of orientations on mechanical properties of the microsized specimens of pure Cu was magnified with reduction of the number of grains inside a sample [15]. This size effect leads to scatter of experimental data of mechanical properties and can be quantified by the ratio between the specimen size: S and average grain size: G, denoted as S/G [19,20]. In this case, S/G ratio of single crystal specimen is "1". The deformation behavior of a single grain is strongly dependent on crystallographic orientation. On the other hand, the deformation of the bulk polycrystalline materials with large S/G ratio is isotropic because orientations of crystals are evenly and randomly distributed within the specimen. When the number of grains within a specimen becomes lower, it means with a S/G ratio smaller than 30, the effect of orientations of each grain and geometry of them becomes stronger in deformation behavior and the S/G ratio of the micropillars in this study was below 30. Thus the yield strength can be different in each pillar. In specimen C, the grains at the top of the pillar were relatively large and yielded in single grains thus having low yield strength and limited work hardening. The detailed effects of the crystal orientation and grain boundary on the grain deformation are discussed below.

The S_E-S_E curves of the microcompression pillar also show work hardening during plastic deformation. Especially in the compression test, the pillar deformed into the shape of a barrel, which is another cause of the work hardening behavior in the engineering stress–strain behavior. In the microtensile study [15], typical serrations during the plastic deformation were observed in the S_E-S_E curve. However, in the compression test, serrations were not observed and the curve is similar to the results obtained with a bulk specimen. The SEM image in Figure 1 showed that the microcompression specimen has smaller and more slip steps than that of the microtensile specimen.

Flat and heterogeneous deformation without any load drop in the microcompression specimen indicates that each grain deformed with small and many slip system activities. On the other hand, in the microtensile test, the deformation was homogeneous and the grains deformed with large slip, and this behavior led to large load drops during the plastic deformation.

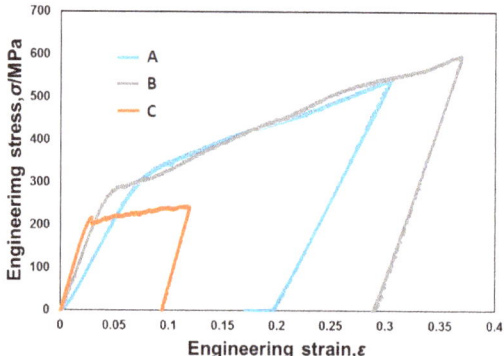

Figure 2. Engineering stress–engineering strain curves of microcompression tests for Cu-Ni-Si alloy specimens A, B and C.

3.2. Work Hardening Behavior of MicroCompression Test

Figure 3 shows work hardening behavior of specimen A in the microcompression test and indicates the enhanced work hardening. The work hardening rate was calculated by the following equation:

$$\Theta = \Delta\sigma/\Delta\varepsilon \tag{1}$$

using the true stress ε; and true strain σ; which were reduced from the engineering stress and the engineering strain observed, respectively. In the compression test, repetition of large bursts and drops could not be observed unlike the microtensile test reported in a previous study [15]. The steady state of work hardening rate curve observed at about 600 MPa may have been caused by the precipitation. The precipitates of δ-Ni$_2$Si forms as disc on (110) matrix plane and the orientation relationship between Ni$_2$Si and matrix is (110)Cu//(001) Ni2Si, [001]Cu//[010] Ni2Si [12,13]. They are reported to work as strong obstacles against the dislocation motion [12,13]. Nonshearable obstacles facilitate the dislocation multiplication by the Orowan mechanism, and generated dislocations are accumulated on the strong grain boundary which may lead to the enhanced work hardening in the Cu-Ni-Si alloy resulted in the steady state of work hardening rate curve.

Figure 4 shows three SIM images of specimen A taken from different directions. In the middle section of the edge part, complex distorted deformations with many shear lines were observed in Figure 4a. Figure 4b,c shows the side view of the distorted deformations, which are from the side plane of the pillar with a square cross section. These findings imply strain concentrates at the grain boundaries and/or near the grain boundaries, and the grain axis rotated gradually during the deformation. Figure 5 shows the crystal characteristic of the Cu-Ni-Si alloy specimen A after the compression test. The color in Figure 5b represents the orientations in inverse pole figure (IPF) against pillar axis shown in the inset. The orientations of grains α and β are plotted in IPF shown in Figure 5c and d. Grain α and β had near <110> and <111> orientations at surface, respectively. Both grains deformed heterogeneously and grain β showed gradual rotation with many slip lines in the side plane. This gradual rotation could come from the different compressive strength values of two or more grains with strong grain boundaries in the pillar. δ-Ni$_2$Si forms as disc on (110) matrix plane and the orientation relationship between Ni$_2$Si and matrix is (110)Cu//(001) Ni$_2$Si, [001]Cu//[010] Ni$_2$Si [12,13].

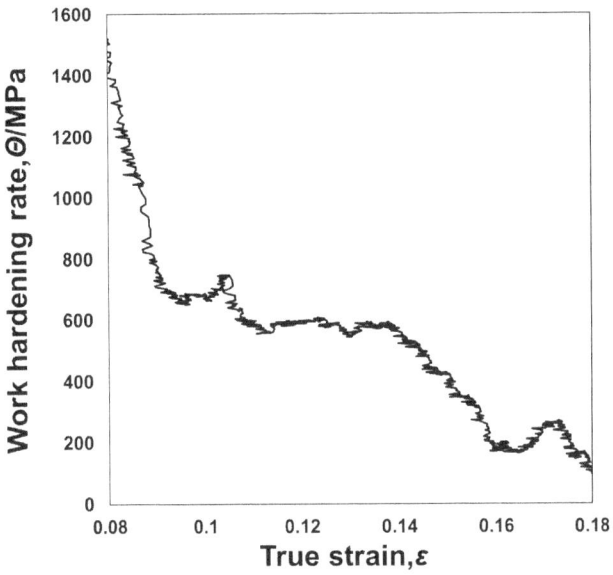

Figure 3. Work hardening behavior of Cu-Ni-Si alloy on microcompression test.

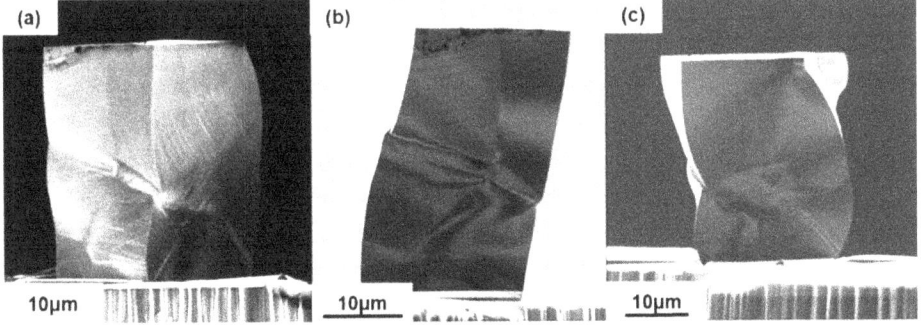

Figure 4. Scanning ion microscope (SIM) images of sample A in Cu-Ni-Si alloy specimens after microcompression test, observed taken from the front (**a**) and sides (**b**,**c**).

Figure 6 shows EBSD mapping of the longitudinal section (Figure 5a) of the deformed specimens. The change in color inside the grains shows gradual rotation of the crystal orientation and the orientation near grain boundaries not determined due to the distorted crystals by high density of dislocation at the boundary, which indicates high strength of the grain boundary in this alloy. In the compression test, stress concentrates at grain boundaries, and, near the grain boundaries, the grain axis rotated gradually. This result suggests that the grain boundary could be strengthened by the δ-Ni_2Si precipitates. The precipitates are reported to have formed in a shape of disk lying on {110} planes and are not shearable by dislocations, thus, they work as strong obstacles against the dislocation motion [12,13].

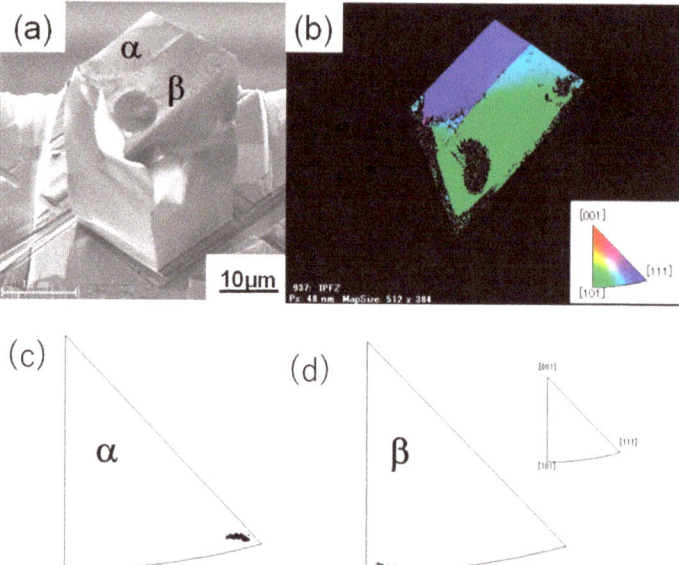

Figure 5. Crystal characteristic of the Cu-Ni-Si alloy specimen A after the compression test observed by (**a**) SIM and (**b**) electron backscattering diffraction (EBSD) image (ND) with inverse pole figures (Z axis) of grains α and β shown in (**c**,**d**) respectively.

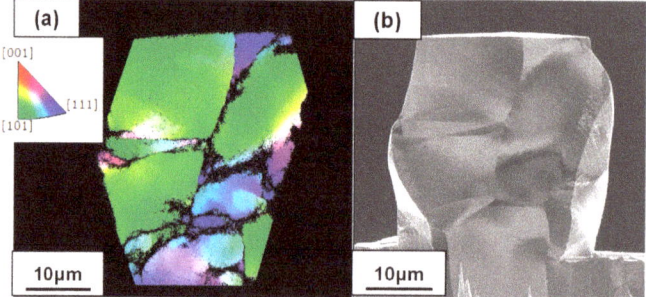

Figure 6. Crystal orientation analysis on longitudinal section of Cu-Ni-Si alloy specimen (correspond to Figure 5a) after the compression test observed by (**a**) EBSD mapping and (**b**) SIM image.

4. Conclusions

Microcompression tests of Cu-Ni-Si alloys were conducted. The mechanical tests of three micropillars showed different yield strength values ranged from 200 to 300 MPa, and this difference is considered to be contributed by the size effect. All of the work hardening curves exhibit plateau regimes, which is similar to the bulk specimen, and the specimen deformed heterogeneously. In the microcompression test, the SIM images showed small slip steps, which mean heterogeneous deformation by less restriction between the indenter and the specimen. EBSD analysis indicated a typical gradual rotation of crystal orientation and high accumulation of dislocation at the grain boundaries which imply enhanced dislocation multiplication at precipitates by the Orowan mechanism. The study showed precipitation effectively strengthened the metals and enhanced work hardening even in the microspecimens with few grains inside. The findings indicate the practical applications of precipitation-hardened materials for microcomponents used in MEMS.

Author Contributions: Conceptualization, S.Y., T.-F.M.C.; Methodology, S.Y., A.U., T.N.; Validation, C.-Y.C., E.K.; Formal analysis, S.Y.; Investigation, S.Y., T.-F.M.C.; Resources, A.A., T.S.; Data curation, S.Y., T.N.; Writing—original draft preparation, S.Y., Writing—review and editing, T.N., T.-F.M.C., M.S.; Visualization, S.Y., Supervision, T.-F.M.C., T.S., M.S.; Project administration, H.H., M.S.; Funding acquisition, T.S., H.H., M.S. All authors have read and agreed to the published version of the manuscript.

Funding: This research was funded by the Grant-in-Aid for Challenging Research (Pioneering) (JSPS KAKENHI Grant number 20K20544) and CREST Project (JST CREST Grant Number JPMJCR1433) by the Japan Science and Technology Agency. This work was performed under the support of the Science and Technology Agency (STA) of the Japanese government.

Conflicts of Interest: On behalf of all of the co-authors, the corresponding author states that there is no conflict of interest.

References

1. Greer, J.R.; De Hosson, J.T.M. Plasticity in small-sized metallic systems: Intrinsic versus extrinsic size effect. *Prog. Mat. Sci.* **2011**, *56*, 654–724. [CrossRef]
2. Bei, H.; Shim, S.; Miller, M.K.; Pharr, G.M.; George, E.P. Effects of focused ion beam milling on the nanomechanical behavior of a molybdenum-alloy single crystal. *Appl. Phys. Lett.* **2007**, *91*, 111915. [CrossRef]
3. Lilleodden, E. Microcompression study of Mg (0 0 0 1) single crystal. *Scr. Mater.* **2010**, *62*, 532–535. [CrossRef]
4. Jang, D.; Greer, J.R. Size-induced weakening and grain boundary-assisted deformation in 60 nm grained Ni nanopillars. *Scr. Mater.* **2011**, *64*, 77–80. [CrossRef]
5. Nagoshi, T.; Mutoh, M.; Chang, T.F.M.; Sato, T.; Sone, M. Sample size effect of electrodeposited nickel with sub-10 nm grain size. *Mater. Lett.* **2014**, *17*, 256–259. [CrossRef]
6. Zhang, Z.; Shao, C.; Wang, S.; Luo, X.; Zheng, K.; Urbassek, H.M. Interaction of dislocations and interfaces in crystalline heterostructures: A review of atomistic studies. *Crystals* **2019**, *9*, 584. [CrossRef]
7. Huang, K.; Marthinsen, K.; Zhao, Q.; Logéa, R.E. The double-edge effect of second-phase particles on the recrystallization behaviour and associated mechanical properties of metallic materials. *Prog. Mater. Sci.* **2018**, *92*, 284–359. [CrossRef]
8. Uchic, D.; Dimiduk, D.M.; Florando, J.N.; Nix, W.D. Sample dimensions influence strength and crystal plasticity. *Science* **2004**, *305*, 986–989. [CrossRef] [PubMed]
9. Kulkarni, K.M.; Kalpakjian, S. A study of barreling as an example of free deformation in plastic working. *J. Eng. Ind.* **1969**, *91*, 743–754. [CrossRef]
10. Kiener, D.; Grosinger, W.; Dehm, G.; Pippan, R. A further step towards an understanding of size-dependent crystal plasticity: In situ tension experiments of miniaturized single-crystal copper samples. *Acta Mater.* **2008**, *56*, 580–592. [CrossRef]
11. Lee, H.J.; Choi, H.S.; Han, C.S.; Lee, N.K.; Lee, G.A.; Choi, T.H. A precision alignment method of micro tensile testing specimen using mechanical gripper. *J. Mater. Process. Technol.* **2007**, *187*, 241–244. [CrossRef]
12. Lockyer, S.A.; Noble, F.W. Precipitation structure in a Cu-Ni-Si alloy. *J. Mater. Sci.* **1994**, *29*, 218–226. [CrossRef]
13. Araki, A.; Poole, W.J.; Kobayashi, E.; Sato, T. Twinning induced plasticity and work hardening behavior of aged Cu-Ni-Si alloy. *Mater. Trans.* **2014**, *55*, 501–505. [CrossRef]
14. Krupinska, B.; Rdzawski, Z.; Krupinski, M.; Pakieła, W. Precipitation Strengthening of Cu–Ni–Si Alloy. *Materials* **2020**, *13*, 1182. [CrossRef] [PubMed]
15. Yanagida, S.; Araki, A.; Chang, T.F.M.; Chen, C.Y.; Nagoshi, T.; Kobayashi, E.; Hosoda, H.; Sato, T.; Sone, M. Deformation behavior of pure Cu and Cu-Ni-Si alloy evaluated by micro-tensile testing. *Mater. Trans.* **2016**, *57*, 1897–1901. [CrossRef]
16. Nagoshi, T.; Chang, T.F.M.; Sone, M. Evaluations of mechanical properties of electrodeposited nickel film by using micro-testing method. *Mater. Trans.* **2016**, *57*, 1979–1984. [CrossRef]
17. Mutoh, M.; Nagoshi, T.; Chang, T.F.M.; Sato, T.; Sone, M. Micro-compression test using non-tapered micro-pillar of electrodeposited Cu. *Microelectron. Eng.* **2013**, *111*, 118–121. [CrossRef]
18. Zhang, G.P.; Takashima, K.; Higo, Y. Fatigue strength of small-scale type 304 stainless steel thin films. *Mater. Sci. Eng. A* **2006**, *426*, 95–100. [CrossRef]

19. Chan, W.L.; Fu, M.W.; Lu, J.; Liu, J.G. Modeling of grain size effect on micro deformation behavior in micro-forming of pure copper. *Mater. Sci. Eng. A* **2010**, *527*, 6638–6648. [CrossRef]
20. Chen, X.X.; Ngan, A.H.W. Specimen size and grain size effects on tensile strength of Ag microwires. *Scr. Mater.* **2011**, *64*, 717–720. [CrossRef]

Publisher's Note: MDPI stays neutral with regard to jurisdictional claims in published maps and institutional affiliations.

© 2020 by the authors. Licensee MDPI, Basel, Switzerland. This article is an open access article distributed under the terms and conditions of the Creative Commons Attribution (CC BY) license (http://creativecommons.org/licenses/by/4.0/).

Article

Computational Modeling of Dislocation Slip Mechanisms in Crystal Plasticity: A Short Review

Khanh Nguyen [1,*], Meijuan Zhang [2], Víctor Jesús Amores [1], Miguel A. Sanz [1] and Francisco J. Montáns [1,3,*]

[1] E.T.S. de Ingeniería Aeronáutica y del Espacio, Universidad Politécnica de Madrid, Plaza Cardenal Cisneros 3, 28040 Madrid, Spain; victorjesus.amores@upm.es (V.J.A.); miguelangel.sanz@upm.es (M.A.S.)
[2] Imdea Materials Institute, Tecnogetafe, Eric Kandel Street 2, 28906 Getafe, Spain; meijuan.zhang@imdea.org
[3] Department of Mechanical and Aerospace Engineering, Herbert Wertheim College of Engineering, University of Florida, Gainesville, FL 32611, USA
* Correspondence: khanhnguyen.gia@upm.es (K.N.); fco.montans@upm.es (F.J.M.)

Received: 9 December 2020; Accepted: 30 December 2020; Published: 4 January 2021

Abstract: The bridge between classical continuum plasticity and crystal plasticity is becoming narrower with continuously improved computational power and with engineers' desire to obtain more information and better accuracy from their simulations, incorporating at the same time more effects about the microstructure of the material. This paper presents a short overview of the main current techniques employed in crystal plasticity formulations for finite element analysis, as to serve as a point of departure for researchers willing to incorporate microstructure effects in elastoplastic simulations. We include both classical and novel crystal plasticity formulations, as well as the different approaches to model dislocations in crystals.

Keywords: crystal plasticity; dislocation slip; phenomenological model; physics-based model

1. Introduction

Analysis of the plastic deformations of metals is very important in many aspects of the design of metallic goods, for example, in the design of the alloys which will be used in the components, in their manufacturing process (e.g., shape forming), and in the behaviour of the material in service, both to sustain the ordinary actions and to behave in a controlled, predictable way (fracture, fatigue, crash-worthiness, etc.) [1,2].

Plastic slip is the most common plastic deformation mechanism in crystalline solids, e.g., most of the metals. This slip takes place as a plastic shearing in specific planes and directions governed by the arrangement of the atoms in a regular crystal structure [3]. This arrangement produces anisotropic elastic and plastic properties depending on the packing of atoms into planes. Compact planes slide in particular directions, a movement which is facilitated by the presence of dislocations (defects in the crystal arrangement, e.g., by the absence of atoms in the structure, line defects). Plastic slips in crystals is a plastic process that does not change the volume. However, for materials whose dominant deformation mechanism is slip, it is usually the origin of yield stress and hardening in macroscopic plasticity. Then, the study of crystal plasticity is fundamental to understand in a rational way the macroscopic behavior of the polycrystal, and how different aspects of the microstructure affect this behavior and the observed critical quantities like yield stress, elastoplastic anisotropy, anisotropic local and global deformation, hardening, thermal properties, etc. [4]. In the late 20th and beginning of the the 21st Century, after the work of Pierce et al. [5], thanks to improved computational power, crystal plasticity finite element simulations have received an important boost because these multiscale procedures, in which the studied phenomena has a length scale of µm, can now be addressed computationally. Then, computational homogenization schemes [6] are a typical procedure to bring the microscale complexity to the macroscale in a simple, standarized way. Thus, current computational power increasingly facilitates this type of analysis, bringing classical materials science to the engineering office [7].

Whereas the scale at which dislocation movements takes place is much smaller than the continuum scale (or precisely because of that, which facilitates the requirement of separation of scales), many effects may be studied at a mesoscale if we consider that a given stress integration point represents a homogenization of the grain, or part of the grain, so crystal plasticity computations give an implicit Representative Volume Element (RVE) of the dislocation theory, being the direction of the Burgers vector the sliding direction, and its modulus the computed sliding times the length of the RVE [8,9].

The choice of crystal plasticity theory as the simulation tool for different inelastic phenomena depends itself on the objectives of such simulation. Several aspects of the behavior of materials are easily extracted from this type of simulations. Examples are texture evolution (e.g., anisotropy symmetries evolution as a result of the evolution of crystallographic planes and grain structure), effects of grain size, effects of twinning or impurities, speed of deformation, influence of grain shape (e.g., after cold forming), etc. Furthermore, damage in these materials due to void nucleation in grain boundaries has also been studied through crystal plasticity simulations (e.g., [10,11], see also [12]). The coupling between the crystal plasticity and phase-field damage is also proposed to study the fracture [13,14]. There are other options, mainly computational, to model plasticity-based phenomena at different scales, specially through the combination of one scale with a larger one in which the desired effect needs to be accounted for. Examples are atomistic models [15,16], dislocation dynamics models [17], and continuum models [18,19]. However, crystal plasticity is probably the most used framework for incorporating the grain and crystal structure as well as the inherent effects of dislocations in the continuum modelling. Then, in this paper after a brief review of the physical aspects, we perform a review of the main ingredients and formulations employed today in crystal plasticity simulations. We finish with some demonstrative simulations as examples of the possibilities of crystal plasticity. The purpose of the review is to serve as an introductory compact presentation of current mostly used formulations for researchers (e.g., continuum mechanics ones) becoming involved in the subject or willing to incorporate texture effects in advanced simulations, for example, by computational homogenization. It is by no means a complete review, so many important works and formulation aspects, are omitted. Advanced long reviews taylored for researchers in materials science are available elsewhere [20–22].

2. Mechanism of Dislocation Slip: Physical Aspects

Crystals have different structural arrangements of atoms, see e.g., Figure 1. Directions and planes are labeled according to Miller indices with reference to the unit cell edge vectors e_i. For example, direction $s_g = 1/\sqrt{2}(-1e_1 + 1e_2 + 0e_3)$ in a face-centered cubic (FCC) is referred to direction $[\bar{1}10]$ (in brackets, using integers and with possible minus signs over the number), and the plane perpendicular to $m_g = (1e_1 + 1e_2 + 1e_3)/\sqrt{3}$ is referred to as plane (111) in the present FCC case; see Figure 2. Equivalent families of directions and planes (physically equivalent, obtained by rotations/symmetries) are referred to as $\langle 110 \rangle = [\bar{1}10], [1\bar{1}0], [110]$ and $\{111\} = [111], [\bar{1}11], [1\bar{1}1], [11\bar{1}]$, respectively for a FCC. In the case of hexagonal crystals, four indices are often employed to allow permutations when referring to families, one index along each possible edge direction, being the last one the height c direction, Figure 1. The prismatic planes of HCP are (0001) and $\{1\bar{1}00\} = (10\bar{1}0), (01\bar{1}0), (\bar{1}100)$, i.e., $(hkl) \to (hkil)$ with $i = -h - k$ (not normal), whereas the directions conversion is $(uvw) \to (UVTW)$ with $U = (2u - v)/3, V = (2v - u)/3$,

$T = -(u+v)$, $W = w$, e.g., the edges are the permutations of $[2\bar{1}\bar{1}0]$ and $[0001]$. The FCC arrangement (Figure 2a) has atoms at the vertices and center of the faces of an ideal hexahedron, resulting in a dense arrangement, and is typical of specially ductile materials (e.g., Al, Cu, Au, Ag). FCC crystals have four $\{111\}$ planes with three $\langle 110 \rangle$ directions in each plane, therefore in total have 12 slip systems, referred to as $\{111\}\langle 110 \rangle$. Body centered cubic crystals (BCC), (Cr, Fe, W) have atoms in the vertices and in the center of the hexahedron, and usual slip planes in the $\{110\}$ family and direction $\langle \bar{1}11 \rangle$ (secondary slip systems may become active for some materials and temperatures). Hexagonal close-packed arrangements have planes $\{0001\}$, $\{10\bar{1}0\}$, $\{10\bar{1}1\}$ with directions $\langle 11\bar{2}0 \rangle$, although the basal plane slip $\{0001\}$ is favored for high height/base ratios (e.g., Mg, Co, Zn), and the other ones for lower ratios (e.g., Be, Ti, Zr); these materials have a more limited slip.

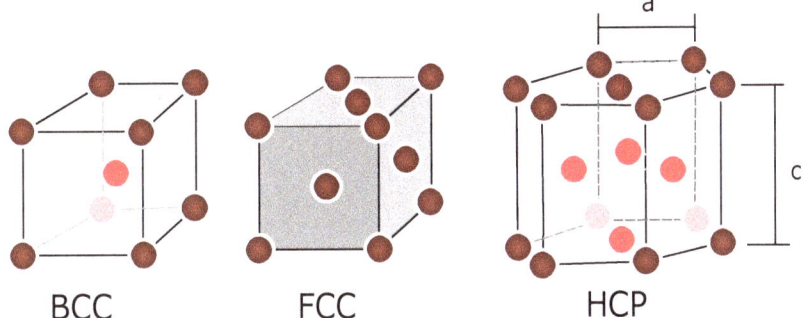

Figure 1. Three typical crystal structures in metals. BCC = Body-Centered Cubic. FCC = Face-Centered Cubic. HCP = Hexagonal Close-Packed. The form ratio for HCP is c/a, ideally $c/a = 1.632$, but for example, cadmium has $c/a = 1.886$ and berilium has $c/a = 1.586$ [23].

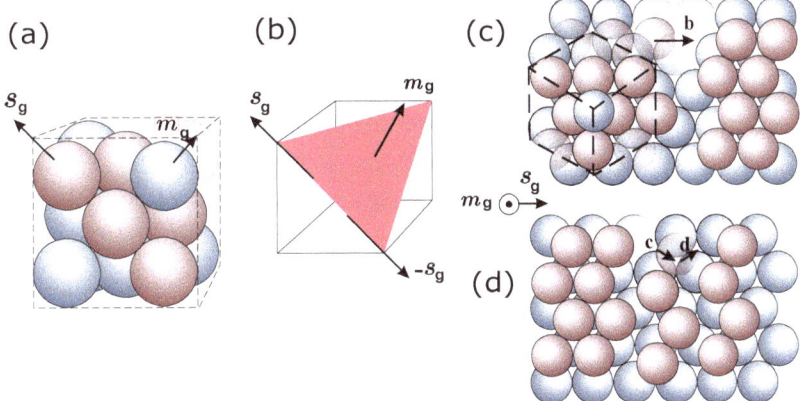

Figure 2. (a,b) Slip plane m_g and direction s_g in a FCC crystal. (c) Total dislocation in a FCC; $b = \frac{1}{2}[\bar{1}10]$ is the Burgers vector. (d) Partial dislocation in a FCC lattice, where the dissociation is given as $b = c + d = \frac{1}{6}[\bar{1}2\bar{1}] + \frac{1}{6}[\bar{2}11]$. Adapted from [23,24].

Dislocations are crystal defects which result, for example, from the absence of some atoms in the otherwise regular arrangement, see Figure 2c,d. Microscopically, slip is a result of the glide of many dislocations, facilitated by the stress field generated around them. This glide does not happen simultaneously in all the grain, but occurs in a worm-like movement, see Figure 3. The glide motion of a dislocation means that the dislocation moves in the plane which contains both its line and the Burgers vector b (Figure 2c). Dislocations and their associated local stress fields interact, blocking or further facilitating their movement. The accumulation of hundreds or even thousands of these displacements form a slip band. These slip bands can be easily observed on the surfaces of polished specimens which have been subjected to plastic deformation. Dislocations, slip bands and slip displacements are small with respect to the size of the grains, so even when the grain microstructure is considered, this plastic slip can be treated as homogeneous from a macroscopic standpoint [25]. The dislocation mechanisms can be homogenized in the form of statistical dislocation populations, and therefore, can be embedded in state evolution constitutive equations in a crystal plasticity theory.

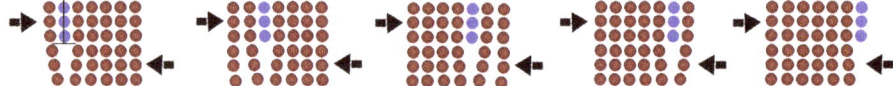

Figure 3. Progressive movement of a dislocation resulting in shear plastic deformation.

For a certain crystal, a slip system is determined by the normal direction of the slip plane m_g (which we can call the plane "direction") and the slip direction s_g in the crystal coordinate system. In Figure 2a,b, the plane formed by red atoms is a typical slip plane in a FCC crystal. At the atomic level, the forming of slip requires some breaking and reforming of dislocation-stressed interatomic bonds. This can be interpreted as a plane of atoms sliding over another parallel one on the slip plane, with the sliding direction as the slip direction. This is a volume-preserving shearing process that is typically not affected by the hydrostatic stress [26]. Slip planes m_g are planes with the highest packing or atom density, whereas the slip directions s_g are usually directions in these slip planes in which the atoms have the closest packing, see again Figure 2. The dyad $s_g \otimes m_g$ constitutes the slip system and $\dot{\gamma}_g$ is the slip (shearing) rate. Sliding may occur in the positive $+s_g$ and negative $-s_g$ directions (hence a FCC crystal may have $4 \times 3 = 12$ slip systems).

Other important quantities of a slip system are the resolved shear stress (RSS) and the critical resolved shear stress (CRSS). The former is the shear stress in the plane and direction of the slip system, and the latter is the critical value for which sliding occurs. For most crystals, the CRSS values of a certain slip system for both "+" and "−" directions can be assumed to be the same [25].

The CRSS of a slip system depends on many factors, among them previous slips and the crystal structure itself, because they play an important role due to the difficulty in breaking and reforming interatomic bonds. Also, an increase in the temperature (increasing vibration) or decrease in the strain rate (facilitating adaptation) increase the thermal activation of atoms and allow dislocations to overcome obstacles more easily. This results in an effective lower CRSS [27–29]. Of course, the modification of the material composition by adding other elements increase in general obstacles (breaking crystal regularity) and could interrupt the glide of the dislocations as the case of BCC iron micropillar [30]. Dislocations pile-up occurs as a cluster of dislocations which are unable to move past the grain boundary. Each successive dislocation will apply a repulsive force to the dislocation incident with the grain boundary. These repulsive forces act as a driving force to help overcoming the energetic barrier for diffusion across the boundary, such that additional pile up causes dislocation diffusion across the grain boundary, allowing further plastic deformation in the material. Decreasing grain size decreases the statistical amount of possible pile up at the boundary, increasing the amount of applied stress necessary to move a dislocation across a grain boundary. The higher the applied stress needed to move the dislocation, the higher the macroscopic yield stress [27]. This effect is well known as the Hall-Petch effect/law [31,32]. Of course, given a grain size, these effects deppend on the amount of statistical dislocations and other deffects.

In essence, slip plane m_g, slip direction s_g and the CRSS τ_g^c are crucial features of a slip system. According to the Schmid law, they determine which slip systems are "active" (sliding) [33]. The crystal structure is anisotropic, and the elasticity and plastic slip are also anisotropic. Furthermore, different slip systems can have different CRSS, which further determine the anisotropy of the material. This holds especially if the material is a single crystal or a polycrystal with a strong texture. Only when the material is a polycrystal that has a large number of different crystallographic orientations, the material can be considered as statistically isotropic on average; and von Mises plasticity is adequate. If the crystallographic orientation has a defined statistical distribution, continuum anisotropic models using Hill's or Barlat's yield functions with anisotropic or isotropic elasticity may also be adequate in some cases. However, note that in continuum plasticity tracking the macroscopic evolution of such anisotropy is not simple [34], which is a good reason to resort to crystal plasticity.

Rate-dependent crystal plasticity modeling is all about computing the shear amount of the activated slip systems from which the overall plastic deformation follows. The key point becomes how to describe/model the evolution of the CRSS of slips. The "hardening" evolution of the CRSS of slips is caused microscopically by the overcoming obstacles during slip gliding and from the interaction with other dislocations, as well as from the generation and annihilation of dislocations. As mentioned, temperature and strain rate are related to the thermal activation, so they are also related to the easiness for a dislocation to glide [27]. Different types of obstacles (as grain boundaries, solutes, and precipitates or other phases) may hinder gliding [30]. Of course, some factors changing the dislocation density or which may influence the interactions among dislocations may contribute to work hardening. An important one is the already pre-existing dislocation density. As well known, this could be reduced by heat treatments like annealing. Active secondary slip systems can cause multiplication of dislocations. Slip transfer at grain boundaries and cross slip can cause annihilation of dislocations. Stress concentrations (mainly at grain boundaries) can even cause the nucleation of dislocations. In practice, the different cross-interactions between dislocations which increase the CRSS are accounted for by means of the latent hardening [35–37].

As explained below, to prescribe the evolution of the CRSS of the slip system, different models have been proposed in the literature by relating its value and evolution to different variables during deformation. One widely used category follows a "phenomenological" approach [5,38–40]. They relate the evolution of the CRSS with the shear rate of all the activated slip systems. Another popular type are the "physical" models, which are dislocation-density based. This type of model relates the evolution of the CRSS with the evolution of the dislocation density of all the slip systems, e.g., the Taylor law [41], the early models [42,43]. This approach is extended to account for more detailed factors that influence hardening, e.g., the annihilation of dislocations [44]; the latent hardening [45–47]; the grain size effect [48]; the slip transfer at the grain boundary [49]; the influence of strain gradient [50]; the incorporation of phase transformation [51].

Dislocation slip mechanisms are not the only ones present in crystals producing permanent deformations. Aside, there are other types of mechanisms as displacive (difusionless) transformations and twinning. The former is a rearrangement of the atoms of a crystal structure to form another structure without a relevant change in volume. The most important case appears in carbon steels (e.g., martensitic transformations). Another plastic mechanism is twinning deformation, which produces mirrored structures. The treatment of these additional mechanisms are out of the scope of this review.

3. Continuum Framework of Crystal Plasticity

3.1. Kinematics

As mentioned, the structure of a crystalline solid at the μm scale may be understood, and hence it is often modelled, as an aggregate of volume elements (finite elements), or lattice blocks (voxels). This constitutes the reference "undeformed" lattice, which contains dislocations (hence microscopically deformed); Figure 2. At the continuum scale (e.g., at a stress integration point in a finite element, computed from the reference coordinates X, the updated coordinates $x = X + u$, and displacement field u as $F = \partial x/\partial X = I + \nabla_X u$ with ∇_X being the reference-coordinates gradient), the deformation can be considered as composition of two basic mechanisms that take place simultaneously: (1) slip between lattice blocks (as an homogenization of dislocation slips), followed by (2) elastic lattice distortions which include (3) lattice rotations (local or "elastic" and rigid-body) that also arise under a general deformation gradient F. Rotations produce observable phenomena of geometric origin, such as the geometric softening of crystals due to their reorientation [52] (similar to structural P-δ effects). In consequence, at large deformations, Kröner and Lee [53] introduced a multiplicative decomposition of F into plastic slips and crystal "elastic" deformations [54,55]. Whereas F is a compatible deformation gradient (X-integrable from the potential $u(X)$ or F^{-1} is x-integrable from $u(x)$), it is decomposed into two (local, incompatible) deformation components as:

$$F = F_e F_p \tag{1}$$

where the plastic deformation gradient F_p is irreversible and transforms the reference configuration into a fictitious local state, termed intermediate configuration. This ("elastically unloaded") intermediate configuration is "stresses-free" and is considered to maintain an "undeformed" slipped lattice. In turn, the elastic deformation gradient F_e is the deformation considered reversible, and includes rigid body rotations; namely after an overimposed rotation Q we have $F^+ := QF = QF_e F_p = F_e^+ F_p$. The possible indeterminacy of the intermediate configuration respect to arbitrary rotations of the type $F_e Q^T Q F_p$ has involved several discussions (see e.g., [56–58]). In summary, specially in the case of crystal plasticity, F_p is continuously time-integrated, so it is always uniquely defined. Remarkably, the conservative/dissipative character of the two gradients give an important physical difference in their polar decompositions: whereas $F_e = R_e U_e$ is meaningful in the sense that R_e are path-independent rotations which in principle do not enter the constitutive equations and U_e are path-independent stretches that may be used in a stored energy as state variables, $F_p = R_p U_p$ is physically meaningless because the crystallographic slip does not change the orientation of the crystal lattice even though F_p include mathematically a rotational part; stretch and rotations are related by the isoclinic flow and must be considered in an incremental setting. Indeed, this is an argument to build the formulation in terms of elastic measurements, avoiding also weak-invariance issues [59,60].

As mentioned, whereas F is compatible, the gradients F_e and F_p are incompatible, meaning that for general non-homogeneous deformations we have

$$\bar{b} = -\oint_\Gamma d\xi = -\oint_\Gamma F_p \cdot dX = -\oint_\Gamma dX \cdot F_p^T \neq 0 \tag{2}$$

where $d\xi$ is the transformation in the intermediate configuration of the material differential dX, and Γ is the transformation of a close circuit C in the reference configuration, i.e., $\oint_C dX = 0$. Even though the gradients are to be interpreted in a continuum sense (at a scale much larger than that of the dislocations), $\bar{b}$ represents the resultant Burgers vector (net Burgers vector of the dislocations tansversing Γ); it may be conceptually interpreted as the Burgers vector b (the sign may be reversed) when one dislocation is present;

i.e., $\bar{b}$ has a modulus proportional to the amount (density) of dislocations; see an ilustration of the concept in Figure 4 motivated in the typical Burgers circuit, and a discussion of the continuum/crystallographic views in [61]. To the last integral in Equation (2), the generalized Stokes theorem may be applied

$$\oint_\Gamma dl * \mathcal{A} = \iint_{S_\Gamma} (ds \times \vec{\nabla}) * \mathcal{A} \tag{3}$$

where S_Γ is a surface with support on Γ, dl is a line differential of Γ, and ds is a surface differential of S_Γ. The operation $*$ is any operation consistent with the physical object $\mathcal{A}$ (vector or tensor of any order), and $\vec{\nabla}$ indicates that the operator operates on $\mathcal{A}$. Then

$$\bar{b} = \oint_\Gamma d\mathbf{X} \cdot \mathbf{F}_p^T = -\iint_{S_\Gamma} (ds \times \vec{\nabla}_X) \cdot \mathbf{F}_p^T = -\iint_{S_\Gamma} (\nabla_X \times \mathbf{F}_p^T)^T \cdot ds \tag{4}$$

where the last identity is inmmediate using index notation. In dislocation density based models, the Kröner-Nye tensor $\boldsymbol{\alpha}$ is handy [62,63], usually defined as the tensor field $\boldsymbol{\alpha}(X)$ such that

$$\bar{b} = b\hat{b} = \iint_{S_\Gamma} \boldsymbol{\alpha} \cdot ds \Rightarrow \boldsymbol{\alpha} = -\frac{1}{b}(\nabla_X \times \mathbf{F}_p^T)^T \tag{5}$$

Of course, because the deformation gradient F is compatible, so $\oint_c F_e F_p \cdot dX = \oint_c dx = 0$, the same integral may be obtained from the spatial configuration and $d\xi = F_p \cdot dX = F_e^{-1} \cdot dx$; i.e., $\boldsymbol{\alpha}(x) = -1/b(\nabla_x \times F_e^{-T})^T$. For the case of infinitesimal deformations, splitting the local displacement into elastic and plastic local (incompatible) parts ($u = e_e + u_p$), we immediately obtain $\boldsymbol{\alpha} = -\frac{1}{b}(\nabla \times (\nabla u_p)^T)^T = \frac{1}{b}(\nabla \times \nabla(u_e)^T)^T$, where we used the small strains approximation $F_e^{-1} \simeq I - \nabla u_e$. Note that for homogeneous elastic or plastic deformations $\boldsymbol{\alpha} = 0$ (dislocations are not geometrically needed because gradients are compatible). Indeed, Nye related them with the curvature in the deformation field.

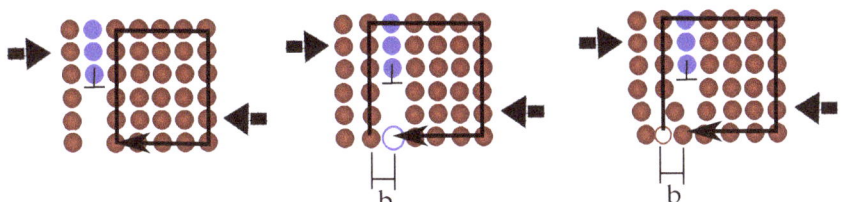

Figure 4. Ilustration of the concept of incompatibility through a moving dislocation. (**Left**) circuit is closed, locally compatible. (**Center**) plastic flow (e.g., a dislocation movement) produces an incompatible field, because the circuit is not closed by the amount b. (**Right**) Elastic deformations bring the field compatible.

The multiplicative decomposition is accepted in crystal plasticity, but at a single integration point in a finite element program, the decomposition assumes that the represented volume is fully transversed by dislocations. However, in reality, this assumption is not fulfilled; in general, there is still a dislocation content between neighboring material points if both experience different slip system activity from nonhomogeneous deformations, even if a full traverse by dislocations or a (statistical) homogeneous population is considered for individual material points. Much research is ongoing in the seek of the best formulation to take into account these effects [64–73].

The constitutive law can be obtained from the Helmholtz free energy density expressed in the intermediate configuration as proposed by [8,69,74]. The free energy density can be decomposed into the elastic and plastic parts [8]. The elastic part is recoverable through elastic strain unloading, whereas the plastic part needs some plastic deformation to recover it (accounts for example, for kinematic hardening).

$$\Psi = \Psi_e(F_e) + \Psi_p(F_p, q) \qquad (6)$$

where q is a set of internal variables. The second Piola-Kirchhoff stress tensor in the intermediate configuration then is obtained, for example, using Green-Lagrange strains, as

$$S^{|e} = \frac{d\Psi_e}{dE_e} \qquad (7)$$

being E_e the Green-Lagrange elastic strain tensor and $S^{|e}$ its work-conjugate stress tensor (both living in the intermediate configuration). Note that the strain energy may be written in terms of any arbitrary deformation measure, and the derivative of the energy respect to that measure gives the work-conjugate stress measure in the same configuration. If the elastic strain energy Ψ_e is defined by a quadratic potential of the Green-Lagrange elastic strains, $S^{|e}$ can be obtained as a linear function of E_e,

$$S^{|e} = \mathbb{C} : E_e = \frac{1}{2}\mathbb{C} : (F_e^T F_e - I) \qquad (8)$$

where $\mathbb{C}$ is the fourth-order elastic tangent moduli tensor of the single crystal.

Considering the decomposition in Equation (1), the velocity gradient L can similarly be decomposed:

$$L = \dot{F}F^{-1} = \underbrace{\dot{F}_e F_e^{-1}}_{L_e} + F_e \underbrace{\dot{F}_p F_p^{-1}}_{L_p} F_e^{-1} = L_e + F_e L_p F_e^{-1} \qquad (9)$$

being L_e and L_p the elastic and plastic velocity gradients. It is noted that the total deformation gradient is decomposed multiplicatively, whereas the decomposition of total velocity gradient is additive. Furthermore, both L and L_e are defined with respect to the current configuration. The plastic velocity gradient L_p, however, lives in the intermediate configuration, so it requires a mapping (push forward) from the intermediate to the current configuration, as shown by Equation (9).

Rice [54] assumed that for multislips, the plastic velocity gradient L_p is the sum of all crystallographic slip rates:

$$L_p = \sum_g \dot{\gamma}_g s_g \otimes m_g \qquad (10)$$

where γ_g is the internal variable that accounts for the accumulated plastic slip in each glide (slip) system g; vectors m_g and s_g are, respectively, unit normal vector to the slip plane and the unit vector in the slip direction that define the same glide system g in the reference configuration. An important aspect here is that the addition in Equation (10) is not exact in discrete algorithms, because the order in which the different slips take place changes the final result. The reason is that after one slip has taken place, L_p for the next one would be in a different configuration; see Figure 5. However, the approximation is considered acceptable in the literature given that steps are typically small. Indeed, recently we have proposed a different framework in terms of corrector rates [75], which results immediately from the chain rule applied to $F_e(F, F_p) \equiv FF_p^{-1}$, namely

$$\dot{F}_e = \left.\frac{\partial F_e}{\partial F}\right|_{\dot{F}_p = 0} \dot{F} + \left.\frac{\partial F_e}{\partial F}\right|_{\dot{F}=0} \dot{F}_p =: {}^{tr}\dot{F}_e + {}^{ct}\dot{F}_e \qquad (11)$$

where $^{ct}\dot{F}_e$ is the continuum corrector rate of the elastic gradient, and $^{tr}\dot{F}_e$ is the continuum trial rate of the elastic gradient; see further details about the elastic corrector rates framework in [75–78]. In this case, Equation (9) may be written as

$$L_e = L - F_e L_p F_e^{-1} = {}^{tr}L_e + {}^{ct}L_e \qquad (12)$$

where $^{tr}L_e$ is the elastic rate with the plastic flow frozen (the concept of a partial derivative), and $^{ct}L_e$ is the elastic rate with the external power frozen. Then, as an alternative to Equation (10), the following addition takes place in the same trial-frozen configuration:

$$-{}^{ct}\tilde{L}_e = {}^{ct}\dot{F}_e \, {}^{ct}F_e^{-1} = \sum_g \dot{\gamma}_g s_g \otimes m_g \qquad (13)$$

As a key difference, this poses an evolution on the elastic deformation gradient instead of on the plastic one, facilitating by construction the weak-invariance property.

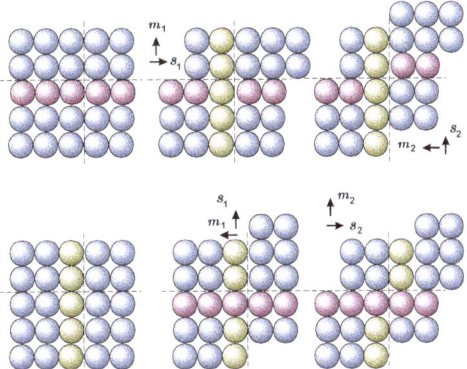

Figure 5. Comparison of the result obtained from two consecutive slips performed in different order.

3.2. Constitutive Models

3.2.1. Phenomenological Models

In crystal plasticity, the phenomenological models can be considered as the first approach adapted from the framework of continuum mechanics, and are dominant in the literature. This approach uses mostly the resolved shear stress (RSS) τ_g on a glide system g, as material state variable. With $S^{|e}$ denoting the second Piola-Kirchhoff stress tensor in the intermediate configuration, the RSS on glide system g, is typically defined as:

$$\tau_g = F_e^T F_e S^{|e} : (s_g \otimes m_g) \qquad (14)$$

where $F_e^T F_e S^{|e}$ is the non-symmetric Mandel stress tensor. In the case of infinitesimal elastic strains which typically hold in crystal plasticity, Equation (14) can be approximated as:

$$\tau_g \approx S^{|e} : (s_g \otimes m_g) \qquad (15)$$

An alternative is to use the recent formulation based of elastic corrector rates, which do not involve the Mandel stress tensor, but the symmetric generalized Kirchhoff one [75,79–81]. This formulation has general validity (not only for infinitesimal elastic strains). In general, the stress is defined as the derivative of the stored energy respect any convenient strain measure, for example, the logarithmic (Hencky) elastic strains $H_e := \frac{1}{2}\log(F_e^T F_e)$ in the intermediate configuration, namely

$$T^{|e} := \frac{d\Psi(H_e)}{dH_e} \text{ such that } \mathcal{D} = -\dot{\Psi}|_{\dot{F}=0} = -T^{|e} : {}^{ct}\dot{H}_e \tag{16}$$

where $\mathcal{D}$ is the plastic dissipation and ${}^{ct}\dot{H}_e$ is the corrector rate of $\dot{H}_e$ (the one with the external power frozen). Plastic slip occurs in this glide system when τ_g reaches a critical level τ_g^c (CRSS) that is typically considered as a function of the total accumulated plastic slip $\gamma = \int |\dot{\gamma}| \, dt$ and its rate $\dot{\gamma} = \sum |\dot{\gamma}_g|$,

$$\tau_g^c = f(\gamma, \dot{\gamma}) \tag{17}$$

This CRSS value can be interpreted as the hardened yield stress of the slip system. In inviscid plasticity $\dot{\gamma} > 0$ just when $|\tau_g| > \tau_g^c$. In rate dependent plasticity, the slip rate in each glide system is then formulated as a function of the RSS τ_g and the CRSS τ_g^c (e.g., as their ratio),

$$\dot{\gamma}_g = f(\tau_g, \tau_g^c) \tag{18}$$

Much work was performed to establish the specific form of Equation (18), e.g., [5,54,55,82,83], who have formulated a power-law function for slip rate as a common choice (other options are possible):

$$\dot{\gamma}_g = \dot{\gamma}_0 \left(\frac{|\tau_g|}{\tau_g^c}\right)^{1/m} \operatorname{sign}(\tau_g) \tag{19}$$

being $\dot{\gamma}_0$ and m material parameters that quantify the reference shear rate and the rate sensitivity of the slip mechanism, respectively. τ_g^c is also referred to as strength or hardness of the slip system. In the above equation $\dot{\gamma}_g$ can take both signs, while in the original work of Pierce et al. [82] the positive and negative slip directions are considered. The sensitivity exponent and the reference shear rate may include the influence of the temperature, for example, as [84,85]

$$m = \frac{kT}{\tau_f \mu b^2 d} \text{ and } \dot{\gamma}_0 = \rho^m b^2 f_D \exp\left(\frac{-\Delta G}{kT}\right) \tag{20}$$

where k is the Boltzmann constant, T the absolute temperature, μ the shear modulus, τ_f is the average strength of the forest, d is the distance to overcome an obstacle, ρ^m is the density of the mobile dislocations (see below), ΔG is the stored Gibbs free energy when a dislocation bypasses an obstacle, and f_D is the Debye characteristic frequency of the crystal, given by

$$f_D = \left(\frac{3N}{4\pi V}\right)^{\frac{1}{3}} v_s \tag{21}$$

where N/V is the number of atoms per volume and v_s is the speed of sound. Typical values are [85] $d \sim 4b$, $\Delta G \sim \mu b^3/5$, $m \sim 0.005$, $\dot{\gamma}_0 \sim 10^{-16} \div 10^{-6}$ [s^{-1}].

For the hardening evolution of the CRSS τ_g^c in each glide system, a hardening law is typically defined. A formulation proposed by Peirce et al. [82] for isotropic type hardening is often used. It includes the explicit contributions of the occurring slip in the same glide system (self-hardening) and of the cross-effect from slips on all other glide systems (latent hardening):

$$\dot{\tau}_g^c = \sum_j h_{gj} |\dot{\gamma}_j| \tag{22}$$

Here h_{gj} is a matrix of hardening coefficients, fitted/determined empirically to capture the micro-mechanical interaction between the different slip systems [82]:

$$h_{gj} = h(\gamma)[q + (1-q)\delta_{gj}] \tag{23}$$

The material parameter q differentiates latent hardening and self hardening of the glide systems; typically in an interval $[1, 1.4]$. For FCC single crystals, a typical choice for the hardening function $h(\gamma)$ is [82]:

$$h(\gamma) = h_0 \operatorname{sech}^2 \left(\frac{h_0 \gamma}{\tau_s - \tau_0} \right) \tag{24}$$

being h_0 the initial hardening rate. τ_0 and τ_s denote initial and saturation values for the CRSS.

We note that Equations (19), (22) and (24) are not the only ones proposed in the literature. Some researchers [86–91] modify Equation (19) to take into account the cyclic deformation by introducing a phenomenological backstress term to formulate the kinematic hardening. Other authors [92,93] use classical hardening laws from continuum plasticity such as Voce's hardening law to avoid vanishing hardening rates though the introduction of a limiting slope.

3.2.2. Physics-Based Models

In contrast to the phenomenological models, the physics-based ones use physical (or physically motivated) measures, as well as principles and important physical insight from microstructure information and its evolution. Examples of that information are dislocation density, precipitate morphology, grain size and shape, etc. Nevertheless, only a few of these aspects have been introduced into physics-based models. In following subsections, we present more in detail some predominant models.

(a) Dislocation density-based models

There are 10^6 to 10^{12} dislocations in every cm^2 of material. The amount is variable and strongly depends on the processing history, in particular deformation and thermal treatments, possibly on microstructure [27,94]. Whereas dislocations are mostly responsible for plastic deformation, crystal plasticity models are usually developed at a different scale. Then, the different scales are considered throughout the formulations in terms of statistically stored dislocation (SSD) densities in the crystal plasticity models. At a given time, there are two types of SSD: mobile (glissile) and immobile (sessile), with respective densities ρ_g^m and ρ_g^{im}; in total $\rho_g = \rho_g^m + \rho_g^{im}$. These densities represent the equivalent length of dislocations (Burgers vectors) along a direction per unit volume, e.g., with units [m/m^3] = [m^{-2}]. Mobile dislocations are the ones producing effective slips, but during deformation a change of status of dislocations is possible [95,96] which means a transfer between ρ_g^m and ρ_g^{im}. This decomposition has been proposed to develop a different treatment needed for both components, because mobile dislocations are the source of plastic deformation and ductility; whereas immobile dislocations block motion, strengthening or hardening the material. The laws of evolution for mobile dislocations require a coupling with the ones for immobile dislocations. The evolution laws for dislocation density is typically based on the following Kocks-Mecking model [95,97,98]:

$$\dot{\rho}_g = \frac{1}{b_g}\left[\frac{1}{l(\rho_g)} - 2\gamma_c \rho_g\right]|\dot{\gamma}_g| \qquad (25)$$

This law includes both storage and recovery effects (in a similar layout to the Armstrong-Frederick rule in hardening) that determine the variation of the dislocation density with plastic deformation. The first addend in the brackets accounts for the storage of dislocations $1/(b_g l(\rho_g))$; the dynamic recovery by the term $2\gamma_c \rho_g$ is associated with the annihilation of stored dislocations. The function $l(\rho_g)$ is interpreted as immobilization of mobile dislocations at obstacles after having travelled a characteristic (statistical average) distance. b_g is the Burgers vector modulus of slip system g. The term γ_c is the critical annihilation distance. Some modifications of Equation (25) are possible; refer, for example, to [99,100].

Further evolution laws for metals of different crystal structures is proposed in [101] and enhanced by subsequent works [102–105]. An example is the following which is a linear superposition of various mobile and immobile mechanisms:

$$\dot{\rho}_g^m = \left[h_g^s(\rho_g^{im}\rho_g^m)^{-1} - h_g^{m0}\rho_g^m - h_G^{im0}\sqrt{\rho_g^{im}}\right]|\dot{\gamma}_g| \qquad (26)$$

$$\dot{\rho}_g^{im} = \left\{h_g^{m1}\rho_g^m - h_g^{im1}\sqrt{\rho_g^{im}} - h_g^r\rho_g^{im}\exp\left[\frac{\Delta G}{kT}\left(1 - \sqrt{\frac{\rho_g^{im}}{\rho_{sat}^{im}}}\right)\right]\right\}|\dot{\gamma}_g| \qquad (27)$$

where ρ_{sat}^{im} is an experimentally determined saturation for immobile dislocation densities, The coefficients h_g^s, h_g^{m0}, h_G^{im0}, h_g^{m1}, h_g^{im1} and h_g^r depend on the degree of activity of the specific dislocation mechanisms g.

Regarding the evolution of the slip rate, in dislocation-based models the equation proposed by Orowan [106] is typically used instead of Equation (19) as it establishes the connection between the plastic slip rate on a given glide system $\dot{\gamma}_g$ and the mobile dislocation density, i.e., it creates a bridge to convert a continuum mechanical term into the physics of dislocations:

$$\dot{\gamma}_g = \rho_g^m b_g v_g \qquad (28)$$

where ρ_g^m is the mobile dislocation density, b_g is the Burgers vector modulus of g and v_g is the average velocity of mobile dislocations that depends on the details of the evolving dislocation mechanism. During slip, depending on whether thermal activation can contribute to get over the obstacle, the barriers to dislocation movement can be referred to as temperature-dependent or independent.

Therefore, the average dislocation velocity is also temperature-dependent [38,68,99,107–109]. Generally, the average dislocation velocity v_g can be expressed by the typical thermodynamic relation [110]:

$$v_g = \begin{cases} 0 & \text{if } 0 \leq |\tau_g| \leq \tau_g^c \\ \lambda_g f_D \exp\left(-\frac{\Delta G}{kT}\right) & \text{if } \tau_g^c < |\tau_g| \end{cases} \qquad (29)$$

where λ_g is the average distance between the obstacles in the glide system g (compare with Equations (19) and (20)). During slip, due to the crystalline structure, there is a local stress field associated to the distortion around each dislocation core. The Gibbs free energy in presence of a general array of obstacles in the slip plane can be formulated as [107]

$$\Delta G = \Delta F \left[1 - \left\langle \frac{|\tau_g| - \tau_g^c}{\tau_t} \right\rangle^p \right]^q \qquad (30)$$

being ΔF the activation free energy necessary to overcome the obstacles without the aid of an applied shear stress, τ_t the strength of the barrier at 0 K, p and q the parameters that are function of distance propagated by the dislocation, and $\langle \cdot \rangle$ is the Macaulay brackets.

Furthermore, it is well known that a crystal always contains a number of immobile dislocations, so if a mobile dislocation keeps moving, it interacts with those immobile dislocations and also with the local stress fields around those.

Immobile dislocations therefore represent obstacles to mobile dislocations, and their density is a key factor determining the threshold plastic stress (CRSS, τ_g^c) required for dislocation motion within a glide system. In addition, the value of τ_g^c also depends on a variety of factors such as: point defects, crystal structure, temperature, and other internal variables [27]. Taylor [111] determined this threshold stress for a single slip to occur:

$$\tau_g^c = \mu b_g \sqrt{\rho_g^{im}} \qquad (31)$$

where μ is the shear modulus. For several slip interactions and to include the influence of the temperature, Elkhodary et al. [112] proposed a modification of Equation (31):

$$\tau_g^c = \left(\tau_g^0 + \mu \sum_{i=1}^{nss} b_g \sqrt{a_{gi} \rho_i^{im}} \right) \left(\frac{T}{T_0} \right)^{-\xi} \qquad (32)$$

being τ_g^0 the initial slip resistance on glide system g, nss the number of slip systems and a_{gi} the Taylor coefficient that represents the strength of interaction between glide systems. T, T_0 are the current and reference temperature, respectively. ξ is the thermal softening exponent.

(b) Geometrically necessary dislocations (GND) models

One of the best known frameworks in crystal plasticity for FE analysis is the one from Kalidindi, Bronkhorst and Anand [39]. Here, the constitutive laws can be totally expressed from the loading history at a stress integration point and are defined as local models. However, experiments show that the mechanical behavior of single crystals is size-dependent and the classical continuum models of plasticity cannot reproduce the details of the behaviour of the material at such small scales, important in nanoindentation [113–117], bending tests for small-scale beam [118–123] and micropillar compression [124]. Hall [31] and Petch [32] introduced the grain size dependence of the flow stress through an empirical relation. Basically, smaller grains result in larger yield stresses. Thereafter, numerous studies have demonstrated that this strengthening effect is due to the fact that in the surroundings of the grain boundaries there is a higher volume fraction of heterogeneous plastic deformation [62,125–127], hence more apparent in smaller grains. Typically, the nonhomogeneuos plastic deformation takes place and it may affect strain gradients and orientation near the boundary. These gradients are associated to the geometrically necessary dislocations (GNDs) [125]. The GND concept appears in nonhomogeneous deformations and is incorporated in the constitutive framework by Nye's dislocation density tensor explained above, see Equation (5). For a given slip system g, given that the Schmid dyad remains constant in the intermediate configuration, we have for that system $\dot{F}^T = F_p^T L_p^T = \dot{\gamma}_g F_p^T \cdot m_g \otimes s_g$ and

$$\dot{\alpha} = -\frac{1}{b}(\nabla_X \times \dot{F}_p^T)^T = -\frac{1}{b}\nabla_X \times (\dot{\gamma}_g F_p^T \cdot m_g \otimes s_g)^T = -s_g \otimes \underbrace{\left[\frac{1}{b}\nabla_X \times (\dot{\gamma}_g F_p^T \cdot m_g)\right]}_{\dot{\rho}_g^{GND}} =: -s_g \otimes \dot{\rho}_g^{GND} \quad (33)$$

where ρ_g^{GND} is the vector of geometrically necessary dislocations for slip system g, proportional to the Burgers vector. This vector can be decomposed in screw and edge dislocations. The screw dislocation is in the direction of the slip direction, whereas the edge dislocations may be decomposed in the direction perpendicular to the slip plane m_g and in the direction in that plane which is perpendicular to the slip direction, i.e., in direction $t_g := m_g \times s_g$. These density rates are

$$\dot{\rho}_{gs}^{GND} = \dot{\rho}_g^{GND} \cdot s_g \quad \text{and} \quad \dot{\rho}_{gm}^{GND} = \dot{\rho}_g^{GND} \cdot m_g \quad \text{and} \quad \dot{\rho}_{gt}^{GND} = \dot{\rho}_g^{GND} \cdot t_g \quad (34)$$

which can also be written as [73,128]:

$$\dot{\rho}_{gmt}^{GND} = \frac{1}{b}\nabla \dot{\gamma}_g \cdot s_g \text{ (edge)}; \quad \dot{\rho}_{gs}^{GND} = \frac{1}{b}\nabla \dot{\gamma}_g \cdot (s_g \times m_g) \text{ (screw)}; \quad (35)$$

The integration of GNDs into the constitutive model can be introduced by different ways in crystal plasticity. In some models, GNDs are added to the statistically stored dislocations (SSDs) and also contribute to hardening by the Taylor hardening relation [129,130]. In the case that the density of GNDs in each system is explicitly resolved, the critical resolved shear stress τ_g can be formulated as [131]:

$$\tau_g = \mu b \sqrt{\sum_g (a_{kg}^{SSD} \rho_g^{SSD} + a_{kg}^{GND} \rho_g^{GND})} \quad (36)$$

where ρ_g^{SSD} is the SSDs density in system g; a_{kg}^{SSD} and a_{kg}^{GND} are coefficients that define the latent hardening interactions among the different glide systems. To obtain the evolution of GNDs, it is necessary to determine the Nye tensor or the slip rate at the integration point level. This task is typically done through an extrapolation of the internal variables from the integration points to nodes in each element [132,133], or alternatively by computing the gradients based on recovery techniques [134]. Another approach to include GNDs in high order models. The most popular model of this type was proposed in the works [65,72,135], and thereafter extended in many works [67,136–141]. Here, the plastic slip in each system γ_g is treated as a kinematic variable in the constitutive model. Another tensorial quantity defined as geometric-dislocation tensor G is proposed to measure GNDs and can be expressed in terms of pure dislocations.

$$G = b \sum_\kappa^K \rho_\kappa l_\kappa \otimes s_\kappa \quad (37)$$

Here ρ_κ is a signed density and $l_\kappa \otimes s_\kappa$ is a dislocation dyad, where l_κ is the line direction (either s_k or $m_k \times s_k$). The relevant aspect in the model is that it focuses on the work for each independent kinematic process that is energy-balanced, and therefore, micro-forces conjugated with slip and its gradient are properly considered in the principle of virtual power. Moreover, for constitutive dependencies on the geometric dislocation tensor, G is decomposed in a typical elastic-strain energy and a specific so-called defect energy Ψ_G. A simple defect energy is defined in a quadratic form of the type [65]:

$$\Psi_G = \tfrac{1}{2}\lambda |G|^2 \quad (38)$$

where λ is a scalar material parameter. The micro-force τ_g gives the following viscoplastic yield conditions (Equation (7.19)) [65]:

$$\tau_g = \underbrace{H(\gamma_g)\sigma_g}_{\text{Phenom. slip hardening}} + \underbrace{J^{-1}\mathbf{S}_g \cdot (\mathbf{T}^{|G}\mathbf{G}^T + \mathbf{T}^{|GT}\mathbf{G}) - \nabla \cdot [J^{-1}F_e(\mathbf{m}_g \times \mathbf{T}_G\mathbf{s}_g)]}_{\text{+ additional hardening due to the energy stored by the GNDs}} \quad (39)$$

where $H(\gamma_g)$ is a function characterizing rate-dependence, σ_g a slip resistance, $\mathbf{S}_g = \mathbf{s}_g \otimes \mathbf{m}_g$ is the Schmid tensor, $\mathbf{T}^{|G} = \partial \Psi_G / \partial \mathbf{G}$ is defined as the thermodynamic defect stress. In Equation (39), there are two different hardening mechanisms. The first addend is purely phenomenological without backstress effects, whereas the second one is purely energetic from restrictions imposed by the thermodynamical framework, and furthermore, it represents a backstress on the slip system g.

(c) Continuum Dislocation Dynamic Models

The GND models presented above are in general based on the Kröner-Nye dislocation density tensor that can be used for modelling size effects. However, those models present a fundamental weakness when only GNDs are considered: after averaging, the dislocation density quantities only account for the GNDs, while all moving dislocations (redundant and macroscopically cancelling SSDs) must be taken into account. In this sense, GND models typically need some phenomenological assumptions in order to account for the contribution of the geometrically redundant dislocations for the deformation process. Such GND models only can provide a closed and kinematically consistent description of the dislocation dynamics if all dislocations share the same line direction [142]. In order to overcome this weakness, recent density-based theories of dislocation dynamics are developed by different researchers [142–150] which contribute to a new family of physics-based plasticity model named continuum dislocation dynamics (CDD).

Here, as an example of such theories, we focus on the more recent CDD model introduced by Hochrainer [142,147,148,150]. This model is based on the description of connected dislocation lines by a generalized dislocation density tensor, i.e., modifies the definition of the dislocation density tensor. As mentioned above, models based on the Kröner-Nye tensor normally average over the line directions of all dislocations contained in a volume element. However, in the CDD models the dislocation density tensor can distinguish a priori dislocations by their line direction before any averaging is introduced, which permits the description of the kinematics of very general systems of dislocations. The modification of the dislocation density tensor can be achieved by defining it in a higher-order configuration space [142, 147,148,150]. For the configuration space $\mathbb{R} \times \mathbb{R} \times \mathbb{S}$ (where $\mathbb{S}$ is the orientation space $[0, 2\pi]$), the modified dislocation density tensor takes the form

$$\alpha^{II}_{(r,\phi)} = \rho_{(r,\phi)} \mathbf{L}_{(r,\phi)} \otimes \mathbf{b} \quad (40)$$

where (r, ϕ) defines a point in $\mathbb{R} \times \mathbb{R} \times \mathbb{S}$, $\rho_{(r,\phi)}$ is the scalar dislocation density at this point (r, ϕ), $\mathbf{b}$ is the Burgers vector, and $\mathbf{L}_{(r,\phi)}$ defines the generalized line direction in the higher-order configuration space that depends on the canonical spatial line direction $\mathbf{l}_{(\phi)} = (\cos \phi, \sin \phi)$ and the curvature $k_{(r,\phi)}$. The evolution of this tensor can be obtained by the following relation [142,147,148]

$$\partial_t \alpha^{II}_{(r,\phi)} = -(\mathbf{V}_{(r,\phi)} \times \alpha^{II}_{(r,\phi)}) \times \nabla \quad (41)$$

being $\mathbf{V}_{(r,\phi)}$ the vector of the generalized velocity in $\mathbb{R} \times \mathbb{R} \times \mathbb{S}$, that is perpendicular to the generalized line direction. By $\partial_t(\bullet)$ we mean partial derivative of $(\bullet)$ respect to t. This generalized velocity consists of spatial component $v = v(r, \phi)v(\phi)$, where $v(r, \phi)$ is the scalar velocity of dislocation segments at point

(r,ϕ) and $v(\phi) = (\sin\phi,\cos\phi)$, and an angular component $\theta((r,\phi))$ that indicates the rotation velocity of these segments. The dislocation density $\rho_{(r,\phi)}$ and the curvature $k(r,\phi)$ then can be formulated from Equation (41) by assuming only glide dislocation movements [147]

$$\partial_t \rho = -(\nabla \cdot (\rho v) + \partial_\phi(\rho\theta)) + \rho v k \tag{42}$$

$$\partial_t k = -vk^2 + \nabla_L(\theta) - \nabla_V(k) \tag{43}$$

This approach allows to address physically relevant microplasticity problems with simple deformation geometries [142,147,148], however the main problem found with this approach is the large computational cost, since there is a large number of degrees of freedom required for conducting simulations in a higher dimensional state space. In order to reduce the dimensionality, a Fourier expansion can be used for the dislocation density and curvature functions, and the expansion coefficients then are used as variables of a simplified CDD model [148]. The total dislocation density ρ^t is defined as the zeroth-order expansion coefficient of the dislocation density function $\rho_{(r,\phi)}$,

$$\rho_r^t = \int_0^{2\pi} \rho_{(r,\phi)} d\phi \tag{44}$$

The total geometrically necessary dislocation density is defined as $\rho_G = \sqrt{(\kappa^1)^2 + (\kappa^2)^2}$, where κ^1 and κ^2 are the first order terms of the Fourier expansion of $\rho_{(r,\phi)}$ in the material point (r,ϕ), which are the components of the dislocation density vector $\kappa_{(r)} = (\kappa^1, \kappa^2, 0)$,

$$\kappa^1 = \int_0^{2\pi} \rho_{(r,\phi)} \cos\phi d\phi, \quad \kappa^2 = \int_0^{2\pi} \rho_{(r,\phi)} \sin\phi d\phi \tag{45}$$

From $\kappa_{(r)}$, the classical Kröner-Nye tensor is recovered as $\alpha = \kappa_{(r)} \otimes b$. Under the assumption that the scalar dislocation velocity and curvature fields are independent of the line orientation, the evolution laws for the total dislocation density ρ^t and the dislocation density vector κ are given by

$$\partial_t \rho^t = -\nabla \cdot (v\kappa^\perp) + v\rho^t \bar{k}$$
$$\partial_t \kappa = -\nabla \times (\rho^t v m) \tag{46}$$

where $\kappa^\perp = (\kappa^2, -\kappa^1, 0)$, m is the glide plane normal, and $\bar{k}$ is the average curvature. A phenomenological evolution equation is normally adapted for the average curvature [148].

3.3. Numerical Implementation Aspects

The finite element method (FEM) is the widely preferred method for computational simulation of solids, and that includes crystal plasticity. Finite element programs typically have a driver subroutine for material models, and commercial ones allow users to implement their own material models via user subroutines. In a typical analysis, the load is applied incrementally in "time" steps, and at each time increment the global equilibrium of the structure is reached by means of an iterative process using a global nonlinear solver (typically based on global Newton-Raphson methods). Once the problem is solved at time t, the crystal model must provide two important quantities at each material point for the solution for the next time increment: (i) the stress at time $t + \Delta t$ and (ii) the tangent moduli tensor $\mathbb{C} = d\mathbf{S}/d\mathbf{E}$ also at time $t + \Delta t$ ($\mathbf{S}$ is the second Piola-Kirchhoff stress tensor and $\mathbf{E} = \frac{1}{2}(\mathbf{F}^T\mathbf{F} - \mathbf{I})$ is now the Green-Lagrange strain tensor, both in the reference, undeformed configuration).

3.3.1. Integration Algorithm

Usually, an elastic-predictor/plastic-corrector algorithm is used for the integration of the constitutive equation. In principle, all data at previous time t is known and the strain increments at time $t + \Delta t$ are given by the element subroutine calling the material one, so one can start "elastically"-predicting any of the involved quantities. Subsequently, the prediction is corrected by solving the nonlinear equations in residual form, typically using a Newton-Raphson scheme. The second Piola-Kirchhoff stress S [39], the slip rate $\dot{\gamma}_g$ [82], the elastic deformation gradient F_e [151], the plastic deformation gradient F_p [152] or the plastic velocity gradient L_p [153] are suggested as starting point in the numerical implementation. An example of elastic predictor/return-mapping algorithm for single crystal models is summarized in Algorithm 1. In the case of viscous-type rate constitutive equations, since no yield function is employed, no check for plastic slip is needed (there is always a plastic slip, even if it is very small). However, convergence problems are common if the rate sensitivity parameter m (see Equation (19)) is too small (as to approach the rate-independent solution), so it is common to increase this parameter artificially just to overcome these numerical problems. On the other hand, in the rate-independent case, non-uniqueness of solution issues may arise (the final stress is unique, but not the active set of slip systems).

Algorithm 1 Typical structure of an elastic-predictor/plastic-corrector return-mapping algorithm

---

Require: Given the state variables at time t and the total strain at time $t + \Delta t$.

1: Elastic predictor: evaluate the trial state
$$^{tr}F_e = {}^{t+\Delta t}_{0}F \cdot {}^{t}_{0}F_p^{-1}; \ {}^{tr}S|^e = \partial \Psi / \partial^{tr}E_e; \ {}^{tr}\tau_g = {}^{tr}F_e^T {}^{tr}F_e {}^{tr}S|^e : (s_g \otimes m_g); \ {}^{tr}\gamma = {}^{t}\gamma$$

2: Plastic slip check
   IF ${}^{tr}\tau_g \leq \tau_g^c({}^{tr}\gamma)$ **THEN**
      Elastic step: set ${}^{t+\Delta t}(\bullet) = {}^{tr}(\bullet)$ and EXIT
   ELSE
      Plastic slip step: Proceed to step 3
   ENDIF

3: Return mapping
   i. Use the standard or exponential backward-Euler for integration of evolution equation
   ii. Local Newton iterations: plastic corrector
       Solve iteratively the residues of the variables used in prediction
       $${}^{t+\Delta t}R = 0$$
       using the Newton-Raphson method
   iii. Update the state variables. Compute the elastoplastic tangent matrix during this phase.

4: EXIT

By using the Newton-Raphson method to obtain the solution, it is necessary to compute the local Jacobian matrix and its inverse. The dimension of this matrix is equal to the dimension of the residues vector, i.e the number of independent variables (and equations) that is used as predictor, for example, six variables for S, eight for F_p, etc. This dimension can be a relevant number if the slip rates are chosen (for example, 48 × 48 for BCC crystals). Taking into account that this system is solved several times until local convergence for each global iteration and at each integration point, it requires some efforts in reducing the number of active slip systems [8], because this task may be the most computational demanding in solving the structure.

3.3.2. Type of Elements

In most crystal plasticity simulations, tri-linear elements are used, because they are easily generated and are inexpensive from a material standpoint (typically $2 \times 2 \times 2$ integration points are employed). However, these elements can not describe strain gradients within one element due to the fact that this type of elements use linear interpolation functions for the displacements. Furthermore, if they employ standard formulation, they are well-known to lock for the incompressible case typical in plasticity. Linear elements are especially not sufficient to capture accurately strong strain gradients. Higher-order elements should be used in such cases with either reduced integration or a formulation to alleviate locking. Typical elements include mixed u/p formulations, incompatible models formulations, or reduced integration. Details on these formulations can be found in [2].

In cases with GNDs-based models, the situation becomes more complicated when those models include strain gradients. The formulations for standard elements are only continuous in the displacements (C^0-continuity) and gradients are often undefined. Enhanced element formulations have been proposed to be used to overcome this problem [67,154]. However, in the case of complex loadings the complexity in defining boundary conditions increases for such element formulations. Furthermore, these element formulations are computationally very costly, and lack of robustness at large strains, so they are often applied to two dimensional problems at small strains. Therefore, the standard elements are still used and the necessary strain gradients are derived based on recovery techniques [134].

4. Searching the Active Set of Slip Systems

The question whether a specific slip system is activated is decided by a yield criterion mathematically expressed in terms of one inequality per slip system. It essentially states that plastic slip occurs if the RSS reaches the CRSS subject to the additional constraint that only slip rates in the positive and negative direction of the RSS are permitted. The overall constitutive material response can be determined once the set of active slip systems is known. However, it is not possible, in general, to know which systems will be active for the final converged deformation. Noteworthy, for multisurface plasticity models such as the Tresca and Mohr-Coulomb models there are robust algorithms to search the active set of yield surfaces. Those algorithms have been formulated based on the geometric representation in the principal stress space. Nonetheless, in the present crystal plasticity case, its application is more complex because the set of active slip systems that satisfies the discrete plastic consistency may not be unique and many combinations of plastic multipliers may exits for a given set of active slip systems, which provide the same incremental plastic deformation gradient [5,155]. If the set of Schmid tensors $S_g = s_g \otimes m_g$ is linearly dependent in the space of deviatoric tensors, non-uniqueness in the solution is possible. In 3D there are 8 linearly independent tensors that generate the deviatoric tensors space. If it happens that the total number of active slip systems is larger than eight, then the Jacobian matrix of the return-mapping system of equations becomes singular. Therefore, an additional condition has to be formulated to be able to select the set of active slip systems which appears in reality. Typically a Moore-Penrose (minimum norm) solution is chosen, although other options are possible [156].

4.1. Rate-Dependent Approach

One of the preferred solutions to overcome the problems related to non-uniqueness are viscoplastic (rate-dependent) formulations. These have been introduced firstly in the works [82,83,157], in which the slip rate is given employing a typical power law function as shown in Equation (19). This approach is then popularly used not only in continuum crystal plasticity models [86–91,93] but also in the homogenization of polycrystals [158–164] or in multiscale modelling of polycrystals [165–168]. As highlighted by [82], this rate-dependent model allows a more accurate large strain formulation for polycrystals. In particular, it can obtain a good prediction of texture and specially obtain its dependence on both strain-hardening/latent-hardening and strain rate sensitivity. But much of the choice of the rate-dependent formulation is to avoid the non-uniqueness problem of rate-independent models, a formulation which can be recovered by using reasonably low values for the rate-sensitivity constant (i.e., a high exponent $1/m$ in the slip law of Equation (19)). However, in this case, the resulting set of equations that are integrated can become extremely stiff as the rate-sensitivity constant reaches a very small value, so a robust formulation for those cases is not simple. These issues have been addressed and tackled in the work of Cuitiño and Ortiz [8].

4.2. Rate-Independent Approach

Borja and Wren [155] proposed a completely different approach. Restricted to the infinitesimal theory, a fairly robust algorithm (the ultimate algorithm) for the selection of the active slip systems was developed within the stress-update procedure for rate-independent single crystal models. The basic idea for determining the set of active slip systems is to start the application of the predictor/return-mapping scheme with some trial set of active slip systems, and then, to generate a sequence of sets of active slip systems as the Newton-Raphson iterations for obtaining the solution are applied. At the end of this root-finding process, the algorithm will converge to a state when the discrete complementary condition is satisfied. The set of active slip systems is the converged set and is unique. Furthermore, to avoid the possibility that the Jacobian matrix can be singular, Borja and Wren proposed the triangular factorization of the Jacobian matrix after the redundant equations are eliminated. This algorithm has been extended recently for the finite strain framework by Borja and Rahmani [169]. Miehe also used this algorithm in the finite strain context [156,170], but introducing the concept of exponential map integrators of [171] into the field of single crystal simulation.

5. Examples of Applications

5.1. Engineering Forming Problems

The above formulations have been employed in a wide number of problems where, for example, texture and its evolution may be important. These have been compared to experimental data and to continuum models to show comparatively enhanced predictions. A typical example is a cylindrical cup deep drawing adopted to investigate the texture effects on the localization of the deformations and in earing [172]. In order to do it, the texture components were incorporated into the crystal plasticity model by introducing the measured orientation distribution function (ODF) at each integration point [173]. Three typical crystal orientations $\{111\}\langle 110\rangle$, $\{113\}\langle 110\rangle$ and $\{001\}\langle 110\rangle$ were selected from the dominant textures in three steel sheets (mild steel Deep-Drawing Quality DDQ dual-phase steel—DP600Ze and high-strength steel CP800). The results obtained from simulations are compared to experimental results. Figure 6 shows the computed results compared with one obtained in experiments for the high-strength steel (CP800) material. The predicted and measured thickness strain distributions (very important in deep drawing), along the rolling direction are similar (see Figure 6c). The thinning of the material appears at the hemispherical bottom region, where a high strain localization can also be observed at the bottom region. Figure 6a,b show the localized necking precluding fracture in numerical and experimental results, respectively. The expected four ears at 45° to the rolling direction are observed. With the fitted simplified texture, the texture component-based crystal model give already results very close to the experimental ones, demonstrating the value of this type of formulations. Furthermore, the results obtained in the study [172] also confirmed that the more the γ fiber texture components, the better is the drawability of bcc steel sheet. The γ fiber also gives rise to four ears at 0 and 90° to rolling direction.

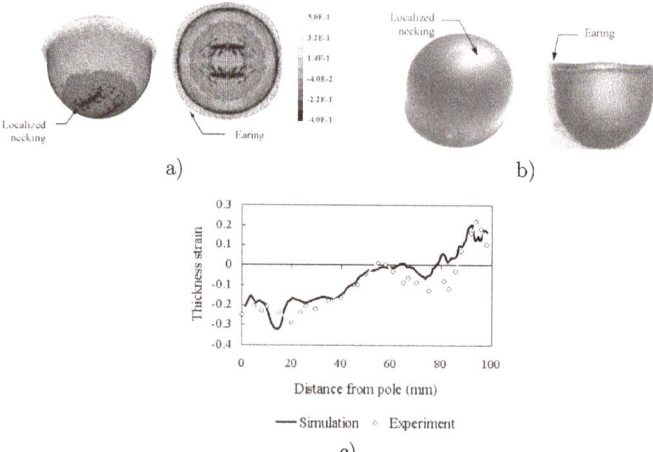

Figure 6. Comparison of the cup drawing between numerical simulation and experiment for CP800 material: (**a**) predicted thickness strain distribution and earing at punch travel 10 mm; (**b**) photographs of experiment specimen; (**c**) thickness strain distribution along the rolling direction. Reproduced from [172], under permission.

The same simulation benchmark was also employed in [174] but using a new concept for representing textures by mapping small sets of mathematically compact spherical Gaussian texture components on the integration points [175–177]. Figure 7 shows the comparison between the computed and experimental results for cube-textured aluminium sheet after cup drawing. In Figure 7a, the result obtained using the texture component-based crystal model is compared with the experiment and also with continuum-based elastoplasticity obtained by use of a Hill-48 yield surface. Two texture components and a random scattering background component are used for fitting the texture component. These results show a better fit of the texture component-based crystal model than that obtained with the classical Hill-based model. Figure 7b,c show simulation results for different approximations using a volume fraction and using two rolling texture components with a random texture component.

Using the elasto-plastic self-consistent homogenization scheme in the finite element framework, Zecevic and Knezevic [178] explored its potential in simulation and prediction of springback of cup drawing from an AA6022-T4 sheet, see Figure 8; only a quarter is modelled due to symmetries. The springback is obtained by eliminating the contact between the blank and other parts when the punch reaches a travel distance of 12 mm. The von Mises stress contours at the end of drawing and springback are shown in Figure 8a,b, while the dimensional changes are depicted in Figure 8c,d.

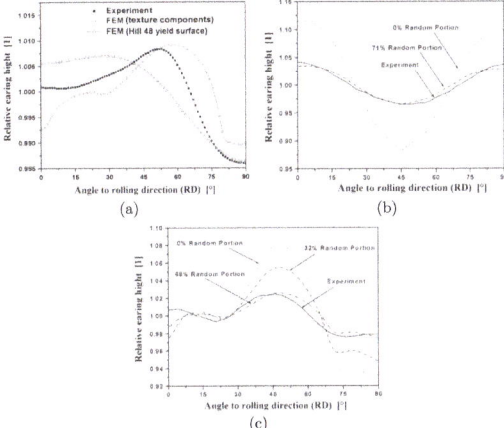

Figure 7. Experimental and simulation results of earing profile after cup drawing for aluminium sheet: (**a**) comparison of results between the texture component-based crystal model and Hill 48 yield surface model and experiment; (**b**) simulation and experimental results in which the texture is approximated using the volume fraction; (**c**) simulation and experimental results in which the texture is approximated using two rolling texture components and a random texture component. Reproduced from [174], under permission.

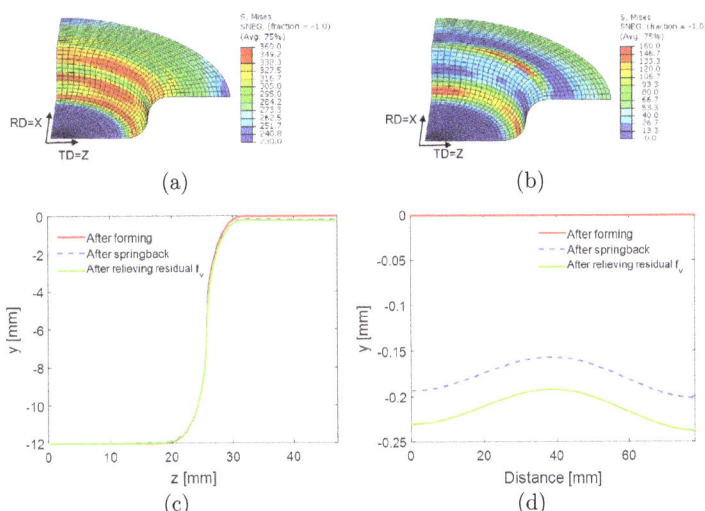

Figure 8. Cup drawing and springback simulation of AA6022-T4 sheet: (**a**) von Mises stress contours at the end of forming; (**b**) von Mises stress contours after springback relaxation; (**c**) shape change after springback of y-z profile; (**d**) shape change after springback of the circumference of the down cup flange. Reproduced from [178], under permission.

5.2. Virtual Material Testing

5.2.1. Texture Evolution

Obviously one of the simulations that can be performed with crystal plasticity models is the analysis of texture evolution and, hence, the evolution of anisotropy. In Figure 9 we show the texture evolution in a copper polycrystal cubic specimen in a tensile test, see full details about the constitutive equations (similar to the avove-presented phenomenological ones) and the material parameters employed in the simulation in [75]. For this simulation we used 512 elements with an initial random texture with 50 orientations. In the same figure, the initial mesh and the deformed mesh is shown. In the middle and lower rows of the figure, the pole figures showing the evolution of texture for directions (111) (middle row) and (100) (lower row) are shown. This evolution has been computed in Figure 9 using both the classical Kalidindi et al. framework and the novel framework employing elastic corrector rates, being the results very similar for this case (see again details in [75]).

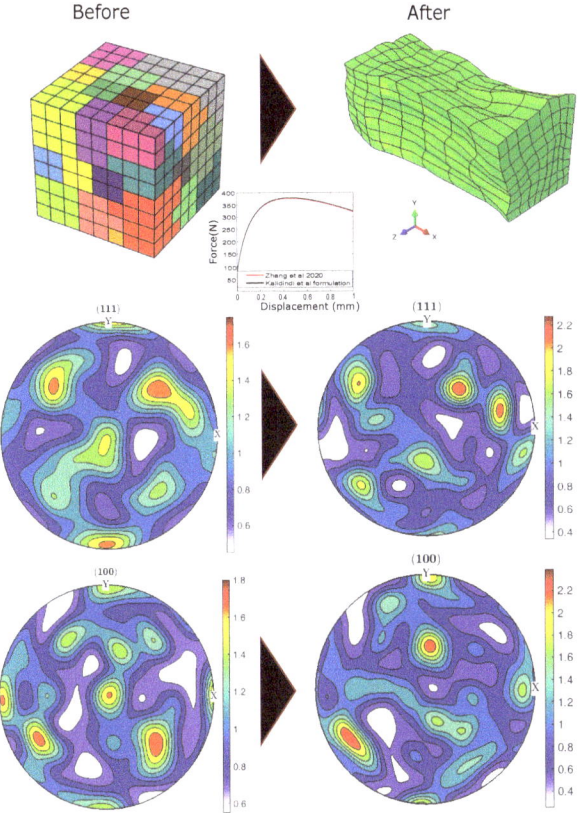

Figure 9. Evolution of texture in copper under a tensile test. Upper row contains the initial and final mesh, where colors correspond to different orientation. Middle row is shows the texture pole plots evolution for direction (111) and the lower row for direction (100). Adapted from [75], under permission. A Matlab toolbox for pole figures, and documentation, is available in https://mtex-toolbox.github.io/.

5.2.2. Impact Test

Taylor impact tests were originally devised to obtain the dynamic yield strength of a material at moderate strain rates, but now such tests are used frequently to validate the constitutive model for the simulation of plastic deformation. Figure 10a shows simulations of Ta cylinder by using the viscoplastic self-consistent (VPSC) model in the finite element framework. The simulation was performed by [179] and compared with the experimental results reported in [180,181]. Figure 10a shows the equivalent plastic strain rate contours that develop in the cylinder during the simulation. It can be seen that the dramatic change in the strain rate is observed over the course of simulation. At the beginning of the impact, the highest equivalent plastic strain rate develops at the center of the foot of the cylinder. As the impact progresses, the high strain rate zone moves away from the foot to the other side of the cylinder as a consequence of the plastic wave propagation. The side profile and footprint obtained in the simulation after the impact are compared with the experimental ones and depicted in Figure 10b. The highly heterogeneous plastic deformation is observed at the center of the foot. The deformation in lateral directions is non-uniform as shown in the ellipsoidal shape of the footprint.

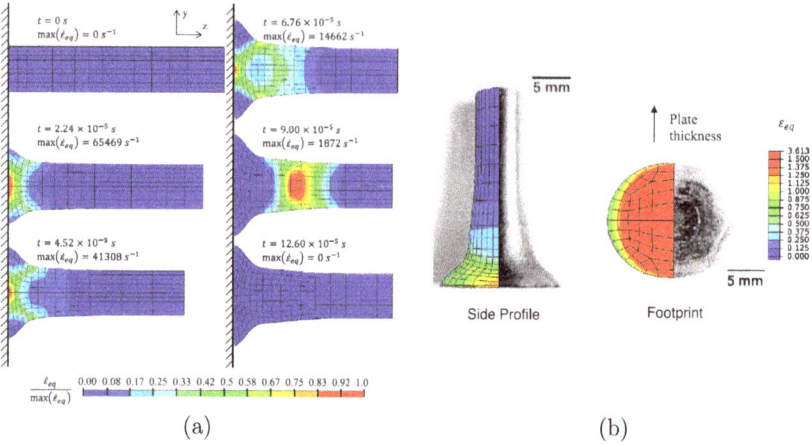

Figure 10. Simulation results of Taylor impact test for Ta cylinder: (**a**) equivalent plastic strain rate contour at different time step; (**b**) comparison of the measured and simulated cylinder geometry after the impact with the superimposed contours of equivalent plastic strain. Reproduced from [179], under permission.

5.2.3. Rolling Test

Using the multiscale framework, a typical rolling test is normally carried out to explore the mechanical behavior of polycrystals subjected to non-homogeneous stress states and examine heterogeneous texture development through the thickness of the sample during the process [165,182]. The rolling simulation that has been done in [182] is here recalled. The material of rolling plate is uranium α-U (see [182] for detailed mechanical and texture properties). The initial texture was represented using 1600 individual orientations. Each orientation was assigned to two elements and distributed over their eight integration points. The sample was rolled between two cylinders leading to 60% reduction in thickness during five steps. The geometry of the rolled plate at each step is depicted in Figure 11a, together with the corresponding contour of the equivalent plastic strain. Due to the highly anisotropic nature of the uranium α-U, the strain is developed heterogeneously through the thickness. Similarly the heterogeneous texture that can be appreciated in Figure 11b. According to [182], the determination of the texture gradients can help with understanding recrystallization kinetics and the final recrystallized microstruture. Consequently, the texture gradients can affect recrystallization by altering crystallographic orientation, grain boundary and dislocation densities.

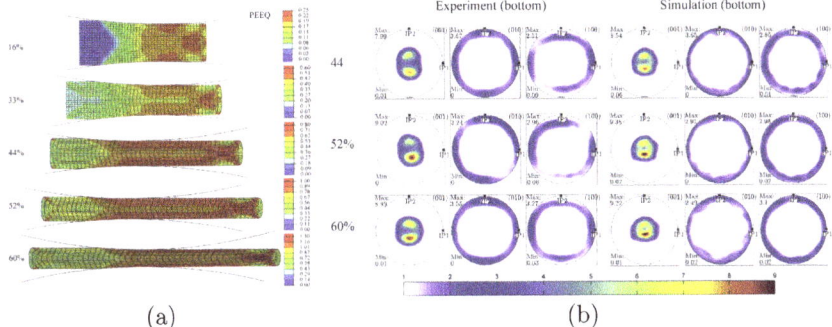

Figure 11. Simulation results of rolling test for the uranium α-U: (**a**) equivalent plastic strain contours at different step; (**b**) comparison between measured and predicted evolution at a polycrystalline material point near the bottom surface of the plate. Reproduced from [182], under permission.

6. Conclusions

In the 21th century, computational crystal plasticity modelling has achieved a high degree of maturity. Thanks to the current computational power and to the algorithms with increased efficiency as multiscale ones based on the FFT, it is becoming an alternative to classical continuum plasticity for engineering problems. Crystal plasticity simulations allow for incorporating many ingredients of the structure of the material as grain size, dislocations, texture, etc. In this paper we have made an overview of different aspects of the present techniques employed in computational finite element simulations of crystal plasticity, as well as some applications of such simulations.

Author Contributions: Conceptualization, K.N. and F.J.M.; methodology, K.N. and F.J.M.; investigation, K.N., M.Z., V.J.A., M.A.S. and F.J.M.; writing—original draft preparation, K.N., M.Z., V.J.A., M.A.S. and F.J.M.; writing—review and editing, K.N., M.Z., V.J.A., M.A.S. and F.J.M.; funding acquisition, F.J.M. All authors have read and agreed to the published version of the manuscript.

Funding: Partial financial support for this work has been facilitated by Agencia Estatal de Investigación of Spain, Grant No. PGC-2018-097257-B-C32.

Conflicts of Interest: The authors declare no conflict of interest. The funders had no role in the design of the study; in the collection, analyses, or interpretation of data; in the writing of the manuscript, or in the decision to publish the results.

References

1. Wagoner, R.; Chenot, J. *Metal Forming Analysis*; Cambridge University Press: Cambridge, UK, 2001.
2. Bathe, K.J. *Finite Element Procedures*, 2nd ed.; Klaus-Jürgen Bathe: Watertown, MA, USA, 2014.
3. Balluffi, R. *Introduction to Elasticity Theory for Crystal Deffects*; World Scientific: Hackensack, NJ, USA, 2017.
4. Kocks, U.; Tomé, C.; Wenk, H.R. *Texture and Anisotropy: Preferred Orientations in Polycrystals and Their Effect on Paterials Properties*; Cambridge University Press: Cambridge, UK, 1998.
5. Peirce, D.; Asaro, R.J.; Needleman, A. An analysis of nonuniform and localized deformation in ductile single crystals. *Acta Metall.* **1982**, *30*, 1087–1119. [CrossRef]
6. Miehe, C.; Schröder, J.; Schotte, J. Computational homogenization analysis in finite plasticity Simulation of texture development in polycrystalline materials. *Comput. Methods Appl. Mech. Eng.* **1999**, *171*, 387–418. [CrossRef]
7. Horstemeyer, M. *Integrated Computational Materials Engineering for Metals*; Wiley: Hoboken, NJ, USA, 2012.

8. Cuitiño, A.M.; Ortiz, M. Computational modelling of single crystals. *Model. Simul. Mater. Sci. Eng.* **1992**, *1*, 225–263. [CrossRef]
9. Fohrmeister, V.; Díaz, G.; Mosler, J. Classic crystal plasticity theory vs crystal plasticity theory based on strong discontinuities. Theoretical and algorithmic aspects. *Int. J. Numer. Methods Eng.* **2019**, *117*, 1283–1303. [CrossRef]
10. Kim, J.; Yoon, J. Necking behavior of AA 6022-T4 based on the crystal plasticity and damage models. *Int. J. Plast.* **2015**, *73*, 3–23. [CrossRef]
11. Aim, U.; Siddiq, M.; Kartal, M. Representative volume element (RVE) based crystal plasticity study of void growth on phase boundary in titanium alloys. *Comput. Mater. Sci.* **2019**, *161*, 346–350.
12. Noell, P.; Carroll, J.; Hattar, K.; Clark, B.; Boyce, B. Do voids nucleate at grain boundaries during ductile rupture? *Acta Mater.* **2017**, *137*, 103–114. [CrossRef]
13. Miehe, C.; Welschinger, F.; Hofacker, M. Thermodynamically consistent phase-field models of fracture: Variational principles and multi-field FE implementations. *Int. J. Numer. Methods Eng.* **2010**, *83*, 1273–1311. [CrossRef]
14. Grilli, N.; Koslowski, M. The effect of crystal anisotropy and plastic response on the dynamic fracture of energetic materials. *J. Appl. Phys.* **2019**, *126*. [CrossRef]
15. Kohlhoff, S.; Gumbsch, P.; Fischmeister, H. Crack propagation in BCC crystals studied with a combined finite-element and atomistic model. *Philos. Mag. A* **1991**, *64*, 851–878. [CrossRef]
16. Shilkrot, L.E.; Miller, R.; Curtin, W. Coupled atomistic and discrete dislocation plasticity. *Phys. Rev. Lett.* **2002**, *89*, 25501. [CrossRef] [PubMed]
17. Shiari, B.; Miller, R.; Curtin, W. Coupled atomistic/discrete dislocation simulations of nanoindentation at finite temperature. *J. Eng. Mater. Technol.* **2005**, *127*, 358–368. [CrossRef]
18. Kochmann, J.; Wulfinghoff, S.; Ehle, L.; Mayer, J.; Svendsen, B.; Reese, S. Efficient and accurate two-scale FE-FFT-based prediction of the effective material behavior of elasto-viscoplastic polycrystals. *Comput. Mech.* **2018**, *61*, 751–764. [CrossRef]
19. Valdenaire, P.L.; Le Bouar, Y.; Appolaire, B.; Finel, A. Density-based crystal plasticity: From the discrete to the continuum. *Phys. Rev. B* **2016**, *93*, 214111. [CrossRef]
20. Roters, F.; Eisenlohr, P.; Hantcherli, L.; Tjahjanto, D.; Bieler, T.; Raabe, D. Overview of constitutive laws, kinematics, homogenization and multiscale methods in crystal plasticity finite-element modeling: Theory, experiments, applications. *Acta Mater.* **2010**, *58*, 1152–1211. [CrossRef]
21. Roters, F.; Eisenlohr, P.; Bieler, T.; Raabe, D. *Crystal Plasticity Finite Element Methods in Materials Science and Engineering*; Wiley-VHC: Weinheim, Germany, 2010.
22. Zhuang, Z.; Liu, Z.; Cui, Y. *Dislocation Mechanism-Based Crystal Plasticity*; Academic Press: Cambridge, MA, USA, 2019.
23. Abbaschian, R.; Abbaschian, L.; Reed-Hill, R. *Physical Metallurgy Principles*; Cengage Learning: Stamford, CT, USA, 2009.
24. Borja, R.I. *Plasticity Modeling and Computation*; Springer: Berlin, Germany, 2013.
25. Hosford, W.F. *The Mechanics of Crystals and Textured Polycrystals*; Oxford University Press: Oxford, UK, 1993.
26. Dunne, F.; Petrinic, N. *Introduction to Computational Plasticity*; Oxford University Press: Oxford, UK, 2005.
27. Hull, D.; Bacon, D. *Introduction to Dislocations*; Butterworth-Heinemann: Oxford, UK, 2011.
28. Zheng, Z.; Balint, D.; Dunne, F. Rate sensitivity in discrete dislocation plasticity in hexagonal close-packed crystals. *Acta Mater.* **2016**, *107*, 17–26. [CrossRef]
29. Testa, G.; Bonora, N.; Ruggiero, A.; Iannitti, G. Flow Stress of bcc Metals over a Wide Range of Temperature and Strain Rates. *Metals* **2020**, *10*, 120. [CrossRef]
30. Rogne, B.R.S.; Thaulow, C. Strengthening mechanisms of iron micropillars. *Philos. Mag.* **2015**, *95*, 1814–1828. [CrossRef]
31. Hall, E. The Deformation and Ageing of Mild Steel: III Discussion of Results. *Proc. Phys. Soc. Sect. B* **1951**, *64*, 747–753. [CrossRef]
32. Petch, N. The Cleavage Strength of Polycrystals. *J. Iron Steel Inst.* **1953**, *174*, 25–28.
33. Schmid Erich, W.B. *Plasticity of Crystals with Special Reference to Metals*; F.A. Hughes: London, UK, 1950.

34. Montáns, F.; Bathe, K. *Computational Plasticity*; Chapter Towards a Model for Large Strain Anisotropic Elasto-Plasticity; Springer: Berlin, Germany, 2007.
35. Park, S.J.; Han, H.N.; Oh, K.H.; Raabe, D.; Kim, J.K. Finite element simulation of grain interaction and orientation fragmentation during plastic deformation of BCC metals. *Mater. Sci. Forum* **2002**, *408–412*, 371–376. [CrossRef]
36. Sachtleber, M.; Zhao, Z.; Raabe, D. Experimental investigation of plastic grain interaction. *Mater. Sci. Eng. A* **2002**, *336*, 81–87. [CrossRef]
37. Ma, A.; Roters, F.; Raabe, D. On the consideration of interactions between dislocations and grain boundaries in crystal plasticity finite element modeling—Theory, experiments, and simulations. *Acta Mater.* **2006**, *54*, 2181–2194. [CrossRef]
38. Kothari, M.; Anand, L. Elasto-viscoplastic constitutive equations for polycrystalline metals: Application to tantalum. *J. Mech. Phys. Solids* **1998**, *46*, 51–67. [CrossRef]
39. Kalidindi, S.; Bronkhorst, C.; Anand, L. Crystallographic texture evolution in bulk deformation processing of FCC metals. *J. Mech. Phys. Solids* **1992**, *40*, 537–569. [CrossRef]
40. Kalidindi, S.R. Incorporation of deformation twinning in crystal plasticity models. *J. Mech. Phys. Solids* **1998**, *46*, 267–290. [CrossRef]
41. Taylor, G. The mechanism of plastic deformation of crystals. *Proc. R. Soc. A* **1934**, *165*, 362–387.
42. Basinski, Z. Thermally activated glide in face-centred cubic metals and its application to the theory of strain hardening. *Philos. Mag.* **1959**, *4*, 393–432. [CrossRef]
43. Kuhlmann-Wilsdorf, D. A new theory of workhardening. *Trans. Metall. Soc. AIME* **1962**, *224*, 1047–1061.
44. Essmann, U.; Mughrabi, H. Annihilation of dislocations during tensile and cyclic deformation and limits of dislocation densities. *Philos. Mag. A* **1979**, *40*, 731–756. [CrossRef]
45. Essmann, U.; Mughrabi, H. Multislip in FCC crystals: A theoretical approach compared with experimental-data. *Acta Metall.* **1982**, *30*, 1627–1637.
46. Lavrentev, F.; Pokhil, Y. Relation of dislocation density in different slip systems to work-hardening parameters for magnesium crystals. *Mater. Sci. Eng.* **1975**, *18*, 261–270. [CrossRef]
47. Bertin, N.; Tomé, C.; Beyerlein, I.; Barnett, M.; Capolungo, L. On the strength of dislocation interactions and their effect on latent hardening in pure Magnesium. *Int. J. Plast.* **2014**, *62*, 72–92. [CrossRef]
48. Arturo Rubio, R.; Haouala, S.; LLorca, J. Grain boundary strengthening of FCC polycrystals. *J. Mater. Res.* **2019**, *34*, 2263–2274. [CrossRef]
49. Haouala, S.; Alizadeh, R.; Bieler, T.; Segurado, J.; LLorca, J. Effect of slip transmission at grain boundaries in Al bicrystals. *Int. J. Plast.* **2020**, *126*, 102600. [CrossRef]
50. Ma, A.; Hartmaier, A. On the influence of isotropic and kinematic hardening caused by strain gradients on the deformation behaviour of polycrystals. *Philos. Mag.* **2014**, *94*, 125–140. [CrossRef]
51. Ma, A.; Hartmaier, A. A study of deformation and phase transformation coupling for TRIP-assisted steels. *Int. J. Plast.* **2015**, *64*, 40–55. [CrossRef]
52. Asaro, R.J. Micromechanics of Crystals and Polycrystals. *Adv. Appl. Mech.* **1983**, *23*, 1–115. [CrossRef]
53. Lee, E.H. Elastic-Plastic Deformation at Finite Strains. *J. Appl. Mech.* **1969**, *36*, 1–6. [CrossRef]
54. Rice, J. Inelastic constitutive relations for solids: An internal-variable theory and its application to metal plasticity. *J. Mech. Phys. Solids* **1971**, *19*, 433–455. [CrossRef]
55. Hill, R.; Rice, J.R. Constitutive analysis of elastic-plastic crystals at arbitrary strain. *J. Mech. Phys. Solids* **1972**, *20*, 401–413. [CrossRef]
56. Simo, J. *Handbook of Numerical Analysis VI*; Chapter Numerical Analysis and Simulation of Plasticity; Elsevier: Amsterdam, The Netherlands, 1998; pp. 183–499.
57. Hashiguchi, K. Multiplicative Hyperelastic-Based Plasticity for Finite Elastoplastic Deformation/Sliding: A Comprehensive Review. *Arch. Comput. Methods Eng.* **2019**, *26*, 597–637. [CrossRef]
58. Gurtin, M.; Fried, E.; Anand, L. *The Mechanics and Thermodynamics of Continua*; Cambridge University Press: Cambridge, UK, 2010.
59. Shutov, A.; Pfeiffer, S.; Ihlemann, J. On the simulation of multistage forming processes: Invariance under change of the reference configuration. *Mater. Sci. Eng. Technol.* **2012**, *43*, 617–625.

60. Shutov, A.; Ihlemann, J. Analysis of some basic approaches to finite strain elastoplasticity in view of reference change. *Int. J. Plast.* **2014**, *63*, 183–197. [CrossRef]
61. Hartley, C.; Mishin, Y. Characterization and visualization of the lattice misfit associated with dislocation cores. *Acta Mater.* **2005**, *53*, 1313–1321. [CrossRef]
62. Nye, J.F. Some geometrical relations in dislocated crystals. *Acta Metall.* **1953**, *1*, 153–162. [CrossRef]
63. Kröner, E. Benefits and shortcomings of the continuous theory of dislocations. *Int. J. Solids Struct.* **2001**, *38*, 1115–1134. [CrossRef]
64. Forest, S.; Barbe, F.; Cailletaud, G. Cosserat modelling of size effects in the mechanical behaviour of polycrystals and multi-phase materials. *Int. J. Solids Struct.* **2000**, *37*, 7105–7126. [CrossRef]
65. Gurtin, M.E. A gradient theory of single-crystal viscoplasticity that accounts for geometrically necessary dislocations. *J. Mech. Phys. Solids* **2002**, *50*, 5–32. [CrossRef]
66. Evers, L.P.; Brekelmans, W.A.; Geers, M.G. Scale dependent crystal plasticity framework with dislocation density and grain boundary effects. *Int. J. Solids Struct.* **2004**, *41*, 5209–5230. [CrossRef]
67. Evers, L.P.; Brekelmans, W.A.; Geers, M.G. Non-local crystal plasticity model with intrinsic SSD and GND effects. *J. Mech. Phys. Solids* **2004**, *52*, 2379–2401. [CrossRef]
68. Ma, A.; Roters, F. A constitutive model for fcc single crystals based on dislocation densities and its application to uniaxial compression of aluminium single crystals. *Acta Mater.* **2004**, *52*, 3603–3612. [CrossRef]
69. Han, C.S.; Gao, H.; Huang, Y.; Nix, W.D. Mechanism-based strain gradient crystal plasticity—I. Theory. *J. Mech. Phys. Solids* **2005**, *53*, 1188–1203. [CrossRef]
70. Kuroda, M.; Tvergaard, V. Studies of scale dependent crystal viscoplasticity models. *J. Mech. Phys. Solids* **2006**, *54*, 1789–1810. [CrossRef]
71. Gurtin, M.E.; Anand, L.; Lele, S.P. Gradient single-crystal plasticity with free energy dependent on dislocation densities. *J. Mech. Phys. Solids* **2007**, *55*, 1853–1878. [CrossRef]
72. Gurtin, M.E. A finite-deformation, gradient theory of single-crystal plasticity with free energy dependent on densities of geometrically necessary dislocations. *Int. J. Plast.* **2008**, *24*, 702–725. [CrossRef]
73. Kuroda, M.; Tvergaard, V. On the formulations of higher-order strain gradient crystal plasticity models. *J. Mech. Phys. Solids* **2008**, *56*, 1591–1608. [CrossRef]
74. Grilli, N.; Koslowski, M. The effect of crystal orientation on shock loading of single crystal energetic materials. *Comput. Mater. Sci.* **2018**, *155*, 235–245. [CrossRef]
75. Zhang, M.; Nguyen, K.; Segurado, J.; Montáns, F. A multiplicative finite strain crystal plasticity formulation based on additive elastic corrector rates: Theory and numerical implementation. *Int. J. Plast.* **2020**, *137*, 102899. doi:10.1016/j.ijplas.2020.102899. [CrossRef]
76. Latorre, M.; Montáns, F. A new class of plastic flow evolution equations for anisotropic multiplicative elastoplasticity based on the notion of a corrector elastic strain rate. *Appl. Math. Model.* **2018**, *55*, 716–740. [CrossRef]
77. Zhang, M.; Montáns, F. A simple formulation for large-strain cyclic hyperelasto-plasticity using elastic correctors. Theory and algorithmic implementation. *Int. J. Plast.* **2019**, *113*, 185–217. [CrossRef]
78. Nguyen, K.; Sanz, M.A.; Montáns, F. Plane-stress constrained multiplicative hyperelasto-plasticity with nonlinear kinematic hardening. Consistent theory based on elastic corrector rates and algorithmic implementation. *Int. J. Plast.* **2020**, *128*, 102592. [CrossRef]
79. Caminero, M.; Montáns, F.; Bathe, K.J. Modeling large strain anisotropic elasto-plasticity with logarithmic strain and stress measures. *Comput. Struct.* **2011**, *89*, 826–843. [CrossRef]
80. Sánz, M.; Montáns, F.; Latorre, M. Computational anisotropic hardening multiplicative elastoplasticity based on the corrector elastic logarithmic strain rate. *Comput. Methods Appl. Mech. Eng.* **2017**, *320*, 82–121. [CrossRef]
81. Latorre, M.; Montáns, F. Stress and strain mapping tensors and general work-conjugacy in large strain continuum mechanics. *Appl. Math. Model.* **2016**, *40*, 3938–3950. [CrossRef]
82. Peirce, D.; Asaro, R.J.; Needleman, A. Material rate dependence and localized deformation in crystalline solids. *Acta Metall.* **1983**, *31*, 1951–1976. [CrossRef]

83. Needleman, A.; Asaro, R.J.; Lemonds, J.; Peirce, D. Finite element analysis of crystalline solids. *Comput. Methods Appl. Mech. Eng.* **1985**, *52*, 689–708. [CrossRef]
84. Fivel, M.; Tabourot, L.; Rauch, E.; Canova, G. Identification through mesoscopic simulations of macroscopic parameters of physically based constitutive equations for the plastic behavior of FCC single crystals. *J. Phys. IV* **1998**, *8*, 151–158.
85. Groh, S.; Maron, E.; Horstemeyer, M.; Zbib, H. Multiscale modeling of the plasticity in an aluminum single crystal. *Int. J. Plast.* **2009**, *25*, 1456–1473. [CrossRef]
86. Meric, L.; Poubanne, P.; Cailletaud, G. Single Crystal Modeling for Structural Calculations: Part 1—Model Presentation. *J. Eng. Mater. Technol.* **1991**, *113*, 162–170. [CrossRef]
87. Cailletaud, G. A micromechanical approach to inelastic behaviour of metals. *Int. J. Plast.* **1992**, *8*, 55–73. [CrossRef]
88. Hu, Z.; Rauch, E.F.; Teodosiu, C. Work-hardening behavior of mild steel under stress reversal at large strains. *Int. J. Plast.* **1992**, *8*, 839–856. [CrossRef]
89. Hasija, V.; Ghosh, S.; Mills, M.J.; Joseph, D.S. Deformation and creep modeling in polycrystalline Ti-6Al alloys. *Acta Mater.* **2003**, *51*, 4533–4549. [CrossRef]
90. Venkatramani, G.; Ghosh, S.; Mills, M. A size-dependent crystal plasticity finite-element model for creep and load shedding in polycrystalline titanium alloys. *Acta Mater.* **2007**, *55*, 3971–3986. [CrossRef]
91. Cruzado, A.; LLorca, J.; Segurado, J. Modeling cyclic deformation of inconel 718 superalloy by means of crystal plasticity and computational homogenization. *Int. J. Solids Struct.* **2017**, *122–123*, 148–161. [CrossRef]
92. Tome, C.; Canova, G.R.; Kocks, U.F.; Christodoulou, N.; Jonas, J.J. The relation between macroscopic and microscopic strain hardening in F.C.C. polycrystals. *Acta Metall.* **1984**, *32*, 1637–1653. [CrossRef]
93. Bassani, J.L.; Wu, T.Y. Latent hardening in single crystals. II. Analytical characterization and predictions. *Proc. R. Soc. Lond. Ser. A Math. Phys. Sci.* **1991**, *435*, 21–41. [CrossRef]
94. Polmear, I.J. *From Traditional Alloys to Nanocrystals*; Butterworth-Heinemann: Burlington, MA, USA, 2006.
95. Estrin, Y.; Mecking, H. A unified phenomenological description of work hardening and creep based on one-parameter models. *Acta Metall.* **1984**, *32*, 57–70. [CrossRef]
96. Estrin, Y. Dislocation-Density–Related Constitutive Modeling. In *Unified Constitutive Laws of Plastic Deformation*; Krausz, A., Krausz, K., Eds.; Academic Press: Cambridge, MA, USA, 1996; pp. 69–106. [CrossRef]
97. Mecking, H.; Kocks, U.F. Kinetics of flow and strain-hardening. *Acta Metall.* **1981**, *29*, 1865–1875. [CrossRef]
98. Devincre, B.; Hoc, T.; Kubin, L. Dislocation mean free paths and strain hardening of crystals. *Science* **2008**, *320*, 1745–1748. [CrossRef]
99. Cheong, K.S.; Busso, E.P. Discrete dislocation density modelling of single phase FCC polycrystal aggregates. *Acta Mater.* **2004**, *52*, 5665–5675. [CrossRef]
100. De Sansal, C.; Devincre, B.; Kubin, L. Grain size strengthening in microcrystalline copper: A three-dimensional dislocation dynamics simulation. *Key Eng. Mater.* **2010**, *423*, 25–32. [CrossRef]
101. Estrin, Y.; Kubin, L.P. Local strain hardening and nonuniformity of plastic deformation. *Acta Metall.* **1986**, *34*, 2455–2464. [CrossRef]
102. Kameda, T.; Zikry, M.A. Three dimensional dislocation-based crystalline constitutive formulation for ordered intermetallics. *Scr. Mater.* **1998**, *38*, 631–636. [CrossRef]
103. Rezvanian, O.; Zikry, M.A.; Rajendran, A.M. Statistically stored, geometrically necessary and grain boundary dislocation densities: Microstructural representation and modelling. *Proc. R. Soc. A Math. Phys. Eng. Sci.* **2007**, *463*, 2833–2853. [CrossRef]
104. Shi, J.; Zikry, M.A. Grain-boundary interactions and orientation effects on crack behavior in polycrystalline aggregates. *Int. J. Solids Struct.* **2009**, *46*, 3914–3925. [CrossRef]
105. Shanthraj, P.; Zikry, M.A. Dislocation density evolution and interactions in crystalline materials. *Acta Mater.* **2011**, *59*, 7695–7702. [CrossRef]
106. Orowan, E. Zur Kristallplastizität. I: Tieftemperaturplastizität und Beckersche Formel. *Z. Phys.* **1934**, *89*, 605–613. [CrossRef]

107. Kocks, U. *Thermodynamics and Kinetics of Slip, Volume 19 of Progress in Materials Science*; Pergamon Press: Oxford, UK, 2003.
108. Rodríguez-Galán, D.; Sabirov, I.; Segurado, J. Temperature and stain rate effect on the deformation of nanostructured pure titanium. *Int. J. Plast.* **2015**, *70*, 191–205. [CrossRef]
109. Shahba, A.; Ghosh, S. Crystal plasticity FE modeling of Ti alloys for a range of strain-rates. Part I: A unified constitutive model and flow rule. *Int. J. Plast.* **2016**, *87*, 48–68. [CrossRef]
110. Cailard, D.; Martin, J. *Thermally Activated Mechanisms in Crystal Plasticity*; Elsevier: Amsterdam, The Netherlands, 2003.
111. Taylor, G.I. Plastic strain in metals. *J. Inst. Met.* **1938**, *62*, 307–324.
112. Elkhodary, K.; Lee, W.; Sun, L.P.; Brenner, D.W.; Zikry, M.A. Deformation mechanisms of an Ω precipitate in a high-strength aluminum alloy subjected to high strain rates. *J. Mater. Res.* **2011**, *26*, 487–497. [CrossRef]
113. Stelmashenko, N.A.; Walls, M.G.; Brown, L.M.; Milman, Y.V. Microindentations on W and Mo oriented single crystals: An STM study. *Acta Metall. Mater.* **1993**, *41*, 2855–2865. [CrossRef]
114. Wang, Y.; Raabe, D.; Klüber, C.; Roters, F. Orientation dependence of nanoindentation pile-up patterns and of nanoindentation microtextures in copper single crystals. *Acta Mater.* **2004**, *52*, 2229–2238. [CrossRef]
115. Zaafarani, N.; Raabe, D.; Singh, R.N.; Roters, F.; Zaefferer, S. Three-dimensional investigation of the texture and microstructure below a nanoindent in a Cu single crystal using 3D EBSD and crystal plasticity finite element simulations. *Acta Mater.* **2006**, *54*, 1863–1876. [CrossRef]
116. Zaafarani, N.; Raabe, D.; Roters, F.; Zaefferer, S. On the origin of deformation-induced rotation patterns below nanoindents. *Acta Mater.* **2008**, *56*, 31–42. [CrossRef]
117. Sánchez-Martín, R.; Pérez-Prado, M.T.; Segurado, J.; Bohlen, J.; Gutiérrez-Urrutia, I.; Llorca, J.; Molina-Aldareguia, J.M. Measuring the critical resolved shear stresses in Mg alloys by instrumented nanoindentation. *Acta Mater.* **2014**, *71*, 283–292. [CrossRef]
118. Motz, C.; Schöberl, T.; Pippan, R. Mechanical properties of micro-sized copper bending beams machined by the focused ion beam technique. *Acta Mater.* **2005**, *53*, 4269–4279. [CrossRef]
119. Weber, F.; Schestaknow, I.; Roters, F.; Raabe, D. Texture Evolution During Bending of a Single Crystal Copper Nanowire Studied by EBSD and Crystal Plasticity Finite Element Simulations. *Adv. Eng. Mater.* **2008**, *10*, 737–741. [CrossRef]
120. Kiener, D.; Grosinger, W.; Dehm, G.; Pippan, R. A further step towards an understanding of size-dependent crystal plasticity: In situ tension experiments of miniaturized single-crystal copper samples. *Acta Mater.* **2008**, *56*, 580–592. [CrossRef]
121. Gong, J.; Wilkinson, A.J. A microcantilever investigation of size effect, solid-solution strengthening and second-phase strengthening for a prism slip in alpha-Ti. *Acta Mater.* **2011**, *59*, 5970–5981. [CrossRef]
122. Zhou, C.; Beyerlein, I.J.; Lesar, R. Plastic deformation mechanisms of fcc single crystals at small scales. *Acta Mater.* **2011**, *59*, 7673–7682. [CrossRef]
123. Norton, A.D.; Falco, S.; Young, N.; Severs, J.; Todd, R.I. Microcantilever investigation of fracture toughness and subcritical crack growth on the scale of the microstructure in Al2O3. *J. Eur. Ceram. Soc.* **2015**, *35*, 4521–4533. [CrossRef]
124. Raabe, D.; Ma, D.; Roters, F. Effects of initial orientation, sample geometry and friction on anisotropy and crystallographic orientation changes in single crystal microcompression deformation: A crystal plasticity finite element study. *Acta Mater.* **2007**, *55*, 4567–4583. [CrossRef]
125. Ashby, M.F. The deformation of plastically non-homogeneous materials. *Philos. Mag.* **1970**, *21*, 399–424. [CrossRef]
126. Evers, L.P.; Parks, D.M.; Brekelmans, W.A.; Geers, M.G. Crystal plasticity model with enhanced hardening by geometrically necessary dislocation accumulation. *J. Mech. Phys. Solids* **2002**, *50*, 2403–2424. [CrossRef]
127. Gao, H.; Huang, Y. Geometrically necessary dislocation and size-dependent plasticity. *Scr. Mater.* **2003**, *48*, 113–118. [CrossRef]
128. Kuroda, M. *Plasticity and Beyond: Microstructures, Crystal-Plasticity and Phase Transitions*; Chapter on Scale-Dependent Crystal Plasticity Models; CSIM Courses and Lectures; Springer: Berlin, Germany, 2014.

129. Fleck, N.A.; Muller, G.M.; Ashby, M.F.; Hutchinson, J.W. Strain gradient plasticity: Theory and experiment. *Acta Metall. Mater.* **1994**, *42*, 475–487. [CrossRef]
130. Gao, H.; Huang, Y.; Nix, W.D.; Hutchinson, J.W. Mechanism-based strain gradient plasticity—I. Theory. *J. Mech. Phys. Solids* **1999**, *47*, 1239–1263. [CrossRef]
131. Cheong, K.S.; Busso, E.P.; Arsenlis, A. A study of microstructural length scale effects on the behaviour of FCC polycrystals using strain gradient concepts. *Int. J. Plast.* **2005**, *21*, 1797–1814. [CrossRef]
132. Busso, E.P.; Meissonnier, F.T.; O'Dowd, N.P. Gradient-dependent deformation of two-phase single crystals. *J. Mech. Phys. Solids* **2000**, *48*, 2333–2361. [CrossRef]
133. Dunne, F.P.; Rugg, D.; Walker, A. Lengthscale-dependent, elastically anisotropic, physically-based hcp crystal plasticity: Application to cold-dwell fatigue in Ti alloys. *Int. J. Plast.* **2007**, *23*, 1061–1083. [CrossRef]
134. Han, C.S.; Ma, A.; Roters, F.; Raabe, D. A Finite Element approach with patch projection for strain gradient plasticity formulations. *Int. J. Plast.* **2007**, *23*, 690–710. [CrossRef]
135. Gurtin, M.E.; Needleman, A. Boundary conditions in small-deformation, single-crystal plasticity that account for the Burgers vector. *J. Mech. Phys. Solids* **2005**, *53*, 1–31. [CrossRef]
136. Yefimov, S.; Van Der Giessen, E.; Groma, I. Bending of a single crystal: Discrete dislocation and nonlocal crystal plasticity simulations. *Model. Simul. Mater. Sci. Eng.* **2004**, *12*, 1069–1086. [CrossRef]
137. Bardella, L. A deformation theory of strain gradient crystal plasticity that accounts for geometrically necessary dislocations. *J. Mech. Phys. Solids* **2006**, *54*, 128–160. [CrossRef]
138. Bayley, C.J.; Brekelmans, W.A.; Geers, M.G. A comparison of dislocation induced back stress formulations in strain gradient crystal plasticity. *Int. J. Solids Struct.* **2006**, *43*, 7268–7286. [CrossRef]
139. Borg, U.; Niordson, C.F.; Kysar, J.W. Size effects on void growth in single crystals with distributed voids. *Int. J. Plast.* **2008**, *24*, 688–701. [CrossRef]
140. Bardella, L.; Segurado, J.; Panteghini, A.; Llorca, J. Latent hardening size effect in small-scale plasticity. *Model. Simul. Mater. Sci. Eng.* **2013**, *21*. [CrossRef]
141. Niordson, C.F.; Kysar, J.W. Computational strain gradient crystal plasticity. *J. Mech. Phys. Solids* **2014**, *62*, 31–47. [CrossRef]
142. Sandfeld, S.; Hochrainer, T.; Gumbsch, P.; Zaiser, M. Numerical implementation of a 3D continuum theory of dislocation dynamics and application to micro-bending. *Philos. Mag.* **2010**, *90*, 3697–3728. [CrossRef]
143. Groma, I. Link between the microscopic and mesoscopic length-scale description of the collective behavior of dislocations. *Phys. Rev. B Condens. Matter Mater. Phys.* **1997**, *56*, 5807–5813. [CrossRef]
144. El-Azab, A. Statistical mechanics treatment of the evolution of dislocation distributions in single crystals. *Phys. Rev. B Condens. Matter Mater. Phys.* **2000**, *61*, 11956–11966. [CrossRef]
145. Acharya, A. A model of crystal plasticity based on the theory of continuously distributed dislocations. *J. Mech. Phys. Solids* **2001**, *49*, 761–784. [CrossRef]
146. Acharya, A.; Roy, A. Size effects and idealized dislocation microstructure at small scales: Predictions of a Phenomenological model of Mesoscopic Field Dislocation Mechanics: Part I. *J. Mech. Phys. Solids* **2006**, *54*, 1687–1710. [CrossRef]
147. Hochrainer, T.; Zaiser, M.; Gumbsch, P. A three-dimensional continuum theory of dislocation systems: Kinematics and mean-field formulation. *Philos. Mag.* **2007**, *87*, 1261–1282. [CrossRef]
148. Hochrainer, T.; Zaiser, M.; Gumbsch, P. Dislocation transport and line length increase in averaged descriptions of dislocations. *AIP Conf. Proc.* **2009**, *1168*, 1133–1136. [CrossRef]
149. Xia, S.X.; El-Azab, A. A preliminary investigation of dislocation cell structure formation in metals using continuum dislocation dynamics. *IOP Conf. Ser. Mater. Sci. Eng.* **2015**, *89*. [CrossRef]
150. Hochrainer, T. Thermodynamically consistent continuum dislocation dynamics. *J. Mech. Phys. Solids* **2016**, *88*, 12–22. [CrossRef]
151. Sarma, G.; Zacharia, T. Integration algorithm for modeling the elasto-viscoplastic response of polycrystalline materials. *J. Mech. Phys. Solids* **1999**, *47*, 1219–1238. [CrossRef]
152. Maniatty, A.; Dawson, P.; Lee, Y. A time integration algorithm for elasto-viscoplastic cubic crystals applied to modelling polycrystalline deformation. *Int. J. Numer. Methods Eng.* **1992**, *35*, 1565–1588. [CrossRef]

153. Roters, F.; Diehl, M.; Shanthraj, P.; Eisenlohr, P.; Reuber, C.; Wong, S.L.; Maiti, T.; Ebrahimi, A.; Hochrainer, T.; Fabritius, H.O.; et al. DAMASK: The Düsseldorf Advanced Material Simulation Kit for modeling multi-physics crystal plasticity, thermal, and damage phenomena from the single crystal up to the component scale. *Comput. Mater. Sci.* **2019**, *158*, 420–478. [CrossRef]
154. Arsenlis, A.; Parks, D.M.; Becker, R.; Bulatov, V.V. On the evolution of crystallographic dislocation density in non-homogeneously deforming crystals. *J. Mech. Phys. Solids* **2004**, *52*, 1213–1246. [CrossRef]
155. Borja, R.I.; Wren, J.R. Discrete micromechanics of elastoplastic crystals. *Int. J. Numer. Methods Eng.* **1993**, *36*, 3815–3840. [CrossRef]
156. Miehe, C. Exponential map algorithm for stress updates in anisotropic multiplicative elastoplasticity for single crystals. *Int. J. Numer. Methods Eng.* **1996**, *39*, 3367–3390. [CrossRef]
157. Asaro, R.; Needleman, A. Texture development and strain hardening in rate dependent polycrystals. *Acta Metall.* **1985**, *33*, 923–953. [CrossRef]
158. Castañeda, P.P. The effective mechanical properties of nonlinear isotropic composites. *J. Mech. Phys. Solids* **1991**, *39*, 45–71. [CrossRef]
159. Lebensohn, R.A.; Tomé, C.N. A self-consistent anisotropic approach for the simulation of plastic deformation and texture development of polycrystals: Application to zirconium alloys. *Acta Metall. Mater.* **1993**, *41*, 2611–2624. [CrossRef]
160. Lebensohn, R.; Uhlenhut, H.; Hartig, C.; Mecking, H. Plastic flow of γ-TiAl-Based polysynthetically twinned crystals: Micromechanical modeling and experimental validation. *Acta Mater.* **1998**, *46*, 4701–4709. [CrossRef]
161. Masson, R.; Bornert, M.; Suquet, P.; Zaoui, A. Affine formulation for the prediction of the effective properties of nonlinear composites and polycrystals. *J. Mech. Phys. Solids* **2000**, *48*, 1203–1227. [CrossRef]
162. Lebensohn, R.A.; Tomé, C.N.; Castañeda, P.P. Self-consistent modelling of the mechanical behaviour of viscoplastic polycrystals incorporating intragranular field fluctuations. *Philos. Mag.* **2007**, *87*, 4287–4322. [CrossRef]
163. Bellec, V.; Bøe, R.; Rise, L.; Thorsnes, T. Influence of bottom currents on the Lofoten continental margin, North Norway. *Rend. Online Soc. Geol. Ital.* **2009**, *7*, 155–157.
164. Lebensohn, R.A.; Zecevic, M.; Knezevic, M.; McCabe, R.J. Average intragranular misorientation trends in polycrystalline materials predicted by a viscoplastic self-consistent approach. *Acta Mater.* **2016**, *104*, 228–236. [CrossRef]
165. Segurado, J.; Lebensohn, R.A.; Llorca, J.; Tomé, C.N. Multiscale modeling of plasticity based on embedding the viscoplastic self-consistent formulation in implicit finite elements. *Int. J. Plast.* **2012**, *28*, 124–140. [CrossRef]
166. Knezevic, M.; Crapps, J.; Beyerlein, I.J.; Coughlin, D.R.; Clarke, K.D.; McCabe, R.J. Anisotropic modeling of structural components using embedded crystal plasticity constructive laws within finite elements. *Int. J. Mech. Sci.* **2016**, *105*, 227–238. [CrossRef]
167. Zecevic, M.; Beyerlein, I.J.; Knezevic, M. Coupling elasto-plastic self-consistent crystal plasticity and implicit finite elements: Applications to compression, cyclic tension-compression, and bending to large strains. *Int. J. Plast.* **2017**, *93*, 187–211. [CrossRef]
168. Knezevic, M.; Beyerlein, I.J. Multiscale Modeling of Microstructure-Property Relationships of Polycrystalline Metals during Thermo-Mechanical Deformation. *Adv. Eng. Mater.* **2018**, *1700956*, 1–19. [CrossRef]
169. Borja, R.I.; Rahmani, H. Discrete micromechanics of elastoplastic crystals in the finite deformation range. *Comput. Methods Appl. Mech. Eng.* **2014**, *275*, 234–263. [CrossRef]
170. Miehe, C. Multisurface thermoplasticity for single crystals at large strains in terms of Eulerian vector updates. *Int. J. Solids Struct.* **1996**, *33*, 3103–3130. [CrossRef]
171. Weber, G.; Anand, L. Finite deformation constitutive equations and a time integration procedure for isotropic, hyperelastic-viscoplastic solids. *Comput. Methods Appl. Mech. Eng.* **1990**, *79*, 173–202. [CrossRef]
172. Nakamachi, E.; Xie, C.L.; Harimoto, M. Drawability assessment of BCC steel sheet by using elastic/crystalline viscoplastic finite element analyses. *Int. J. Mech. Sci.* **2001**, *43*, 631–652. [CrossRef]
173. Helming, K.; Eschner, T. A new approach to texture analysis of multiphase materials using a texture component model. *Cryst. Res. Technol.* **1990**, *25*, K203–K208. [CrossRef]

174. Raabe, D.; Roters, F. Using texture components in crystal plasticity finite element simulations. *Int. J. Plast.* **2004**, *20*, 339–361. [CrossRef]
175. Zhao, Z.; Roters, F.; Mao, W.; Raabe, D. Introduction of a texture component crystal plasticity finite element method for anisotropy simulations. *Adv. Eng. Mater.* **2001**, *3*, 984–990. [CrossRef]
176. Raabe, D.; Zhao, Z.; Roters, F. A finite element method on the basis of texture components for fast predictions of anisotropic forming operations. *Steel Res.* **2001**, *72*, 421–426. [CrossRef]
177. Raabe, D.; Sachtleber, M.; Zhao, Z.; Roters, F.; Zaefferer, S. Micromechanical and macromechanical effects in grain scale polycrystal plasticity experimentation and simulation. *Acta Mater.* **2001**, *49*, 3433–3441. [CrossRef]
178. Zecevic, M.; Knezevic, M. Modeling of Sheet Metal Forming Based on Implicit Embedding of the Elasto-Plastic Self-Consistent Formulation in Shell Elements: Application to Cup Drawing of AA6022-T4. *JOM* **2017**, *69*, 922–929. [CrossRef]
179. Zecevic, M.; Knezevic, M. A new visco-plastic self-consistent formulation implicit in dislocation-based hardening within implicit finite elements: Application to high strain rate and impact deformation of tantalum. *Comput. Methods Appl. Mech. Eng.* **2018**, *341*, 888–916. [CrossRef]
180. Maudlin, P.J.; Gray, G.T.; Cady, C.M.; Kaschner, G.C. High-rate material modelling and validation using the Taylor cylinder impact test. *Philos. Trans. R. Soc. A Math. Phys. Eng. Sci.* **1999**, *357*, 1707–1729. [CrossRef]
181. Maudlin, P.J.; Bingert, J.F.; House, J.W.; Chen, S.R. On the modeling of the Taylor cylinder impact test for orthotropic textured materials: Experiments and simulations. *Int. J. Plast.* **1999**, *15*, 139–166. [CrossRef]
182. Zecevic, M.; Knezevic, M.; Beyerlein, I.J.; McCabe, R.J. Texture formation in orthorhombic alpha-uranium under simple compression and rolling to high strains. *J. Nucl. Mater.* **2016**, *473*, 143–156. [CrossRef]

© 2021 by the authors. Licensee MDPI, Basel, Switzerland. This article is an open access article distributed under the terms and conditions of the Creative Commons Attribution (CC BY) license (http://creativecommons.org/licenses/by/4.0/).

MDPI
St. Alban-Anlage 66
4052 Basel
Switzerland
Tel. +41 61 683 77 34
Fax +41 61 302 89 18
www.mdpi.com

Crystals Editorial Office
E-mail: crystals@mdpi.com
www.mdpi.com/journal/crystals

www.ingramcontent.com/pod-product-compliance
Lightning Source LLC
LaVergne TN
LVHW070125100526
838202LV00016B/2231